Comprehensive Logistics

Timm Gudehus · Herbert Kotzab

Comprehensive Logistics

Springer

Dr. rer. nat. Dr.Ing. habil. Timm Gudehus
Strandweg 54
22587 Hamburg
Germany
www.TimmGudehus.de

Prof. Dr. Herbert Kotzab
Business School Copenhagen
Dept. Operations Management
Solbjerg Plads 3
2000 Frederiksberg
Denmark
hk.om@cbs.dk

ISBN 978-3-540-30722-8 e-ISBN 978-3-540-68652-1
DOI 10.1007/978-3-540-68652-1
Springer Dordrecht Heidelberg London New York

Library of Congress Control Number: 2008940418

Cover design: eStudio Calamar S.L.

Printed on acid-free paper

Springer is part of Springer Science+Business Media (www.springer.com)

Preface

This book presents the scope, variety and importance of modern logistics. It deals with all relevant aspects: space and time; systems, structures and processes; networks and supply chains; economics and technology; micrologistics and macrologistics; intralogistics and extralogistics; planning, scheduling and control; management, organization and operation. The tasks and goals of modern logistics and the options for actions are thoroughly investigated and explained. Current logistic trends and fashions are critically discussed. Strategies and methods for planning, scheduling and operating logistic networks and systems are developed. The book contains rules, algorithms and formulas for solving actual problems. Their application is demonstrated by examples from business practice.

The work is based on the well-known German reference-book "Logistik" by *Timm Gudehus*, which was published in several editions (Gudehus 1999/2000/2003/2005/2007/2010). After translation by *Herbert Kotzab*, the content was thoroughly revised for the international edition by both authors. We have reformulated and supplemented the whole text and updated figures and tables.

The most difficult task when writing the international edition turned out to be finding the right terminology for the multitude of logistic subjects. As logistics is a relatively young discipline, in English and American, even more than in German, there are several expressions with similar meanings, for example, inventory, stock, store, storage and warehouse. To avoid misunderstandings, we selected the most common expressions, defined the terms as precisely as necessary and used them consistently. The terms can be found via the index, which makes this book as well a dictionary of modern logistics.

Throughout the years, many clients and colleagues, scientists, students, readers, consultants, managers and other people have contributed to this book. We thank them all for listening, fruitful discussions, critical remarks and constructive proposals. Our sincere gratitude goes to *Hitesh K. Gadhia* in Copenhagen for proofreading and many helpful comments. We express our thanks to the *Kühne School of Logistics and Management* in Hamburg for transferring figures and tables into English, and to *Miebach Consulting* in Frankfurt for revising cost figures and technical data.

Most of all, we would like to thank our wives. Without their patience and understanding this book would not exist.

June, 2009
Hamburg
Copenhagen

Timm Gudehus
Herbert Kotzab

Contents

Introduction

The history of logistics goes far back. Since people began to cooperate, *operative logistics* has been practiced under different names: conveying, lifting and transport; buffering and storing; handling, packing and stacking; carrying, shipping and traveling. Cars, trucks, trains and ships were invented, cranes and handling devices were developed, and stores, silos and depots were erected. Roads, channels, railways and ports have been built all over the world.

The *logistic service providers* of the past were the *merchants* of Venice, Florence, London and the *Hanseatic League*, the *couriers* and *postal services*, the *carriers*, *shippers* and *forwarders* and the *operators* of stores and market places. More than 150 years ago, these logistic entrepreneurs already procured goods from all over the world, moved huge quantities around the globe and delivered letters over long distances on the next day.

Impetus and innovations of *modern logistics* – besides the new name – are the increasing multitude of technical solutions and the larger capacities, higher speeds, cheaper energy, improved performances and extended services. The most important development, however, is the growing *integration* of the single logistic activities. This was stimulated by new information and communication systems and by modern process control technique. Integrated systems and extended networks have been set up, which supply companies and consumers quick, cheap and reliable within broad variety of goods and services. Today logistics is the backbone of global trade.

Since mid 20th century, scientists have observed and described this development. They noticed the importance of *planning*, *organization* and *scheduling* for the *supply management* and named the discovered field of activities '*logistics*' (Baumgarten 1981, 1999; Kapoun 1981; Krampe 2000; Morgenstern 1955; Morphy and Woods 1996, 2004; Plowman 1962; Schulte 1995; Weise 1998).

Theoretical logistics evolved from the *science of war* (Jomini 1881), *engineering* (Andler 1929; Harris 1913; Maister 1976), *business planning*, and *economics* (Christopher 2003; Forrester 1961; Lambert 1997; Langford 1999; Pfohl 1990). It is pursued under different names: *operations research* (OR) (Churchman 1961; Ford and Fulkersen 1962; Domschke 1985, 1990, 1995; Gudehus 1975, 1976; Schneeweiß 1981), *material handling* (Arnold 1995, 2004; Bahke 1973; Jünemann 1963; Meller et al. 2004; Miebach 1971), *traffic theory* (Gudehus 1975; Harders 1968; Wehner 1970), *transport economy* (Aberle 2003; Ide 1991), *material*

Fig. 1 Historical development of transport and logistics

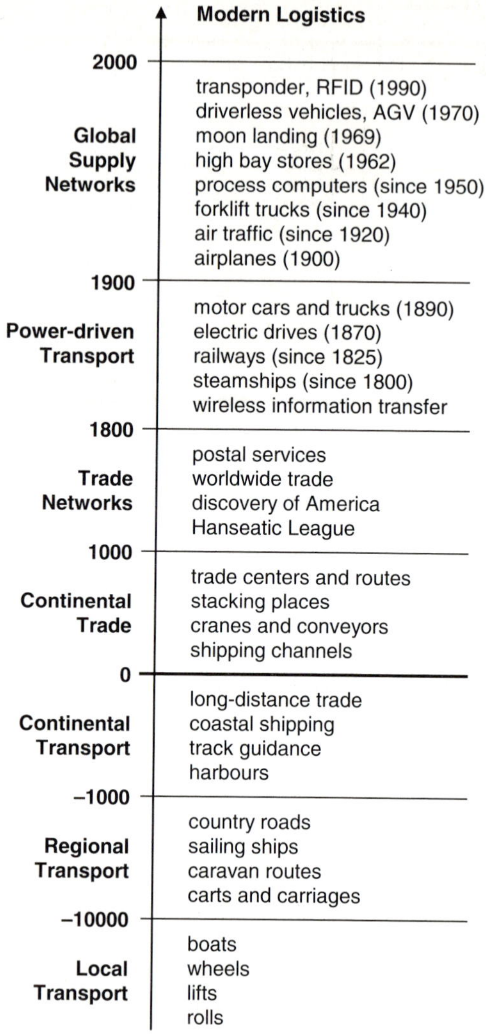

management (Soom 1976; Inderfurth 1999; Tempelmeier 1999), and *supply chain management* (Bretzke 2005, 2008; Cooper 1997; Kotzab 1999, 2001; Kuhn 2002; Schönsleben 1998; Simchi-Levi et al. 2008).

Up to now, logistic scientists primarily study the historically grown systems and current practices, describe techniques and options, and develop solutions for selected problems. However, in order to cope with fast changing requirements and to use innovations more efficiently, it is necessary to convert logistics from a descriptive *empirical science* into a normative *analytical science*. This book supports and promotes this change. It presents the *principles*, *strategies* and *operations* of

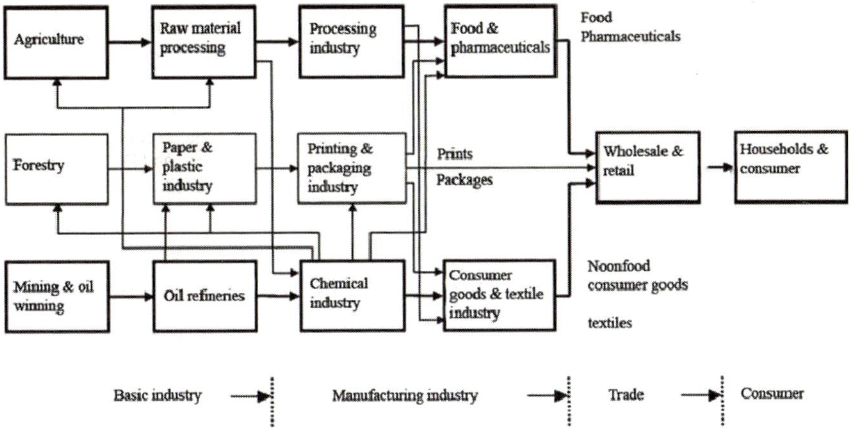

Fig. 2 Logistic networks and supply chains for consumables

logistics and develops the organizational, technical and economic options for the systematic planning of efficient logistic systems and for the solution of practical tasks.

The book starts with the tasks, objectives and opportunities of logistics. Along with the introduction into the *structures* and *processes* of logistic systems, the terminological foundations for *analytical logistics* are laid. Analytical logistics aims to develop general *rules* and *methods* for planning, scheduling and solution of specific tasks and *algorithms* for mathematical modeling and optimization of logistic processes and networks. Further topics are *strategies* and *decision rules* for the operation of logistic systems.

Many companies consider their logistic problems as unique and use company- and business-specific terms. Logisticians in theory and practice and of different industries prefer their own ideas and do not care much for the solutions of others. However, who analyzes the logistics of companies in different industries and countries, soon realizes that the same principles are valid everywhere and similar methods are applicable to solve logistic problems. Therefore, this book abstracts from specific industries, countries and technologies.

As the technical aspect or economic aspect alone veils the view of the whole and hampers many options of logistics, the common specialization on *technical logistics* and *business logistics* in science and teaching are detrimental. Technology, economics, engineering, informatics and other areas contribute to *logistics*, which is an *interdisciplinary science*. Hence, this book presents the organizational, technical and economic aspects of logistics with same the weight.

As for Operations Research (OR), the mathematical basis of logistics is *arithmetic*, *algebra* and *analysis*, in particular *probability theory*, *statistics*, *queuing theory* and *theory of graphs* (Biggs 1999; Gross 1998; Ferschl 1993; König 1936; Kreyssig 1975). The special OR-methods for solving cutting-, transport-, location-, sequencing- and other problems will be presented here as far as necessary

(Churchman 1961; Domschke 1985, 1990, 1995; Müller-Merbach 1970). The same holds for the basics of economics, engineering and informatics (Mankiw 2003; Samuelson 1995).

The principles, strategies and formulas described in this book were developed for business practice. They have been successfully applied in several projects to solve real problems. Although the problems presented here originate from practice, the theoretical solutions will be developed first. Possible applications are illustrated by examples and selected business cases afterwards.

The book presents options for action, offers methods for solution and optimization and supports decision making in logistics. It contains rules and tools for logistic planning and consulting, helps to avoid common errors and indicates the dangers of standard programs and methods. Results are *design rules*, *operating strategies* and *scheduling principles* as well as general formulas for the *computer-aided design*, *dimensioning, simulation* and *optimization* of logistic networks, systems and supply chains.

In the *first part* of the book the *principles, methods and strategies* of logistics are developed. It starts with a delimitation of the objectives and tasks, followed by the configuration, structure and organization of logistic processes and performance systems. Topics of the next chapters are *planning and realization*, *potential analysis* and *strategies*. The *economic basics* of logistics are presented in the chapters on *logistic costs* and *logistic pricing*. The chapter *time management* gives an analysis of the role of time for logistics and develops strategies for time-scheduling. This is followed by a chapter on *random processes* and *dynamic forecasting*. The *demand forecast* is starting point for *order scheduling* and *inventory management*, presented in the subsequent chapters.

Logistic units are the smallest elements passing through logistic chains. Their functions and determining factors are explained in a separate chapter, closing with the *logistic master data* required for order processing and process optimization. The subsequent chapter presents the *limit performance laws* and *queuing effects*. They are decisive for dimensioning systems and networks and for assessing capacities and performances. The last chapter of the first part outlines the relations and potential conflicts between *sales*, *purchasing* and *logistics*.

Topics of the *second part* of the book are the *systems, networks and operations* of logistics. It begins with a general introduction into *logistic networks* and their *subsystems*. The following chapters present the most important subsystems, which are the *storing systems*, *commissioning systems* and *transport systems*. They start with the definition of the functions and the specification of the performance requirements of the respective system. Then the elements, structures and processes of the systems are analyzed. The system analysis and a classification of the possible solutions lead to *selection rules* and *design options* to fulfill the system-specific requirements.

The next chapter shows how the optimal layout and dimensions of a *logistic hall* can be found and how the subsystems are optimally arranged. The dimensioning methods and arrangement strategies can be used in the *layout planning* of transshipment buildings, logistic centers and factory halls but also for allocating of buildings on a site. The resulting plants and logistic centers are the sources, nodes and sinks of the logistic networks of industry, retailers and logistic service providers. In a subsequent chapter the fundamentals of *production logistics* are outlined.

The central chapter of the second part presents methods, strategies and tools for the optimal design, selection and operation of *supply chains* and *logistic networks*. They are fundamental for *supply chain management* (SCM). The next chapter explains the advantages and disadvantages of *outsourcing* logistic performances and analyzes the peculiarities of *logistic service providers*. The last chapter discusses the role of *people in logistics*.

This book gives a comprehensive and consistent presentation of all relevant areas of modern logistics. Both parts and the single chapters of the book depend on each other and are linked by cross references. However, each chapter is understandable in it self and can be read separately.

Through the extensive *keyword index*, the book becomes a dictionary of modern logistics. In order to enable easy searching via the index, new terms and key words are written in *italics*. Bullet points (•) emphasize important definitions, strategies and statements. Indication arrows (►) mark general principles and fundamental rules.

Many *figures* and *tables* illustrate the text and support the understanding. The *data tables* with target values are useful for model calculations and as reference. To simplify programming, the formulas are written mostly in EXCEL-spread-sheet notation on one line with slanting fraction strokes.

Functional, safe and elegant bridges and buildings are the result of consequent application of the rules and laws of static and mechanics. Correspondingly, in logistics holds:

► Efficient, effective and competitive logistic networks and systems can be achieved only if the rules and laws of modern logistics are known and correctly applied.

The aim of this book is to present the knowledge and to establish the basics for this purpose. Further goals are to support the general understanding of logistics, to stimulate thoughts and ideas, and to offer new impulses for research and development.

Part I
Principles, Strategies and Organization

Chapter 1
Tasks and Aspects of Modern Logistics

Consumers and companies need products, material and other physical objects at a time when and a place where they are generally not produced. This leads to the *4R-task of operative logistics*:

- Logistics has to provide the ***r**equired quantities* of goods most efficiently at the ***r**ight place* in the ***r**ight order* at the ***r**ight time*.

The tasks of *analytical logistics* are to develop and organize optimal processes, structures, systems and networks for the operative logistics. The main tasks of *logistic management* are to execute the orders and to fulfill the requirements of consumers and companies at lowest costs with adequate quality.

Objects of logistics are physical goods such as raw materials, preliminary products, unfinished and finished goods, packages, parcels and containers or waste and discarded goods. Also, animals and even people can be logistic objects, which need special care and service.

The *sources,* i.e. the *suppliers* or *senders* of logistic objects, are plants, factories, storages and warehouses of *producers, wholesalers*, and *logistic service providers*. The final *sinks*, i.e. the *destinations* or *receivers* at the end of the *logistic chains*, are the department stores, markets and sales outlets of retailers and the *points of consumption*. Not only the final customers, but also their suppliers are *receivers* of goods and products from other, upstream sources. In addition, producers, retailers and consumers are sources of empties, waste, used-up goods and other materials that must be removed. These sources, sinks and intermediate stations, linked by transport elements, make up a *logistic network*.

Logistics in a narrow sense takes the locations of sources and sinks as fixed and the ordered, produced and consumed quantities as given. Under this aspect, logistics has to execute only the *basic functions* shown in *Fig. 1.1*:

transport to bridge space
handling to adjust quantities
storing to bridge time
commissioning to fill orders
(1.1)

In addition related services are executed.

T. Gudehus, H. Kotzab, *Comprehensive Logistics*,
DOI 10.1007/978-3-540-68652-1_1, © Springer-Verlag Berlin Heidelberg 2009

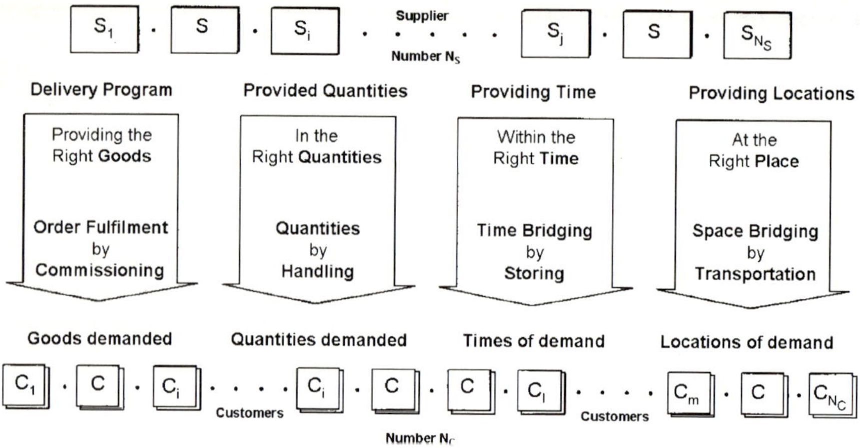

Fig. 1.1 Functions and tasks of operative logistics

The technology and intrinsic processes of mining, cultivation, production, manufacturing and bottling are not part of logistics. Logistics has to *supply* these processes with materials and parts, to *distribute* the produced goods and to *remove* the arising waste and residual material. This implies more than just transport, storing and handling. It means:

- Logistics has to manage physical goods in *space* and *time* in order to execute *orders*.

Logistic systems, which execute only the basic tasks (1.1), are special *performance systems*. Other performance systems, which execute also non-logistic tasks, are the subject of *logistics in a general sense*. In this sense, logistics has to design, set up, operate and optimize systems, which generate physical goods and immaterial services. These tasks overlap with *production planning*, *process technology*, *industrial engineering*, *operations research*, *informatics* and other fields of technology and economics.

Logistics in the broadest sense includes also *purchase* and *sales*. Sales representatives and purchasing managers negotiate the *terms of business* and the *prices* of goods and services. They initiate the supply processes and link the logistic chains between companies, customers and consumers.

Modern logistics is an interdisciplinary *applied science*. It uses the knowledge of other sciences for which logistics in turn is an *ancillary science*. This holds also for *informatics*, which deals with storing, processing, transferring and providing data and information. Although informatics plays an important role for the whole economy, it is for logistics, as for other disciplines, a mean to an end but not an end in itself.

In this chapter the elements and functions and the general processes, structures and organization of performance systems and logistic networks will be analyzed. In more detail, the functions and effects of *logistic centers* are explained. The last sections deal with *network management* and the *task of logisticians*.

1.1 Systems and Networks

A *system* is a set of *elements* connected by certain *relations*. *Networks* are systems, where the elements are stations or nodes and the relations are material or information flows between these elements. A *logistic system* is called *logistic network*, if the distances between the elements are far longer than the extension of the elements. This holds generally for *external logistic systems*. For *internal logistic systems* it depends on the distance of the observer.

The term *performance system* is a generalization of the term *machine system*. Therefore, many definitions and principles of the *theory of machine systems* and of *systems analysis* are transferable to performance and logistic systems (Bertalanffy 1968; Churchman 1970; Hubka 1973; Lenk and Ropohl 1978). A *machine system* executes *production orders*. It processes, transforms, handles and moves physical objects by a certain *technique*. Automatic machine systems operate *independent of persons* in a *deterministic* manner. They have only limited degrees of freedom. Examples of such machine systems are printing machines, chemical plants, and fully automatic assembly lines. Special logistic machine systems are *automatic high bay stores* (HBS), *sorter systems* and *automatic guided vehicle systems* (AGV).

Besides many analogies, there are fundamental differences between machine systems and performance systems. *Performance systems* depend on people and execute varying *customer orders* and *service orders*. Examples of *technical performance systems* are plants and factories but also logistic centers, traffic networks and company logistic networks.

Initiated by randomly arriving orders with different content, a performance system operates with *processes*, *throughputs* and *lead times*, which vary *stochastically* and are generally order-dependent. In order to cope with a varying demand, performance systems have several modes of operation and are partly de-centrally organized. While the *kinetic chains* of a machine system are defined by the *structure*, the *performance chains* of logistic systems and networks depend on *structure and strategies*. Therefore, a central task of *analytical logistics* is to develop strategies for the design and the operation of performance systems.

The functions of a performance system are determined by the requirements. Design, dimensioning and optimization of a system depend on the *customer* who specifies performances, outputs and objectives, defines the interfaces and frame conditions, and requires *quality* and *quantities*. For this purpose, the customer can choose between *result specification*, *functional specification* and *technical specification*:

- A *result specification* defines only the output and allows a variety of solutions. The methods, technology, structures and processes remain open.
- A *functional specification* determines processes and methods, thereby limiting the number of possibilities and solutions.
- A *technical specification* prescribes materials, elements and structure of the system in addition to methods and processes.

The goals and competencies of the customer and the type of the system determine the kind of the specification. For machine systems, a functional specification is supplemented by a technical specification of the critical elements. For internal logistic and performance systems, a result specification is combined with a functional specification of the critical processes. A pure result specification is adequate for external logistic systems.

1.2 Tasks and Objectives of Logistics

Every logistic task has certain aims, concerns a limited area and deals with defined aspects. The most general aspects of logistics – corresponding to *macroeconomics* and *microeconomics* (Mankiw 2003; Samuelson and Nordhaus 1998) – are *macrologistics* and *micrologistics* (Ihde 1991).

The aim of *macrologistics* is to ensure the efficient supply of consumers, companies and state with goods and to organize the traffic flows between sources and destinations within a region, a country and around the globe. This is aimed independent of the ownership of the goods, sources and sinks. In order to achieve the optimal economical development of a country, besides capable institutions and suitable laws an efficient *logistic infrastructure* is necessary.

The aim of *micrologistics* is to supply – based on private orders, agreements and contracts – companies and consumers with the required goods most efficiently and to cover the *mobility demand* of individuals. For this purpose, companies and logistic service providers plan, set up and operate logistic systems and networks. The task of micrologistics is to realize and operate logistic systems and to manage transport chains and supply networks in order to fulfill the expectations of customers and to ensure the optimal development of a company.

The main area of micrologistics is *company logistics*. As shown in *Fig. 1.2* it comprises internal logistics and external logistics. *Internal logistics*, also called *indoor logistics*, *material handling* or *Intralogistics*, connects the receiving docks, internal sinks and sources, and the shipping docks of the same site, which can be a logistic center, transshipment point, plant or market. *External logistics* or *Extralogistics* connects the shipping docks of one or several locations with the receiving docks of other locations.

Depending on the direction of the material flows, it is common to distinguish between *inbound logistics* or *procurement logistics*, *outbound logistics* or *distribution logistics* and *reverse logistics* or *disposal logistics*. *Procurement logistics* focuses on the supply of goods from the sources to the companies. *Distribution logistics* deals with the delivery of goods from the companies to the recipients.

Hence, procurement and distribution are two different aspects of the same task. Their objectives are determined either by the customer or by the supplier. From the customer's point of view, the supplier's distribution system is part of his own procurement system. Likewise, for the supplier the customer's procurement system is part of his distribution system.

Reverse logistics is the chronological reversion of supply. Tasks are to collect, transport, store and recycle or dispose production residues, consumer waste, packaging material, empties, depleted goods and worn out material. Special areas of

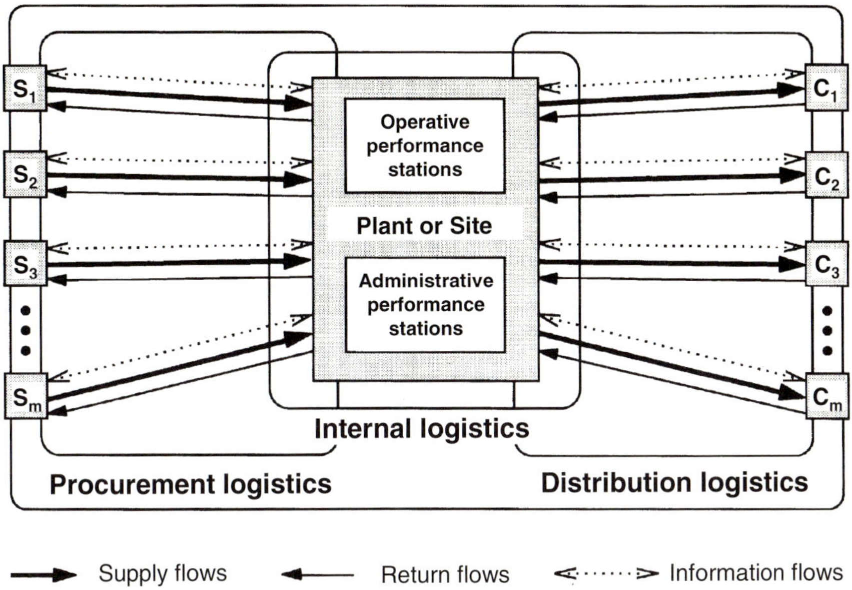

Fig. 1.2 Areas of company logistics
S_i: suppliers C_j: customers

reverse logistics are *garbage collection*, *waste management* and *empties logistics* (Dekker et al. 2004; Murphy 1996/2004).

Traffic networks and *transport systems* convey physical goods and people from departure points to destinations. The *nodes* or *stations* of a traffic network do not keep stocks. Their functions are junction and diversion respectively buffering, sorting, dispatching and other *transshipment services*.

The primary *objectives of company logistics* are:

performance
service quality (1.2)
cost efficiency

These are also the objectives for planning, realization and management of other performance systems. Their contents and priorities depend on the general goals of the company and the specific tasks of a project.

1.3 Structures and Processes

Performance systems are networks of stations generating certain services and performances. As sketched in *Fig. 1.3*, *material flows* and *data flows* pass through this network. Apart from the production and administrative processes within the stations, any performance system is a logistic system.

As in *hydrodynamics*, performance and logistic systems can be looked at from a stationary point of view under the *structural aspect* or from a dynamic point of

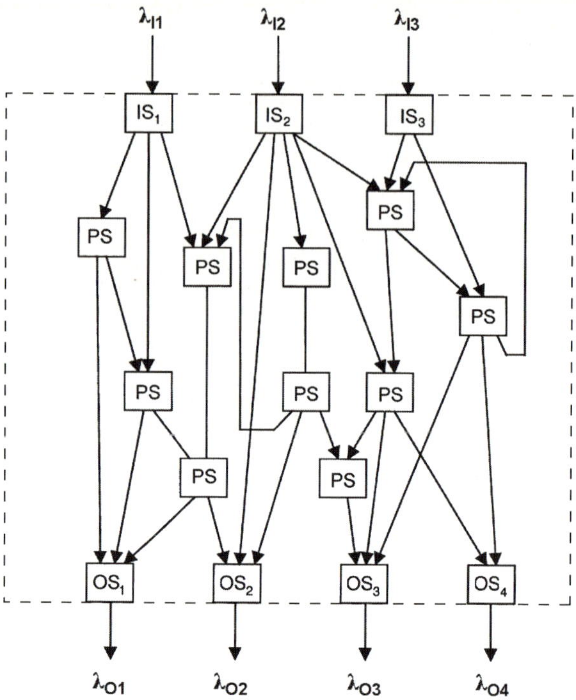

Fig. 1.3 General structure of a performance or logistic system

PS: performance stations
IS_i: incoming stations — OS_j: outgoing stations
λ_{Ii}: incoming flows — λ_{Oi}: outgoing flows
→ material or data flows — - - - border of the system

view under the *process aspect*. Some problems, such as process optimization, can be solved easier from the dynamic point of view. For other problems, such as design of systems or networks, the stationary aspect is opportune. For the tasks of modern logistics, both aspects must be considered. Hence, a logistician has to think in processes, structures and systems.

1.3.1 Structural Aspect

Under the structural aspect, a stationary observer analyzes structure, functions, capacities and performances of a system or network. From this point of view, the task of logistics is *system optimization* (Gudehus 1975/II; Lenk and Ropohl 1978):

- The logistic system or network has to be designed, dimensioned, organized and operated in order to fulfill the requirements most efficiently under specified restrictions.

The first step of system optimization is a *structure analysis* or *potential analysis*. By this analysis the stations and the configuration of the system are scrutinized in order to find out how well they execute the orders and how far they operate efficiently (see

Sect. 4.4). The assessment includes the material flows and data flows between the stations.

However, looking at a logistic system only from a stationary point of view involves the danger to lose sight of the processes. This is avoided by the *Chandler-principle* (Chandler 1962):

- Structure follows processes.

That means, processes and strategies should be developed before the structure is designed. However, this is only partly feasible, since the processes are depend on the structure of the system.

1.3.2 Process Aspect

Under the process aspect, an observer follows the flows of goods and data on their way through the system. The observer scrutinizes the sequences of the activities in the logistic chains and the time consumption in the stations. The task of logistics from a dynamic point of view is *process optimization*:

- Out of a variety of possibilities, the most efficient processes and logistic chains meeting the requirements must be selected, designed and combined.

The first step is a *process analysis* aimed to recognize the efficiency of the stations in the performance chains (see *Sect. 4.3*). This analysis assesses whether the system meets the objectives of the customers, the specifications of the orders and the expectations of the recipients.

The basic *principle for process design* is:

- Only if all relevant processes in a system or network are known, it is possible to dimension the stations, to fix the connections, to calculate the costs and to reach the overall optimum.

A purely process-oriented approach often ignores *competing processes* and misses possible *synergies*.

1.3.3 Dynamic Network Aspect

Under the influence of *operations research*, theoretical logistics has dealt for a long time primarily with the optimization of *structures* and *networks* for *stationary flows* (Bucklin 1966; Churchman et al. 1957; Daganzo 1999; Domschke 1985, 1995; Domschke and Drexel 1995; Ford and Fulkerson 1962; Müller-Merbach 1970). The stochastics and dynamics of the flows in the supply chains have been widely disregarded.

Supply chain management (SCM) focuses on the processes in the logistic chains from the suppliers of the suppliers, through the company to the customers of the customers (Christopher 1992; Cooper et al. 1997; Kuhn and Hellingrath 2002; Murphy and Wood 2004; Schönsleben 1998; Scott et al. 1991; Stadtler 2005). The pure process aspect of SCM, however, neglects the structures and the interactions between competing supply chains. SCM therefore often misses the synergies resulting from the multi-use of networks and resources (Bretzke 2008).

The performance requirements are generally *stochastic* and *vary in time*. Therefore, the central task of *company logistics* and *network management* is to design and to optimize processes *and* structures in order to obtain the optimal system for a stochastically fluctuating and dynamically changing demand (see *Sects. 1.9* and *2.9*). To support network management in this task, analytical-normative logistics has to develop *strategies for* planning, scheduling and operation of *dynamic systems*.

1.4 Elementary and Compounded Performance Stations

Performance systems as well as logistic systems consist of *elementary performance stations*. They are connected and arranged into *compounded stations*, *function modules*, *production plants*, *logistic centers* or *organizational units*.

A general performance station as shown in *Fig. 1.4* is defined as follows:

- A performance station produces tangible and intangible goods due to orders or commands, using material and resources, such as persons, areas, buildings, machines and equipment.

The basic task of economic performance stations is to generate useful results and to create value at lowest costs. For *controlling*, all performance stations are *cost centers*, but not all cost centers defined by controlling are performance stations.

Several performance stations located in the same site can be organized as a *function module*, *profit center* or another kind of *performance area* as shown in *Fig. 1.5*. Performance *areas* that produce similar goods or cover a specific part of a performance chain can be delimited and organized as an *organizational unit*.

Organizational units are plants, companies and business units where a *manager* or a *third party* is responsible for output, performance, quality and costs (see

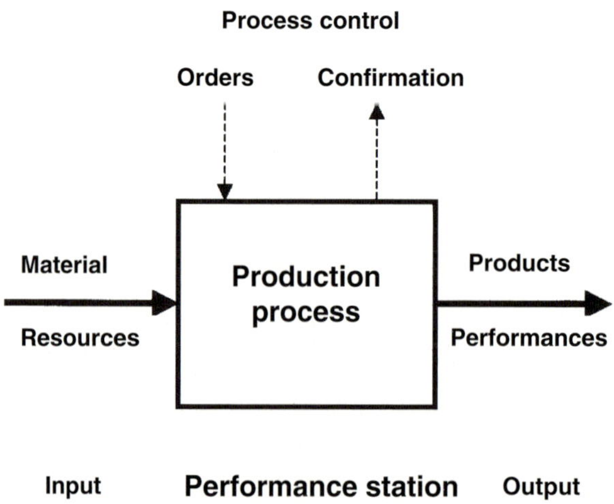

Fig. 1.4 Input and output of a performance station

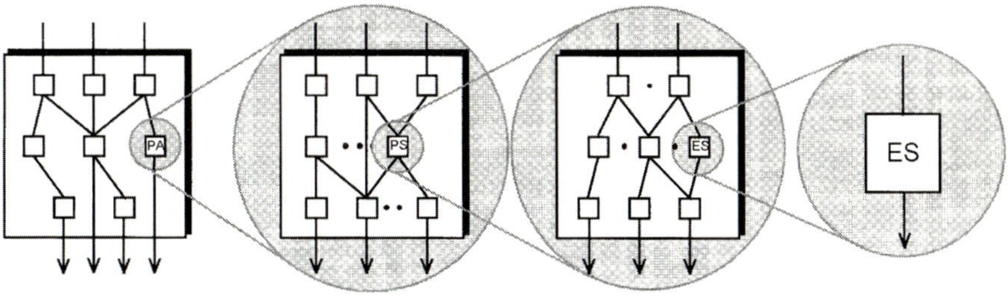

Fig. 1.5 Bundling and unbundling of stations and organization units

Chap. 22). In order to *call for tender* and to contract a *third party provider*, appropriate performance stations must be selected and delimited in an organizational unit, which generates well specified services (see *Chaps. 7* and *22*).

The kind of the output, e.g. the properties of the products or the transport length and storing time, defines the *performance type*. Additional requirements are delivery times, storing conditions, safety regulations and quality requirements. For systems analysis and design, it is advisable to classify the performance stations corresponding to function, output or other features into different categories.

1.4.1 Output of Performance Stations

The output of a performance process can be tangible or intangible:

- *Tangible outputs* are physical objects such as material, buildings, industrial products, consumer goods or in general *products* that result from an extraction, production, manufacture, refinement, machining, assembling or filling process.
- *Intangible outputs* are the results of a transformation of space, time or information of physical and/or immaterial objects, such as rearranging, stacking, packing, coding, handling, transport or storing.

A process with a tangible output is called *production* or *manufacturing process*. Processes with intangible output are called *performance* or *service process*. However, the differentiation between a production process and a performance process depends only on the point of view or on the ownership of the goods. Refinement, assembling, bottling and packaging are considered part of the production process as long as these processes occur within the same company with own material. The same processes are called performance processes, or e.g. in textile industry *passive subcontracting*, if a third party executes them with external material.

From the process aspect, the distinction between production process and performance process is not of much use, as the physical subject of the performance process carries with it the results of the consecutive process steps.

Output and *throughput* of performance stations are measured in *performance units*. Performance units for tangible outputs are the *measure units* for *weight* [kg; t], *length* [m, km], *area* [m^2] and *volume* [*l*, m^3] and the *number of pieces* or *load units* [LU]. Measures for intangible outputs are either the measures of the tangible goods undergoing the operation, or *activity units* such as *orders*, *positions* or defined scopes of services.

Special measure units for the basic *logistic performances* (1.1) are:

- *Transport performance units*: ton-kilometer [t-km], load-unit-distance [LU-km], pallet-kilometer [pal-km], passenger-kilometer [pass-km]
- *Handling performance units*: handling-unit [HU], load-unit [LU]; storage-unit [SU], handled item, piece
- *Storage performance units*: load-unit-time [LU-day], pallet-day [pal-d], square meter-month [m^2-month], car-parking-hour [car-hour]
- *Commissioning performance units*: order [Ord], position [Pos], pick unit [Pick], collection unit [CU = box, pallet, container, case]

1.4.2 Types of Performance Stations

The capability of a performance station is determined by its *limit performances*, i.e. by its maximal input, output and/or throughput. They depend on the number of functions a station can execute. Depending on the number of functions, monofunctional and multifunctional stations can be differentiated:

- *Monofunctional performance stations* can only execute one of process kind.
- *Multifunctional performance stations* can execute several different processes in *parallel* or in *sequence*.

The output of a performance station is either of *direct use* for the business, as for instance the result of an assembling line, or of *indirect use*, such as the services of a repair and maintenance station. *Internal performance stations* are located within the operating sites, building, or plants belonging to the company. *External performance stations* are located outside of the operating site or operated by a *service provider*. Depending on the objectives and required level of detail, it is either necessary to look at the elementary stations, or more advantageous, to consider combined stations (see *Fig. 1.5*):

- *Elementary* or *irreducible stations* cannot be separated without loosing their function. At one time they can execute only one specific process.
- *Combined* or *compounded stations* consist of parallel or serially arranged elementary stations. They can execute different processes simultaneously.

Another differentiation of performance stations results from the objects of the process:

- In *operative stations* physical objects are produced, processed, transformed, stored, moved or handled.
- In *administrative stations* orders, data or information are produced, processed, stored or transmitted.

In some cases an operative station executes also administrative tasks with order documents and accompanying information.

In a *transformation station* the incoming objects such as material, goods, parts or semi-finished products are transformed into other products and loose their identity, e.g.:

- *Converting stations* convert raw material in a chemical, physical or technical process into synthetic materials.
- *Manufacturing stations* form, combine and connect input material and parts into modules, semi finished or finished goods.
- *Assembling stations* erect and assemble vehicles, machines, constructions or buildings from parts, components and modules.
- *Filling stations* bottle or fill goods into bottles, cans, bags, containers or packages and generate *article units* or *primary package units*.
- *Packaging stations* are filling packages, parcels or boxes with article or logistic units and generate *secondary* or *tertiary package units* or *load units*.
- *Dismantling centers* take apart and sort discarded products for a further recycling process.

Service stations are operative performance stations where the incoming objects keep their identity, e.g.:

- *Control stations* identify, check, code and control physical objects without further change.
- *Treatment stations* execute any kind of service on material objects without changing them physically.
- *Repair stations* fix and mend defective products, transport means or used equipment.
- *Logistic stations* handle, sort, buffer, store and move physical objects, goods, products or load units without changing their properties.

Logistic service stations and pure logistic networks execute only logistic functions.

1.4.3 Key Figures of Performance Stations

A performance station can be characterized by the following key figures:

Services	types of orders service characteristics SC_i functions F_α transformation processes
Objects	properties of the incoming and outgoing physical objects type of incoming and outgoing data and information
Times	operating times, running times, working hours, processing and throughput times (1.3)

Static Limit performances	buffer and storage capacities for tangible objects memory capacities for data and information
Dynamic Limit Performances	limit performance of production μ_P limit performance of throughput μ_{ij}
Resources	areas and space equipment, machines and facilities conveyors and transport means personnel
Relations	location of the station operational and organizational relations links and interfaces to other stations

The *partial utilization* ρ_{ij} [%] of a performance station is determined by the *partial input, output* and *throughput rates* λ_{ij} [PU/TU], measured in *performance units* PU per *time unit* TU, related to the respective *partial limit performances* μ_{ij} [PU/TU], which are determined by the *cycle times* τ_{ij} [TU/PU] and *throughput times* T_{ij} [TU/PU] (see also *Fig. 1.6*). If operating times, machine running times and working hours are known and the scheduling and operating strategies are given, the required personnel, the number of equipment and the number of transport means can be calculated from the above key figures of the station by the formulas and algorithms which will be derived later in this book.

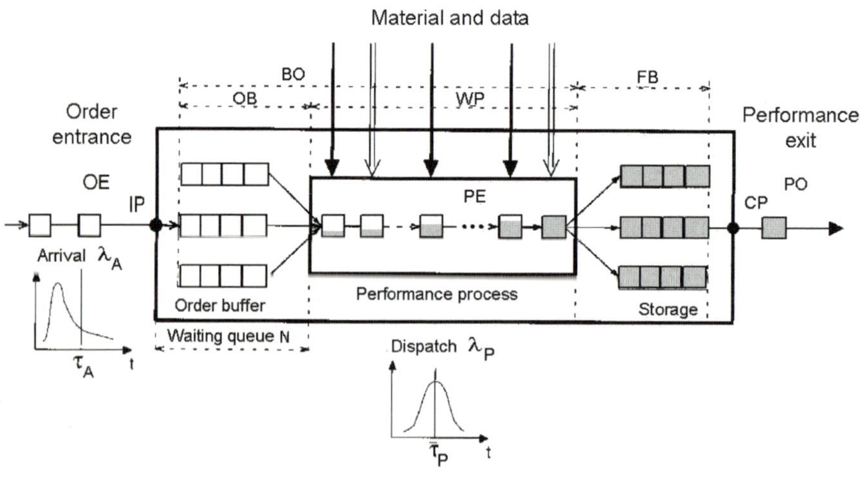

Fig. 1.6 Performance process and key figures of a performance station

IP: identification point CP: control point

$\lambda_A = 1/\tau_A$: input rate = 1/mean arrival time length

$\mu_P = 1/\tau_P$: limit performance = 1/minimal average cycle time

$\lambda_O = 1/\tau_O$: output rate = 1/mean leaving time length

OE: order entry PE: performance execution PO: performance output

OB: order buffer WP: work in progress FB: finished goods buffer

BO = OB + WB: back orders

1.5 Structures of Logistic Networks

Logistic systems and networks are performance stations linked by connections which transfer goods and information flows (see *Fig. 1.3*). The operative stations process, buffer and handle incoming flows of goods and convert them into outgoing flows. The administrative stations generate and process information and data that initiate or accompany the material flows of goods through the network and within the operative stations.

Storage systems, *commissioning systems* and *handling terminals* are special logistic systems that execute only one or two of the basic logistic functions (1.1) (see *Chaps. 16, 17* and *18*). *Transport systems* and *traffic networks* bridge spatial distances. They consist of *transport connections* and *transport nodes*. A transport connection links a departure station directly with an arrival station. By a transport node incoming transport flows are joined and/or branched into outgoing flows (see *Sect. 13.2.6*).

The number of intermediate stations determines the *stage degree* of a logistic chain:

- An *N-stage logistic chain* consists of N transport sections connected by N-1 *intermediate stations*.

In *Operations Research*, the stages of a chain or network are called *echelons*. In the *automotive industry* they are the upstream and downstream *tiers* as shown in *Fig. 1.15*.

The structure of a logistic network is defined by the *structure parameters*:

- number, locations and functions of the *sources* or *delivery points*
- number, positions, locations and functions of *intermediate stations*
- number, locations and functions of the *sinks* or *receiving points*

Intermediate stations can be *transport nodes*, *transshipment points* (TSP), *stores* or multifunctional *logistic centers*.

Some of the structure parameters, such as the locations of suppliers and customers, are *fixed points* which cannot be changed in short time. Other parameters, e.g. the number and locations of intermediate stations, are variable and can be used as *design parameters*. For known performance requirements and restrictions, it is possible to optimize a logistic system or network by varying these parameters.

A further option for extended networks is to connect sources and sinks with high exchange flows directly, while letting weaker flows of goods run over two, three or more stages. This procedure results in a *hybrid network* (Bretzke 2008) with *mixed structure*, which is a *superposition* of systems with different stage degree (see *Figs. 1.15*, *21.22* and *21.24*)

In order to explain the structural possibilities and to discuss the basic characteristics of logistic networks, it is useful to investigate networks with only one stage degree. In practice, generally hybrid networks with mixed stage degree must be considered (see *Chap. 21*).

1.5.1 Single-Stage Networks

As shown in *Fig. 1.7* a single-stage transport network has only direct connections by *unbroken transports* between sources and sinks.

As long as the transport means are fully utilized by the consignments to be transported from a source to a sink, a *non-stop delivery* by a *one destination transport* is most efficient.

If a source supplies several sinks in a surrounding *service area* with smaller consignments, it is opportune to execute *mixed destination transports* on longer *delivery routes*. Smaller consignments from many sources in a surrounding area are collected in a *mixed source transport*, also called *milk run* (see *Fig. 21.5*).

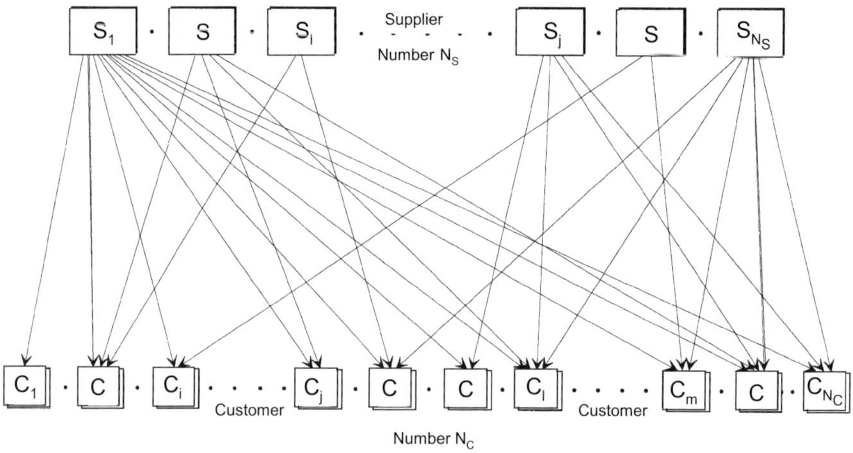

Fig. 1.7 Single-stage transport network for direct delivery
S_i: supplier C_j: customer

1.5.2 Two-Stage Networks

In a two-stage network, sources and sinks are separated by one intermediate station. That can be a stockless *transshipment point*, a *store* or a *logistic center*.

A two-stage transport network is opportune, if a small number of delivery points has to be supplied from many far-off sources, provided direct transports are not efficient. The transshipment points then are located as *collection stations* in the centers of the *source areas*.

If only if a small number of sources has to supply many, widely spread and far-off recipients, it is opportune, if direct transports are not efficient, to convey the goods via transshipment points that serve as *distribution stations* and are located in the center of the *destination area*.

1.5.3 Three-Stage Networks

When supplying many recipients from many far distant sources with small quantities, a three-stage network structure can be opportune. The connections between the sources and sinks in a three-stage network are twice broken. This is possible by passing through two consecutive transshipment points of a *freight network* as shown in *Fig. 1.8* , or through a *logistic center* followed by a transshipment point as shown in *Fig. 1.9*.

The collection stations are located in the areas of the sources, the distribution stations in the areas of the recipients. In a transshipment point as well as in a logistic

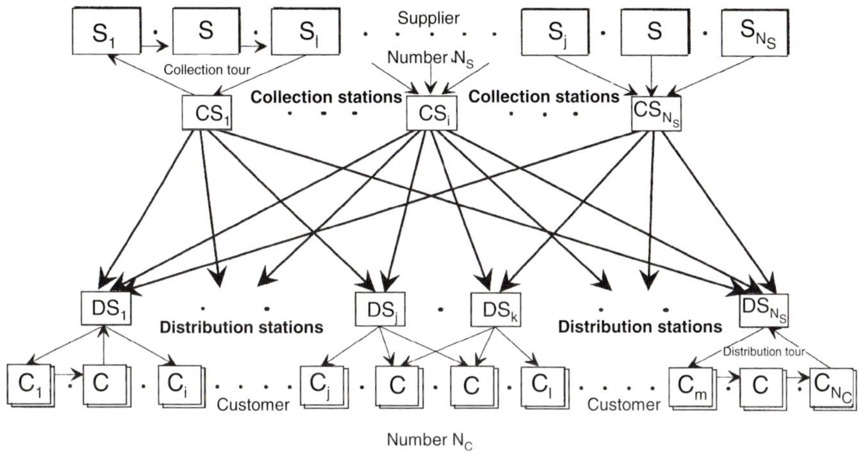

Fig. 1.8 Three-stage freight network with transshipment points for collection and distribution

S_i: suppliers C_j: customers CS_i: collection stations DS_k: distribution stations

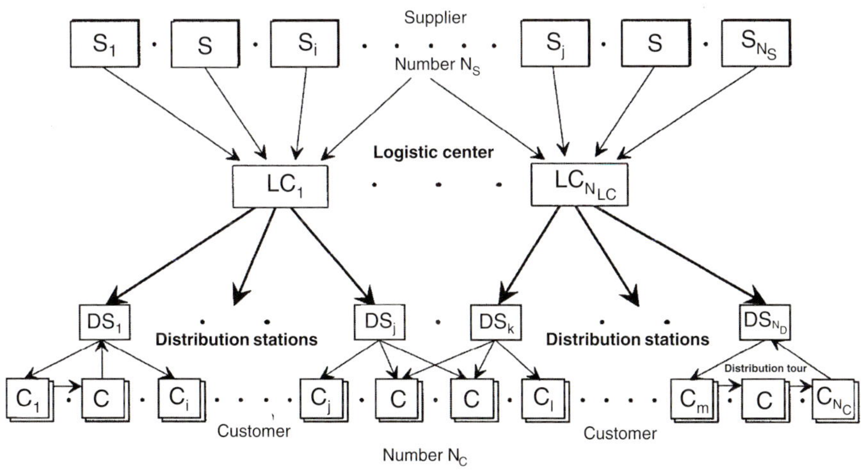

Fig. 1.9 Three-stage logistic network with logistic centers and distribution stations

LC_n: logistic centers

center, the handling of the goods for freight consolidation can be organized as *cross-docking* of complete load units without sorting or as *transshipment* which implies sorting of single packages. *Main leg transports* bridge the long distances between the collection and distribution stations.

The transport costs in a three-stage freight network can be considerably reduced by load consolidating in the transshipment points, selection of most economic transport means, combining outbound and return freights in the *main carriages,* and by optimal collection and delivery tours.

A further possibility to minimize logistic costs and to improve the logistic service is to allocate *additional services* in the *transshipment centers*. In a partly centralized organization, the functions of several storekeeping stations are bundled as shown in *Fig. 1.9* into one or a few *logistic centers* without increasing the number of stages.

1.5.4 Multi-Stage Networks

In a multi-stage network, the connection between sources and sinks is broken more than twice. Four-stage logistic networks as shown in *Fig. 1.10* result if one or more *logistic centers* with central stocks and additional services are inserted between the collection and distribution stations. Multi-stage *global freight networks* emerge in the *multimodal transport* when using air or sea transport for the long distances, rail transport for medium distances and road transport for local pickups and deliveries (see *Figs. 21.4* and *21.21*).

In most cases, the logistic network of a company is a *superposition* of networks with different stage-degree. Logistic chains with different stages and deviating distances, throughput time and performance costs connect the sources and sinks. From these, the *scheduler* has to select the optimal chain that fulfills the required delivery times at lowest costs.

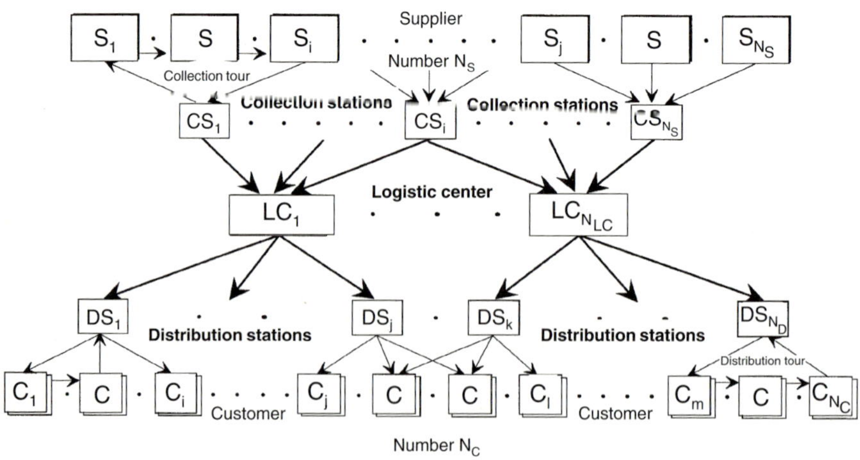

Fig. 1.10 Four-stage network with collection and distribution stations and several logistic centers

1.6 Functions of Logistic Centers

In a *logistic center* several logistic functions are bundled as shown in *Fig. 1.11*. It offers a wide range of services and consolidates procurement and distribution flows in order to reduce handling and transport costs.

An *open logistic center* consists of several buildings or handling places surrounded by traffic areas and directly connected to external roads, railway, waterways or airports. Examples for open logistic centers are railway stations, airports, harbors and seaports comprising the sites of several logistic service providers. Other examples are *city logistic centers* located at the periphery of metropolitan areas and big cities. They are a combined source and sink for *city logistics* that aims to consolidate freights into and out of the service areas (Taniguchi 2001).

A *confined logistic center* is made up of performance stations and functional modules located in a separate building or site. Generally, confined logistic centers are connected directly to the external road network. In some cases they have a link also to the railway net, a waterway or an airport.

A confined logistic center can be the operating site of a manufacturer, wholesaler or retailer or of a logistic service provider. Examples are *consolidation centers* (CC), *distribution centers* (DC) and *shipping centers* (SC), *local warehouses* (LWH) and *central warehouses* (CWH), *transshipment points* (TSP), *regional distribution centers* (RDC) or *freight terminals* (FT).
Most logistic centers can execute the *basic operative logistic functions*

transfer of goods for many suppliers and customers
storekeeping for one or more suppliers (1.4)
commissioning for many customers

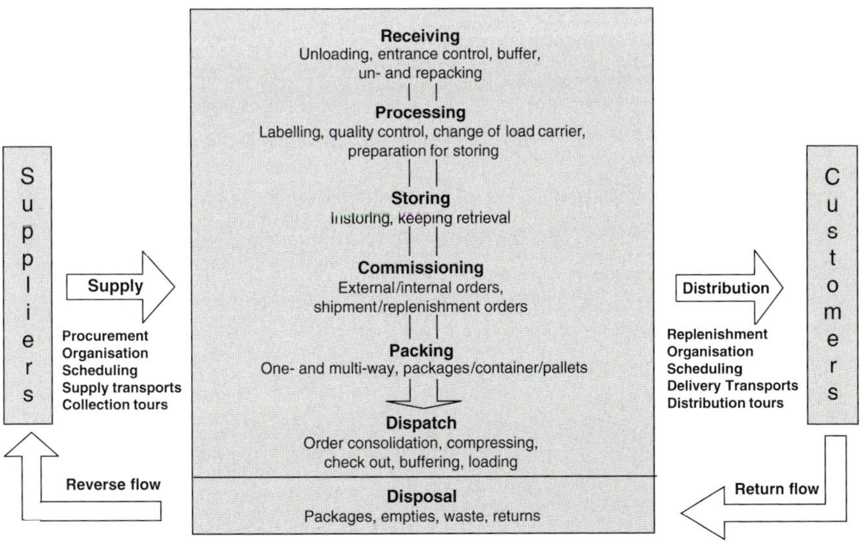

Fig. 1.11 Functions of a logistic center

For this purpose, they have the *standard functional areas*

receiving area
storage systems
commissioning systems
internal transport systems (1.5)
sorting systems
dispatch area

Many logistic centers offer *additional services* or *added values*, such as (Stock and Lambert 2001; Murphy 2002):

quality control
bottling and packaging
wrapping and unwrapping
building up and braking down load units
cutting and weighting (1.6)
assembling
repair and maintenance
handling of returns and reclamation
handling of empties

These services are executed in special functional areas such as *quality control, return reception, reclamation* or *repair and maintenance stations*.

Apart from the operative functions and services, logistic centers generally execute *administrative services* such as

scheduling of external transports and freight
inventory management and stock controlling
customs clearance
invoicing (1.7)
order scheduling
data processing

The operative areas (1.5) of a logistic center with the standard functions (1.4) and the different *internal logistic chains* are shown in *Fig. 1.12*. A logistic center with a wide scope of *added values* such as (1.6) and (1.7) is also called *competence center*.

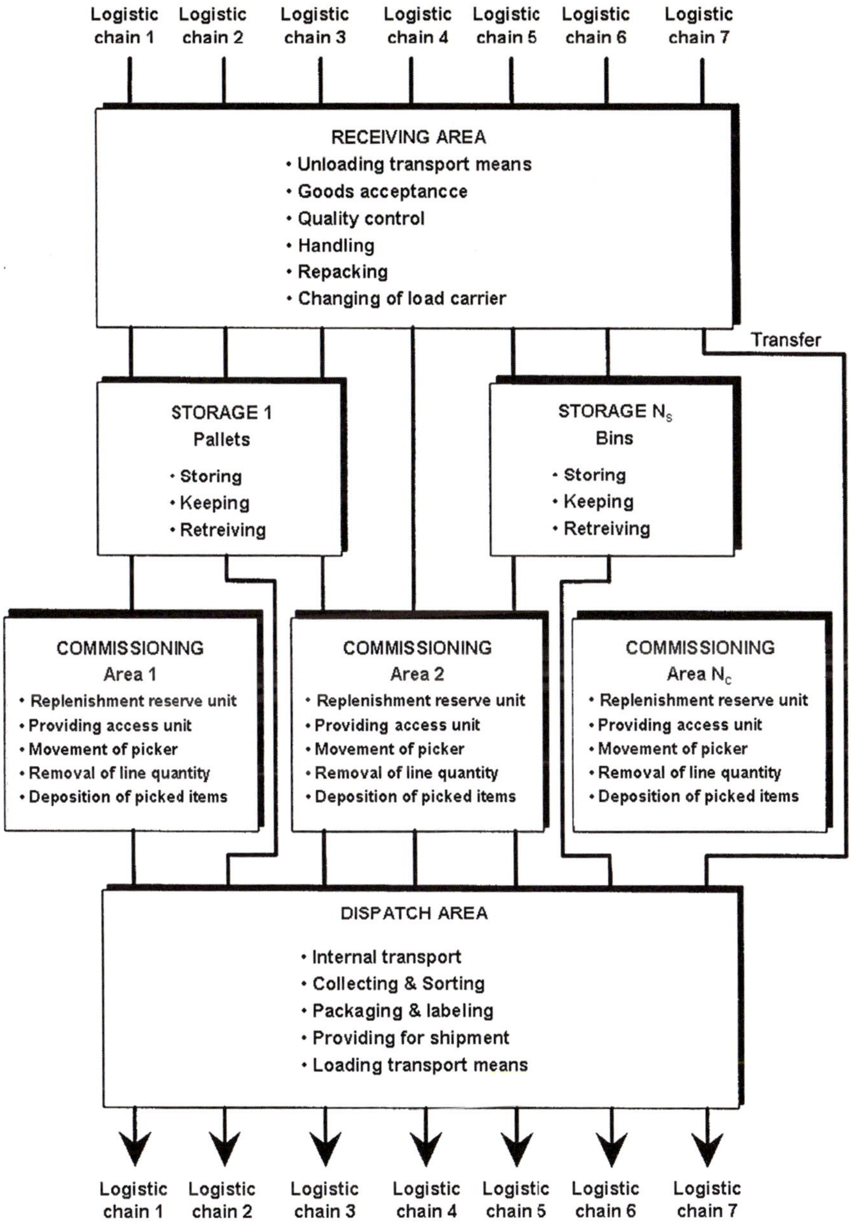

Fig. 1.12 Operative performance areas and internal logistic chains of a logistic center

1.7 Process Chains and Logistic Chains

A chronological sequence of operations executed in a spatial chain of organization units and performance stations, which results in a product or service of certain value, is called *process chain*, *performance chain* or *value creation chain*. Depending on whether the operations take place in operative or administrative stations and whether they refer to tangible or intangible objects, a performance chain is a *logistic chain*, *information chain* or *order chain* (see *Fig. 1.13*):

- A *logistic chain* is a sequence of operative stations passed by material objects. In- and outgoing objects of logistic chains are material, goods or load units that change in time, space and sequence during the process. The flow of goods through a logistic chain is called *material flow*.
- An *information chain* is a sequence of stations passed by data or information. The in- and outgoing objects of an information chain are intangible. The flow through an information chain is called *information flow* or *data flow*.

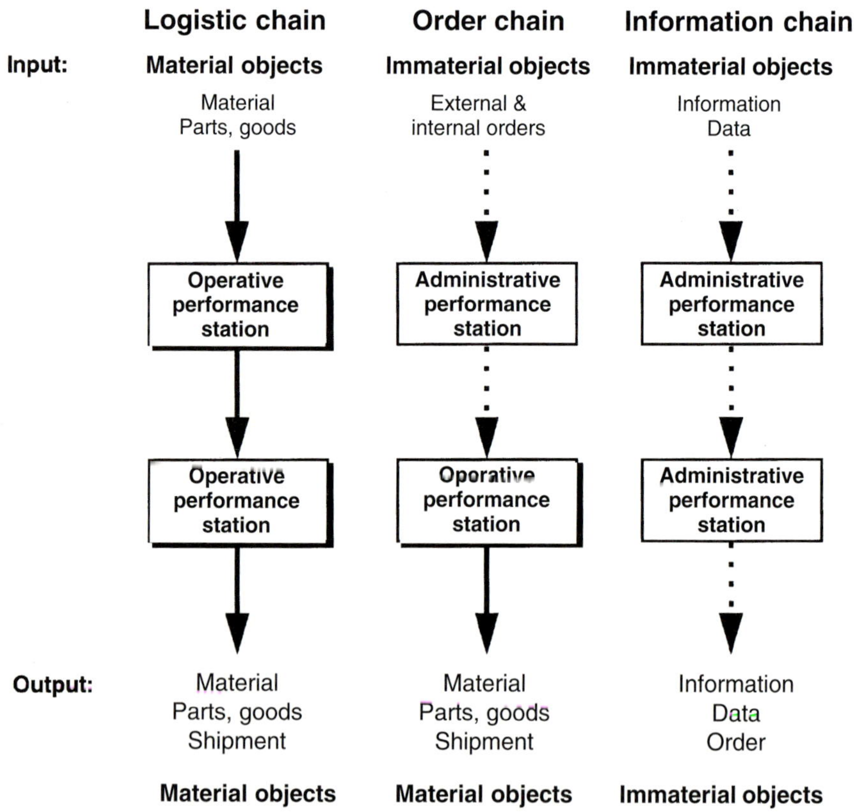

Fig. 1.13 Logistic chains, information chains and order chains
→ material flows ---> data flows

- An *order chain* is a sequence of administrative and operative stations passed through first by orders and later by the generated objects. The administrative stations accept, process and control the orders. The following operative stations execute the production or service orders.

The incoming objects of an order chain are immaterial orders. The outgoing objects are either tangible, such as products or load units, or immaterial such as services or information (see *Fig. 3.7*). A logistic chain reflects the complete delivery process from a supplier to a customer. The order chain describes the order process from the customer and back to the customer. In many cases the logistic chain is triggered by an order chain and goes along with an information chain. Information chains and logistic chains meet at so-called *identification points* (IP) and *control points* (CP) (see *Sect. 2.6*).

In many cases, for the execution of the same order, several order chains are possible. It is the task of supply chain management to identify, depending on the kind and content of the orders, the most favorable process chains and to design them according to the requirements.

A *complete performance chain* includes all performance stations between a source and a sink. It can be split up into external and internal logistic chains:

- *External logistic chains* consist of stations outside a company and of transport connections between shipping locations, transshipment points, logistic centers and destinations.
- *Internal logistic chains* consist of stations and connections inside a company, station, site, transshipment point or logistic center.

Design, set-up and optimization of external logistic chains will be outlined in *Chap. 21*. *Internal logistic chains* are the possible process paths of physical goods from the entrances through the operative stations (1.5) to the exits of a plant or site. For a logistic center, they are shown in *Fig. 1.12*. Internal logistic chains start in the *goods receiving area* with the following handling and administrative activities:

unloading of arriving vehicles
receiving of goods
quality control (1.8)
unpacking and repacking
building up storage units

For *transfer goods* and *transit units* that are not kept on stock but only transshipped, the processes in the receiving area are followed by a *direct transport* via a *bypass* to the dispatch area. This transfer of arriving goods on unchanged pallets is called *crossdocking* (see *Sect. 21.1.3*).

If the incoming goods are kept on stock, the processes in the receiving area are followed by a transport to the storage area, where the three *steps of the storage process* are executed (see *Chap. 16*):

in-storing of the storage units
keeping the storage unit on a store place (1.9)
retrieval of the storage unit

If not only full loads units but also *single items*, *mixed load units* and *sorted load units* are required, the storage process is followed by *sorting* and *order picking* which is called also *commissioning*. The *steps of the commissioning process* are (see *Chap. 17*):

replenishment of reserve units
providing of access units
movement of the order picker (1.10)
gripping of the order line quantity
deposition of the picked items

Internal logistic chains end with the transport to the *dispatch area* where the following activities are executed:

sorting and consolidating
packing and labeling
compressing and sealing
preparing for shipment (1.11)
final control and checkout
transport to the loading dock
loading of the waiting vehicles

In the stations of a logistic center the above activities are executed in parallel or successively. The single activities, the allocation to the different stations, the combination of these stations and their position within the logistic center differ from case to case.

If for the different *article clusters* several parallel storage and commissioning systems exist as shown in *Fig. 1.12*, for some articles more than only one internal logistic chain is possible. To find out the most efficient internal logistic chain, appropriate *selection strategies* are needed.

1.8 Effects of Logistic Centers

The total *logistic costs* for delivering goods via a logistic center are caused by the following *partial costs*:

- *transport costs* for the shipment from the sources to the center
- *interest costs* for the inventory capital
- *performance costs* for the functions within the logistic center
- *distribution costs* for the delivery from the center to the recipients

These partial costs depend on the degree of consolidation of procurement, inventories, functions and distribution. A further factor of influence is the *number of logistic centers* between delivery points and receiving locations.

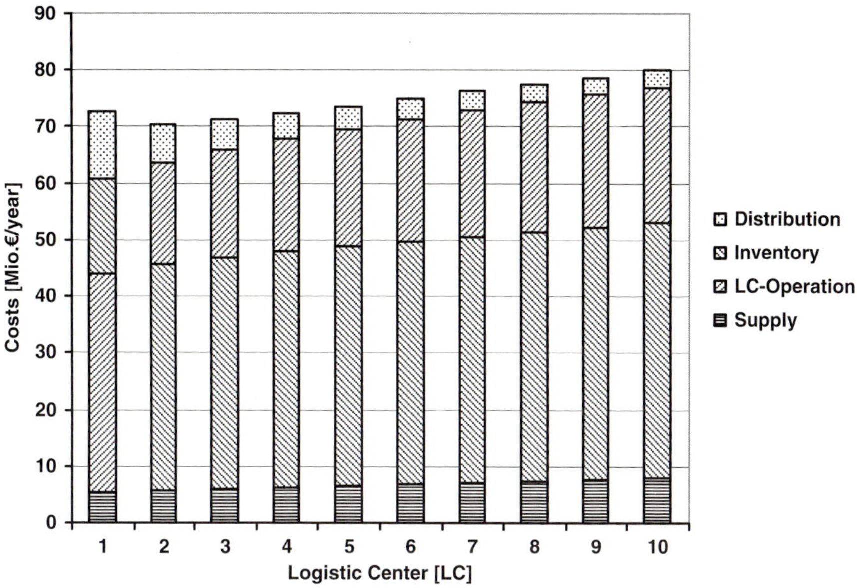

Fig. 1.14 Dependence of the total logistic costs on the number of logistic centers
procurement network of a German retailer with 250 outlets
network structure see *Fig. 1.9*, supply chains see *Fig. 21.23*

The partly opposite influences of the partial costs lead to the *optimization rule*:

- For any given set of requirements and restrictions, exists an *optimal number of logistic centers*.

One example is presented in *Fig. 1.14*. It shows the optimization of the *procurement network* of a *department store retailer* with a structure as outlined in *Fig. 1.9*. The total costs first decrease with decreasing number of logistic centers, until a flat minimum is reached for the *optimal number of logistic centers*. For the example the optimal number is 2. The total costs for only one logistic center increase as the distribution costs increase extremely. In this case, it was possible to reduce the total logistic costs by about 12% and to improve service level and logistic quality by consolidating the functions, stocks and flows of goods from formerly 10 regional warehouses in 2 logistic centers.

1.8.1 Procurement Consolidation

The consolidated shipment of many small procurement orders in large load quantities via a logistic center reduces the costs for the suppliers. It simplifies scheduling and improves the utilization of production facilities. The costs for order processing, operations, storing, commissioning and dispatch also decrease.

By procurement consolidation, the total logistic costs can be reduced in a range from 2 to 5% of the purchase prices, depending on the type and value of the goods and on the depth of manufacturing. The negotiation of more favorable terms of delivery, such as quantity discounts, just-in-time-delivery or the use of standard load units reduces the costs further and contributes to improving the competitiveness of all participants in the supply chains.

1.8.2 Consolidation of Inbound Flows

The consolidation of many smaller shipments from one supplier to many customers into a few shipments to one logistic center reduces the number of inbound transports without changing the transport frequency. At the same time the quantity per shipment increases.

The utilization of standard load units such as pallets or containers is improved by higher filling rates. Also, the utilization of the transport means is improved. Vehicles with larger capacity, such as semi-trailers, trailer trains, swap trailers or wagons, and transport networks with lower specific costs, such as railways, can be used. In addition, efficient loading techniques reduce costs.

The mean travel distance of the inbound flows of goods can be minimized by locating the logistic center in the *transport gravity center* of the shipment points (see *Sect. 18.10*). Compared with the mean travel distance of direct deliveries, the travel distance of consolidated inbound flows can either be reduced or at least kept constant. From this follows the *rule*:

- The costs of the inbound flow of small shipments from many sources can be reduced by consolidation via a logistic center.

As shown in *Fig. 1.14*, the costs for the inbound flows cause between 5 and 10% of the total logistic costs. The relevant costs are reduced up to 25% by replacing 10 regional distribution centers by 2 logistic centers.

1.8.3 Consolidation of Stocks

By consolidating the stocks of the same goods from many regional warehouses in a central store, either the total stock level can be reduced remarkably while maintaining the service level, or the service level can be improved without changing the stock level. This, however, holds true only with optimal *inventory management* (see *Chap. 11*).

According to the *square root law of stock centralization* (Maister 1976), which will be proved in *Sect. 11.10*, the stock levels of two regional warehouses with the same throughput can be reduced by a factor $1/\sqrt{2} = 0.71$, i.e. by about 30% if centralized in one logistic center. At the same time the inventory turnover is increased by the inverse factor of the stock reduction. That means by consolidation of two equal stocks of the same range of articles in one central store the *inventory turnover* can be increased by the factor $\sqrt{2} = 1.41$. From this follow the *rules of stock centralization*:

- Stocks, inventory capital and interest costs can be reduced by a smaller number of storage locations.
- Space demand for storeplaces and storage costs decrease with increasing storage capacity.

- Centralization of many local stocks in a few logistic centers reduces the inventory turnover and hence the costs for in- and out-storing and handling.

As explained in *Chap. 11*, these effects are achievable for articles with a continuous and sufficiently predictable demand. Other stocks that cannot be influenced by scheduling, such as buffer stocks or the stock for *sales promotions*, reduce the effects of stock consolidation.

For the above example the interest costs for the inventory capital are in the range between 25% and 35% of the total logistic costs. By 2 logistic centers instead of 10 regional warehouses, inventory capital and interest expenses are reduced by 20%. The costs for running the two logistic centers further decrease by the smaller space needed for the stock, by the increased inventory turnover and by other consolidation effects such as the lower specific costs of an automatic high bay store.

1.8.4 Bundling of Functions

Provided the logistic center is optimally designed, dimensioned and organized, the bundling of the logistic functions (1.4), (1.5), (1.6), (1.7), (1.8), (1.9), (1.10) and (1.11) has the following positive effects:

- increased efficiency of operations and administration
- applicability of more efficient technique
- reduced share of partly filled load units and storage units
- use of standard load units with optimal dimensions
- decreasing handling costs
- lower costs due to the reduced number of storeplaces
- chances to balance and compensate peaks of demand
- opportunity to operate facilities round-the-clock
- reduced administration costs by modern IT
- lower total overhead costs for the smaller number of locations

The most interesting contributions to the savings are the decreasing investment with increasing storage capacity and the reduction of handling cost rates with increasing turnover.

For example, an *automatic high bay store* with capacity for more then 10.000 pallets and high throughput rates which is operated round-the-clock has lower storage cost rates by a factor of 3 than a conventional store served by forklift trucks. Automatic high bay stores are only one example for efficient technology and modern logistic systems. Their potential can only be captured within large logistic centers (see *Chaps. 16, 19* and *21*). Another possibility of efficient consolidation of functions is the transfer of order picking from regional warehouses or from the delivering vehicles to a logistic center. Normally, the *general rule* holds:

- The costs of internal logistics decrease with the number of logistic centers.

The extent of cost savings by consolidation of functions in logistic centers varies from case to case. In the above example, the internal logistic costs, which amount here to 50–60% of the total logistic costs, were reduced by 15% by consolidating the functions from 10 regional warehouses to 2 logistic centers. In other cases the savings were even higher.

1.8.5 Consolidation of Distribution

By consolidating many single shipments to customers or outlets in a few big shipments via a logistic center, the delivery transports can be reduced substantially. At the same time the transport quantities increase.

As for the inbound flow, standard load units and vehicles with large capacities can be used for the long-distance transport from a logistic center to distribution stations, outlets and key accounts. By this means, the transport capacities are better and more evenly used and the share of *less-than-car-loads* and *less-than-truck-loads* is reduced.

The *area coverage*, i.e. the delivery from the regional distribution stations to the single customers, can be executed by local *freight forwarders*. Logistic centers that are located near metropolitan areas can also participate in the *city logistics*. The consolidation of delivery tours for many companies leads to additional optimizations, which consequently relieves regional and local traffic (Bock et al. 1996).

These effects of a consolidated distribution and better area coverage are partly compensated with decreasing number of logistic centers by the longer distances to the customer locations. If the long-distance transports are not executed by rail, logistic centers can cause higher traffic on the roads.

Therefore, the *general rules* hold:

- With decreasing number of logistic centers, the delivery frequencies can be reduced and more efficient transport means used.
- Longer travel distances in extended distribution areas cause higher distribution costs although the transport frequency is decreasing with the number of logistic centers.

Figure 1.14 shows the dependency of distribution costs on the number of logistic centers for the considered department store. The costs for distributing to the single department stores from only 2 logistic centers instead of 10 regional warehouses increase by the factor 3. This leads to a contribution of the distribution costs to the total logistic costs of about 14% instead of 4%. The savings generated by 2 logistic centers are partly compensated by the higher distribution costs as shown by the total cost curve in *Fig. 1.14*.

1.8.6 Additional Effects and Potentials

The implementation of logistic centers with high throughput and central stocks leads to attractive saving potentials compared to regional logistic stations. Practice shows that the cost savings by consolidation of transports and functions are often higher than in the above example.

The savings depend critically on the total structure of the logistic network, on the right design and selection of the supply chains and on the proper layout and competent management of the logistic centers. Methods, strategies and solutions for this purpose will be developed throughout this book.

The employment of a competent *logistic system provider* for the operation of a logistic center and for the execution of the inbound and outbound transports can lead to further cost savings and service improvements. A system provider has the option to achieve additional *synergy effects* by using a logistic center for several customers (see *Chap.* 22).

1.9 Network Management

In war times holds (Gardiner 2006): "The scope and success of all *military operations* are completely dependent on what can be achieved logistically". The same is true in peace times:

- Scope and success of business operations critically depend on what can be achieved logistically.

That is determined by the quality of the network management. The logistic network of a company, institution or household is part of an extended network that goes far beyond their direct influence. Therefore, it has to be decided first where to set the limits of the own logistic network. The objectives of *network management* result from the kind of business and the type of the logistic network in which a company operates. This can be a temporary, permanent, flexible and a combined network.

1.9.1 Temporary Logistic Networks

Temporary networks are set up for a limited time in order to serve a temporary demand. Examples of temporal and spatial limited networks are the logistic networks of building sites, exhibitions, fairs, events and development projects:

- The management of temporary logistic networks is task of *project logistics.*

Project logistics is the core competence of companies that are specialized in the execution of major projects in alternating locations. Examples are *building site logistics* of corporate building groups (Kulick 1981), *plant logistics* of engineering companies, *object logistics* of event organizers and the *disposal logistics* of demolition and salvage companies. Central tasks of project logistics are *development* of a temporary logistic network, *assignment* of specialized service providers, e.g. for furniture, heavy load or bulk goods, and *management* of the logistic network and sites.

If the projects are nonrecurring events, such as relocation of an office, trade fair participation or a building project, it does not make sense for the company, to have an own project logistics department For these special purposes, qualified *project logistic providers*, such as *moving companies* or *building site providers*, are available on the market.

1.9.2 Permanent and Flexible Networks

Permanent networks are set up for unlimited time in order to serve a long lasting demand. The continuity of the demand and the size of the orders determine the network:

- *Permanent* or *fixed logistic networks* consist of logistic stations at fixed locations, such as receiving stations, transshipment points and logistic centers, that are connected with each other by a permanent transport net.

Many *logistic network provider*, e.g. freight forwarders, railways, postal, courier, express and parcel services and airlines, are owner of a permanent logistic network. The procurement networks of retail companies with dedicated logistic centers and regional warehouses are also fixed logistic networks (see *Sect. 21.11*).

As in electricity industry, a permanent logistic network is opportune for a continuous basic demand, since with permanent high utilization of the installed limit performances and capacities, the performance costs of a fixed network become minimal.

However, the flexibility of a fixed network is generally rather low. Hence, for seasonal or stochastic peak demand that exceeds by far the basic demand, a flexible network is opportune:

- *Flexible* or *virtual logistic networks* are networks with changing stations, varying transport links and altering partners involved.

The operating costs of a flexible network are generally higher as for a fixed network. Operators of flexible networks, who own no transport means and have no permanent operation sites, are the so-called *Fourth Party Logistic Service Providers* (4PL) (see *Sect. 22.3.3*).

According to the demand, a fixed regional or national network can be connected with flexible local or global networks thus resulting in combined networks:

- *Combined networks* consist of a number of fixed stations connected by regular main transports in combination with flexible local networks and spontaneous relation transports.

The procurement and distribution networks of large car manufacturers, chemical companies and consumer goods producers with plants and suppliers around the globe are such combined networks.

As an example *Fig. 1.15* shows the logistic network of a car manufacturer whose module suppliers are located close to the assembly plant. The *procurement network* of the factory ranges upstream from 2nd and 3rd *tier suppliers* of parts and components to the 1st tier of the module suppliers. The *distribution network* reaches downstream from the end of the assembling line and a central warehouse for accessories and spare parts over transshipment points to the car dealers all over the world.

Medium sized companies with plants in one continent, e.g. in Europe, need a fixed continental network, e.g. a *Euro-logistic network*, combined with a flexible international network in the other parts of the world.

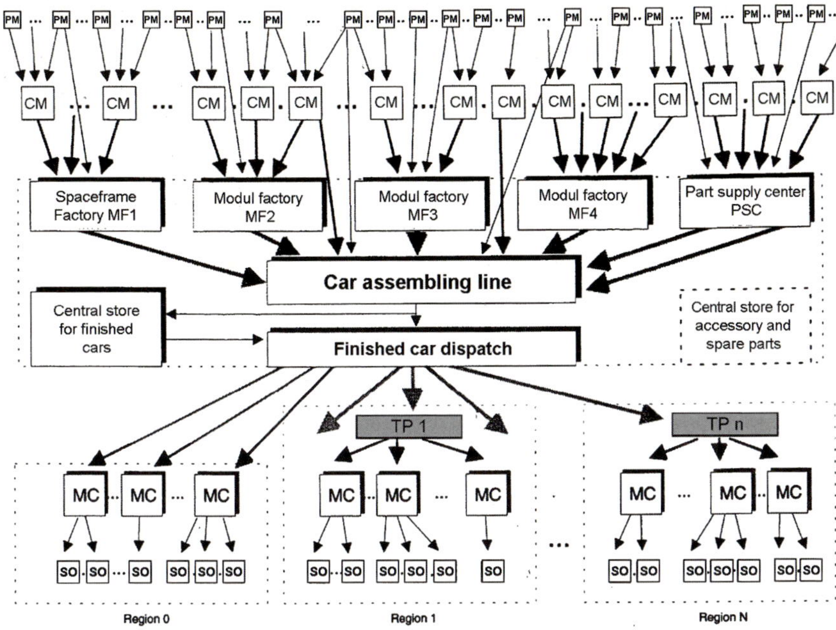

Fig. 1.15 Logistic network of a car manufacturer (SmartCar)

PM: parts manufacturer (3rd tier)	CM: component manufacturer (2nd tier)
MF: module factory (1st tier)	CA: car assembly;
CW: central warehouse for parts	TP: transshipment points
MC: market center	SO: sales outlet

Also *global logistic service providers*, such as international freight forwarders, airlines and shipping lines, operate with combined logistic networks. They connect their own fixed global network with flexible local networks of contract partners or *sub contractors*. By this means, they can offer a complete logistic network that covers the whole globe.

1.9.3 Objectives of Network Management

Logistics is unlimited. Therefore, it is necessary to set borders, to define links and interfaces, to set goals within the limited network, to decide on the objectives, to allocate the execution of the orders and to control the results. These are the tasks of *network management.*

Who has to solve logistic problems must look beyond the borders of the own logistic network. In order to avoid barriers, a logistician must know at least the first stages of the *downstream chains* of the customers and the last stages of the *upstream chains* of the suppliers. Barriers have to be transformed into connections that enable uninterrupted flows of goods and information.

Company logistics includes *supply chain management* and *network management*. Depending on the corporate objectives and the logistic network, the general *tasks of company logistics* are:

planning and forecasting of demand
development of strategies
design of network structure and standard supply chains
planning and implementation of own logistic systems (1.13)
synchronization and adjustment of interfaces and links
scheduling of orders and stocks
logistic controlling and consulting

The special tasks of logistics differ from company to company. In case of logistics being a *core competency*, company logistics should be organized as independent corporate unit on the same level as production, finance, administration, purchasing, and sales (see *Sect. 2.9*). Temporary tasks, such as development of strategies, configuration of the network or planning of a project, may be partly delegated to an experienced *consultant*.

1.9.4 Future Tasks of Logistics

After industrialization of production and manufacturing, which has started in the early 19th century, is widely finished today, it is the challenge for the 21st century to industrialize the performance processes.

The implementation of fixed transport networks and the building of large logistic centers will come to an end in the next future. In the densely populated and highly industrialized countries space for new roads, railroads, airports and logistic operations becomes scarce. Therefore, the goal of the future is to manage *flexible networks* by using the existing limited resources in the best possible way.

Logistic research should investigate the necessary *legal frame conditions* and develop suitable *strategies* for this challenge. This research is an important field of activity for *analytical-normative logistics* as well as for international institutions such as the EU and the OECD (see *Chap. 22*).

1.10 Task of Logisticians

The word *logistics* has three basically different meanings:

- *Practitioners* regard logistics as the activities necessary to set up and operate transport, storage, traffic, and handling systems and networks.
- *Planners* understand logistics as the design, dimensioning and optimization of logistic networks, processes and systems.
- *Theorists* see logistics as investigation of practices, search for principles, examination of options, and development of strategies, algorithms and rules for planning, set up and operation of systems and networks.

Many people are not aware of these differences (Murphy 2004). However, without distinction between the different meanings of the term logistics, there is a risk of contradictions and drawing pointless conclusions (Popper 1971). This explains many misunderstandings and contradictions between *logisticians*.

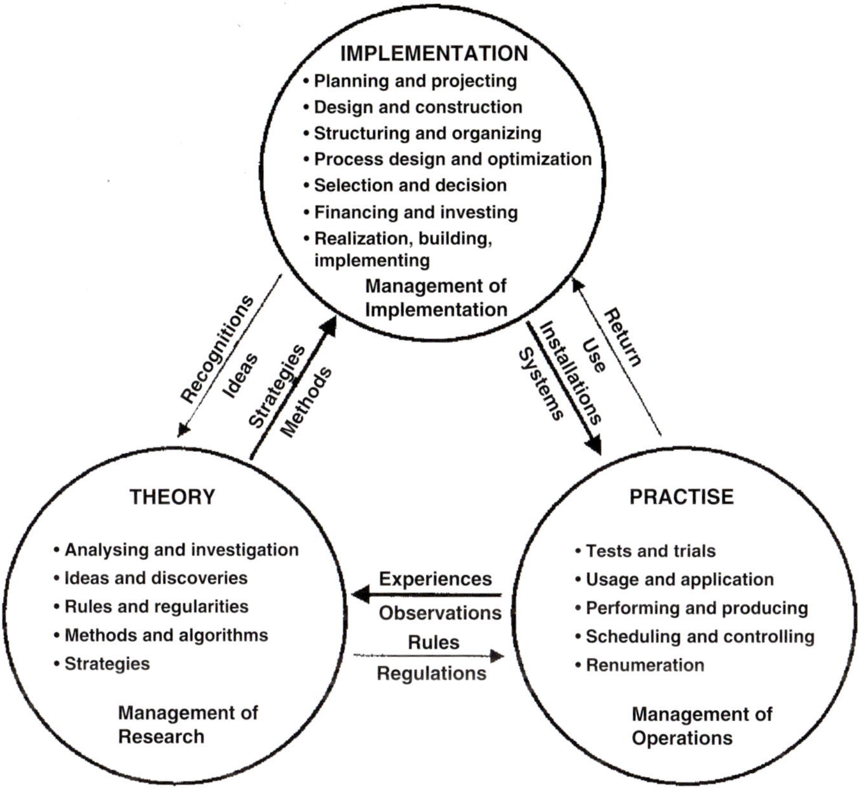

Fig. 1.16 Tasks of logisticians and relations and tensions between theory, implementation and practice in logistics

The more logistics is expanding and specializing, theory, planning and practice diverge. This development can be observed in other disciplines too. *Figure 1.16* shows the tasks, responsibilities and links of the three fields of logistics.

According to the different tasks of logistics, there are *strategic logisticians* or *theoreticians*, *implementing logisticians* or *planners* and *operative logisticians* or *practitioners*, who respectively focus on theory, implementation or business practice. From the most competent and successful members of these groups, come the *logistic managers* or *supply chain managers*. Managers set the goals, decide upon suggested solutions and determine the course for new developments and conceptions.

Conditions for the success of strategic, implementing and operative logisticians are mutual respect, knowledge of the objectives and contributions of the others and a joint orientation towards practical benefits for customers, company and society.

1.10.1 Strategic Logisticians and Theorists

The objective of strategic and theoretical logisticians is to *contribute to practical use* through analysis of the basic principles of logistics and by developing innovative logistic concepts and strategies. This group consists of academics at institutes and universities and of consultants, strategists, organizers and systems analysts.

Prerequisites for a successful strategic logistician are analytical thinking, openness about new ideas, creativity and judgment. *Analytical-normative logistics* requires knowledge of the methods and strategies of logistics and business administration, sufficient command of arithmetic, algebra, probability theory and statistics and some know how in *operations research*. A further necessity is to know the realities and requirements of the daily business.

Many theorists are *generalists*. They tend to set up tautological terms and definitions, to develop models out of touch with reality and to make considerations with little relation to practice. Logistics as *applied science* is justified by its applicability in practice. Hence, the measure for strategic logisticians is their contribution to practical use.

Whoever asks "What is logistics?" or "What do we mean by Supply Chain Management?" will produce a number of new terms but will not create practical benefit. Only those who asks "What are the objectives and tasks of logistics?" and "How can we solve this problem?" will come to beneficial answers and create the necessary terminology for the specific purpose (Popper 1971).

Some theorists follow the principle "why look for a simple answer if a complex one is possible?" This group wants to use a smart algorithm, to test a new OR-technique, to do *chaos research* or to apply *complexity theory* even though no relevant practical improvements can be achieved. Instead, the challenges for theorists are to understand a complex logistic system, to find out its rules and to make the system controllable (see also *Sect. 10.5.5*) (Popper 1971).

1.10.2 Implementing Logisticians and Planners

Implementing logisticians and planners have the objective to develop, construct, plan, organize, program, and set up machines, sites, systems and networks that *bring about practical benefit*.

First of all, the planners and project managers of companies and general contractors for storage, conveyor, transport and logistic systems belong to this group. Others are the developers, planners, programmers and systems engineers working for the suppliers of machines, equipment, vehicles, control systems and software.

Prerequisites to success for an implementing logistician are constructive thinking, organizational competencies and a profound knowledge of the possible solutions, the specific circumstances of the project and the necessities of operative logistics.

Planners and implementing logisticians tend towards specialization and over-engineered solutions. Their position is often quite thankless. In case of success, it was due to the initiative of the client or his consultant. In case of failure, it was the planner's or the executing company's fault.

1.10.3 Operative Logisticians and Practitioners

Operative logisticians and practitioners have the objective to *generate permanent benefits* by putting the best solution into action. Schedulers, plant managers, operators and users of logistic centers and logistic systems belong to this group. Most of them work for *logistic service providers*.

Also *logistic managers* are in many respects practical logisticians. They manage logistic operations or a logistic network and take care for the sustainable competitive advantage of the company.

Operative logisticians need the ability to think practically. They have to have a solid knowledge of the techniques and capacities of their equipment, sites and systems. Operative logisticians in management position must be capable to organize processes and to decide under uncertainty.

Practitioners schedule incoming orders, operate computers and lead business plans to success. Not all plans survive the contact with reality, not all software is suitable in practice. The schedulers make decisions, on which the daily profit or loss of orders and in the long run of the whole business depends. Therefore, they should be well educated and trained (Gardiner 2005 and 2002; Murphy 2004). To keep them motivated, their work and contribution has to be appreciated by the management (see *Sect. 23.3.2*).

Practitioners that have been working only in one company or one specific industry for a long time tend to get routine-blinded. This often leads to arrogance towards theorists who vice-verse are inclined to look condescendingly at practitioners.

1.10.4 Specialists and Generalists

Most results, options and solutions of logistics are quite trivial or straightforward. However, they are multifaceted and numerous. Therefore, if not proceeding systematically, complexity causes confusion.

In logistics the wheel is often reinvented and old wine is sold in new bottles. This happens to beginners when they start to think about the causes, effects and possibilities of logistics. However, some experienced specialists and generalists also tend to reinvent the wheel.

Specialists are experts in a narrow area. They know all about their specific field. However, they also recognize the limits and barriers. Specialists think in construction and technology, in experience and examples, in programs and computers or in profit and cash flow.

Due to many reservations, some specialists are incapable of making decisions. They lose the overview and tend to overestimate partial aspects. Who is lost in the thicket of numbers and the multitude of technical details, does not see the wood for the trees and fails to find the best solution.

Generalists are familiar with many subjects. They keep an eye on a very broad field, recognize correlations and think in systems. Generalists are often more decisive and willing to take risks than specialists. Due to a lack deeper knowledge and limited know-how, generalists tend to underestimate problems, do not see realization barriers and are unaware of innovative solutions. Some generalists replace know-how by belief. For instance, by uncritical belief in ISO-regulations they become unrealistic "ISOterics".

Without knowing the principles and realities and without keeping critical details in mind, there is the danger to fantasize. *Visionaries* tend to develop "intergalactic" solutions of no practical benefit. Such detached generalists consider a high number of simple relations as "highly complex". They call the network of relations between suppliers, customers and service providers "virtual enterprise" or reinvent the work-shop production by calling it "fractal factory".

Other theorists and strategists consider only one aspect of logistics: *Just in Time, Kanban, Outsourcing, Benchmarking, IT, PPS/MRP/ERP/APS, Supply Chain Management, RFID* and *e-Logistics*. Each of these fashionable terms claims to solve just about all tasks. However, logistics has many facets and should be considered under all relevant aspects.

Good logisticians are specialists in one or two fields and generalists in all other areas of importance for logistics. They follow the *hawk principle*:

- The logistician stays above theory and practice and observes with sharp eyes structures, processes and connections. If a certain field shows progress, the view is focused, good solutions are assessed, details analyzed and useful ideas are captured from the lowland of theory and practice.

The permanent change between *top-down* to *bottom-up* and vice verse widens the competencies. By this process, the logistician gains further abilities to solve actual problems.

Who has no distance, does not see the whole. Who only focuses on details, can not understand the relations. A system is more than the sum of its elements, but the function of the total system can depend on one element only (Lenk and Ropohl 1978). This holds especially for *bottlenecks* that arise in practice everywhere, but are not noticed immediately (Goldratt 2002; Gudehus 1975/I). Bottlenecks are decisive for the capability and efficiency of performance, production and logistic systems, and therefore are a central topic in this book.

1.10.5 Theory and Practice

Apart from the contributions of *Operations Research*, logistics is in many areas a *skill* based on experiences and experiments. Most practical innovations in logistics are still found by *trial and error*. Some practitioners emphasize this proudly. Faced with a theoretically based proposal, they object: "That might be right in theory but is of no use in practice".

More than 200 years ago, the German philosopher *Immanuel Kant* wrote an article about this objection (Kant 1793). He criticized theoreticians, who never will become practical, and told the practitioners: "Nobody can disregard theory without being an ignorant in his field".

Tensions have always existed between theory and practice and will remain in future. For logistics, they are reflected in the *tension triangle of logistics* shown in *Fig. 1.16*. Without these tensions, neither progress in practice and nor discoveries in theory are possible.

One goal of this book is to develop a practice-oriented *theory of logistics*. Without unnecessary complexity and abstractness, it should help to a better understanding. Analytical-normative logistics offers *rules* for solving problems, develops *methods* for attaining logistic objectives and designs *strategies* for the planning and operation of logistic systems.

Kant also has said: "nothing is more practical than a good theory". Applied to logistics and business practice, follows:

- Nothing is more useful than a good strategy.

Hence, the development of strategies of practical use is another central focus of this book.

Chapter 2
Organization, Scheduling and Control

The *organization* of performance stations and networks, the *scheduling* of orders, inventories and resources, and the *control* of the processes determine the productivity and efficiency of companies and logistic systems.

The common terms *organization*, *planning*, *scheduling* and *control* have different meanings in business and engineering, in production, logistics and informatics as well as in theory and practice. The same holds for the term *management*, which involves planning, scheduling and control, but comprises in practice far more (Christopher 1992; Cooper et al. 1997; Schönsleben 1998). As well the names and abbreviations of IT systems for planning, control and scheduling are misleading (Fritsche 1999; Scheutwinkel 1999; Scott et al. 1991; Stadler et al. 2007). In order to avoid any misunderstandings, these central terms are defined in the following precisely.

- *Planning* is the optimal selection, dimensioning and organization of resources, stations, networks and processes necessary to fulfill *future performance requirements*.

Planning generally deals with inaccurate information, mean values and uncertain expectations. *Production planning* takes care of the capacities and performances of the resources to fulfill the requirements and demand of future periods. *Business planning* is carried out periodically on a monthly, quarterly or annual basis. *Strategic planning* covers five years, ten years or even longer periods of time. Other kinds of planning are *project planning* for a specific project and *marketing planning* of sales promotions.

- *Scheduling* is the *disposition* of orders and quantities for *actual requirements* and the time-phased *allocation* of the resulting internal orders to the available resources.

The orders are either *external orders* from customers or *internal orders*. Internal orders are derived from short term *demand forecasts*, generated by project planning or come from other departments of the same company. Scheduling requires reliable data and accurate information. The *scheduling procedure* is either triggered

T. Gudehus, H. Kotzab, *Comprehensive Logistics*,
DOI 10.1007/978-3-540-68652-1_2, © Springer-Verlag Berlin Heidelberg 2009

by a relevant *event*, e.g. by an incoming order or a change of a resource, or started *periodically* in short periods of time, e.g. hourly, daily or weekly (see *Sect. 10.6*).

Scheduling is called also *detailed planning*, *short term planning*, *order planning* or *supply chain event management*. Those terms however are misleading as they lack clear distinction from planning and control.

- *Control* is the *steering, guiding* and *adjustment* of performance stations and transport means with the aim to execute given orders within due dates.

Process control triggers, guides and adjusts the processes of *machine systems*, for example of an *automatic guided vehicle system* (AGV) or an automatic *high bay store* (HBS). *Business control* leads and controls the procedures and actions in *performance stations* and *business units* where people are involved.

The term *organization* stands for both, for the action of *organizing* and for the result of this action (Ackhoff 1970; Drucker 1988; Probst 1992; Weber 1997):

- *Organization as action* is the *analysis* and *design* of the *relational structures* and of the *consecutive processes* within a system of people, stations and other elements with the aim to achieve certain tasks.
- *Organization as result* are formal *regulations*, *advices* and informal *rules* for people, for the *operation* of system elements, and for the *relations* between the elements.

Organizing is a central task of *planning*, which is topic of the next chapter. The organization of a performance or logistic system results from the following *organizational options of action*:

design of processes and structures
number and functions of the organization levels
centralization or decentralization
process organization and structure organization (2.1)
configuration of hardware and software
standard software or individual software
organization of the scheduling procedures

This chapter presents and explains the different organizational options of action and their results. It offers recommendations for the *organization of company logistics* and explains the basic *principles of dynamic scheduling*.

2.1 Orders

All business transactions and processes in a production plant, a logistic system or a supply network are initiated by *orders*. Each order includes delivery requirements, operational directives and logistic prescriptions:

- The *delivery requirements* are given by the *order lines* or *positions*. They indicate which quantity of what article, product or service is required.
- The *operative directives* specify what and how to produce, to deliver and to accomplish. They are directives for production, assembling and handling.

- The *logistic prescriptions* determine the *times* when and the *locations* where to pick up, deliver and provide the required quantities.

In order to ensure a correct order execution in due time, the delivery requirements and logistic prescriptions must provide the following information:

- *order lines* or *positions* identifying the article or product
- *order quantity* giving the required amount or number of article units
- *addresses* of the pickup station and/or the delivery station
- *time information* about the due dates of pick up, dispatch and delivery

External or internal orders, which determine the events within a performance, production or logistic system, are:

delivery orders
production orders
handling orders
dispatch orders (2.2)
pickup orders
transport orders
procurement orders

For *order scheduling* it is necessary to differentiate between

- *single position orders* concerning only one article
- *multi position orders* referring to several articles
- *single piece orders* requiring only one article unit
- *multi piece orders* demanding more than one article unit

External orders are placed by customers and schedulers of other companies. *Order scheduling* converts, bundles, and divides the external orders into *internal orders*, which initiate the processes within the operative performance stations and the transports between them. Internal orders addressed to the stations of a logistic systems are:

replenishment orders
pickup and delivery orders
transport orders
in-storing and out-storing orders
supply and retrieval orders (2.3)
handling orders
sorting orders
packaging and bottling orders
loading and unloading orders
control and checking orders

2.2 Order Management and Logistic Scheduling

Each company and operative unit needs an *order management*. It has to identify, check and prepare the incoming orders, to process and convert them into time-phased internal orders and to control their correct execution. Order management includes commercial, technical and logistic tasks (Darr 1992):

- *Commercial order processing* accepts, checks and confirms incoming external orders, invoices the executed orders and performs additional administrative activities.
- *Logistic order processing*, also called *order scheduling* or *logistic scheduling*, converts commercially accepted orders by splitting, clustering, parts listing and other *scheduling strategies* into internal orders, which are transferred to the performance stations, and controls their execution.
- *Technical order processing*, also called *production scheduling*, prepares the technical execution of the orders and assigns them to machines, working places or other stations of the performance network.

Commercial order processing is normally the task of the *sales back office*. Technical order processing is in most companies part of *production planning and control*. The responsibility for the logistic scheduling is not always clearly determined and distributed differently.

After order execution, the resulting products, delivered goods and executed services are charged to the customer's account (see *Chap. 7*). Additional *administrative services* in connection with the order processing are:

providing information about delivery ability, status and due dates
tracking and tracing of articles and shipments
inventory management and scheduling of supplies (2.4)
invoicing, encashment and reminding

More and more *logistic system providers* execute these administrative services for their clients (see *Sect. 22.3*).

Logistic order processing is either centralized or decentralized. It can be the job of the sales department, the production management or of a separate *order center*. There is no universal rule for the organization of company logistics and order processing. It depends on the size, the number of locations, the kind of products and services, and on the distribution channels of a company. Nevertheless there are general *organizational rules* and *strategies*, which are applicable for the organization of logistics, order processing and other operations (see *Chap. 10*).

The goal of order scheduling is to ensure the punctual and efficient execution of the current orders by the available performance stations and resources. This can be achieved by

- *Sourcing strategies* such as *make-or-buy*, *source-to-order*, *source-to-stock*, *make-to-order* or *make-to-stock*
- *Time strategies* in order to keep the required delivery dates and times, and at the same time, to leave sufficient time for operating strategies
- *Allocation strategies* for the cost efficient allocation of orders to the available performance stations and resources keeping the requested delivery dates
- *Inventory strategies* for storekeeping articles to ensure the required *ability to deliver* at lowest cost

Besides the make-or-buy decision, which depends primarily on business policy, *sourcing strategies* are mainly influenced by logistics (Baumgarten 1992). The possibilities for the *outsourcing of logistics* and the aspects regarding the employment of *logistics service providers* are discussed in *Chap. 22*.

Chapter 8 will present *time strategies* and criteria for the decision between make-to-order and make-to-stock. *Chapter 10* describes the possibilities and strategies of *order scheduling*. They are related closely with *production planning* and the *operating strategies* of performance stations. In *Part 2* of this book, it will be shown how specific *operating strategies* for storage, commissioning and transport systems and for supply networks can be derived from the general results and rules of *Part 1*. Topics of *Chap. 11* are *inventory strategies* for optimal replenishment and storekeeping. In *Sects. 11.2* and *11.12* the general criteria for make-to-order or make-to-stock will be supplemented by special criteria for storable goods.

2.3 Process Organization and Structure Organization

Corresponding to the *process aspect* and the *structure aspect*, the process organization and the structure organization must ensure the efficient, correct, complete and punctual execution of the orders (Bowersox et al. 2007; Hammer and Stanton 1999):

- *Process organization* regulates the consecutive *operations* and *activities* on orders, information, and material passing through the *performance chains*. It directs and controls the flows of *orders, data, material* and *products*.
- *Structure organization* regulates the *relations* between people, performance stations and system elements. It specifies the functions, tasks and dependencies of the elements and determines the *organizational structure* of the system.

In order to make it controllable, the organization structure of extended systems should be *hierarchic* (Ackhoff 1970; Beckhard 1969; Blau et al. 1971). In a *hierarchical organization* the tasks, responsibilities and objectives are allocated to different *organization levels*.

Each level reports to the next higher level and receives instructions from there. It gives internal orders and instructions to the next lower level. The general tasks and characteristics of a typical hierarchical organization with three levels are shown in *Table* 2.1.

2.3.1 Strategic Level

The *strategic level, planning level* or *administrative level* of the organization is responsible for business policy, development of strategies, determination of products and services, business planning and *central services*. At this level, general agreements with customers and procurement contracts with suppliers are negotiated. The aim of the strategic level is to ensure the capability of the company or system to fulfill the *requirements of the future*.

Planning and other activities at the strategic level are based on uncertain information, on expectations and on medium to long range forecasts. To reach the corporate goals proper *planning, scheduling* and *operation strategies* are needed.

Table 2.1 Tasks and characteristics of organization levels

Strategic Level	
Tasks	• Corporate planning • Strategy development • Program planning • Marketing, purchasing, sales • Corporate finance and accounting • Human resources • Controlling
Characteristics	Orders of top management Market requirements Application of planning strategies Uncertain information Long decision times (days or weeks)
Tactical Level	
Tasks	• Order scheduling • Order management • Production scheduling • Work preparation/process planning • Inventory management • Supply scheduling • Manufacturing resource planning • Order tracking and tracing • Controlling of operative processes
Characteristics	External orders Rules and instructions from administrative level Application of scheduling strategies Relatively definite information Medium processing times (hours or days)
Operational level	
Tasks	• Initiation of processes • Controlling of single activities • Process control • Assurance of execution
Characteristics	Internal or direct customer orders Instructions from scheduling level Application of operative strategies Well defined orders, precise information Short reaction times (seconds, minutes or hours)

2.3.2 *Tactical Level*

At the *tactical level* or *scheduling level* external customer orders are converted into internal orders for the operative units applying the rules and instructions of the strategic level. The tactical level also generates and transfers *procurement orders* to the suppliers.

The tactical level manages and controls orders, inventories and resources. It handles current orders and requirements following *scheduling strategies*. The scheduling is based on definite information and short term forecasts. The tactical level is concerned with the *efficient use of the available resources*.

2.3.3 Operational Level

The *operational level* or *executing level* is responsible for the efficient order execution and concerned with the secure and proper operation of the resources. For this purpose, it has to coordinate the single performance stations. The operational level manages and controls the execution of the orders transferred from the scheduling level or placed directly by external customers.

Both, *business control* and *process control*, need *operating strategies* that ensure the efficient and correct execution of the orders and a high availability of the resources. The activities in the operational level are based on well defined orders and precise information.

Within a flexible organization the hierarchies are not strictly separated and not realized in a rigorous manner. In complex systems and larger companies it is opportune to subdivide the three organizational levels further or to install them parallel in different business units.

2.4 Organization Principles

For the division of the tasks and responsibilities between the organization levels, the following *organization principles* are helpful. They are based on theoretical considerations and have been confirmed in practice.

Top-down instructions and bottom-up confirmations and reports lead to close interrelations between the organizational levels. In order to achieve fast feedback and quick goal adaptation under changing conditions, it is necessary to delegate the decision-making to the lowest possible organization level. This leads to the

- *Principle of Delegation*: Decision making should be as local as possible and only as central as useful and necessary.

Disregarding this principle causes hindering decision procedures and long *reaction times* and results in a lethargic and inefficient organization. The application of this principle to the organization of planning and scheduling leads to the

- *Principle of Subsidiarity*: Planning and scheduling should be as local as possible and only as central as useful and beneficial.

Due to this principle, a central planning department or an inter-company order center should only take over tasks, which cannot be executed by the local units or single companies and which are of advantage for the cooperating companies (Brucks 1997) (see also *Sect. 21.17*).

A flexible, powerful and customer oriented company observes the following *organization principles*:

- *Self Control*: Instructions, responsibilities, quality control and remuneration have to be organized in such a manner that a performance station executes the orders in its own interest efficiently, correctly and in due time.

- *Result orientation*: Each performance station is responsible for its contribution to the business and must know the required results and all related processes.
- *Clear instruction sources*: No station should receive instructions for the same task or for the same order from more than one side.
- *Information discipline*: Each station must receive and deliver all information needed respectively by itself and other stations for the correct, complete and punctual order execution.
- *Check and Balance*: Reciprocal control of information by the consecutive performance stations ensures high process quality.
- *Limited Span of Control*: No organizational unit or performance station should have more functions and responsibilities than a qualified manager can cope with.

The development of *standard procedures* for the regular business processes is the first step when designing a new organization. The next steps are allocation of functions to the single stations and adaptation of the different processes.

The *irregular processes*, necessary to cope with incorrect data, errors, quality faults or failures of customers, suppliers, performance stations, resources or employees are often ignored or neglected. The development of *emergency procedures* and *failure strategies* is generally more difficult than the organization of the regular processes. However, as safeguard against *Murphy's law* "If anything can go wrong, it will", an organization should observe the

- *Safety Principle:* By *flexible precautions* for coping with errors, failures and irregularities the organization must be completed and made safe.

Complex systems with many closely linked subsystems and performance stations are difficult to control. They are at the same time clumsy and fragile. Even by the most advanced *simulations*, such systems cannot be improved substantially. However, as outlined in *Sect. 13.5*, from *queuing theory* follows the

- *Decoupling Principle:* By inserting adequate *material buffers* or *order buffers*, a logistic system can be separated into decoupled subsystems and stations, which are not critically affected by tailbacks, feedbacks and interruptions.

If this principle is observed, the decoupled subsystems and performance stations can schedule independently and execute the orders they receive from outside the company, from adjacent stations or from an order center. For example, the production stations can autonomously schedule their resources and the material necessary to execute the given orders (see *Chap. 20*).

The application of modern IT-systems for data exchange and data processing, as well as the electronic access to the master data of central databases open up new possibilities for planning and scheduling. However, the above principles still hold. Extended networks and complex systems can be kept under control only with the help of the *delegation principle*, the *subsidiarity principle* and the *decoupling principle* (Heymann 1997).

A *potential analysis* as described in *Chap. 4* reveals whether the organizational principles are observed in a company. Their disregard indicates *organizational weak points* which can be eliminated within short time.

2.5 Software Levels and Computer Configuration

Today, *software* assists planning, scheduling and control, and executes many tasks without people. Electronic *databases* store, process and administrate the business data and logistic master data. Besides tapping in data and parameters and reading of information, staff members execute difficult procedures and make critical decisions that cannot be left to the computer.

Corresponding to the three hierarchical levels of business organization, software systems can be organized in *software levels*:

1. *Planning software* is applied for corporate planning, accounting, administration of human resources and for other tasks at the strategic level. Special software modules for the *Advanced Planning and Scheduling* (APS), *Enterprise Resource Planning* (ERP) and *Supply Chain Management* (SCM) determine the capacities, the resources and their optimal allocation for the future demand (Fritsche 1999; Kuhn and Hellingrath 2002; Meyr 2004; SAP 1994, 2000; Stadtler and Kilger 2007; Scheutwinkel 1999).
2. *Scheduling software* for *Network Resource Planning* (NRP), *Material Requirement* or *Resource Planning* (MRP I/MRP II) and *Production Planning and Control* (PPC) manages inventories, processes current orders, allocates stocks, generates replenishment orders, registers the utilization of transport means, plants and other resources, calculates routes and registers the performances.
3. *Control software* is steering, guiding and adjusting the workflow, operations, transport and material handling in and between the different performance stations and subsystems. It controls the proper execution of internal orders.

Following the principles of *delegation* and *subsidiarity* the different software modules can be assigned to a real or virtual hierarchy of administrative, scheduling and control computers. The computer hierarchy is generally designed as *client-server-system*:

- The *host computer* is the *central server* of a company and performs strategic planning, executes central tasks and keeps *master data*.
- *Client stations* in the departments and performance stations enable the local data in- and output and support local scheduling, planning and control.

For the specific tasks of logistics, either *special software modules* such as *Merchandise Information Systems* (MIS), *Warehouse Management Systems* (WMS) or *Transport Management Systems* (TMS) can be used or *standard software modules* such as the SAP modules SD for *Sales and Distribution*, MM for *materials management* and PP for *production planning* (SAP 1994 and 2000). Both, quality and capability of standard software as well as special software differ very much. They depend on the design of the software, on the programmed strategies and algorithms and on the application. Unqualified software or incorrect parameters lead to wrong results and can cause severe damages (Dittrich et al. 2000; Gudehus 2002). Therefore, one should assess any software critically before applying.

Today in most companies host and client computers and other peripheral devices are connected by an internal data network, the *Intranet*. In order to avoid double

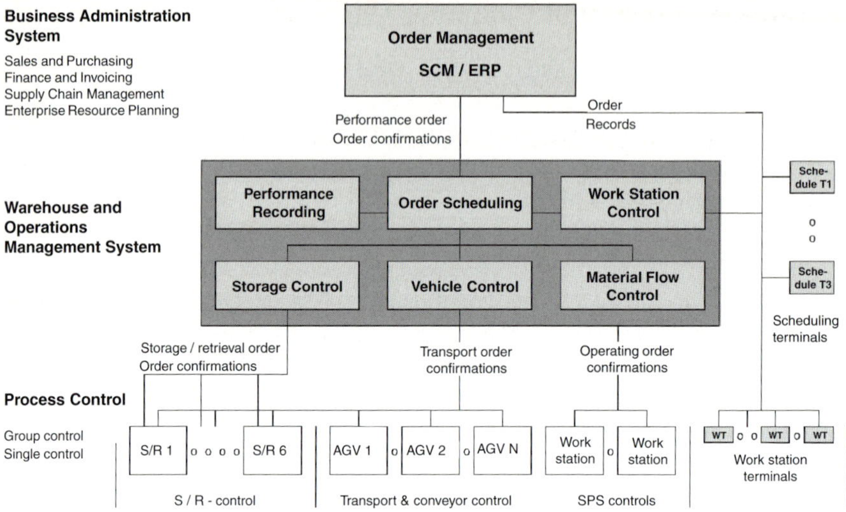

Fig. 2.1 Computer hierarchy and organization of a logistic center

files and inconsistencies of data, all software modules and local stations should have access to only one *database* and operate with the same master data. For this purpose standard *database software* solutions are available.

A basic problem that is often not addressed and solved when designing IT-systems is caused by the *response times* and *processing times*. Due to the *queuing laws* of *Sect. 13.5* the response times for data entries, inquiries and commands increase over-proportional with growing numbers of users and contacts. Response times lasting only a few seconds under normal circumstances, can increase up to several minutes and more in phases of frequent use, if the system capacity is insufficient.

These inefficiencies are not recognized until the IT system is in use. Therefore, the hardware or at least the software modules for time critical processes should be decoupled from the IT system for slower planning and administrative processes by subordinated *process computers* and special software modules such as *Production Planning and Control Systems* (PPC), *Warehouse Management Systems* (WMS) or *Transport Management Systems* (TMS).

If a technical system is operated *real-time*, the critical elements and subsystems must be controlled by *local control units* operating below a central process computer. As an example, *Fig.* 2.1 shows the organization structure of a *logistic center* with an IT-hierarchy designed due to these principles (see also *Fig. 18.6*).

2.6 Data Flow and Information Flow

The incoming orders trigger the processes in production, performance and logistic systems and generate *information flows*. Additional data and information flows are necessary to trigger and control the material flows through the logistic chains. The exchange of instructions and feedbacks between the performance stations and the computer network cause further *data flows*.

When an item, such as a product, parcel or load unit, enters a performance station, an order with instructions of what has to be done with it is needed. These instructions accompany the item or can be transferred in advance before it arrives. In any case, the entering goods must be identified, checked and inspected at an *identification point* (I-point).

When leaving a station, the outgoing good has to pass a *control point* (C-point), where the correct execution of the order is checked and the *logistic data* are captured. For example, the last C-point at the end of a supply chain for consumer goods is the *point of sales* (POS).

In order to avoid multiple data processing, the C-point of the delivering station should be identical with the I-point of the subsequent station. The merging of I-points and consecutive C-points are interesting saving potentials within many *intra-organizational performance chains*. A merger of I- and C-point is feasible to a certain extend also within *inter-company logistics chains*, if *electronic data interchange* (EDI) between delivery and receiving stations is possible (Forster 1995).

Accompanying information is necessary in order to identify the item, to control the process and to secure the indisputable *transfer of risk*. It allows for *tracking and tracing* of single articles, load units and shipments through the logistic chains. The *accompanying logistic information* includes

- *identification information* to identify the article, package or load unit
- *sender information* to specify the point of origin and the sender
- *destination information* determining the point of destination or the receiver
- *control information* to indicate the delivery chain and intermediate stations

In addition to this information the *postage* or *freightage* is indicated on letters, parcels and other shipment units.

Documentation and provision of the accompanying information can be carried out in different ways (Arnold 2002):

- *Complete information* in plain writing or as a printed bar code *directly on the article unit*, package or load unit. The information can be read by a person, a scanner or by another device at the I- and C-points
- *Complete information on a code carrier*, such as a *label* or a programmed *transponder* attached to the article, package or load unit
- *Identifying information* on the item or a code carrier. Other information is contained in *accompanying papers* or *shipment documents*
- *Identifying information* without accompanying papers. After identification at the I-point or C-point further information is generated by computer

Coding, labeling, reading and scanning require time and generate costs which sum up in the total chain to non-negligible values. Many coding and identification systems are standardized (Arnold 1995; Kotzab 1997). The most applied coding system is the *International Article Number* (IAN), earlier called *European Article Number* (EAN), which is based on standardized bar codes and numbering systems (CCG 1994, Kotzab 2005).

Nowadays, *paperless transmission* of data, orders and other information is possible by *Electronic Data Interchange* (EDI) or by *Internet*. For the EDI standardized information transaction sets are needed. In Europe EDIFACT, ODETTE, SEDAS, CEFIC or ANSI ASC X.12 are common standards (CCG 1993, 1995). However, they are only valid for special industries and their wider application was partly inhibited by the Internet.

The introduction of the *eXtensible Markup Language* (XML) has offered new possibilities for transmitting and exchanging order, data and information (Kotzab 2005). XML was introduced by the *World-Wide Web Consortium* in 1998. It is recommended as being the standard for exchanging information over the web and can be seen as a further development of the *Hypertext Markup Language* (HTML), that also allows computers to process the content of websites (www.w3c.org/XML). XML can therefore be characterized as an open and flexible standard to store, publish and exchange information (Daum et al. 2001, p. 18).

2.7 Potentials of Information Technology for Logistics

Modern *information and communication technology* (ICT) offers new possibilities and saving potentials for logistics (Scheer 1998). However, they also imply a danger for misinterpretation, exaggerations and misuse.

The application of modern ICT in logistics, somewhat misleading also called *e-logistics* (Straube 2004), offers the following potentials:

- The *transaction costs* for orders, data and information are reduced substantially by *EDI* and *Internet*.
- Based on *article data, inventory information and orders*, *efficient scheduling* strategies can be realized and quick decisions are possible.
- EDI enables the *advanced shipping notification* of the receivers.
- *Electronic ordering systems*, *order acknowledgement* and *invoicing* speed up and simplify the order processing between industry and retailers.
- *Errors are reduced*, *response times are shortened*, and multiple data collection is avoidable by integrated C- and I-points.
- *Continuous Replenishment Programs* (CRP) of the manufacturers enable automated replenishment based on agreed delivery abilities and lead times.
- *Demand forecast* can be improved by computer integrated merchandising systems, which connect widely distributed and far away *point of sales* (POS) with the production and replenishment systems.
- The actual information about locations and loads of transport units, send via satellite, per Internet or by EDI, enable *dynamic transport scheduling* and effective control of transport fleets.
- *Advanced* application, allocation and operation *strategies* can be realized (see *Chap.* 10).

- The order and load information gathered on the I- and C-points can be used for *logistic controlling* and for reimbursement of logistics services (see *Chap.* 7).
- *Tracking and tracing* of shipments can be realized, e.g. with the help of *transponders* and *RFID* (Finkenzeller 2002; Shephard 2004).
- Based on the current utilization of resources and networks intelligent *booking systems* and *pricing models* are possible.

The application of ICT in logistics has just started. The consequent realization of these potentials and further possibilities will influence the development of logistics (Scheer 1999; Kuhn et al. 2002; Straube 2004).

2.8 Risks of Information Technology in Logistics

Although the potentials of ICT application in logistics are considerable, some expectations are too optimistic and many announcements are exaggerating (Kuhn et al. 2002; Schmidt 2000; Straube 2004).

The dangers and risks are often ignored. More errors, mistakes and failures than thinkable can be observed in practice. Negative consequences of the naïve application of ICT and Internet in logistics, of *telematics* and *e-logistics* are (Dittrich 2000; Fritsche 1999; Gudehus 2002/2006; Murphy et al. 2004):

- Although the *software* for planning, scheduling and business processes is updated permanently, in many respects they still *do not keep the promises*. Often the software development is driven more by informatics then by logistics. Some software modules are based on improper assumptions and do not cope with the requirements of business practice.
- The efforts for *adapting standard software* and the time needed for the *implementation* are often underestimated. That is why *customization* of software and *adaptation* to the specific operational needs can lead to a never ending story and harm daily business.
- Instead of realizing the necessary set up, *trade off implementation* is accepted, which leads to *unacceptable simplifications*. The software then hinders daily business instead of optimizing and disburdening it.
- *Logistic master data* are often incomplete or improper structured. Many standard programs lack important logistic master data such as dimensions and weights of article units, packages and load units (see *Sect.* 12.6).
- The offer of too many forecast methods, scheduling algorithms and free parameters causes incorrect applications. The user is swamped with the selection of the proper algorithm and the adjustment of the parameters. Therefore, in practice, dispatchers and schedulers disregard many possibilities of the software.
- *Insufficient data administration* and software maintenance lead to *false results*.
- *Long reaction times* for response, data processing or printouts hinder work flow and reduce efficiency.
- *Batch processing* causes long waiting times and delays the work within the performance stations. It also affects the reaction times for other users.

- *Multiple data registration* of the same information at consecutive I-points and C-points causes unnecessary effort and costs.
- Because certain *logistic master data* are not available, many optimization strategies cannot be realized.
- The program produces *incorrect outputs* because wrong parameters have been set, false assumptions have been made or the programmed algorithms are unknown to the user.
- Incorrect or too simplified scheduling algorithms generate *insufficient proposals*.
- Computers, Internet and software might be *misused* for computer games, private e-mails or other non-business purposes.
- Over*complicated pricing models* for passenger fares, freight rates and logistic services are installed on the computer, which are of doubtful use for customers, sometimes even for the selling company (see *Chap. 7*).
- The potentials of IT induce *exaggerated controls*, lead to *information overload* and stimulate *overstated graphic design* on the display or of print outs. Beautiful, colored or three dimensional figures, diagrams and animations without any use only generate costs and inefficiency.
- In other cases *results and data are insufficiently presented and poorly visualized*. Badly arranged displays, overloaded spreadsheets, endless print outs and statistical junk impede a useful application.
- *Unnecessary data* and *outdated information* are registered, administered, stored and exchanged.
- Parcels, packages and other logistic units carry *too many* or *unnecessary labels*, markings or codes.

In order to avoid the last of these problems the following *IT design principles* should be observed:

- Only as many data and as much control as really necessary
- Instead of plenty information only necessary information
- The user, not the programmer determines the information to be offered

For the effective application of ICT in logistics, *experience*, *sense of proportion* and *judgment* are needed. The most important issue is to focus all IT-activities on the interest of the customer and the goals of the company. The *golden rule of IT* "garbage in, garbage out" has to be kept in mind also for logistics.

The receiver of a punctual, complete and error free shipment is not interested, which tour it took and what stations it has passed on its way. Even for the disappointed customer the recent location of the missing good, the reasons of a delay or the last checking point are of minor interest. The customer wants to receive the expected parcel and ordered goods at the agreed date. That means:

- Identification, communication and information systems are only means to an end and not an end in itself.

They are important tools for generating efficient logistics services, but do not create value for a customer as long as he does not want and notices the additional value. This holds in particular for the application of *transponders*, *smart labels* and *RFID* in logistics (Finkenzeller 2002; Shephard 2004). The most promising areas of transponders for electronic coding and of RFID for the collection of data are the *logistics of empties*, the *object identification* within delivery chains and the *electronic KANBAN* (see *Sect. 12.7*).

2.9 Organization of Company Logistics

Due to its cross-functional character, company logistics should be a separate organizational unit which cooperates closely with sales, production, purchase and other departments of the company (see *Chap. 14*).

The two central responsibilities of *company logistics* are *strategic logistics* and *operative logistics*. The medium and long-term oriented *strategic logistics*, also called *system management* or *network management*, includes *logistic controlling* and *logistic planning*. The short term oriented *operative logistics*, also called *system operation* or *network operation*, consists of *logistic scheduling* and *logistic operations* (Chopra and Meindl 2007; Heymann 1977).

An approved *organization structure* for company logistics is shown in *Fig.* 2.2. In small and medium sized companies the logistic tasks are more concentrated and certain hierarchical levels do not exist. In big companies with many locations further differentiation, specialization and decentralization are opportune. In any organization one must keep in mind: All performance stations of company logistics should be primarily *service stations* focusing on the requirements of the customers and the benefits the company.

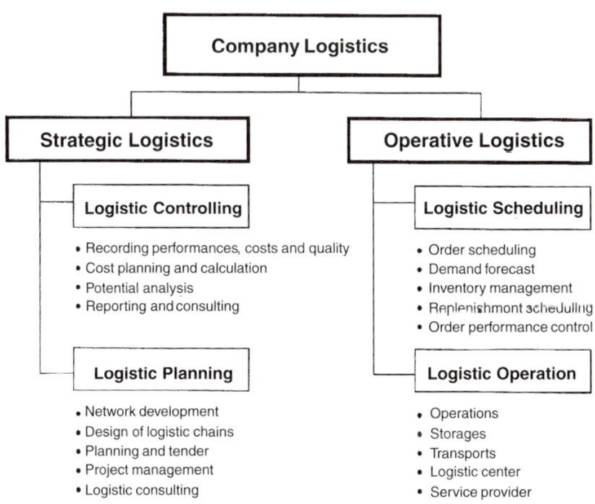

strategic logistics = system management = network management
operative logistics = system operation = network operation

Fig. 2.2 Organization, areas and tasks of company logistics

2.9.1 Logistic Controlling

Logistic controlling monitors the efficient execution of logistic services within the given budgets. It informs the planning department, order scheduling, logistics operations and other business units about the performed services and the logistic costs and prices. Common logistic controlling methods and tools are *logistic cost accounting* and *logistic reporting* (Horváth 1992; Küpper 1991; Lambert and Burduroglu 2000; Seuring 2006; van Amstel and Farmer 1990; Weber 1993).

If some logistic services have been outsourced, the logistics controlling has to develop the *remuneration system* for the *logistic service provider* (see *Sect. 7.3*). Also monitoring of the market prices for logistic services and controlling of the correct billing are tasks of logistic controlling.

2.9.2 Logistic Planning

The overall goal of logistics planning is to guarantee the competitiveness of the company by efficient logistic services. This involves:

- development of company logistics
- delimination and design of the logistic network
- design and optimization of the supply and delivery chains
- planning and implementation of logistic centers and logistic systems
- design and optimization of internal logistic processes
- selection of qualified logistic service providers
- organization of scheduling procedures
- determination of logistic quality levels
- selection and implementation of demand forecast

For these tasks, logistic planning must participate in the design of the information and communication systems and in the data processing of the company.

Both, *logistic controlling* and *logistic planning*, consult other departments of the company, in some cases also customers and suppliers, in all questions of logistics. In particular, the sales and the purchase department must be informed about the possibilities, services and costs of logistics (see *Sect. 14.8.4*).

2.9.3 Logistic Scheduling

Logistic scheduling or *order scheduling* prioritizes the commercially accepted orders and transforms them into internal orders.

The internal orders are transferred to the relevant operations and performance stations (see *Chap. 10*). Further tasks are *time management* and *scheduling* of replenishments, inventories and resources within the operative logistic stations of the company. This includes demand forecasts, inventory control, update of reorder points and safety stocks (see *Chaps. 8* and *11*). Logistic scheduling also monitors whether the involved operative performance stations execute the orders completely and correctly in due time. This requires close cooperation with sales, production, purchasing and external suppliers.

Many schedulers in lower hierarchical position are servants of the sales department or underlings of production. Due to lack of competence or courage some schedulers try to please everybody. Their decisions often do not match the expectations as they are torn between the front lines. Other reasons for poor scheduling are lacking interest of top management for scheduling and missing appreciation for qualified schedulers.

Schedulers and *dispatchers* are the *strategists of the daily business* as they decide on the efficient use of resources and determine the performance of a company. However, they can also cause unnecessary costs and losses. Top management consequently has to implement and ensure an independent, strong and competent scheduling.

2.9.4 Logistic Operation

Logistic operation is responsible for the *effective competitiveness* of all logistic activities of the company. This includes the management of human resources and production facilities within the logistics operations of the company, such as internal transport and stores, distribution centers and the own truck fleet.

If certain operative logistic functions such as the storing of finished products, the operation of a logistics center, the external transport or the complete distribution have been outsourced, the activities of the logistic operations are limited to *system management*, *coordination* and *monitoring* of the performances of the *logistics service providers* (see *Chap. 22*).

2.10 Organization of Scheduling

From the raw material sources to the end users, different supply chains compete against each other (Christopher 2003; Covinato 1992; Kuhn and Hellingrath 2002). Within these supply chains, only companies survive that execute the customer orders in short lead times at minimal costs. This requires *efficient scheduling*.

The supply chains must permanently be adjusted as competition changes and new products are launched. The flows of goods must be adapted to the varying demand of companies and customers. Incoming inquiries and orders have to be immediately processed. As will be shown in *Sect. 10.6*, immediate reaction towards changing requirements can be achieved by *dynamic scheduling*. Therefore, the dynamic scheduling of orders and inventories in supply and delivery chains is the keystone of network management.

Based on a medium term forecast, the scheduler decides which articles should generally be delivered *ex stock*. The criteria for this key decision are *service*, *delivery time* and *opportunity costs*. By the same criteria the scheduler decides whether an actual order should be executed – complete or in parts – from an anonymous stock or whether it should be directly produced in the own factory or specifically procured from a supplier.

The reorder points of storekeeping articles are dynamically calculated in order to achieve minimum costs in a self-regulated manner. Dynamic scheduling ensures market consistent delivery times and a cost efficient ability to deliver. It avoids too high as well as insufficient inventories.

Characteristics, principles and steps of the dynamic scheduling are (Gudehus 2002/2006):

1. *Correct task sharing between scheduling and planning*
   - scheduling in short time intervals for short terms
   - planning in longer time intervals for medium and long terms
   - scheduling of the current demand
   - planning of the long term and project demand
2. *Right organization of scheduling* (*Sect. 10.6*)
   - local scheduling of the single performance stations
   - central scheduling for delivery chains and networks
   - coordination between intra- and inter-company scheduling
   - observation of the principles of subsidiarity and decoupling
3. *Short term dynamic forecast and medium term rolling forecast* (*Chap. 9*)
   - dynamic forecast of short term demand
   - rolling forecast of medium and long term demand
4. *Service dependent decisions about storekeeping articles and order articles*
   - time opportunity of storekeeping (see *Chap. 8*)
   - cost opportunity of storekeeping (see *Sect. 11.12*)
   - storekeeping decision for finished goods and preliminary products
   - repeated update of the storekeeping decision
5. *Current order scheduling* (see *Sect. 10.6*)
   - decision on delivery from stock or production/procurement on order
   - synchronized logistic scheduling and production scheduling
   - optimal consolidation of supply and dispatch
6. *Dynamic inventory scheduling* (see *Sect. 11.13*)
   - selection of the optimal replenishment strategy
   - current calculation of cost efficient reorder point
   - self-regulating assurance of the required stock availability
   - correct procedure of inventory scheduling
7. *Correct and complete master data and cost rates* (see *Sect. 12.6*)
   - regulation of responsibility for the provision of master data
   - calculation of adequate cost rates for logistic services
   - provision of current market prices for logistic services
   - binding responsibility for data input and updating
8. *Task sharing between scheduler and program*
   - scheduling of normal demand by scheduling program
   - scheduling of irregular demand and critical orders by persons

For a broad range of articles and a high number of incoming orders it is essential to assist and disburden the schedulers by advanced *scheduling software*. The scheduling of normal orders and regular replenishments can be performed by program. The scheduler then has time to focus on irregular and express orders, on the implementation and updating of master data as well as on the monitoring of delivery dates and the adherence to requirements.

An efficient, correct and self-controlling support of a scheduler or dispatcher by a program is possible only if the program operates with the right *strategies*, *forecasting methods* and *algorithms*. *Scheduling parameters* such as smoothing factors, current demand, reorder points, safety stocks and replenishment quantities, have to be calculated by the program dynamically from the incoming orders and the actual article stocks. Purchasing, inventory and replenishment strategies and the selection of load and transport units must be adapted regularly to the current demand.

The more standard procedures are executed by computer, the more schedulers can focus on other activities. Hence, the higher the number of employees engaged with scheduling, the higher the potential savings by computerized scheduling. Wherever *remote scheduling* by computer is possible, such as in local production facilities, purchasing departments, sales departments or retail outlets, full time scheduling by persons is no longer necessary. Skilled employees and executives can perform the remaining scheduling tasks in addition to their other activities.

Only in larger business units, an *order center* with a small number of highly qualified schedulers is necessary in addition to local scheduling. Their job is to assist the remote scheduling, to control critical parameters and to adapt the scheduling programs. They harmonize the local scheduling procedures and results with production planning, customers and suppliers.

Also for the inter-company *supply chain management* a central planning and scheduling unit might be necessary. This can be set up by *cooperation* between the planning divisions and order centers of the independent companies or by establishing an inter-company *planning and scheduling center*.

However, in many cases, the information for inter-company planning and scheduling is not available. Only a few companies are willing to hand over unlimited information on sales, orders, stocks and resources to suppliers or customers (Murphy et al. 1996/2004). The readiness for information exchange depends on the additional value of inter-company scheduling compared with the independent scheduling of the single companies. That means:

- Central planning and scheduling can be only realized, if the costs for all participants are reduced or at least kept.

Central scheduling can be an advantage for *sales promotions* of consumables, in *bottleneck situations* and for *storekeeping articles* which are delivered from a *central store*. With exception of these cases, approved *selection rules* and *application criteria* for central planning and scheduling are not known up to now. Their development and evaluation are still open tasks for *logistic research* (see *Sect. 21.17*).

Chapter 3
Project Planning and Realization

Tasks of *project planning* are the design and organization of a system, which fulfills certain requirements under given restrictions at lowest costs. This includes the selection and dimensioning of equipment, resources and other elements, the connection of these elements to performance chains and the design of logistic networks. The tasks of *project realization* are the scheduling of the implementation, the construction and manufacturing of the system elements, the build-up of the whole system and finally the start-up and tests. Both, planning as well as realization, need qualified *project management* (Nicholas 1990; Pintot and Slevin 1987; Lock 1996).

This chapter will provide insight into the possibilities, objectives, tasks and procedures of the planning and realization of logistic projects. After explanation of the *possibilities of action* and the *objectives of logistics*, the *performance requirements* and *restrictions* are discussed. The following sections present the *means for specifying* systems and processes, *computer tools* for planning and optimization, and methods for the *selection of the best solution*. The last section gives a survey on the role of *technology in logistics*.

3.1 Possibilities of Action

In order to plan successfully, one needs to know the *objectives*, *requirements* and *restrictions* of the specific project, as well as the *possibilities of action*. The basic *options for logistics* are (see *Fig. 3.1*):

- *Organizational options*: design of *processes* and *structures*; development, selection, and combination of *strategies* and variation of the *strategy variables*; *linking* of performance stations and *integration* of subsystems to *networks*; *production* or *procurement* either *on order* or *on stock*
- *Technological options*: selection, invention, construction and improvement of machines, devices and transport means; layout of sites, halls and buildings; dimensioning and optimization; specialization, mechanization, automation; application of process control and IT.

T. Gudehus, H. Kotzab, *Comprehensive Logistics*,
DOI 10.1007/978-3-540-68652-1_3,

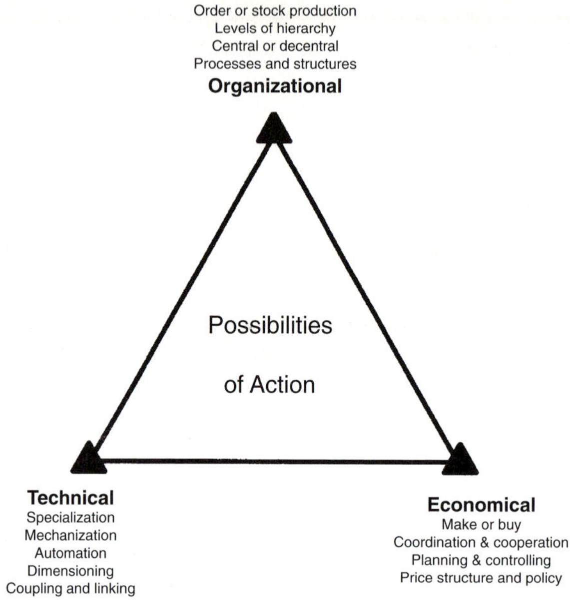

Fig. 3.1 Possibilities of action in logistics

- *Economical options*: *make or buy*; *cooperation* and *alliances*; *synergies* and *economies of scale*; *cooperative planning and scheduling*; setting of *prices* and design of *pricing models*; *utility-dependent remuneration*

The practical use of these options depends on the situation of the company and on the particular project. Before planning a new system, one tries first to adapt, improve or extend the existing systems. Only if the required performance can no longer be achieved within the existing structures at competitive costs, a new system will be envisaged.

However, in order to exploit all possibilities, it is advisable to permanently develop new ideas and concepts. The existing systems should be compared with the own potentials of the company, not only against *key performance indicators* or *benchmarks* of other companies (see *Sect. 4.5*).

Performance, *quality* and *costs* can be improved substantially only by breaking down grown structures, i.e. by *reengineering* and *business processes redesign* (Hammer and Champy 1993; Scheer 1984; Schönsleben 1998). For this purpose, a *potential analysis* is advisable, which points out the *deficiencies* of the existing systems. It reveals, whether it is sufficient to adapt and optimize the present processes and structures or whether it is necessary to design and realize a new system (see *Chap. 4*).

If market requirements and general conditions are changing, the company is forced to rethink its organizational and technical possibilities. In successful companies, rationalization, optimization and redesign are a permanent process. The success of this process, in Japan called *Kaizen*, depends not only on the employees' involvement and motivation but also on the readiness of the top management for change. However, motivation and involvement are not sufficient. Further prerequisites are *competence* and *experience* (see *Chap. 23*).

The *phases of planning and realization* of a logistic project are shown in *Fig. 3.2*. The indicated durations of the single phases are based on the experience from many

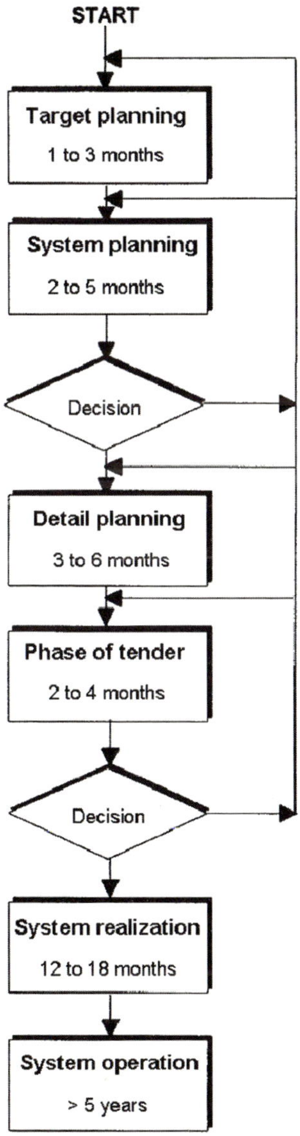

Fig 3.2 Steps of planning and realization of logistic systems

realized projects (Kleiber 1992). Crucial factors for the *duration of the planning* are the *competence* of the project team and the *decision willingness* of the management. The *duration of the realization* and the *success of the project* depend on the *qualification*, the *performance readiness* and the *experience* of the project manager, the contractors and the operators (see *Sect. 13.2*). In addition, the current economic situation influences the realization time.

3.2 Planning Phases

In order to achieve the goals in limited time and to ensure that no useful option is ignored, a systematic planning procedure based on proven methods is necessary. The *phases of project planning* are shown in *Figs. 3.2* and *5.5*. They comprise *target planning*, *system planning*, *detail planning* and *tendering* (Baumgarten 1999; Gudehus 1973a; Kleiber 1992; Pradel 2005). As planning and optimization are *iterative processes*, the steps and loops of these phases must be passed until all requirements are realized.

3.2.1 Target Planning

The general *steps of target planning*, also called *demand planning, goal planning*, *pre-planning* or *basic planning*, are:

evaluation of the demand
definition of the objectives
derivation of the tasks
documentation of processes
determination of functions (3.1)
data collection and analysis
identification of restrictions
quantifying the performances

The result of this starting phase is a *documentation* of the goals and objectives of the project, and of all relevant prerequisites for the following planning phases. The *demand record book* is presented to top management or the principal to authorize the results as basis for the further planning.

3.2.2 System Planning

Depending on the project, whether it is a subsystem, building, logistic center, transport system or a logistic network, *system planning* is also called *material-flow-planning, material-handling planning, blueprint planning*, *system design*, *layout planning* or *concept development* (Baumgarten et al. 1999; Gudehus 1973a; Kleiber 1992). The general *steps of system planning* are:

analysis of requirements and restrictions
segmentation of articles and orders (see *Sect. 5.6*)
specification of logistic units (see *Chap. 12*)
development of strategies (see *Chap. 5*)
design of processes and structures
conception of possible system solutions
static and dynamic dimensioning and optimization (3.2)
layout development (see *Chap. 19*)
design of organization and process control
blueprint planning of halls and buildings
budgeting of investment and operating costs
selection of the optimal solution

As explained in *Sects. 16.8, 17.7, 18.3, 21.9* and *10.10,* the general steps of system planning have to be adapted to the specific systems and projects.

The result of system planning is a *planning report*. In this report the recommended solution is presented in the form of blueprints, diagrams, tables and descriptions. It includes a calculation of the necessary personnel and equipment, a budget of the investment, a rough calculation of the *operating costs*, the *possible stages* and a *time schedule for the realization*. The planning report is the basis for the decision as to whether a project should be realized or not.

3.2.3 Detail Planning

After the decision to realize the project, a detailed planning is necessary in order to work out the solution more precisely and to get the approval of authorities. In this phase, besides logisticians, experts from other disciplines are involved, such as traffic planners, computer specialists, architects, engineers and economists. The *stages of detail planning* are:

updating of planning data
architecture of the buildings
detailed planning of logistic system modules
engineering of constructions and technical systems
planning of the building installations
design of load carriers (3.3)
specification of master data
design of process control
specification of computer hard- and software
approval by technical, state and other authorities
actualization of investments and operating costs
mile-stones and time schedule for the realization

Results of detail planning are *requirement record books*. They specify by descriptions, drawings, tables and plans the required functions, performances, capacities, qualities and further details. The requirement record books for the whole system, subsystems and critical elements are the basis of the *call for tenders*.

3.2.4 Tendering

The goal of a call for tenders is to find the right partners for the realization of a project and/or for operating the system. The general *steps of tendering* are (see also *Fig. 22.2*):

determination of the procedure
selection of qualified suppliers and contractors
preparation of bidding documents
conception of performance and quality remuneration
draft of price blank forms (3.4)
working out and submission of offers by the bidders
evaluation and comparison of the offers
negotiations with selected bidders
contract negotiation and conclusion

At the beginning, it is necessary to decide on the modalities of the tender. It can be a *performance tender* for a service offer, a *functional tender* for a general contractor or specified *single tenders* for the different elements, subsystems or performance packages. At the end, either several *single suppliers* for the part systems or *service providers* for performance packages or only one *general contractor* or *system provider* for the whole system will be awarded with the realization and /or the operation (see *Chap. 22*).

3.3 Realization Steps

The realization of a logistic project comprises the following steps:

project management
time-control, performance-control, cost control (3.5)

execution planning
building application and approval (3.6)

site development
building of traffic areas and outdoor facilities (3.7)

foundation work
erection of the buildings
installation of house automation and technique (3.8)
construction and production of system elements
delivery and assembling of the logistic modules

duty record book for computer hard- and software
procurement and installation of the hardware (3.9)
programming and implementation of the software

test runs of subsystems and total system
acceptance of partial and total performances (3.10)
start-up of the complete system

recruitment of employees
training and instruction (3.11)

Some of these steps are executed in parallel. After the successful tests of the functions, performances and availability the realization phase ends with handing over the completed system to the user (see *Sects. 13.7* and *13.8*).

3.4 Logistic Objectives

The objectives of logistics result from the demands of society, the goals of the economy and the objectives of the company. In times of war, *military goals* dominate logistic objectives (Jomini 1881).

The *objectives of company logistics* are:

achievement of *performance*
securing of *quality* (3.12)
minimization of *costs*

They result from the economic goal of any company to achieve high profit. The correct·execution of orders with adequate quality is a prerequisite for continuous turnover at good prices. In combination with minimal costs, this assures competitiveness and sustainable high profits.

3.4.1 Humanitarian Objectives

Humanitarian objectives of logistics, as well as of technology are:

- maximal safety for people
- reliable supply of vitally important goods
- relief of people from heavy manual work
- easing of work by ergonomic design
- elimination of primitive routine work
- forecasts of driving times, congestion and redirections of traffic
- affordable, frequent and area-wide transport connections
- shortest availability of emergency vehicles, police and fire brigades
- quickest possible care of sick and injured persons

The extent, to which the different humanitarian objectives have to be met, is determined by law and legal norms, by factory inspection boards, safety and health control and by company regulations.

3.4.2 Ecological Objectives

Ecological objectives are important for all logistic systems. For *reverse logistics*, they are crucial. The ecological objectives of logistics are:

- avoidance and reduction of waste
- small pollution
- reduction of noise
- protection of nature (3.13)
- minimal use of material and resources
- minimal energy consumption
- reduction of area usage

Many ecological objectives, such as minimal consumption of energy, material and area, are compliant with the economic objectives. Other objectives that can only be achieved with additional effort and expenses are prescribed by law (see *Chap. 23*).

3.4.3 Performance Objectives

The central objectives of performance are:

- execution of orders
- observance of due dates and delivery dates
- cope with the required throughput (3.14)
- storing the resulting stocks
- performance of additional services

Benchmarks for observing these objectives are the specific performance requirements. The requirements for each of the objectives (3.14) have to be quantified before planning and realization and must be updated permanently during operations (see *Sect.* 3.6.).

3.4.4 Quality Objectives

Dependent on the subject, it is necessary to differentiate between *product quality, service quality* and *performance quality.* Product quality is achieved through technical operations in a factory for physical goods. Service quality depends on the scope of services. Performance quality is essential for performance systems, in particular for logistic systems.

The basic features of performance quality are *availability*, i.e. the probability of finding a system capable to perform a certain operation, and *reliability*, i.e. the probability that the operation is executed without failures.

The *availability* is the relation of the time of correct operating capability of a system to the total *operating time* T_{tot}, which is the sum of *capability time* T_{cor} and *failure time* T_{fail}:

$$\eta_{avai} = T_{cor}/T_{tot} = T_{cor}/(T_{cor} + T_{fail}). \quad (3.15)$$

Fig. 3.3 Objectives of logistics

In order to measure the *reliability* of a system for a certain function, the number of *correctly executed functional events* n_{cor}, the number of *failed events* n_{fail} and the total number of *events* $n_{tot} = n_{cor} + n_{fail}$ are counted within an operating time of sufficient length. The *measured reliability* is the relation of the number of correct events to the total number of events:

$$\eta_{rely} = n_{cor}/n_{tot} = n_{cor}/(n_{cor} + n_{fail}). \tag{3.16}$$

The measurement can be restricted to partial functions, to special aspects of the performance or to other features of the operation. Availability and reliability are linked by the performance rate λ and the *Mean Time Between Failure* MTBF = T_{fail}/n_{fail}. The dependency of the availability on performance rate, MTBF and reliability will be investigated in *Sect. 13.6*. There also the availability and reliability of performance chains and systems will be derived from the respective properties of the elements.

The most important *quality objectives of logistics* and their features are:

- *performance availability* η_{PA}
  availability to deliver storekeeping articles (see *Sect. 11.8*) (3.17)
  availability to execute production orders

- *shipment quality* η_{SQ}
  completeness
  intactness (3.18)
  faultlessness

- *delivery reliability* η_{DR}
  observance of due dates (3.19)
  observance of pickup and delivery dates

The performance availability (3.17) is defined by relation (3.15). The shipment quality or shipment reliability (3.18) and the delivery reliability (3.19) are given by relation (3.16). Together the quality objectives (3.17), (3.18), and (3.19) determine the service level:

- The *service level* η_{SL} is the probability that the orders are executed and delivered *complete* and *correct* in *due time*.

The service level – also called *logistic quality* – is the product of performance availability, shipment quality and delivery reliability:

$$\eta_{SL} = \eta_{PA} \cdot \eta_{SQ} \cdot \eta_{DR} \tag{3.20}$$

For instance, with an availability of a storekeeping article of 98.0%, a shipment quality of 99.0% and a delivery reliability of 95.0% a service level of $\eta_{SL} = 0.98 \cdot 0.99 \cdot 0.95 = 92.2\%$ is achievable.

A specific quality objective for the *order management,* an *order center* or a *call center* is the

- *readiness to inform* about delivery ability, dates, status, destination and arrival of orders and shipments.

This quality feature of company logistics depends on the number of correctly answered inquiries for information and is measured correspondingly by relation (3.16). A further important logistic quality objective is the

- *flexibility to perform* under changing requirements, seasonal fluctuations and changes of the assortment.

The flexibility of a system or an organization cannot be measured directly, but may affect the service level severely.

The *benchmarks* for achieving the quality objectives are the *quality standards*, which have been agreed with customers or determined by top management, government or market. *Quality defects* are unacceptable deviations from the quality standards. They are collected in *default statistics* and related to the agreed standards.

3.4.5 Cost Minimization

The overall goal of a company is to reduce costs without affecting performance and quality. Possibilities to reduce logistic costs are (see also *Sect. 6.9*):

- avoidance or reduction of handling and transport
- optimization of inventory (see *Chap. 11*)
- effective use of infrastructure, areas, buildings and transport routes
- maximal utilization of capacities, transport means and networks
- increasing the performance of bottlenecks (see *Chap. 13*)
- improvement of the data and information flow (see *Chap. 2*)
- efficient use of personnel
- optimal use of time (see *Chap. 8*)
- employment of logistic service providers (see *Chap. 22*)

(3.21)

The success of these and other options is measured by the achieved reduction of the *operating costs* in relation to the necessary *investment* (see *Sect. 5.1*).

3.4.6 Goal Conflicts

Many logistic goals and objectives are incompatible. The top management, not the logisticians must solve these goal conflicts by prioritizing the different objectives (Churchman 1961). Logisticians can reveal the conflicts and support the decision by a *utility value analysis*, i.e. by weighting the objectives corresponding to their utility value.

However, the utility value analysis often conceals unsolved conflicts. In order to identify the real and to avoid ostensible conflicts, the needed functions must be specified, the required performances be quantified and the quality expectations be determined.

If the functions, performances and quality expectations are known, it is possible to check, whether the cost objectives are compatible with them or not. In case of incompatibility, the performance and/or quality requirements can be adapted to meet the cost guidelines. In some cases, also the frame conditions must be questioned and eventually changed as far as possible.

3.5 Frame Conditions

Frame conditions are *constraints*, *restrictions* and other *restraining conditions* for a project and the company. They determine the *limitations* and the *fix-points* for planning and operation and reduce the possibilities of action. Frame conditions for logistic systems are:

- *Spatial restrictions*

The locations of the sources and sinks are fixed or restricted to certain areas. The space for storing and the widths, heights and paths for transports are limited.

- *Temporal restrictions*

Operating times and shift schedules are predetermined. Timetables are set. Processing steps require certain *cycle times*. The durability of perishables is limited. Legal restrictions of the working hours must be respected.

- *Technical constraints*

Properties of the goods and load units, limited storage capacities, speed and capacity of transport means, limit performances of transport elements or the interfaces to adjoining systems reduce the applicable techniques, load carriers, transport means or traffic routes.

- *Structural constraints*

The existing internal or external infrastructure limits or affects the possible solutions. The *logistic infrastructure* is given by transport nets and traffic routes and by the locations of transshipment points and logistic centers.

- *Organizational conditions*

Prescribed processes, unavailable data, limited information, existing coding systems, computers, standard software, unchangeable business strategies and the organization of the company prevent certain solutions.

- *Economical restrictions*

Operating costs have to be calculated with rates for depreciation, interest, human resources and other cost factors, which are prescribed by the company. Performances and prices for outsourced services are more or less given by the market. The capital resources of a company are limited. The *return on investment* (ROI) is limited.

- *Safety requirements*

Internal and external safety requirements for people and goods are binding. Access to goods and availability of critical articles must be ensured. Deprivation of valuable, periled or dangerous goods by theft, deterioration, casualties and fire should be averted. The probability of lengthy breakdowns and unacceptable side effects must be minimized.

- *Competitive conditions*

Services, performances and quality standards are determined by the market. Lower costs and prices of a competitor can set the agenda for the company's logistics.

- *Legal restrictions*

Laws, regulations, tariffs, norms, rules and ecological restrictions are coercively to be considered.

These frame conditions limit the possible solutions. Some restrictions are *exclusion criteria* or *knock out (KO) criteria*, which eliminate certain solutions. In order to avoid working on unsuitable solutions, it is advisable to list all identifiable *KO-criteria* before planning.

Not all frame conditions are really fixed. In some cases it is possible to enable an interesting solution by elimination of a restriction and to win more than the costs for changing the obstacles.

3.6 Performance Requirements

The performance requirements are either primary or secondary requirements. *Primary performance requirements* are determined by customer demand or by top management. *Secondary performance requirements* are derived from the primary requirements.

In order to avoid misunderstandings between economists, logisticians and engineers, *monetary* or *commercial quantities* and *non-monetary* or *technical quantities* must be distinguished carefully. These are often mixed up by using the same word for different subjects, such as the term *inventory* for the *stock value* and for the *stock quantity*, or of different words for the same subject, e.g. the terms *sales* and *shipments* for the number of article units sold in a certain period. With known prices and costs, the monetary requirements of logistics can be calculated from the non-monetary requirements.

The *primary requirements of logistics* are:

3.6.1 Product Attributes

properties of products and articles
measures and weights of article items
measures, weights and content of the sales units (3.22)
purchase prices, production costs or sales prices

3.6.2 Order Requirements

type of orders to be fulfilled
number of orders per period
number of positions per order (3.23)
number of items or sales units per position
number of performance units per order line

3.6.3 Time Requirements

pickup dates
shipment dates (3.24)
delivery dates or times

If the *logistic units* and the *packaging hierarchy* are known, the product requirements, order requirements and time requirements can be converted into order quantities, performance rates and material flows (see *Chap. 13*). This results in the first category of *secondary requirements of logistics*, the

3.6.4 Flow Requirements or Dynamic Demand

performance rates [PU/PE]
piece flows [items/PE]
volume flows [m^3/PE] (3.25)
weight flows [t/PE]
value flows [€/PE, $/PE...]

The first four of these requirements are *non-monetary quantities*. They determine the capacity, performance and dimensions of the stations and the structure of a logistic system. *Value flows* are the piece flows times the price per piece. They are *monetary quantities*. *Sales* and *turnover* are the value flows of all articles sold or delivered in a certain *period* [PE] of time. The monetary values are needed e.g. to calculate the *economic order quantities* (EOQ) (see *Chap. 11*).

The stochastic fluctuations of the piece flow determine the *safety stocks* necessary to assure a required *availability* of the storekeeping articles (see *Sect. 11.8*). From the safety stocks and the strategies of inventory management result the second group of secondary requirements, the

3.6.5 Inventory Requirements or Static Demand

number of storekeeping articles
stock per article [AU/article] (3.26)
inventory per article [€/article, $/article ...]

The article stock, i.e. the stored quantity of an article, is a *non-monetary quantity*, whereas the *inventory*, i.e. the *stock value* of an article is a *monetary quantity*. The inventory is the stored quantity times the monetary value of the article, which can be the purchase price, the production value or the sales price. The different monetary values used in practice are misleading. Therefore, the monetary value, on which the inventory is based, should be stated when inventory figures are reported.

Holding inventory of articles and buffering quantities of goods is not an end in itself, but a means to fulfill certain objectives. Stock level and buffer quantities are important parameters for planning and optimizing storage systems, logistic processes and value chains. The planning of the assortment, the determination on in-house and out-house production and the delimitation of the storekeeping articles are important strategic decisions. They have to be made before planning and should be permanently revised during operations.

The buffer quantities of *non-storekeeping products* result from the strategies and the synchronization of the production and delivery steps (see *Chaps.* 8 and 10). The optimal stocks of *storekeeping articles* result from the minimization of inventory and performance costs (see *Chap. 11*). Therefore, the cost optimal stock is normally not the minimal possible stock. *Just-In-Time* and *JIT-procurement* without intermediate stocks at the supplier or the customer and without buffer at the point of consumption, are generally not cost optimal.

3.6.6 Time Dependency

Characteristics of most performance requirements are short term *stochastic fluctuations*, which are the result of randomly incoming orders or demand, and medium- and long-term *systematic changes*, that result from daily, weekly and annual changes in production and consumption.

As an example for systematic changes, *Fig. 3.4* shows the seasonal change of the *total shipments* and the *inventory level* for the storekeeping articles in the logistic center of a retail company (see also *Fig. 9.9*). Due to stochastic fluctuations and other unpredictable influences, the performance requirements of the future can never be determined precisely. Experience has shown that the *error range* is at least ±5%. Hence, it is unnecessary to apply formulas or algorithms and to perform calculations or simulations with accuracy higher than the error of the input data.

3.7 Determination of Planning Data

The basic *planning data* for a project are the quantified performance requirements and frame conditions. They are either pre-set by the top management or derived from the expected sales values and planned production for a certain operating time until the planning horizon.

As planning and realization of a logistic system lasts at least one or two years, it is not advisable to plan a new system for a *planning horizon* shorter than 5 *years*. If possible, a planning horizon of *10 years* should be considered. In addition it is advantageous to develop a *stage concept* for the flexible realization of the whole system in economical steps with increasing demand.

It is common practice of many planners, when outlining the capacity for a new storage system to simply extrapolate the actual inventory up to the planning horizon with sales augmentation multipliers. As shown in *Sect. 11.9*, this approach can lead to excessive storage capacities and may disregard important potentials for inventory reductions.

Other planners derive the future stock level from the planned sales and demand by turnover factors. However, turnover factors from the own business or from other companies are risky and often misleading. These *key performance indicators* include specific prerequisites and are generally not transferable. In particular, if the prices and the basis of the cost calculations are unknown, *benchmarks* from other companies are dangerous (see *Sect. 4.5*).

Planning and scheduling of the stocks within the supply chains should be based on the non-monetary requirements *measured in article units*, not in monetary values. The future demand can be derived from the demand of the past by *programme planning* and *forecasting* of the *real sales* without price changes.

The non-monetary flow requirements (3.25) and the number of load units, bins or pallets in a store must be calculated from the throughputs and stocks, the sizes of article units and from the capacities of the load carriers. As shown in *Chap. 12*, the

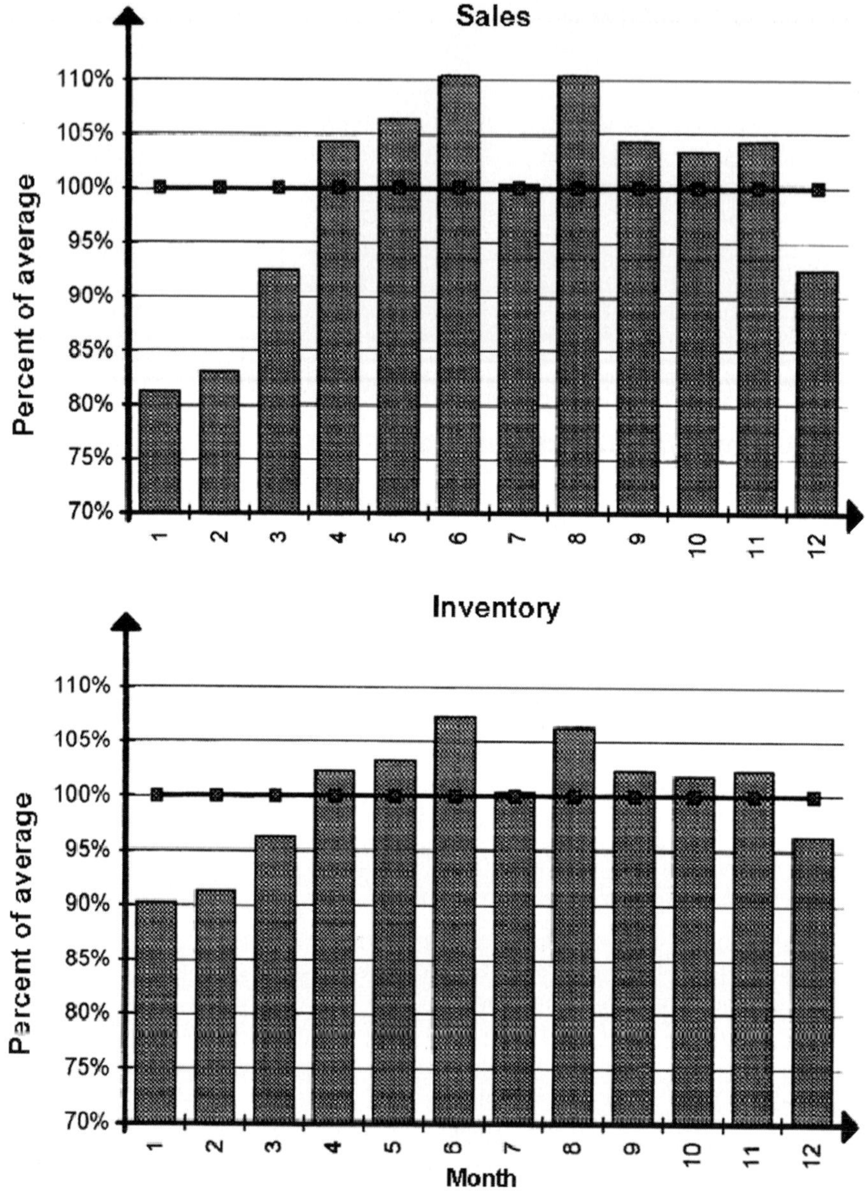

Fig. 3.4 Seasonal deviations of sales and inventories from annual average for storekeeping articles in a retail logistic center

number of load units can be minimized by appropriate *packing* and *filling strategies*. This leads to the *load unit planning rules*:

- The extrapolation of the number of existing load units by growth factors results in too high stock and throughput data.
- By optimal selection, design and allocation of the load carriers, stocks and throughputs of the load units can be minimized.

The most reliable starting information for the determination of planning data are the orders and article data of a past business period. They should be available in the order management system of the company for at least a full accounting year. However, important *logistic master data* such as measures and weights of the article units are often missing. If not available in product specifications and catalogs, the missing data must be required from the suppliers or measured directly in the production or warehouse.

By the mathematical methods described in *Chap. 9*, the monthly demand of standard articles and services can be forecasted quite reliably for the coming years from the consumption or shipments of the past, as long as these show a recognizable systematic behavior. That means (Fisher 1997):

- The future demand for *standard articles* and *standard services* can be *forecasted* by mathematical methods with relatively high accuracy and reliability.

This reduces the sales risks but also affects the profits as most companies prefer to offer low-risk standard articles and services. To keep the business and to replace expiring articles, it is inevitable to offer also new products, promotion goods, fashion articles and special services. The future demand of special articles and services cannot be derived from the past, but must be *estimated* based on experience, market analysis, and comparison with the life cycles of comparable products. This means:

- The future demand of *special articles* and *special services* can only be *estimated* with low accuracy and high risks.

However, the risks are compensated by higher profits, as only a few companies dare to develop and offer special articles and services.

Any determination of planning data involves the danger of producing graveyards of irrelevant data. Other data, that are important for planning and optimization, are missed later. This leads to the *planning data recommendation*:

- As few planning data as possible, only as many as absolutely necessary.

Before starting the planning, it is essential to decide, which data of what accuracy are really necessary, where they are available or from where they can be derived. If certain data are not directly available, it is possible in many cases to calculate them from known data by *correlation factors*.

3.8 Presentation of Systems and Processes

In order to make the structures and processes of a system transparent, it is necessary to separate the operative and the administrative performance stations and to scrutinize their relationships. In addition, for each performance station, the *key performance indicators* (1.3) must be specified. Also the incoming and outgoing material flows and data flows should be known. The results of this *input-output-analysis* can be presented either as *block diagram*, as shown in *Fig. 1.6*, or in a *spreadsheet program*.

The spatial, temporal and logical relations between the performance stations and the flows within the system can be depicted as

structure chart
process chart (3.27)
flow chart

Each of these three *system charts* shows another aspect of the system. Only together they give a complete picture. However, the different aspects should be kept apart and not presented in one blended picture.

The input-output-analysis of the performance stations and the compilation of the system charts are helpful tools for a comprehensible presentation of the structures, processes and relations and help to recognize deficiencies. Years of logistic consulting revealed the *complexity rule*:

- Complex material flows and data flows, unclear flow charts and several process chains for the same result indicate potentials for improvements.

The system charts (3.27) allow finding *bottlenecks*, identifying *weak points* and eliminating *flaws* of the system. These and other *tools for system visualization* are useful also for the design of optimal process chains and logistic systems (Scheer 1984; Krallmann 2002).

3.8.1 Structure Charts

A *structure chart* or *structure diagram* is a picture of the *spatial structure* of a logistic system. The *standard symbols* of structure charts as well as of process charts are:

- bold bordered rectangles representing operative stations
- fine bordered rectangles representing administrative stations
- thick directed lines standing for material flows
- dotted directed lines symbolizing data and information flows

Material flows run from one operative performance stations to another. Information flows connect administrative with operative or with administrative performance stations.

A typical structure chart with material and data flows shows *Fig. 3.5* for the logistic structure of a bottling company. This was first established for a business case in the *chemical industry* and has later been applied in a project for a *beverage company*. Further examples for structure charts and process charts, which have been worked out for other projects, are presented and described in *Sects. 21.11* and *21.12*.

A *quantified structure chart* indicates the material and information *flow rates* as well as the *stocks* and the *buffers* before and within the stations. If the breadths of the lines are drawn proportional to the flow rate, the structure chart is called *Sankey-diagram* (Schulte 1995; VDI 3300 1959).

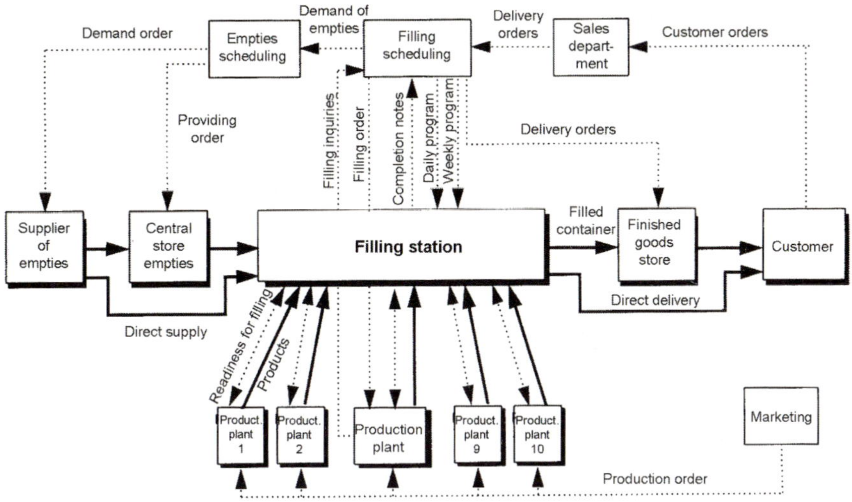

Fig. 3.5 Structure chart of a bottling company
bold framed rectangles: operative performance stations
fine framed rectangles: administrative performance stations
bold directed lines: flow of materials and goods
dotted directed lines: flow of order, data and information

3.8.2 Process Charts

A process chart shows the *chronological sequence* of the consecutive performance stations for a business process. The presentation of a process chain follows the path of a selected object through a system. Any possible flow of the object generates a different process chain. For example, the different order chains of the bottling company with the structure chart *Fig.* 3.5 are presented in *Fig. 3.6*.

Which objects are selected and what processes are analyzed depends on the specific problem. For example, for the business process *customer delivery*, the specific objects are the *customer orders* and the processes are the *order processes*.

The objects passing an *order chain* are immaterial orders, which are first processed in administrative stations and later executed in operative stations resulting finally in a product, a service or another kind of executed performance.

The objects passing a *logistic chain* are material objects. These can be consignments to be dispatched, bottles or cartons to be filled, loose tobacco to be transformed into cigarettes or full pallets passing a logistic center.

3.8.3 Flow Charts

Flow charts present the *chronological sequence* and *logical connection* of the single process steps. *Figure* 3.7 shows the *standard symbols* that are used for system analysis and process design (DIN 66001 1993; Slack et al. 2004), i.e.:

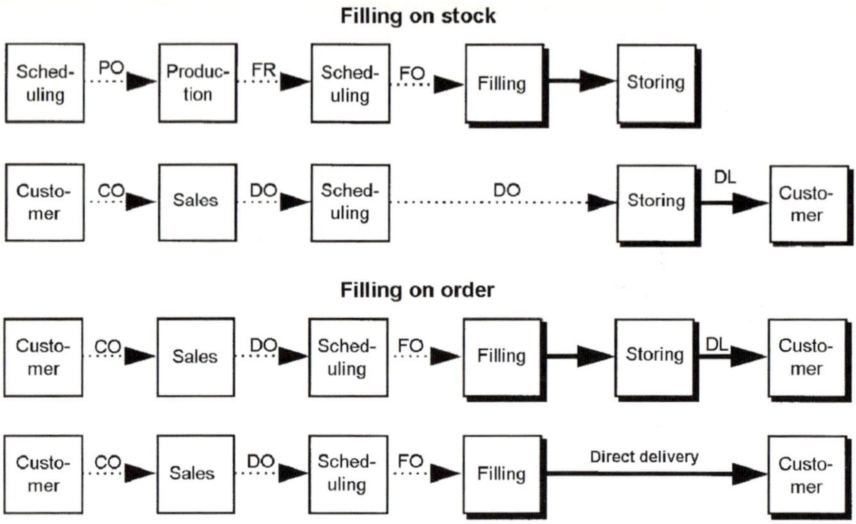

Fig. 3.6 Order chains of a bottling company

CO customer order	FR filling request
DO delivery order	FO filling order
PO production order	DL goods delivery

- *Rectangles* depicting actions that are started by an incoming information or object and which release objects or information
- *Rhombi* depicting a conditional branching dependent on a decision or a comparison of information

The rectangles are marked with the activities. The rhombi contain the branching conditions. Directed arrows connect the rectangles and rhombi, corresponding to the chronological process and logical connection.

Sequences of activities containing a longer series of single actions and internal decisions are depicted by rectangles with a bold or double side frame. The internal structure of such a *sub-process* can be presented in a separate chart, where the *incoming interface* is marked by a circled I and the *outgoing interface* by a circled O. By this means the complex processes of larger systems are made transparent.

3.9 Selection of the Best Solution

To select the best solution, all suitable solutions have to be assessed, evaluated and compared. Methods of selection are *feasibility study*, *performance analysis*, *economic evaluation* and *utility value analysis*. These methods can be applied during the planning process on subsystems and at the end to the complete solution.

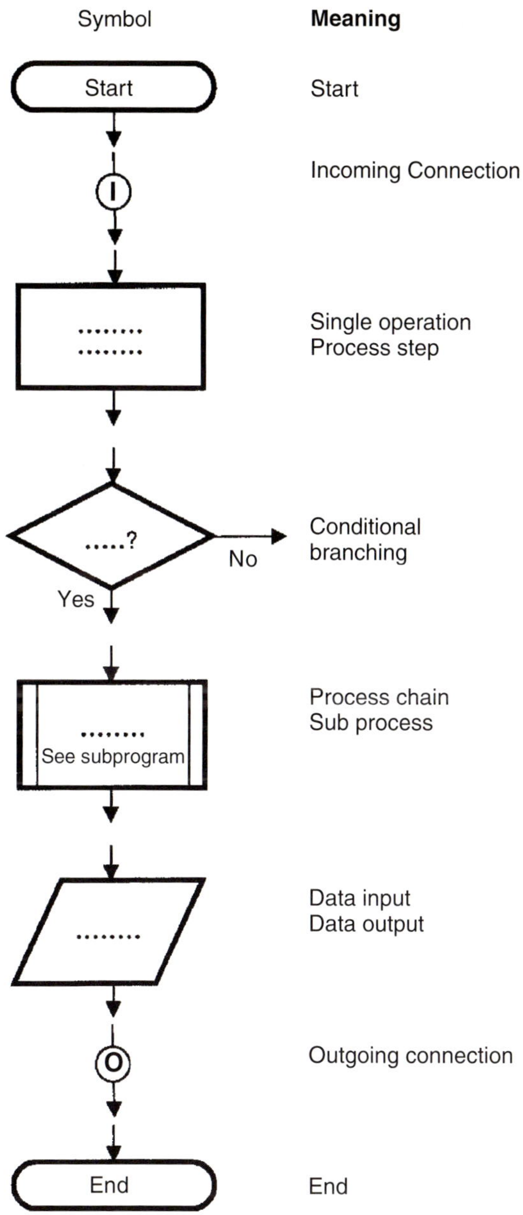

Fig. 3.7 Standard symbols of flow charts according to DIN 66001

3.9.1 Feasibility Study

A feasibility study investigates the suitability and the feasibility of principally possible solutions. This involves checks of

- technical feasibility
- fulfillment of the performance requirements
- compliance with the frame conditions
- realization within the given time

Goals of a feasibility study are the selection of realizable solutions and the elimination of inappropriate solutions.

3.9.2 Performance Analysis

Feasible solutions generally have different *limit performances* and *reserve capacities*. In the performance analysis, the limit performances, reserve capacities and the *flexibility* of the feasible solutions are assessed and compared. The goal is to select the most effective solutions (see *Sect. 13.7*).

3.9.3 Economic Evaluation

In an economic evaluation, the investments and the operating costs of the feasible, effective and most flexible solutions are compared with each other. Besides the *investment* and *operating costs*, the comparison can include the *performance costs* and the *return of the investment* (ROI) (see *Sect. 5.1*). The goal is to identify and to select the solutions with the lowest operating costs at maximum capacity, which can be realized within a given investment budget. The economic evaluation may stop the realization of a project or a plan if the budget is not kept or the ROI benchmark cannot be achieved (see *Sect. 6.9*).

3.9.4 Utility Value Analysis

In many cases, several solutions exist that fulfill all performance requirements, but differ in flexibility, mechanization, automation, and other features. For comparing the non-monetary features of different solutions, a *utility value analysis* or other *scoring methods* are suitable. Afterwards, in a *cost-benefit analysis* the resulting *utility values* are related to the cost differences of the relevant solutions (Churchman 1961; Zangemeister 1972). The steps of a utility value analysis are:

1. Development of a catalogue of independent, non-monetary *ranking criteria* for all relevant properties of the systems to be compared (see *Table 3.1*)
2. Selection of a suitable *evaluation scale* to assess the degree of fulfillment of the evaluation criteria (see *Table* 3.2)
3. Derivation of the *weights* for the ranking criteria from their relative importance and utility for the company
4. *Scoring of the evaluation criteria* for all relevant solutions based on the degree of fulfillment of the ranking criteria (see *Table* 3.3)

Table 3.1 Ranking criteria for comparing logistic systems

Ranking Criteria	Determining Factors
Service Quality	Kind and frequency of failures Ex-stock availability Delivery times and due-date reliablility Consequential costs of failures
Availability	Reliability MeanTime between failure (MTBF) MeanTime to restore (MTTR) Redundancies Costs of interruptions
Flexibility against	Variations of throughput and inventory Variations of programme and articles Change of logistic units Alterations of order structure Change of locations and number of sources and sinks
Flexibility by	Reserves of performance and capacities Adaptability, modularity, extendibility Specialization, universality, usability
Compatibility	Connection ability Interface requirements Fulfilment of frame conditions
Manpower	Workforce Qualification Specialists and know-how
Technical Risks	Breakage, damages, deterioration, rejects Shrinkage, theft, burglary Fire, water or accidents Accidents and safety for workers Consequential damages
Cost Risk	Exceeding budgets Consequence of bad planning Risks of forecast and usage Variability of cost factors and fixed costs

5. *Calculation of the utility value* or *total score* by summation of the partial scores resulting from multiplication of the weights with the single scores
6. *Sensitivity analysis* of the utility value by varying the weights of the most important ranking criteria

The scoring should be performed by several members of the planning team and by competent representatives of the management and the later user. Resulting are the *utility values* or *total scores* for the solutions to be compared.

Table 3.2 Possible evaluation scales for the ranking criteria

Positive criteria Ranking	**Negative criteria** Ranking	**Ranking scales** Mark	 Points	 Ranks
Excellent Optimal fulfilment	Minimal Negligible	1	8 to 10	2 ++
Good Better than standard	Minor Tolerable	2	6 to 8	1 +
Sufficient Up to standard	Average Fairly acceptable	3	4 to 6	0
Just Sufficient Less than standard	Below Average Just acceptable	4	2 to 4	–1 –
Insufficient Almost not acceptable	Too high Almost intolerable	5	1 to 2	–2 – –
Insufficient Out of standard	Unacceptable Intolerable	6	0 K.O.	x

Table 3.3 shows the partial scores and the resulting utility values of three different storage and order picking systems for palletized goods, which are described in detail in *Chaps. 16* and *17*. The automated *high bay store* with dynamic commissioning has the highest total score of 2.90. This has to be compared with the total score of 2.80 for the *narrow aisle channel store* and with the total score of 2.55

Table 3.3 Utility value analysis of three different storage and commissioning systems

Ranking criteria	Weight	**Solution 1** Fork Lift Truck Store with static article access **FTS**	**Solution 2** Narrow Aisle Store with static article access **NAS**	**Solution 3** High Bay Store with dynamic article access **HBS**
Service quality	15%	3	2	2
Availability	15%	2	2	3
Flexibility	25%	2	3	4
Compatibility	10%	2	4	3
Manpower	15%	4	3	2
Technical risk	5%	4	3	2
Cost risks	15%	2	3	3
Utility value	100%	**2.55**	**2.80**	**2.90**

Solution 1: truck operated store with conventional order picking
Solution 2: narrow aisle channel store with conventional picking
Solution 3: automated high bay store with dynamic order picking

for the *fork lift truck store*, both with conventional commissioning. In this case the most favorable solution is the truck operated storage system with conventional order picking.

The utility value analysis should be applied with care and judgment. The example *Table 3.3* shows how close the utility values of different solutions can be, as the sums of the evaluated criteria often compensate the differences. Even minor alterations in the weights or of the evaluations can change the ranking.

Even more dangerous methods are *cost-benefit-analysis* (Christopher 2003; Frazelle 2001) and *balanced scorecard* (Kaplan 1996). Here the non-monetary utility values or total scores are related to cost differences. As long as the score of a solution is better than the score of another solution with higher or equal costs, the decision is clear. However, if a better score is connected with higher costs, the decision stays open, since it is impossible to determine the adequate cost difference for a difference of the scores. That leads to the *rule*:

- Utility value analysis and cost-benefit analysis support the selection of the best solution, but generally cannot replace the final decision.

The utility value analysis is also applicable for the comparison of different offers in the *tendering phase* or for the selection of a suitable organizational solution. However, in any case the final decision has to be taken by the top management.

3.10 Planning and Optimization Tools

Nowadays, computer tools are of high value and inevitable for planning, optimization and scheduling of logistic systems and performance chains. Their objectives and applications are manifold: extensive calculations with mass data; implementation of complicated formulas and algorithms; examination of dependencies on several parameters; system optimization; sensitivity analysis; system and process simulation; support and shortening of planning and scheduling.

However, for many reasons an uncritical application of computer tools can lead to wrong or useless results: The tool may be over simplified, too universal, too specific, too complex, opaque, inflexible, over accurate or in some cases simply wrong. Some programmers do not consider the consequences of assumptions made only in order to simplify the programming. Others change algorithms, strategies or structures without authorization.

If experienced practitioners and analysts access the output and reject implausible results, only the programming and computing effort is in vain, whereas the blind application of false results can lead to enormous damages (see *Sect.* 2.8). The following *rules for tool-programming* help to avoid these dangers:

- as simple but also as realistic as possible
- as specific as necessary but also as universal as possible
- the results should not be more accurate than the input values
- user-friendly and comprehensible program configuration
- consistent documentation of program steps, formulas and algorithms
- plausibility checks of the results by control calculations
- extensive tests and debugging by calculations with real data

(3.28)

Many programs and software tools have been developed based on the algorithm and formulas presented in this book. They are applied for

collection and calculation of planning data
demand forecast and development of scenarios (*Chap. 9)*
order scheduling and production planning (*Chap. 10*)
selection and allocation of load carriers and transport means
performance calculations for machines vehicles, and personnel
dimensioning and optimization of storage and commissioning systems
calculation of limit performances (*Chap.13*)
calculation of queuing effects (*Sect. 13.5*)
dimensioning and optimization of transport systems (*Sect.18.8*)
calculation of investment and operating costs
scheduling of inventory and replenishment (see *Table*11.7)
model calculation for remuneration systems (*Chap. 7*)
calculation of cost rates, prices and tariffs
calculation of order-, customer- and article-specific logistic costs
dimensioning and optimization of logistic centers
selection of optimal transport chains (*Sect. 21.10*)
determination of transport optimal locations (*Sect. 18.10*)
optimization of network structures (*Chap. 21*)
tour planning and optimal routing (*Sect. 18.11*)
simulation of dynamic scheduling in supply networks
selection and optimization of logistic chains (*Chap. 21*)
optimization of logistic systems and networks

(3.29)

Computer tools for many logistical applications are *spreadsheet programs* such as MS-EXCEL and MS-ACCESS. In order to generate customized programs for specific projects, modularized standard programs can be combined and linked.

As an example from a consulting project, *Fig.* 3.8 shows the structure and the influence factors of a process model for the costs of the possible *supply chains* between industry and sales outlets.

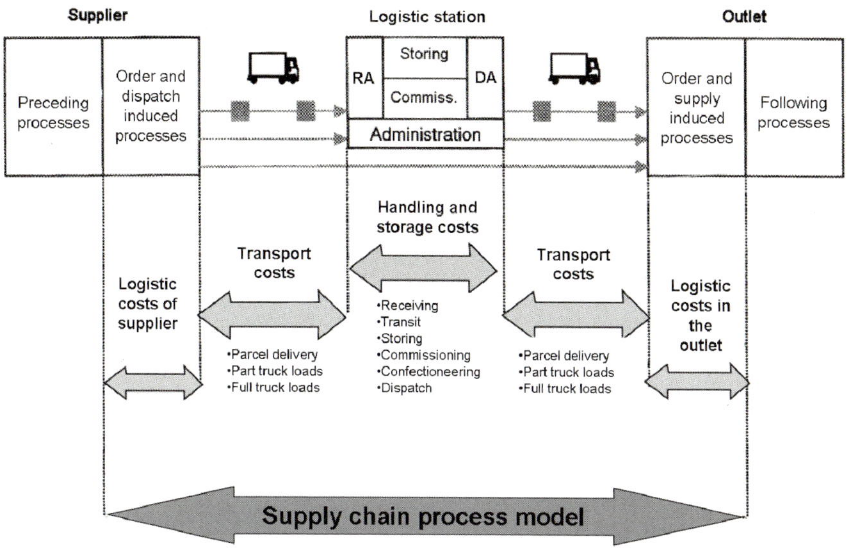

Fig. 3.8 Process cost model for selecting the optimal delivery supply chains between industry and retail

3.11 Technique and Logistics

Technique enables the solution of logistic tasks and the achievement of logistic objectives. Therefore, technology, i.e. technical know-how, is essential for logistics. Who is not interested in technology and ignores the technical possibilities will never be a competent logistician.

3.11.1 Technical Development and Possibilities

The technological development and the application of technique in logistics evolved in *historical phases*, which correspond to the *technical possibilities of action*. These are:

1. *Mechanization*: Certain functions, such as handling, moving and lifting, which have been performed by persons or have technically not yet been possible, are realized by inventing, redesigning or further developing mechanical devices, machines and transport means.
2. *Performance improvement*: The performance of a machine, device or transport mean is improved by expanding the capacity, increasing the speed or acceleration, or by simplifying the mechanical process.
3. *Cost reduction*: The manufacturing costs of machines, devices or transport means are reduced by simpler construction, modular configuration, cheaper material and economies of scale. Operating costs and performance costs are diminished by reduced deterioration, easier maintenance and longer *total machine running time*.

4. *Quality improvement*: Service quality, user convenience, operating safety, required space, reliability and availability are improved.
5. *Automation*: Electronics and mechanical devices guide and control the movements of robots, machines, devices and transport means and make them independent from persons.
6. *Linking*: Transport elements, devices and machines are connected to transport, production or logistic chains. Production, packaging, bottling or assembling lines are built up by linking conveyors, vehicle systems, storage systems, and robots with machines.
7. *Networking*: Several transport chains, parallel production lines, internal and external transport means, and stores are integrated to production, transport, and performance systems and connected to logistic networks.
8. *Process control*: Chains and networks are controlled by computers and management software in order to operate the system or network with high availability at minimal costs.

The most important criteria for the application of technique in logistics are the *operating costs* and the *performance costs*. Logistic costs can be reduced in various ways such as decreasing capital expenditures, increasing performance at equal labor utilization or decreasing labor utilization at equal performance. The most effective way is to increase performance with decreased labor utilization.

Costs can be further reduced by diminishing the area demand, e.g. by implementing a *high bay store* instead of a conventional storage hall, or by higher utilization of the resources. Another possibility is the better utilization of a limited room, for instance by implementing a *flow rack store* or an overhead *monorail system*. Improvements in quality are of economic interest only if they reduce costs, increase the demand or achieve higher prices.

3.11.2 Directives for Logistic Technology

The application of technique in logistics generally increases the fixed costs, as the invested capital raises interest and depreciation. This leads to the *fixed-cost-dilemma of logistics* discussed later in *Sect. 6.8* and the following *directives for technique in logistics*:

- The more mechanization and automation, the more necessary it is to use the facilities and resources intensively and permanently throughout the year.
- External transport means with larger capacity and higher investment, such as container ships, air planes and railways, but also large vehicle fleets require day-and-night operation.
- High performance systems, such as automatic sorters, automatic guided vehicles or dynamic picking systems, are efficient only if operated in multiple shifts on more than 250 days a year.
- Capital intensive storage systems, like automatic high bay stores, require a high filling degree of the store places all year round.

From these directives, follow the *application rules*:

- Conventional transport and warehouse equipment with low mechanization and automation are more flexible and more cost efficient if low performances and small capacities are required or if the utilization varies strongly.
- The performance costs of highly mechanized and automated systems can be lower by more than factor 2 than the costs of conventional systems, as long as the performance and capacity requirements are high and the utilization is continuous.
- The application of technique in logistics becomes more efficient with increasing *centralization* of functions and consolidation of transports and stocks.

This explains why large companies in the *automotive*, *chemical* and other *industries* have pushed automation and mechanization of their logistic systems for more than 50 years. Also *logistic network providers* such as parcel distributors, airlines, railways and shipping companies have invested with great success in technique.

Big retail companies followed with a time lag of about 20 years compared to their industrial counterparts by taking over and centralizing their procurement logistics. They enforced the application of technique after building logistic centers and central warehouses. Today smaller logistic service providers and medium-sized industrial companies must as well mechanize and automate their logistic systems in order to operate more resourcefully.

3.11.3 Challenges for Manufacturers and IT-Suppliers

Consequences from the above criteria and prerequisites for the successful application of technique in logistics are the following *tasks for manufacturers*:

- *improving the reliability and availability* of machines, devices, vehicles and systems
- *prolongation of total running times* of machines, devices, transport means and sorters
- *increasing capacity* and *limit performance* without higher investment
- *reducing operating costs* while keeping performance and quality
- *higher utilization for the customer* of newly developed machines and other technical products
- *standardizing* of elements and *modularization* of systems in order to ease and to accelerate assembling, maintenance and repairs
- development of more *flexible* handling devices, robots, sorting and picking systems

The following *requirements* are especially addressed *to the suppliers of IT-equipment and systems*:

- enabling *paperless logistic processes*
- *cheaper* and *simpler coding systems*
- *automated attachment of labels and codes*
- *automatic high performance measurement of weight and size*
- *inexpensive* multi-vendor capable *information transfer*

A still unsolved problem is the fully automatic *picking out of a box* with mixed, unordered and non-cubic items that would replace *manual gripping* (see *Sect. 17.13*).

The *harmonization* of the dimensions of logistic units, load carriers and transport units as well as the *standardization* of logistic master data, codes, information interchange and interfaces are well known examples of successful cooperation between technology, informatics and logistics. *Norms* and *standards* enable the synchronization of the procurement and supply processes in inter-company logistic chains. They also help to achieve the objectives of *Efficient Consumer Response* (Corsten 2004; Kotzab 2005). However, standardization has its price:

- Premature standardization and overregulation cause inflexibility and solidification and inhibit further development.

3.11.4 Technical Innovation in Logistics

Those that concentrate too early on technical solutions, lose sight of the strategies, processes and structures. This leads to the *planning recommendation*:

- Successful planners first consider strategies, processes and structures, and afterwards the appropriate technology.

Those that do not know enough about the technical possibilities may overlook appropriate, efficient and approved techniques or propose unrealistic solutions.

Good and feasible ideas as well as excellent technical solutions are rare. Besides some pioneer areas, where in a few years many new solutions spring up like mushrooms, the innovation of technique in logistics is lower and slower than generally expected and claimed. The time span between a good idea and the first successful realization is often many years long. For example, the development of the highly efficient *automated mini-load systems* (AMS) for dynamic order picking of small items goes back to the seventies of the last century (see *Fig. 17.34*) (Gudehus 1977). It took more than 15 years before their advantages were noticed and appreciated by potential users. Today more and more AMS are installed with great success.

The explanation for this delay is that too many companies are risk-averse when it comes to introducing a new technique. Technical innovation needs not only creative engineers but also risk-taking entrepreneurs.

Chapter 4
Potential Analysis

The task of a *potential analysis* is to compare the services and performances of a company with the external and internal requirements. During a *logistic audit*, the capabilities, capacities and performances of the logistic stations, the network structure and the processes are scrutinized and evaluated (Straube and Gudehus 1994). *Goals* of such an analysis are:

- discovery of *potential areas* for most effective improvements
- identification of *weak points* and *possibilities of action*
- *assessment* of possible *improvements* and *savings*

For illustration, in *Fig. 4.1* the potential areas of the logistic network between manufacturers and retailers for *consumer goods* are shown.

Short term results of a potential analysis can be proposals for optimizations, improvements and elimination of weak points. Medium and long term results are new concepts, suggestions for planning and well defined projects.

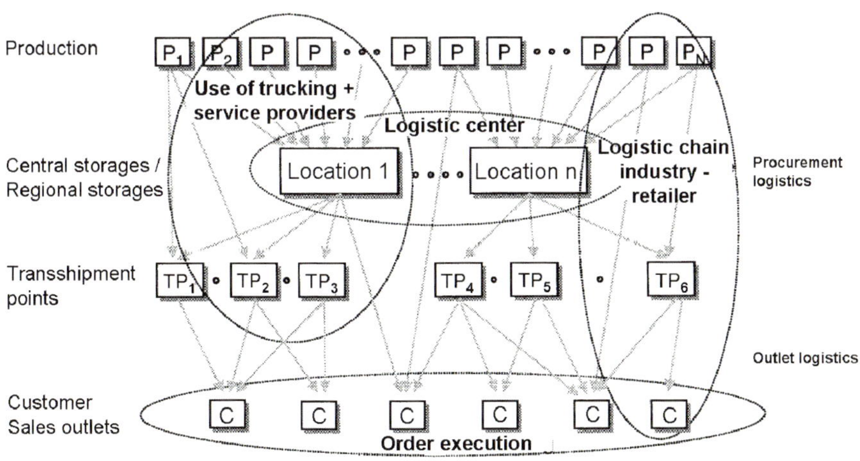

Fig. 4.1 Potential areas of improvements in the logistic network between manufacturers of consumer goods and retailers

T. Gudehus, H. Kotzab, *Comprehensive Logistics*,
DOI 10.1007/978-3-540-68652-1_4, © Springer-Verlag Berlin Heidelberg 2009

Depending on the present situation of the company, the potential savings range between 10 and 20% of the total logistic costs. In specific areas and under certain circumstances the potentials might even be higher. With an operating margin between 1 and 3% and logistic costs being 10% of the total costs of the company, a reduction of the logistic costs by 10% results in a profit improvement between 25 and 100% (Shapiro 1984).

Depending on the size of the company, a potential analysis takes between 4 weeks and 3 months. The expenses for such an analysis by a competent consultant are comparably low. The return by the achieved savings and improvements is generally less than one year.

A potential analysis is not only relevant for assessing possible savings. It also offers ideas for improving services, quality and competitiveness. However, a potential analysis cannot replace planning and optimization. The results of a potential analysis enable the top management to decide, which projects might be most successful and should be planned and realized.

In this chapter the steps of a potential analysis are described. These are:

requirement analysis
performance analysis
process analysis (4.1)
structure analysis
benchmarking

The single steps involve several *check lists* for examining the different potential areas and finding possible weak points.

4.1 Requirement Analysis

First step and basis of any potential analysis is the critical assessment of those logistic performances and services that are required by customers, markets, sales and other departments of the company. The analysis examines the questions:

- Do the requirements towards logistics correspond to the overall goals of the company?
- Is the present cost-benefit-ratio adequate in order to fulfill these requirements?
- Is the prioritization correct? Are the important market segments adequately addressed? Do the most profitable customer groups get the best service?
- Is the present programme of products, articles and services adequate?
- Does the product range include unprofitable articles or services that can be eliminated without affecting the competitiveness (see *Sect. 5.8*)?
- To what extent can logistic services and quality be improved or reduced and what are the consequences of such measures?

Sales departments are inclined to require exorbitant services, as long as the costs involved are unknown. In many cases they demand a *permanent delivery availability* of 100%, while an *average availability* of 98% is sufficient (see *Sect. 11.8*). In other cases, very short delivery times or *24-hour-delivery* are required for all customers, though only a few customers ask for such services. Instead of focusing on reliable delivery dates, which can be achieved at low costs by efficient processes, shorter delivery times and express delivery are demanded. To decide about such demands, the *adequacy principle* is helpful:

- The costs for any service improvement have to be measured against the achievable additional sales and profit.

If the extra costs for special services are explicitly charged, e.g. in forms of an *express surcharge* or a *packaging fee*, many customers will disclaim such additional services.

Results of the requirement analysis are recommendations for a *balanced programme* of products and articles, for *adequate logistic services* and for *differentiated quality standards*.

4.2 Performance Analysis

The task of the performance analysis is to examine at which costs and with which quality the operative and administrative performance stations of procurement, production, distribution and sales fulfill the requirements. In addition, the economic value added by the involved stations is scrutinized (Axson et al. 2003).

For this purpose, the *key performance indicators* (1.3) of the single stations and the order, material and information flows between them are determined, depicted and analyzed. This *input-output-analysis* shows, at which costs and with which input of resources the orders are executed. By this means the weak points of the company logistics are determined and first possibilities to cure them can be developed.

Many potential analyses carried out by experienced consultants have shown that in particular the following *weak points* impede smooth processes and increase the performance costs:

4.2.1 Narrow Pass Stations

Narrow pass stations – generally called *bottlenecks* – operate in peak times above 95% of the installed *limit performance*. They cause long *queues* and *waiting times* for the incoming orders or objects and limit the output of a total system.

In many cases long delivery times can be reduced effectively by increasing the limit performance and capacity of only one or a few critical narrow pass stations. Therefore, it is highly advisable to look out and to take care of the bottlenecks in the system first (Goldratt 2002).

4.2.2 Wide Pass Stations

Wide pass stations are stations, which operate for longer periods of time below 70% of the installed limit performance. Even in peak times they never reach their limits. Wide pass stations are often overstaffed and cause exceeding costs. They do not add sufficient value to the business processes and are typical *waste stations*.

Wide pass stations can be removed by adjusting staff, resources and capacities, by reorganization or by disintegration and assignment of their functions to other stations. In many cases these measures do not only improve the cost situation but motivate the staff in other areas.

4.2.3 Failure Stations

Failure stations are stations with *availability* far below 90%. Such stations block upstream performance stations by frequent or longer lasting interruptions. They can cause the under-utilization of downstream stations, longer delivery delays and the excess of due dates (see *Sect. 13.6*).

The availability of a failure station can be improved by training and qualification of staff, by increasing the quality of the resources, by removing the most common breakdown causes and by organizing a *maintenance service*. These measures can be realized with comparable low efforts. The positive effects of removing a failure station are often remarkable.

4.2.4 Redundancy Stations

Redundancy stations have the same functions as one or more other stations, which execute the same orders on the same kind of objects successively or in parallel.

Redundancy is necessary to ensure the provision of services in cases of breakdown, which occur with high probability and cause major damages. In such an event it is possible to switch to a redundancy station.

However, it is advisable to check, to what extent the offered redundancy is really necessary. Quite often it turns out, that *part redundancy* is sufficient instead of *full redundancy* (see *Sect. 13.6*).

4.2.5 Delay Stations

Delay stations frequently or extremely exceed the required throughput times and completion dates. They put a risk on delivery dates and cause additional costs at other stations.

Delay stations are often bottlenecks or failure stations. However, delays can also be caused by unexpected work load, lack of parts, material and auxiliary supplies or by poor management. These causes can be removed by proper planning, efficient scheduling, reliable supply, adequate buffers and better management.

4.2.6 Fault Stations

Fault stations cause serious errors with unacceptable frequency. Defects and quality deficiencies do not only affect performance and costs. They also initiate delays, disturbances, inefficiency, rework, efforts and costs in the succeeding stations and other performance chains. This can lead from temporary irritations to the permanent loss of customers.

Fault stations can be removed by elimination of the error sources, by training and qualification of staff, by improved management or by specification of *quality standards*. To keep the quality standards, *incentives* or *bonus schemes* are helpful. Also *total quality management* (TQM), which implies a *quality securing system*

(QSS), helps to comply with the quality standards of the market (Dahlgaard et al. 1997; DIN EN ISO 9000 1994).

4.2.7 Main Cost Stations

Main cost stations cause the highest operating costs. In logistics they are those performance stations with the highest contribution to the total logistic costs. For example, the main logistic cost stations of a *retail company* as described in *Sect. 1.8* are the *sales outlets*, which generate between 45% and 60% of the total logistic costs, followed by *freight and transport*, which – depending on the terms of delivery – cause between 25% and 35%, and the *logistic center*, that contribute between 10% and 25% to the total logistic costs.

Naturally, the main cost stations offer the largest potentials for savings. Cost reductions can be achieved by improved organization, by rationalization, mechanization and automation, by application of IT and by other means, which will be developed and described later (see e.g. *Sect. 6.9*).

4.3 Process Analysis

Tasks of the process analysis are to assess the *order processes* from customer to customer and to evaluate the *logistic processes* from the suppliers to the receivers of the goods (Baumgarten 1993; Dahlgaard et al. 1997). For this purpose a complete documentation and a critical review of the present order and logistic chains are necessary (see *Sect. 3.8*).

As all logistic considerations should start with the customer, the analysis of the order chains begins at the order acceptance followed by order processing, procurement, production and distribution. The last step is the handover of the required objects to the customer.

Whereas the *order chains* are scrutinized *along the order flow*, the *logistic chains* are analyzed *against the flow of goods*. This helps to observe the *value contribution* and *customer orientation* of the stations participating in the supply processes.

During the process analysis, which is basis of any *process modeling* (Scholz-Reiter 1999), the following subjects and questions are examined:

4.3.1 Logistic Units (see Chap. 12)

- Which *load carriers* and which *logistic units* are used in the internal and external logistic chains?
- Are the *dimensions* and *capacities* of the logistic units harmonized?
- Are the load carriers used with high *filling degree*?
- Is the *range of load carriers* sufficient and appropriate?
- Who is responsible for the *packaging hierarchy*?
- Who decides, based on what *criteria*, about *selection*, *application*, *introduction* and *elimination of load carriers*?
- Is the *logistics of empties* properly organized?

4.3.2 Logistic Master Data (see Sect. 12.6)

- Does a well organized *logistic data base* exist?
- Are the *logistic master data* complete, up to date and correct?
- Who is responsible for the *management of logistic data*, i.e. for their registration, updating, correctness and completeness?
- Who defines and *norms* the logistic master data of the company?
- Who ensures the *compatibility* between the company's logistic data and the corresponding data of suppliers and customers?
- Are the *possibilities* of the logistic master data fully realized?
- Is the *logistic data exchange* between internal stations, suppliers and customers organized correctly?
- Is the *technical data exchange* provided professionally and efficiently, e.g. by *EDI* or by *Internet*?

4.3.3 Time Management (see Chap. 8)

- Are the present *delivery times* and the adherence to the *delivery dates* in line with the market?
- How well are the *throughput times* kept in the performance chains?
- Are the *temporal options of action* effectively used?
- Are the *operating times* of consecutive and parallel stations *synchronized*?
- Are the lengths of *planning periods* and of *scheduling periods* adequate and correctly adjusted?
- Are the *throughput times* of orders and material too long?
- Are the relevant *time strategies* known and appropriately applied?
- Does an efficient *time management* exist?
- Where are the time-killing *bottlenecks*, *failure stations* and *delay stations*?

4.3.4 Costs (see Chaps. 6 and 7)

- Are the *performance costs* in the single sections of the logistic chains known and are they appropriate?
- Does an effective *logistics controlling* exist?
- Who examines the *appropriateness* of the *cost rates* for the internal and of the *prices* for the external logistic performances and services?
- What are the *specific logistic costs* per article unit?
- What are the logistic costs per category, of a supplier or for a customer?
- Where and how can logistic costs be saved without reducing service and performance?
- What are the main *logistic cost drivers*?
- Which are the *waste stations*, *redundancy stations* and *main cost stations*?

4.3.5 Stores and Stocks (see Chap. 11)

- Are the *stores* and *buffers* at the different stages of the logistic chains necessary and appropriate?
- Who decides, based on which criteria, on the *storekeeping articles* of the assortment?
- Who determines the *availability to deliver* of the storekeeping articles?
- Are the *safety stocks* sufficient to ensure efficient utilization, uninterrupted performance and market-driven customer delivery?
- Is the *safety stock level* derived from *benchmarks* or do the *safety stocks* of the single articles result from the required *availability*?

4.3.6 Quality (see Sect. 3.4)

- Are the required services provided with *adequate quality*?
- Does a *quality management and securing system* exist?
- Are the internal and external *quality defects* regularly recorded?
- Which quality deficiencies are fined by what *penalties*?
- Are the processes sufficiently *flexible* for the changing demands of customers and for the *dynamic markets*?
- Which are the *failure*, *delay* and *fault stations*?

4.3.7 Interfaces and Connections

- How good is the co-operation between internal and external performance stations?
- Are the *dimensions and capacities of* the exchanged *load units* harmonized?
- Are the flows of material and information between different performance stations free of *interruptions* and *disturbances*?
- How efficient is the *exchange of information* and the *communication* along the logistic and performance chains?

4.3.8 Scheduling and Process Control (see Chaps. 2, 10 and 11)

- Are the available resources and capacities used efficiently?
- Are the *right strategies* applied for order scheduling, inventory management and stock replenishment, and for production scheduling?
- What *tools*, *methods* and *programs* are used for scheduling, process control, information exchange and controlling and are they applied correctly?

4.3.9 Logistic Chains (see Chap. 21)

- What are the present logistic chains of procurement and distribution?
- Are the logistic chains *complete* and *optimally* selected for the different flows of material, parts and finished goods?

- What are the *strategies* and *criteria* for selecting a logistic chain?
- Are the *possibilities of action*, such as optimal *bundling* and *sequencing*, known and efficiently used?

4.3.10 Make or Buy (see Chap. 12)

- Which parts of the logistic and performance chains are *core competencies* of the company?
- Which services and performances can be efficiently outsourced to *suppliers, system providers* or *logistic service providers*?
- Who decides, based on what *criteria, about make-or-buy*?

Results of the process analysis are recommendations for optimizing the processes, for the most efficient use of the own resources and for the *outsourcing*. In addition, the economic benefits of the different possibilities of action are estimated. A process analysis also offers insights for staffing and for improvements in the different performance stations.

4.4 Structure Analysis

After analysis of the requirements, performances and processes, the *structure analysis* examines whether the given system structures satisfy present and future demands. Based on the results of the analysis it is possible to decide, which service, quality and cost improvements can be achieved by changing the present structures or by setting up new systems.

For this purpose the *structure diagrams* of the logistic network of the company and of all subsystems of interest have to be worked out. Some weak points and possible improvements can already be recognized by this analysis. Performance stations that receive orders for the same job from more then one source indicate conflicts. Flows of the same material from the same source through different logistic chains to the same destination offer potentials for improvements (see *Sect. 3.8*).

Task of a structure analysis is to examine the following subjects and questions:

4.4.1 Location of Stations

- Are the plants, warehouses, logistic centers, transshipment points, delivery stations and stores located at optimal places?

4.4.2 Allocation of Functions

- Are the functions, tasks and stocks rightly allocated amongst plants, warehouses, logistic centers and transshipment points?

4.4.3 Degree of Centralization (see Sect. 1.8)

- What functions should be executed *centrally* and which should rather be executed *locally*?
- What is the *optimal number* of plants, warehouses, logistic centers, delivery stations and sales outlets?

- How should they be allocated and connected with each other?
- What are the possible improvements of costs and performance by *consolidating* local stocks and functions *in a logistic center*?

4.4.4 Stages of Supply Chains (see Sect. 1.5)

- What is the optimal number of stages for the procurement and distribution?
- Where are avoidable handling or transaction activities?
- Is the selection between direct delivery and delivery via transshipment points or logistic centers based on the right *criteria*?

Results of the structure analysis are recommendations for *improving* the structure as well as suggestions for a *redesign* of the whole logistic network or of subsystems. The structure analysis also offers proposals for centralizing or localization of functions and stocks, as well as assessments of the achievable improvements of costs, service, performance and competitiveness.

4.5 Benchmarking

Benchmarks are reference values. Correspondingly, *benchmarking* means the comparison of costs, performance, quality and other *key performance indicators* (KPI) of several companies, business units or stations with similar tasks and functions. It can also include a comparison of the operation methods, organization and strategies (Axson et al. 2003; Camp 1989; Leibfried et al. 1992; Reider 1999; Röder 1998; Watson 1993).

For benchmarking it is essential that the business units, whose tasks, functions, costs and key performance indicators are compared, comply sufficiently. Relatively small differences between companies, plants or stations can lead to quite different key performance indicators, which are neither better nor worse. For example, due to the different sizes and values of the articles an external *logistic benchmarking*, especially between companies in unlike industries, is generally misleading.

Three different kinds of benchmarking are possible: *external benchmarking*, *internal benchmarking* and *analytical benchmarking*. The necessary effort for them and their usefulness are quite different.

4.5.1 External Benchmarking

External benchmarking *compares the key indicators* of the performance stations *of different companies*, that have similar functions, such as plants, warehouses or logistic centers (Axson et al. 2003; Reider 1999; Röder 1998).

A common mistake of external benchmarking in logistics is the separate comparison of single indictors such as *stock levels* or *delivery ability* without taking into account the impact and relations between them. According to the *best practice value* from other companies, the management might set a limit for reducing the inventory in the own company. However, by doing this, they unconsciously reduce their delivery ability or increase the total costs. On the contrary, the postulation for an improved delivery ability in order to achieve best practice might drive up stock levels and costs.

Other misleading benchmarks are the *logistic costs in relation to turnover*, as companies define and record their logistic costs differently (see *Chap. 6*). The average value of full pallets with the same size but different content varies considerably. Therefore, the logistic costs in relation to the value of the content can differ by a factor of 10 or more, although the logistic costs per pallet are equal.

Also the results of *interviews* and *trend polls* can be misleading, as the specific circumstances and goals of other companies are generally unknown. The answers reflect the opinion, competence and intentions of the persons, who filled out the question forms, but do not give deeper insight into the performance and strategies of the companies. Even if all participants would answer honestly and correctly, it remains open, whether the company has chosen the right strategy or copies only general trends and fashions.

These objections hold as well for surveys exclusively conducted by external consultants for leading companies within a certain industry. The leaders are often neither willing to present their key performance indicators nor to make their strategies public. Furthermore, they also do not know, whether there might be even better strategies.

Those who only follow trends cannot be better than average and will make the same mistakes as others. Following a general trend does not allow the right strategy for a company to be identified. In order to become the *best in class* and to have a lead over the competition, a company must develop its own strategies.

4.5.2 Internal Benchmarking

Internal benchmarking *compares the performance indicators* of operative or administrative stations *within the same company*, which have similar tasks and functions (Camp 1989; Watson 1993). It can only be carried out in larger companies with several similar performance stations.

Internal benchmarking allows for the checking of how far the objects, tasks and functions of the stations to be compared are equal or different. The performance indicators can be defined and measured everywhere in the same manner. Therefore, only internal benchmarking really shows, how well the plants, warehouses, logistic centers and stores fulfill their jobs and why they differ in costs and performance.

Larger differences in costs and performance indicate potential improvements, which can be realized immediately by transferring the *internal best practice* to the other performance stations.

4.5.3 Analytical Benchmarking

Analytical benchmarking compares the key indicators of an existing performance area with the key performance indicators of an optimally planned and organized performance area, which fulfils the same functions, executes the same orders and provides the same throughput.

Analytical benchmarking generally requires a far higher effort than external or internal benchmarking. It requires the development of own strategies and the planning of company specific solutions. However, only analytical benchmarking allows recognizing the options of a company, accessing the changes for the better, calculating the costs and investment and planning accurate actions to achieve the options.

From a medium-term perspective, analytical benchmarking is the only promising way for companies that want to improve against external competitors *and* against the best present internal performance. To achieve sustainable competitiveness, a company should not rely on "Me too" or "Best Practice".

4.5.4 Options of Action and Strategies

Those who have no goal, cannot find the right way to success. Before starting a project or doing things for the sake of doing something a potential analysis should be carried out, since:

- Only by a *potential analysis* the really important goals can be identified and the most profitable projects be defined.

Potential analyses and benchmarking reveal goals, options of action and possibilities for savings and improving the company logistics.

However, a potential analysis does not offer solutions. Only *strategies*, i.e. effective procedures to reach a certain goal, help finding the optimal solution.

Chapter 5
Strategies of Logistics

Strategies are logical procedures to reach certain *goals*. Suitable strategies are prerequisites for the effective use of resources and often the simplest mean to improve efficiency and to reduce costs. In logistics, the conception of new strategies and the analysis of their effectiveness are as important as the development of new techniques and the design of new systems.

Goals and strategies are often confused or equated. In order to develop a strategy the goal must be well defined (Chandler 1962; De Wit and Meyer 2005; Goldratt 1984). However, a goal is not a strategy. Without a suitable strategy it remains open, whether a certain goal can be achieved at all. On the other hand, to aim at a wrong goal leads to useless results. For example, the reduction of stocks can result in higher total costs due to insufficient availability to deliver. Hence, before developing a strategy, relevance and compatibility of the goals should be checked.

Corresponding to the field of application, strategies can be classified into:

- *solution and optimization strategies* for the design of new and the optimization of existing systems
- *utilization and allocation strategies* for the utilization of planned and existing systems
- *scheduling and operating strategies* for the operation of existing systems.

A certain goal can often be achieved by different strategies. If a goal is described in qualitative terms, only a comparison of the relative effectiveness of different strategies is possible. If *target functions* or *target figures* are given, the effectiveness of a strategy can be quantified by the strategy effect:

- The *strategy effect* is the measure of the goal achievement.

The strategy effect depends on the performance requirements and restrictions and on the strategy variables:

- *Strategy variables* are free parameters that can be varied within certain limits to optimize the strategy effect.

With many strategies, the effect becomes maximal for an *optimal value* of the strategy variable. For an *overdrawn strategy*, the variable has a value beyond the

T. Gudehus, H. Kotzab, *Comprehensive Logistics*,
DOI 10.1007/978-3-540-68652-1_5, © Springer-Verlag Berlin Heidelberg 2009

optimum, where the effect decreases. In such cases, the deviation from the optimum can be corrected by a *counter strategy*.

Within the hierarchical levels of logistic networks, systems and organizations, many different strategies are applicable. Some of them are incompatible. Their interfering effects spoil each other. Up to now, there is little known about the interference and the mutual compatibility of strategies.

Strategies are determined by the goals, restrictions and required performances. Optimal networks, systems and processes can be designed only after suitable strategies have been developed. If goals, restrictions or requirements are changing, the strategies must be scrutinized and adapted or other strategies have to be developed.

This chapter defines the *objectives* and *target functions* and presents the *basic strategies* of logistics. From these, general *solution and optimization strategies* are derived. The development of *utilization and allocation strategies* and of *scheduling and operating strategies* will be presented in the following chapters.

5.1 Target Functions and Target Figures

Many logistic objectives can be quantified by a *target function* or a *target figure* (see *Sect. 3.4*). By a suitable strategy, the target figure can either be minimized or maximized.

The most important objectives of a company are generally *monetary targets*, such as , *investment*, *costs rates* and *return on investment*. *Non-monetary targets*, such as *performances increase*, *service improvement* or *quality assurance*, affect, influence or limit the monetary targets.

5.1.1 Operating Costs

The primary objective of planning and optimization is to reduce the operating costs. They are measured in *monetary units* [MU = €, $, £...] per *period* [PE = day, weak, month, year]. The operating costs are a function of the *throughput* and *performance rates* λ_i, the *restrictions* r_j and the *strategy variables* x_k:

$$K_{op} = K_{op}(\lambda_i; r_j; x_k) \qquad [\text{MU/PE}]. \qquad (5.1)^1$$

An important restriction and in some cases a *KO-criterium* is the limited availability of capital. As only solutions with investment lower than the maximal available investment capital I_{max} are realizable, all solutions have to meet the *investment restriction*:

$$I \leq I_{max} \qquad [\text{MU}]. \qquad (5.2)$$

Both, investment and costs depend on the *logistic units* [LU] or *performance units* [PU] and on the strategy variables. Within logistics, in many cases this dependency is a discontinuous *step-function* of the variables. Due to *whole-number effects*, logistic costs can change extremely, when altering a certain parameter (see *Chaps. 6* and *12*).

---

[1] In order to differ clearly between costs and capacities, in this book operating costs are denoted by K, specific cost rates by *k* and capacities by C.

5.1.2 Performance Costs

The *performance costs* or *cost rates* are the target function for optimizing an order, logistic or performance chain. They are the sum of the partial operating costs $K_{op}(\lambda)$ [MU/PE] of all stations involved in the performance process for a certain performance related to the throughput or performance rate λ [PU/PE]

$$k = K_{op}/\lambda \qquad [\text{MU/PU}]. \tag{5.3}$$

Generally, a company can influence its external logistic costs, i.e. the procurement and distribution costs, only to certain extent. How far this is possible depends on the *terms of delivery* and on its market power. All external logistic costs, which can be changed by negotiation, even if they are part of the purchasing prices, should be taken into account for system design and optimization.

5.1.3 Payback Time and ROI

As investments, savings and revenues are uncertain and risky over longer times, many companies require a short payback time. A short payback time limits business risks and prioritizes alternative investment plans.

The *payback time* of an investment is the time until the investment is amortized by the achieved savings and profits. The *return on investment* (ROI) of a solution S_1 with investment I_1 [MU] and operating costs K_1 [MU/year] compared to an initial solution S_0 with investment I_0 and operating costs K_0 is given by:

$$\text{ROI} = (I_1 - I_0)/(K_1 - K_0) \qquad [\text{years}] \tag{5.4}$$

The required ROI or *payback time* depends on the business policy and profit situation of a company and on the conditions of the capital markets. In many companies the *maximal payback time* is set between 3 and 8 years.

Short term oriented companies favor short payback times rather than sustainable cost reductions. This policy is questionable, since these companies tend to execute only low budget projects that promise quick but mostly small profit improvements and cost savings. They postpone higher investments and larger projects, which enable minimal operating costs and maximal profit in the long run.

5.1.4 Non-monetary Target Figures

Most non-monetary target figures of logistics result from the goals of performance increase, service improvement and quality securing.

Targets of performance, which have to be *minimized*, are

personnel
transport means
storage space
path lengths (5.5)
total network length
transit-, throughput- or cycle times

Targets of performance, that have to be *maximized*, are:

performance and utilization of personnel
performance of transport means
limit performances of a network
utilization of transport paths and networks (5.6)
filling degree of load carriers and transport means
utilization of store capacities
utilization of devices, machines and plants

A higher performance or throughput generally requires additional resources whereas a better utilization is to certain extent achievable without investment. Therefore, strategies aiming at higher utilization are generally means for cost reduction.

The specific *targets of quality* are (see *Sect. 3.4.4*):

readiness to deliver
availability and reliability
completeness of orders
punctuality (5.7)
rate of failure and claims
shipment quality
accident rate

Quality target figures should be improved until they reach a required *quality standard*. Improvements beyond these standards may increase total costs significantly. Hence, standards for the quality requirements (5.7) are restrictions when optimizing costs and performances.

5.1.5 Restrictions

When minimizing or maximizing a target value, all limitations and restrictions have to be considered. Besides the restrictions from quality requirements, they result from the *frame conditions* of *Sect. 3.5*.

Some frame conditions are determined by *minimal or maximal values* such as

minimal necessary storage time
maximal tolerable storage time
longest tolerable delivery time (5.8)
technical operating lifetime

Before starting planning, the contribution of these target values to the overall business goal has to be examined carefully. This prevents of developing sub-optimal solutions, which have to be omitted later (Churchman et al. 1961).

5.2 Clustering, Sequencing, Securing

Many strategies can be derived from the basic strategies *clustering*, *sequencing* and *securing* and their counter strategies *separating*, *re-arranging* and *slacking* (Gudehus 1992). These basic strategies can be applied widely in logistics and other business areas. They are general strategies to govern the *complexity* of problems, systems and organizations (Simon 1962).

The interrelation of the three basic strategies and their main objectives are shown in *Fig. 5.1*. They are connected with each other by stress forces, which result from their limited compatibility.

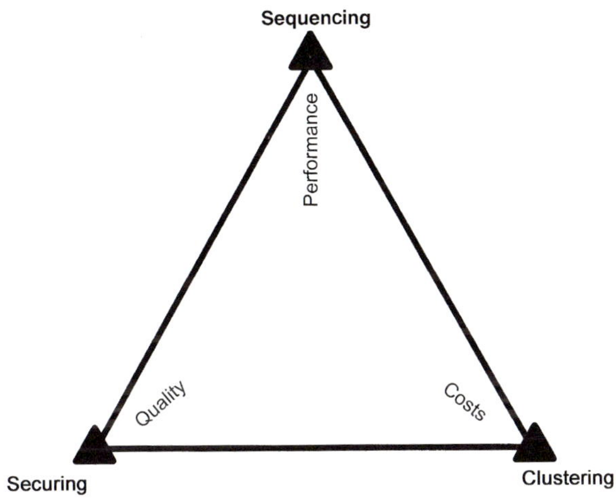

Fig. 5.1 Basic logistic strategies and their objectives

5.2.1 Clustering

Orders, shipments or goods as well as transport flows, stocks, functions and processes can be bundled due to certain criteria into spatial or temporal clusters (Collingwood et al. 2004; von Neumann and Morgenstern 1994). *Cluster strategies* or *bundling strategies* normally aim for cost reductions. Typical *cluster strategies* of logistics are:

- clustering of assortments, orders or services (see *Sect. 5.6*)
- *ABC-classification* (see *Sect. 5.8*)
- bundling larger quantities of goods by *load carriers* (see *Chap. 12*) (5.10)
- consolidation of functions and stocks in a *logistic center* in order to utilize *economies of scale* (Nowitzky 2003) (see *Sects. 1.8, 3.10, 6.9, 11.10, 16.4, 17.14* and *18.11*)
- bundling of services in logistic centers (see *Sects. 1.5 and 1.8*)
- generation of collective-, series-, large-scale and batch-orders
- production and replenishment in *optimal lot sizes* (see *Chap. 11*)

- consolidation of small single shipments, e.g. *less than truckload* (LTL) into larger collective shipments, e.g. *full truckloads* (FTL) in order to reduce shipping costs (see *Sect. 21.2.4*).
- *categorization* of merchandise, customers or suppliers

All clustering strategies imply the question, to which end which elements should be consolidated. The number $N_{part}(n)$ of possibilities to cluster n elements equals the number of *partitions* of an integer n into a sum of numbers. For example, 4 orders can be clustered in 5 different ways corresponding to the five decompositions 1+1+1+1, 1+1+2, 2+2, 3+1, 4 of the integer 4. Diagram *Fig. 5.2* shows the dependency of the number of partitions on the number of elements n. Keeping in mind the logarithmic scale, one can read from this diagram that the number of partitions increases almost exponentially with the number of elements. For 40 elements the number of partitions is already 37,338 (Biggs 1999; Hardy and Ramanujan 1918).

To find out the optimal clustering of more than 10 elements by *trial and error* or by *simulation* is practically impossible, because the number of possibilities is too high. Optimal clustering of elements requires a *cluster strategy* that is aiming for a defined objective and enables the selection of the elements for the different clusters by a program based on an *algorithm*.

Cluster strategies are often quite simple and can be realized with low organizational effort. Compared to sequencing strategies, cluster strategies normally need longer preparation time and advanced technology.

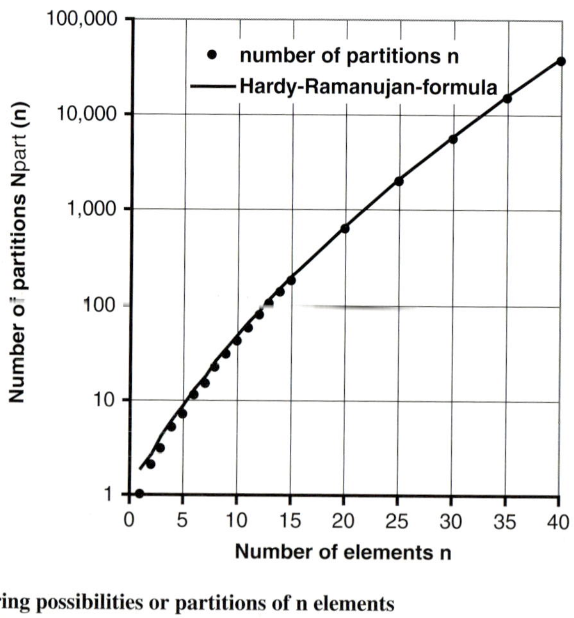

Fig. 5.2 Clustering possibilities or partitions of n elements

Points: exact values calculated by recursion formula of *Biggs* (Biggs 1999)
Curve: Hardy-Ramanujan-Formula $N_{part}(n) = \exp(\pi\sqrt{2n/3}) / (4n\sqrt{3})$

5.2.2 Sequencing

Orders, shipments, process chains and procedures, logistic units, load carriers, logistic units, function modules or equipment can be arranged in a *spatial order*, located in an *optimal layout*, sequenced in a *priority order* or put on a *chronological scale* (Baker 1974).

The *criteria of sequencing* depend on the objectives. Sequence strategies should generally achieve performance improvements. But also cost reductions are possible by sequencing. Typical *sequence*, *arranging* and *ordering strategies of logistics* are:

- *Pareto- or Lorenz-sequencing* (see *Sect. 5.8*)
- *packing strategies* and *filling strategies* (see *Chap. 12*)
- *routing and transport strategies* (see *Sect. 18.11*)
- *sequence optimization* (Müller-Merbach 1970) (5.11)
- *connecting strategies* and *layout strategies* (see *Sect. 19.9*)
- *just in time* (JIT) and *just in sequence* (JIS)
- *priority rules* (see *Sect. 10.4*)

Sequence strategies aim for a reduction of proportionate setup times, volume losses and space demand. Furthermore, they are applied in order to shorten transport paths, running times or throughput times and to improve the utilization of stores, transport means or other resources. Sequencing strategies can also help to reduce stocks and to improve efficiency.

As for the clustering strategies, it is necessary to define, which elements should be sequenced or arranged for which purpose. The number of sequencing possibilities equals the number of *permutations* of n elements. This is given by the factorial $N_{perm}(n) = n! = 1 \cdot 2 \cdot 3 \cdot 4 \cdots n$. The dependency of the number of permutations on the number of elements is shown in *Fig. 5.3*. It increases more than exponentially with the number of elements and much faster than the number of partitions. For example, the number of permutations of 8 elements is 40,320 while the number of partitions is only 22.

Optimal sequencing of even a small number of elements is therefore only possible by a *sequence strategy* which is aiming for a defined objective. Then the optimal sequencing of elements is possible by a programmable algorithm.

Sequence strategies and their algorithms are generally more complex than cluster strategies and cluster algorithms (Baker 1974; Curchman 1961; Domschke 1990, 1995; Müller-Merbach 1979). However, today's computer capabilities allow the application of such algorithms with a justifiable effort. Compared to cluster strategies, the implementation of sequencing strategies does not require costly technique.

5.2.3 Securing

The design and control of systems, processes and organizations are not only determined by cost targets, performance requirements and service demand but must also achieve safety. In order to fulfill the various safety needs, *securing strategies* have

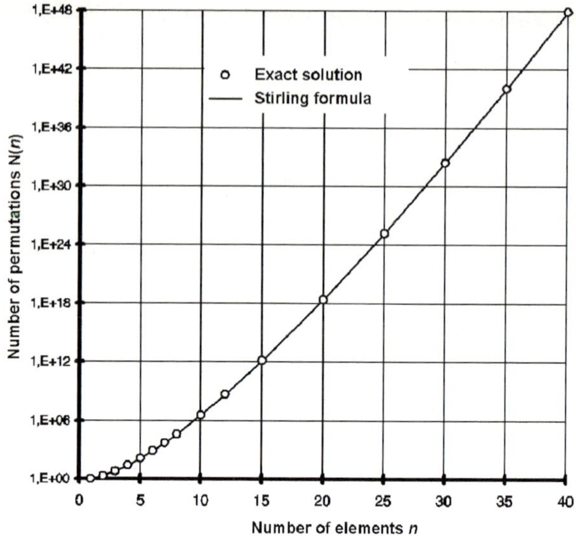

Fig. 5.3 Sequencing possibilities or permutations of n elements

Points: exact values given by $N_{perm}(n) = n!$
Curve: *Stirling-Formula* (Kreyssig 1995)

to be developed. They should ensure performance and availability with a sufficient level in case of break downs, errors or unscheduled changes of demand. Securing strategies are primarily directed at quality objectives. However, in many cases they also affect performances, capacities and costs. Typical *securing strategies of logistics* are:

- *regulation and controlling processes*
- *prevention strategies and safety regulations*
- *decoupling of stations and subsystems* (see *Sect. 13.5*)
- *emergency and breakdown strategies*
- *tracking and tracing*
- *safety chains* for hazardous or valuable goods
- *check and balance*
- *quality securing* and *controlling* (5.12)
- *redundancy* and *universality*
- *buffer stocks* and *safety stocks*
- *performance and capacity reserves* (*Sect. 16.1.3*)
- *production and procurement in advance*
- *time buffers*
- *fire protection*
- *escape routes, stairs and doors*

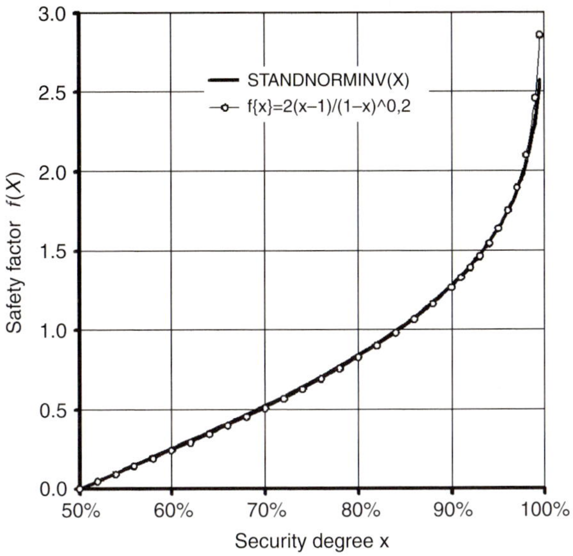

Fig. 5.4 Dependency of the safety factor on the required safety level
Dots: inverse standard normal distribution
Curve: approximation formula (11.33)

The costs of most securing strategies increase over proportional with the required safety degree. That makes extreme safety very expensive. Total safety requires total control. This is neither feasible nor necessary.

The correlation between *safety level* and securing effort is given by the *safety factor* $f_S(\eta)$ as shown in *Fig. 5.4*. For a quantity with randomly varying values, which are normally distributed, the safety factor times the standard deviation gives the value, which is not exceeded with probability η (see *Sect. 9.3.5*). This probability is the required safety level.

The safety factor is used e.g. to calculate the *breath reserve* of the store capacity, which limits the risk of an overflow of the store (see *Sect. 16.1.3*). It also determines the *safety stock* of articles, the *delivery ability* and the *time buffer* for ensuring a requested *punctuality* (see *Sect. 11.8*).

The dependency *Fig. 5.4* of the safety factor on the safety level shows that the securing efforts exceed all limits, when the requested safety approaches 100%. Hence, absolute safety, e.g. the requirement of 100% delivery ability or 100% punctuality, is unaffordable for stochastically varying demand and performances.

5.2.4 Combination Strategies and Counter Strategies

In order to achieve more than one objective, the basic logistic strategies – clustering, sequencing and securing – can be combined. However, the combination of two strategies reduces the effect of the single strategies, if they are not fully compatible.

If clustering, sequencing and securing are overdrawn, *counter strategies* can be applied. These are:

- *separation* or *disintegration* of shipments, orders, stocks or functions: single order processing instead of batch order processing; small batches instead of large batches; single transports instead of consolidated transports; small deliveries instead of bulk delivery; single ordering instead of collective ordering; specialization instead of standardization; localizing instead of centralizing; delegation instead of concentration.
- *reordering, rearrangement, relocation* and *redistribution* of orders, stocks, stations or functions.
- *slackening*, *loosening* and *deregulation* by reduction of excessive control, securing efforts, safety stocks and redundancies.

The three basic strategies *clustering*, *sequencing* and *securing* primarily aim at the three elementary goals *costs*, *performance* and *quality* (see *Fig. 5.1*). Hence, one could expect, the optimal solution of a given logistic task is achievable by combining the basic strategies and their counter strategies. However, finding the optimal solution is a far more difficult task.

5.3 System Strategies

In addition to the basic logistic strategies and their combinations, the optimal design and scheduling of whole systems, larger organizations, business networks and multistage supply chains require more sophisticated *system strategies*. These strategies concern several stages of a supply chain or network and make use of the options and possibilities of modern *information and communication technology* as described in *Sect. 2.7*.

Well approved *system strategies and central strategies* in *logistics* are:

- *real stock-centralization* for all suitable articles in a central store or logistic center
- *virtual stock-centralization* by simultaneous *central scheduling* of local stocks and *procurement consolidation* for several demand stations, e.g. for the outlets of a department store
- *spatial postponement* and/or *temporal postponement* of storekeeping and/or customizing
- *advanced production* for sales promotions or expected bottleneck phases
- *optimal stock location* of purchased material and parts, of primary and intermediate products as well as of finished goods and merchandise in the different stations of the supply chains
- controlled, smoothed, clocked or *throttled throughput* of parts, orders or shipments through multi-stage delivery, supply and performance chains
- *assignment policies* for *goods of short supply* and *bottleneck strategies* for seasonal peaks of demand
- division and assignment of orders to parallel performance stations or performance chains

- *advanced scheduling* and *adaptive planning* (APS) of upstream performance stations based on direct information from *points of sales* (POS)
- *load bundling* and *transport consolidation* of shipments from suppliers located within the same area (*milk run*)
- *shipment consolidation* by bundling single deliveries of several delivery stations to full truck loads
- *freight consolidation* of shipments for different recipients within the same area
- *selection of the cost-optimal freight chain* or *delivery chain* (see *Chap. 21*)

The implementation of system strategies for an extended network requires an *order center* and *central scheduling*. Central scheduling, though, does not only provide advantages but is generally connected with certain drawbacks. Central scheduling and control of activities and processes affects the motivation of the people in the local stations. It diminishes personal responsibility and individual initiative and reduces *self-regulation*, *flexibility*, and *efficiency* of local stations. Excessive centralization increases the *fault liability* of a system.

The potential savings and improvements of *part-network-strategies* for limited part-nets can sometimes be quantified or at least be estimated. Evaluation and calculation of the effects of complex system strategies, aiming for the optimization of a whole network with large extension, such as the network of a car manufacturer shown in *Fig. 1.15*, are not possible up to now (Inderfurth 1999).

The positive effects of system strategies are often overestimated, e.g. the saving potentials of using instantaneous information from the points of sales in the upstream stations of the supply network. In some companies, central scheduling has been forced by top management to build up huge stocks and to place excessive procurement orders based on *speculation* or *balance sheet policy*. Such behavior causes the so called *bullwhip effect* and may later result in large *stock revisions* (Forrester 1958; Lee et al. 1997).

The highest danger of central scheduling and planning is the application of system strategies, where the positive effects are not sufficiently ensured and the negative consequences are unknown or ignored. In order to avoid these deficiencies, it is highly recommendable to estimate in advance all possible consequences of central strategies.

5.4 Methods of Solution and Optimization

Planning, optimization and scheduling of networks, systems and processes requires adequate *methods* (Kirsch 1973). The methods as described in the following subsections have proven to be of high value for logistics. However, planning techniques, methods and instruments can never replace creativity, intuition, experience and judgment. Only knowledge of real life business, sufficient *know-how* of logistics technology and experience of good and bad solutions in practice guides successful planning and allows evaluating possible solutions, no matter how they have been generated (see *Sect. 21.9*).

5.4.1 Analytical System Design

During the *iterative steps of the planning- and optimization process* shown in *Fig. 5.5* suitable solutions for the subsystems are developed, selected and dimensioned, and combined into a complete solution. In these steps, *selection, design and planning rules* are applied. For dimensioning and optimization, the analytical relations between the design parameters and the required properties of the system must be known. In addition general calculation formulas and proven algorithms are extremely helpful.

If performance, store capacity and costs depend on several parameters and strategy variables, the functions, relations and performance of a subsystem or the whole solution should be reproduced as *mathematical model* in a computer program. The program calculates the target values and can be used to generate the *optimal solution* by variation of strategy variables and design parameters. The *analytical simulation* on a computer is based on functional relationships, algorithms and formulas (see *Sect. 3.9*).

5.4.2 Optimization by OR-Methods

The general goal of *Operations Research* (OR) is to find the optimal solution for a defined problem with given target function systematically out of a number of possible solutions. *Total enumeration*, *Branch & Bound* (B&B), *simplex method* and *Linear Programming* (LP) are well known OR-methods for finding exactly the optimal solution. *Heuristics* such as the *Add- and Drop Algorithm* or the *Gradient Search Method* start with an analytically constructed *opening solution* and find approximate solutions in shorter time (see *Sect. 18.11*).

Several logistic problems, such as optimizing sequences, routes, packaging and locations, can be solved by approved and efficient OR-methods. Other OR-methods are helpful for the optimization and simulation of stochastic and dynamic systems (Churchman et al. 1961; Domschke 1985, 1990; Kirsch 1973, Müller-Merbach 1970).

In logistics, the relations between performance requirements, parameters and target values are often non-linear or whole-numbered. Many problems are of dynamic nature and have to cope with stochastic fluctuations. Here, *OR standard procedures* aimed at solving static and linear problems are of limited use. In addition, it is often difficult to understand the dependency of a mathematical solution which has been generated by sophisticated OR-methods on the requirements.

Many OR-experts prefer to focus on model building, classification of problems and mathematical methods. Only exceptionally they discuss the characteristics of the underlying problem and the business consequences of a mathematically optimal solution. On the other hand, in some cases, OR-models are too simplified or based on assumptions that are out of touch with reality.

5.4.3 Tests by Digital Simulation

The *digital simulation* or *stochastic simulation* reproduces the performances and other key data of the elements, their relations and the whole structure of a given system by a *simulation model* in a computer program (Kuhn et al. 1993). A random generator creates incoming orders and logistic units with assumed time sequence, frequency and distribution. For the generated input flows, which pass the stations and transport connections of the system, the program calculates throughput times, waiting queues, blockages, buffers and stocks and their temporal variation by counting the passing, incoming, waiting and outgoing units at the single performance stations and connections, for subsystems and for the whole system.

Digital simulation is an *experiment* with the help of a *mathematical model* of a system, which has already been planned or exists in practice. Where real field experiments are impossible or too expensive, model experiments are suited to test the limit performances of a system and/or of theoretical predictions. However, model experiments, such as digital simulation, can not replace the analytical system design. By digital simulation alone, no general laws and functional connections are derivable (Hollingdale 1976; Nersesian and Swartz 1996; Flores and Whybark 1986).

Digital simulation is a suitable tool for examining functionality, capacity and temporal behavior of complex systems, which have been developed and optimized by analytical methods. It helps to find out whether a designed system meets systematically varying and stochastically fluctuating demand. That means:

- Digital simulation of an analytically constructed and optimized solution increases planning safety. It allows the examination of the dynamic system behavior with different operating strategies for stochastically and/or systematically varying demand.

However, a digital simulation does not answer the questions, why a system works, whether the same performances could not be achieved by a simpler system and by which strategies a system can be optimized.

5.5 Solution and Optimization Procedure

Within the different phases of planning, described in *Sect. 3.2* and shown in *Fig. 3.2*, the construction, design and optimization of subsystems, whole systems and supply chains are performed in an iterative *solution and optimization procedure* with increasing detail. This procedure includes the following ten steps presented in *Fig. 5.5*:

1. Compilation of Requirements

specification of functions
determination of performance requirements (5.13)
comprehension of frame conditions

2. Analysis of Options

systems elements and design parameters
strategies and strategy variables (5.14)

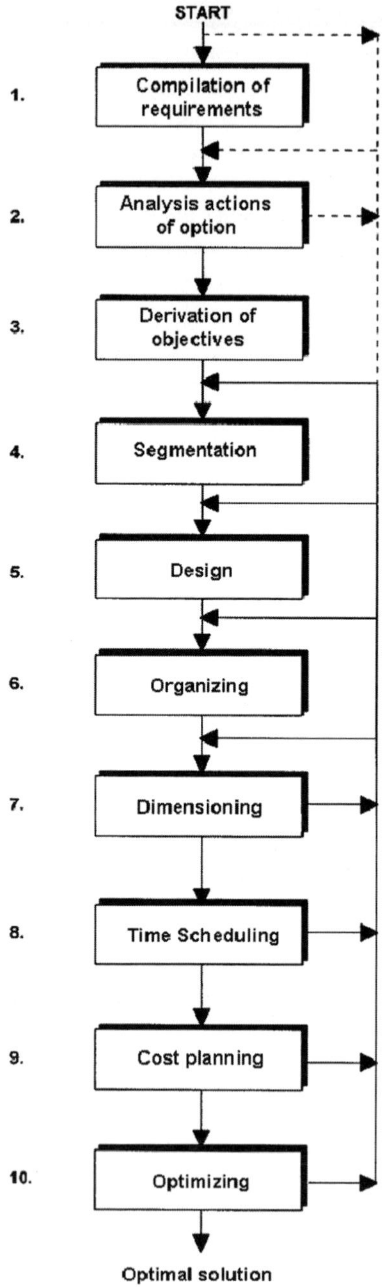

Fig. 5.5 Steps of the solution and optimization procedure

3. Derivation of Objectives

minimal and maximal values
figures to be maximized (5.15)
figures to be minimized

4. Segmentation

clustering and arrangement of articles
consolidation and sequencing of orders
consolidation and sequencing of shipments (5.16)
assignment of load carriers and transport means

5. Design

selection of system elements
combination of subsystems and subnetworks
selection and design of processes and performance chains (5.17)
linking of subprocesses to whole processes

6. Organizing

organization of departments, scheduling and controlling
development of assignment, scheduling and operation strategies (5.18)
design of hardware and software configurations

7. Dimensioning

calculation and determination of dimensions, capacity and number of load units
dimensioning of storage modules, handling stations, transport elements, areas and buildings (5.19)
determination of capacity, limit performances and speed of conveyor and vehicle systems
calculation of the number of vehicles, conveyors, handling, picking and packing stations, storage systems and other equipment

8. Time Planning

synchronization of operating times, hours of work and shift schedules
set up of time tables and schedules
calculation of throughput and delivery times (5.20)
determination of standard dispatch and delivery times

9. Cost Planning

specification and evaluation of investments
calculation of cost rates for internal performances
request of prices for external logistic services (5.21)
calculation of total logistic costs and cost rates

10. Optimization

optimization of target values by variation of the free parameters
comparison and evaluation of possible solutions
selection and suggestion of the best solution
decision by the management

(5.22)

Competing solutions, which achieve the goals in different ways at similar operating costs, can be compared by an *utility analysis* as explained in *Sect. 3.11*. This only concerns solutions that are not discarded before due to *K.O.-criteria*, such as missing minimal requirements, fixed frame conditions or indispensable goals. Monetary target values, such as investment or operating costs, should not be included in the utility analysis as points of an evaluation cannot be converted into monetary values in an objective way.

5.6 Segmentation and Classification

Segmentation strategies are special *cluster strategies* whose variables are the *classification criteria*.[2] Segmentation and classification of orders, goods, shipments and services into *categories* and their assignment to the possible *classes* of load carriers, storage systems and transport means are the first important steps of planning logistic systems and of scheduling process chains. Only few logistic managers and planners are aware of this.

5.6.1 Segmentation of Assortment

The central task of *category management* is segmentation of the range of products and merchandize according to sales aspects, costs or other criteria. This determines the scale of articles and services to be offered on the market.

Typical *logistic assortment classifications* are:

- storekeeping articles and order procured or produced articles
- own products and third party products
- articles with continuous demand and with temporary demand
- standard articles and promotion articles
- classifications due to physical properties, such as size and weight
- classification of articles based on throughput, sales or inventory into *A*-, *B*- and *C-articles* (see *Sect. 5.8*)

5.6.2 Classification of Orders

- categorization of orders based on value, quantity or volume into *small orders*, *normal orders*, *big orders* and *bulk orders*

---

[2] In ancient times the *theory of classes of elements and their relations* has been part of *formal logic*, which was founded by *Gottfried Wilhelm Leibnitz* (1646–1716) and called *class logistics*.

- classification of orders into *single item orders* and *multiple item orders* or into *single piece orders* and *multiple piece orders*
- segmentation of orders based on their priority or urgency in *rush orders* or *emergency orders*, *standard delivery orders* and *fixed date orders*
- categorization of orders into *standard orders* and *special orders* due to their complexity, handling effort or the competence required for the processing

5.6.3 Segmentation of Stocks

- categories with similar storage requirements such as *high value articles*, *safety items*, *cooled articles*, *small parts* or *bulky items*
- classification based on the kind of *load carrier* to be used for storaging such as *containerized goods*, *palletized goods*, and *bin-articles*
- *stock splitting* according to the number of containers, pallets or bins necessary to store the stock of an article
- assignment of the different stock classes to storage systems such as *block place store*, *standard rack store* or *dynamic store* (see *Chap. 16*)

The segmentation of storekeeping articles into classes to be stored in different types of load carriers and the assignment of these classes to different storage systems are important steps of storage planning. The consequences of a wrong segmentation can be severe. For this task, *classification criteria* and *assignment strategies* are needed, which will be developed in *Chaps. 12* and *16*.

5.6.4 Segmentation of Shipments

- classification of shipments due to their urgency into *express*, *due date* and *standard delivery time* shipments
- differentiation of the consignments into *single item* and *multiple item* shipments or into *hazardous goods*, *heavy goods* and *valuable cargo*
- classification due to size in *parcel shipment*, *break bulk shipment*, *partial load consignment* and *full load consignment*
- classification corresponding to the transport mode into *road*, *railway*, *waterway*, *sea* and *air freight* (see *Sect. 21.2*)
- distinction between *homogenous shipments* consisting of load units of the same kind and *mixed shipments* containing load units of different kind

5.6.5 Classification of Transports

- freight differentiation based on the *load carriers* to be used for the transport such as bins, boxes, pallets, containers or swap bodies
- distinction between *irregular transport orders*, that occur spontaneously with varying quantities, and *regular transport orders*, which come in regularly
- classification of the cargo based on the *physical properties* such as *gases*, *liquids*, *bulk cargo*, *letters*, *hazardous goods*, *fresh products*, *chilled goods*, *furniture* and *bulky or heavy goods*

- classification by the *transport modes* into road, rail, water and air transport
- assignment of the different classes of shipments to the *transport vehicles* such as car, truck, van, trailer and tractor trailer, inland, coastal and sea ships, or cargo and passenger airplanes
- division into *single-modal transports* using only one transport mean and *intermodal transports* or *combined transports* using different transport means in the consecutive steps of a *multi-modal transport chain* (see *Sect. 21.1.8*)
- differentiation due to *areas of origin and destination*, such as export, continental and overseas, countries and shipment areas, collection tours, delivery routes and fixed relation transports
- separation based on *individual transport time* and *scheduled standard running times* into rush orders, expedited transport, normal, scheduled or planned transport, and time-phased pickup or delivery

The assignment of transport orders to specific load carriers and transport means, as well as the selection of the *traffic mode* and the separation into areas of origin and destination, are crucial activities of planning and optimization of transport systems and supply chains. The necessary *selection strategies* and *assignment strategies* and formulas for calculating the strategy effects will be developed in *Chaps. 18* and *21*.

5.7 Specialization and Universality

The separation of orders, articles, goods, shipments or functions into clusters and their assignment to classes of load carriers, storage systems, transport means and technologies, lead to the *problem of specialization or universality* with the basic questions:

- To what extent is it opportune to use *different logistic units* and *specialized* storage, transport and handling systems for the different classes?
- To what extent is it better to employ *standard logistic units* and *universal systems* and *technologies*?

Individually designed load carriers and specialized handling, storage, commissioning and conveyor systems are opportune only in larger logistic centers. Specialized transport units and vehicles are economical only for a continuous flow of high volumes of a special good. When frequent changes of flows, demand or goods are expected, the *principle of maximal flexibility* must be considered:

► Only as many different logistic units, storage systems, performance stations, transport means and technologies as absolutely necessary and as few standard units, multi purpose systems and universal technologies as possible.

Specialization and universality are related with the *range of assortment*, *management of variants* and *spectrum of services*. They imply the questions:

- What is the optimal range of merchandize articles for a *retailer* and what is the optimal number of products and variants for a *manufacturer*, in order to fulfill the market requirements and to generate maximal profit?
- What is the optimal spectrum of services for a *logistic service provider*, in order to be competitive at lowest costs with highest profit?

The present assortment range, number of variants and service spectrum has to be scrutinized permanently and adapted to changing demand, especially before starting a major investment project. For this purpose, *company logistics* must know all costs which are caused by procurement, production and distribution of the products and services of the company.

5.8 ABC-Analysis

ABC-analysis is a well known tool for structural analyses. It is recommended by many theoreticians and consultants. However, ABC-analysis is often misused and overestimated. The benefits of ABC-analysis are limited and the danger of drawing wrong conclusions from it is high. However, if applied with care, ABC-analysis can contribute to solve certain problems (Flores and Whybark 1986).

5.8.1 Pareto-Classification and Lorenz-Curve

A *Pareto-classification* is the ordering of a number of objects with respect to decreasing or increasing quantities of a certain property or attribute. In *economics*, the ordering due to *de*creasing quantities of a property, such as *income* or *wealth* of *households*, is usual. In *logistics*, an ordering of the objects due to *in*creasing quantities is common. Examples for *objects* with total number N are:

orders
articles
shipments (5.23)
customers
consumers

Examples for the *properties* or *attributes*, which sum up in total to M, are:

throughput or sales quantities [SU/year]
sales volume [€, $, £.. per year]
stock quantity [pieces, items, SU or LU]
inventory [€, $, £..]
profit margin [€, $, £.. per SU]
number of positions [Pos/Order] (5.24)
order quantity [pieces, items, SU or LU per order]
purchasing power [€, $, £.. per customer and year]
lead time [days per order]
delivery time [days per shipment]

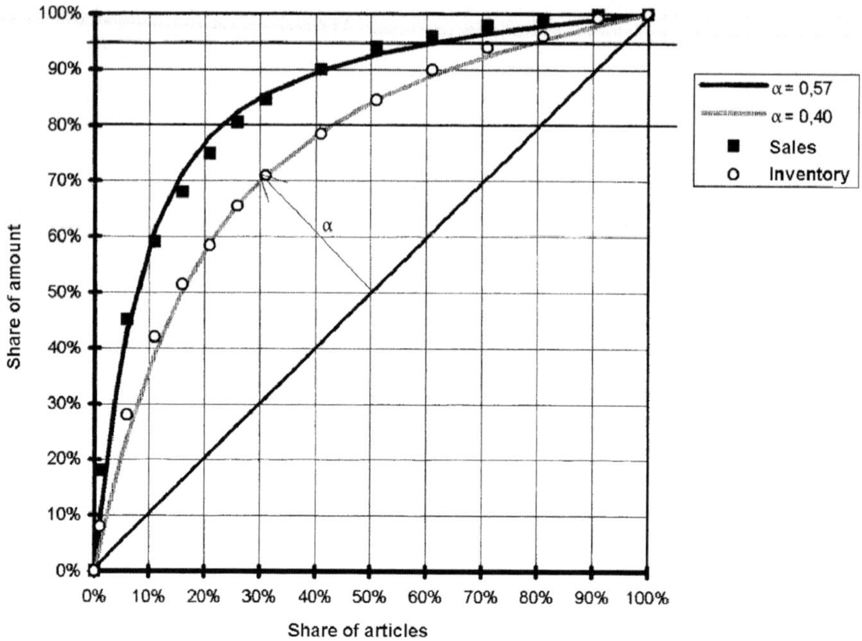

Fig. 5.6 Lorenz curves of sales volume and inventory of storekeeping articles in the logistic center of a merchandiser

Curves: parameterization by function (5.29)
Squares, circles: results of a sales and inventory analysis

The *Pareto-classification* can be visualized by the *Lorenz-curve*.[3] The relative share ρ_N of a total number N of objects is plotted on the horizontal axis, while the relative share $\rho_M(\rho_N)$ of the objects of the total quantity M as function of ρ_N is plotted on the vertical axis. *Figure 5.6* presents the Lorenz curves of the sales volume and the inventory of the storekeeping articles in the logistic center of a department store (see *Sect. 1.6* and *Fig. 3.4*).

5.8.2 ABC-Classifications

For a given set of objects (5.23) and properties or attributes (5.24), two different versions of ABC-classification can be found in logistics. This is quite misleading and causes confusion. The first version is the regular ABC-classification, where the objects are categorized due to their share of the property:

- In a *regular ABC-classification* the total number N of objects is divided due to their increasing shares of a property or attribute into classes of A-, B- and C-objects with the numbers N_A, N_B and N_C in a fixed proportion

[3]The Lorenz-curve is named after *M.O. Lorenz*, who in 1905 suggested this curve in order to visualize *V. Pareto's* income distribution

$$N_A : N_B : N_C = \rho_{NA} : \rho_{NB} : \rho_{NC}. \tag{5.25}$$

The quantities or shares of these classes of the total quantity M of the property or attribute is given by the relation

$$M_A : M_B : M_C = \rho_{MA} : \rho_{MB} : \rho_{MC}. \tag{5.26}$$

Very common in logistics and elsewhere are the fixed classes:

$$N_A : N_B : N_C = \rho_{NA} : \rho_{NB} : \rho_{NC} = 5\% : 15\% : 80\%. \tag{5.27}$$

In *Table* 5.1, the results of the ABC-analysis of the sales volume of a retailer are shown. In this case the standard classification (5.27) of the objects results in the shares 39%: 33%: 23% of the total sales volume.

For an *inverse ABC-classification,* the roles of the objects and the quantities of the property are interchanged. In logistics, the same classes for the regular ABC-classification are common for the irregular ABC-classification, but now for the property shares:

$$M_A : M_B : M_C = \rho_{MA} : \rho_{MB} : \rho_{MC} = 5\% : 15\% : 80\%. \tag{5.28}$$

As long as the Lorenz-curve is symmetric with respect to the falling diagonal of the diagram, the regular ABC-classification and the irregular ABC-distribution result in the same curves. For example, in the case of the retail-assortment the inverse ABC-classification into the standard classes (5.28) of the sales volume results in the shares 38%: 39%: 23% of the number of articles. These are almost the same shares as for the regular ABC-classification. However, in practice many Lorenz-distributions are not symmetric. Therefore one must be critically aware, which version of the ABC-classification is applied.

The results of the ABC-analysis of the retail-assortment, as presented in *Table 5.2*, also reveal that the so called *Pareto- 80:20-rule*, claiming that 80% of the objects represent 20% of the property, does not hold in general. In this example it is fairly true for the sales volume, where 20% of the articles make up 77% of the sales amount. It is definitely wrong for the inventory. Here 20% of the articles cover only 58% of the inventory. That means:

► The often stated 80:20 rule generally does not hold in practice.

Table 5.1 Regular ABC-analysis of objects and their attributes

	Objects **Articles**		Attribute Sales		
Class	Share	Number	Share	Amount	Unit
A-objects	5%	1,500	39%	24.6	Mio.items/a
B-objects	15%	4,500	38%	23.9	Mio.items/a
C-objects	80%	24,000	23%	14.5	Mio.items/a
Sum	100%	30,000	100%	63.0	Mio.items/a

Objects: storekeeping articles *Attributes*: sales amount

Table 5.2 Irregular ABC-analysis of the attributes and their objects

	Attribute **Sales**			Objects **Articles**	
Class	Share	Amount	Unit	Share	Number
A-objects	80%	50.4	Mio.items/a	23%	6,900
B-objects	15%	9.5	Mio.items/a	39%	11,700
C-objects	5%	3.2	Mio.items/a	38%	11,400
Sum	100%	63.0	Mio.items/a	100%	30,000

Attributes and objects see Table 5.1

This is one of the reasons, why ABC-analyses are often misleading. Another reason for misuse and misunderstandings is the fact that many different ABC-classifications are possible. A Lorenz-curve depends on the kind of objects, which are analyzed, and on the kind of properties, which are measured. By the combination of the listed examples of objects (5.23) and properties (5.24), one can derive many different Lorenz-curves. In addition, the 20% best selling articles may not be the 20% articles with the highest throughput. The 20% articles with the highest stock value are not necessarily the 20% with the highest consumption, although with optimal scheduling of the replenishment, a correlation between inventory and consumption exists (see *Chap. 11*).

Before using the results of an ABC-analysis from past periods, one has to consider the *rule of experience*:

- The runners of yesterday may be the sleepers of tomorrow and vice versa.

If this experience and the other implications of an ABC-analysis are not kept in mind, incorrect conclusions may be drawn.

5.8.3 Parameterization of Lorenz-Curves

The ABC-analysis is the first step for many clustering and sequencing strategies. The achievable objective and the application should be known before carrying out an ABC-analysis, as this may be time consuming and costly. The selection of the objects and properties to be analyzed are determined by the purpose of the analysis.

Neither the standard classification (5.27) nor the limitation to three classes ABC is necessary. For many purposes they can even be inadequate. The restriction to three classes with a fixed ratio limits the possibilities of the optimization. From this follows the general *principle of cluster analysis*:

- The number of object classes, their ratio and the attributed properties of a Lorenz-analysis are freely definable parameters and can be used as strategy variables for optimizations.

For optimization and calculations, a suitable parameterization of the Lorenz-curve is needed. A helpful result of many article analyses (performed for trade and industry) and mathematical considerations is the following parameterization of Lorenz-curves:

$$\rho_N = 1 + \left((2 - (1-\alpha)^2) \cdot \rho_N - \sqrt{4 \cdot \rho_N^2 - 4 \cdot (1-\alpha)^2 \cdot \rho_N^2 + (1-\alpha)^4} \right) \Big/ (1-\alpha)^2. \tag{5.29}$$

This function is a 45° rotated hyperbola with the *inequality parameter* or *Lorenz-parameter* α which corresponds to the so-called *Gini-coefficient*. It ranges between 0 and 1 and is proportional to the distance of the Lorenz-curve from the ascending diagonal $\rho_M = \rho_N$. For an equal distribution the Lorenz-parameter α is 0 and the curve becomes the diagonal $\rho_M = \rho_N$. For an extreme unequal distribution, the Lorenz-parameter α is close to 1 and the curve runs up very steeply and approaches fast to the horizontal line $\rho_M = 1$. The Lorenz-parameter is determined by only one point of the corresponding ABC-distribution.

5.8.4 Applications and Risks

ABC-analysis and Lorenz-curves can be applied for:

- focusing activities on the most important objects
- test of the effects of an action for a small number with high weight
- specialized solutions for the ABC-classes of the objects
- reduction of the positions of series orders (see *Sect. 17.12*)
- assortment reduction by elimination of C-articles
- suppression of A-objects and favoring of B- or C-objects

Many applications imply risks which quite frequently cause unsatisfactory results. Such *risks of ABC-analyses* are:

- wrong conclusions from the A-objects for the total number
- disregard of the effects of the B- and C-objects
- inattention of the temporal changes of the ABC-classification
- improper interpretation of a given distribution

In the following section, several examples for logistic ABC-analyses are given. An application which demonstrates the limits of the ABC-analysis is the *fast-mover strategy* intended to reduce the mean traveling times in storage and commissioning systems (see *Sect. 17.6*). A positive example is the order consolidation for large scale production (see *Sect. 17.12*).

5.9 Logistic Article Classifications

The first goal of the logistic analysis of an assortment is to support the sales department when shaping and developing the range of goods and services to be offered on the market (see *Sect. 14.3*). The second goal is to enable optimal strategies and solutions for company logistics.

The first step of the *logistic article analysis* is to classify the range of products and services into different categories, based on sales-oriented and logistical characteristics and objectives. The following *logistic classification criteria* for assortments, articles and products are of particular importance:

- *storability*: storable and non-storable items
- *distribution areas*: local, regional and national assortments; European, USA and International products (see *Sect. 9.7*)
- *movement rate*: A-, B- or C-articles with high, medium and low sales or consumption
- *demand characteristics*: X-articles with *regular*, Y-articles with *irregular* and Z-articles with *sporadic* demand (see *Fig. 9.8*)
- *value classes*: high-value, medium-value and low-value goods
- *application purpose*: capital goods, wearing parts, spare parts, durable and non-durable consumer goods, food and non-food
- *application range*: standard parts, standardized products and made-to-order, customized or tailored products
- *durability*: perishables, short-living (non-durable), long-living (durable), non-perishable and long-lasting goods
- *temperature and freshness*: classification, e.g. of *dairy products* (DP) or *food*, based on *temperature* and *maximal date of expiry* (MDE) or *best before date* (BBD) such as:

$$\begin{array}{lll} \textit{frozen food}: & \text{temperature } -18 \text{ to } -5\,(-1)^\circ\text{C} & \text{MDE 180 to 360 days} \\ \textit{fresh food}: & \text{temperature } +1 \text{ to } +4\,(+7)^\circ\text{C} & \text{MDE 3 to 30 days} \\ \textit{durables}: & \text{temperature } +10 \text{ to } +20^\circ\text{C} & \text{MDE} >360 \text{ days} \end{array} \quad (5.30)$$

- *life cycle*: perennial, annual, seasonal, fashion goods
- *out of stock costs*: loss of sales, profits and profit margins, costs for compensatory replenishment, downtime and standstill costs, penalties
- *type of packaging*: loose or bulk goods, packaged goods in bags, boxes, bottles, barrels, blister packs, trays, parcels, containers
- *variant variety*: *one variant goods* and *multi variant goods* in different sizes, colors, materials, qualities, components or designs
- *composition structure*: *one component articles* and *multi component articles*, such as sets, partsystems, sub plants, whole systems and plants
- *stage of production*: *primary products* or *end products* and *secondary products*, such as modules, parts and raw material.

For production, assembly, bottling and packaging of multi component and multi variant articles a *bill of materials* (BOM) is needed. It includes the composition of the components, their quantities and their origin. For material, parts and incoming products destined for different end products, *lists of application* are necessary. With the help of these lists, the demand of secondary products can be calculated from the demand of the primary products.

Logistic article classifications are necessary and useful for calculating supply quantities and safety stocks, for selecting storage, handling and transport systems, for order scheduling and for production planning as well as for the design and optimization of logistic chains, however. However, one should always keep in mind:

▶ Logistic article classification is never final and has to be scrutinized and updated regularly.

As already stated, this holds especially for ABC-analyses based on

sales volume and sales value
stock quantity and inventory (5.31)
profit margin and profit

In *Fig. 5.6*, the Lorenz-curves resulting for the sales volume and the inventory of a *department store* have been presented. Other Lorenz-curves, which result for the sales and inventory of a *computer wholesaler*, are shown in *Fig.* 5.7. For both cases the parameterization (5.29) has been calculated.

The Lorenz-parameter for thes inventory distribution of the department store assortment has been calculated with $\alpha_I = 0.40$. It is noticeably lower than for the sales, where it is $\alpha_S = 0.57$. The findings for the computer assortment are the Lorenz-parameter $\alpha_I = 0.84$ for the inventory and the Lorenz-parameter $\alpha_S = 0.80$ for the sales volume. Both values are far higher than the corresponding parameter

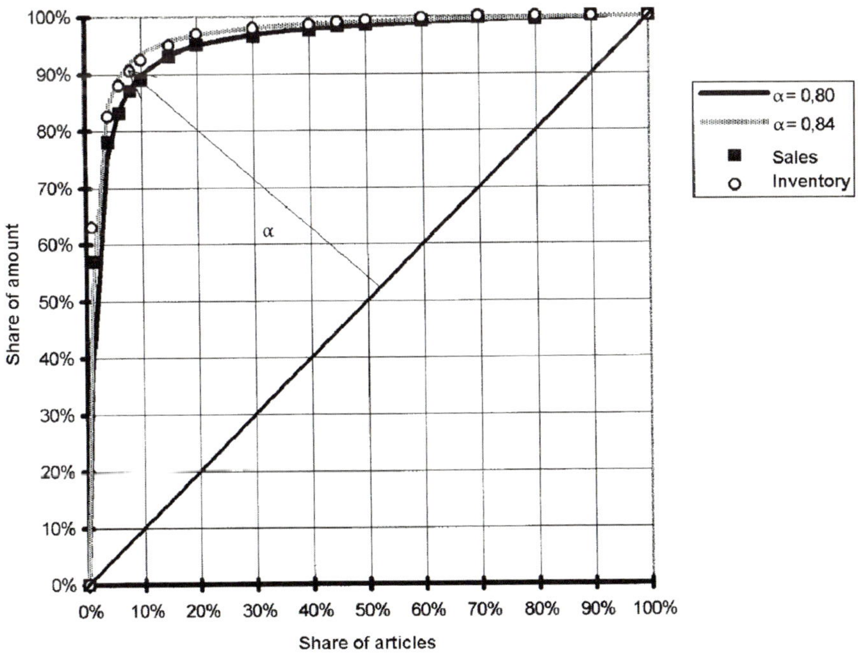

Fig. 5.7 Lorenz curves of sales and inventory of a computer wholesaler
Curves: parameterization by function (5.29)
Squares, circles: results of a sales and inventory analysis

Table 5.3 Service classification of the products of a machine manufacturer

		Service level				
Abbr.	**Service Classes** Characteristics of production	Standard delivery time DT	Ex-stock availability SA	Due-date reliability DR	Position service level $(SA)^m x DR$	Order service level $(SA)^{mxn} x DR$
S	**Store keeping articles**					
SA	store keeping standard articles	1 working days up to dispatch	98.0% per article	99.0% for shipment	97.0%	91.3% Pose/order 4.0
CA	confection articles from stock	2 working days up to dispatch	99.5% with stock keeping components	98.0% confection + shipment	95.8% 4.5 parts/article	91.6% 3.0 Pose/order
O	**Order manuctured articles**					
AO	standard assembling to order from preproduces parts	5 working days up to dispatch	99.0% with store keeping parts	97.0% Assembly + shipment	92.2% 5.0 parts/article	87.7% 2.0 Pose/order
PO	standard production to order including production of parts	10 working days up to dispatch	99.0% with store keeping parts	95.0% production + assembly + shipment	94.1% 6.0 parts/article	84.2% 2.0 Pose/order
SO	individual production of specialities prototypes test articles	>20 working days up to dispatch	98.0% with store keeping parts	92.0% planning + production + shipment	90.2%	90.2%

values of the department store. Even worse: in the second case the Lorenz-parameter for the inventory is greater than for the sales.

With optimal scheduling, the stock of an article with regular demand increases proportionally to the square root of the sales (see *Sect. 11.10*). From this *square root law of logistics* follows the general *inventory rule*:

- ▸ If storekeeping articles with regular demand are scheduled optimally, the Lorenz-parameter of the inventory is lower than the Lorenz-parameter of the sales.

A deviation of the Lorenz-curves for sales and inventory from this rule, as observed for the computer assortment, indicates either a high share of *promotion articles*, *non-optimal scheduling* or different prices for sales and inventories (see *Sect. 11.9*).

Since the assortment of the computer company had been running out of hand, a *differential analysis* was performed. This led to a reduction of the assortment by about 20% of the articles. In addition, the introduction of optimal scheduling led to enormous stock reductions, especially for the A-articles.

The delivery programme of a manufacturer should be differentiated into *service classes*, in order to specify and keep the service levels and delivery times for the storekeeping and non-storekeeping products. For the different classes, customers can be guaranteed standard delivery times and standard service levels. The *service classification* depends on the technology of production, the sources of supply and the business policy. As an example, *Table* 5.3 presents the service classification of a machine manufacturer.

The stated target values for delivery times, punctuality, delivery ability and stock availability are results of a detailed analysis of the *capabilities of production* and a consequent *standardization of processes*. The *service standards* depend on company, business and competition.

A similar service classification is advisable for the assortment of a *retailer*. The merchandise is divided into storekeeping articles and customer specific procured articles. This classification depends on the lead times and service levels of the sources of supply, the manufacturers and the wholesalers.

Segmentation, clustering and classification of articles, orders or customers never is an end in itself. It is a first step for developing strategies and depends on the specific objectives. The economic objective of any business is to achieve sustainable profits by maximizing revenue and minimizing costs. Articles, orders and customers should consequently be differentiated according to their profit contribution into

profitable articles
contributing articles (5.32)
losscausing articles

The classification according to profit is especially helpful for setting up service classes. The most profitable articles should have highest delivery ability and stock availability, the most profitable orders should be executed with best punctuality and the most profitable customers should receive the best service.

Chapter 6
Logistic Costs and Controlling

Logistic costs are defined differently in companies. In many cases, the reported logistic costs of companies even within the same business differ more than justified by their operations. Some companies do not count interest and depreciation on inventories as logistic costs. Others include the distribution costs of their suppliers or the purchasing costs. In some cases, even the purchase value of the procured goods is included in the logistic costs (Baumgarten et al. 1993; Gudehus and Kotzab 2004; Weber 2002).

Another problem, which arises not only in logistics, is *costing and pricing of intangible goods*. Intangible goods, such as logistic services, provide immediate utility and are generally not storable. Therefore, the conventional methods of accounting, costing and pricing, which have been developed for tangible goods, are of limited value for logistics (Cooper 1992; Horvàth 1999; Johnson 1987).

More appropriate for the calculation of *performance costs* are *process-related cost accounting* and *activity based costing*. However, in logistics as well as in other areas of business, the definition and calculation of process costs differs (Bragg 2001; Hicks 2002; Horvàth 1999; Pohlen and LaLonde 1994; Poist 1974). This holds especially for the performance costs of *multifunctional logistic systems*, for *pricing of integrated performances* and for the consideration of *fixed costs*.

As long as they are defined, measured and calculated differently, logistic costs, cost rates and prices cannot be compared. Any *benchmarking* based on such doubtful indicators is misleading (see *Sect. 4.5*). Hence, reported *market volumes* and *market shares* of logistics are at best educated guesses (Baumgarten et al. 1993; Kille and Klaus 2007; Müller-Steinfahrt 1998).

The situation in *logistic controlling* and *supply chain controlling* is even worse (Cooper and Kaplan 1998; Manrodt et al. 1999; Seuring 2006). Only a minority of companies records and monitors logistic costs separately and continuously (Weber 2002). Whereas in industry the total logistic costs range between 5 and 15% of turnover, in trade companies they make up between 10 and 25% of turnover (Baumgarten et al. 1993; Gudehus 1999/2007). For retailers, logistic costs can use up more than one third of the profit margin. Despite this, it is still the exception for retailers to record and monitor the logistic costs from the ramp of the suppliers to the point of sales.

T. Gudehus, H. Kotzab, *Comprehensive Logistics*,
DOI 10.1007/978-3-540-68652-1_6, © Springer-Verlag Berlin Heidelberg 2009

Logistic controlling does not only include calculation, budgeting and recording of costs, but also the monitoring of performance and quality. Controlling should consult management in the planning, set up and operation of optimal systems. To enable this, it has to determine and specify for improvements in service, performance, quality, and costs (Cooper 1992; Darkow 2001; Horvàth 1999; Johnson 1987; Weber et al. 1993, 2002).

Where and with what accuracy *costs*, *performances* and *quality* should be recorded and monitored depends on the contribution of logistics to the *value creation*, on the *core competencies* and *objectives* of the company, and on current projects. In logistic controlling, as in other areas, *less is more*: it is better to control a small number of meaningful *key performance indicators* (KPI) with adequate accuracy in longer time spans, than to monitor all possible performance, quality and cost data with high precision permanently without knowing the demand for these information (Manrodt et al. 1999). For controlling, not the precision of the performance and cost data, but their practical use and application are decisive.

In this chapter, the *logistic costs* are consistently defined, the fundamental issues of *logistic costing* are presented, and practicable methods for the calculation of *use dependent cost rates* are developed. This includes a discussion of the *fixed-cost dilemma of logistics*, the relationship between logistic costs and performance rates and the most effective *options for reducing logistic costs*. Using the results of this chapter, in the following chapter *cost-based prices* and *pricing systems* for logistic performances and services are derived.

6.1 Cost Accounting and Performance Costing

Corresponding to the stationary or *structural aspect* and to the dynamic or *process aspect*, two different types of accounting are necessary. *Cost accounting* for longer periods keeps a stationary point of view, while *performance costing* for shorter periods reflects the dynamic perspective.

6.1.1 Cost Accounting

Cost accounting is a system of full costing from a stationary point of view. It is required for *investment decisions* and *business planning* and for the *profit and loss accounting* (P&L) of a company (Bragg 2001; Wöhe 2000). Topics of cost accounting are the *operating costs*.

When talking about costs, one has to differentiate carefully between past and current costs and expected or planned costs:

- The *past* and *current operating costs* K_{op} [€/PE][1] are the *accumulated costs* for a former, respectively a current *accounting period* [PE = 1 month, 1 quarter, 1 year], in which the recorded *flows* λ_i [PU_i/PE] of different goods or services G_i, $i = 1,2 .. N_G$, with *performance units* [PU_i] have been generated or processed.

[1] Here and in the following € represents the *monetary unit* [MU] which could as well be \$, £, or another currency unit.

- The *expected* or *planned operating costs* K_{op} [€/PE] are the costs for a future *planning period* [PE] with expected, respectively planned *flows* λ_i [PU_i/PE] of goods and services G_i.

For the production of physical goods, the *performance units* are the different *product units* or the standard *measuring units* for weight, length and volume. For the generation of intangible goods, they are *operation units, task units* or *service units*, which are appropriate to measure elementary or compounded *operations, tasks* and *services*.[2]

The logistic costs are defined as follows:

- The *logistic costs* K_{Log} [€/PE] are the total *operating costs* of a single *logistic performance station*, a *logistic profit center*, the *logistic network of the company* or of a *logistic service provider*.

For non-logistic companies, *logistic cost accounting* has to register, calculate, monitor and report all logistic costs, which are caused by the business between the dispatch ramps of the suppliers and the receiving ramps of the customers. However, in some cases this might be difficult or even impossible, for example if the procurement costs are included in the purchase prices.

Because distribution concerns the own products and is crucial for the competitiveness, companies generally know their distribution costs far better than their procurement costs. In order to avoid double counting in *econometrics of logistics*, the following *definition of company logistic costs* is necessary:

- For *market surveys* and inter-company comparisons, the total *company logistic costs* comprise all logistic costs between the receiving ramps of the company and the receiving ramps of their customers.

Due to this definition, procurement costs are part of the logistic costs of the supplying companies. In addition, expenses for external logistic services should be taken as part of the logistic costs of the companies.

6.1.2 Performance Costing

Performance costing is a flow related costing from a dynamic point of view. It is a generalization of *product costing* and *activity based costing* (ABC) (Cooper and Kaplan 1998; Hicks 2002; Horvàth 1999; Mayer 1991; Weber 1993).

The performance costs that arise from the execution of logistic tasks and services have to be known in order to *optimize process chains*, to select *optimal delivery chains*, to *schedule cost-efficiently* and to calculate *cost-based prices*, which determine the allocation of the resources for the whole economy (see *Sect. 7.6*).

The performance costs of a *mono functional performance station*, which can execute only one operation, task or service, are the logistic operating costs K_{Lop} [€/PE] of a period divided by the performance rate λ [PU/PE] of this period. For

---

[2] One has to be aware that the word "performance" has a threefold meaning: 1. the process of the generation of products or the execution of an operation, 2. the results of this process, 3. the rate in which the results are generated.

a *multifunctional performance station*, which can be a network, system or complex service station, that executes different *tasks* or *services* TS_i with the *partial performance rates* λ_i [PU_i/PE], $i = 1,2 \ldots N_{TS}$, it is necessary to separate the total operating costs K_{op} according to *utilization* into a sum of *partial operating costs* $K_i(\lambda_i)$:

$$K_{op} = \sum_i K_i(\lambda_i) \qquad [€/PE]. \tag{6.1}$$

The *partial performance costs* k_i [€/PU_i] can be calculated by relating the partial operating costs $K_i(\lambda_i)$ to the partial performance rates λ_i:

$$k_i = K_i(\lambda_i)/\lambda_i \qquad [€/PU_i]. \tag{6.2}$$

All costs caused directly by the performance flows are *variable costs*. They can be split up into a sum of partial variable costs due to the different tasks and services causing them. All other costs which are not directly caused by performance flows are *fixed costs*. Their splitting into partial fixed costs is possible by the *utilization related fixed cost allocation* as described in *Sect. 6.7*.

6.2 Logistic Cost Calculation

As the general cost calculation of a company, the logistic cost calculation comprises *standard cost calculation*, *accompanying cost calculation* and *final cost calculation* (Horváth 1999; Weber 2002; Wöhe and Döring 2008).

6.2.1 Standard Cost Calculation

Subjects of *standard cost calculation* or *planned cost calculation* are the future operating costs for an existing or a planned system. Results are *standard logistic costs* and *target performance costs*.

Standard cost calculation is necessary for *investment decisions*, for *planning* systems, processes and projects, for *cost accounting* and *benchmarking* of future periods and for the *calculation of prices and tariffs*.

6.2.2 Accompanying Cost Calculation

Accompanying cost calculation aims for a continuous control of all costs caused by the execution of logistic *tasks and services* during the current accounting period. The result of accompanying cost calculation is information for management about the current *costs* and *utilization of resources*.

Knowing the costs and the utilization of the resources allows initiating appropriate measures for reducing costs, adaptation of resources and improving capacity utilization in due time. The results of the accompanying cost calculation can be used also for invoicing and compensation of *logistic service providers*, if *costs-based prices* have been agreed (see *Chap. 7*).

6.2.3 Final Cost Calculation

Subjects of *final cost calculation* or *post calculation* are the operating costs of closed periods in the past. The *real logistic costs* and *cost rates* can be compared with the respective *target values* and *benchmarks*. This allows conclusions for standard costing and pricing.

Most important causes for deviations of real costs from the target values in logistics are:

- *Cost factors*, especially personnel costs, have been planned, assumed or expected too high or too low.
- *Utilization* of resources, such as transport means, storage systems, machines, and production facilities, has been planned or expected falsely.
- *Empty runs* of transport means and *filling degrees* of transport and load units were incorrectly planned.
- The actual *utilization structure* of the logistic system differs from the anticipated structure.

The first two reasons for differences between real and target costs are normally caused by the *planner* and the *operator* of a logistic system. A too high share of empty runs and bad utilization of storage capacities is in many cases also the result of unqualified planning or poor scheduling. However, this can be caused also by a *user*, who changed transport relations, *demand structure* or stock levels. An insufficient utilization can also be initiated by a wrong demand forecast or false information from the customers.

For a *dedicated logistic system*, which is used for a longer period of time by one or a small number of companies based on individual *contracts*, the users must bear the risk of changing demand and the cost differences resulting from a deviating utilization of the ready held resources. Final cost calculation for dedicated logistic systems can be used for the utilization based allocation of surpluses or additional costs to the different users (see *Sect. 7.5.8*).

For a *multi-user logistic system*, where tasks and services are offered on the market and used only for shorter periods of time by many different customers, the risk for changing demand and insufficient utilization is born by the *logistic service provider*. This risk is compensated by the chances for higher profit from better utilization or favorable demand structure. Furthermore, the service provider can influence the demand by his sales efforts and by offering *utilization dependent prices*. For multi-user logistic systems the structure and utilization risk are incorporated in the prices (see *Chap. 7*).

6.3 Components of Logistic Costs

The total logistic costs are a sum of *specific logistic costs*, *additional logistic costs* and *administrative costs*:

- *Specific logistic costs* are all costs of a performance station, a profit center or a company, which are caused by executing the genuine operative logistic tasks *transport*, *handling*, *storing* and *commissioning*.
- *Additional logistic costs* are caused by executing *additional operative tasks* which are directly connected with the genuine logistic tasks, such as *packing*, *labeling*, *loading and unloading*, *quality control* or *handling of empties*.

- *Administrative logistic costs* are costs for related *administrative services*, such as *scheduling*, *quality management* and *controlling*, which go along with the execution of logistic performances and additional services.

Costs for non-logistic tasks, such as research and development, construction, production, assembling, marketing, sales and general administration, are not part of the logistic costs. Also, the costs for buying and procuring merchandise, parts, material and equipment are not logistic costs as long as they are not directly caused by the execution of logistic tasks and related services. For instance, the costs for packing *sales units* are production costs, whereas the costs for packing material, pallets, bins and load carriers are *material costs of logistics*.

When designing and optimizing company logistics as well as when scheduling orders and inventories, it is necessary to keep in mind that many logistic activities also have an effect on non-logistic costs and revenues. They influence *setup costs*, *out-of-stock costs*, *disruption costs* and *ordering costs* as well as *prices*, *profit margins* and *turnover*. Hence, logisticians always have to bear in mind the *economic principle*:

► Logistic activities as all other activities in the company should maximize the difference between revenues and costs at lowest capital investment.

6.3.1 Elements of Logistic Costs

The following elements of logistic costs should be registered separately (Horvàth 1992; Mayer 1991; Weber 1993):

- *Personnel costs*: wages for workers and salaries for employees with logistic responsibilities, including personal taxes, vacation, illness, absence, etc.
- *Space and area costs*: Depreciation and interest for the owned assets and buildings, rents and leasing fees for external buildings, halls and areas, including related heating, climate, maintenance and surveillance costs.
- *Route and network costs*: Depreciation and interest for own and fees for external driveways, routes, roads, highways, railroads and transshipment points.
- *Costs for logistic equipment*: Depreciation, interest and operating costs for own as well as rental fees and leasing costs for external logistic equipment such as racks, forklifts, transport means, cranes, conveyors and handling equipment, control systems and process computers, including the equipment-caused energy, cleaning, repair and maintenance costs.
- *Load carrier costs*: Depreciation and interest for own as well as rental fees and leasing costs for external load carriers, such as pallets, bins, barrels, racks, cassettes and containers, including the costs for cleaning, repair, maintenance and *empties management*.
- *Logistic material costs*: Expenditures for packing material, transport packing, load securing, labels and other material, which is needed in order to perform logistic tasks and services.
- *Logistic IT-costs*: Depreciation, interest and operating costs for own IT-systems as well as costs for external IT-systems as far as used for logistic purposes.

- *Third party logistic expenses*: Freights, rental fees and other expenses for logistic service providers.
- *Taxes, duties and insurance fees*, which accumulate during the execution of logistic tasks and services, as far as related to logistic purposes.
- *Planning and project costs*: Depreciation and interest on activated expenses for planning, project management and implementation accumulated up to the start of the economic utilization of a logistic system.
- *Inventory holding costs*: Interest and write offs on all stationary and moving inventories, in stocks, on buffer places and in transport.

In some companies the inventory holding costs include only the interest caused by the capital commitment. *Obsolescence costs* as well as *write-offs* due to non-marketability, deterioration or stock decline are often neglected. However, the *write-offs* on inventories of fashion, perishable, high value or electronic goods can be as high as or even higher than the interest.

6.3.2 Types of Logistic Costs

Depending on the function of a station or site, on the kind of tasks and services, or on the responsibility, logistic costs can be differentiated into:

- transport, storing, handling and commissioning costs
- procurement, distribution and disposal costs
- operative and administrative costs
- intralogistic costs and extralogistic costs
- direct and indirect logistic costs
- internal or own costs and external or third party costs

According to the definition of *intralogistics* in *Sect. 15.1*, *intralogistic costs* are all logistic costs that arise between the receiving and dispatch ramps of the plants, stations, logistic center and other locations of the company. Correspondingly, *extralogistic costs* are the sum of all procurement, intermediate transport, distribution and disposal costs of the company. The *procurement costs* arise between the dispatch ramps of the suppliers and the receiving ramps of the company. *Intermediate transport costs* are caused by transports between different locations of the company. The *distribution costs* arise between the dispatch ramps of the company and the receiving ramps of the customers.

Direct logistic costs are the operative and administrative costs of all departments and stations which execute separately registered logistic tasks and related services. *Indirect logistic costs* are the costs of administrative stations, which are only indirectly concerned with logistic tasks, such as the departments for human resource, accounting, planning, sales or general management.

With exception of the service providers, whose central business is logistics, the execution of logistic tasks and services does not directly generate sales or other revenues. If the sales of the company are not directly generated by logistic services, it is sufficient to take into account only the direct logistic costs when calculating the logistic costs of products, articles, orders or customers. Only logistic service providers charge their customers by prices or cost rates, which include also indirect costs.

Depending on the *terms of delivery*, the procurement costs may be part of the distribution costs of the suppliers and are included in the purchasing prices. The same holds for the distribution costs, which might be part of the procurement costs of a customer, who picks up his shipments at the dispatch ramp.

The terms of delivery have been agreed 1936 by the *international chamber of commerce* in Paris and standardized in the *International Commercial Terms*, the so called *Incoterms*. Typical terms are *ex work* (EXW), *free on board* (FOB), *cost, insurance and freight included* (CIF) and *free to door*. Not only the terms of delivery, but also the logistic system and the size of the orders, the kind of the shipments, the *delivery times* and *service level* affect the logistic costs of a company as well as of its suppliers and customers.

6.3.3 Fixed and Variable Logistic Costs

In order to calculate performance cost rates, the operating costs have to be separated into fixed costs and variable costs:

$$K_{op} = K_{var} + K_{fix} \tag{6.3}$$

The *variable costs* K_{var} are costs which change with throughput and performance rates and can be avoided as long as there is no demand. The *variable logistic costs* are made up by:

- *expenses for variable human resources*, i.e. for workers and employees as far as hours of work and payment are determined by the performance rate
- *material consumption costs* of the logistic stations as far as these are directly caused by the generated logistic tasks and services
- *renting and leasing costs for transport means*, such as cars, vehicles and other mobile equipment and devices, as far as they are related to utilization
- *energy costs* for fuel, gas, electric power, illumination, heating and air conditioning of areas, buildings, transport means and other logistic equipment
- *repair and maintenance costs* of the transport means, load carriers, vehicles, conveyors, handling installations and logistic equipment
- *road fees and network prices*
- *use-related service expenses*
- use-related *taxes, fees* and *insurance* rates
- *use-dependent depreciation costs for dynamic system elements*, such as transport means, load carriers, conveyors, handling installations and other equipment
- *throughput-dependent inventory costs*, i.e. interest and write off on the use-related share of the total inventory

Depreciation costs, which depend on the current performance rate, are variable costs and not part of the fixed costs. Also, the costs for repair and maintenance which vary generally with the utilization are not fixed costs. It is misleading to assume that these cost factors are independent of utilization and to calculate them for all periods with a constant percentage of the total investment.

Some parts of the variable costs are not as adaptable to a varying demand as generally assumed. For example, in a company with flexible working times, but annual working time contracts, an employee gets guaranteed a minimal number of working hours, which have to be paid even if they have not been used productively.

The *fixed costs* K_{fix} are the remaining costs, which amount independently whether products or services are generated or not. Main elements of *logistic fixed costs* are:

- *time-dependent depreciation* of the investment for *static system elements*, such as areas, buildings, plants, store places, routes and transport networks
- *expenses for fixed human resources*, such as *management* and *standby staff*, which are present independent of demand
- *interest rates* for the total average capital employed in logistics
- *fixed rental fees and leasing costs*
- *fixed personnel costs* for a basic staff of workers and employees
- *permanent expenses* for external services independent of performance
- *fixed taxes, duties, insurance fees* and *tariff rates*
- *depreciation and interest* on activated expenses for planning, project management, implementation and consulting
- *permanent inventory holding costs*, i.e. interest and write off on the share of the total inventory, which is independent of throughput.

In practice the separation of fixed costs and variable costs can be quite difficult. For example, the wear and tear of transport routes is proportional to the intensity of the traffic. Consequently, this part of the network costs should not be taken as fixed but as part of the variable costs (see *Sect. 18.12*). The separation of *dynamic system elements* with use-dependent depreciation and the *static system elements* with time-dependent depreciation is explained in detail for storage, commissioning and transports systems in *Chaps. 16, 17* and *18*.

6.4 Depreciation and Interest

The operating costs of technical systems are to a large extent determined by the depreciation and interest on the invested capital. Different methods are legally permitted and in use for the calculation of depreciation and interest. The applied method depends on the objectives, such as taxation, funding, balancing, costing, pricing or investment decisions (Wöhe and Döring 2008).

For investment decisions as well as for costing and pricing the calculation with *use-dependent depreciation* and *interest on average bounded capital* is appropriate. Besides its simplicity, the advantages of this method are constant cost rates until the end of the technical lifetime even for varying utilization.

6.4.1 Use-Dependent Depreciation

If Λ [PU] is the total *utility stock*, measured in performance units PU, of a *dynamic system element* with *investment* I [€] the *use-dependent depreciation* is

$$k_{\text{dep}} = \text{I}/\Lambda \qquad [€/\text{PU}]. \tag{6.4}$$

For machines and technical devices, such as cranes, conveyors or sorters, the performance unit is the *running hour* and the utility stock is given by the *total running time* or *technical lifetime* Λ_T [h]. For cars, trucks and other vehicles, the performance units are *km* or *miles* and the utility stock is the *total mileage* Λ_M [miles; km]. For example, using formula (6.4) for a truck-trailer with purchase price I = 100,000 € and a *total mileage* of $\Lambda = 1{,}200{,}000$ km results in a depreciation of 8.33 €-Cent/km.

Information about the technical lifetime or mileage can be obtained from the manufacturers. These values are guaranteed by manufacturers, provided maintenance and repair are performed according to their recommendation. *Table 6.1* presents *standard values* for the *technical lifetimes* of selected intralogistic equipment and the resulting *depreciation times* for different utilization. These values are derived from information of manufacturers and experiences of logistic projects.

If $\lambda(t)$ [PU/PE] is the actual performance flow in period t, the *use-dependent periodic depreciation* is:

$$\text{K}_{\text{dep}}(t) = \lambda(t) \cdot \text{I}/\Lambda = \lambda(t) \cdot k_{\text{dep}} \qquad [€/\text{PE}]. \tag{6.5}$$

With the help of formula (6.5) the use-dependent periodic depreciation of *dynamic system elements* can be calculated also for a varying performance $\lambda(t)$. If for example, a truck-trailer has been driven 150,000 km in a year, the periodic depreciation for that year is 12,500 €. If it has been driven 80,000 km, the periodic depreciation is only 6,667 €.

If a system element is *multifunctional* and performs different partial operations, tasks or services, the *partial performance rates* λ_i and the corresponding *partial utility stocks* Λ_i have to be inserted in formulas (6.4) and (6.5). In this way result the *partial depreciations* $k_{\text{dep i}}$ and the use-dependent *partial periodic depreciations* $\text{K}_{\text{dep i}}(t)$

A *linear depreciation* over the whole lifetime, independent of the actual utilization, results in cost rates and prices, which are higher than the real costs for times with low demand, and in cost rates and prices, which do not cover the real costs in times with high demand. Too high prices which are charged to customers or offered on the market prevent a better utilization in times with low demand. On the other hand, lower, but insufficient prices stimulate increasing demand in times where the utilization is already high and generate further loss.

Only if the utilization of a device, machine or facility with *investment* I [€], total *lifetime* T_{LT} [PE] and a *residual value* R [€] is constant for the whole time, the *standard formula for linear depreciation* is correct:

$$\text{K}_{\text{dep}} = (\text{I} - \text{R})/\text{T}_{\text{LT}} \qquad [€/\text{PE}]. \tag{6.6}$$

Table 6.1 Standard values for technical lifetimes of intralogistic equipment and depreciation times for different utilization at 8 hours per shift and 250 days per year

Equipments facilities	Utility stock		Depriciation time			
	Minimal running time	Time unit	1-Shift 2,000	2-Shift 4,000	3-Shift 6,000	h/a
Storage equipment						
Narrow aisle RS unit	30,000	h	15	8	5	a
Rack serving unit	40,000	h	20	10	7	a
Cranes	60,000	h	30	15	10	a
Material handling equipment						
Fixed fork trucks	15,000	h	8	4	3	a
Fork lift trucks	20,000	h	10	5	3	a
Tow trains	30,000	h	15	8	5	a
AGV-Systems	40,000	h	20	10	7	a
Conveyers systems						
Transfer trolley	40,000	h	20	10	7	a
Vertical transfers	40,000	h	20	10	7	a
Parcel sorter	40,000	h	20	10	7	a
Electric monorail	50,000	h	25	13	8	a
Chain conveyers	60,000	h	30	15	10	a
Continuous conveyers	60,000	h	30	15	10	a
Equipment						
Elevators	60,000	h	30	15	10	a
Handing roboter	40,000	h	20	10	7	a
Weighting stations	20,000	h	10	5	3	a
Racks						
Tray racks	15–20	a	–	5–10	–	a
Gravity racks	60,000	h	20	10	7	a
Mobile racks	40,000	h	20	10	7	a
Rotary racks	30,000	h	15	8	5	a
Computer systems						
Hardware	5–10	a	–	5–10	–	a
Software	3–5	a	–	3–5	–	a
Load carrier						
Euro pallets	3–5	a	–	3–5	–	a
Bins	5–10	a	–	5–10	–	a
iso-container	7–12	a	–	7–12	–	a
Buildings						
Storage halls	25–40	a	–	25–40	–	a
Silo buildings	10–20	a	–	10–20	–	a
Building facilities						
Air condition systems	15–20	a	–	15–20	–	a
Sprinkler systems	20–30	a	–	20–30	–	a
Heating installations	20–30	a	–	20–30	–	a
Electric installations	20–30	a	–	20–30	–	a

This formula can be applied for *static system elements*, for example for storage racks, ground areas and buildings. For *land* the deprecation is 0, if the residual value can be assumed to be equal to the purchase price.

For a dynamic system element with utility stock Λ [PU], which is performing with a constant rate λ [PU/a], the total life time is $T_{LT} = \Lambda/\lambda$ [PE]. Inserting this into relation (6.6) results in relation (6.5), if the residual value $R = 0$.

Even if tax regulations or balance policy allows a progressive or decreasing depreciation, investment decisions, costing and pricing should be based on the use dependent depreciation. Otherwise wrong decisions, incorrect cost rates and misleading price incentives would result.

For example, in Germany tax regulations allow writing off a silo constructed high bay store like a machine over a period of 10 years. However, statistics of existing high bay stores indicate total lifetimes of 20 years and more provided the system is properly maintained. Hence, it would be wrong to choose a truck operated store instead of a high bay store only because tax regulations allow to write off a conventional storage building over 25 years and to write off the high bay silo over 10 years.

In logistics, most important for the decision between alternative investments are the *operating costs* calculated with *use-dependent depreciation* and *average interest costs* (see *Table 16.6*). Additional criteria are the *return of investment* of the different solutions (see *Sect. 5.1*).

6.4.2 Average Interest Costs

The interest costs are the product of the *current investment value* I_{act} [€] and the *interest rate* i [%/a]. The current value of a large number N_A of assets, such as buildings, machines, equipment and vehicles, which have been invested in different times with *start values* I_k [€], $k = 1,2, \ldots N_A$, and which are expected to have the *residual values* R_k [€] at the end of the respective life time, is the sum of the time average values $(I_k - R_k)/2$ of the single assets:

$$I_{act} = \sum_k (I_k - R_k)/2 \quad [€]. \tag{6.7}$$

From this consideration results the *average interest calculation principle* (Wöhe and Döring 2008; Gudehus 2005):

- For costing and pricing and for investment decisions, the interest costs are given by the product of the average interest rate and the average time value

$$k_{int} = i \cdot (I - R)/2 \quad [€]. \tag{6.8}$$

From this principle the following *calculation rules* can be derived:

- For machines, transport means and other equipment with residual value 0, interest costs can be calculated on half of the start investment value.
- For land and other assets, which keep their value over time, the interest costs must be calculated on the full purchase value.

When – deviating from these rules – interest costs are calculated on the declining current value, just as for the profit and loss statement, the performance costs become time dependent. In the beginning they are higher than at the end of the utilization.

If cost rates and prices are calculated with the fiscally accepted *degressive depreciation* and *decreasing interest costs*, they are far higher for a newly invested facility than for an old facility, which executes the same task and services. This calculation punishes the users at the beginning and rewards the users at the end of the depreciation time. Customers that had been attracted by low prices at the end of the depreciation time will drift away due to the high prices of a new installation, if they have alternatives. Companies with this kind of calculation generally postpone a new investment for too long.

6.5 Performance Units and Performance Flows

The logistic costs depend on the *kind* and on the *quantity* of the executed task and services. The kind of tasks and services is specified by certain *characteristics and features*. The periodical *quantity* is given by the *throughput* or *performance rate* λ [PU/PE], which is the number of *load units* or *performance units* PU passing respectively leaving the station or system during period PE.

6.5.1 Characteristics, Features and Performance Units

The result of a task or a service and the circumstances and frame conditions of their performance are generally specified by certain characteristics and features. The specific *characteristics and features* of logistic tasks and services are:

- physical properties of logistic units, such as measures, volumes and weights
- required delivery times and delivery dates
- delivery ability, punctuality and shipment quality
- included services and tasks of performance packages
- terms of delivery, storing instructions, and safety requirements

The performance units for logistic tasks and services can be:

- *measuring units*, such as *weights* [kg, t], *lengths* [m] and *volume* [l, m^3]
- *number of logistic units*, such as *article units* [AE] or *load units* [LU]
- *processing units* for tasks or services, such as *orders*, *positions* or *shipments*

Based on these general performance units are the special performance units for the basic logistic tasks *transport*, *storage* and *handling*:

- *transport-distance performance units* [TPU]: load-unit-distance [LU-km], transport- unit-kilometer [TU-km], volume-kilometer [m^3-km], ton-kilometer [t-km]
- *handling-performance units* [HPU]: article units, packages, pallets and other items, which are handled, moved and stored separately
- *in-storing* and *out-storing units* [I-SU and O-SU]: storage units, bins, pallets, load units [LU]

- *store-place-occupation-time units* [SU-t]: load-unit-storage-time [LE-day], such as pallet-day or car-parking-day, article-storage-time [AE-days], volume-storage-day [m^3-days], area-occupation-months [m^2-months]
- *waiting-time units* [WTU]: taxi-minutes; person-hours; setup-hours

These special *logistic performance units* are applied for measurement, costing and pricing of single and compounded performances.

6.5.2 Single and Compound Performances

The calculation of cost rates and performance prices requires a differentiation between single and compound performances:

- *Single performances* [SP] are elementary *tasks* or *services* and can be measured by only one elementary performance unit.
- *Compound performances* [CP], *complex tasks* and *service packages* consist of a number N_{CP} of single or *partial performances* SP_r, r = 1, 2,. . . . . .N_{CP}, which are executed in parallel or in sequence to achieve a required result.

In order to measure compound performances, generally several elementary performance units are necessary.

Logistic service providers often call their compound or *integrated services* "products". For instance, a typical product of a warehousing company is *storing pallets*. Storing comprises the dynamic tasks of *in-storing* and *out-storing*, which are measured by the number of *pallet-in* and *pallet-out* moves, and the static task of *storekeeping*, which is measured by *pallet-days*.[3]

In many cases, the contained tasks and services of a compound performance include further *sub-performances*. An example for a compound intralogistic-performance is the *commissioning* of orders with a certain number of positions per order and several picking units per position. A compound extralogistic-performance is the *shipping* of a number of single packages, which have to be palletized and are transported via one or two transshipment points to a given destination.

In general, the customer is not interested, which single tasks are necessary in order to execute his order. First of all, customers expect a *performance result*, which adds value. For the *pricing* it is therefore necessary to differentiate between the final performance results, which are ordered and noticed by the customers, and the single part-performances, which are necessary to achieve the final result.

The results of logistic performances are intangible, such as a consignment prepared for shipment on the dispatch ramp or a delivered shipment at the receiving ramp. The input for these results are generally single tasks and services, but can also include physical goods needed for the performance, such as packing or load securing material.

[3]This differentiation of *storing* is well known by most warehousing specialists but unknown to or neglected by many economists, theorists, consultants and managers.

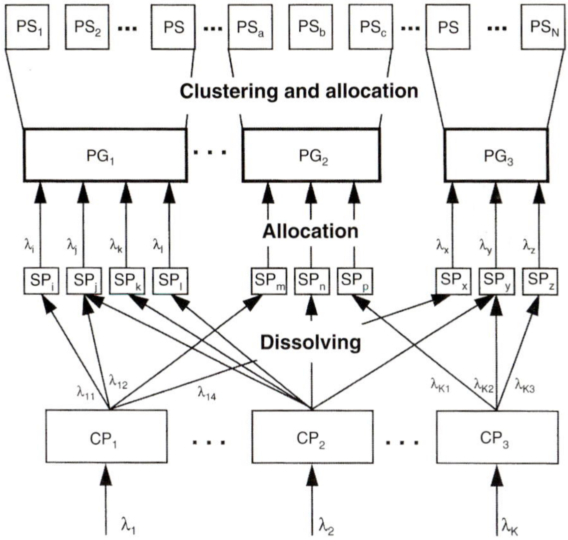

Fig. 6.1 **Dissolving compound performance (CP) into single performances (SP) and allocation to performance groups (PG) consisting of clustered single performance stations (PS)**

6.5.3 *Induced Sub-Performances*

A *performance flow* λ_{CP} [CP/PE] of a compound performance or integrated performance CP, which contains the *sub-performances* SP_r, $r = 1,2...N_{CP}$, with the *content quantities* m_r [PU_r/CP], induces the *sub-performance flows*:

$$\lambda_\tau = m_r \cdot \lambda_{CP} \quad r = 1, 2, \ldots\ldots N_{cp} \qquad [PU_r/PE]. \tag{6.9}$$

If a *sub-performance* SP_r includes n_{rs} *sub-sub-performances* SP_{rs} , $s = 1,2...N_{SPr}$ with the content quantities m_{rs} the sub-performance flow (6.9) induces the *sub-sub-performance flow*:

$$\lambda_{rs} = m_{rs} \cdot m_r \cdot \lambda_{CP} \qquad [PU_{ru}/PE]. \tag{6.10}$$

The vector of *content quantities* (m_1, m_2,..., m_N) and the dissolution of a compounded performance into sub-performances, as shown in *Fig. 6.1*, correspond to the *bill of material* of production planning. The number of sub-performances depends on the kind of compound performance, on the *order scheduling* and on the capacity and the filling degree of the load carriers and transport means.

If the transport means and load carriers differ along the logistic chain, it is advisable to relate the logistic costs to the most elementary logistic units, which flow unchanged through the logistic chain from the origin to the destination. The performance costs for those parts of the total logistic chain, which are passed by load units filled with other units, are allocated to the elementary logistic units (see *Sect. 12.2*).

6.5.4 Partial Performances and Utilizations

In a certain period PE a *multifunctional station* or a *logistic system* can execute different tasks or services with the *partial performance flows* λ_i [PU_i/PE]. The *performance vector*

$$\lambda = (\lambda_1; \lambda_2; \lambda_3; ....; \lambda_N). \quad (6.11)$$

reflects the *performance structure* or *demand structure* of a multifunctional performance station or logistic system.

Each of the possible tasks or services alone can be executed by the station or system up to a certain *limit performance* μ_i [PU_i/PE]. These limit performances are determined by the process, the technique, the organization and the capability of the personnel (see *Chap. 13*).

Corresponding to the performance vector (6.11), the capability of a multifunctional performance station can be represented by a *limit performance vector*:

$$\mu = (\mu_1; \mu_2; \mu_3; ......; \mu_N). \quad (6.12)$$

The utilization of a station with performance vector (6.11) and limit performance vector (6.12) is given by the *utilization vector*

$$\rho = (\rho_1; \rho_2; \rho_3; ......; \rho_N), \quad (6.13)$$

which is made up by the *partial utilizations*

$$\rho_i = \lambda_i/\mu_i \quad [\%]. \quad (6.14)$$

The *utilization vector* reflects the *utilization structure* of a multifunctional station or logistic system. A facility, station or system generally executes the single tasks and services with different partial utilizations (6.14).

When calculating cost rates and prices, it is necessary to differentiate whether a performance station or a logistic system can execute the partial performances *independently* at the same time or only *competitively*. Competing partial performances, such as in-storing and out-storing, can only be executed as long the device, e.g. the storage equipment, does not execute other possible partial performances. For stations with *competitive partial performances*, the sum of the partial utilizations (6.14) can never exceed 100%. This leads to the *limit performance laws* of *Sect. 13.4*.

6.6 Cost Centers and Cost Drivers

Any performance station generates costs and can therefore be defined as a separate *cost center*. The costs of a performance center depend on the *kind* of tasks and services, the installed *limit performances* and the executed *performance rates*. Any performance unit that influences the operating costs is a *cost driver*.

The number of elementary performance stations and their different operations is in most cases so large, that a differentiated registration of all performance flows for all stations is a complex and costly task. This task can be reduced by *clustering* several elementary stations into a small number of *performance groups* and by *bundling* single tasks and services into *performance packages* (see *Fig. 6.1*). Depending on the requirements, this approach allows either a quite *differentiated costing and pricing* or a simple *inclusive costing and pricing* (Horvàth 1999; Lewis et al. 1956; Weber 1993).

6.6.1 Differentiated Costing and Pricing

In order to reduce complexity, all stations, which perform similar tasks and services for the same kind of orders with the same cost drivers, are clustered into *cost centers*. All *relevant performance units* are registered and invoiced separately, irrespective of their contribution to the total costs.

Logistic service providers and internal *profit centers* can efficiently execute differentiated costing and accounting by electronic data collection and computers. The individually recorded performancc units of the intralogistic tasks and services are recorded by a *warehouse management system* (WMS) and the units of the extralogistic tasks and services by a *transport management system* (TMS). Both systems are linked with an ERP-system, which also registers the cost rates, prices and executed orders.

After a closed *billing period*, the ERP-system calculates the charges to the different users or the *invoices* for the customers by multiplying the registered performance rates with the *cost rates*, respectively with the agreed *prices* (see *Sect. 7.4*). At the end of an *accounting period*, the ERP system can consolidate the orders, costs and performances and calculate the logistic costs per customer, article, supplier or for other categories of interest.

Differentiated costing and pricing has the advantage that the users of a system are directly charged with the costs, caused by their individual demand, and for the resources, which have been used up by them. In this way, cost effective changes of the demand structure are promptly passed on to the users. This stimulates the users in a self-controlled manner to keep their demand and the most expensive requirements under control. Vice versa, the operator of the system, in his own interest, will adapt personnel, resources and capacities to a changing demand in short time.

Differentiated costing and pricing avoids conflicts, disputes or claims due to structural changes. Its application is advisable especially for the remuneration of system providers, who operate – based on a longer lasting contract – a *dedicated logistic system* for only one or for a small number of customers (see *Sect. 7.4* and *Chap. 22*).

6.6.2 Inclusive Costing and Pricing

For inclusive costing and pricing, as many performance stations as useful are consolidated to a small number of *cost centers*. The single tasks and services are clustered into a minimal number of *standard services*. For these service packages only the *main performance units* are registered, which cause the highest costs and are the *leading cost drivers*.

The costs for minor performances, which go along with a standard service, are not separately registered. They are incorporated in the cost rates respectively in the prices for the standard services. Since the accuracy of most costs calculations is limited by the *fixed-cost dilemma*, it is generally justified to incorporate all tasks and services, which cause less then 5% of the total costs. This is sufficient as long as the *structure relation* of minor performances to the main performances remains. Hence, the basic *condition for inclusive pricing* is:

▸ The *demand structure* should not change by more than ± 10% during the time, for which fixed *standard prices* have been agreed.

In any case, it is necessary to control from time to time the rate of the included services and the demand structure. Major changes of the demand structure require an adaptation of cost rates and prices.

Standard costing is relatively simple and effortless and therefore appropriate for charging the logistic performances of an *internal profit center* to the participants within the company. An external *logistic service provider*, who wants to apply inclusive pricing, should carefully specify and communicate his standard services, the included services and the demand structure on which the pricing is based. For the case, that major changes of the demand structure happen, additional agreements are necessary. Under these conditions, inclusive pricing is applicable for any kind of *logistic standard services*, in particular in *extralogistics*. Examples are the *list-prices* and *tariffs* for letter, parcel, pallet and container shipments.

The extreme are *all-inclusive prices*, such as day rates for car rental, and *flat rates*, such as railway passes for trains and season tickets for ski-lifts. They are easy to handle for both sides. However, their disadvantages are the missing relation to the individual utilization of the resources.

6.7 Performance Cost Rates

Ideally, the users or customers of a performance station or system should be charged with the variable costs they have caused by their orders and with the fixed costs according to proportionate utilization. Then, orders will only be placed and performances are required if really needed and economically justified. Otherwise the *principle of economic misallocation* will become effective:

▸ Tariffs, fees and prices independent of costs and utilization generate in the short run excessive demand, misuse and waste, while in the long run, the resources will be used inappropriately and inefficiently.

The total costs of a performance station or system include – besides other operating costs – the sum of the use-dependent depreciation and the average interest costs. In order to calculate performance cost rates (6.2) for mono- and multi-functional stations and compounded systems, the total operating costs have to be split according to relation (6.1) into the partial variable costs and fixed costs for the different kinds of performances.

From the performance cost rates with the help of formula (7.1) of the next chapter, cost-based performance prices can be calculated. Provided they are accepted by the customers, *cost-based prices* ensure, that the *turnover* covers variable costs, depreciation and interest, overhead costs and risks, and generates appropriate profit.

6.7.1 Splitting of Variable Costs

As long as the variation of the total operating costs with the partial performance flows λ_i is linear, the variable operating costs can be split up into a sum

$$K_{var} = \sum_i K_{vari}(\lambda_i) \qquad [€/PE] \qquad (6.15)$$

of *partial variable costs*

$$K_{var\,i} = k_{mar\,i} \cdot \lambda_i \qquad [€/PE]. \qquad (6.16)$$

The proportionality factors $k_{mar\,i}$ are the *partial marginal costs*. For a steady cost dependency they are the partial differentials of the total operating costs:

$$k_{mar\,i} = \partial K_{op}/\partial \lambda_i \qquad [€/PU]. \qquad (6.17)$$

The marginal costs $k_{mar\,i}$ depend on the characteristics and features of the respective tasks or services and on the costs directly caused by their execution.

The linear dependency (6.16) is valid only as long as the operating costs are a steadily differentiable function of the flow. However, due to *whole number effects* of indivisible load units, transport means and other resources, in logistics the cost dependency is often a *step function*. In these cases, it is sufficient to approximate the step function by a continuous function and to calculate with the mean dependency (see *Fig. 12.10* and *Sect. 12.5*).

6.7.2 Splitting of Fixed Costs

The fixed costs are independent of the current throughput and performance rates. They are determined by the investment for fixed resources and by other costs factors, which enable the station to execute required tasks and services up to certain limit performances. Hence, the fixed costs depend on the installed *limit performances* μ_i [PU_i/PE]:

$$K_{fix} = K_{fix}(\mu_1; \mu_2; \mu_3; \ldots\ldots; \mu_N) \qquad [€/PE]. \qquad (6.18)$$

They can be separated in an objective manner into a sum of *partial fixed costs*

$$K_{fix} = \sum_i K_{fix\,i} \qquad [€/PE]. \qquad (6.19)$$

by the rule of *utilization dependent fixed cost allocation*:

- The fixed costs of *multifunctional elements* are allocated to the different tasks and services according to the partial *utilization* (6.14).

Due to this rule, the fixed costs of multifunctional dynamic elements are allocated due to the partial *utilization* of the installed limit performances. The fixed costs of static elements, such as spaces, areas, buildings, offices or store places, are allocated to the executed tasks and services in relation to the partial *occupation* of their space or place capacity. The expenses for fixed human resources are allocated due to the partial use of their time.

Other kinds of fixed cost allocations are *action caused* or *fair fixed cost allocation* (Horvàth 1999; Weber 2002). However, fixed costs are not *caused* by any actions and *fair* can be defined in many ways, whereas the utilization structure can be determined in an objective manner and measured beyond dispute.

6.7.3 Calculation of Performance Costs

Inserting the relations (6.16) and (6.19) into the sum (6.3) and of this into (6.1) gives the *performance dependency of partial operating costs*:

$$K_i(\lambda_i) = K_{fix\,i} + k_{mar\,i} \cdot \lambda_i \qquad [€/PE]. \tag{6.20}$$

This relation, in combination with relation (6.2), leads to the *calculation formula for partial performance costs*:

$$k_i(\lambda_i) = k_{mar\,i} + K_{fix\,i}/\lambda_i \qquad [€/PU]. \tag{6.21}$$

The relations (6.20) and (6.21) mean:

- ▸ Operating costs increase proportional with throughput and performance rate, while performance cost rates decrease inversely proportional with throughput and performance rate.

With definition (6.14) of the partial utilization, relation (6.21) can be converted into the formula for the *utilization dependency of cost rates*:

$$k_i(\rho_i) = k_{mar\,i} + k_{fix\,i}/\rho_i \qquad [€/PU]. \tag{6.22}$$

Herein

$$k_{fix\,i} = K_{fix\,i}/\mu_i \qquad [€/PU] \tag{6.23}$$

are the partial fixed costs per performance unit at maximal performance $\lambda_i = \mu_i$, i.e. at *full utilization* $\rho_i = 100\%$. For example, *Fig. 16.29* shows the utilization dependency of the performance costs for different storage systems.

Due to relation (6.20), the operating costs increase by the marginal costs if an additional performance unit is produced. This leads to the *marginal cost law*:

- ▸ Only if the achieved cost rate or price per unit is higher than the marginal costs, a user or customer contributes to covering the fixed costs.

To ensure the long term existence of a company, not only the variable costs but also the total fixed costs and a sufficient profit must be redeemed. Hence, for price negotiations, it is necessary to bear in mind that a company only survives, if the prices cover the full costs and generate adequate profits (see *Sect. 7.3*).

However, in business practice the full costs cannot be achieved in all times. In logistics this is generally caused by underutilization of storage facilities, low filling degrees of transport means and free return capacities. As long as surplus storage and free cargo space are available, the prices are determined on the markets by *demand and offer* and not only by costs (Gudehus 2007) (see *Sect. 7. 7*).

6.7.4 Cost Rates for Compound Performances

The cost rates for *compound performances* and *integrated services* of logistic chains and networks result with the following *compound performance calculation rule*:

► The cost rate k_{CP} for a compound performance CP, which is generated sequentially or simultaneously by single performance stations, is the sum of the *content quantities* m_r of the sub-performances SP_r multiplied by the partial cost rates k_r of the stations involved, i.e.

$$k_{CP} = \sum_r m_r \cdot k_r \qquad [€/PU]. \tag{6.24}$$

From these cost rates, *cost-based prices* can be calculated by adding provisions for overhead, risks and profit. A formula for the price calculation and indications for the order of magnitude of overhead rates, risk surplus charges and profits in logistics are given in *Sect. 7.2*.

6.7.5 Investment Decisions

In non-logistic companies, most of the investments in logistics aim either for capacity expansion or for cost reduction. The performance costs generally decrease with increasing capacity and limit performance, at low values progressively and for higher values slower. However, this *economy of scale* is limited. At a certain *optimal size* and *optimal performance* the performance costs reach a *minimal value*. For larger sizes, they no longer decrease. As will be shown in *Sect. 19.11*, the performance costs of *logistic centers* even increase, when the optimal throughput and size are exceeded (Nowitzky 2003).

Since extensive mechanization and automation require higher investments and increase the fixed costs, the economy of scale for *high-tech solutions* is higher than for *low-tech solutions*. As shown in *Figs. 16.26, 16.27* and *16.28* for 4 storage systems with different mechanization and automation, at low performances the performance costs of a high-tech solution are generally far higher and at high performances substantially lower than for the low-tech alternatives.

This leads to the question: What is the best technique and the optimal degree of mechanization and automation for an investment project? It can only be answered by planning and optimizing the different technical solutions and calculating and comparing their *investment*, operating costs, *performance cost rate* and *return of investment* (ROI). If these economic key data are known for the feasible solutions, the decision results from the *investment rules*:

► Economically optimal is the solution with *highest capital earning value*.

► Of two solutions with same capital earning value, the one with lower performance costs and shorter ROI is preferable, as it involves *lower risks*.

The economic opportunity of investment alternatives can also be assessed by comparing the flow and capacity dependency of the relevant performance cost rates and the required investment.

6.8 Fixed Costs Dilemma and Utilization Risk

The utilization dependency of the performance costs (6.22) leads to the general *fixed-cost dilemma*:

- A system planned and set up for a certain performance flow generates fixed costs independent of the utilization.

The fixed-cost dilemma is caused by the necessity to provide fixed resources and infrastructure for the expected demand in advance. The higher the investment for installed limit performances, the more critical becomes the fixed-cost dilemma. Consequences are:

- The performance costs of a system decrease with growing utilization of the installed limit performances, increasing filling degree of load units and transport means, and decreasing number of empty runs.
- Approaching 100% utilization, maximal filling degrees and vanishing empty runs, the performance costs reach the *full-utilization cost rates*.
- Cost rates and performance prices hold only for the utilization, filling degrees and share of empty runs anticipated in the standard cost calculation.
- The *accuracy of calculated cost rates* and performance prices decreases with increasing variation of the expected demand.

Due to the fixed-cost dilemma, investors tend to dimension the capacities of a new logistic system as narrowly as possible and to avoid additional investment for higher capacities. Planners and general contractors though have a tendency to invest more, in order to reduce the prospective operating costs, and to install higher *reserve capacities* for unexpected variations and increases of demand, in order to avoid the later reproach of *misplanning*. This goal conflict between investors and planners can be solved by the following *planning and dimensioning rules*:

1. A *start solution* is designed that fulfills the mean performance requirements and frame conditions at lowest investment.
2. *Alternative solutions* are developed, which allow reductions of operating costs but require larger investment and generate higher fixed costs.
3. Solutions with higher fixed costs are opportune only if the *payback period* of the higher investment is shorter than the time, for which the planned utilization is sufficiently ensured.
4. *Logistic centers* and stations are planned with *modular design* for the demand of a *planning period* of at least 5, better 10 years, and set up in *extension steps* covering the demand of the next 2 or 3 years (see *Chap. 19*).
5. With exception of the storage capacities, the *limit performances* should be sufficient to serve the mean annual demand within the normal *operating time*, which has to be fixed as to leave a sufficient *time reserve* for flexible adaptation to the expected peak demand (see *Sect. 8.3*).
6. The dimensioning of the *storage capacity* has to take into account the annual peak stock. If outsourcing of the peak store place demand is possible, the storage capacity can be dimensioned for the mean annual stock.
7. *Transport means* and *mobile equipment* are bought, leased or rented only as far as necessary for the peak demand of the next following year.

The utilization dependency of the performance costs and the fixed-cost dilemma are often not sufficiently taken into account. For example, a reduced performance demand resulting from transport consolidation or inventory optimization does not increase the profit as long as the free resources are not used otherwise and contribute to the fixed costs by new revenues.

Another example of neglecting the fixed-cost dilemma can be found in the *waste disposal industry*. Due to successful waste prevention, the upcoming waste quantities are steadily decreasing. This causes the operators of waste depositories to raise the price per ton in order to cover the fixed costs. The higher prices force even more customers to reduce their waste or to use alternatives. In the end, such an *ill-advised price policy* ruins the entire business. The same can happen in *public transport* or in other business with high fixed costs.

In order to ensure the competitiveness of the company, logistic service providers should apply the following *costing and pricing rules*:

- Performance costs should to be calculated for the realistically expected utilization of the resources.
- *Performance prices* are calculated from the performance costs by adding a *utilization risk surcharge* which covers the expected utilization risks.

Overcapacities in the market, which may be caused by competition or decreasing demand, become evident, if the utilization sinks irreversibly below 80%. In such a situation, the *economic lifetime* of the resources concerned becomes shorter than the *technical lifetime*. If this holds on for longer than a year, an *exceptional depreciation* of the asset value is necessary. By this adaptation of the asset value, the prices can be decreased and the situation may improve, as long as the prices are higher than the marginal costs. This policy is the only way to survive if the utilization decreases persistently. If the demand is further dwindling and the market prices decrease below marginal costs for a long time, it is inevitable to shut down capacities or to close the business (Gudehus 2007 p. 290).

On the other hand, if the demand is increasing, producer and service provider have the chance to increase the prices noticeably. As long as the resources are in short supply or lacking, the market prices can increase in excess of the full costs and will generate *windfall profits* for all suppliers, whose prices are not bound.

Customers and suppliers can secure themselves against the risk of demand-driven short term price fluctuations in both directions by closing a *long-term contract*, which regulates *remuneration* and *performance prices* for at least one year and specifies the modalities for *price adjustments* (see *Chaps. 7* and *22*).

6.9 Options for Reducing Logistic Costs

According to their premises and effects, the options for cost reductions and savings in logistics can be differentiated into:

- invest-free and invest-effective measures
- performance-neutral and performance-affecting savings

- short-term, medium-term and long-term measures
- profit-improving and profit-neutral measures
- organizational, technical or economic possibilities

Performance-affecting measures are cost reductions, which either prolong delivery times or reduce the service level. Measures that reduce the expenditures of the company are directly profit-improving. An example is the employment of a service provider with lower prices than the own cost rates or the prices of the former provider.

The organizational, technical and economic possibilities for cost reductions should be carefully examined because they are often interdependent. They can enhance, but may also compensate and exclude each other.

6.9.1 Organizational Cost Saving Measures

Organizational cost saving measures are of particular interest as they can often be realized in short time without considerable investment and have direct effect on the net income. The most effective cost savings potentials result from the *cluster*, *sequencing* and *allocation strategies* described in *Sect. 5.2*, e.g.:

- ▸ The spatial and temporal *bundling* and *consolidation* of orders, shipments, flows of goods, inventories, functions and processes improves the *utilization* of the installed limit performances of the resources and the *filling degree* of load units and storage capacities.
- ▸ The spatial and temporal *sequencing, ordering* and *allocation* of orders, shipments, inventories, capacities and processes increases the *efficiency* of performance stations, improves the performance of resources and personnel, eliminates partly filled logistic units and transport means and avoids empty runs.

However, many of these strategies elongate lead times, delivery times, and throughput times (see *Sect. 8.11*).

Other examples for organizational saving potentials are *elimination of non-value adding activities*, especially in the administrative areas, *simplification of the organization* and *improvement of demand forecasts* (see *Chap. 9*).

6.9.2 Economic Cost Saving Measures

Economic cost saving measures and possibilities for increasing the net income in logistics are:

- ▸ Development of *innovative services*, which improve competitiveness, increase sales and allow higher prices (see *Sect. 7.7*)
- ▸ *Reduction* of the supplier base, the assortment or the variety of variants
- ▸ *Logistic discounts* for ordering unbroken, full or larger packages, load units, pallets, container or transport units (see *Sect. 7.6*)
- ▸ *Economy of scales and high-tech solutions* enabled by increasing sales or by consolidating the demand of several companies

- *Concentration on core competencies* and *outsourcing of non-core activities*, e.g. by employment of external logistic service providers for selected tasks and services (see *Chap. 22*)
- *Cooperation* within the logistic chain between manufacturers, suppliers and retailers, in order to achieve an *Efficient Consumer Response* (ECR)
- Negotiation and selection of the most favorable *terms of delivery*
- *Cost-based prices*, *tariffs* and *discounts* (see *Chap. 7*)
- Evaluation, selection and assignment of the *most cost efficient logistic chains* for procurement and distribution

6.9.3 Technical Cost Saving Measures

The most important technical cost saving measures and possibilities for logistics are (see also *Sect. 3.11*):

- *Mechanization and automation* in order to substitute manpower by machines and control devices
- *Improvement of limit performances*, for example by higher speed and acceleration, shorter downtimes, improved handling and shorter driving times
- *Introduction of innovative technologies*, such as a *automatic high-bay stores* (HBS) instead of a conventional forklift truck stores or *automatic guided vehicles* (AGV) instead of man-operated cars and trucks
- *Larger plants and logistic centers* with modern techniques
- *High capacity load units*: As long as the filling degree is sufficient and the expenses for consolidation and break-up are lower than the savings in handling, storing and transport, larger load units are more efficient (see *Chap. 12*)
- *Larger transport means*: High capacity trucks and wagons, longer trains, big container ships and wide-bodied aircrafts
- *Standardization of logistic units and equipment*: Standard logistic units and equipment reduce operating cost through the whole logistic chain
- Standardization, synchronization and acceleration of data and information exchange
- Standardization and synchronization of the operational processes in the whole logistic chain

These and further possibilities and potentials of *technology in logistics* are described in more detail in *Sect. 3.11*.

6.9.4 Economies of Scale and Principle of Critical Mass

Technical cost saving measures require additional investments and cause higher depreciation and interest. Larger investments become profitable only when the demand is sufficient and the utilization reaches a certain critical level. On the other hand, significant improvements of efficiency and cost reductions are only achievable by *mechanization*, *automation* and *advanced technology*, which need higher investments.

For example, *Fig. 16.21, 16.22, 16.3, 16.24, 16.25, 16.26, 16.27, 16.28* illustrate how the performance costs decrease with increasing demand for 4 storage systems with differing technique and investment. With increasing capacity, the storage throughput costs can be decreased by more than a factor of two. For highly used capacities above 10,000 pallet places, the high-tech solutions are increasingly profitable. This shows the general *principle of critical mass* in logistics:

- Networks, logistic centers, high speed transport means, big load units and high-performance technology become opportune when the demand exceeds a *critical mass* of freight volume, stock level and/or material flow.

In many cases considerable organizational and technological efforts are necessary in order to reach the critical mass and to obtain the benefits of economies of scale. Substantial financial funds have to be dedicated and risks must be borne. However, when the critical mass is exceeded and the aimed cost reductions are reached, a self multiplying process starts: Due to reduced costs, lower prices can be offered. This increases demand and orders and improves the utilization. By this, the costs decrease further, and so forth (Gudehus 2007 p. 290).

The principle of critical mass and its momentum have already been observed by *logistic entrepreneurs* more than 100 years ago, who built up railway networks, post organizations, harbor centers and trade networks. Nowadays, global logistic service providers, airlines, shipping companies and service contractors are fighting for critical mass. The same happens with *e-commerce*, *retail companies* and *IT-service providers*.

6.9.5 Influence Factors on Logistic Costs

Normally, transport, handling and storage costs for physical goods are far higher than the IT costs for processing data and information. This fact leads to the general *logistic cost rule*:

- The logistic costs are determined primarily by the *flows* and *stocks* of the *physical goods* and by their *weight*, *volume* and *properties*.

Apart from interest and insurance fees, logistic costs neither depend on the value of the goods nor on the turnover. Whoever uses the relation of logistic costs to turnover or to unit costs of the articles as *benchmarks*, does not understand the dependencies of logistic costs. By using such ill-defined key figures for the *pricing*, bulky and heavy articles will be sold at too low prices and generate losses. On the other hand, orders for small articles with to too high prices may be lost. Other mistakes result if calculating *optimal lot sizes* and *economic order quantities* based on storage costs, which depend only on the value and not also on the volume and weight of the storekeeping articles (see *Sect. 11.7*).

Additional influence factors on logistic cost rates and performance prices are the *duration of storing*, the *transport distances*, the *technology*, the *utilization* of the resources and – last but not least – the *market situation*. Due to this multiplicity of influencing factors, cost rates and performance prices for the same task or service can differ extremely without being incorrect. This is demonstrated e.g. by the different cost rates for storage systems. *Benchmarking* in logistics must be aware of these facts (see *Sect. 4.5*).

When planning a new project, economic calculations have to be performed by using *preliminary cost factors* and *cost rates*, which are taken from comparable projects, roughly estimated, or inquired from suppliers and logistic service providers. Some *standard values* for logistic costs and prices can be found in this book. The planning and optimization of a logistic network, logistic center or system based on preliminary input data result in preliminary dimensions, capacities, limit performances, stock levels and flow values. With this information more precise cost factors and cost rates can be inquired or calculated. If the start cost values differ too far from the corrected values, the optimization and calculation must be repeated using the improved values. By this procedure, planning becomes an *iterative process* as shown in *Figs. 3.2, 5.5* and *15.1*.

Chapter 7
Logistic Pricing and Marketing

For many companies, logistics is only a *cost factor*, although it can be also a *value driver*. With the help of professional *marketing strategies* based on competitive quality and customer oriented pricing, logistics can generate additional revenues and higher profit (Andraski and Novack 1996; Shapiro 1984; Weber 1993).

In a *free market economy*, logistic service providers are relatively free of legal constraints to set up their own price schemes. In some areas and industries, however, the freedom of pricing is misused. This especially happens at times with *bottlenecks* and in markets with *monopolists*. In other areas, pricing is restricted, for example in *public transport*, where the government regulates many tariffs (Berry and Yadav 1997; Gudehus 2007; Simon 1996).

A company that wants to outsource logistic tasks and services can influence the pricing significantly. By specifying the *remuneration scheme* and requiring cost- and value-related prices in the *tender documents* the offers become comparable and a fair contract can be negotiated. In order to establish a long-term cooperation, *logistic system providers* and *contractors* offer fair pricing schemes in their own interest (see *Sect. 22.4*). Also *logistic service providers* and *integrators* are forced to keep *fair pricing principles*, if they intend to build up a reputation as trustable partners. Long-term contracts with fair and consistent remuneration protect service providers and customers alike against unexpected changes. They are the most reliable basis for investment decisions.

After analyzing the negative consequences of common pricing practices in logistics, *general principles for fair pricing* will be presented in the first section of this chapter. From these principles a standard *performance and quality remuneration scheme* for logistic services is developed, which has been applied successfully in practice. In the following section, the steps to establish a *project specific remuneration scheme* are outlined. For the provider and the customer, revenues and profit critically depend on *marketing and pricing strategies* which are described in *Sect. 7.7*. In the last section the role of *economics* for *logistics* is discussed.

T. Gudehus, H. Kotzab, *Comprehensive Logistics*,
DOI 10.1007/978-3-540-68652-1_7, © Springer-Verlag Berlin Heidelberg 2009

7.1 Pricing Principles

Negative consequences of *unfair pricing* and evidence for the misuse of unlimited market freedom can be observed in many areas. In logistics these are:

- *Specifications* and *price lists* that are incomplete, difficult to understand or not accessible for the customers.
- Performance prices that are not cost-based and not use-related.
- A comparison of prices for the required performances is hardly possible.
- Required performances are linked with others that are not needed.
- Only undifferentiated, all-inclusive prices are offered.
- The price covers only the core-performance. Services, which are inevitable related with the core-performance, are charged extra.
- Price changes are unfounded and happen more often than justified.
- Due to *discounts*, *all-inclusive prices*, complicated *price models*, *subscriptions* and *free extras*, the customer is not able to calculate the prices for single performances, in particular if the future demand is not known exactly.
- *Business conditions* and *terms of trade* restrict the *obligation to fulfill* and to keep *quality standards*.

Unfair pricing is quite common on logistic markets: misleading freight tariffs, which are still based on outdated tariff models; cartages or other fees that the receiver has to pay afterwards; oblique prices for storage services; different pricing schemes of parcel distributors; *monopoly prices* for letter mail, which cross subsidize parcel post; *customer cards* and *special prices* of railway companies; cross subsidy of railway passenger fares by excessive network tariffs for cargo trains; confusing tariffs, fees, mileage programs and overbooking practices of *airlines* (see *Sect. 22.4*).

To find out the complete, correct and lowest price, the customer is troubled with lengthy inquiries and analyses of confusing price lists. Such pricing methods are misleading, unacceptable and sometimes even illegal. They annoy the customers and make them suspicious. Service providers who try to increase profit by such practices harm themselves in the long run, as their profit and loss accounting is confused by prices and revenues, which are not related to costs and utilization. In addition, they lose reputation and customers.

Prices, which are independent of costs and not related to the utility value for the customer, generally cause waste and misuse of resources. Cross subsidized sectors of the business will not be rationalized. Business opportunities are not realized due to overrated prices. An example is the shifting of cargo transport from the railways to the roads where excessive network prices for cargo trains subsidise the passenger fares (see *Sect. 21.14*).

Service providers who offer performance prices, which are transparently related to costs and utility value, can gain additional market shares. Transparent pricing also forces competitors to correct and adjust their price structure. An US-based airline doubled its market share within short time due to stable, uncomplicated and attractive prices that did not include unwanted services. This airline became one of the most profitable airlines in the USA. Similar strategies have been successfully applied by some telecommunication companies and parcel distributors (Berry and Yadav 1997).

In order to secure competition and functioning markets, the following *principles of fair pricing* ought to be obligatory in logistics, as well as in other businesses (see *Chap. 23*):

1. Performance prices should be easy to understand and transparent, based on costs and related to utility value of the services.
2. Cross subsidized *dumping prices* are unacceptable. Under normal circumstances, at 100% utilization the sales prices should cover full costs.
3. All standard logistic tasks and services have to be specified in a *catalogue*.
4. The single performance prices and the *terms of payment* must be documented in *price lists*, which are available for all customers.
5. Performance prices should not include unnecessary or unwanted services.
6. The sales price must include all tasks, services and sub-performances which are inevitably connected with the offered performance.
7. Additional and special services, expensive extras and insurance fees can be ordered separately at adequate prices.
8. Prices for a longer lasting demand should be valid for longer time, if possible at least for a year, and only be changed if justified by cost or structure alterations.
9. *Reimbursements*, *discounts* and *sliding-scale prices* are only acceptable for pre-defined quantities or for a long-term obligation of a customer.
10. The controlling of invoices must be uncomplicated. Invoices should only contain executed performances at the agreed prices.
11. Cost intensive *extra services*, such as *express delivery*, are charged separately only to those customers that have ordered them.
12. Performance prices include the obligation to perform as specified within the agreed *quality standards*. Disregarded performance generation will be reimbursed. Deviations from quality standards are fined.

These principles of fair pricing do not impede a competitive, innovative and efficient service provider to offer profitable prices. The above US-examples have shown how fair pricing creates additional competitive advantages. Fair pricing principles are major preconditions for a positive development on the logistic markets and of the whole economy (Gudehus 2007).

7.2 Performance Costs and Prices

The remuneration for performances and the calculation of performance prices depend on the legal relation between user and provider. Companies, whose core business is not logistics, normally incorporate the logistic costs in the overhead surcharges. However, in order to keep control and to minimize the logistic costs, it is necessary to establish an *article- and order-specific logistic accounting* and to calculate the article- and order-specific logistic costs separately:

- The *article-logistic costs* and the *order-logistic costs* are the sum of all logistic costs per article unit respectively per order which are caused by procurement, production and distribution.

From the quantities and costs of the logistic services and performances, which are required to procure, produce and distribute the article units, the article-logistic costs are calculated with the help of the rule for compounded performances (6.24) by multiplying the relevant quantities with the cost rates.

Logistic tasks and services that have been performed by an internal logistic department are charged to the other divisions of the company with *target performance costs* which are calculated as described in the previous chapter. For this purpose the logistic departments, such as a *logistic center* or a *shipping department*, must be organized and operated as *profit centers*.

A *logistic service provider* calculates from the *cost rates* k_i [€/PU] for the different tasks and services, which have been calculated for a planned utilization, the *performance prices* P_i [€/PU] to be offered on the market. In order to cover sales and administrative overheads and to compensate failure and utilization risks, the cost rates are increased by *surcharges* and *profit margin*:

$$P_i = (1 + p_{sas} + p_{risk} + p_{uti} + p_{prof}) \cdot k_i \qquad [€/PU]. \qquad (7.1)$$

The performance price calculated for a planned utilization without profit is the *full cost price*. The surcharges are necessary to ensure that the *total revenues*

$$R_{tot} = \sum_i \lambda_i \cdot P_i \geq K_{tot} \qquad [€/PE] \qquad (7.2)$$

generated by the expected *performance demand* λ_i [PU/PE] cover the *total costs* K_{tot} including all risks and generate adequate profit R_{tot}-K_{tot}.

7.2.1 Sales and Administration Surcharge

The *sales and administration surcharge* p_{sas} (SAS) has to cover all expenses for sales, administration and overhead that have not been taken into account separately in the calculation of the cost rates.

Smaller logistic service providers, such as freight forwarders facing strong competition, calculate with SAS between 10 and 15% depending on the market segment. Bigger logistic corporations, such as railroad companies, shipping companies and airlines, have sales and administrative overheads between 20 and 50%, sometimes even higher. For a logistic service provider with a contract lasting longer than five years, a sales and administrative overhead surcharge between 8 and 10% is generally sufficient. For long relationships between service providers and customers, sales surcharges are unnecessary.

7.2.2 Failure Risk Surcharge

The *failure risk surcharge* p_{risk} has to cover all risks and warranties which result from failures to perform and from not keeping an agreed quality standard.

If no *penalty* is paid to the customer, the risk surcharge is determined by the statistical mean *reclamation costs*. If a failure of performance or quality is penalized, the risk surcharge depends not only on the reclamation costs, but also on the kind of the performance, the probability of failures and the height of the fines.

An agreement upon such fines should be based on measurable failures, in order to make the risk surcharge calculable. If e.g. in case of a failure to keep a certain delivery date within the agreed tolerance, the customer receives as refund the double performance price and if the probability of such an event is 0.7%, the failure risk surcharge is $p_{\mathrm{risk}} = 2{\cdot}0.7\% = 1.4\%$. The failure risk surcharges for other penalized qualities can be assessed similarly.

Well managed logistic service providers who ensure common quality standards generally calculate with a total failure risk surcharge between 3 and 5%.

7.2.3 Utilization Risk Surcharge

The *utilization risk surcharge* p_{uti} protects a logistic service provider against the risk of *underutilization* compared with the planned utilization. The utilization risk surcharge depends on the predictability of the demand for the period, for which the prices are fixed, and on the *fixed cost share* of the performance costs.

If a customer contractually ensures a minimal utilization or allows an adjustment of the performance prices to the utilization as described in *Sect. 7.5.9*, no utilization risk surcharge is necessary.

7.2.4 Profit Margin

The *profit margin* p_{prof} depends on the *price policy*, which has to meet market demands, competition and customer benefits (see *Sect. 7.7*). The profit margin of a *monopolistic provider* is capped only by the *utility value* of the offered services for the customer. In this case the profit margin is the difference between realized sales revenues and total costs including sales, overheads and risk surcharges. A *competing provider* has to set a lower profit margin limit, which covers the general business risks and generates sufficient profit. In the long run, a logistic service provider who operates on free markets can only survive with a profit margin of at least 3%.

The profit margin for innovative performances can be much higher, as long as there is a competitive advantage, which offers additional value to the customers. The expectation of a high profit that covers the innovation risk and pays back the development costs in short time is the main motivation for innovations.

7.2.5 System Management Fee

The total surcharge of a logistic system provider, who offers and executes integrated tasks and services based on a long-term *contract*, has to cover all the above surcharges. In addition, a system provider will charge a *system management fee*:

- The *system management fee* is the bonus that a customer has to pay for *outsourcing* the management efforts and business risks related with the execution of system performances.

It corresponds to the *general contractor surcharge* for a turnkey project. In markets with strong competition a system management fee on top of the other surcharges is generally not achievable, as many customers argue that the system provider is compensated by better *purchasing margins* and *economy of scale savings* (see *Chap. 22*)

According to the above values, the total surcharge of a system service provider who is secured against the utilization risk, is in a range between 15 and 20%. If a logistic service provider bears the utilization risk, the system management fee increases by the utilization risk surcharge.

7.3 Objectives of Remuneration Schemes

Any long-term contract with a logistic service provider must regulate the remuneration. The more complex the tasks and services and the higher the performance rates are, the more important is a clear regulation of the remuneration (see *Chap. 22*).

A good *performance and quality remuneration scheme* aims at the following *objectives* and has to fulfill certain *requirements*:

- The remuneration should stimulate the service provider in a self-regulating manner to perform efficiently and correctly.
- It must be clear, feasible, and in accordance with the above *principles of fair pricing*.
- Procedure, dates and acceptable reasons for price adjustments, such as costs or demand changes or *rationalization*, must to be agreed.
- Demand structure, performance rates and bearing of the utilization risk have to be clarified and agreed upon in advance.
- Performance rates, failures and quality defects should be registered with minimal effort based mainly on otherwise registered data.
- Invoicing and payment should be feasible by computer with the help existing standard software.

An effective remuneration scheme makes detailed controlling and operative management by the customer unnecessary. Alone the provider, not the customer, is responsible for management of the operations and controlling of performance. The customer can confine to examining invoices and reclaiming eventual failures and quality deficiencies.

7.4 Standard Remuneration Scheme

The basic concept of a performance and quality remuneration scheme, which fulfils the above objectives and has been approved in business practice, is presented in *Fig. 7.1*. This standard scheme can be adapted to project specific conditions and circumstances. Its procedure is quite simple:

- All tasks, services and *standard performances* SP_i, $i = 1,2,...N_{SP}$, are specified and documented in a *performance catalogue*.
- The agreed cost-based *performance prices* P_i [€/PU] for the standard performances SP_i with *performance units* PU_i are documented in a *price list*.
- Intolerable *quality defects* QD_r, $r = 1,2,...N_{QD}$, are specified in a *quality defect catalogue*.
- For the different quality deficiencies, *defect penalties* D_r [€/QD] based on appropriate *defect units* DU_r are determined in a *list of penalties*.
- The *remuneration* R [€/PE] for *period* PE is the sum of the products of the *performance rates* λ_i [PU_i/PE] multiplied by the corresponding performance prices

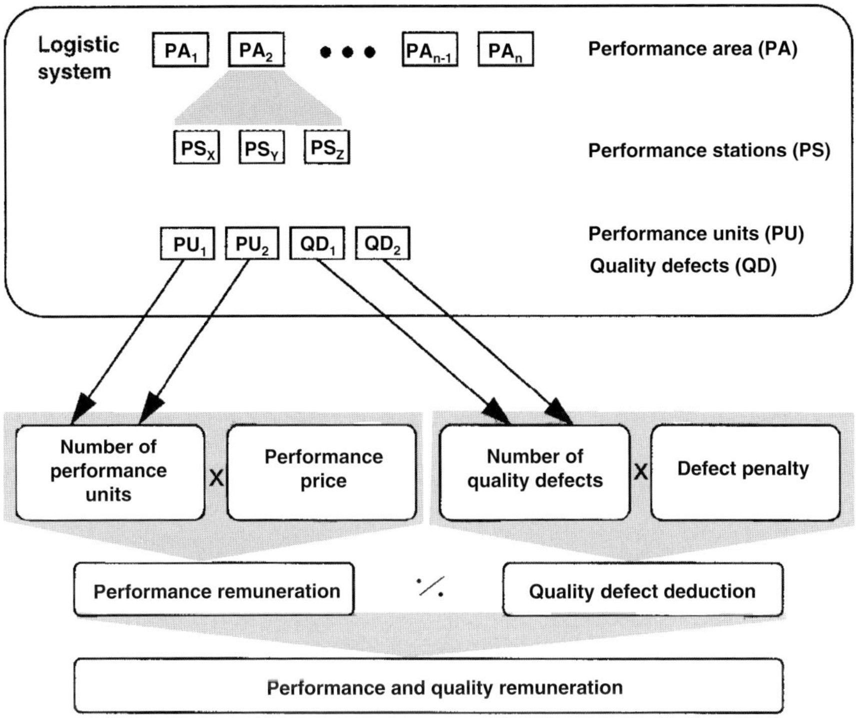

Fig. 7.1 Concept of the standard performance and remuneration scheme

P_i [€/PU] minus the sum of the products of the *failure rates* σ_I [QD_r/PE] multiplied by the corresponding defect penalties D_r [€/QD]:

$$\mathbf{R} = \sum_r \lambda_i \cdot \mathbf{P}_i - \sum_i \sigma_r \cdot \mathbf{D}_i \qquad [€/\mathbf{PE}]. \tag{7.3}$$

Orientation values for cost-based performance prices of internal logistic tasks and services are listed *Table 7.1*. For external logistics, they are presented in *Table 7.2*. These values have been calculated by the methods and rules outlined in *Chap. 6* for a standard utilization of 90%. For the special tasks of storing, commissioning, transport and freight, the calculation of cost rates and performance prices will be described in more detail in *Sects. 16.13, 17.14, 18.12* and *21.14*.

The reduction of the remuneration by penalties forces a provider to focus not only on a high performance, but also on the observance of the agreed quality standards. Not only cost-efficient operation, but also the avoidance of quality defects will improve the bottom line of the provider. In this way, the above performance and quality remuneration scheme becomes *self-regulating*.

No human built system is free of defects, not even the transport to the moon. No *zero-defect-scheme* can alter this fact. Hence, a company should set limits for the *tolerable quality defects*, which may be 1%, 1‰ or 1×10^{-6}.

General *3-sigma* and *6-sigma schemes* require a quality level of 3 times, respectively 6 times the standard deviation σ of a stochastically varying target value (Goldsby 2005). As shown in *Fig. 9.6*, the *safety factor* 3 corresponds to a safety level of 1‰. This might be adequate for consumer goods. In *informatics*, 3-sigma or 1‰ are far too low. Here 6-sigmas are required. In logistics, 1‰ is generally too tight. Business experience has proven (see *Sects. 3.3.4* and *17.4*):

- For *performance availability*, *punctuality* and *delivery reliability*, a quality level between 1% and 5% is sufficient and affordable.
- For *shipment quality* and *order picking quality*, a level between 0.1 and 1.0% is necessary and feasible.

Table 7.1 Standard performance prices for intralogistic tasks and services

FERFORMANCE AREA **Type of performance**	**Performance unit**	**PU**	**Performance price**	**Price unit**
GOODS RECEIVING	Receiving controls			
Unloading				
pallets out of truck	Pallet	Pal	1.00	€/Pal
packages out of truck	Package	Pac	0.05	€/Pac
STORAGE	Strong addicted to capacity and service requirements			
In-storing	Pallet	Pal	1.00–3.00	€/Pal
from goods receiving	Bin	Bin	0.15–0.25	€/Bin
Store keeping	Pallet-day	Pal-Cday	0.10–0.20	€/Pal-Cday
in a storage place	Bin-day	Bin-Cday	0.02–0.04	€/Bin-Cday
Out-storing	Pallet	Pal	1.00–3.00	€/Pal
providing in despatch area	Bin	Bin	0.15–0.25	€/Bin
GOODS DESPATCH				
Order picking	Inclusive pallet replenishment			
packages on pallets	Package	Pac	0.15–0.25	€/Pac
items into bins	Item	Item	0.02–0.05	€/Item
Loading	With control of good issue and load securing			
pallets into truck	Pallet	Pal	1.20–1.50	€/Pal
packages into truck	Package	Pac	0.05–0.10	€/Pac
EXTRA SERVICES				
Supply scheduling	Supply order	SOrd	3.00–5.00	€/Sord
Internal transport	Distance 50–100 m			
from pick up point	Pallet	Pal	1.00–1.50	€/Pal
to destination	Bin	Bin	0.08–0.15	€/Bin

Orientation values for sufficient demand based on costs level 2006

Load unit	**Mean weight**	**Mean dimensions**
Pallet (CCGI)	500 kg	1,200 × 800 × 1,050 mm
Container	25 kg	600 × 400 × 300 mm
Bin	5 kg	10 *l*/barrel
Article unit	0.5 kg	1 *l*/item

When setting the quality level, one has to take into account that the quality securing costs within the own company, as well as the risks allowances of an external service provider increase with the required quality level infinitely. Total quality or zero-defect as well as 100% punctuality, 100% availability or 100% safety are impossible, unaffordable and unnecessary (see *Sect. 5.2.3*).

7.5 Project Specific Remuneration Schemes

For *contract logistic projects*, the remuneration scheme must be designed individually. This can be achieved by adapting the standard remuneration scheme to the specific demand and circumstances as follows:

Table 7.2 Performance prices for extralogistic tasks and services

PERFORMANCE AREA **Type of performance**	**Performance unit**	**PU**	**Price**	**Price unit**
DIRECT TRANSPORT	Full and partially loaded trucks			
Distance up to 250 km	Transport order	T-Ord	70.00	€/T-Ord
	Intermediate stop	I-Stop	20.00	€/I-Stop
	Transport distance	T-km	1.20	€/T-km
Distance over 250 km	Transport order	T-Ord	120.00	€/T-Ord
	Intermediate stop	I-Stop	20.00	€/I-Stop
	Transport distance	T-km	0.90	€/T-km
LOCAL AREA TRANSPORT	Inclusive transshipment without haulage to TSP			
Delivery Transport	Delivery stop	D-Stop	5.00–8.00	€/D-Stop
Delivery tours from TSP	Pallets	Pal	20.00–30.00	€/Pal
to destination	Cargo	Cgo	6.00–8.00	€/100 kg
Pick up transport	Pick up stop	P-Stop	5.00–7.00	€/P-Stop
Collection tours from collection points	Pallets	Pal	18.00–25.00	€/Pal
to TSP	Cargo	Cgo	4.00–6.00	€/100 kg

Orientation values for sufficient demand based on costs level 2006
The prices for area transports depend on the size of the area
Transport unit: trailer or trailer truck with 2 swap-bodies
Pallets dimensions: 800 × 1,200 mm, height <1,500 mm,
weight: mean 500 kg/pal, max 700 kg/pal
General cargo: single items and standard parcels

7.5.1 Specification of Tasks and Performance Rates

In the first step is clarified, what tasks, services and compound performances have to be provided at what *rates*. The precise definition and specification of the required *kinds of performances* and the quantification of the expected *performance rates* are basis of any remuneration scheme.

For this purpose, a qualified *target planning* is necessary, as described in *Sects. 3.2, 3.6* and *3.7*. The results have to be documented in a *catalogue of requirements and performance rates*.

7.5.2 Specification of Performance System

In the second step, the system is specified, which will execute the required functions, services and performances.

For this purpose, the performance stations, staff, resources and equipment and the *limit performances* are described and evaluated as far as necessary. After this, suitable *cost centers* are formed and *cost drivers* are defined as described in *Sect. 6.6*.

7.5.3 Definition of Performances and Performance Units

In the third step is determined, where and how the required performances will be executed. As outlined in *Sects. 6.5* and *6.6* and shown in *Fig. 6.1*, the required tasks and performances are split up into sub-performances and basic-performances and allocated to the executing performance stations. For the basic-performances, suitable *performance units* are selected.

Performances of successive stations of a performance chain, which relate to the same performance unit, are integrated in *compound performances, standard performances* or *primary performances*. The expected *performance structure* is documented as described in *Sect. 6.6*.

7.5.4 Agreement of Contract Period and Invoicing Period

Contract periods in logistics range from 1 year to 10 years (see *Sect. 22.4*). For *transport and freight*, the contract period is generally 1 year, for *storage tasks and services* at least 3 years and for *customized system performances*, which are connected with high customer specific investments, 5 to 10 years and longer.

The *planning period* for costing and accounting is generally the business year or the calendar year. As far as possible, the prices should be kept constant for the planning period. In alignment with the monthly cost and activity accounting, for the remuneration an *invoicing period* of 1 month is adequate in most cases. The longest meaningful invoicing period is 1 year, the shortest 1 week. Also quarterly invoicing may be adequate.

The contract partners also must agree upon the *terms of payment*, which could be *credit memo procedure, debit advice* or issuing of an *invoice* with an allowed time for payment.

7.5.5 Calculation of Cost Rates and Performance Prices

According to the procedure of the previous chapter, the specific cost rates for the different standard and compound performances are calculated with the help of relation (6.21). From these cost rates *cost-based performance prices* are calculated with the help of relation (7.1) by adding appropriate surcharges for sales, administration, risks and profit.

The surcharges and the profit margin result from negotiations between customer and provider. They can be used for an *open book calculation* after the best provider with the most favorable prices has been selected during a *bidding process* as described in *Chap. 22*. After agreement about surcharges, profit margin and *start prices* the open book is closed. The remuneration is based only on the agreed performance prices. They are documented in *price lists*, as shown in *Table 7.1* and *Table 7.2*, which are valid in combination with the performance specifications and the frame conditions.

7.5.6 Definition of Quality Defects

If quality defects are penalized, the quality standards and their measurement must be specified. When requiring a certain quality level, one has to keep in mind that the risk surcharge increases with the number and the precision of the quality standards. As described in *Sect. 3.4.4.*, typical quality defects and deficiencies of logistic performance systems are:

- *delivery insufficiencies*

  delayed pickup
  delayed or unpunctual delivery (7.4)
  failure to comply with agreed delivery dates

- *shipment insufficiencies*

  incompleteness
  quantity deviations
  damages (7.5)
  loss and shrinkage
  wrong articles

- *scheduling insufficiencies*

  unavailability of storekeeping articles
  insufficient delivery ability (7.6)
  insufficient ability to perform

An external service provider can only be made responsible and fined for the unavailability of stocks, if he is responsible for inventory scheduling. *Outsourcing of inventory scheduling*, however, is risky and can cause excessive stocks.

After specifying the quality deficiencies, customer and provider also have to agree upon the *measurement* and *registration* of defects and the settlement of *complaints* (see *Sect. 22.4*).

7.5.7 Agreement upon Defect Penalties

For any deviation in quality, a certain *defect penalty* has to be agreed. Its height is generally related to the performance price. For example, the penalty is double or triple the regular price for the corresponding performance (see e.g. the *German BGB Sect.* 431 (3)).

The settlement of *direct damages*, such as loss or damage of cargo, has to be regulated independently of the defect penalties. It can also be covered by an appropriate *insurance*.

Normally, a customer should not expect from a service provider liability for *consequential damages*. They are difficult to assess in advance and therefore extremely expensive or impossible to insure.

7.5.8 Regulation of Utilization Risk

To secure a logistic service provider against the risk of *underutilization*, three different regulations are possible:

- *Fixed cost remuneration*: A *basic fee* is charged that covers all fixed costs. The remuneration is calculated with performance prices that take into account only the *variable costs*. If there are several users, the basic fee for each one is proportional to the individual share of utilization.
- *Utilization guarantee*: The customer guarantees a *minimal utilization* and takes over the non-covered fixed costs, if the minimal utilization is not met. For exceeding utilization, over-charged fixed costs are refunded.
- *Price adjustment*: The current remuneration is calculated with *target prices* $P_{i\,plan}$, which are based on the planned performance rates $\lambda_{i\,plan}$. At the end of the business year, *adjusted performance prices* $P_{i\,real}$ are calculated for the real *performance rates* $\lambda_{i\,real}$, keeping the other parameters of the pre-calculation constant. In case of lower rates than planned, the service provider is credited, in case of higher rates, the service provider is debited with the difference

$$\Delta = \sum_i \lambda_{i\,real} \cdot (P_{i\,real} - P_{i\,plan}) \qquad [€/a] \qquad (7.7)$$

Any of these fixed cost regulations stimulates the users to give reliable forecasts of their demand and to use the contracted resources as intensive as promised.

7.5.9 Regulation of Price Adjustments

Each contractual partner has the right to require in due time an adjustment of the prices. The acceptable reasons for price adjustments, such as changes of cost factors or performance structure, are documented in the contract.

If the acceptable reasons and the procedure of price adjustments are not agreed in advance, disputes about price requirements can cause premature termination of the contract. This happens more often than generally known.

7.5.10 Documentation

The regulations of the remuneration scheme should be completely documented and signed by both parties. This includes:

specifications of tasks, services and performances
lists with the actual performance prices
calculation schemes for cost rates and prices
agreed surcharges and profit margin (7.8)
quality defects and defect penalties
terms of payment

For big projects and high performance rates, it is advisable to develop a *price calculation program* and make it part of the contract. Such a program enables *sensitivity analyses* of the consequences of parameter changes and price calculations for different *utilization scenarios*. It can also be used for the calculation of the annual *fixed cost remuneration* and for the adjustment of the current prices to cost and structure changes.

7.6 Logistic Tariffs and Discounts

The standard remuneration scheme can also be used to develop pricing schemes and tariff systems for *compound logistic services* which are offered on the market, such as letter postage, parcel services, freight forwarding, railroad transport, container services and storing. On the market, however, arises the problem that sales prices and pricing schemes should be related also to the *utility value* for the customers and cannot be based only on the *cost drivers of the service provider*. As shown later, this conflict is solvable, if the prices are related to cost drivers of the producer which are also value drivers of the customers.

7.6.1 Logistic Tariffs

A *cost-based and value-dependent price* for an executed order is calculated from a basic *order tariff* and from specific *performance tariffs*:

- The *order tariff* is charged per order, per shipment or for another *basic unit* and covers all basic tasks, services and performances.
- The *performance tariffs* are charged for the specific performances and cover all costs related to the specific *performance units*.

With these tariffs the price for an executed order is:

$$\textbf{\textit{Price}} = \textbf{\textit{basic unit}} \times \textbf{\textit{order tariff}} + \textbf{\textit{specific units}} \times \textbf{\textit{performance tariff}} \quad (7.9)$$

The order tariff for *handling orders*, such as packing, loading, unloading or sorting, includes all costs caused by the order. The performance tariff covers all costs that depend on the number of items handled, transported or sorted.

Storage orders are charged cost-based with an *in- and out-storing-tariff* for in- and out-storing of the single load units and with a *store-place-tariff* that covers the costs for keeping a load unit for a certain time (see *Sect. 16.13*). For *commissioning*, a cost based tariff is more difficult (see *Sect. 17.14*).

Alternative cost-based *transport tariffs* for the use of a *transport unit* [TU] or for conveying a number of load units [LU] are (see *Sect. 18.12*):

- *basic tariff* [€/TU or €/LU] and *distance tariff* [€/TU-km or €/LU-km].
- *relation tariffs* [€/TU or per €/LU] for defined *transport relations* $A_i \rightarrow B_j$.
- *zonal rates* [€/TU or per €/LU] for defined *distance zones*.

Freight orders are charged with an order-related *basic tariff* that covers roll-up and provision of transport means and the use of transshipment points, a *handling tariff* for sorting, loading and unloading, and a *distance tariff* for the transport of the load unit over a certain distance (see *Sect. 21.15*).

7.6.2 Logistic Discounts

In business practice, logistic discounts are often granted as certain percentages of the performance price or of the total turnover (Chopra and Meindl 2006). Economically such *flat rate discounts* do not make much sense.

Logistic discounts are opportune only, if they correlate with the expected savings or stimulate higher utilization. For example:

- *Logistic order discounts* are granted in order to stimulate the ordering of *complete logistic units,* such as full packages, article-pure pallets or completely filled transport units.
- Logistic *quantity discounts* or *volume based discounts* are granted in order to improve *supplier loyalty* and to ensure high utilization of the resources. They depend on the number of load units, transports or performance units ordered by the same customer within longer period of time.

In order to stimulate the intended behavior, the logistic discount must be related to the cost savings they generate. The appropriate values for the different discounts depend on the special circumstances.

Quantity discounts are granted in order to reduce the costs for logistic services. The customers should be aware, that quantity discounts generate larger order quantities and higher stocks. Hence, in order to avoid increasing costs, the discounts must be taken into account when calculating the *economic order quantity* (see *Chap. 11*). Larger order quantities stimulated by logistic discounts can contribute to the so called *bullwhip effect* (Forrester 1958; Lee 1997).

7.7 Marketing and Pricing Strategies

Prices are the result of *price building processes*, which are the central part of the *marketing process*. Price building is a competition between *offerers* who want to sell and *customers* that demand a certain good or service. The marketing process ends with the *contract price* at which the seller hands over a good or performs a service for the buyer.

From the offerer's perspective, marketing is a *sales process* that includes, besides pricing, other processes such as promotion, advertising and explaining the offer. From the customer's perspective, marketing is a *purchasing process* that, besides negotiation, includes inquiring, examining and selecting potential suppliers. *Specification* of the goods and performances and *negotiation* of the prices and *terms of business* are parts of the marketing processes of both parties.

In most markets, several pricing processes happen simultaneously and consecutively and influence each other. The same product, service or performance, generally has different prices on separate markets and at different times. The *market price* is the *mean value* of all prices that have been paid on a defined market within a certain time period. It is a statistical value that depends on the circumstances and the number of selling events. For many goods and services, the market price is unknown or can only be estimated. That holds in particular for logistic services.

The duration of the price building process depends on the frame conditions. Price building happens in very short time if an offered *fix-price* is just accepted, and lasts longer if prices and discounts are negotiated. The negotiations following a written price offer may take days and weeks. Tariff negotiations of interest groups last even longer. The resulting sales prices depend on the general *market rules*, the *market constellation*, the *costs* and on the *marketing and pricing strategies* of buyers and sellers (Gudehus 2007).

A central part of the market rules are the *price building rules*. On *logistic markets*, the price building rules are often incomplete, changing, partly unknown and unfair. Existing rules are not kept in any case since they cannot be controlled. Consequences are limited competition, inefficient markets and arbitrariness up to fraud. This causes high *transaction costs*, unfair prices and a *misallocation of resources*.

7.7.1 Price Building Process and Rules of Fair Trade

The single price building process, as shown in *Fig. 7.2* is a contest between seller and buyer with uncertain result. The seller aims for maximal revenues at lowest costs, the buyer for maximal benefits at lowest price. As in any competition, the marketing process and price building are restricted by *market rules*. Some rules are set by law and cannot be modified. Other rules, in particular for standard products on consumer markets, are generally known. Additional regulations can be agreed upon individually between seller and buyer.

In order to ensure fair trade and efficient competition, the following *rules of fair trade* should be obligatory for all participants:

- All rules must be available and accepted by the actors before trading.
- During the trade, no rule should be unilaterally changed, added or canceled.
- All participants must follow the rules without exception.
- For all participants hold the same criteria for market admission and exit.
- An actor who meets the admission criteria should not be hindered.
- Each participant must accept leaving the market without success.
- No actor is forced to participate in a certain trade.
- No one should deliberately take advantage of an emergency situation.
- If rules are imprecise or lacking, the price building process is interrupted until the actors have agreed upon the regulation.

Up to now these rules do not hold on all markets. They are only partly secured by law. Consequences are, e.g. the above mentioned unfair practices on logistic markets.

Disputes are settled by a *neutral authority* that has been accepted by the participants before trading has started. This can either be a consultant, an expert, an official institution or the *chamber of industry or commerce*. An *arbitral court* or a court of justice is only needed for difficult cases of high value.

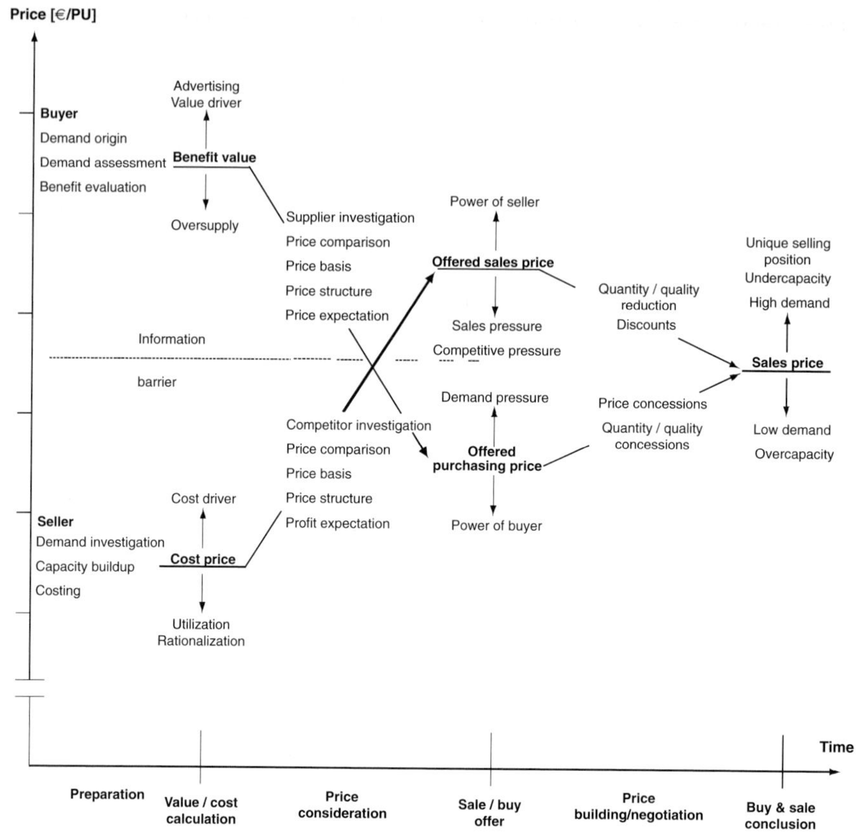

Fig. 7.2 Process and main influence factors of price building by negotiation

Consumers do not have such arbitration authorities. The protection of consumers and other small market participants is a governmental task. Some rules against unfair competition in logistics have been outlined in *Sect. 7.1*.

7.7.2 Logistic Pricing Units and Price Structure

The price P [MU/PU] of a tangible or intangible good is measured in *monetary units* [MU] per *pricing unit* [PU]. Whereas the money units are given indisputably by the legal currency, the *pricing units* are generally free. There determination requires special considerations and an agreement between buyer and seller. The pricing bases for *tangible goods* are generally known, often standardized and for certain goods even regulated by law. The pricing bases for *intangible goods*, such as logistic tasks and services, are often unknown and generally unregulated.

The offered price of the seller must cover the purchasing, production and provision costs and ensure sufficient profit. For this purpose the seller calculates the cost rates from the operating costs for an expected or planned sales volume and adds the necessary surcharges and aimed profit margin as explained above.

For logistic pricing, a differentiation between internal and external cost drivers is necessary. *Internal cost drivers* of the service provider are related to his internal processes and procedures. These are e.g. internal load and transport carriers, handling processes, vehicle movements, tracking and tracing, data processing, and other internal activities. *External cost drivers* are the freight units, load units and performance units, which are related to the customer orders.

The customer has insight and is interested only in external cost drivers, which are determined by his *requirements*. The requirements of the customer result from the intended use of the good or service. The monetary *utility value* is the highest purchasing price the buyer is willing to pay. Just as the seller keeps his costs and profits secret, the buyer keeps the highest acceptable purchasing price to himself. This mutual behavior causes – as indicated in *Fig. 7.2* – an *information barrier* between buyer and seller.

The *utility value* of a good, service or performance depends on *value drivers* that can either be economically calculable or are soft and incalculable. *Universal value drivers* are the size, quantity and number of performance units, by which the economic benefit is measured. *Special value drivers* are properties, which are of value only under certain circumstances or only for some customers. *Soft value drivers* are incalculable and related to personal preferences or taste, such as appearance, brand, entertainment value of the good or service, or as prestige and reliability of a supplier.

For the same logistic service or performance, often different pricing bases are possible. This holds for freight services and in particular for warehousing services, where the pricing basis is not standardized. Hence, buyer and seller must agree upon a pricing basis which is efficient and objectively measurable. From the relation between cost drivers and value drivers follows the *price bases rule*:

► Only objectively measurable external *cost drivers* and *value drivers* should be used as pricing bases.

For simple standard performances it is sufficient to use the main cost driver as single pricing base. For compound performances and integrated services a single pricing unit is not sufficiently related to the utility value for the costumer and the utilization of the resources. In spite of this, many *integrators* offer *all inclusive prices*, *flat rates* or *package prices*, in order to suppress transparency, comparability and competition. This can be prevented by the *principle of price differentiation*:

► Compound performances and integrated services should be charged by *differentiated prices,* which are based on *costs* and related to *utility value*.

Cost-based prices for compound performances are derived from the performance cost rates as explained in *Sect. 6.7*.

For a customer, differentiated prices improve *price transparency* and *comparability*. The costs are allocated to the users according to the orders and to the utilization of resources. The seller is protected by differentiated prices against the risk of *structural changes* after closing the contract. Differentiated, cost-based and value-related prices are also of advantage for the whole economy, as they favor contractors with the best offers.

However, there exists a *goal conflict* between differentiated pricing and efficient accounting and invoicing. For example *commissioning* has so many cost drivers, that a remuneration based on all of them would be very complicated. Therefore, it is necessary to limit the price bases to a small number of *main cost drivers* which are equal to the *main value drivers* of the customers. Other cost drivers are incorporated in the main cost drivers as outlined in *Sects. 6.5* and *6.6*.

The general conflict can be solved in an objective way without affecting competition by a *standardization of price bases*:

- For customary *standard goods and performances*, a neutral authority should determine as many differentiated price bases as necessary for transparency, cost dependency, competition and control, and as few as possible, in order to enable efficient accounting and invoicing.

Such a standardization of the price structure is legal, as long as it concerns only the price bases and not the height of prices. The standardization of pricing and performance units is necessary in order to enable price assessments and to improve the fairness for consumers and smaller market participants. In addition, standardized services, performance units and price bases reduce the *transaction costs* and are prerequisites for efficient price building (Gudehus 2007).

7.7.3 Price Building Types

Basically, five different price building types can be found on the markets (Gudehus 2007). Their features, advantages and disadvantages are presented in *Table 7.3*.

Most simple are

- *Externally Fixed Prices*: The sales price is fixed by the producer, provider, retailer or the state for a longer time and not negotiable.

Table 7.3 Characteristics and effects of the basic price building types

Price building type	Reaction to demand/cost	Price building time	Price building effort	Effect of competition	Price fluctuations
Externally fixed prices	Months to years	Minimal	Low	Low	None
Binding sales prices	Days to weeks	Minimal	Low	High	Small
Negotiated prices	Hours to days	Medium	High	High	High
Auction and stock exchange	Minutes to hours	Short	Average	Very high	Very high
Cost plus pricing	Years	Periods	High	None	None

In logistics there are several examples for externally fixed prices: postal prices for standard letters; fares for public transports and taxis; ticket prices and flat rates of airlines and railroad companies.

The advantages of externally fixed prices are stability and predictability of the market prices and the elimination of tedious price negotiations. Disadvantages of externally fixed prices are that they generally act to the advantage of one side, mostly for the sellers, and to the disadvantage of the other side, often the customer. Externally fixed prices are generally adjusted quite fast to higher costs or when demand increases. They are reduced only slowly when demand decreases and are rarely adjusted to decreasing costs.

More flexible than external fixed prices are

- *Binding Sales Prices*: The sales prices are announced in advance to the customers and are binding for the seller.

The advantage of binding sales prices and *list prices*, compared to externally fixed prices is that they are generally more flexible and faster adjusted to changing costs and demand. Increasing demand, under-capacity and lack of resources generate increasing sales prices, while they are lowered if demand is falling. Costs increases lead to increasing sales prices only as far as demand and competition allows them. Rationalization or improved technology enable price reductions, stimulate sales and increase profits.

Further advantages of binding sales prices are small price fluctuations and the elimination of price negotiations. They are appropriate for mass products and standard services, which are required by many anonymous buyers in small quantities. Examples in logistics are passenger transport and parcel services.

If the announced or listed prices are not binding they are

- *Negotiable Prices*: Prices are not binding and can be negotiated.

Depending on the *market power* of the actors, the negotiated sales price lies somewhere between the *monetary utility value* of the buyer and the *full cost price* of the seller. Negotiable prices react immediately on *offer and demand* and the competitive situation. However, the negotiation process may last long and requires experience and competency of the actors.

Written price requests are most important for the procurement of a high demand, valuable products and complex systems. Long-term contracts for transports, freight demand, warehousing and system services are generally closed by negotiated prices at the end of a *tendering process* (see *Chap. 22*).

Time consuming price negotiations are avoided by auctions and stock exchange:

- In a *seller auction*, interested sellers offer a *sales price* to a buyer and the lowest quoted price wins the bid. In a *buyer auction*, interested customers bid a *purchase price* and the highest price offer wins the bid.
- On a *stock exchange* the quantities and prices of prospective customers and sellers are brought together. From the demanded and quoted quantities and prices, the *sales prices* and *sales quantities* are calculated with the help of an *algorithm* which achieves certain aims.

Aims of the *broker institution* can be maximization of transfer quantities, total turnover, profit of sellers or buyers, or a minimal or maximal market price. Different aims lead to different results for the actors (Gudehus 2007). Advantages of auctions and stock exchanges are fast and efficient price building, low transaction costs and immediate reaction to offer and demand. The main disadvantages are highly *volatile market prices* and the unknown aims of the institutions, who organize auctions and stock exchanges. Volatile prices complicate forecasting and affect planning reliability.

Due to the *Internet*, the marketing and price building at stock exchanges and auctions are nowadays quite simple, cost efficient and fast. The reach of the Internet has increased the number of market participants. This explains the high expectations towards *e-commerce*, *e-logistics*, *Internet-based auctions* and *Internet-based stock exchanges* (Grieger 2006). However, auction and stock exchange are suitable markets only for goods and services with well defined quantity and quality.

Examples for auctions in logistics are *transport exchanges* for road and shipping capacity and *freight exchanges* for bulk cargo, such as ore and coal, and for standard cargo, such as ISO-containers or EURO-pallets. In order to allow for efficient Internet-based *auction pricing* for other logistic services, such as warehousing, handling services or transshipment, it is necessary to standardize content and pricing of these services. Standardized information, specification and pricing are keys for further success of *e-commerce* and *e-logistics*.

Logistic service providers quite often propose

- *Cost-Plus-Pricing* or *Open Book Compensation*: The provider calculates the prices by adding the agreed profit margin to the accounted costs which have been caused by the services and performances for the customer.

Cost-plus-pricing is common for *public orders*. It is not a free price building process but a risk-free transfer of total costs plus profit to the customer. It prevents a contractor to reduce costs and improve efficiency, stimulates to generate more services than needed and generates additional costs. After longer collaboration, disputes arise between contractor and customer about the adequacy of the charged costs (see *Chap. 22*), if not a consistent remuneration scheme based on cost-related prices has been agreed in advance.

7.7.4 Market Constellations

Besides costs and pricing strategies, the market constellation determines the result of the pricing process shown in *Fig. 7.2*. The resulting sales price depends on the pressure to buy and on the pressure to sell.

The *pressure to buy* increases with dwindling stocks, the acuteness of demand and the danger of lacking supply. In order to use these circumstances to their advantage, sellers investigate the demand situation of their customers. The *pressure to sell* increases with growing stocks, declining utilization and dwindling liquidity. Therefore, buyers try to find out stock levels, utilization of resources and financial situation of the sellers.

In general, buyers and sellers are simultaneously or consecutively involved in several price building processes. This causes a *competition pressure*, which depends on the market constellation. On logistic markets as on most other markets, the basic market constellations are:

$$\begin{aligned}
&\textit{no competition}: \text{ one seller and one buyer}\\
&\textit{seller monopoly} \text{ one seller and many buyers}\\
&\textit{seller oligopoly}: \text{ few sellers and many buyers}\\
&\textit{buyer monopoly}: \text{ many sellers and one buyer}\\
&\textit{buyer oligopoly}: \text{ many sellers and few buyers}\\
&\textit{complete competition}: \text{ many sellers and many buyers}
\end{aligned} \qquad (7.10)$$

On logistic markets, numerous monopolies and oligopolies can be found. In many countries still *letter mail monopolies* and *railway monopolies* exist. Examples of oligopolies and cartels are the world wide *shipping conferences* and *airline cooperatives* that determine prices and capacities and *airport operators*, who set *airport charges*. Also the state, as owner of the public transport network, is acting as a monopolist and sets *network prices*, such as *car tax*, *petrol tax*, *toll* and *Maut*.

In addition, there are *effective monopolies* of market dominating companies and *oligopolies* in sub-segments of logistic markets, which allege performance standards and set price levels, such as the IATA *International Air Transport Association*. For many logistic markets there is a lack of knowledge on the competitive situation.

Outsourcing of logistic services can lead to the situation, that a buyer monopolist faces a seller monopolist. This is the case when a logistic service provider is committed by a long-term contract and high investments to a customer, who himself depends on the provider. The mutual dependency eliminates the free price building on the market. The negative consequences of outsourcing and *inter-company SCM* for competition, prices, flexibility and innovation are widely ignored so far (Bretzke 2005).

7.7.5 Pricing Strategies

By appropriate *pricing strategies* the actors try to influence the final price to their advantage. Theory and experience reveal the *basic market law* (Gudehus 2007; Mankiw 2003; Wöhe 2000; Smith 1776; Schneider 1968):

- Market prices tend to increase with growing demand and decreasing offers and to decrease with falling demand and growing offers.

In combination with the considerations of *Sect. 6.7*, the general market law leads to the following *pricing strategies* (Gudehus 2007):

- If competition is low and the market is unsaturated, the sales prices can be increased up to the *utility value* of the customers.
- In saturated markets with high competition and low purchasing pressure, the sales price is lowered to the *full-costs price* until 100% utilization is reached.
- If the capacities are under-utilized for longer time, the sales prices must temporarily be reduced, in the worst case down to the *marginal costs*.

The ongoing price experiments of *network operators* and *integrators*, such as postal services, railroad companies, airlines and telecommunication companies, show that the problem of pricing has not been resolved yet for network operators with high fixed costs. Up to now, no generally valid solution is known, that ensures in a self-regulating way fair pricing and high utilization of fixed resources. A possible solution of this problem could be the

- *Fixed Costs Pricing Strategy*: In times of low utilization, the price for goods and services produced with high fixed costs is decreased to the *full-cost price* for 100% utilization. For times of higher utilization, the price is increased as high as sufficient to compensate for the losses from underutilization in low demand periods and to achieve sufficient profit for the whole business year.

The fixed cost price strategy results from the above pricing rules and the *fixed costs dilemma* discussed in *Sect. 6.8*. It offers a fair solution of the utilization dilemma for sellers and buyers. If with the fixed cost pricing strategy over longer time a profitable business is impossible, there is no further need for the offered resources. In this situation, the investment has to be written off step by step until the sales price reaches the marginal costs. If for longer time not even the marginal costs are covered by the prices, the operation must be closed down.

7.7.6 Marketing Strategies of Sellers

The marketing strategies of sellers aim to maximize the total profit by achieving the optimal sales price for a maximal sales volume. *Long term strategies* are the *market positioning strategies* (Wöhe 2000; Porter 1980):

- *Value Leadership*: All efforts are focused on the value of the offer for the customer. Marketing and R&D analyze all *value drivers* of a product or service in order to gain as many *unique selling points* as possible and to reduce competition. This allows prices above full costs, even in times of low demand.
- *Cost Leadership*: All efforts of R&D and production are focused on creating a range of highly standardized products and services at lowest costs. Prices marginally below the prices of all competitors ensure full utilization and generate reasonable profit.
- *Flexibility Leadership*: A relatively broad variety of products and services is modularized in such a way that, with highly flexible production facilities, a quick adaptation to changing demand and cost adjustments are possible. This allows flexible and customized pricing.

Some companies try to combine these basic marketing strategies. However, quite often they fail, since it is difficult to save costs in periods of high revenues and to keep service and quality in periods of low revenues.

In logistics, *systems service providers* generally follow the value leadership strategy. Examples for cost leadership can be found in the markets for parcel services, container forwarding and bulk goods transport. Flexibility leadership is shown by courier service providers, trucking operators, tramp shipping companies or universal carriers.

Besides the medium-term strategies of *customer attraction* and *demand investigation*, suitable short term *price investigation strategies* of sellers are:

- *Competition price investigation*: Before delivering the own price quotation, the seller tries to find out the quoted prices of the competitors.
- *Expectation price investigation*: Before a price is offered, the seller tries to find out the utility value and the price expectations of the customer.

The calculation of the full-cost prices for different utilization scenarios is the most important step of the pricing. Based the full-cost prices for 100% utilization, the following *price differentiation strategies* are possible (Ekelund 1970; Simon et al. 2003):

- *Regional Price Differentiation*: If the buying power deviates, the market is split up in price regions, where different *local prices* are offered.
- *Temporal Price Segmentation*: If the demand is time dependent, different *time-based prices* are offered.
- *Customer Group Segmentation*: Depending on different demand or value drivers, customers are grouped into *customer categories*, which are served with different prices using separate *sales channels*.
- *Product and Service Differentiation*: Different sizes, measures, packaging, design and fitting as well as differentiation by service, quality, added values or special ambience, enable different *services* and *quality category prices*.
- *Quantity Differentiation*: Quantity discounts are offered for bigger orders according to the savings in order processing and by higher utilization.

Some segmentation and differentiation strategies are in conflict with the above *principles of fair pricing*, e.g. if price differences do not correlate with benefits for the customers.

For the opening of the *information barrier* between seller and buyer, two different *price information strategies* are possible: In an *open price strategy*, the sales price is shown to all market participants, also to the competitors. In a *covered price strategy*, the sales price is offered only to customers.

Too frequent, extreme or unjustified *price changes*, varying *rebates* and high *discounts* make customers skeptical towards the seller and the value of the offer. In consequence, many customers shift their purchases, choose other suppliers or even cease to buy. Therefore, *price continuity* as an important issue.

7.7.7 Marketing Strategies of Buyers

Marketing strategies of buyers aim to maximize the benefits at lowest price. If the buyer's demand compared to the capacities and stocks of the seller is small, the buyer has only restricted options of action. In this situation, *counter strategies* help coping with a dominating seller. Counter strategies for a weak buyer are *refusing* unnecessary features, add-ons and compound products or services. Other counter strategies are *supplier exploration* and *evasion* by shifting, reduction or abandonment of the demand. In logistics, typical evasion strategies are switching to another transport mean or delivery chain (see *Chap. 21*).

Buyers with high demand can pursue own strategies. Most important are the *demand consolidation strategies*:

- *Internal Demand Consolidation*: The total demand of all departments in a company for the same product or service is consolidated and purchased together from only one supplier.
- *External Demand Consolidation*: Several companies or households consolidate their demands and purchase the total demand from one supplier.
- *Temporal Demand Consolidation*: The single or collective demand for a longer time is consolidated and purchased from one supplier.

A typical example of internal demand consolidation in logistics is the *consolidation of shipments*. By this means, either full truck loads instead of less than truck loads are achievable or switching from parcel expedition with high rates to pallet shipment with lower rates is possible (see *Sects. 21.13* and *21.14*).

Examples for external demand consolidation are the *purchasing associations* of small shops and the *cooperation* of independent companies. They act as *purchase agent* and *logistic service provider* for their members. A *wholesale company* operating its own central stores plays a similar role for its outlets. Both, cooperation and wholesale are examples for *inter-company SCM*.

The consolidation of demands improves purchasing power, helps against market segmentation and avoids expensive multi-stage distribution. Demand consolidation also improves the seller's costs situation. In order to achieve these advantages, the following *buying strategies* can be pursued:

- *Multiple Sourcing*: The total demand of a particular product or service is divided amongst several suppliers.
- *Single Sourcing*: The total demand of a particular product or service is allocated to one supplier with best conditions at lowest prices.
- *Consolidated Sourcing*: Suitable parts, articles or performances are consolidated to tender packages in order to be purchased from a small number of *key suppliers* with best conditions at lowest prices.
- *Total procurement*: All parts, articles or performances needed for a longer period of time or for a big project, are tendered together and assigned to a *main contractor* or *service provider* with best conditions.

Criteria for selecting the right buying strategy are *total price*, *purchasing costs*, *integration costs*, *delivery time* and *independence*.

Multiple sourcing is unavoidable, if the demand exceeds the capacities of one supplier. It has the advantage of reducing the dependency on only one supplier. By single sourcing the lowest prices are achievable due to *economies of scale*. However, single sourcing increases the dependency on the supplier. Consolidated sourcing and total procurement require *buying competence* and *integration competence*.

7.7.8 Negotiation Strategies

A price negotiation starts with the announcement of a *start price* either by the seller or the buyer. If the seller has announced the start price, he first tries to convince the customer of the value of the offer.

After this opening phase, possible *negotiation strategies* for both sides are:

wait and see
ask for the price expectation of the other side
concessions towards quantity, quality and service
play along
incremental offering of discounts
immediate offer of the last price (7.11)

Business negotiations are like playing poker. It is the aim of *game theory* to develop, select and combine negotiation strategies, in order to achieve the best result (von Neumann and Morgenstern 1944). However, game theory is not sufficient. Negotiations require knowledge of human nature and behavior, intuition, psychology and tactics and the assessment of the expectations and situation of the other side. Last but not least, professional experience and judgment are essential for negotiations.

7.8 Economics and Logistics

Ultimate purpose of the economy is to supply the society with the required goods and services at lowest costs. Primary goal of a company is to achieve sustaining high profits by maximal revenues at lowest costs. This leads to the overall *goal of logistics*:

- The logistic services and performances, needed by private households and companies, have to be provided and executed at lowest costs and to be marketed with optimal profit.

The tasks and objectives of *technology* and *economics* in logistics are closely related. Tasks of technology are to enable, simplify and improve logistic performances. Its objectives are efficiency of performance and quality of services (see *Sect. 3.11*). Tasks of economics in logistics are evaluation of technical solutions, organization of administrative processes, and development of business strategies for marketing, costing and pricing.

In order to solve these tasks successfully, economists and technicians must keep in mind that logistics is *interdisciplinary*. The separation of *economical logistics* and *technical logistics* hampers many advantages and potentials.

7.8.1 Contributions of Economics to Macrologistics

Economists can contribute to macrologistics by investigating the following *topics* and *problems*:

- segmentation and quantification of the *logistic markets*
- *effects of logistics* on national and international economy

- mechanisms and conditions of logistic *pricing*
- fair and effective *price structures* and *pricing principles* in logistics
- effects of *price structures* on the utilization of resources
- economic consequences of *cross price subsidies* and *allowances*
- *network economy* in traffic and logistics
- consequences of the *fixed cost dilemma*
- limits of *economies of scale* in logistics
- *econometrics* of costs, prices and performances in different logistic markets
- investigation and comparison of *logistic costs* of trade and industry
- *performance and cost comparisons* of logistic service providers
- potentials, consequences and limits of *make or buy* and *outsourcing*
- potentials and limits of *inter-company supply chain management*
- tasks, power and consequences of *associations* and *institutions* in logistics
- purchasing strategies and supply chain management of *private households*
- options and necessities for political action in logistics (see *Chap. 23*)

Some of these macrologistic topics are well established disciplines, such as *transport economy* (Aberle 1996), *transport management* (Ihde 1978) and *waste management*. Other important areas offer promising research opportunities, such as *logistic market segmentation* (Klaus 2006), *econometrics in logistics*, *network economy* and *logistic pricing*.

Transport and freight performances are still measured in undifferentiated tonkilometers. The effects of shipment sizes and dispatch types on transport costs and freight tariffs are seldom discussed in economic textbooks. The economic aspects of *Intralogistics* are almost totally neglected by economists. It is remarkable, that the frame conditions for and pricing of intralogistic performances are almost unknown, although the market for these services is fast growing.

7.8.2 Contributions of Economics to Micrologistics

Some economists view logistics and supply chain management as new business approaches. They claim SCM and logistics to be synonymous to *consumer orientation* of all operations and to *thinking in processes and networks*. To them, supply chain management means inter-organizational *collaboration* and *holistic approach* (Cooper 1997; Pfohl 1990; Straube 2004; Schönsleben 1998; Weber 2002). Despite these noble claims, up to now only few contributions of economics to company logistics are of practical use or at least enlightening (Bretzke 2008).

If economists really want to contribute to practical use, they should solve real problems and answer the following questions of *company logistics*:

- standardization of *logistic accounting* and *logistic costs* (see *Sect. 6.3*)
- standard methods for *use-dependent depreciation* (see *Sect. 6.4.1*)
- potentials and limitations of *universality* and *diversification* in logistics
- delimitation and allocation of *fixed costs* (see *Sect. 6.7*)

- *standard methods for costing and pricing* of single, special, compound and integrated logistic tasks and services
- practicable solutions for *fixed cost compensation* (see *Sect. 7.5.9*)
- standard procedures for the calculation of *order-logistic-costs* and *article-logistic-costs* (see *Sect. 6.2*)
- *bill of quantities* and *price lists* for standard logistic performances
- *standard tender documents* for *contract logistics*
- *consequences for management decisions* of the *fixed cost dilemma* (see *Sect. 6.8*), of *multi-user synergies* and of *economies of scale in logistics* (see *Sects. 6.9* and *19.11*)
- investigation of direct and indirect *logistic revenues*
- *cost-based and value-related prices* and *incentive pricing* in logistics (see *Sect. 7.7*)
- further development of *logistic marketing*
- potentials, challenges and limits of horizontal and vertical *inter-organizational supply chain management* (Bretzke 2005, 2008)

In modern textbooks on company logistics, these questions and problems hardly occur (Christopher and Peck 2003; Cooper et al. 1997; van Hook and Harrison 2004; Langford 1999; Pfohl 1990; Kuhn and Hellingrath 2002; Schönsleben 1998). Practically useful information, feasible solutions, general rules and helpful recommendations are rare. With the exception of *Operations Research* (Churchman 1970, 1961; Domschke 1985, 1995, 1992; Domschke and Drexel 1990; Inderfurth 1994, 1999; Müller-Merbach 1970), there is still a lack of methods to solve problems of practical importance for company logistics.

Optimization of logistic chains, *efficient order scheduling* and *optimal inventory management* are not feasible without *use-related cost rates* and *cost-based prices*. Inter-organizational supply chain strategies cannot be assessed, if the added value compared to optimal *bilateral strategies* cannot be calculated and objectively measured (Otto and Kotzab 2002; Gudehus 2002/6). Only cost based and value related prices are incentives for the efficient use of resources.

7.8.3 Challenges for Economists in Logistics

Many management textbooks and publications of economists on logistics start with extensive definitions and interpretations of the term *logistics* and propagate new terms without practicable benefit. They present the *history of logistics* and describe established business practices. Endless surveys investigate the *fashions* and *trends* of logistics in academia and practice. However, with response rates far below 20%, the results of such surveys lack validity.

In former times, economists obviously contributed more to logistics. An example still valid today is the *theory of optimal location* of the German economists *J.H. von Thünen* (1783–1850) and *W. Launhardt* (1832–1918). They invented the *Thünen-circle* for the maximal distribution area and the *Launhardt-cone* of the increasing distribution costs around a production site (Schneider 1968 S.77ff).

Nowadays, preferred topics of scientific economists in logistics seem to be *logistic management* for top executives and *logistic controlling*. Some publications on logistics promise huge cost saving potentials, demand new research projects and offer unfounded speculation as *visions*. Instead of that, companies need feasible costing and pricing methods, applicable remuneration schemes, reliable information about logistic markets and approved marketing and pricing strategies.

This and the previous chapter contain solutions and offer suggestions for economical problems of practical relevance in logistics. The solutions have been developed for consulting projects and were successfully implemented in business practice. Economists are invited to check and improve them and to create better solutions.

Most important tasks for economists in logistics are to develop strategies, to solve relevant problems, and to give feasible advice of practical use. This opens challenging opportunities for economists in research and education (see also *Sect. 18.14*).

Present and future challenges are *demand saturation* and *bottlenecks* in the high developed societies. Bottlenecks are depleted raw material resources, limited energy supply, scarcity of land and limited capacity of traffic routes, which cannot be further expanded. Also, saturated markets at the end of the delivery chains are a special kind of *bottlenecks*. Due to these bottlenecks, the economy can no longer rely on expansion. Economists, engineers and logisticians together must find solutions, which ensure reliable and efficient supply of the demanded goods and services despite of the bottlenecks.

Chapter 8
Time Management

The *locations* of the stations and their *distances* define the structure, whereas the *dates* and *times* determine the flows within a logistic network. Processes are specified by locations and times. Hence, time is the fourth dimension of logistics.

During the last centuries the general awareness of time has changed (Landes 1983). In logistics, time became an important aspect by the *Just-In-Time-movement* (Hopp and Spearman 2000; Kotschek 1992; Schniederjans and Olsons 1999; Schönsleben 1998; Voss and Robinson 1987; Wildemann 1987). Today, *delivery times* and *punctuality* are recognized as important competition factors.

Nevertheless, in practice *delivery times* are still not binding, imprecise or not kept reliably (Stalk 1988). Contrary to the principle "punctuality before speediness" (see page 301) many companies exaggerate *Just-In-Time* or *24 h-delivery*, but neglect *punctuality*. Others see short delivery times as expensive means of marketing and ignore them as chance to reduce costs (Vorkuta and Lummus 2000; Zibell 1990). Up to now, the options of time, the influences on costs and delivery times, and the potentials of time scheduling have hardly been realized (Gudehus 2003/2006).

In this chapter, after specifying the characteristic *dates* and *time spans* of logistics, the composition, influence factors and consequences of *lead times*, *delivery times* and *cycle times* are analyzed. Based on this analysis *time strategies* will be developed and recommendations for *time management* and *order scheduling* are derived.

8.1 Time Points and Time Spans

Time points are *moments*, specified by a *date* referring to a *time zero* t_0, which can be the Birth of Christ or the start of a calendar year, business period or operating day. Customary dates to fix time points are:

$$\begin{array}{l} \text{calendar year (CY)} \\ \text{calendar month (CM)} \\ \text{calendar week (CW)} \\ \text{calendar day (CD)} \\ \text{time of day (TD)} \\ \text{scaling periods } (PE_t,\ t = 0,1,2\ldots) \end{array} \tag{8.1}$$

T. Gudehus, H. Kotzab, *Comprehensive Logistics*,
DOI 10.1007/978-3-540-68652-1_8,

Points of time for planning, project management and scheduling are:

- *Start dates* t_{start}: starting time points, such as the starting date of a project, the pickup time for a consignment, a departure date or an arrival date.
- *End dates* t_{end}: closing times or completion dates; finishing dates of projects; delivery dates; arrival times; order execution dates; expiry and durability dates.
- *Milestones* t_{mile}: intermediate start and/or end dates of part or projects, subprocesses; completion dates of the sections of a certain process
- *Decision dates* t_{dec}: *point of no return* for the release of a project or an investment or the decision date about an action, e.g. placing an order

Time points are either *temporal fix-points*, or *temporal action parameters* which can be varied and used for optimizations in planning and scheduling.

Time spans are *time intervals* between two points in time. A time span can be a *single time length* T_i, a *mean time* T_m, a *minimal time* T_{min} or a *maximal time* T_{max}. If not indicated otherwise, in this book time spans T denote *mean times*. *Time intervals* τ, *time spans* T, and *time requirements*, are measured in *time units* [TU]. Customary *time units* or *periods* [PE] are:

second [s]
minute [min]
hour [h]
day [d] (8.2)
week [w]
month [m]
year [a]

Together with the *length units* m, km, miles, the time units are the basis for the measurement of *speeds* v [m/s; km/h; miles/h] and *accelerations* b [m/s^2].

8.1.1 Logistic Time Spans

The most important time spans in logistics are:

- *Time units* [TU] as standard for the measurement of *flows, performances, throughput* and *rates* of events, output or arrivals
- *Planning period* which is the time span between a certain calendar date and the *planning horizon*
- *Scaling periods* of equal length to structure longer periods, e.g. a longer operating time, a business year or a planning period
- *Cycle times* of cyclical operations, e.g. process-cycles, grip-cycles, in- and out-storing-cycles, moves of a turn-table or crane-cycles
- *Clock times* τ_C of recurring events, such as passing, servicing, production, filling and controlling of discrete units
- *Event times* T_E of single events, e.g. processing time, travel time, driving time, transfer time, basic time, performance time, waiting time, holding time, setup time and storing time

- *Lead times* T_{LT} such as delivery time, transport time, through-put time, order execution time, material lead time and replenishment time
- *Product lifetimes* T_{Prod} and *life cycles* T_{LC} of finished goods
- *Technical lifetimes* T_{Tech}, i.e. technically possible running times of machines, transport means and equipment
- *Economic lifetimes* T_{Econ}, i.e. economically useful operation times of machines, transport means, plants, sites and buildings

The minimum of technical and economic lifetime is the *utility time* of an investment, equipment or facility. The utility time multiplied with the limit performance gives the *utility stock*, which determines as explained in *Sect. 6.4* the operating and performance costs.

Specific *logistic time spans* are transport times and storing times:

- The *transport time* is the total time between takeover of a freight or load and handover of the consignment at the destination.
- The *storing time* is the total time between deposition of an article unit or a load unit on a store place and its retrieval from that place.

Both, storing time and transport time may be limited by the *maximal shelf-life* of products and merchandize or by the order requirements. The transport time is determined by *transport scheduling* and *transport strategies* and can be calculated from the technical data of the transport means, the number and lengths of the stops, and from the route length. The storing time for a single unit of a storekeeping article is unpredictable, if it has been produced or replenished anonymously. The *mean storing time* of storekeeping articles depends on demand, sales or consumption and is determined by *replenishment strategies* and *inventory management* (see *Chap. 11*).

8.1.2 Definition of Time Units

Some time spans, e.g. event times and performance times, are fixed by the requirements or frame conditions. Other time spans, such as product lifetimes, running times and annual cycles, are determined by nature, markets or technique. However, many time spans, such as delivery times, storing times and scheduling cycles, are variable to a certain extent. Others, e.g. planning times and scaling periods are free parameters. The determination of variable and free *time parameters* is an important task of time management, since these parameters can be used to optimize lead times and minimize costs.

The appropriate definition of *time units* for measuring time spans and specifying dates depends on the required *accuracy*. Too short time units pretend unachievable exactness and cause unnecessary efforts. Too long time units elongate lead times and affect punctuality. If delivery times are measured in weeks, deviations of several days are generally accepted. If they are specified in days, deliveries will not be precise on the hour. From this follows the *time unit rule*:

▸ The appropriate length of the time units for measuring dates and time spans ranges from 1% to 5% of the considered lead and process times.

The appropriate time unit for *cycle times* and *clock times* is generally the *second*, as already seconds of an operation, which recurs with high frequency, have great influence on the required number of workers, transport means or equipment. Therefore, *cycle times* of cranes, forklift trucks, storage and retrieval units and other intralogistic means are normally specified in *seconds*.

Internal transport times are generally measured in *minutes*. External *delivery times* and *travel times* on the road, by train or by airplane are given in *hours*. *Sea transport times* are usually stated in *days*.

The appropriate time unit as standard for measuring internal performance rates, material flows, order arrivals and data flows is the *hour*. External flows of orders, transports, freights or traffic are generally related to a *day* (see *Chap. 9*).

8.1.3 Planning Periods

The *long-term planning* of a company covers a planning period of at least 5 years and can reach up to 10 or 20 years. For the planning of traffic networks, public authorities operate with planning periods of 20 to 50 years. The planning period for *medium-term planning*, such as business-, sales- and production- plans, ranges from 1 to 3 years.

The longest lead times of a business determine the length of *short-time planning*, i.e. for the *forecast* of the actual demand and the *scheduling* of orders and resources. The short time planning periods in the chemical industry, of many machine manufacturers and of large producers of consumer goods are months and quarters. Smaller enterprises, such as handicraft businesses or repair services, with delivery times from some days up to a few weeks, operate with short time planning periods of weeks or months. *Logistic service providers*, such as forwarders, parcel couriers, railways, airlines and postal services, have planning periods of a day or week, in order to cope with the fast fluctuating demand.

8.1.4 Scaling Periods

For the planning of systems and cost calculation, as well as for forecasting, scheduling and remuneration of performances, it is necessary to split a longer planning or business period into a sequence of *scaling periods* PE_t, $t = 1,2, N_{PE}$. The *scaling periods* PE_t are intervals of a defined *length*, which are measured in time units (8.2), and counted by a time index $t = 1, 2, 3 \ldots$ from a certain *starting date* (8.1) (see *Chaps. 9, 10* and *11*).

The proper length of the scaling periods results from the above *time unit principle* and depends on the specific task. Quite often, the period length is kept by tradition or adopted from other planning processes. However, by doing this, valuable options of time may be omitted.

Depending on the required accuracy, the scaling periods are segmented in *subperiods* by scaling- and sub-scaling periods, e.g.:

- *calendar* year or business with sub-periods

$$\begin{array}{ll} \textit{quarters} & QU_t,\ t = 1, 2, 3, 4 \\ \textit{calendar months} & CM_t,\ t = 1, 2, \ldots, 12 \\ \textit{calendar weeks} & CW_t,\ t = 1, 2, \ldots, 52 \\ \textit{calendar days} & CD_t,\ t = 1, 2, \ldots, 365 \end{array} \tag{8.3}$$

- *months* segmented into

$$\textit{days of month} \quad MD_t,\ t = 1, 2, \ldots, 28(29)/30/31 \tag{8.4}$$

- *weeks* divided into

$$\textit{week days} \quad WD_t,\ t = 1, 2, \ldots, 7 \tag{8.5}$$

- *days* split into

$$\textit{operating hours} \quad OD_t,\ t = 1, 2, \ldots, 24 \tag{8.6}$$

The length of the scaling periods determines the accuracy and the reaction time of forecasts (see *Sect. 9.8*). Also the accuracy of scheduled dates and the *punctuality* of deliveries depend on the scaling periods. For example, the timetables of railways are scaled in hours and segmented in minutes, in order to ensure punctuality to the minute.

In some companies the length of scheduling cycles for orders and inventories is equal to the length of the scaling periods of the medium-term planning. Hence, orders are collected and scheduled only once per day, week or even month. Correspondingly, stocks are checked and replenished only daily, weekly or monthly. However, as outlined in the *Sects. 10.6* and *11.11*, lead times and inventories increase with the length of the scheduling cycle. A consequence is the *rule of cyclical scheduling*:

► Lead times and inventories can be reduced by shorter scheduling cycles.

For example, some years ago, a big chemical company changed from monthly to weekly scheduling and achieved by this means inventory reductions of several hundred million Euros.

The possibility to reduce lead times and to improve performance rates by shorter cycles or clock times is used in *informatics* more than in logistics. Whereas in informatics the computer clock-cycles have been continuously increased over the last forty years with the result of higher capacities and shorter response times, the freedom to determine the lengths of periods and scheduling cycles is still not consequently used in logistics.

8.2 Operating Time and Working Time

The next step of time management after fixing the planning and scaling periods is the determination of the operating times and working hours. The operating time scheme is given by a sequence of *operating times* $T_{op}(t)$ with fixed or variable length and start dates $t = 1, 2, \ldots N_{Top}$. Customary operating time schemes are:

- *business calendar* with *operating weeks* and *operating days*

$$N_{fd} = 250 \text{ to } 300 \quad \text{days/year}$$
$$N_{ow} = 50 \text{ to } 52 \quad \text{weeks/year} \qquad (8.7)$$
$$N_{wfd} = 4 \text{ to } 7 \quad \text{days/week}$$

- *working shifts* per week or per day

$$N_{wshift} = 5 \text{ to } 28 \quad \text{shifts/week}$$
$$N_{dshift} = 1 \text{ to } 4 \quad \text{shifts/day} \qquad (8.8)$$

- *operating hours* per week, per day or per shift

$$N_{h\,ow} = 28 \text{ to } 168 \quad \text{hours/week}$$
$$N_{h\,od} = 8 \text{ to } 24 \quad \text{hours/day} \qquad (8.9)$$
$$N_{h\,os} = 5 \text{ to } 10 \quad \text{hours/shift}$$

The business calendar with the *operating time scheme* must be harmonized with the *working time scheme* for the personnel. The working time scheme consists of general *working hour regulations* and a factory-specific *deployment plan*. The working hour regulations specify the annual, weekly and daily working obligations of a *full time worker* (FTW) or of a *part time worker* (PTW) and their holiday entitlements. The deployment plan determines the weekly, daily or shift presence of the single worker.

The relation of the *operating hours* T_{op} to the *total time* T_{tot} gives the *time efficiency* of the operating time scheme:

$$\eta_{Teff} = T_{op}/T_{tot} \quad [\%]. \qquad (8.10)$$

For example, the time efficiency of the common operating time scheme of a week with 5 working days, 2 shifts per day and 7 hours per shift is $\eta_{Teff} = 5\times2\times7/(7\times24) = 41.7\%$. With this time scheme, a factory executes orders and generates output during less than 42% of the time. In 58% of the total time, the resources remain unused.

8.2.1 Operating Time Strategies

For the operation of a single station or a whole business, the following *operating time strategies* are possible:

- *Demand-related operation*: Starting point and duration of the operation depend on the current demand. Transports are executed whenever passengers or freights orders come up. Orders are accepted during 7 days around the clock. Shipments and deliveries are possible at all times.
- *Plan-based operation*: Operations are executed at regular times following a *timetable*. Fixed *time schedules* regulate the transports. A logistic center or factory has planned *operating times*. Orders are only accepted during certain acceptance times. Shipments and deliveries are executed at certain dates. Shops, outlets and stores have fixed *shop hours*.

A plan-based operation is feasible and sufficient for a constant or cyclic demand, which can be reliably planned or forecasted. In these cases, workload, time demand

and operating hours are scheduled in advance in accordance with the expected demand (see *Sect. 9.9*). For example, in public transport the time tables for busses and trains are adjusted to the hourly and daily demand.

8.2.2 Regulation of Working Times

The determination of the working time is an entrepreneurial decision with severe consequences, since:

▸ The working time regulation of a company determines the ability to perform, flexibility, delivery times and utilization of resources.

Principally, starting dates and operating times are within certain limits free parameters, which allow fulfilling a temporal and quantitative demand as efficient and flexible as possible. However, business practice must respect many *regulations*, which restrict the options, such as

- contractual working hours and holiday entitlements
- public and legal holidays
- working regulations for holidays, night shifts and overtime
- industry specific limitations for machine running times
- sales and opening time regulations for retailers
- restricted driving times for commercial transports
- worker participation rights for working hours and factory holidays

In Europe and other parts of the world, legal and contractual working time regulations have limited the options of time management considerably. This has caused the loss of global competitiveness of certain businesses in some countries, e.g. of the textile industry, which almost closed down in USA and Europe.

However, global competitive pressure has induced *deregulation* and loosening of too strict operating and working time regulations. Nowadays, operating and working times are more driven by market and customer requirements. Only companies who are able to adapt the operating times to the necessities of the markets will remain competitive.

8.3 Adaptation, Synchronization and Flexibility

Adaptation and synchronization of operating times and scheduling periods are *time strategies* that reduce lead times and delivery times, improve efficiency and save costs. In order to realize these time strategies, flexible working times are necessary.

8.3.1 Adaptation of Operating Time

Depending on demand, the length of the operating time is extended or shortened by adapting the number of working days or shifts or by varying the number of working hours per day or the number of working hours per shift. Possible *achievements by flexible operating times* are:

- The daily performance of stations with technically fixed *limit performance* can be adapted to a varying demand.
- Waiting times caused by lacking orders are minimized; the working time of personnel is more efficiently used.
- Driving times of transport means are adaptable to a varying transport demand.
- Production to stock becomes avoidable as far as it has been practiced only to keep the personnel busy.

Flexible operating times are especially important for *service providers*, as services cannot be produced on stock and service demand can vary extremely. Also manufacturers, e.g. the *automotive industry*, have created the *breathing factory* with demand-dependent operating times (see *Sect. 10.5.1*).

Disadvantages of demand-related operating times, especially when the demand fluctuates extremely, are *vacant costs* for resources, which are required only in peak times. Therefore, in businesses with extreme seasonal variations of demand, *preproduction on stock* is still necessary and longer delivery and service times are unavoidable in high season.

8.3.2 Synchronization of Start Times

The start times of the daily or weekly operation of parallel or consecutive stations can be synchronized and harmonized. This allows for

- *reduction of lead times* as orders can be executed without delay by the consecutive stations of an order chain
- *shortening of transport* and *delivery times*, since longer idle periods in the transshipment stations are avoided
- *minimization of waiting times* of workers after start of the operation by shifted start times

For example, the operation of a *commissioning system* should not start before the order center has prepared a sufficient number of picking orders. If *24-hour-service* is required, a logistic center must enable shipment of the executed orders also after the operating time in the commissioning ended. Synchronized collection tours, main runs and dispatch tours are prerequisites for short delivery times of *freight forwarders*.

8.3.3 Flexible Working Times

Labor contracts, which regulate only the total annual and the mean weekly working hours, permit temporarily reduced or elongated working times and enable demand-related shift lengths and operating times.

For the organization of flexible working times hold the following *experience rules*:

- ▸ Operating times of administrative departments without direct customer contact should follow the working times in the operative stations.
- ▸ Up to several days, increasing demand and time pressure in stations with limited personel improve the performance rate.

This holds especially for administrative departments. However, in the long run work overload affects performance and quality. Hence, for administrative stations, staffing for the average demand is generally sufficient, as long as the peak demand does not last longer than a few days.

8.4 Order Lead Time of Single Stations

The *order lead time* T_{OLT} of a single station is the time span from the entry of an order until its completion (Hopp and Spearman 2001). It is generally the sum of *waiting time* T_W, *setup time* T_S, *performance time* T_P and *ripening time* T_R:

$$T_{OLT} = T_W + T_S + T_P + T_R \tag{8.11}$$

The different contributions to a lead time, cycle time or clock time can either be *measured* in practice or *determined analytically* by special methods, such as *Work Factor* and *MTM* (*Method of Time Measurement*). Mean storage cycle times, picking times and travel times can also be *calculated* with the help of *formulas*, which will be presented in the *chapters 16*, *17* and *18*.

8.4.1 Setup Times

The *setup time* T_S is the time required for preparing the performance station to execute an order and to clear it for the next order. Typical setup times are

order acceptance time
preparation and clearing times
material provision times
conversion and switch over times
in- and out-storing times (8.12)
basis times of commissioning
loading and unloading times
data processing times

For the calculation of the total cycle time of a performance station, preparation and clearing times must be taken into account only, as far as these activities happen before or after order execution. *Shadow strategies* shift a part of the setup time into the *time shadow* of other activities. This *parallelization* reduces lead times and costs (see *Sect. 8.11*).

8.4.2 Performance Times

The *performance time* or *order execution time* T_P is the time that a ready station needs to execute the order-specific tasks or services. Performance times of *operative stations* are:

$$\begin{array}{l}\text{production cycle times}\\ \text{processing times}\\ \text{assembly cycle times}\\ \text{service times}\\ \text{picking and handling times}\\ \text{walking and travel times}\\ \text{bottling, filling and packing times}\end{array} \tag{8.13}$$

The *minimal order execution time* of a station with *limit performance* μ [PU/PE] for orders with quantity m [PU/Ord] is:

$$T_{P\,min} = m/\mu \quad [PE]. \tag{8.14}$$

Performance times of *administrative* and *creative departments* are:

$$\begin{array}{l}\text{data processing times}\\ \text{scheduling times}\\ \text{planning times}\\ \text{construction times}\\ \text{development times}\end{array} \tag{8.15}$$

Generally, these times cannot be calculated precisely in advance and must be estimated.

8.4.3 Ripening Times

Order related processes which take place after the performance process has been executed are *ripening times* T_{ripe}, such as

$$\begin{array}{l}\text{drying times}\\ \text{curing times}\\ \text{hardening times}\\ \text{deposit times}\\ \text{ripening times}\\ \text{fermentation times}\end{array} \tag{8.16}$$

Similar to the setup times, these times only must be taken into account, as far as they do not happen parallel to other processes. This can as well be achieved by *shadow strategies* and *parallelization*.

8.4.4 Minimal Order Lead Time

The sum of minimal setup time, minimal performance time and minimal ripening time without waiting times, is the *minimal order lead time*:

$$T_{OLT\,min} = T_{S\,min} + T_{P\,min} + T_{R\,min}. \tag{8.17}$$

By insertion of the relation (8.14) into (8.17) follows with the *basic lead time* $T_{LTo} = T_{S\,min} + T_{R\,min}$ the *dependency of lead time on order quantity*:

$$T_{OLT}(m) = T_{LTo} + m/\mu \quad [PE]. \tag{8.18}$$

This dependency has to be taken into account for the calculation of *economic order quantities* (EOQ) and *safety stocks*, if the stock is replenished by a supply station with limited performance (see *Chap. 11* and *Sect. 20.3*).

8.4.5 Waiting Times

The *waiting time* T_W is the difference between total order lead time and minimal order lead time. Contributions to and causes of waiting times are

- *dead times* due to lacking data, information, decisions or instructions
- *downtimes* caused by disruptions, failures or lacking personnel
- *correction times* for errors and quality defects
- *congestion times* caused by the occupation of the station with precedent orders (see *Sect. 13.5.3*)
- *blocking times* resulting from a tailback of a subsequent performance station (see *Sect. 13.5.4*)
- *procurement delay times* caused by too late arrival of material that is sourced or produced for this stage
- *interruption times* causes by a breakdown or an out-of-stock-situation in a precedent performance station (see *Sect. 13.6*)
- *buffer times* that result from regular scheduling and order collection in order to optimize the utilization of the resources

Principally, dead times, downtimes and correction times are manageable. They can exceed the productive performance times, if the equipment is maintained insufficiently or if the operation is managed badly. Dead times, down times and correction times of well run operations should be much shorter than the process times.

Irregular loss times as well as *regular buffer times* can add up to *total loss times*, which are much longer than the minimal lead time (8.17), and elongate order lead times severely. Therefore, keeping the irregular loss times under control and optimizing the regular buffer times are central tasks of time management.

8.5 Lead Times of Performance Chains

The *order lead time* of a performance chain is the time between order entry into the chain and leaving of the last order result at the end of the chain. Orders can be sales, production, shipment, delivery, replenishment, transport, dispatch, haulage or other kinds of performance orders. Accordingly, the order lead time is called *delivery time*, *production time*, *transport time*, *service time*, *travel time* or *replenishing time*.

For example, the *replenishing time* is the total time between placement of the replenishment order and deposition of the ordered quantity on the store place. For *inventory scheduling* it is necessary to differentiate between the *initial replenishment time*, that includes setup of the order process and preparation of the first production, and the *regular replenishment time* for an established supply process. Normally, the regular replenishment time is significantly shorter than the initial replenishment time.

8.5.1 Processing of Compound Orders

Compound orders are multi-position orders, production orders and assembly orders which require several articles, several products and/or different steps of operation. As shown in *Fig. 8.1*, compound orders and induced suborders pass a network of consecutive and parallel administrative and operative stations. The orders generate information, material and performance flows in the parallel performance chains of the network.

The different performance chains converge into performance stations, which complete the order step by step. At the end, the final product, the ordered performance or the order result is completed and/or delivered. The last station can be the

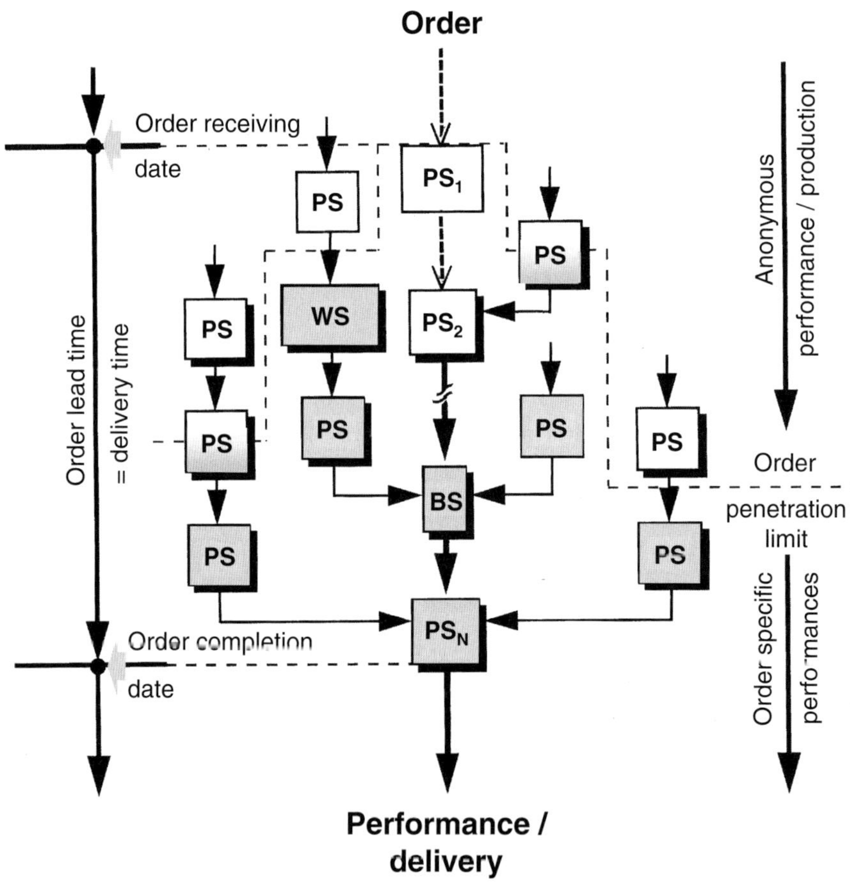

Fig. 8.1 Order execution tree with parallel performance chains

PS_i: performance stations BS: bottleneck station WS: waste station
→: primary performance chain = critical path
→: secondary performance chains or supply chains
- - →: order penetration limit

final assembly line of a production site, the goods dispatch dock of a company or the destination point of a shipment.

8.5.2 Decoupling Stations and Order Penetration

Generally, in the execution of orders for more than one article or for an integrated service or complex product several performance chains are involved. Each performance chain has a separate *entry station* whose process is triggered by an internal suborder.

The *order specific chains* start either with a *production station* with *free capacity* or with a *storage station* with a *free stock* from which the required order quantity can be taken. As long as the free capacity, respectively the free stock is sufficient to cover the demand for more than just one order, the production or storage station is a *decoupling station*. A decoupling station separates the preceding stations from the time critical order process.

As indicated in the *order execution tree* of *Fig. 8.1*, the decoupling stations are the *order penetration points* which define the *order penetration limit* (Chopra and Meindl 2007). The *upstream stations* before the order penetration limit *produce on stock*; the *downstream* stations *produce to order*.

8.5.3 Primary and Secondary Performance Chains

The order chain with the longest lead time is the *time critical performance chain* or *primary performance chain*. The primary performance chain determines the total order lead time. The other, uncritical performance chains are *secondary performance chains* or *supply chains*.

By reduction of the lead times or by insertion of a decoupling station, a primary chain can become a secondary chain. This can also happen, if the lead times of a secondary chain are extended or if a process in a secondary chain does not start on time.

For complex performance networks, it is necessary to specify all performance stations and their possible connections with the help of appropriate software, such as *MS-Project*. The program can determine the time critical performance chain or *critical path* by a *network planning algorithm*, such as *PERT* or *CPM*, from the relations and the process times for the single activities. The program can also find out the *bottleneck stations*, if all *limit performances* are known. Such programs are applied e.g. for big building projects, scheduling of complex IT-systems and erection of large construction sites.

8.5.4 Order Lead Times in Networks

The *total order lead time* T_{OLT} for a compound order executed by a performance network is the sum of the *partial order lead times* T_{OLTi} of all performance stations PS_i, i = 1,2, ...N of the main performance chain:

$$T_{OLT} = \sum_i T_{OLT\,i}. \tag{8.19}$$

The *minimal order lead time* $T_{OLT\,min}$ for the chain is the sum of the minimal order lead times of all performance stations of the primary performance chain:

$$T_{OLT\,min} = \sum_{i} (T_{Si\,min} + T_{Pi\,min} + T_{Ri\,min}). \tag{8.20}$$

This is the sum of the minimal setup times, performance times and ripening times of the time critical process chain.

8.5.5 Decoupling and Postponement

Designing a performance network, dimensioning of the stations and positioning of the decoupling stations are important steps of *process development* and *system planning*. For this purposes the following *network design rules* are useful:

- *Inserting decoupling stations*: Order specific performance chains can be shortened, order lead times be reduced and punctuality be improved by decoupling stations.
- *Spatial postponement*: The closer the decoupling stations are located to the end of the order execution tree or chain, the shorter become the lead times.
- *Temporal postponement*: The later material, parts and final products are allocated to customer orders, the shorter are the lead times and the smaller is the *risks of misallocation*.

However, when shortening the primary performance chain, one always has to examine whether a secondary performance chain will not become time critical.

In the automotive industry lead times are minimized by *individualization* of the cars at the latest possible stage. The module and part suppliers are located close to the factory. Their production is based on a *rolling sales plan* and triggered by the current assembly process. The parts and modules are delivered *just-in-sequence* (JIS) and *just-in-time* (JIT) to the assembly line. By these means the *ex-works lead time* of a car can be reduced to a few days.

Preconditions are sales driven production planning and a demand that does not exceed the assembly capacity for longer time. The probability that the rolling planning for parts and modules meets the real demand decreases with increasing range of variants. Hence, essential for the achievability of short lead times by decoupling and postponement are consequent *variant management* and *predictable demand* (see Sect. 9.9).

8.6 Material Lead Time

The *material lead time* is the time between the date when the material enters a performance chain and the date when it leaves it. It is the sum of *flow times* and *waiting times*. Long material lead times tie up capital, cost interest, cover space, block store places and involve risks.

The material flow for *order-specific material* is initiated by an *external order*. A basic provision to keep order lead times is the *material lead time condition*:

► The material lead time for order specific material must be shorter than the required order lead time minus the necessary downstream order execution time from the consumption station to the exit.

If this condition is not fulfilled, shorter order lead times are only possible by storekeeping of all materials with too long procurement times.

The total inventory costs increase linearly to the material lead time. As value is added step by step, fixed capital, interest costs and inventory risks increase during the order passes the performance chain. This can be minimized by the *stock location strategy*:

- ▸ If storekeeping is opportune, the material should be stored at an early stage of low value close to the start of the performance chain.

However, this strategy conflicts with the above postponement strategy. The conflict between low costs and short lead times is solvable only for the single case.

The delivery times for finished goods, parts or modules can be shortened by a *pre-procurement* or *make-to-stock strategy*. With these strategies, *bottlenecks* are avoidable and larger order quantities with lower proportionate procurement and setup costs are achievable. On the other hand, they cause higher storekeeping costs. This goal conflict leads to the *task of manufacturer inventory management* (see *Chap. 11*):

- ▸ The stock of goods must be kept on a level that ensures the required *delivery ability* and *lead times* at minimal costs.

This task corresponds to the *goal of retail inventory management*:

- ▸ The article stocks in the outlets and in the upstream stores should ensure a competitive availability at minimal costs.

The sourcing of non-order-specific material is based on sales forecasts or on production plans and initiated by an *internal order*. This material is purchased or produced in advance and put on stock until it is required by an *external order*. The difference between material lead time and delivery time is the *material waiting time* for orders, which result in *buffering times* and/or *storing times*.

8.7 Time Scheduling of Single Stations

The goal of time scheduling is to keep promised delivery dates with a required punctuality at lowest costs. The achievable *punctuality* η_{DD} [%], i.e. the probability that a delivery date DD is kept, is determined by the *stochastic variations* of the demand, the *lead time fluctuations* within the critical order chain and by the *time strategies* of scheduling.

The lead time of a single *performance station* is extended against the planned lead time by *stochastic waiting times* which are caused by fluctuations of the order flows and/or the performance times. The stochastic waiting times grow with the utilization, which depends on the number of incoming orders (see *Sect. 13.5*). The overlaying distributions of the waiting times and the performance times cause a right tilted lead time distribution as shown in *Fig. 8.2* for a single station. From a known lead time distribution the *X% punctuality-lead-time* (XLT) which is kept with *probability* X can be read of. To keep a required delivery date DD with probability X the order execution must start before the latest *start date*:

$$SD_{max} = DD - XLT \geq OED \tag{8.21}$$

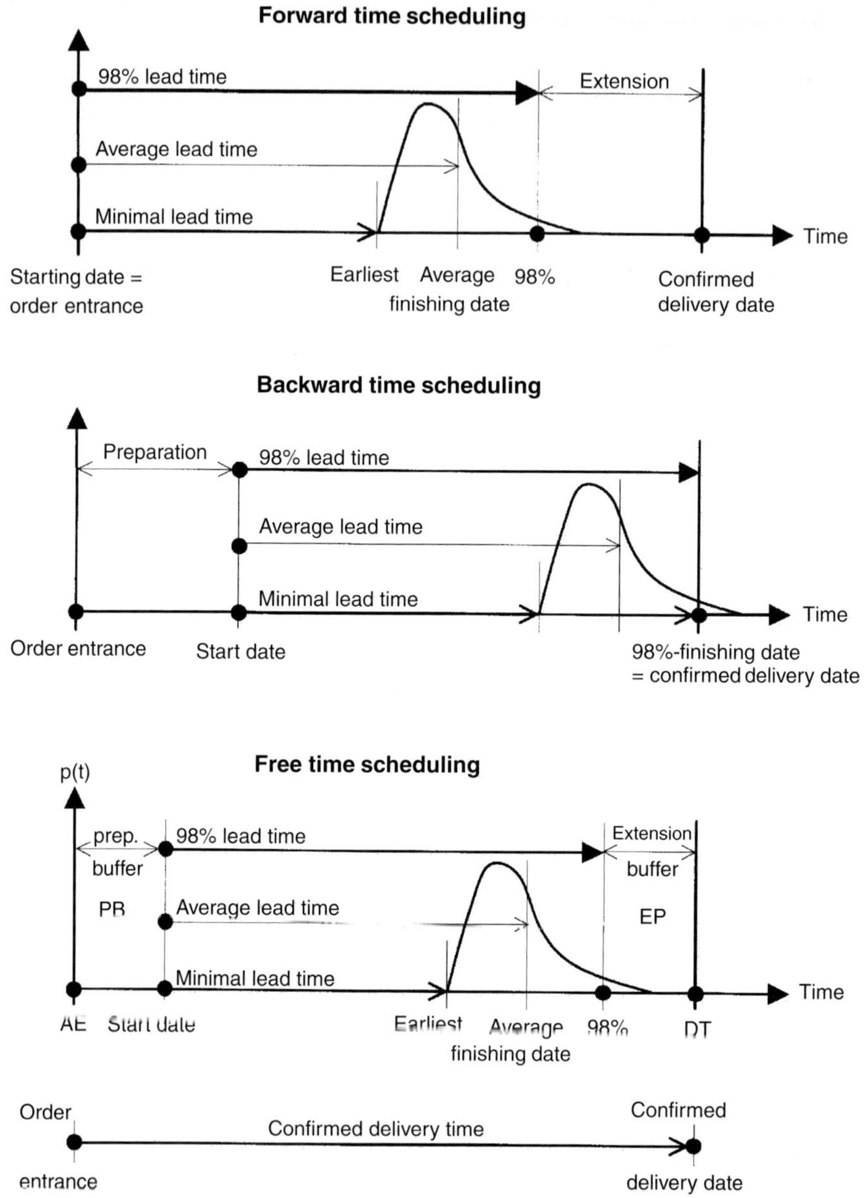

Fig. 8.2 Time strategies and punctuality for a performance station
p(t): probability density of the lead time distribution

The time difference DD-OED between a required *delivery date* DD and the *order entry date* OED is a *time buffer* which ensures punctuality. The earlier an order is started, the longer is the time buffer and the higher is the probability to keep the delivery time. However, one must keep in mind:

- For stochastically varying demand and/or fluctuating limit performances *100% punctuality* is neither achievable nor affordable.

As shown in *Fig. 8.2*, time difference DD-OED much longer than the X%-lead-time offers several options for time management, such as *forward scheduling*, *backward scheduling* and *free scheduling*.

8.7.1 Forward Scheduling

With the forward scheduling strategy, the order execution starts as soon as the performance station is available. The incoming orders are executed in the chronological order of arrival due to the principle *first come first served (FCFS)*.

After finishing the order, the result is kept for the *post buffer time* PoBT = DD-OED-XLT until the delivery is due. This requires buffer- or storeplaces.

8.7.2 Backward Scheduling

With the backward scheduling strategy, the order execution starts at the latest possible start date (8.21). Several orders are executed in the chronological order of the latest start dates. Before starting an order, it is kept in an *order-buffer* for the *pre-buffer time* PrBT = DD-OED-XLT. The *order result*, which has been finished *just in time*, leaves the station immediately.

Advantages of backward scheduling are no stocks, buffers and storeplaces on the output side. However, these advantages face the risk of missing the deadline.

8.7.3 Free Scheduling

With a free scheduling strategy, the order execution starts at an *intermediate date* ID between the order entry date OED and the latest possible start date (8.21).

The orders are kept in an *order buffer* OB for a *pre-buffer time* which can be selected within the range 0 < PrBT < DD-OED-XLT. The result of the order execution is kept until delivery in a buffer or storeplace for the *post buffer time* PoBT = DD-PrBT-OED-XLT.

The *order buffer* OB(t) is time dependent and varies stochastically, if the *order entry rate* OE(t) and/or the *production capacity* PC(t) are time-dependent (see *Sect. 10.5*). In the order buffer several single orders can be collected in order to execute them most efficient as a *series order*. Start dates, sequence and content of the series orders are *strategy parameters* of free scheduling, which can be used to minimize costs (see *Chap. 10 and 20*).

8.8 Time Scheduling of Performance Chains

If an order passes a multi-stage chain of stations, either the single performance stations operate with *local scheduling strategies* or an *order center* schedules the *intermediate start dates* SD_i for the *performance stations* PS_i due to *central scheduling strategies*. Here, again forward, backward or free scheduling can be applied.

8.8.1 Delivery Dates and Punctuality

If the difference OED-DD between order entry date OED and delivery date DD is longer than the minimal order lead time (8.20), it is possible to keep the delivery time with required reliability by adjustment of the intermediate start dates SD_i. This also allows for achieving cost savings and further logistic goals.

For example, *Fig. 8.3* shows a consecutive *backward scheduling* in a performance chain with and without pre-buffer times. The adherence to the partial closing dates is assumed to be $X_i = 98\%$. If XLT_i are the partial *X%-punctuality lead times* of the performance stations PS_i, the intermediate start dates SD_i must fulfill the conditions:

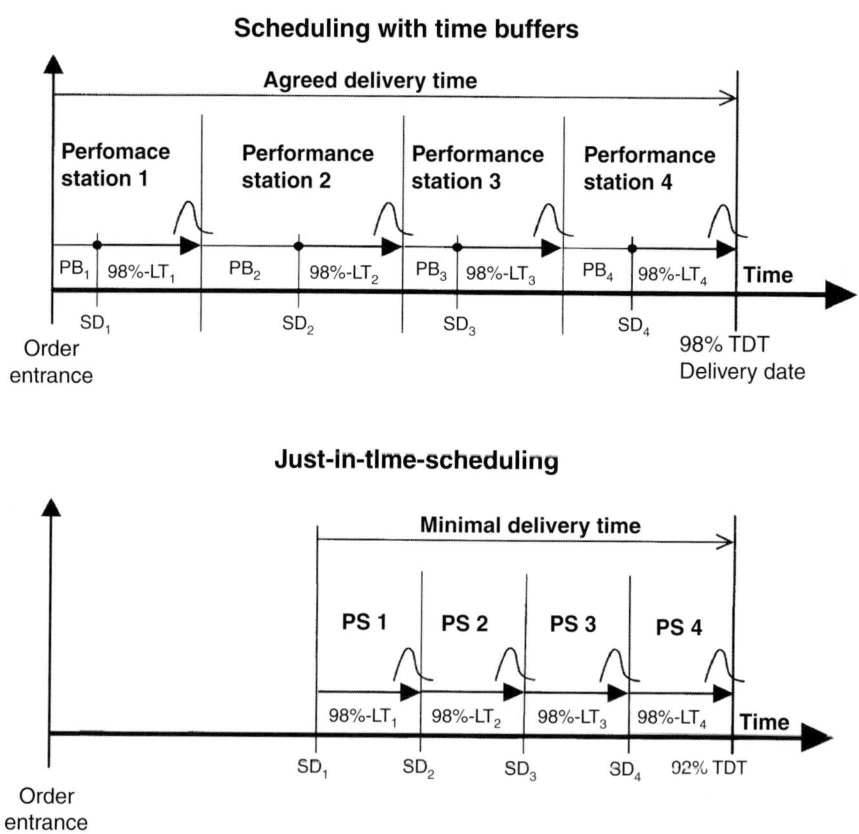

Fig. 8.3 Backward scheduling in a performance chain with and without time buffers

PS_i performance stations SD_i partial start dates
TLT total lead time LT_i partial lead times
PrBT pre-buffer times

$$SD_i + XLT_i \leq SD_{i+1} \quad \text{for } i = 1 \ldots N. \tag{8.22}$$

SD_1 is the start date of the first performance station PS_1 and SD_N the start date of the last station PS_N.

From the *product rule of probability theory* follows, that the probability X [%] to keep the delivery date of a performance chain is the product of the probabilities X_i [%] to the keep the intermediate lead times, i.e. the *punctuality condition*:

$$\prod_i X_i = X_1 \cdot X_2 \cdots X_N \leq X \tag{8.23}$$

The punctuality condition (8.23) implies the *punctuality rule*:

▸ A required punctuality of X for the order delivery necessitates a mean punctuality $X_i \geq X^{1/n}$ of the n performance stations PS_i of the order chain.

For instance, if the punctuality for an order chain of 3 stations, should be 97%, the punctuality of each station must be better than $0.97^{1/3} = 0.99 = 99\%$.

For an order entrance date OED, a required delivery date DD can be kept with probability X only if the partial X%-lead times fulfill the *lead time condition*:

$$\sum_i XLT_i \leq DD - OED = OLT. \tag{8.24}$$

If it is impossible to fulfill both conditions (8.23) and (8.24), the required order delivery date OLD cannot be kept with punctuality X. In this case, the customer must accept either a later delivery date or lower punctuality. As long as both conditions (8.23) and (8.24) are fulfilled with unequal sign < there is freedom for one of the following *scheduling strategies*.

8.8.2 Central Scheduling by Push Principle

As shown in the top of *Fig. 8.4*, an *order center* collects the external orders and divides them into internal *suborders*. Based on appropriate *allocation strategies*, which are outlined in *chapter 10*, the suborders are allocated and send to the first, second and all following performance stations of the order chain (see *Sects. 2.2., 2.3* and *14.3*). The order center prescribes also start dates and pre-buffer times for the single performance stations.

The single stations wait for orders from the order center and send the order results immediately to the next station. The suborders and intermediate results push the single stations. That means, the whole order chain is driven by the *push principle*. The intermediate results are stored, if at all, in the receiving stations.

8.8.3 Central Scheduling by Pull Principle

The external orders are also collected and divided into suborders by an order center. However, with this strategy, the suborders are simultaneously allocated to all performance stations of the order chain, with the provision that the order result is sent to the subsequent performance station only on request.

Starting at the final performance station, the orders together with the intermediate results are subsequently *pulled* through the whole order chain. Hence, it is driven by

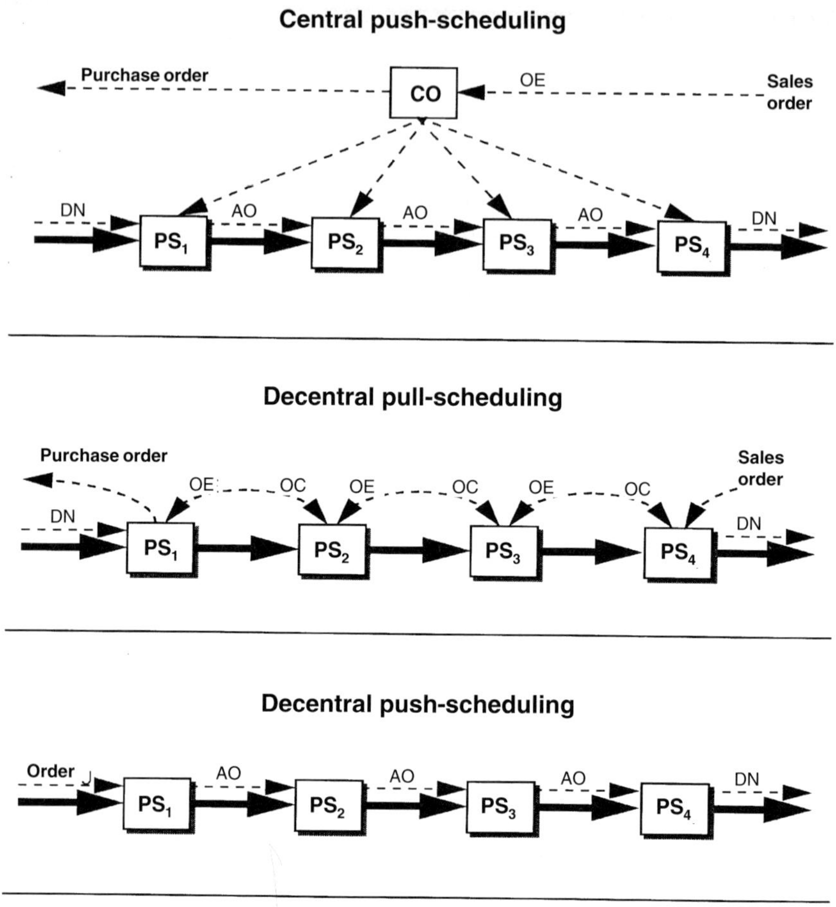

Fig. 8.4 Decentral (local) and central scheduling by pull and push principle

→: material flow	- - →: information flow
PSi performance stations	CO: central order scheduling
OE: order entry	OC: order confirmation
AO: accompanying order	DL: delivery note

the *pull principle*. The intermediate order results, such as semi-finished products, are stored in the delivery stations.

The order center controls the adherence to all dates, sources the required material, provides the stations with resources and manages the inventories.

8.8.4 Central Bottle Neck Scheduling

Again, the external orders are collected and processed by the order center. In a first step, the order center allocates the suborders of appropriate order series to the *bottleneck stations* of the performance network in order to achieve the optimal utilization of the most critical *limit performances*. Here the *processing strategies* of *Chap. 10* can be applied.

In a second step, the orders for the remaining performance stations of the primary and secondary order chains are scheduled. The stations downstream of the bottleneck station generally receive their orders by the push principle, the upstream stations by the pull principle.

Central bottleneck scheduling is a common strategy of many *production planning systems* (PPS) and *Enterprise Resource Planning* software (ERP) (see *Chap. 20).*

8.8.5 Local Scheduling by Pull Principle

Local scheduling does not need an order center. If driven by the pull principle, the last performance station or exit station, that finalizes the complete order, receives directly the incoming external orders. This station immediately schedules the start date and the pre-buffer time for that part of the order that is executed by itself.

From the external orders, *part orders* with certain delivery dates are derived and sent to the preceding performance stations. They schedule the incoming orders in the same way and hand the resulting part orders to the upstream stations until the *order penetration limit* is reached (see *Fig. 8.4* middle).

After checking its delivery ability, each station of the network reports its scheduled delivery date to the downstream station. Having received all *order confirmations* and scheduled dates from the upstream stations, the exit station schedules the binding delivery date and informs the external customer.

Any performance station can individually decide on material sourcing, holding of inventory and the use of its own resources. The performance stations also react to breakdowns and interruptions in their own responsibility.

In the *order preparation phase*, the single performance stations *pull* the order confirmations from the preceding stations. In the following *order execution phase*, the partial order results are *pulled* by the following station. By doing this, adjacent performance stations establish *customer-supplier relationships* and monitor each other. The performance chain is driven by the *pull principle.*

Kanban, the Japanese word for *sign-card*, is a simple realization of the *local pull principle* (Im and Schonberger 1988; Ohno 1988). A station reports its current demand by providing an emptied bin with a card at a *pickup point*. Alternatively, the Kanban card which indicates type and replenishment quantity of the required article is placed on separate board. The supplier collects empty box and card, after serving the last order by placing a refilled box, thus receiving the next order (see *Sect. 12.7*).

The local pull principle can also be found in *assembly lines*, where many variants of a final product are mounted in shortest time at lowest costs from a relatively small variety of parts and modules. This is typical for *automotive*, *household appliance*, *consumer electronics* and *computer industry*.

8.8.6 Local Scheduling by Push Principle

As shown in the bottom of *Fig. 8.4*, local scheduling by the push principle is similar to the local pull principle. In this case, however, the external order with accompanying parts enters the first station of the main performance chain. The *order parts* may be parcels or consignments in combination with a transport order or preliminary products, material or parts for a production or assembly line.

The *order entry station* schedules its own pre-buffer and start date, gives an *advanced notice* to the next performance station and passes the order result together with an *attached order* to it. The customer or recipient receives the result, e.g. a parcel, shipment or finished product, together with a *delivery note* from the last performance station of the order chain. By this means, the whole chain is driven by the *push principle.*

The *local push principle* is typical for haulage and dispatch orders and applied by *postal services, parcel service providers* and *freight forwarders.* The strategy is also suitable for a *job-shop production.*

8.8.7 Central Scheduling Versus Local Scheduling

The aim of *central scheduling* is to achieve an overall optimum of lead times and costs by applying all possible time options and scheduling strategies. Central scheduling becomes active, if interruptions, breakdowns or delays happen or when rescheduling is necessary. It schedules not only the customer orders, but also the stocks and transports in the whole performance network.

An order center causes costs and needs additional processing time. Further disadvantages of central planning and scheduling are the reduced commitment and responsibility of people in the local stations. That may cause lower efficiency, less motivated employees and delays, especially when unexpected breakdowns happen.

The advantage of *local scheduling* is the motivation of personel by responsibility and empowerment. Under normal conditions, with highly motivated local people the whole order process runs *self-regulated* without additional costs and time losses for central planning and scheduling. A pure local scheduling, however, has the following disadvantages:

- Single performance stations may not be prepared for unexpected high demand and variations of order contents. This results in longer order confirmation and order lead times.
- New products and changing requirements need additional time for preparation and harmonization
- Performance stations located close to the order entry tend to *self optimization* and leave less time for the other stations.

These drawbacks and problems of pure local scheduling are manageable under the following conditions:

- The *reaction times* of the single stations to inquiries are *not too long.*
- The single *order lead times* are relatively *short* and *stable.*
- The *demand* is sufficiently *predictable* so that each performance station is able to forecast its own demand for materials and resources.
- The orders refer to *standard performances* and *standard products* with well-defined procedures and only modestly fluctuating lead times.
- Bilaterally harmonized *scheduling rules* for buffer times and start dates help to avoid time losses.

If one of these conditions for local scheduling is not accomplishable, a *combination of local and central scheduling* may achieve most of their advantages and avoid the disadvantages. In business practice, various combinations of central scheduling and local scheduling in connection with pull and push principle can be found.

Generally, it is opportune that the *order center* is responsible for overall *system strategies* and *cooperation rules* and for medium- and long-range *resource planning*. *Logistic controlling* can ensure optimal utilization of installed limit performances and adherence to delivery dates and costs.

8.9 Just-in-Time

Essentially, *Just-In-Time* (JIT) is backward scheduling of a performance chain without time buffers between the stations. As shown in *Fig. 8.3*, the results of the executed part orders are passed to the subsequent station *just in time* (Schniederjans 1999; Schönsleben 1998; Voss and Robinson 1987; Wildemann 1987; Zibell 1990).

Advantages of JIT, which can be applied to the main performance chain as well as to a whole supply network, are:

- minimal total order lead times
- no buffer places and no inventories

These advantages, however, face the following *disadvantages* of JIT:

- A minimization of costs by using time buffers for optimal utilization of capacities and limit performances is not possible.
- The probability of keeping the delivery date decreases with the product $\eta = \eta_1$ x η_2 x η_3 x .... η_N of the partial punctuality η_i of the N stations.

The total punctuality for the example shown in *Fig. 8.3*, with four stations in a main performance chain with a single punctuality of 98%, is only $98\%^4 = 0{,}98^4 = 0.922 = 92.2\%$.

Therefore, real Just-In-Time can only be successful under the conditions:

- All order specific stations operate with *high punctuality*.
- The potential cost savings by order consolidation are low.

These JIT-conditions are fulfilled by assembly lines for single products with constant throughput, small lead time fluctuations, short setup times and high availability of the participants. Accordingly, JIT can be found in the final assembly of cars, home appliances or computers. In these and other industries, by JIT combined with *Kanban* the lead times have been reduces dramatically.

In other industries, that have irregular throughput rates and fluctuating lead times, JIT has not been as successful. Either stocks and buffers could not be eliminated completely or the operation turned out to be too expensive. Even companies with JIT often keep buffers and accept deliveries not on the minute, but on the hour or on the same day.

Today, JIT has lost its dominant importance for logistics. All attempts for a light *JIT-philosophy* by slackening supply dates and allowing transit buffers result in the same task as *logistics,* namely to provide the *right good at the right time at the right place.* Nevertheless, the JIT-movement made logisticians aware of time and stimulated many improvements (Stalk 1988; Wolff 1994; Zibell 1990).

8.10 Strategies for Lead Time Reduction

Besides Just-In-Time, many other time strategies can achieve reductions of lead times and delivery times. Corresponding to their cost implications, they can be differentiated into *cost reducing*, *cost neutral* and *cost increasing strategies.*

Ranked according to their cost implications, the most important strategies to reduce lead times and delivery times are:

- *Elimination*: The elimination of unnecessary stages of an order chain, the reduction of idle periods and holding times, as well as the elimination of non value adding activities reduce lead times and costs. Avoidable holding times and processes can especially be found in administrative performance stations, such as waiting for information, carrying on with file cards although data bases are existent, multiple registration of the same data and successive receiving and dispatch inspections (see *Sects. 2.6* and *2.7*)
- *Clearing of disturbances*: Disturbances within the performance chains can often be cleared and eliminated in short time. This reduces fluctuations of lead times and improves punctuality.
- *Simplifying*: Simpler procedures and processes reduce lead times and costs.
- *Standardizing*: Definition and implementation of standard processes for the core activities enable cost savings and shorter lead time. As well, standardization of parts, modules, products and load carriers generates cost savings and lead time reductions.
- *Sequencing*: Free scheduling, in combination with optimal sequencing strategies, enables time reductions, improved utilization and cost savings.
- *Timing*: The agreement of binding dates between all participants of the critical performance chain is beneficial for the punctuality.
- *Synchronizing*: The start dates of parallel stations are harmonized and for successive stations are mutually shifted (see *Sect. 8.3*). Synchronization is a cost efficient strategy that improves the lead time for all orders. Well known examples for synchronization in logistics are train timetables, progressive signal system in road traffic and shelf replenishment in retail outlets before the sales time.
- *Flexibility*: By flexible resources, personnel and operating times and by pools of reserve personnel longer waiting times in peak phases can be avoided. In spite of the higher costs, flexibility is often the only way to reach competitive lead times.
- *Parallelizing*: Big orders are separated into part orders, which are executed simultaneously in parallel performance stations. Other kinds of parallelization are setup and ripening processes in the time shadow of core processes. Generally parallelizing is cost neutral and can be combined with *specialization.*
- *Decoupling*: Delivery times can be drastically reduced by inserting decoupling stations, in order to shorten the time critical order chain.

- *Postponement*: By shifting the decoupling stations closer to the end of the order network, the parts and products can be allocated later to a customer order. By this mean, delivery times can be reduced and misallocation be avoided.
- *Flip-Flop*: The set up of two stations with the same function, where one station alternately operates while the other station is prepared, helps to avoid interruptions and to reduce lead times.
- *Prioritizing*: The simplest prioritization strategy is to execute *rush orders* first. As long as the share of rush orders does not exceed 5%, the lead times for rush orders can be extremely reduced. However, rush order prioritization elongates the lead times of normal orders and causes additional costs. This holds especially if rush orders are absolutely prioritized and the execution of normal orders is interrupted for urgent orders.
- *Splitting*: Execution of many customer specific single orders instead of a small number of large series orders, breaking up larger orders into smaller suborders, direct delivery of single shipments instead of bulk sending, and small lot sizes are effective means to reduce lead times. These *splitting strategies*, however, are incompatible with consolidation and may increase costs.
- *Acceleration*: Lead times can be reduced by increasing speeds, clock rates and performance rates of the performance stations. Acceleration, e.g. by raising the driving speed of transport means, increases costs but can also decrease lead times. This may lead to cost savings as resources will be better utilized due do the higher circulation of transport means.
- *Bottleneck elimination*: The elimination of bottlenecks by implementing additional resources is a costly – but in many cases inevitable – strategy to achieve sustainable reductions of lead times for all orders, especially during phases of peak demand. The reduction of waiting times upstream and downstream of the bottlenecks may lead to overall cost savings (Goldratt and Cox 2002).
- *Dynamic Scheduling*: By dynamic scheduling, as outlined in *Sect. 10.6*, either triggered by orders and events or in short scheduling cycles of a day or an hour, lead times can be shortened, delivery dates kept and costs minimized (Gudehus 2003/2007).

In order to achieve the effectiveness of these time strategies, central *time controlling* is recommended (Wolff 1994). However, in many cases local order scheduling, in combination with a self-regulating remuneration, is more effective than central time controlling. As outlined in *Sect. 7.3*, by this means the local stations are stimulated to keep short delivery dates and take care for cost savings and performance improvements.

8.11 Economic Order Lead Time

Lead times and performance costs are interdependent. Extremely short lead times and tight delivery dates imply high hold out costs and prevent cost-saving consolidation strategies. Shorter delivery times than offered by the competitors can generate additional orders and allow higher prices, which may outbalance the additional costs.

Too long lead times and open end dates are also expensive. Long delivery times involve high inventories, require more buffers and storeplaces and cause extra costs. If delivery times are not competitive, sales and revenues dwindle.

The interaction of all these effects results in a dependency between *order execution costs* and delivery times as shown in *Fig. 8.5*. From this idealized curve one can derive that there exists, beside of the technically determined *minimal lead time* (8.20), an *economic order lead time* (ELT), at which the total costs are minimal. However, in practice the cost dependency on the lead time is very difficult to assess, as it depends on many factors, some of which are not quantifiable. The dependency can also be a *step-function* if *strategy changes* happen or *economy of scale effects* occur.

The lead times of industry are generally criticized as being too long. Empirical studies indicate that the sum of all waiting times can be five to ten times longer than the minimal possible lead time. Longer delivery times are not necessarily a result of overloaded capacity and bottlenecks. In many cases they are caused by inefficient or poorly harmonized processes and lacking time management. Longer lead times may even turn up during periods of low occupation, as underemployed personel tends to hold tight to work and extend performance times.

A short-sighted conclusion is, that the most powerful leverage for shorter lead times is reducing waiting times (Wolff 1994; Zibell 1990). However, the biggest share of the waiting times is made up of regular buffer times which are used for cost optimal scheduling. The more the room for maneuver for scheduling is limited

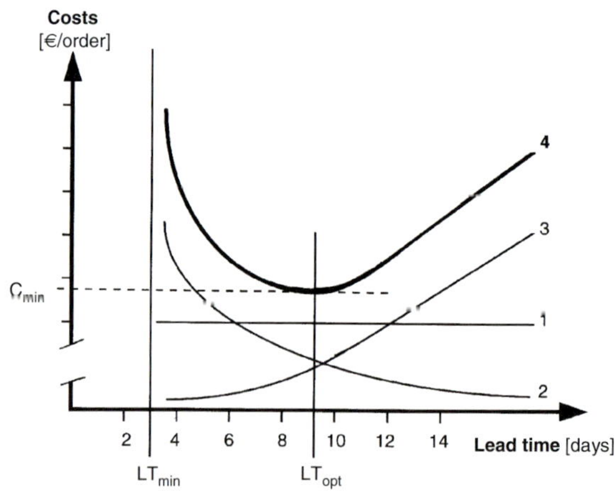

Fig. 8.5 Dependency of order execution costs on the lead time

1. curve: costs independent of the lead time
2. curve: costs increasing with the lead time
3. curve: costs decreasing with the lead time
4. curve: total order execution costs
OLT_{min}: minimal order lead time OLT_{opt}: optimal order lead time

by short lead times and tight delivery dates, the less cost optimization is possible. The demand for shorter delivery times cannot be fulfilled without consequences, as there is a goal conflict between short lead times and minimal costs. By suitable time strategies reduction of costs *and* shorter lead times are achievable only up to a certain extent.

Chapter 9
Random Processes and Dynamic Forecasting

The actors of a free market economy decide autonomously on buying and selling. Companies, consumers and customers of products and services independently place orders at times and with quantities which correspond to their individual demand.

Within the agreed delivery times, producers, suppliers and service providers consolidate, separate and schedule the incoming orders and execute them at optimal times. The output rates of many production and service stations vary randomly. Unpredictable disturbances and breakdowns cause *stochastic fluctuations* of lead times. The uncorrelated behavior of customers and companies and the deviations between ordered and produced quantities cause *random processes* within the whole economy (Churchman et al. 1957; Ferschl 1964).

Examples for random processes and stochastic flows in logistics are the arrival of persons or trucks (Arnold and Rall 1998), the passing by of vehicles on a road (Leutzbach 1956) and the arrival of orders and information. Other examples are delivery processes, production stations, handling activities and service stations with randomly distributed cycle times and lead times. The *dates, clock times* and *quantities* of these events are generally *random variables*.

If the mean rate and the average quantity of the events vary over time, the flows are *instationary* and the processes *dynamic*. The *systematic changes* are hidden by the stochastic fluctuations of the single events and become observable only after several periods. Stochastic and dynamic processes and flows affect *logistics* in many ways. They determine

- possibility and quality of mathematical forecasts
- scheduling of orders, replenishment and inventory
- limit performances of stations and nodes
- queuing effects in logistic networks
- buffer places, congestion lines and store capacities
- adherence to delivery dates
- design of processes and systems

This chapter deals with the characteristics and consequences of *random processes, stochastic flows* and *dynamic forecasting* in logistics.

T. Gudehus, H. Kotzab, *Comprehensive Logistics*,
DOI 10.1007/978-3-540-68652-1_9, © Springer-Verlag Berlin Heidelberg 2009

9.1 Random Processes and Stochastic Flows

A time sequence of events with uncorrelated, randomly fluctuating *clock times* τ and/or stochastically varying *quantities* m is called a *random process*. The resulting *rate* $\lambda(t)$ [PU/TU] of *process units* [PU] per *time unit* [TU] is a *stochastic flow* (Ferschl 1964).

9.1.1 Order Flow, Data Flow and Material Flow

The design, dimensioning, behavior and control of single stations as well of whole logistic systems are determined by order flows, data flows and material flows:

- *Order flows* or *order entry rates* are stochastic flows where the events are the arrival of *orders* for sales units, supply units or *performance units* (see *Fig. 9.1*)
- *Data flows* or *information flows* are flows of incoming or outgoing data sets, records, documents, codes or other *information units*.
- *Material flows* are flows of physical goods, e.g. of transport units, vehicles, persons, article units, packages, freight units or *load units* [LU].

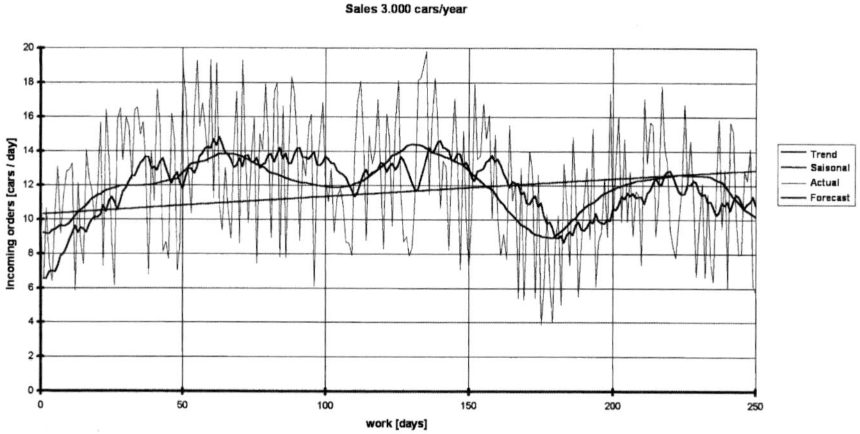

Fig. 9.1 Daily order entry of a car dealer

Orders single orders for passenger cars of the same type
Sales 12 ± 3.5 cars per day

9.1.2 Types of Discrete Flows

Different from the *continuous flows* of liquids or gases, order, data and material flows are *discrete flows* of single *process units* [PU]. The laws governing discrete flows deviate in several respects from the laws of continuous flows.

For a process P with a mean *clock time* or *cycle time* τ_P [TU/E] between successive events E, the *event rate* or *clock rate* is

$$\lambda_E = 1/\tau_P \qquad [E/TU]. \tag{9.1}$$

In some cases, e.g. for a traffic flow, the clock times are *continuous variables* varying within a certain range. In other cases, e.g. for an elevator serving several levels, they are *discrete variables* limited to a finite number of values.

As illustrated in *Fig. 9.2*, the variations of clock times and process quantities determine the type of stochastic flows and random processes.

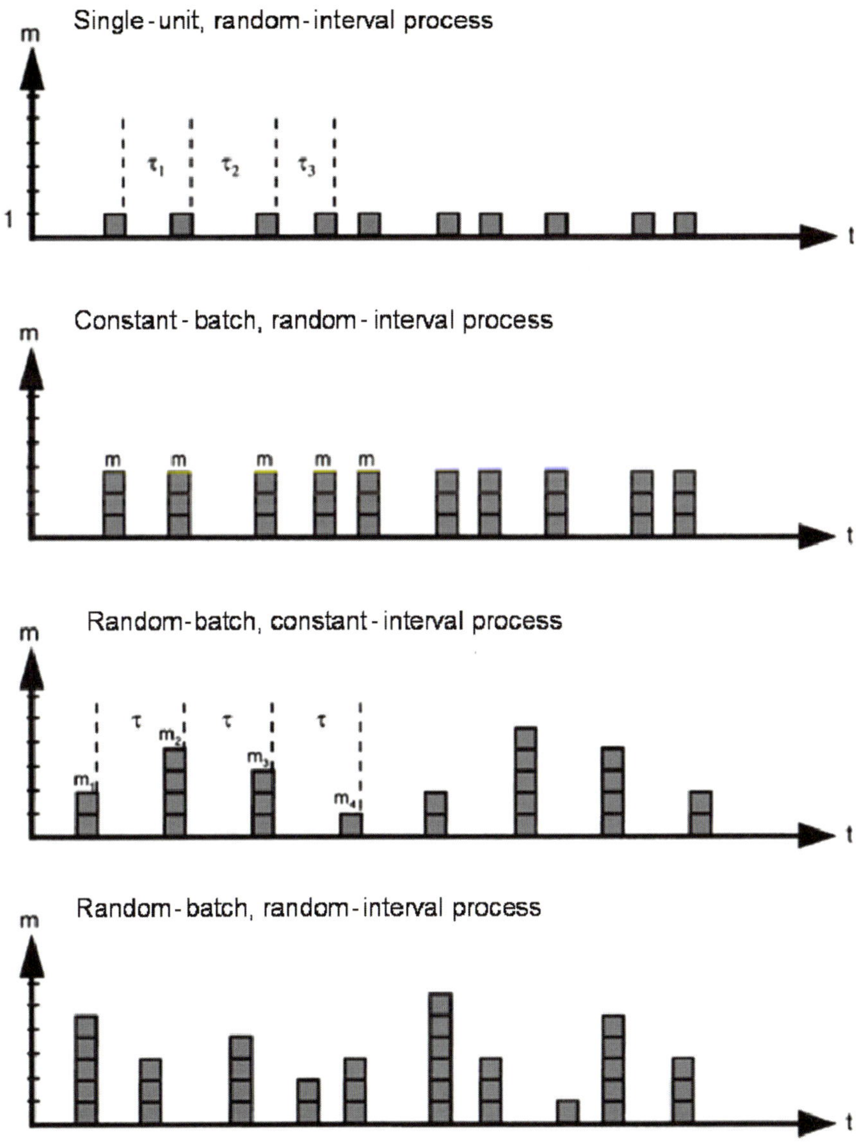

Fig. 9.2 Random processes and stochastic flows

τ_P: clock times
m_P: process quantities

For a process with the clock rate (9.1) and a mean *process quantity* m_P [PU/E] per event the *flow rate, flow intensity* or *throughput* is:

$$\lambda_P = m_P \cdot \lambda_E = m_P/\tau_P \qquad [\text{PU/TU}]. \tag{9.2}$$

The process quantities are either *discrete numbers*, e.g. bits, bites, process units or load units, or *continuous values*, such as the weight or volume of bulk goods. Depending on the variation, the following processes can be distinguished:

1. *Single-unit random processes* and *stochastic flows*: Single units arrive uncorrelated with randomly varying clock times. The process is stochastic in time and constant in quantity with $m = 1$ for all events. The clock rate equals the throughput rate.
2. *Constant-batch random processes* and *stochastic flows*: Batches of equal quantity arrive uncorrelated in randomly varying clock times. The process is stochastic in time with constant quantity $m > 1$. The throughput rate is higher than the clock rate by the factor m.
3. *Random-batch cyclic processes* and *random-batch cyclic flows*: Batches with uncorrelated random quantities arrive in constant cycle times. The process is constant in time and random in quantity.
4. *General-random processes* and *general stochastic flows*: Batches with random quantities arrive uncorrelated in random clock times.

An example for a single unit stochastic flow is the order entry for single cars as shown in *Fig. 9.1*. If stochastically varying numbers of cars are ordered by the different customers, it is a general stochastic flow. Other examples of stochastic flows are the load unit flow resulting from unloading randomly arriving transport means with varying load and the batch-wise passing cars on a road.

9.1.3 Stationary and Dynamic Flows

As long as the mean clock time and the mean process quantity are time independent, the stochastic flow is *stationary*. Seasonal behavior, varying demand, a product life cycle or other influences can change the mean throughput rate over time and cause an *instationary flow* or *dynamic flow*:

$$\lambda_P = \lambda_P(t) \tag{9.3}$$

The time dependency of the flow rate can be caused by changing clock rates $\tau_P(t)$, altering process quantities $m(t)$ or by simultaneous change of both. As shown by *Fig. 9.1*, the systematic long term variations of a stochastic flow are superposed by random short term fluctuations.

Stochastic fluctuations and systematic variations have different effects in logistics:

- Short term *stochastic fluctuations* of order, data and material flows cause *queuing effects* in front of performance stations, determine the level of *safety stocks* and influence the *breathing reserve* of the storage capacity.
- Medium and long term *systematic variations* of the order, data and material flows determine the *limit performances* needed for the peak demand and the *buffer stocks* necessary for balancing changes of demand.

For random processes, the single dates and quantities of the events are principally unpredictable. However, if the probability distributions of clock times and quantities are known, the average effects of a stochastic process are calculable. Also the future mean behavior of a stochastic flow can be derived under certain conditions from the patterns of the past. The mathematical forecasting, however, is made difficult by the superposition of the systematic time dependency and stochastic fluctuations.

9.2 Probability Densities and Time Distributions

The probabilities of events with continuous values are given by a *probability density* (Kreyssig 1975). Times in logistics, such as clock times, throughput times and lead times, only have positive values. Hence, the *time distribution*, i.e. the probability density of stochastically distributed times, differs from zero only for positive arguments. Probability densities and time distributions are defined as follows (Ferschl 1964; Kreyssig 1975):

- The product $w_P(\tau) \cdot d\tau$ of the probability density or time distribution $w_P(\tau)$ with the differential $d\tau$ is the probability that τ has a value between τ and $\tau + d\tau$.

In practice it is difficult or impossible to determine a time distribution or probability density. At best, the mean value and the standard deviation can be measured or estimated. To evade this difficulty, suitable *standard distributions* are used (Kreyssig 1975; Ferschl 1964). The mathematical properties of standard distributions allow the explicit solution of *queuing problems* and other tasks (Dorfwirth 1961; Ferschl 1964; Gudehus 1976/I; Krampe et al. 1973; Schaßberger 1973). Also, the *limit performance* of traffic nodes with priority can be calculated with the help of a suitable standard distribution (Gudehus 1976/II, Harders 1968).

9.2.1 Mean Value and Standard Deviation

For a time distribution $w_P(\tau)$, the probability that the observed times τ are below a certain value T is given by the *distribution function*:

$$W_P(T) = \int_0^T w_P(\tau) \cdot d\tau \tag{9.4}$$

Since all times have positive values, the probability for $\tau \geq 0$ is 1 and the time distribution fulfills the *normalization condition*:

$$W_P(T \to \infty) = \int_0^\infty w_P(\tau) \cdot d\tau = 1 \tag{9.5}$$

The *clock times* τ of a stochastic process P with time distribution $w_P(\tau)$ vary randomly around the *mean value* or *expectation value*:

$$\tau_P = \int_0^\infty \tau \cdot w_P(\tau) \cdot d\tau . \tag{9.6}$$

The *standard deviation* s_τ of the single times τ around the mean value (9.6) is the square root of the *variance*:

$$s_\tau^2 = \int_0^\infty (\tau - \tau_P)^2 \cdot w_P(\tau) \cdot d\tau. \tag{9.7}$$

A dimensionless measure for the random fluctuations of a stochastic process is the *variability*

$$V_P = (s_\tau / \tau_P)^2. \tag{9.8}$$

Another measure is the *variation* $\nu_P = s_\tau / \tau_P$ which is the square root of the variability.

9.2.2 Logistic Time Distributions

Examples for time distributions of random logistic processes are shown in *Figs. 9.3* and *8.2*:

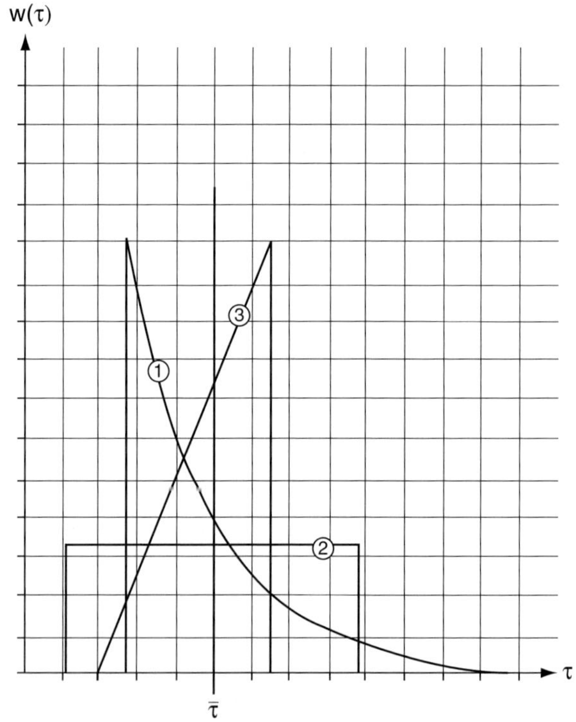

Fig. 9.3 Examples of specific logistic time distributions

Curve 1: modified exponential distribution of passing vehicles in one lane
Curve 2: rectangular distribution of forklift truck service times in a storage area
Curve 3: triangular distribution of stacker crane cycle times in a high bay rack

- The distribution of service times of a storage area by a forklift truck is a rectangular distribution, if all locations are served with same probability.
- Cycle times of a stacker crane serving a high bay rack from one corner have a triangular distribution.
- Time intervals between passing cars in a single lane have a modified exponential distribution.
- Grip times of order picking can be approximated by a normal distribution.
- Passing times through a station have a slightly right skewed distribution which is shifted on the time axis by the minimal passing time (see *Fig. 8.2*).

9.2.3 Exponential Distribution

The simplest standard distribution is the *exponential function* (see *Fig. 9.4*):

$$w_E(\tau, \tau_P) = (1/\tau_P) \cdot \exp[-\tau/\tau_P]. \tag{9.9}$$

The standard deviation s_P of the exponential distribution is equal to the mean value τ_P and the variability is 1. A random process with an exponential time

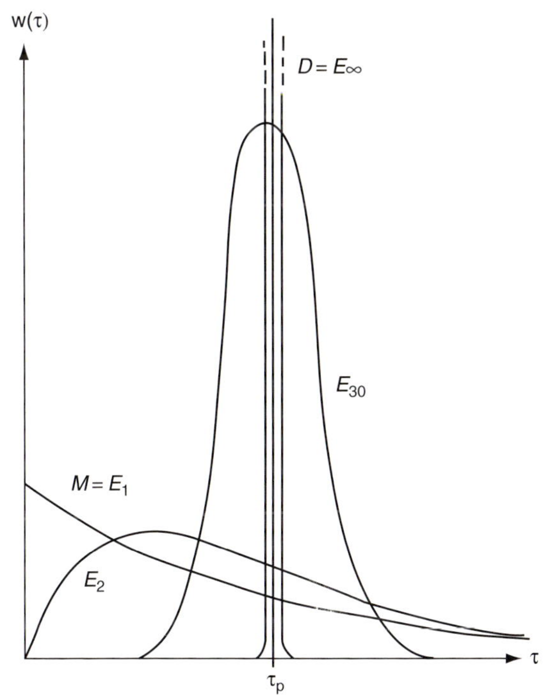

Fig. 9.4 Standard time distributions

$k = 1$	Exponential distribution
$k > 1$	Erlang-distribution
$k \to \infty$	Dirac-distribution

distribution is called *Poisson-process* or *Markov-process* and indicated by an *M* (Ferschl 1964; Krampe et al. 1973; Schaßberger 1973). An exponential distribution can be used for the approximation of a random process, if the times can have any positive value with continuously decreasing probability. This leads to the *recommendation*:

- ▸ If only the mean value is known, it is safest to calculate with an exponential time distribution.

Many observations have shown, that the time distributions of order entries, incoming data, arriving persons or telephone calls are approximately exponentially distributed (Arnold and Rall 1998; Churchman et al. 1957). Also, the waiting queues in front of stochastically operating performance stations are exponentially distributed (see *Fig. 13.21*).

Due to the finite length or caused by the limit performance of preceding stations, moving load units, transport units and vehicles in a single lane have a *minimal time distance* $\tau_o > 0$. The time distributions of such material flows can be approximated by the *modified exponential distribution* with the probability density (Gudehus 1976/I/II):

$$w_E(\tau, \tau_P) = \begin{cases} 0 & \text{for } \tau < \tau_o \\ \exp[-\tau/(\tau_P - \tau_o)]/(\tau_P - \tau_o) & \text{for } \tau \geq \tau_o. \end{cases} \tag{9.10}$$

The zero point of a modified exponential distribution is shifted to the right side of the time axis by the *minimal clock time* τ_o. The standard deviation $s_\tau = (\tau_P - \tau_o)$ is determined by the mean clock time τ_P and the minimal clock time τ_o.

A modified exponential distribution can be applied if the clock times of a stochastic process can take any positive value $\tau \geq \tau_o$ above a minimal value τ_o with continuously decreasing probability. A random process with modified exponential distribution is called *modified Poisson-process*.

For a small minimal clock time, i.e. for $\tau_o \to 0$, the modified exponential distribution evolves into the exponential distribution (9.9). If the minimal clock time approaches the mean clock time, i.e. for $\tau_o \to \tau_P$, it becomes a *Dirac-distribution* of a process with equal clock times. For example, the output of a production with constant clock times and 100% utilization is a *Dirac-flow*.

9.2.4 Erlang-Distributions

The probability distribution of the *k*-Erlang-distribution is (Ferschl 1964):

$$w_k(\tau, \tau_P) = (k/\tau_P)^k \cdot \tau^{k-1} \cdot \exp[k \cdot \tau/\tau_P]/(k-1)!. \tag{9.11}$$

The *Erlang-parameter* k is a positive integer equal to the reciprocal variability:

$$k = 1/V_P = (\tau_P/s_\tau)^2. \tag{9.12}$$

The *mean value* τ_P and the *standard deviation* s_τ determine the Erlang-parameter k completely.

The Erlang-distribution evolves into the *exponential distribution* (9.9), if the variability approaches 1, i.e. for $V_P \to 1$ or $k \to 1$. For very small variability, i.e. for $V_P \to 0$

and $k >> 1$, the Erlang distribution becomes a *standard normal distribution* (9.19). For $k \rightarrow \infty$, it evolves into a Dirac-distribution (see *Fig. 9.4*). That means:

- Erlang-distributions involve the exponential distribution, the standard normal distribution and the Dirac-distribution.

Erlang distributions can be used, if the mean value and the standard deviation of a process are known. This holds in many cases for the cycle times of service or operation stations, such as call centers, repair stations, toll stations, cash desks, information booths or order acceptance offices.

The *distribution of lead times* of an order chain with stochastically fluctuating performance times of the single stations is approximately a *modified Erlang-distribution*. Corresponding to the modified exponential distribution (19.10) it is shifted to the right side of the time axis by the *minimal lead time* (8.20) of the main performance chain (see *Fig. 8.2*).

9.3 Frequency Distributions of Discrete Values

A *frequency distribution* represents the probabilities of stochastic events with a countable number of discrete values. It is defined analogous to the probability density of continuous variables:

- The *frequency distribution* $w_P(x_i)$ of a discrete random variable x is the probability that x has the value x_I.

In the simplest case, the discrete values of the random variable are integers, i.e. $x_i = n_i$ with $i = 1, 2, 3 \ldots$.

Similar to the time distributions, it is generally difficult and in many cases impossible to measure the frequency distribution of a discrete random quantity in practice. A determination of a frequency distribution by *stochastic simulation* is only possible if the time and frequency distributions of the input values are known. If the real distributions of incoming flows and service rates are unknown, they must be approximated by a *standard frequency distribution*.

9.3.1 Mean Values and Standard Deviation

With a frequency distribution $w_P(x_i)$, the probability that x takes a value below the value x_n is given by the *distribution function*:

$$W_P(x_n) = \sum_{i<n} w_P(x_i) \tag{9.13}$$

A frequency distribution fulfills the *normalization condition*:

$$\sum_{i=0}^{\infty} w_P(x_i) = 1. \tag{9.14}$$

The *mean value* or *expectation value* of a discrete quantity with the frequency distribution $w_P(x_i)$ is:

$$m_P = \sum_{i=0}^{\infty} x_i \cdot w_P(x_i). \tag{9.15}$$

The *standard deviation* s_m is the square root of the *variance*:

$$s_m^2 = \sum_{i=0}^{\infty} (x_i - m)^2 \cdot w_P(x_i). \tag{9.16}$$

The *variability* $V_m = (s_m/m_P)^2$ is the square of the *variation* $\nu_m = s_m/m$.

9.3.2 Discrete Quantities in Logistics

Examples for randomly distributed discrete quantities in logistics are:

- quantities per event of a stochastic batch process
- number of events per period of a general stochastic process
- bulk content of batch-wise arriving flows
- contents of shipments
- stocks of articles with regular demand
- article consumption or demand within a certain period
- discrete cycles time of a station or a transport node

Stochastic processes cause queuing effects, such as waiting lines and waiting times in front of single stations and blockage of successive stations. Analytical investigations and simulations have shown (Gudehus 1976/I):

- In first order approximation, the queuing effects depend only on the mean values and the standard deviations of the flows and service times.

The influence of the exact distribution is generally negligible, as long as mean value and standard deviation are correct. Hence, it is generally sufficient to approximate frequency distributions by *standard distributions* with mean values and standard deviations equal to the measured values and with a shape that is in accordance with experience.

9.3.3 Binominal Distribution

The binomial distribution is defined for integers in the range $0 \le x \le N$ and given by the function:

$$w_B(x; N, p) = [N!/((N - x)! \cdot x!)] \ p^x \ (1 - p)^x. \tag{9.17}$$

$[N!/((N - x)! \cdot x!)]$ is the *binominal coefficient* with the factorial $N! = 1\,2\,3 \ldots.. N$.

The binomial distribution is a right-skewed function with *mean value* $m = N \cdot p$, *standard deviation* $s_B = \sqrt{N \cdot p \cdot (1 - p)}$ and *variability* $V_B = (1 - p)/(N \cdot p)$. Its value is the probability that an event P with the *single probability* p appears exactly x-times after N independent trials. For example, the probability that the same number comes out exactly three times when playing five times with a six numbered dice is $w_B(3;5;1/6) = 5.4\%$.

The binomial distribution can be used to approximate the frequency distributions of random integer variables with small mean value x_B and standard deviation $s_B < x_B$. In agreement with observations, mathematical considerations lead to the *application rule*:

▸ The binominal distribution (9.17) can be used to approximate the frequency distribution of a stochastic flow or demand, the contents of shipments, the bulk-lengths of incoming flows and the replenishment quantities of supply orders of many independent customers.

9.3.4 Poisson-Distribution

For very large N, i.e. for $N \to \infty$, and constant mean value m the binomial distribution evolves into the *Poisson-distribution*:

$$w_P(x; m) = (m^x/x!) \cdot \exp[-m]. \quad (9.18)$$

The Poisson distribution has the *mean value* m, *standard deviation* $s = \sqrt{m}$ and variability $V_m = 1$. A general mathematical *theorem* says (Ferschl 1964):

- If the time distribution of stochastic events is exponential, the frequency distribution of the numbers of events in equal time periods is a Poisson-distribution and vice versa.

The Poisson-distribution is useful for the approximation of integer distributions:

▸ If only the mean value of a random integer quantity is known, the assumption of a Poisson distribution gives results on the safe side.

This holds for instance for the distribution of the number of positions per order as shown in *Fig. 9.5*.

9.3.5 Normal Distribution

The most important distribution of statistics and probability theory is the *Gauß-Distribution* or *normal distribution* (Kreyssig 1975). The normal distribution of a value x around a *mean value* m with *standard deviation* s is given by the density function:

$$w_N(x;m,s) = \exp[-(x/s - m/s)^2/2]/\sqrt{2\pi s^2}. \quad (9.19)$$

As shown in *Fig. 9.6*, the normal distribution is symmetric around the mean value m. It has positive values for positive and negative arguments. Hence, the normalization (9.5) must be integrated from $-\infty$ to $+\infty$.

In logistics, the normal distribution can be applied for analyzing random processes with continuous and discrete distributions. Its density function (9.19) evolves asymptotically for large m from the binomial distribution (9.17), as well as for large *k*-values from the Erlang-distribution (9.11).

9.4 Mean Values and Variances in Logistics

With the help of *probability theory* and *statistics*, mean values and variances of logistic quantities can be calculated or at least estimated. For this purpose, the *standard distributions*, the *safety law*, the *theorem of large numbers* and the *law of error propagation* are useful. Many strategies of logistics, such as centralization and safety stocks, follow from these laws.

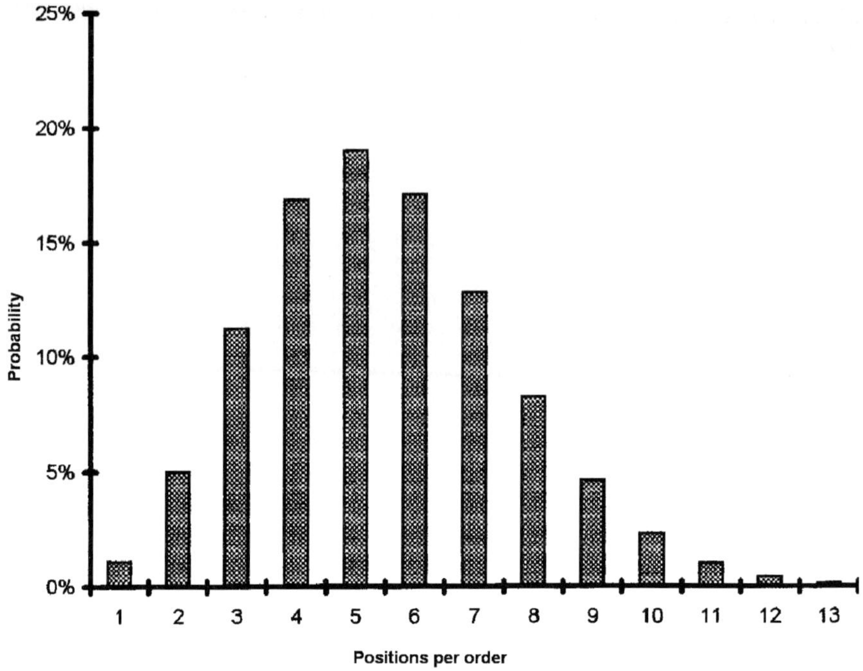

Fig. 9.5 Poisson distribution

Variable: positions per order
Mean value: 5.5 positions per order

9.4.1 General Safety Law

The distribution function of a centered normal distribution with mean value m = 0 and standard deviation s = 1 is the *standard normal distribution* Φ(x). The inverse function is the *inverse standard normal distribution*:

$$f_S(\eta_s) = \Phi^{-1}(\eta_s) \tag{9.20}$$

This function is crucial for the *general safety law*:

► If m is the mean value and s the standard deviation of a normal distributed random variable, its values lay with probability η_s below $m+f_s(\eta_s)\cdot s$.

With the inverse standard normal distribution (9.20), which is given by the MS-Excel-function NORMSINV(η), the *safety factors* of *Table 11.5* and *Fig. 5.4* have been calculated. They determine the *safety stock* for a required *stock availability* (see *Sect. 11.8*) and the *overflow reserve* for the *storage capacity* (see *Sect. 16.1*).

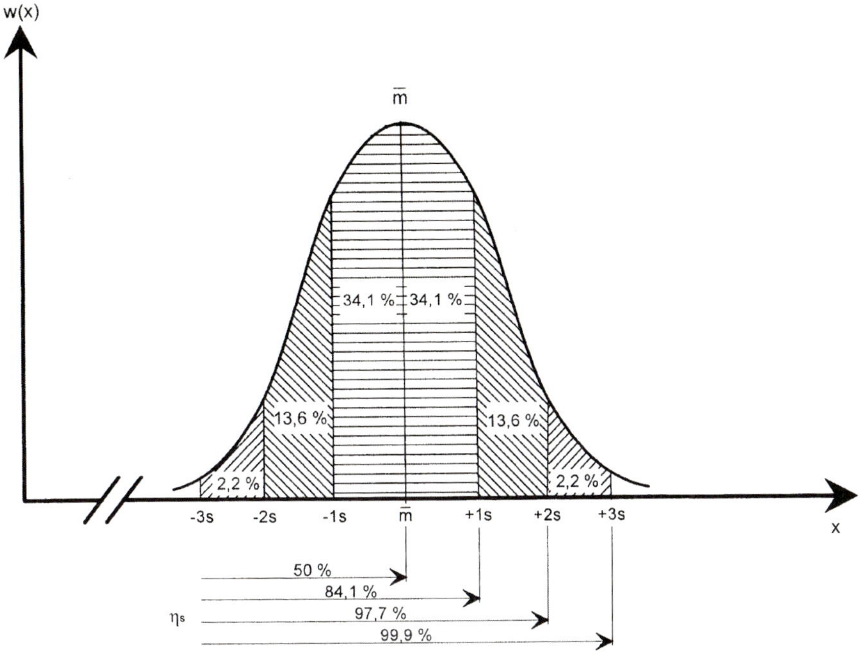

Fig. 9.6 Normal distribution or Gauß-distribution and safety degree
mean value m *standard deviation* s *safety degree* η_s

9.4.2 Theorem of Large Numbers

The *theorem of large numbers* says (Kreyssig 1975):

- The sum value

$$Z = \sum_{i=1}^{N} z_i \tag{9.21}$$

of a large number N of independent random variables z_i with single variances s_i^2 has approximately a normal distribution with variance

$$s_Z^2 = \sum_{i=1}^{N} s_i^2 \tag{9.22}$$

From the law of large numbers follows the *centralization rule of logistics*:

- ▸ The stochastic variation of accumulated performances, throughput rates and demand of many local stations is much smaller than the local variations.

The centralization rule is applied in order scheduling, inventory management, resource planning and in many other areas.

If the variances of the single random variables are all equal to s^2, the variance of the accumulated value is

$$s_Z^2 = N \cdot s^2 \tag{9.23}$$

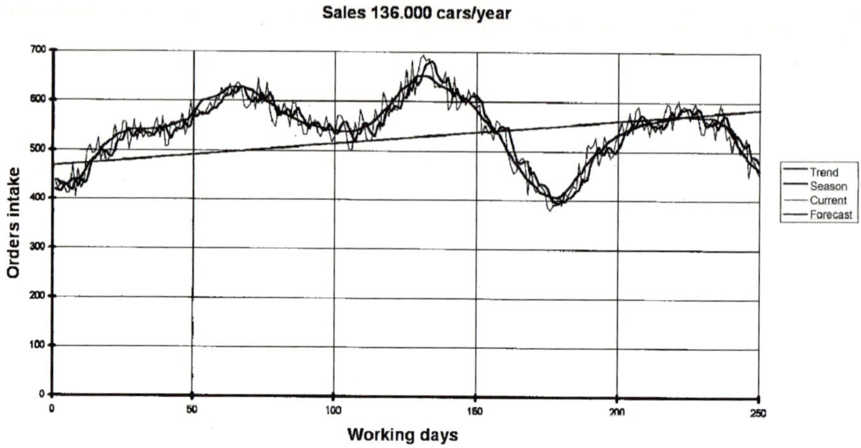

Fig. 9.7 Daily order entry of a car factory from 45 dealers
Total daily demand: 544 ± 23.3 cars/day

If the standard deviation s of all N random variables equals 1, the standard deviation of the sum is $s_Z = \sqrt{N}$. This can be applied in many areas of logistics for the estimation of standard deviations, uncertainties and stochastic errors.

For example, the car dealer of *Fig. 9.1* receives on average $\lambda_m = 12$ orders per day. The single car orders are variables with standard deviation 1. Hence, due to relation (9.23) the daily order flow has the standard deviation $s_\lambda = \sqrt{12} = 3.5$ cars and a variation of 29%. The sum of the sales of 45 car dealers, who sell on average 12 cars per day, gives a mean order entry rate for the car manufacturer of $\lambda_S = 540$ cars per day. From the theorem of large numbers (9.22) follows, that the total sales vary as shown in *Fig. 9.7* with standard deviation $s_{\lambda S} = 23.2$ cars/day, i.e. with variation of only 4,5%.

This example demonstrates the *effect of demand accumulation*:

▸ The accumulation of the small stochastic demand of a large number of local stations leads to a regular demand.

According to this effect, the consolidation of all local orders for nationwide sold articles by an *order center* or a *central computer* has considerable advantages for forecasting and scheduling. The advantages increase with the size of the consolidated sales area.

The fact, that the total demand is more regular and can be forecasted better than the local demand, leads to the *stock centralization rule* (see *Sect. 11.10*)

▸ Several local demand stations can be served more efficiently with higher *stock availability* or lower *safety stock* from a central store.

An application of this rule is the nationwide or global *spare-part distribution* from a central warehouse.

9.4.3 Law of Error Propagation

A generalization of the theorem of large numbers is the *law of error propagation* (Kreyssig 1975):

- The values of a steady differentiable function $F(z_1, z_2, \ldots, z_N)$ of N independent random variables z_i with variances s_i^2 have approximately a *normal distribution* with the *variance*

$$s_F^2 = \sum_{i=1}^{N} (\partial F/\partial z_i)^2 \cdot s_i^2. \tag{9.24}$$

By applying (9.22), (9.23) and (9.24), the variability and the standard deviation of compounded random variables, such as the variability of flow intensity, stock level fluctuations or demand variations, can be calculated. For example, from (9.24) follows:

- A *demand* $\lambda = m \cdot \lambda_R$ resulting from a mean *order entry rate* λ_R with *rate variance* s_R and a mean *order quantity* m with *quantity variance* s_m^2 has the *standard deviation*:

$$s_\lambda = \sqrt{m^2 \cdot s_R^2 + \lambda_R^2 \cdot s_m^2} \tag{9.25}$$

and the *variability*

$$V_\lambda = (s_\lambda/\lambda)^2 = (s_R/\lambda_R)^2 + (s_m/m)^2 = V_R + V_m \tag{9.26}$$

This means, the variability of the demand is the sum of the variability of the order rate and of the order quantities.

Another consequence of (9.24) is the *calculation accuracy rule*:

► The relative accuracy of the sum of many cost components or investment values is higher than the relative accuracy of the single values.

It is therefore possible to calculate a total investment or the total operating costs quite precisely, even if the single investment contributions respectively costs factors are only known roughly. This is of practical importance for *investment decisions*, *costing* and *pricing*, where the total error can be calculated from the estimated errors of the single factors by the law of error propagation.

9.4.4 Approximation and Averaging in Logistics

Scheduling, dimensioning and costing are generally based on the *mean values* of the order entry, order structure, performance rates, throughput and other stochastic values. It is an open question how far this proceeding is justified.

The *arithmetic mean* of the discrete values $F(x_i)$, $i = 1,2...,N$, of a steady differentiable function F(x) of N arguments x_i is:

$$F_m = (1/N) \cdot \sum_{i=1}^{N} F(x_i). \tag{9.27}$$

If the single arguments are distributed around a *mean value* x_m with small differences $\Delta_i << x_m$, i.e. if

$$x_m = (1/N) \cdot \sum x_i \quad \text{and} \quad x_i = x_m + \Delta_i, \tag{9.28}$$

the mean value function (9.27) can be developed into a *Taylor-sequence* of the *differentiations* $F^{(n)}(x) = d^nF/dx^n$ with increasing degree n. The first terms of the Taylor-development are:

$$F_m = (1/N) \cdot \sum_{i=1}^{N} F(x_m) + F^{(1)}(x_m) \cdot \Delta_i + F^{(2)}(x_m) \cdot \Delta_i^2/2 + ...) \tag{9.29}$$

Since the sum of the deviations Δ_i equals zero and higher potentials Δ_i are negligible, the sum (9.29) results in the *approximation rule*:

- As long as the curvature $F^{(2)}(x)$ in the range of the arguments x_i is much smaller than 1, the mean value F_m of a steady differentiable function is approximately equal to the function value at the mean argument x_m:

$$F_m(x) \approx F(x_m) \tag{9.30}$$

Consequences of the approximation rule are the *averaging rules of logistics*:

- ▸ In *first order approximation*, planning, dimensioning, scheduling and costing can calculate with the *mean values* of stochastic variables.
- ▸ By *perturbation calculation*, the influence of the *variances* on queuing, stock overflow, adherence to delivery dates and other key performance indicators can be evaluated (see *Sect. 15.4*).

After separating orders, assortments and flows into logistically similar classes, the calculations can be performed separately for the *homogenous classes* with the mean order rates and material flows (see *Sect. 5.5*).

Condition for the averaging rules of logistics is that the dependency has no leaps. This holds for process and operating costs that depend linear or reciprocally linear on the arguments as long as integer effects are negligible. Leaps in the cost curves that result from *integer effects*, e.g. from integer numbers of load units or transport means, can be smoothed out with a smoothing function (see *Chap. 12*).

9.4.5 Fluctuation Law

It is possible to derive from the theorem of large numbers (9.22) and the law of error propagation (9.24) the following *fluctuation law* (Gudehus 1999; Inderfurth 1999; Schneeweiß 1981; Tempelmeier 1999):

- The number of units of a stochastic demand or flow λ with *flow variance* s_λ^2 fluctuate for *time intervals* T with *variance* s_T^2 around the *mean value*

$$M(T) = \lambda \cdot T \tag{9.31}$$

with the *standard deviation*

$$s_M(T) = \sqrt{T \cdot s_\lambda^2 + \lambda^2 \cdot s_T^2}. \tag{9.32}$$

and the *variability*

$$V_M(T) = (s_M/M)^2 = s_\lambda^2/(T \cdot \lambda^2) + s_T^2/T^2 = V_\lambda/T + V_T. \tag{9.33}$$

The fluctuation law (9.32) will be applied in *Sect. 11.8* to calculate the *safety stock* of an article with stochastic consumption.

If T is the constant length T_{PE} of consecutive *periods* PE, the time variability V_T is zero and M(T) is the flow per period λ_{PE}. This leads to the *rule of periodical fluctuations*:

- The variability of a stochastic flow decreases inverse proportional with the length T_{PE} of the measuring period

$$V_{\lambda^{PE}} = V_\lambda / T_{PE}. \tag{9.34}$$

This rule is of importance for the selection of the appropriate measuring period for scheduling and forecasting (see also *Sect. 8.1.4*).

9.5 Mathematical Forecasting

For planning logistic systems and for scheduling orders and stocks sufficient knowledge of the future *demand* and *requirements* is needed. A *prediction* of the exact figures of a stochastic demand or stochastic flow for a future period is principally impossible. However, under certain conditions the *mean values* can be mathematically forecasted (Makridakis and Wheelwright 1978; Winters 1960).

The mean development of a dynamic flow or demand can be forcasted for a time T, if the systematic changes have a *regular pattern* or *recurring course* that can be observed with sufficient accuracy in the past, provided the pattern remains similar in future (Box and Jenkins 1979; Gardner 1985; Holt 1957; Lütkepohl 1993). If the temporal course has no recurring regularity, i.e. if it changes suddenly and unforeseeably, no mathematical forecast is possible.

9.5.1 Forecasting Conditions

A measure of the quality of a forecast is the forecast error (Kreyssig 1975):

- The *forecast error* of a prediction for the course of a time dependent flow or demand $\lambda(t)$ in a span of N periods $t = 1,2...N$ is measured by the standard deviation of the series of the *predicted values* $\lambda_{Pr}(t)$ from the *observed values* $\lambda_{Obs}(t)$:

$$\Delta_{Pr} = \sqrt{\sum_t (\lambda_{Pr}(t) - \lambda_{Obs}(t))^2/(N-1)}. \tag{9.35}$$

The forecast error is determined by the *stochastic error* given by the variability (9.34), by the *delay error* caused by delayed observation of systematic changes and by the *systematic error* due to unpredictable deviations.

In order to keep the stochastic error small, the length of the periods should be as long as possible. On the other hand, if systematic changes should be observed with small delay, the period must be as short as possible. This leads to the *determining rule for the period length*:

- ▸ The length of the periods should be short enough to observe systematic changes without too much delay, but must be so long that stochastic fluctuations are smoothed out sufficiently.

Consequences of this rule are the *forecasting conditions*:

1. The demand or flow rates have to be known for a statistically sufficient number of periods, in order to observe and calculate the systematic course.
2. The stochastic fluctuations of demand or flow rates must be much smaller than the required forecast accuracy.

The accuracy of mathematical forecasts depends critically on the variability of the stochastic flow or demand. Similar to the spoiling of information flows by noise, high stochastic fluctuations reduce the predictability.

9.5.2 Zero Period Rule

The stochastic behavior restricts the predictability of random events. A measure of the randomness and an indicator for predictability is the *zero period share*, which is the relation of the number of periods without event to the total number of periods. The share of zero periods measured for a long time is the probability that no event occurs in a period.

In many cases the *frequency distribution* $w(\lambda)$ of a stochastic demand or flow λ with variability $V < 1$ is approximately a *binomial distribution* with $p = 1 - V$ and $N = 1/(1 - V)$. For a binomial distribution the zero period probability $w_B(0)$ is calculable with the help of (9.17). The result is the *zero period rule* (Tempelmeier 1999):

► The zero period share of a stochastic demand or flow λ with variability V is approximately

$$w_B(0) = V^{\lambda/(1-V)} \underset{\text{for } V \to 1}{\to} e^{-\lambda}. \tag{9.36}$$

Depending on the zero period share, stochastic events can be differentiated into *XYZ-classes* (see *Fig. 9.8*):

- *X-events* or *regular events* with zero period share below 1%: A forecast of the mean values is possible with high reliability and short delay, provided the pattern of the observed systematic time course of the past is persistent in the future.
- *Y-events* or *irregular events* with zero period share between 10% and 50%: A systematic time course is observable only with delay. A less reliable forecast of mean values is possible, if the observed time course is persistent.
- *Z-events* or *sporadic events* with zero period share above 50%: The systematic variations of the mean values are very difficult to recognize even after many periods; if at all, their future development is predictable only with low reliability and long delay.

Due to relation (9.36) the zero period share decreases exponentially with the flow rate λ which increases with the period length. That leads to the *ABC-XYZ-correlation*:

► The XYZ-classification of stochastic events depends on the period length and correlates with the ABC-classification.

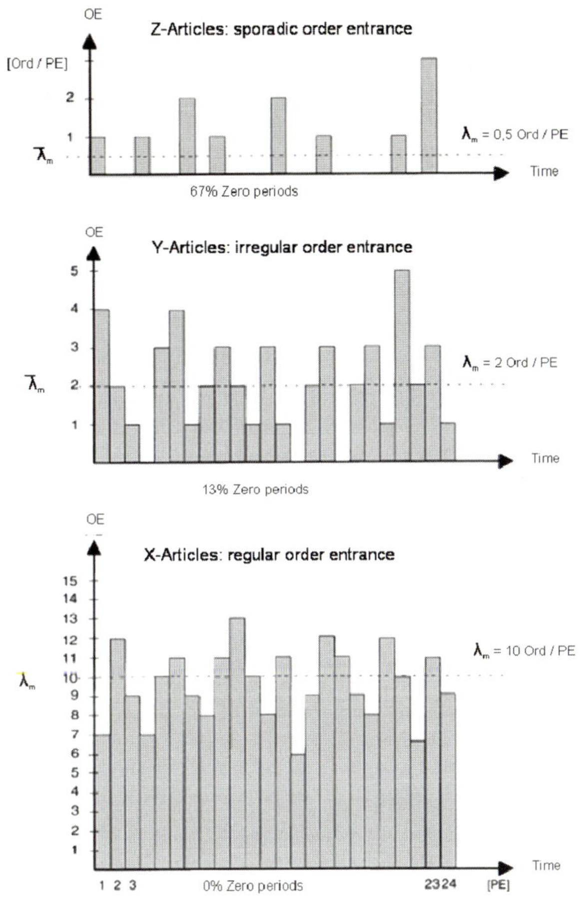

Fig. 9.8 Fluctuation and zero share of periods for XYZ-articles

The ABC-classification of an assortment involves the XYZ-classification and vice versa. Both depend on the period length (Tempelmeier 1999). A consequence is the *classification rule*:

- Period length and the borders between the ABC-classes and XYZ-classes are free parameters, which can be fixed due to the special task.

These rules and correlations are quite unknown. Hence, the practical usefulness of ABC- and XYZ-classifications is often overestimated. Their temporal changes are generally neglected (see *Sect. 5.6*).

9.5.3 Forecasting Methods

All mathematical forecasting methods assume the continuation of a pattern that has been observed for a longer time.

If the systematic course has changed only slowly in the past, it is probable that the mean value of the last periods remains for the next following periods. Under these conditions, the *expectation values* for the next future can be calculated by mathematical smoothing methods from the past values, such as *floating mean* or *moving average* and *exponential smoothing*.

If the systematic course of the events shows a repeating pattern within a certain *cycle time*, a suitable forecasting method is *cycle time forecasting*. The cyclic behavior can be caused by natural or calendar cycles:

- *Natural cycles* are the revolutions of earth and moon, the annual seasons and the harvesting cycles of natural products.
- *Calendar cycles*, such as daily, weekly, monthly and annual cycles, are related to the natural cycles.

A cyclic pattern can be caused also by a cultural, economic or business cycle:

- *Cultural cycles*, such as Christmas, Easter, Ramadan or vacations, are often linked with calendar cycles.
- *Economic cycles* are *trade-cycles*, *market-cycles* and other cycles of demand and supply, whose length and variations are difficult to explain (Gudehus 2007, p. 80).
- *Business cycles* are consequences of timetables, opening and operating times, *catalogue cycles* and periodical activities of the company.

Different from other cycles, business cycles can be changed to a certain extent, i.e. they are *action parameters* for planning and scheduling (see *Chap. 8*).

9.5.4 Floating Mean Value

Floating mean value or moving average forecasts are based on the assumption, that the mean flow, throughput or demand $\lambda(t+k)$ of future periods $t+k$ is equal to the *linear average* of the values in the last n periods:

$$\lambda(t+k) = \lambda_m(t) = (1/n) \cdot \sum_{j=1}^{n} \lambda(t-j) \quad \text{for } k \geq 0 \tag{9.37}$$

The resulting values become dynamic, if the calculation (9.37) is repeated at the end of each period t using the last n observed values.

Since all n past periods are taken into account with equal weight, the floating mean (9.37) reveals time changes with a *time delay* which increases with the *smoothing range* n. Due to the law of large numbers for a stochastic flow with standard deviation s_λ the *stochastic error* $s_{\lambda m}$ of the floating mean (9.37) is:

$$s_{\lambda m} = s_\lambda / \sqrt{n} \tag{9.38}$$

To keep the stochastic error small, the smoothing range must be taken as large as possible. In order to achieve a short reaction time, n should be as small as possible. This reflects the basic goal conflict of mathematical forecasting.

Another disadvantage of the floating mean value forecast is that the computer must keep permanently a large amount of data in the memory buffer.

9.5.5 Exponential Smoothing

Exponential smoothing forecasting avoids the disadvantages of the floating mean (Gardner 1985; Holt 1957; Winters 1960). It assumes, that the mean flow, throughput or demand $\lambda(t+k)$ of future periods $t+k$ equals the *weighted average* of the *predicted mean value* $\lambda_m(t-1)$ and the *observed value* $\lambda(t-1)$ of the last period $t-1$:

$$\lambda(t+k) = \lambda_m(t) = \alpha \cdot \lambda(t-1) + (1-\alpha) \cdot \lambda_m(t-1) \quad \text{for } k \geq 1. \quad (9.39)$$

The *smoothing factor* α is a real number within the interval $0 < \alpha < 1$. The exponential smoothing becomes dynamic by repeating the calculation (9.39) at the end of each period t with the values of the last period.

By successive insertion of the previous values into relation (9.39) results:

$$\lambda_m(t) = \sum_{j=1}^{t} (1-\alpha)^j \cdot \lambda(t-j-1). \quad (9.40)$$

This relation shows, that the observed values $\lambda(t-j-1)$ of the past periods $t-j-1$ contribute to the mean value (9.39) with exponentially decreasing weight $(1-\alpha)^j$. Therefore, the exponential smoothing reacts faster on changes than the floating mean value (9.37), if the smoothing factor α is taken sufficiently close to 1.

Due to the law of large numbers, for a stochastic flow with standard deviation s_λ the *stochastic error* $s_{\lambda m}$ of the exponentially smoothed mean value (9.39) is:

$$s_{\lambda m} = s_\lambda \cdot \sqrt{\alpha/(2-\alpha)}. \quad (9.41)$$

In order to keep the stochastic error small, the smoothing factor α should be taken as close to 0 as necessary. However, with smaller α the *time lag* for observing systematic changes increases. Again, these competing requirements for the smoothing factor reflect the basic goal conflict of mathematical forecasting.

The comparison of the relations (9.39) and (9.41) shows, that the smoothing factor α corresponds to an effective smoothing range $n = (2-\alpha)/\alpha$. For example, a value $\alpha = 0.2$ corresponds to a smoothing range of 9 periods. All values from periods before the effective smoothing range contribute to the exponentially smoothed mean value (9.39) with a weight of 13% (Gudehus 2003/2007, p. 4).

For exponential smoothing, only the forecast value and the observed value of the last period are needed. No further data of the past must be buffered. This makes the computation considerably easier and faster. As outlined in *Sect. 9.9.2*, further improvements are achievable by an *adaptive α-factor*.

9.5.6 Cycle Time Forecasting

If a certain temporal behavior reappears within a *cycle time* T_Z, the cycle time can be divided into N periods with length $T_{PE} = T_Z/N$ and end points $t_{Zo} + t \cdot T_Z/N$ for $t = 1, 2, \ldots, N$. Then the temporal course of a cyclic flow, which varies during the cycle time T_Z around a *mean value* λ_m, can be written with the *cycle weights* $g_Z(t)$ as

$$\lambda(t) = \lambda_m \cdot g_z(t) \qquad [\text{ME/PE}]. \quad (9.42)$$

The cycle weights $g_Z(t)$ of the past are given by the observed periodic values $\lambda_{obs}(t)$ related to the measured *mean value* λ_m of a complete cycle:

$$g_Z(t) = \lambda_{obs}(t)/\lambda_m \quad \text{for } t = 1, 2, \ldots N. \tag{9.43}$$

They are normalized to N over the complete cycle:

$$\sum_{t=1}^{T} g_z(t) = N. \tag{9.44}$$

The observed cycle weights of the last cycles are used to forecast the flow, throughput or demand for the future cycles.

Due to the stochastic fluctuations, also the observed cycle weights (9.43) have a stochastic error. The cycle weights of a single article with low demand may fluctuate so much from cycle to cycle that a reliable determination of the cycle weights is impossible. In such a case, the cycle weights can be taken from the cyclic behavior of a whole *article category* with comparable demand structure, since due to the higher demand of the whole group the stochastic fluctuations of its weights are smaller.

The forecast of the cycle weights for the coming periods can be improved by applying the method of exponentially smoothing to the cycle weights. Such *combined methods* help to reduce the stochastic error.

In a similar manner it is possible to forecast a demand or flow with dynamic behavior that is a *superposition of different cycles*. Examples are the weekly and the monthly cycle shown in the upper diagrams of *Fig. 9.9*, which are superimposed by the daily cycle within a week shown in the last diagram.

9.5.7 Forecasting by Model Functions

Forecasting methods based on model functions try to fit the systematic behavior by a mathematical function, which depends on free parameters. Standard model functions are *power series* in time and *Fourier developments*. Other functions are *life cycle curves*, which result from the analysis of the behavior of comparable articles. In many cases, the following *standard test-function* (9.48) is applicable. The free parameters of the model function can be determined by minimizing the deviation (9.33) between the values of the model function and the observed values with the help of the *least square method* (Kreyssig 1975).

Forecasting by model functions must be applied with care, since the extrapolation of the past functional behavior into to the future is not always justified. A further disadvantage is that it is difficult to separate the stochastic fluctuations from the systematic changes. Instead of using general model functions with parameters, which cannot be explained, it is advisable to apply *exponential smoothing, cycle time method, life cycle curves* or the standard test function (9.48), which depend on understandable influence factors.

9.6 Demand Planning and Forecasting

For planning and forecasting the future demand, it is necessary to differentiate between primary and secondary demand:

- The *primary demand* for products and performances of a company is caused and influenced by *external factors*, such as customer demand, competition, natural cycles, calendar cycles and economic cycles. It can be influenced by the company only to a certain extend via quality, price and performance.

- The *secondary demand* is caused or induced by the primary demand. Depending on the organization and technical relations, the secondary demand can be calculated with the *bill of materials* from the primary demand, if the lead times, stock levels and limit performances of the performance stations for the orders are known.

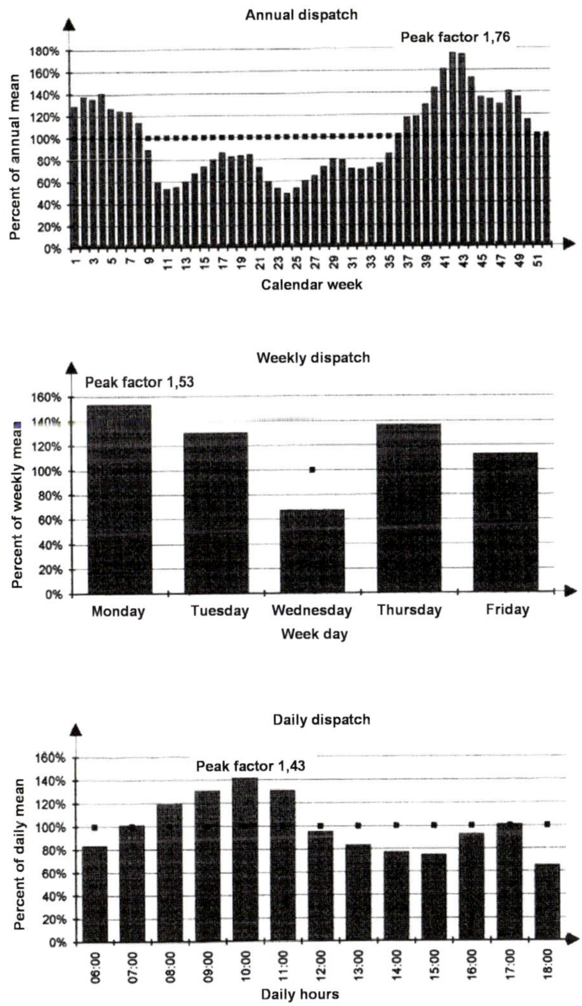

Fig. 9.9 Cycle weights and peak factors of the annual, weekly and daily dispatch of a central store of a retail chain

The secondary demand depends crucially on the scheduling strategies, operating times and capacity planning. For *Materials Requirement Planning* (MRP) and calculation of the secondary demand from the primary demand, nowadays sophisticated *MRP-programs* are available.

9.6.1 Primary Demand

If the predictability conditions are fulfilled, the *primary demand* can be mathematically forecasted to a certain extent. In addition, business activities, such as price changes, promotions, product changes or alterations of services, must be taken into account. The impact of certain activities, such as a *price change*, can be estimated or planned according to experience. The behavior of customers or competitors is far more difficult to predict or even incalculable (Gudehus 2007).

Therefore, one should keep in mind that forecasts of the primary demand are of limited reliability. This holds particularly for *service companies* that cannot produce to stock. They must have highly flexible systems and capacities. Also manufacturers, wholesalers and retailers of consumer goods must be able to cope in short time with changing requirements. Hence, in a saturated society it is better to serve the *consumer markets* due to the *pull-principle* instead of the *push-principle*.

9.6.2 Secondary Demand

The *secondary demand* can principally be planned or calculated from a known primary demand. For example, in *aircraft industry*, *plant building*, *construction business* and *project construction*, the future demand for material and parts is assessed from the actual order stock, the planned completion date for the product or project and the delivery times of parts, modules, components and material.

However, if material and parts are used for different requirements, material demand planning for the whole scope of secondary articles becomes highly complex. This holds in particular for storekeeping articles, raw and auxiliary material, modules and parts that are used for different final products which are sold in many variants on several markets.

It is therefore necessary to assess the efforts and advantages of MRP in comparison to an independent forecast of the secondary demand directly from consumption. Irrespective of whether their demand is primary or secondary, forecasts based on consumption and *dynamic inventory scheduling* are opportune for articles and products which are anonymously sourced or produced to stock, since stocks decouple supply from demand.

9.6.3 Peak Factors

When designing and dimensioning logistic networks, logistic centers and production systems, the demand during *peak times* must be taken into account. If the demand shows a cyclic pattern, the peak time requirement equals the mean demand multiplied by the *relevant peak factor*. The relevant peak factor results from the product of the maximal cycle weight of all cycles that are longer than the required reaction time of the system. This leads to the *dimensioning rule*:

- For planning and dimensioning, the peak factors of all cyclic changes, with cycle times longer than the maximal tolerable lead and delivery times, must be taken into account.

In general, the dispatch areas of a *logistic center* or a *transshipment station* must be capable to react within hours. The required lead times of big logistic centers are generally one or two days. This can be achieved by the following *dimensioning rules for logistic centers* and *transshipment stations*:

- The dimensioning of packing places and dispatch areas and of the equipment and personnel in these areas must take into account the requirements of the peak hour of the annual peak day. This results by multiplying the mean annual demand with the *dispatch peak factor* f_{Disp}, which is the product of *day peak factor* f_{Day}, *week peak factor* f_{Week} and *season peak factor* f_{Sais}:

$$f_{\text{Disp}} = f_{\text{Day}} \cdot f_{\text{Week}} \cdot f_{\text{Sais}}. \tag{9.45}$$

- For dimensioning limit performances, equipment and personnel in the other functional areas the mean daily throughput is relevant. This results by multiplying the mean throughput by the *throughput peak factor* f_{Trough}, which is the product of *week peak factor* f_{Week} and *season peak factor* f_{Sais}:

$$f_{\text{Trough}} = f_{\text{Week}} \cdot f_{\text{Sais}}. \tag{9.46}$$

- For dimensioning *storage capacities* the *stock peak factor* f_{Stock} is needed. Provided the optimal scheduling strategies outlined in *Sect. 11.9* are applied, the stock peak factor is the square root of the throughput factor:

$$f_{\text{Stock}} = \sqrt{f_{\text{Trough}}}. \tag{9.47}$$

For example, *Fig. 9.9* shows the daily, weekly and seasonal course of the requirements of a logistic center of a retailing company with many department stores. The product of the daily, weekly and seasonal peak factors gives a dispatch peak factor $f_{\text{Disp}} = 3.86$. This means, that the performance in the absolute peak hour of the dispatch area is almost four times higher than the annual mean value.

The throughput peak factor can be read off as $f_{\text{Trough}} = 1.76 \cdot 1.53 = 2.69$. From this results a stock peak factor $f_{\text{Stock}} = \sqrt{2.69} = 1.64$, which is noticeably lower than the dispatch and the throughput peak factor. By *pre-production* for the peak demand, the stock peak factor can be leveled out and the storage capacity requirements reduced.

To illustrate the smoothing effect of longer periods on the stock peak factor, *Fig. 3.4* shows the seasonal course of the monthly throughput and mean inventory per month for the same logistic center, for which the diagrams of *Fig. 9.9* hold. The seasonal peak factors of the monthly values are obviously lower than the seasonal peak factors for the weekly values. This is due to the smoothing over of longer periods. The resulting stock peak factor of 1.07 is in approximate accordance with the square root of the throughput peak factor 1.15.

9.7 Test Functions and Scenario Calculations

The daily numbers and the single contents of orders, performance requirements and demand can neither be forecasted accurately nor planned. This holds in particular for *consumer goods*. Therefore, more and more manufacturers and retailers of consumer goods apply the *pull principle*. Similar to the service sector, the consumer goods industry has to adjust to changing market conditions and to flexibly respond to customer requirements. This is meant by *Efficient Consumer Response* (ECR) and *Continuous Replenishment* (CRP) (Kotzab and Bjerre 2005).

In order to plan flexible logistic networks and production systems, which involve high investments and have long realization times, it is essential to develop *scenarios* for different requirements and market conditions. Also the consequences of different *scheduling strategies* should be assessed by scenario calculations.

For *scenario calculations* suitable *test functions* are needed which allow simulating expected demand, incoming orders, material flows or other requirements. Compared to the use of real data of past periods, a suitable test function has the advantage that expected or planned changes can be generated by free parameters.

9.7.1 Standard Test-Function

Test-functions are constructed by multiplication and summation of mathematical functions, which depend on *adjustment parameters*. These *parameters* generate an observed or expected *trend*, *cyclic behavior*, *product life curve* and *stochastic fluctuations*.

From general considerations and many applications results the

- *standard test-function*

$$\lambda_{STF}(t) = m_E(t) \cdot \lambda_E(t) \qquad [PU/PE], \tag{9.48}$$

which is the product of the standard *event function*

$$\lambda_E(t) = ROUND(\lambda_E \cdot g_{Trend}(t) \cdot g_{Cycl}(t) \cdot g_{Dist}(t) \cdot g_{Erand}(t)) \qquad [EV/PE] \tag{9.48a}$$

and the standard *quantity function*

$$m_E(t) = MAX(1; ROUND(m_E \cdot g_{Mrand}(t)) \qquad [PE/EV]. \tag{9.48b}$$

This function has been applied successfully in *theory* to test forecasting methods, scheduling strategies and algorithms and in *practice* to perform potential and scenario calculations for consulting projects (Gudehus 2003/2007).

With the standard test function (9.48) the stochastic *event* E of integer *process units* PU, e.g. the arrival of orders or logistic units, can be generated for any process with a systematic time behavior during N_{PE} periods $t = 1,2,\dots N_{PE}$. The process units may be order lines, article units, sales units, cars, transport means, performance units or load units. The rounding operation in (9.48a) generates integers, which for weak flows are 0 in some periods.

The mean *intensity of the event flow* can be adjusted by the *start value* λ_E of the event function (9.48b). An observed or expected linear trend is modeled by the *trend function*:

$$g_{\mathrm{Trend}}(t) = 1 + c_{\mathrm{Trend}} \cdot t/N_{\mathrm{PE}} \tag{9.49}$$

with the adjustable *trend parameter* c_{Trend}. A cyclic behavior is generated with the *cyclic function*

$$g_{\mathrm{Cyc}}(t) = 1 + (f_{\mathrm{Cyc}} - 1) \cdot \mathrm{SIN}(2\pi \cdot \nu_{\mathrm{Cyc}} \cdot t/N_{\mathrm{PE}}). \tag{9.50}$$

The amplitude of the cycles is adjustable by the *cyclic factor* f_{Cyc} and the frequency by the *cycle frequence* ν_{Cyc} [EV/N_{PE}]. The *distortion function* $g_{\mathrm{Dist}}(t)$

$$g_{\mathrm{Dist}}(t) = \mathrm{IF}(t < t_S;\, 1;\, \mathrm{IF}(t > t_E;\, 1;\, f_{\mathrm{Cdist}}) \tag{9.51}$$

generates a sudden increase or reduction by a *distortion factor* f_{Cdist} between a *start date* t_S and an *end date* t_E. The *random event function*

$$g_{\mathrm{Erand}}(t) = 1 + f_E \cdot (2 \cdot \mathrm{RAND}() - 1) \tag{9.52}$$

calculates with equal probability random values in the interval [$1-f_E$;$1+f_E$]. By the *random event factor* f_E the *variation* ν_E of the event flow can be adjusted. The resulting variation is

$$\nu_E = s_{\lambda E}/\lambda_E = f_E^3/3. \tag{9.53}$$

The standard *quantity function* (9.48a) is the product of the *mean quantity* m_E. per event and the *random quantity function*

$$g_{\mathrm{Mrand}}(t) = 1 + f_m \cdot (2 \cdot \mathrm{RAND}() - 1). \tag{9.54}$$

This function generates stochastic fluctuations of the actual quantities around the mean value by calculating random values in the interval [$1-f_m$;$1+f_m$]. The *random quantity factor* f_m determines the *variation* ν_m of the generated quantities:

$$\nu_m = s_m/\lambda_m = f_m^3/3. \tag{9.55}$$

As an example, *Fig. 9.10* shows the result of the simulated sales for a business year with $N_{\mathrm{PE}} = 250$ working days. The test function (9.48) generates a *general stochastic flow* that varies stochastically around a mean value, which changes systematically with a linear trend and a cyclic pattern.

Another example is shown in *Fig. 9.13* where the same sales function has been suppressed by the distortion function (9.51) to 0 from the 1st to the 60th working day and the sales start suddenly at the 61st day. Both examples, *Fig. 9.10* and *Fig. 9.13*, have been used to test strategies of *dynamic forecasting*, which will be outlined in the next section. Also the strategies of *dynamic order and production scheduling*, which will be explained in *Sects. 10.5* and *10.6*, and the *dynamic inventory scheduling*, as outlined in *Sect. 11.13*, have been developed and tested with the model function (9.48).

9.7.2 Natural Growth Function and Logistic Functions

Instead of the linear trend function (9.49), other *model functions* can be inserted into (9.48a) to simulate the systematic time behavior of a process.

The *natural growth function*

$$g_{\mathrm{NG}}(t) = 1 - \mathrm{EXP}[(t - t_o)/T_G] \tag{9.56}$$

depends on the *growth time* T_G. As shown in *Fig. 9.11*, it ascends at *start time* t_o from 0 with decreasing gradient asymptotically to 1. The growth function (9.56) describes natural processes, such as the growth of grass. It is suitable to approximate *startup processes*, e.g. in the in *car industry*.

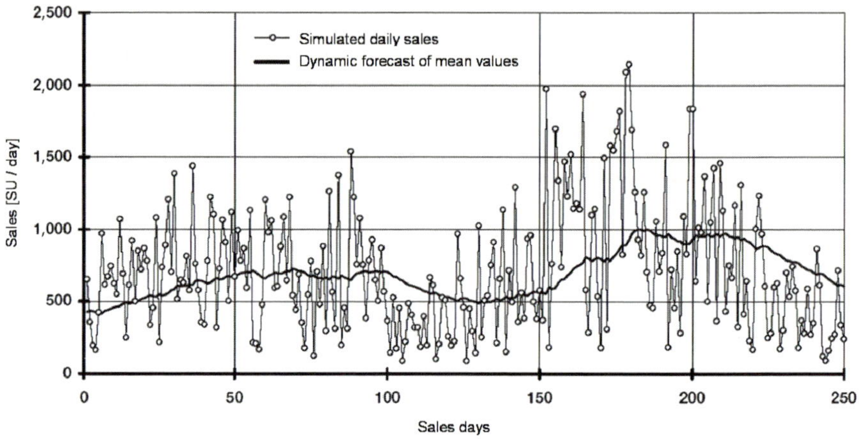

Fig. 9.10 Test of the dynamic forecasting for the sales of an article with stochastically and systematically changing demand

Mean daily sales 700 ± 360 sales units
Trend + 100% p.a.
Seasonality ± 50%

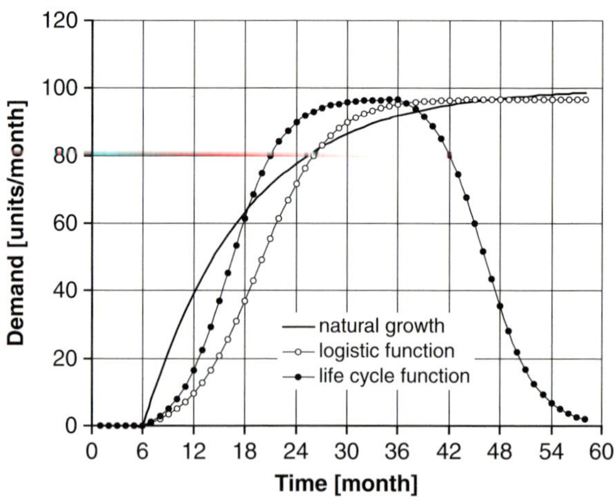

Fig. 9.11 Growth function, logistic function and life cycle function

Quite useful is the *increasing logistic function*:

$$g_L^+(t) = 1/(1 + a^+ \cdot \text{EXP}[-(t - t_o^+)/T_N^+]) - 1/(1 + a^+). \qquad (9.57)$$

It depends on an *increase-time* T_L^+ and an *ascending parameter* a^+, by which the gradient of the curve can be adjusted. *Figure* 9.11 shows that the S-shaped logistic function (9.57) starts slowly at point t_o and increases until point t_o+T_N. From there the curve approaches 1 with decreasing gradient. This is typical for the *demand curve of a new product*, which reaches a *saturation level* after a certain time.

The *decreasing logistic function*

$$g_L^-(t) = (2 + a^-)/(1 + a^-) - 1/(1 + a^- \cdot \text{EXP}[-(t - t_o^-)/T_N^-]). \qquad (9.58)$$

depends on the *decrease-time* T_L^- and a *descending parameter* a^-. It describes the *demand curve for a dying product*. As shown in *Fig. 9.12*, this curve falls from the beginning end of the life cycle t_o^- at first slowly, then faster and after longer time asymptotically down to zero.

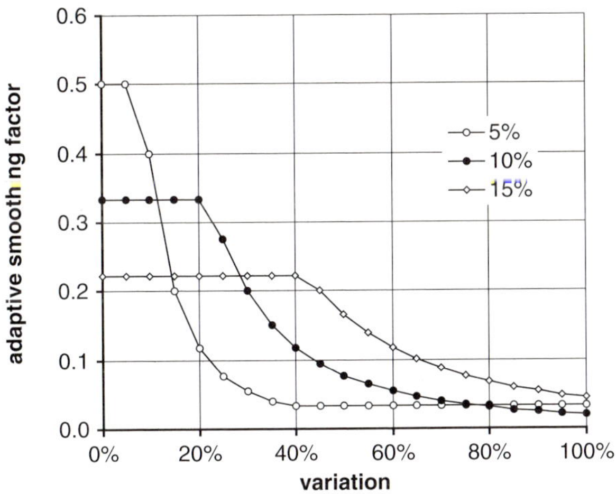

Fig. 9.12 Dependency of the adaptive smoothing factor on the variation

The sales of a product during its life time can be described quite well by the *life cycle function*

$$g_{LC}(t) = g_L^+(t - t_o^+; T_N^+; a^+) - g_L^+(t - t_o^-; T_N^-; a^-). \qquad (9.59)$$

This model function is the combination an increasing and a decreasing logistic function. As shown in *Fig. 9.11*, by the 6 *adjustment parameters* t_o^+, T_N^+, a^+, t_o^-, T_N^- and a^-, different product life cycles can be approximated.

The life cycle function can also be applied to forecast the sales for the *catalog cycle* of a *mail order company*. For this purpose, the 6 adjustment parameters are derived from the sales of previous catalog cycles for articles with similar demand pattern. From cycle to cycle, the experience and reliability of the parameters improve. By this method, it is possible to forecast the sales of established articles for the total catalogue season from the sales of the first couple of days with accuracy better than 5%.

9.8 Dynamic Forecasting

For the *dynamic scheduling* of orders and inventories the mean values of the demand and flows and their variance must be forecasted for a time length, which is at least as long as the replenishment time, i.e. for a couple of days up to several weeks (see *Sects. 10.6* and *11.13*) (Holt 1957; Winters 1960; Tempelmeier 2005).

If order entrance rate and demand do not change too fast, this is possible by *dynamic forecasting* with the algorithms:

- *Dynamic Mean Value Forecast*: The mean value of the future flow or demand is calculated at the end of each period from the demand $\lambda(t-1)$ and the forecasted mean value $\lambda_m(t-1)$ of the last period with an *adaptive smoothing factor* $\alpha_\lambda(t)$ by exponential smoothing

$$\lambda_m(t) = \alpha_\lambda \cdot \lambda(t-1) + (1-\alpha_\lambda) \cdot \lambda_m(t-1). \tag{9.60}$$

- *Dynamic Variance Forecast*: The future variance of the flow or demand is calculated from the demand $\lambda(t-1)$, the forecasted mean value $\lambda_m(t-1)$ and the forecasted variance $s_\lambda(t-1)$ of the last period with the same smoothing factor $\alpha_\lambda(t)$ by exponential smoothing

$$s_\lambda(t)^2 = \alpha_\lambda \cdot (\lambda(t-1) - \lambda_m(t-1))^2 + (1-\alpha_\lambda) \cdot s_\lambda(t-1)^2. \tag{9.61}$$

The time dependent smoothing factor $\alpha_\lambda(t)$ is given by the following formula (9.62). It is determined by the current flow or demand and its variance.

Dynamic forecasting is quite simple, since it needs only the information of the last period. Its algorithms are mathematically proved. Their limitations have been tested in many practical cases (Gudehus 2003/2007).

In order to forecast the sales of a *new article*, the mean demand and its variance must be estimated before the sales starts. These expectation values are taken as start values. As *Fig. 9.13* shows, already a few periods after the first orders arrived, the dynamic forecasting operates *self-regulating*.

For *multi-item orders* with $m > 1$, mean value and variance of *order entrance* λ_E and *order quantities* m can also be forecasted separately by two sets of formulas, corresponding to (9.60), (9.61) and (9.62) with two different smoothing factors α_E and α_m. The separate forecast of order entrance and order quantity is necessary for deciding about *storekeeping* and for judging the influences on the *safety stock* (see *Sects. 11.8, 11.12* and *11.14*).

9.8.1 Adaptive Smoothing Factor

Many forecast programs based on exponential smoothing calculate with a fixed *α-factor*, which must be adjusted by the scheduler. However, nobody can permanently control and adjust this sensible parameter for many articles. Therefore, it is mostly kept fixed for long time although a constant smoothing factor can cause quite wrong results.

This problem is solved by the *adaptive smoothing factor* which is calculated at the end of each period by the formula

$$\alpha_\lambda(t) = 2 \cdot \mathrm{MIN}(\nu_\lambda(t)^2;\nu_{\max}^2) / \left(\nu_\lambda(t)^2 + \mathrm{MIN}(\nu_\lambda(t)^2;\nu_{\max}^2)\right). \tag{9.62}$$

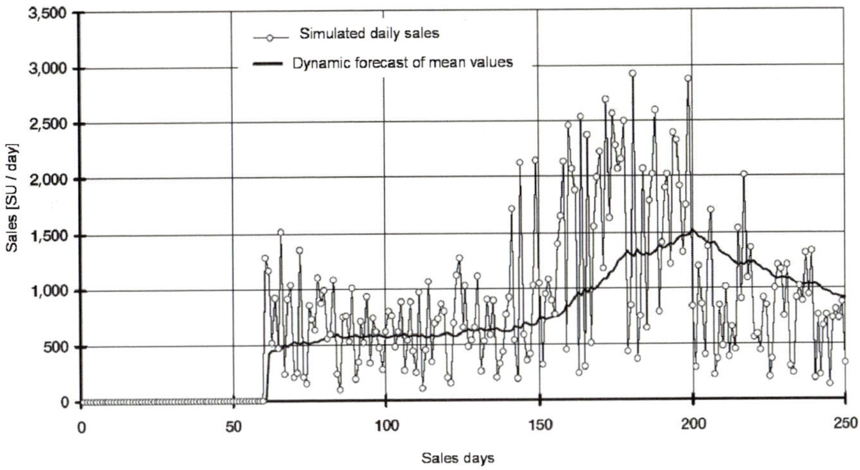

Fig. 9.13 Dynamic sales forecast for an article with abrupt starting demand
dynamic forecast by the algorithms (9.60), (9.61), (9.62), (9.63), (9.64), (9.65)
start of sales at 61st day other parameters see *Fig. 9.10*

The *current variation*

$$\nu_\lambda(t) = s_\lambda(t-1)/\lambda_m(t-1)$$

is determined by the mean value $\lambda_m(t-1)$ and variance $s_\lambda(t-1)$ of the demand, which have been forecasted for the last period.

As shown in *Fig. 9.12*, the adaptive smoothing factor (9.62) changes the *effective smoothing range*

$$n_{\lambda\mathrm{eff}} = (2-\alpha_\lambda)/\alpha_\lambda. \tag{9.63}$$

The smoothing length (9.63) is shortened when the variance decreases and is elongated when the variance increases. Due to the adaptive smoothing factor (9.62) the mean value $\lambda_m(t)$ follows self-regulated the systematic changes as fast as compatible with the tolerable stochastic error.

The stochastic error of the mean value λ_m is limited by the *maximal acceptable variation* ν_{max} in formula (9.26), which is determined by the aim of the forecast. For instance, for inventory scheduling a variance of λ up to 10% is tolerable, since the *economic replenished quantity* and the *safety stock* vary only with the square root of λ. For example, with a tolerance of λ of 10% they vary only by ± 5%.

$$\nu_{max}/n_{min}/n_{max}:\ 5\%/3/60 \quad 10\%/5/120 \quad 15\%\ /8/180$$

In order to keep the effective smoothing range (9.63) under control, it is necessary to restrict the α-factor by a lower limit α_{min} and an upper limit α_{max}. This results in the additional condition for the α-factor:

$$\alpha_{min} \le \alpha_\lambda(t) \le \alpha_{max}. \tag{9.64}$$

For example, if the smoothing should take into account effectively the last 60 periods, due to relation (9.63) the *minimal smoothing factor* α_{min} is 0.033. If the smoothing should take into account at least the past 5 periods, the *maximal smoothing factor* is $\alpha_{max} = 0.33$.

When the demand or flow within the minimal smoothing range has dropped to 0, the adaptive smoothing factor must be set up to the maximal value. This leads to a second condition for the α-factor:

$$\alpha_\lambda(t) = \text{IF}\left(\sum_{j=1}^{nmin} \lambda(t-j) = 0; \alpha_{max} \text{ for the next } n_{min} \text{ periods}\right). \quad (9.65)$$

The summation in (9.65) covers the minimal smoothing length n_{min}, which results from (9.63) with the maximal α-factor. The condition (9.65) is necessary in order recognize immediately the start or restart of the demand. It also ensures the recognition of the end of demand. *Figure 9.13* shows how fast the dynamic sales forecast by the algorithms (9.60), (9.61), (9.62), (9.63), (9.64), (9.65) follows the abruptly starting sales of a new product with first demand on day 61.

9.8.2 Indication of Irregularities

For *dynamic scheduling* it is essential to notice irregular changes as soon as possible. If the scheduler has to observe many articles or events at once, a critical change may be noticed too late.

Due to the safety law of *Sect. 9.4.1*, the probability that the demand or flow event stays in the range $\lambda_a \pm s_\lambda$ of one standard deviation is 85%. Hence, the probability that the value for n succeeding periods stays on the same side of this range is $1 - (1 - 0{,}85)^n$. This leads to the *irregularity indication*:

- ▸ If the demand deviates from the mean value λ_m in 3 or more succeeding periods in the same direction by more than the standard deviation s_λ, the probability of an irregular change exceeds $1 - (1 - 0{,}85)^3 = 99{,}7\%$.

The irregularity indication can be used to generate *warning signals*, which causes the scheduler to investigate the possible reasons for the irregular behavior and to adapt the scheduling appropriately.

9.9 Demand Forecasting in Logistic Networks

The demand of the stations within a multi-stage logistic network can theoretically be calculated from the demand of the final stations, if the *bill of material* of the final and intermediary products, the inventory levels and the resources, as well as the scheduling strategies of all downstream stations, are known. Nowadays, such a *network demand calculation* is principally possible by high performance computers, if the number of final products is not too large and the necessary information is available.

However, with larger numbers of products and extensive bills of material the demand calculation for many-stage networks becomes increasingly complex. Furthermore, for *inter-company delivery chains* the required information is generally not available. Only exceptionally companies disclose information on sales, inventory, resources and business strategies to suppliers or customers. Generally, such information is *business secret.*

This holds in particular for the *price policy*. If sales are receding, a retailer might stimulate demand by a price reduction without informing the supplier who expects further decreasing demand. When sales are booming a price raise is possible, which may affect the further increase expected by the supplier.

9.9.1 Collaborative Forecasting and Scheduling

The willingness of companies to share information depends on the *added value* and advantages of collaborative forecasting, planning and scheduling (Seifert 2002). The main advantages of collaborative forecasting are:

- The knowledge of the cumulative demand of many final stations improves the *predictability* of systematic changes of the demand for all participants.
- Instantaneous information about the demand of the final stations or from the *point of sales* can be used for *Efficient Consumer Response*, i.e. for serving the markets better and faster.

According to the law of large numbers, the variation of the cumulated demand is smaller than the variation of the single demands. Also the seasonal weights can be determined more accurately by using the cumulative demand of many stations for the same article. Hence, the knowledge of the cumulative sales improves the forecast of demand, sales and turnover. Sudden demand changes are recognized faster and *bottlenecks* can be anticipated.

9.9.2 Time Lag of Demand Information

If a supply station is separated from the final consumption stations by one, two or more stages, the scheduling is normally based on the orders from the immediately following stations. Due to the delayed information, the upstream stations can react to a systematic change of the final demand only with *time lag*.

The mean time lag caused by an intermediate storekeeping station equals half of the *stock reach time*. A station, which operates on order, causes a mean time lag equal to the half *consolidation time* of the orders. As indicated by the upper drawing of *Fig. 9.14*, the time lag of the demand increases with the distance of a station from the final stations in a delivery chain.

If all stations of a supply network are linked with the final stations by *EDI* or via *Internet*, it is principally possible to determine the future demand based on the simultaneously transferred final demand information, as shown in the lower drawing of *Fig. 9.14* without delay (Lee et al. 1997). When forecasting the own demand it is, however, necessary to take into account also the scheduling strategies of the downstream stations. Business practice has shown that this information is generally not available. This is the main reason, why many ambitious ECR and CRM projects have failed.

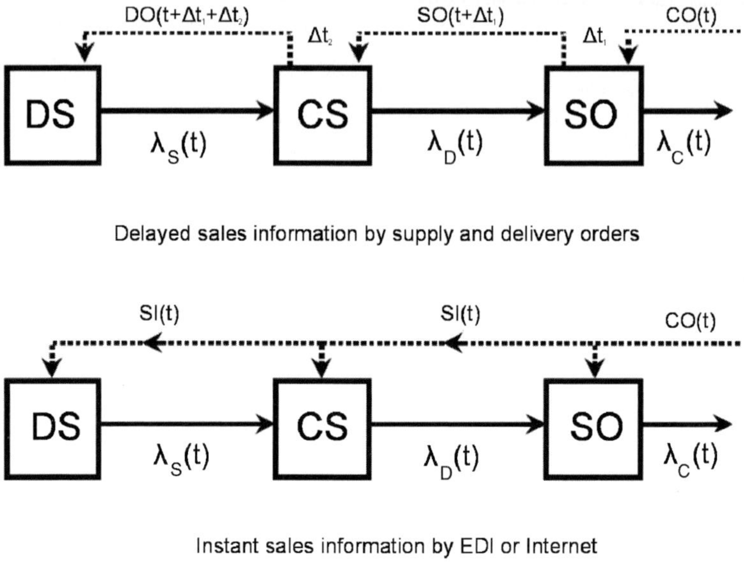

Fig. 9.14 Delivery chain with delayed and instant sales information

DS: delivery station CS: central store SO: sales outlet
DO: delivery orders SO: supply orders CO: customer orders
λ_D: deliveries λ_S: supplies λ_C: customer sales
SI: instant sales information by EDI or Internet

9.9.3 *Effects of Inter-company Supply Chain Management*

Even if the stocks and scheduling strategies of all storage and performance stations, as well as the actual sales of all final stations are known, this would be of rather limited value. Simulations of two-stage supply networks show that forecasts based on the immediate final demand instead on the delayed demand, do not necessarily result in significant stock reductions or cost savings as long as no *bottlenecks* occur (Gudehus 2001).

Theoretical analysis, simulations and practical experience lead to the *supply network rules*:

- ▸ As long as no bottlenecks are to be expected, the upstream stations of a supply network do not need to know immediately the demand of the final consumption stations for their short term scheduling.
- ▸ The knowledge of the cumulative demand of the final consumption stations is helpful for *medium term forecasts* and *resource planning* of the whole supply network and for the provision for *bottleneck phases*.

Who ignores these rules, tends to overestimate the possibilities and potentials of *inter-company supply chain management* (see also *Sect. 21.17*).

Chapter 10
Order Scheduling and Operating Strategies

Before the execution of external orders it is necessary to decide, when, where and in which sequence they should be started. These are the *tasks of order scheduling*:

- In order to execute *external orders* within required times at lowest costs they have to be accumulated, sequenced, divided or bundled into *internal orders* which are allocated to the available performance stations and resources.

Traditionally, order scheduling has been the task of people, the *schedulers*. Nowadays, scheduling is supported and performed to an increasing extent by computer. Schedulers, as well as programs, apply *scheduling strategies* which result from experience or have been systematically developed. Some strategies are documented as *rules* and *working instructions*; others are programmed as *algorithms*. Many strategies only exist in the mind of the scheduler.

The qualification of schedulers, the *added value* they generate, but also the *damage* caused by unqualified schedulers and unsuitable procedures are hardly known. In many, even big companies, only a few people schedule orders and resources. When experienced schedulers retire, often problems arise, as their know-how has not been transferred to the successors.

In order to support and improve scheduling, suitable strategies must be developed, selected, documented and programmed. Good *scheduling programs* perform mathematical forecasts and schedule *standard orders* without people. Relieved from routine tasks, schedulers can focus on special orders, exceptional situations and difficult tasks.

In this chapter, *operating strategies* for the optimal utilization of single and integrated performance and production stations are investigated. They can be distinguished in *time, processing*, *allocation*, *production*, *sequencing* and *dispatch strategies*. These general operating strategies are applicable to all kinds of production, performance and logistic systems, such as the storage, commissioning and transport systems described in *Chaps. 13* to *19*.

Time strategies have already been presented in *Chap. 8*. *Sourcing, inventory and replenishment strategies* will be developed in *Chap. 11*. The combination of operating strategies with time and replenishment strategies results in scheduling strategies for integrated production and performance networks with competing order

T. Gudehus, H. Kotzab, *Comprehensive Logistics*,

DOI 10.1007/978-3-540-68652-1_10,

chains (see e.g. *Fig. 8.1*). Applications of the general scheduling strategies presented in this chapter are *order scheduling* and *production planning* of workshops, bottling and packaging stations, assembly lines and production chains (see *Chap. 20*). Other applications are *administrative stations*, such as *call centers*, offices and *order centers*.

The *qualitative effects* of a strategy with respect to certain objectives, such as utilization, delivery times, stock levels or performance costs, are in most cases relatively easy to assess. Quantifying the *strategy effect* and determining the *optimum of a strategy variable* are more difficult. In many cases these are still unsolved problems. The strategy effects on the total costs for a longer planning period can be calculated only for relative simple systems under very limited conditions (Gudehus 2003/2007; Inderfurth 1994).

Order scheduling strategies are closely related with *production strategies* such as *make-to-stock* and *make-to-order*. In order to demonstrate the connection between *order scheduling* and *production planning*, an algorithm which is used to calculate the order buffer, the stock of finished goods and the lead times for different operating strategies, order flows and limit performances will be presented here. The *model calculations* quantify the effects of different operating strategies.

The relatively simple example of a single production station coupled with a single storage station turns out to be already quite complex. Delivery, production and supply *networks* are of far higher *complexity*. They become manageable by the *decoupling principle* and the *subsidiary principle* of section 2.4. These principles and the scheduling strategies are applied to the *dynamic scheduling* of *instationary order flows* in the last section of this chapter.

The scheduling strategies, the decision rules for *make-or-buy*, the opportunity of *make-to-order* or *make-to-stock* and the general procedures of production planning outlined in *Chap. 20* are the key of *production planning and scheduling* (PPS), of *enterprise resource planning* (ERP) and of *advanced planning systems* (APS) (Frazelle 2001; Gudehus 2003/2007; van Hook 2004; Meyr 2004; Park and Narayan 1997; Scheer 1998; Scheutwinkel 1999; Schönsleben 1998; Stadtler and Kilger 2007; SAP 1994, 2004).

10.1 Performance and Production Structures

The successive actions and part-processes within a single performance or production station are (see *Fig.1.6*):

order entry
buffering of orders
preparation for execution
start of performance (10.1)
order execution and production
buffering or storing of results
dispatch of the results

Buffering or storekeeping of the finished goods is only possible, if the order results are storable.

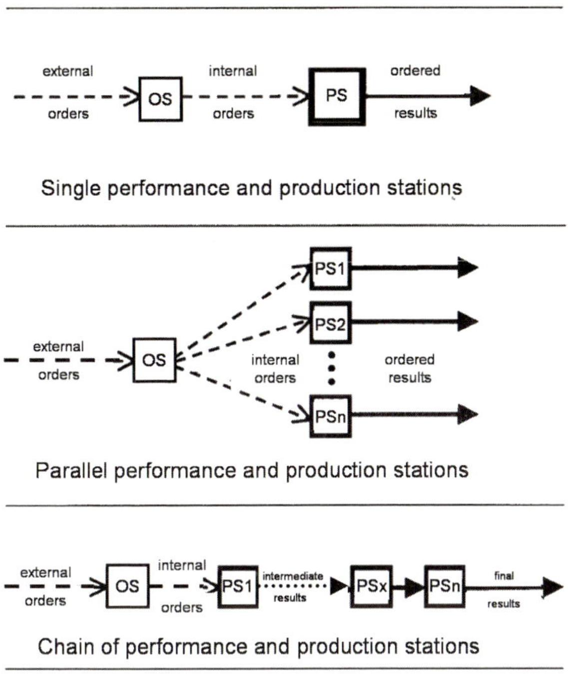

Fig. 10.1 Elementary configurations of performance and production stations
Results: services or products
OS: order scheduling PS: performance or production station

Figures 1.3, 8.1 and *10.1* show how single performance and production stations can be connected, combined and chained to networks of different structure and complexity. Complexity and scheduling of extended networks become manageable by the

- *Decoupling Principle*: All stations and subsystems which are not technically linked should be decoupled by *order buffers* or *material stocks* and as far as possible scheduled and controlled locally.

For systems and networks designed and organized according to this principle, scheduling can be divided into *central scheduling of external orders* for the total system and *local scheduling of internal orders* within the decoupled stations and subsystems (see *Sect. 2.2*).

10.1.1 Single Stations

In the simplest case shown at the top of *Fig. 10.1*, only one performance or production station executes the orders. Examples of single *performance stations* are *bottling stations* for liquids and *packaging stations* for consumer goods (see *Fig. 3.5*). Single *logistic stations* are loading ramps, receiving docks, packing places and transport vehicles.

Scheduling strategies for single stations are combinations of the *time strategies* of *forward*, *backward* and *free scheduling* of *Sect. 8.7* (see *Fig. 8.2*) with the following *processing*, *production* and *sequencing strategies*.

10.1.2 Parallel Stations

If the limit performance of a single station is insufficient for the demand, several stations are necessary, which operate as shown in the middle of *Fig. 10.1* in *parallel*. Which order is executed by which station is regulated by the *allocation strategies* of the following *Sect. 10.3*.

Parallel stations are typical for *workshop production*. Other examples can be found in *intralogistics*, such as docking fronts, packing areas and storage aisles.

10.1.3 Linked Stations

If an order is executed by a chain of closely linked performance and production stations, as shown at the bottom of *Fig. 10.1*, additional options for scheduling are the *time strategies* of *Sect. 8.8*, such as the *push principle*, the *pull principle*, *Just-in-Time* and their combinations.

Linked performance stations are typical for *line production*. Examples are simple assembly lines, cigarette machines linked with packing stations, bottling stations directly followed by packaging and palletizing stations, or sequences of machines and working places. Linked stations in logistics are the freight, supply and delivery chains described in *Chap. 21*.

10.1.4 Performance Trees

Complex orders are executed by a *performance tree* as shown in *Fig. 8.1*, which is a network of secondary performance chains merging at certain stations into the time critical *main order chain*. Examples of performance trees are complex assembly lines for cars or machines and printing lines for newspapers.

In order to regulate the flow from the secondary chains into the main chain, additional *supply* and *replenishment strategies* are necessary.

10.1.5 Performance Networks

A performance and production network consists of a larger number of stations that can execute several different orders at the same time. Examples are *freight networks, airline networks, railway networks* and the *production networks* of car manufacturers (see *Fig. 1.15*).

In many cases the performance network consists of one or several main order chains and many secondary supply chains. If the secondary chains are decoupled by intermediate buffers, the main performance chains can be scheduled by an *order center*. It collects, clusters, sequences, divides and converts external orders into internal orders for the different stations of the main performance chains. For this purpose, suitable clustering, sequencing and allocation strategies are needed.

In order to keep the operation of a *dynamic network* simple and efficient, *self-regulating processes* should be organized. The performance chains and stations upstream of the *order penetration limit* shown in *Fig. 8.1* can be operated by the

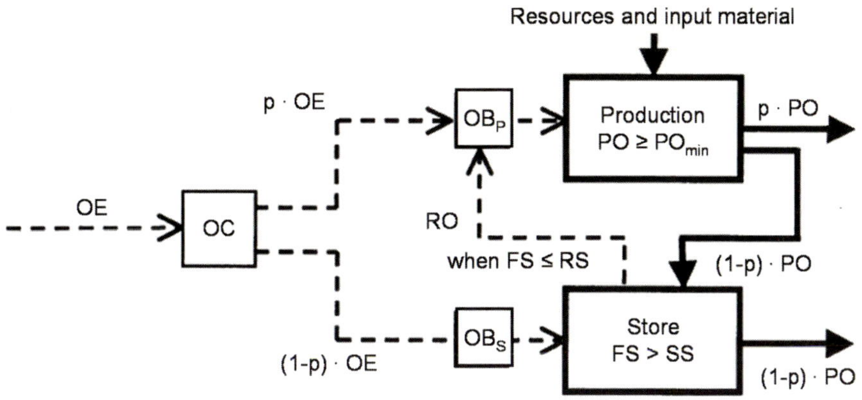

Fig. 10.2 Combined production and storage system

OE:	order entry	OC: order center	RO: replenishment orders
OB_P	production order buffer	OB_S	stock order buffer
PO	production output	PO_{min}	minimal production output
FS:	finished goods stock	RS	reorder stock level
SS	safety stock	p	share of direct production

pull-principle. They receive orders directly from the immediately following stations and generate internal orders, which are routed to the preceding stations. If several orders chains are available for the same order, *selection strategies* and *allocation strategies* are needed. In addition, *coordination* and *synchronization strategies* for local stations and secondary chains are necessary, if their cooperation is not regulated by the pull-principle.

10.1.6 Combined Production and Storage Systems

Combined production and storage systems consist of a production station and a finished goods store as shown in *Fig. 10.2*. They are quite common. Food, beverage, chemicals and drugs are bottled and packed to stock or to order, depending on the current demand. Utility goods, like automobiles, are either assembled on order or to stock. For example, in *USA*, cars are mainly produced to stocks of dealers, whereas in *Germany* car production on customer orders dominates.

The combined production and storage system of *Fig. 10.2* is a two-stage feedback system with the two different order chains shown in *Fig. 3.7*. The *first order chain* runs from an order center via a production order buffer to the production station and from there through a finished goods buffer directly to shipment. The *second order chain* runs from the order center via a stock order buffer followed by a finished goods store, which executes the external orders. If the stock level falls below a certain *reorder stock*, an internal replenishment order is send to the production station.

Corresponding to the two order chains exist two logistic chains: The *first logistic chain* is the main production chain, which receives pre-finished goods, material and resources from secondary supply chains. It ends with direct shipment to the customer. The *second logistic chain* also includes the main production chain, but from there goes to a finished goods store. After a storing time of varying length the goods are shipped due to the customer orders.

10.1.7 Production Strategies and Operating Modes

Basic *production strategies* for a combined production and storage system are:

- *Order-production* or *make-to-order*: All incoming orders go directly to the production, which only executes external orders. With exception of a short term buffer exists no stock of finished goods.
- *Stock-production* or *make-to-stock*: All incoming orders go to the *finished goods store*, where they are executed. The production operates only for internal replenishment orders.
- *Stock-order-production* or *make-to-stock and to order*: A share p [%] of the external orders is executed by the production, the remaining share (1−p) [%] by the finished goods store.

By the *strategy variable* p certain objectives are achievable, for instance a *steady utilization* of production capacity.

These basic production strategies can be combined with the following *operating modes* (Naylor et al. 1997):

- *Continuous production*: If technology forbids interruption or when demand exceeds the limit performance the production operates continuously.
- *Discontinuous production*: If the demand is lower than the limit performance and interruptions are technically tolerable, the production can operate discontinuously.

For a discontinuous production, start and endpoints of the order execution must be scheduled. Possibilities are *immediate*, *scheduled* or *advanced execution*.

The basic production strategies and operation modes, the strategy variables and their qualitative effects on lead times and stock levels are listed in *Table 10.1*.

Table 10.1 Effects of basic production strategies and operation modes

Production Strategy strategy variable	**Lead Times**	**Finished Goods Stock**
Order-Production	− to ++	++
order threshold	dependent on demand	without stock
Stock-Production	++	− to o
economic lot size	independent of utilization	dependent on demand
Continuous Production	− to ++	− to ++
running times	dependent on demand	dependent on demand
Discontinuous Production	++	− to ++
start dates	dependent on utilization	dependent on demand

Optimal achievement of the objective: ++
Satisfactory achievement of the objective: +
No effect on the objective: o
Negative influence on the objective: −

10.2 Processing Strategies

After examining the correctness and completeness of order content and requirements, *order preparation* starts with bundling and/or separating the external orders into internal orders, which can be executed in a closed process. If the execution concerns several secondary chains, the order is separated according to the *bill of material* into suborders for pre-products, components, parts and modules, which must flow from secondary chains *just in time* into the main chain.

Table 10.2 Effects of different processing strategies

Processing Strategy strategy variables	**Process Costs**	**Lead Times**	**Punctuality**
Single Processing none	–	++	–
Batch Processing batch size	++	–	+
Complete Execution none	–	++	+
Partial Execution partitions	++	–	+
Immediate Execution none	–	+	–
Scheduled Execution start date	+	o	+
Advanced Execution start date, lot size	+	++	++
Parallel Execution parallel stations, partitions	–	++	+

Symbols: see Table 10.1

The preparation of the external orders and the separation according to the bill of materials results in *internal orders*, which are transferred to the respective performance chains and stations. There, they are allocated to the available resources and executed following a certain *processing strategy*. *Table 10.2* contains the possible processing strategies and their qualitative effects on costs, lead times and punctuality (Churchman 1961; Schulte 1995).

10.2.1 Single Processing and Batch Processing

Single processing is the simplest strategy. The orders are allocated to a performance station separately and executed independently. Advantages of single processing are short throughput times, the possibility of customer specific production sequences and the immediate execution of *rush orders*. These advantages face the following disadvantages, which generally cause higher costs:

- longer setup times, dead times and walking times
- lower punctuality due to the higher fluctuations of lead times
- inefficient utilization of resources
- lower filling degrees of transport means and load units

If short delivery times are required or if the order buffer does not contain several orders with the same process, single processing is unavoidable.

For *batch processing*, several orders with the same process are bundled and executed as one *batch order*. The *batch size* c_B is called in production *lot size* and in logistics *bulk length* or *series size*. It is a *strategy variable* which can be used to minimize costs. If the setup costs are high, the lot size must exceed a certain minimum, in order to ensure an efficient production.

Batch processing offers several advantages, which increase with the batch size:

- low proportionate setup times, dead times or walking times
- higher performance rates
- high punctuality due to smaller fluctuations of lead times
- better utilization of the resources
- higher filling degrees of load carriers and transport means

The cost savings by these effects are partly compensated by the disadvantages:

- additional organizational efforts
- longer lead times
- limited possibilities to handle rush orders
- final separation and sorting of the results for the external orders

By adjusting the batch size and by sequencing the single orders within the batch as well as of successive batch orders, the required delivery dates can be realized at minimal costs (see *Chap. 20*). In the extreme, the optimal batch length is either $c_B = 1$, which means *single processing*, or it becomes $c_B = \infty$, which means *continuous production*.

10.2.2 Complete Order Execution or Partial Order Execution

Complete order execution means that each order is completely executed in one uninterrupted process. Complete execution is possible without additional organizational effort. The order result is generated completely. A further advantage is that no intermediate storing of part order results is necessary.

In some cases, complete order execution is forced by technology or required by the customer. The complete execution of big orders occupies the performance stations for a long time and may cause extended working times or weekend shifts. After a big order has been finished, a certain *rest time* is left free at the end of a shift or working day, which may be insufficient to complete other orders. These effects cause additional costs.

The complete execution of big orders can be combined with two different *delivery strategies*:

- *Closed dispatch*: The order is executed completely and the whole order quantity is shipped after the order has been finished.
- *Continuous dispatch*: The continuously produced order quantity is filled into load or transport units and shipped immediately after a unit is filled.

Continuous dispatch enables effectively short replenishment times, which are independent of the order size. For high demand this procedure leads to smaller *economic lot sizes* and lower *stocks* (see *Sect. 20.3*). It allows *continuous replenishment* (CRP) and helps to avoid the so-called *bullwhip effect* (Forrester 1961; Gudehus 2006; Lee et al. 1997).

Partial order execution results from separation of a big order into several part orders. *Strategy variables* of partial order execution are the *part order quantities* m_r. They are limited by the minimal and the maximal tolerable production lot size. Partial order execution offers the following *options* and *advantages*:

- preference and insertion of rush orders
- minimal time losses at the end of shifts
- optimization of utilization
- optimal sequencing and mixing with other orders
- maximal filling degrees of load units and transport means

Disadvantages of partial order execution are:

- additional organizational effort
- intermediate buffer space for part order quantities
- final consolidation of the part orders
- longer order lead times if not executed on parallel stations

Often big orders can only be executed in parts, e.g. if the total order quantity exceeds the transport capacity or if it cannot be executed by one station until the required delivery date.

10.2.3 Immediate, Scheduled and Advanced Order Execution

Immediate order execution means that a received single or batch order is executed by the respective performance or production station as soon as capacity is available. Immediate execution implies *first-come-first-go*.

If the demand approaches the limit performance, *queues* pile up in front of the performance and production stations. Their length depends on the fluctuation of the incoming orders and of the process times (see *Sect. 13.5*). The stochastically fluctuating waiting times affect *punctuality*. As long as the limit performance exceeds the demand significantly and the stochastic fluctuations are small, immediate execution enables short delivery times. In order to keep the effects of bundling strategies, immediate order execution should be limited to 5% of all orders.

Scheduled order execution means that the incoming single and batch orders are executed due to one of the *time strategies* of *Sects. 8.7 to 8.9*:

forward scheduling
backward scheduling
free scheduling (10.2)
just-in-time execution
bottleneck scheduling

Just like immediate order execution, just-in-time execution completely determines the sequence of the order execution.

Not all time strategies are compatible with all production strategies. Under certain conditions, forward and backward scheduling can be partly combined with other sequencing strategies. With bottleneck scheduling in combination with other strategies high *flexibility* is achievable.

Advanced order execution means that orders are executed long before delivery date or without external orders. Examples are the execution of transport orders or deliveries before due date. Another kind of advanced execution is *make-to-stock*. Here, the internal orders are anonymous replenishment orders (see *Chap. 11*). As explained in *Sect. 20.4*, the advanced execution of anonymous stock orders is a strategy to avoid *bottleneck phases*.

Goals of advanced order execution are:

optimal utilization
minimal setup and process costs
high delivery ability (10.3)
short and reliable delivery times
avoidance of bottlenecks

These goals are partly incompatible and not achievable simultaneously. Strategy variables of advanced order execution are *starting date*, *order quantity* and *replenishment time* or *lead time*.

10.3 Allocation Strategies

Allocation strategies assign the single orders to the actually available process chains and parallel stations. Goals are:

steady utilization of capacities
maximal utilization of stations
minimal order buffers
minimal waiting times (10.4)
short lead times
lowest stock levels
minimal process costs

In order to achieve these goals, it is necessary to permanently control utilization, order and material buffers, lead times and costs.

Special allocation strategies of logistics are presented in the *Sects. 10.5* and *13.3*. An allocation strategy connected with partial order execution is:

- *Parallel part-order execution*: The part-orders of a separated big order are executed in parallel by the available performance stations or chains.

Strategy variable of parallel part-order execution is the number of employed stations or chains.

Main advantage of parallel execution is the reduction of the lead time, in particular for big orders. In addition, considerable cost savings are possible by this strategy. A disadvantage of parallel execution is the additional effort caused by the distribution of part-orders to parallel stations, coordination and synchronization of their execution, additional setup costs, collection of the resulting parts and buffering them until the last part is finished.

Parallel order execution is well known in production, as well as in logistics. It is extremely useful for very big orders and projects. It is applied also in informatics for high-performance data processing, large calculation tasks or big sorting orders which are executed by parallel computers.

10.4 Sequencing Strategies

As long as several orders are waiting in the order buffer, by applying *sequencing strategies* costs can be reduced, performances be increased, delivery times be shortened and other objectives be achieved. *Strategy variables* are number, sizes and sequences of the *sorted order clusters* (see *Sect. 5.2*) (Müller-Merbach 1970; Schulte 1995; Gudehus 2003/2007).

An example is the so called *pearl-string* of cars with different colors, features and equipment, on the final assembly line of the *automobile industry*. According to the sequence in the pearl-string, the modules and parts for the individual cars are supplied *just-in-sequence* (JIS) and *just-in-time* (JIT).

10.4.1 First-In-First-Out and First-Come-First-Go

With the *first-in-first-out principle* (FIFO), the sequence of the *completion dates* is equal to the sequence of the arrival dates. With the *first-come-first-go principle* (FIGO), the sequence of the *start dates* determines the sequence of the arrival dates.

For a single station and for equal order process times, both strategies are identical. If the process times differ and the orders are executed in parallel, FIFO and FIGO can lead to different sequences.

10.4.2 Urgency Sequencing

The order buffer is segmented due to the urgency of the orders into two or more *priority categories*. Urgent orders are executed before less urgent orders, rush orders before normal orders. In the extreme, each order has an externally required completion date that determines the sequence of the start dates.

10.4.3 Lead Time Sequencing

The orders of the order buffer are sorted after ascending or descending lead times and executed in the resulting sequence.

Table 10.3 Effects of sequencing strategies

Sequencing Strategy strategy variables	**Process Costs**	**Lead Times**	**Punctuality**
First-In-First-Out order completion	–	+	+
First-Come-First-Go order arrival	–	++	+
Urgency Sequencing urgency classes	–	+	++
Lead Time Sequencing as or descending process time	o	+	o
Setup-Cost Sequencing order sequence	++	–	o
Setup-Time Sequencing order sequence	+	+	+
Value Sequencing in or decreasing values	+	–	o
Quantity Sequencing in or decreasing quantities	+	–	o

Symbols: see Table 10.1

10.4.4 Setup Sequencing

The orders are sorted and executed in the sequence with the minimal sum of *setup times* or *setup costs* for the single orders.

For example, the orders of a bottling company, printing company or dyeing mill are executed in a *bright-dark-sequence*. The dark colors follow bright colors in order to minimize *cleaning times* and *maculation.*

10.4.5 Value Sequencing

Orders with high value are executed before orders of low value, or in reverse sequence, low value orders are executed before high value orders.

Make-to-order in decreasing value sequence improves liquidity and saves interest, if the finished orders can be invoiced immediately.

For replenishment orders or orders which can only be invoiced at a later date, the execution in increasing order value reduces inventory capital and interest.

10.4.6 Quantity Sequencing

Orders with big quantities or large number of positions are executed before orders with small quantities or number of positions. Under certain conditions, also the execution in sequences of increasing quantities is opportune.

10.4.7 Effects of Sequencing

The results of a qualitative evaluation of the effects of the various sequencing strategies are shown in *Table 10.3*. The effects of sequencing on throughput and queues are quantified with the help of the *limit performance and queuing laws* of *Sects. 13.4* and *13.6*.

Costs reduction, short delivery times and punctuality can not always be achieved simultaneously by the sequencing strategies: FIGO and urgency sequencing focus on delivery times and punctuality. Setup cost sequencing aims at cost minimization. Setup time sequencing helps to reduce costs and delivery times. Value sequencing can decrease interest costs and improve liquidity.

The effects of sequencing strategies differ greatly. Before implementation, they should be carefully analyzed and, if possible, quantified, since for some strategies, such as time or quantity based sequencing, the effects are rather small.

10.5 Order Production and Stock Production

The decision between make to stock and make to order is generally based on tradition, experience or qualitative considerations and not on objective criteria and systematic analysis. If decision criteria exist, they are applied only to single products or single orders. By the following *algorithm* the effects of *order-production, stock-production* and combined *stock-order-production* can be calculated. The algorithm has been developed for the production and storage system shown in *Fig. 10.2* and first applied in the *automotive industry*. Later it was adapted and applied in the *beverage industry, chemical industry* and *tobacco industry*. An extension to complex production trees and networks is possible.

10.5.1 Order Entry

The decision between order-production and stock-production depends critically on the *order entry rate*

$$\mathrm{OE}(t) \quad [\mathrm{PU/PE}] \tag{10.5}$$

for the required *production units* [PU] per *period* PE. The start dates of the periods t with length $T_{PE}(t)$ are given by the partition

$$\mathrm{SD}_t = [t_o + (t-1)\cdot T_{PE};\, t_o + t\cdot T_{PE}] \quad \text{for} \quad t = 0, 1, 2, \ldots, N_{PE} \tag{10.6}$$

of a considered time span of N_{PE} periods. As outlined in *Sect. 8.2*, the length of the scheduling periods can be an hour, a work day or a week. For the following model calculations, the considered time span is a business year with 250 work days [WD]. The production units are single cars with an order entry (10.5) generated by the *standard test function* (9.48) as shown in *Fig. 9.10*.

10.5.2 Production Capacity and Limit Performance

The *production capacity* $PC = PO_{max}$ [PU/PE] is the maximal possible output of production units per period. It is given by the product of the *maximal operating time* T_{OPmax} [h/PE] per period with the *limit performance* μ_P [PU/h] of the production station:

$$PO_{max} = \mu_P \cdot T_{OPmax} \cdot \quad [\mathrm{PU/PE}] \tag{10.7}$$

A *fix-production* operates independent of the current demand due to a *production plan* with *scheduled operating times* $T_{OP}(t)$. It yields in period t the *production output*:

$$PO(t) = \mu_P \cdot T_{OP}(t). \qquad [PU/PE] \qquad (10.8)$$

In a *flexible production* or a *breathing factory* the production capacity is adjusted to the current *production demand* $PD(t)$ by *flexible operating times*:

$$T_{OP}(t) = PD(t)/\mu_P \cdot \qquad [h/PE] \qquad (10.9)$$

Flexible operating times are achievable as far as compatible with employment laws and process technology by overtime or short time work, variable shift length or multiple shift operation. In the best case, they yield a production output, which is precisely equal to the actual demand, i.e. $PO(t) = PD(t)$.

In order to ensure an efficient production, in most cases a minimal operation time T_{OPmin} [h/PE] must be kept. This leads to a *production threshold* with *minimal production output*:

$$PO_{min} = \mu_P \cdot T_{OPmin} \cdot \qquad [PU/PE] \qquad (10.10)$$

A restart of production after interruption is only efficient when the order buffer has reached a certain *order threshold* or *minimal order buffer* OB_{min} [PU]. Otherwise, the proportionate setup and restart costs are too high.

The production is adjustable to the actual demand until a certain *reaction time* or *freezing time* t_R [PE] before the production starts at period $t + t_R$. When the freezing time has passed, the scheduled sequence of the production cannot be changed any more.

The considered model factory for cars can reach in three 8 h shifts a maximal output of PO_{max} = 750 cars/WD. The reaction time is t_R =1 WD. The production threshold is an 8 h shift operation with minimal output of PO_{min} = 250 cars/WD. The production starts when the order buffer has reached a minimum of OS_{min} = 2,500 cars, which ensures a 1 week two shift operation.

10.5.3 Order Production

If the order buffer at the beginning of a period t is $OB_P(t-1)$ and the current production order entry is $OE_P(t)$, a production with output $PO(t)$ leads to the

- current *production order buffer* at the end of a period

$$OB_P(t) = MAX(0; OB_P(t-1) + OE_P(t) - PO(t)) \qquad [PU]. \qquad (10.11)$$

For a production plan with *scheduled operating times*, the current production output is given by relation (10.8). With *flexible operating times* (10.9) the current production output is kept at maximal output PO_{max}, as long as $t - t_R$ periods ago the order buffer $OB_P(t - t_R)$ was higher than PO_{max}, and is reduced to the order buffer $OB_P(t-t_R)$ as long as it exceeds the minimal output PO_{min}. If the order buffer drops below the minimal output, the production is stopped. It starts again, when the order buffer has reached the *order threshold* OB_{min}. This leads to the algorithm for the

- current *production output* of a flexible production

$$\mathrm{PO}(t) = \begin{cases} \mathrm{MIN}(\mathrm{PO}_{\mathrm{max}};\ \mathrm{OB}_{\mathrm{P}}(t - t_R)) & \text{if } \mathrm{OB}(t - t_R - 1) > \mathrm{OB}(t - t_R) > \mathrm{PO}_{\mathrm{min}} \\ 0 & \text{if } \mathrm{OB}(t - t_R - 1) < \mathrm{OB}(t - t_R) < \mathrm{OB}_{\mathrm{min}} \end{cases} \tag{10.12}$$

For a flexible production, the production lead time LT(t) in period t equals the reaction time t_R plus the time needed to execute the actual order buffer (10.11). If the production has been interrupted, the lead time increases by the waiting time until the order entry OE(t) has accumulated the minimal order buffer. This results in the

- current *lead time* of a flexible production

$$\mathrm{LT}_{\mathrm{P}}(t) = \begin{cases} t_{\mathrm{R}} + \mathrm{OB}_{\mathrm{P}}(t)/\mathrm{PO}_{\mathrm{max}} & \text{if } \mathrm{PR}(t) > 0 \\ t_{\mathrm{R}} + \mathrm{OB}_{\mathrm{P}}(t)/\mathrm{PO}_{\mathrm{max}} + \mathrm{OB}_{\mathrm{min}}/\mathrm{OE}(t) & \text{if } \mathrm{PR}(t) = 0 \end{cases} \tag{10.13}$$

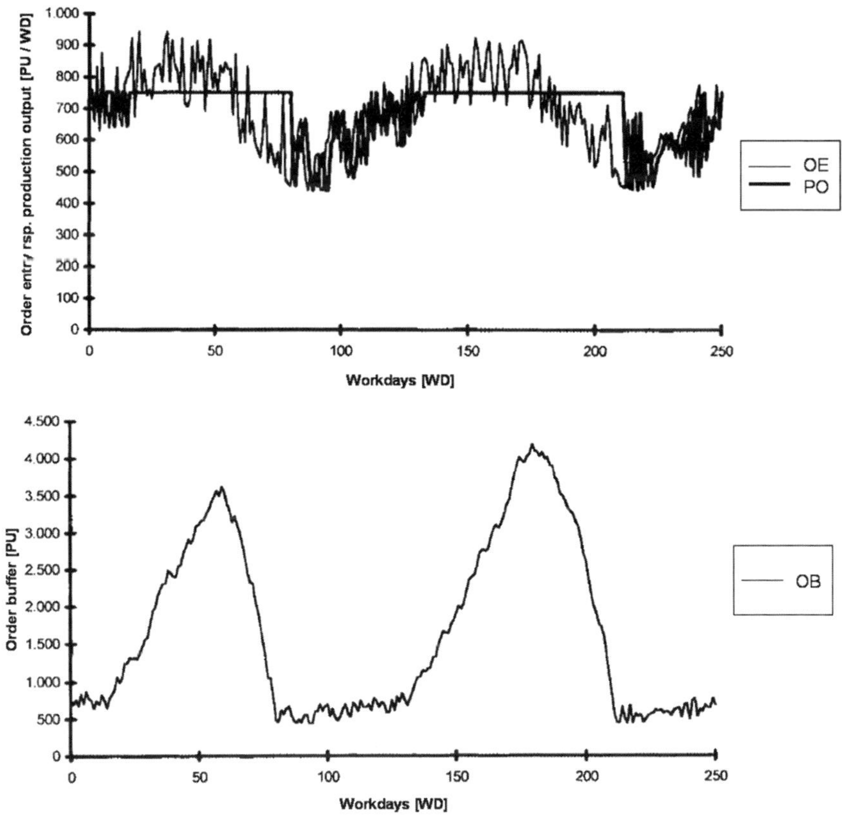

Fig. 10.3 Production output and order buffer for a continuous flexible order production of 175,000 cars per year

OE(t):	order entry	PO(t):	production output
$\mathrm{PO}_{\mathrm{min}}$	250 cars/WD	$\mathrm{PO}_{\mathrm{max}} =$	750 cars/WD
OB(t):	order buffer	$\mathrm{OB}_{\mathrm{min}} =$	2,500 cars
t_{R}:	reaction time 1 WD	average OE:	700 cars/WD

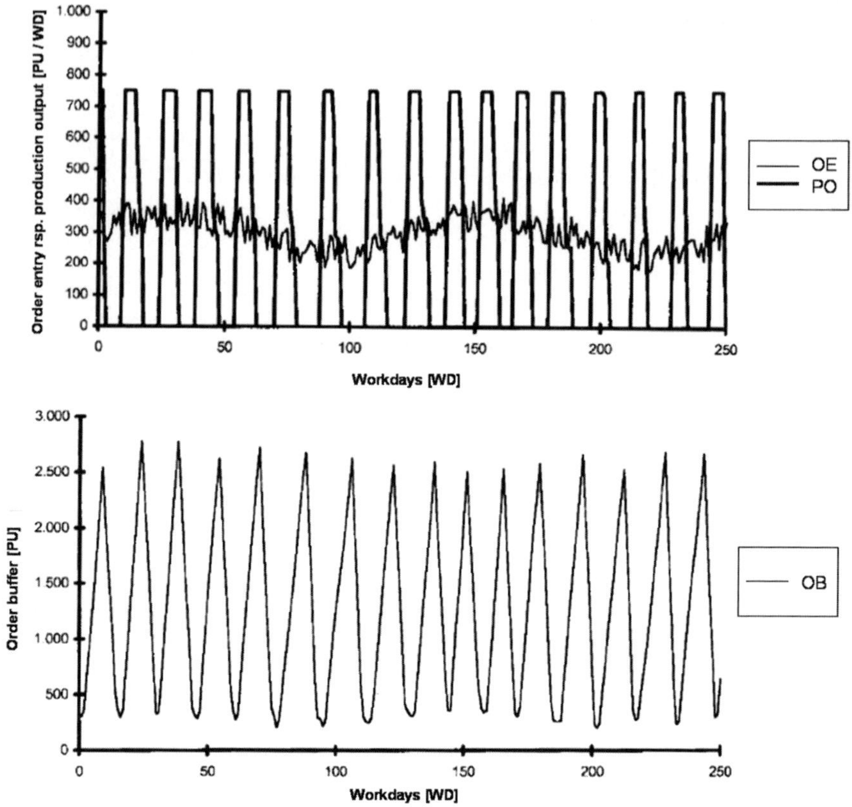

Fig. 10.4 Production output and order buffer for a discontinuous flexible order production of 62,500 cars per year

OE(t):	order entry	PO(t):	production output
PO_{min}	250 cars/WD	PO_{max} =	750 cars/WD
OB(t):	order buffer	OB_{min} =	2,500 cars
t_R:	reaction time 1 WD	average OE:	300 cars/WD

Figure 10.3 shows the results of model calculations for the considered car factory with the above algorithm for a mean demand of 700 cars/WD and *Fig. 10.4* for a mean demand of 250 cars/WD. The upper diagrams present the production output (10.12) and the lower diagrams the order buffer (10.11).

The mean demand of 700 cars/WD of *Fig. 10.3* is only 7% less than the maximal output and so high that even in weaker seasons the order buffer does not fall below the minimal output. Therefore, *continuous production* at full capacity runs from the 20th to the 75th workday and from the 130th to the 223rd workday. In these *phases of over-demand* the order buffer increases up to more than 4,000 cars. In the times of lower demand, the order buffer decreases after the reaction time of one day down to the daily capacity of 750 cars. The *lead times* calculated by relation (10.13) reach the minimum of 2 WD in times of lowest demand and increase up to a maximum of 9 WD in peak times. The annual mean value of the lead times is 6.8 WD. This shows, that the lead times become longer with increasing and shorter with decreasing capacity utilization.

In *Fig. 10.4* the mean demand of 250 cars/WD is far lower than the maximal production capacity of 750 cars/WD. The order buffer repeatedly falls below the minimal production output of 250 cars/WD and the production is interrupted until the order buffer has reached the order threshold of 2,500 cars. The result is a *discontinuous production.*

The *lead times* are minimal 2 WD, maximal 8 WD and in average 3.9 WD. The mean and the maximal lead times of the order production depend on the order threshold OB_{min}, which also determines the length and the lot size of the production phases and the total production costs. That means:

- ▶ The *strategy variable of order production* is the *order threshold*, i.e. the number of orders, which must accumulate to start the production.

The order threshold can be used to minimize the costs at competitive and reliable lead times. The minimal lead time is the sum of the reaction time and the order minimal processing time.

10.5.4 Stock Production

Shortest delivery times are achievable with stock production and delivery ex stock. With this strategy, the production demand PD(t) results from the *replenishment orders* of the finished goods stock, which are determined by the *replenishment strategy*. For optimal inventory management as outlined in *Chap. 11* holds:

- ▶ *Strategy variables of stock production* are *safety stocks* and *replenishment quantities.*

They can be used to minimize the total costs. The safety stock is necessary to ensure a required *stock availability*. The *replenishment quantity* which achieves lowest costs is the *economic order quantity* (EOQ).

The latest date for releasing a replenishment order is reached when the stock is just sufficient for the expected demand until the first supply will arrive in store. This *reorder stock* RS is the sum of the *safety stock* SS and the demand during the *replenishment time* T_{repl}, which is the product $OE_S(t)\cdot T_{repl}$ of *storage order entry* $OE_S(t)$ and replenishment time. The replenishment time is the sum of the reaction time t_R, the production time and the transport time t_{Tr} to the store place.

With *closed dispatch*, the production time for the total replenishment quantity RQ is RQ/PO_{max}. With *continuous dispatch*, the replenishment time is the time until the first unit has been produced and transported to the stock, i.e. the sum of reaction time and transport time. This leads to the

- current *reorder stock*

$$\mathrm{RS}(t) = \begin{cases} \mathrm{SS} + (t_R + t_{Tr} + \mathrm{RQ}/\mathrm{PO}_{max}) \cdot \mathrm{OE_S}(t) & \text{for closed dispatch} \\ \mathrm{SS} + (t_R + t_{Tr}) \cdot \mathrm{OE_S}(t) & \text{for continuous dispatch.} \end{cases} \tag{10.14}$$

The replenishment orders are triggered when the *finished goods stock* FS(t) has reached the reorder stock, i.e. when FS(t) = RS(t). The production starts delayed by the reaction time t_R and runs at maximal capacity PO_{max} until the complete replenishment quantity RQ has been produced. This causes the

- current *production order buffer* for stock production

$$OB_P(t) = \begin{cases} 0 & \text{if } FS(t) > RS(t) \\ RQ & \text{if } FS(t - t_R) = RS(t - t_R) \\ OB_P(t-1) - PO_{max} & \text{if } FS(t - t_R) < RS(t - t_R) \end{cases} \quad (10.15)$$

If operating with *closed dispatch* the complete replenishment quantity is sent to the finished goods store at the end of the production phase leading to the

- current *finished goods stock* for closed dispatch

$$FS(t) = \begin{cases} SS + RQ & \text{if } FS(t - t_R - t_{Tr} - RQ/PO_{max}) \\ & \quad = RS(t - t_R - t_{Tr} - RQ/PO_{max}) \\ FS(t-1) - OE_S(t) & \text{if } FS(t - t_R - t_{Tr} - RQ/PO_{max}) \\ & \quad \neq RS(t - t_R - t_{Tr} - RQ/PO_{max}). \end{cases} \quad (10.16)$$

In addition to the stock in the finished goods store (10.16) during the production time $t_{RQ} = RQ/PO_{max}$ of the replenishment quantity RQ, a *temporary buffer stock* of finished good piles up in the exit of production.

For a stock production with closed dispatch, the upper diagram of *Fig. 10.5* shows the order entrance and the resulting production output for the example of the automotive industry. The lower diagram presents the time dependency of the finished goods stock without the temporary buffer in the exit of production. In this case, the typical *saw-tooth curve* has vertical flanks, which are caused by the arrival of the closed replenishment quantity, and decreasing slopes, which are generated by the continuous order entry for single units.

If operating with *continuous dispatch*, the produced units are sent out immediately or daily to the finished goods store leading to the

- current *finished goods stock* for continuous dispatch

$$FS(t) = \begin{cases} SS + (t - t_R - t_{Tr}) \cdot PO_{max} - OE_S(t) & \text{if } FS(t - t_R - t_{Tr}) \\ & \quad = RS(t - t_R - t_{Tr}) \\ FS(t-1) - OE_S(t) & \text{if } FS(t - t_R - t_{Tr}) \\ & \quad \neq RS(t - t_R - t_{Tr}). \end{cases} \quad (10.17)$$

In this case, no additional buffer of finished goods piles up in the production exit. For the same order entrance as before, the calculated production output and finished goods stock are shown in *Fig. 10.6* for a stock production with continuous dispatch. Again, the time dependency of the stock is a saw-tooth curve, however, due to the continuous supply, the flanks increase after arrival of the first replenishment unit and decrease after the last replenishment has arrived.

The comparison of the algorithm and the results of the model calculations for the two dispatch strategies lead to the following *dispatch rules*:

- Due to the shorter effective replenishment time of continuous dispatch the production starts at a lower reorder stock than for complete dispatch.

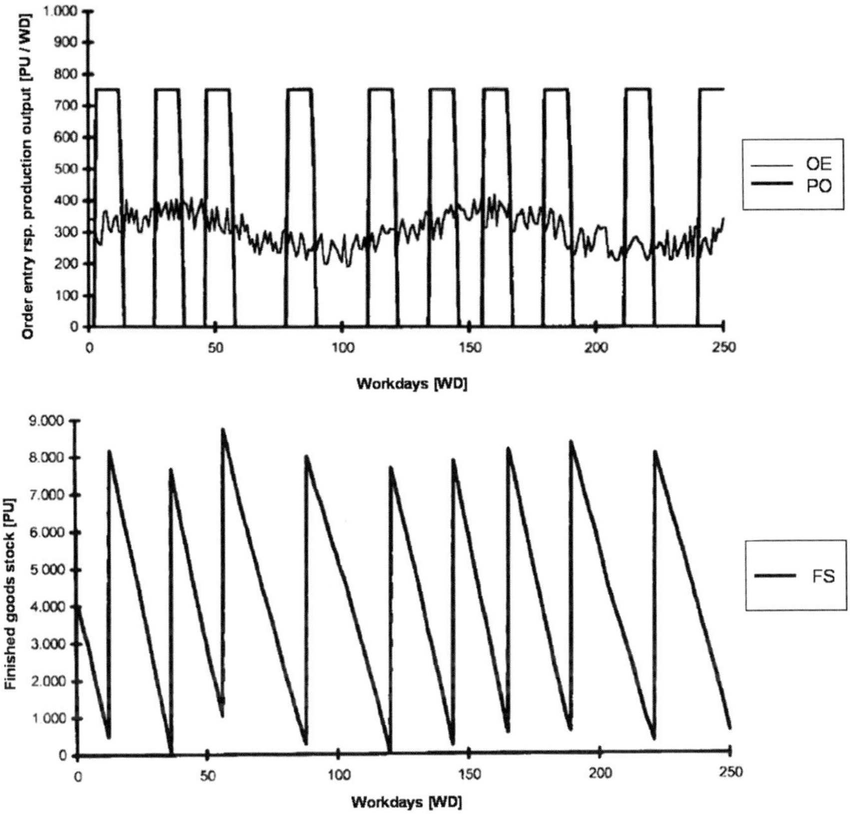

Fig. 10.5 Production output and finished goods stock for stock production with closed dispatch for a demand of 75,000 cars per year

OE(t): order entry	PO(t): production output
PO_{min} 250 cars/WD	PO_{max} = 750 cars/WD
FS(t): finished goods stock	RQ = 8,000 cars
t_R+t_{Tr}: 1 WD	average OE: 300 cars/WD

- The mean stock level for continuous replenishment is lower than for closed replenishment.
- The difference between closed and continuous replenishment diminishes with decreasing replenishment quantity and disappears, if the total quantity are produced in less than a day.

Strategy parameters for the minimization of the transport costs are the capacities of the load units and the transport means (see *Chap. 11*). The transport from production to stock is most effective in full load units, which are transported by completely filled transport means. This leads to the

► *Shipment Rule for Stock Replenishments*: The costs for stock production with high demand are minimal, if the replenishment quantities are integer multiples of the load units and of the transport capacity and are transferred in full load units and transport units immediately from production to the store.

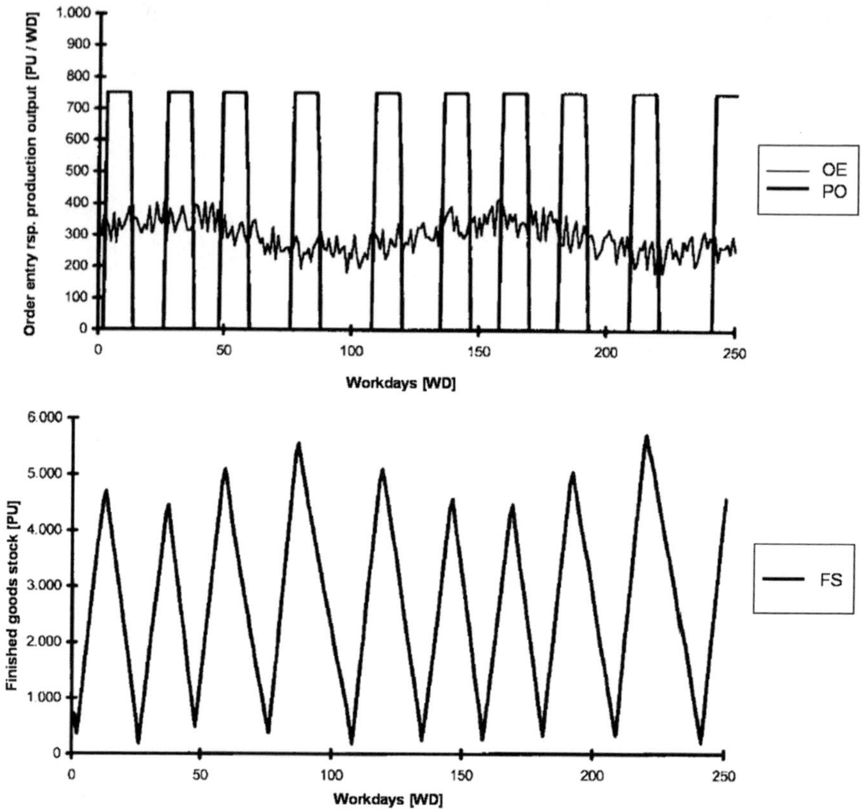

Fig. 10.6 Production output and finished goods stock for stock production with continuous dispatch for a demand of 75,000 cars per year

Parameters: see Fig. 10.4

The delivery times of stock production are the storing times, which are less than a day, as long as the stock is sufficient to serve the incoming orders. That means:

▸ Stock production enables minimal delivery times, if the production output is sufficient and the safety stock is adequate.

If the production is not capable of fulfilling the demand or if the consumption during the replenishment time exceeds the reorder stock level, also with stock production piles up a

- *stock order buffer* at the end of a period

$$OB_S(t) = MAX(0;\ OB_t(t-1) + OE_S(t) - FS(t)) \qquad [PU]. \qquad (10.18)$$

The price for the short delivery times of stock production compared to order production are the additional costs for the transports from production to store, for in- and out-storing, for occupying storeplaces and for the interest and risks of the inventory. Up to a certain level of the demand, the additional costs of storekeeping can be overcompensated by the savings that result from the *economic order quantities* of the production. The *opportunity limit* between stock production and order production will be derived in *Sect. 11.12*.

10.5.5 Combined Stock-Order Production

For a combined stock-order production the external orders are allocated according to specific *selection criteria* directly to production or to the finished goods store. Also for this case the stock order buffer is given by (10.18), whereas the *production order buffer* is:

$$OB_P(t) = MAX((0; OB_P(t-1)+OE_P(t)-PO(t)+IF(FS(t) < RS; RS; 0)). \quad (10.19)$$

In practice different selection criteria are possible. They can be evaluated either by simulation using the formulas (10.7), (10.8) and (10.19), or by analysis of the opportunity limits as outlined in *Sect. 11.12*.

The number of influencing factors, such as size, seasonality or stochastic fluctuation of the order entry, the number of strategy parameters, such as replenishment quantities, production capacity or reaction time, and the number of options, such as the shares p of make-to-order and 1-p of make-to-stock, is very large. The evaluation and discussion of all these issues would go beyond the scope of this book. Interested readers can develop a program using the given algorithm and formula and perform own calculations. Results of such calculations are the diagrams *Fig. 10.3*, *10.4*, *10.5*, *10.6* and *10.7*.

Figure 10.7 shows the results of a simulation for a combined stock-order-production with constant production output, which is equal to the mean order entry rate. External orders are executed from finished goods store, as long as stock is available. If the stock is depleted, external orders are passed directly to production and delivered from there. During times where the order entry is lower than the production capacity, stock piles up. When order entry exceeds production capacity, the stock is stepwise reduced. When the stock has reached 0 an order buffer piles up, that will be reduced as soon as the order entry drops below the production capacity.

The simulations result in the following *production scheduling rules*:

- Big orders are produced more efficiently to order whereas smaller orders are more efficiently delivered ex stock.
- Small rush orders should be delivered completely ex stock, big express orders partly ex stock and partly from production.
- In phases of very high demand, standard orders are most efficiently produced on order and shipped directly from production.
- In phases of low demand, standard orders are more efficiently served and shipped from the finished goods stock.

In the *consumer goods industry*, big orders for *sales promotions* are loaded right after production to trailer skips that have been provided on time. They are then directly shipped via regional stores and transshipment stations to the outlets of the *retailers*, instead of being delivered via a central store (Kotzab 1997).

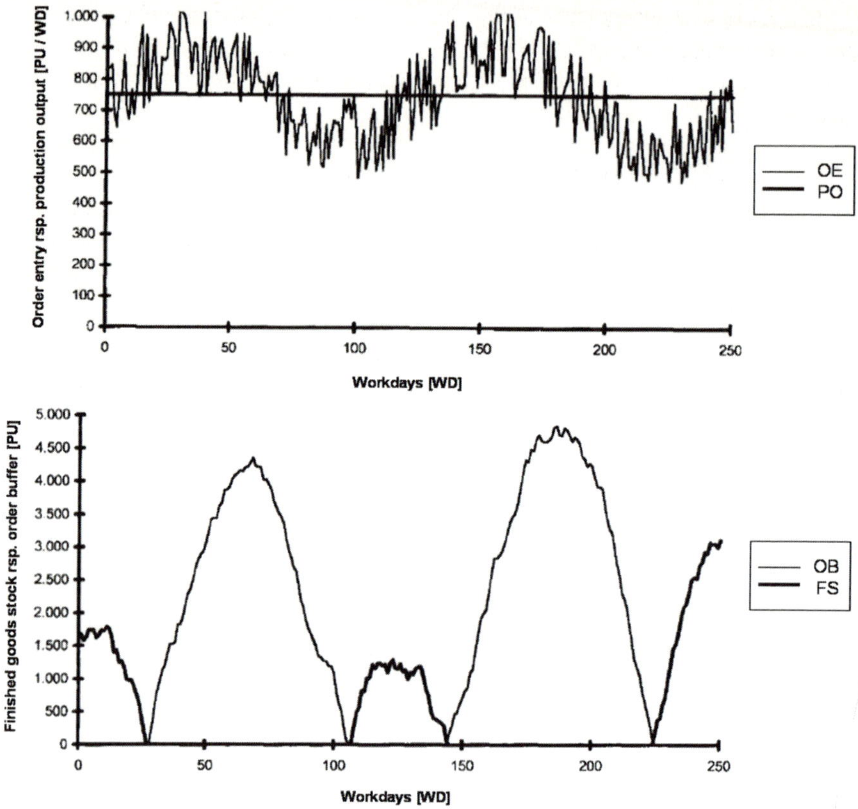

Fig. 10.7 Production output and finished goods stock for stock-order production with complete dispatch for a demand of 187,500 cars per year

OE(t): actual order entry PO = average OE = 750 cars/WD
FS(t): finished goods stock OB(t): order buffer

10.5.6 Governing Complexity

The analysis of the combined production and storage system of *Fig. 10.2* with only two coupled stations has shown how many influences and options must already be considered in this relatively simple case. With the number of coupled stations the possibilities of bundling and allocation increase more than exponentially (see *Figs. 5.2* and *5.3*).

The model calculations also show that the *stochastic simulation* of the relatively simple system is quite complex and does not lead to more insights than the *system analysis*. A simulation can reveal the effects of changes, conditions and influence factors. However, a simulation does not explain the causes of the effects. With the complexity of the system, also the complexity of the simulation model and of the results increases. Therefore, stochastic simulation of a complex system with many options and parameters alone is not sufficient to develop strategies.

The complexity of a *supply network* consisting of a large number of production, performance and storage stations becomes manageable by the *decoupling principle* of *Sect. 10.1* and the *subsidiary principle* of *Sect. 2.4*. Decoupled subsystems can independently schedule the orders they receive from adjoining systems, from an order center or directly from external customers. As long as no *bottlenecks* arise, low costs are achieved, which are close to the theoretical overall minimum if the subsystems aim at reaching the local optimum within their sphere of influence. However, which scheduling strategies are best suited for this aim is still an open question.

10.6 Dynamic Scheduling

To a great extent, conventional scheduling is static. It is generally performed at fixed dates in longer intervals. The basic strategies and their parameters are generally left unaltered for long time (Park and Narayan 1997).

In *processing industries*, like *chemical* or *steel industry*, and in *plant contracting companies* monthly scheduling is still normal. *Machine manufacturers*, producers of *consumer goods* and *wholesaler* generally schedule weekly. Market-oriented manufacturers, even the *automobile industry*, and *logistic service providers* schedule once a day. Some logistic companies, such as *freight forwarders*, *public transport companies* and *airlines*, even schedule on an hourly basis.

Newly incoming or not yet started orders are re-scheduled at the regular scheduling dates taking into account the availability of material and production and the most important changes of resources. However, in most companies the scheduling strategies, such as make-to-stock or make-to-order, as well as the strategy parameters, such as replenishment quantities and safety stocks, are left unchanged over a long time. In this respect, their scheduling remains static even if the scheduling periods are short.

The shorter the scheduling periods, the quicker is the *reaction* and the higher the *punctuality*. The better scheduling strategies and strategy variables are adapted to changing conditions, the higher is the *utilization* of resources and the better are delivery ability and competitiveness. This can be achieved by dynamic scheduling (Gudehus 2003/2007):

- *Dynamic Scheduling* (DS) is performed in short time intervals dependent on the required delivery times and punctuality with strategics and parameters which are adapted to the current situation.

Depending on the trigger for scheduling, a periodic-dynamic scheduling or a event-dynamic scheduling is possible:

- *Periodic-Dynamic Scheduling* (PDS) is performed periodically in short time intervals and takes into account all new and open orders, resource changes and other relevant events of the last period.
- *Event-Dynamic Scheduling* (EDS) is performed whenever a relevant event happens, such as an incoming order, a break down of a machine, a failing transport mean, a setup of a resource, a failure in the execution or a delay of supply.

Event-dynamic scheduling is most flexible. However, the organizational efforts and the time demand increase with the number of events. Therefore, pure event-based scheduling is generally not feasible, not even by high sophisticated computer programs. The goal conflict between *flexibility* and *feasibility* is resolved by combining periodic- and event-dynamic scheduling:

- ▶ Dynamic scheduling is regularly performed in short time intervals of suitable length and repeated when urgent, important or big orders are coming in, critical resources fail or other important events occur.

The most important features of dynamic scheduling, compared to conventional scheduling, are quick reaction to actual events and adaptation of strategies and parameters to changes. For this purpose adequate *adaptation strategies* must be developed.

Chapter 11
Inventory Management

Stocks are necessary to balance temporal deviations between demand and supply and between consumption and production. Buffer stocks enable high utilization by decoupling stations with deviating production and consumption rates. Safety stocks ensure availability when demand varies stochastically and when production or supply are temporarily interrupted or delayed. Supply costs are minimal if the right quantities of the right articles are stored at the right stages of the supply network.

However, inventory ties up capital, costs interest, needs space and is risky. That is why management often requires inventory reductions and high turnover rates without considering the effects on total costs and availability. In booming times on the other hand, stocks often exceed the necessary level. The inappropriate behavior and deficiencies which can be observed in practice result partly from a general lack of knowledge of the methods, strategies and effects of *inventory management* despite the barely manageable number of textbooks and publications on this subject (Chopra and Meindl 2007; Bellmann et al. 1995; Bogeschewsky 1995; Hadley and Whitin 1963; Harris 1913; Inderfurth 1994, 1999; Schneeweiß 1981; Silver et al. 1988; Tempelmeier 1995; Wöhe 2000; Zwehl 1979). Further reasons are incorrect recommendations and formulas of some publications and shortcomings of many scheduling programs (Dittrich et al. 2000; Gudehus 2003, 2007).

Other causes of poor inventory management are

- shared responsibility for storekeeping and costs
- lack of correct selection criteria for storekeeping articles
- excessive requirements for the stock availability
- insufficient, speculative or too optimistic demand forecasts
- unsuitable scheduling strategies and algorithms

The first three points can be avoided by the following *strategic measures for inventory management*:

- ▸ responsibility of the same person for storekeeping and costs
- ▸ optimal selection of storekeeping articles
- ▸ determination of stock availability according to necessity

T. Gudehus, H. Kotzab, *Comprehensive Logistics*,
DOI 10.1007/978-3-540-68652-1_11, © Springer-Verlag Berlin Heidelberg 2009

After deciding at what stage of the supply network which articles with what availability should be stored, and after determining the scheduling parameters, starts the *task of inventory scheduling*:

- Determination of the optimal *reorder points* and *replenishment quantities* in order to ensure the required *ex-stock availability* at *minimal costs*.

  The *steps of inventory scheduling* are:

1. Forecast of article *consumption* or *demand*
2. Calculation of the optimal *replenishment quantity*
3. Calculation of the *safety stock*
4. Determination of the *reorder point*
5. Check and release of *replenishment orders*

These steps can be executed cyclically, weekly, daily or event dependent, e.g. triggered by the incoming orders.

After an analysis of the *functions of stocks*, in this chapter different *replenishment strategies* for storekeeping articles will be developed. From the *logistic cost function* for the process of replenishment and storekeeping, a general formula for the *optimal replenishment quantity* or *economic order quantity* (EOQ) is derived. The *safety stock* is calculated for a given demand from the required *ex-stock availability*. Consequences of the general formulas for replenishment and safety stock are the *square root laws of logistics* and other applications. The chapter closes with an analysis of the *cost opportunity of storekeeping* and a presentation of *strategies for inventory optimization*.

11.1 Functions of Stocks

For inventory management, it is necessary to distinguish principally between the different functions of stocks presented in *Table 11.1*, which are *buffering*, *storing* and *keeping*. In practice, these functions are often mixed up and the same stock can have several functions. Therefore, the terms buffering, storing and keeping are often used synonymously and the transition from buffering to storing and from storing to keeping is sliding or blurred.

Buffering, storing and keeping are not limited to physical goods. *Orders* and *information* can also be buffered, stored and kept. As outlined in the previous chapter, the consolidated execution of collected orders and the selected processing of orders from an *order buffer* enable efficient utilization of capacities and minimization of production costs.

Two basic *strategies of order scheduling* are *procure-to-stock* or *make-to-stock* and *procure-to-order* or *make-to-order*. With make-to-stock the total costs are minimized by bundling many small delivery orders, which are executed from stock, into a small number of bigger replenishment orders, which are produced or procured in advance on stock. An additional strategy for storekeeping articles is, to split the incoming orders into *ex-stock orders*, which are delivered from stock, and *direct orders*, which are delivered directly after order-specific production or procurement (see *Sect. 11.12*).

Table 11.1 Functions of stocks

	Buffering	Storing	Keeping
Functions	**provision**	**keep ready**	**bridge time**
	for consumption	of merchandise	until production,
	production, service,	production factors	transport, delivery,
	control	finished goods	sorting, sales, use
Objectives	high utilization	required availability	optimal batches
	interruption protection	minimal costs	minimal costs
	minimal space	optimal availability	maximal
Demand	**permanent**	**permanent**	**temporarily**
Assortment	minimal	broad	small
Stock level	random variation small	random variation	constant
	mean value	sawtooth pattern	in/decreasing
Storage time	undetermined short	undetermined	predetermined
		medium to long	differing
Scheduling	**self-regulating**	**pull-principle**	**push-principle**
	Kanban/FlipFlop	demand determined	plan dependent
Influences on stock	variance of supply and	consumption	production plan,
level	consumption of supplier	replenishment	sales plan, loading,
	reliability	availability process	tours cycle times
		costs	

11.1.1 Buffering

Buffering is the provision of small quantities of an article for a performance or service station with stationary consumption, such as production, assembling or processing. Purpose of the buffer stock is to ensure *high utilization* of a station with stochastically fluctuating input and/or performance rates.

As shown in *Fig. 11.1*, the buffer stock varies during longer time randomly around a *mean stock level* m_S. The mean stock must be kept at such a level that the probability of an interruption of the supply is small. The waiting times of the articles in a buffer are generally short.

The suitable buffer stock can either be kept *self-regulating* or by *scheduling*. Examples of self regulating *buffers without scheduling* are *waiting queues* in front of a performance station or at the entrances of a *transport node* with random inflow and/or stochastically fluctuating performance rates. Due to the *queuing laws* of *Sect. 13.5*, the mean buffer stock results from the relation and variation of the arrival and performance rates.

Examples of *scheduled buffers* are

- *material buffers* in front of processing, production and working stations, which decouple supply and demand and reduce *down times* caused by lack of material
- *article buffers* in front of picking places in warehouses or in sales counters, and in the shelves of retail outlets, which ensure a required availability

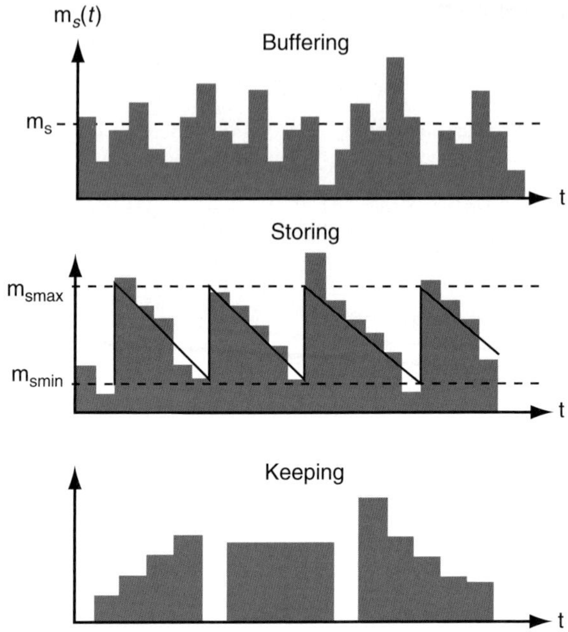

Fig. 11.1 Time dependency of stock for buffering, storing and keeping

$m_S(t)$	current stock
m_{Smax}	average maximal stock
m_{Smin}	average minimal stock or safety stock

The replenishmentof a scheduled buffer is driven by demand and similar to the replenishment of stores for continuous demand. It observes the *pull-principle* with the aim to keep a required availability within the available buffer space.

11.1.2 Storing

Storing is the scheduled keeping of article stocks in order to serve a longer lasting demand. Typical for storing is the *saw-tooth pattern* of the stock level of single articles in the course of time as shown in the *Figs. 10.5*, *10.6*, *11.1*, *11.4*, *11.20* and *11.21*. The stock falls stepwise from a maximal to a minimal value. When the *reorder point* is reached, a *replenishment quantity* is ordered to fill up the stock. The storage time of the single item is not predetermined. Due to the *law of large numbers*, the total stock of many articles fluctuates far less than the stocks of the single articles (see *Sect. 16.1.3*).

Goals of storing are:

- availability of the storekeeping articles with probability equal to the required *ex-stock availability*
- smoothing of random fluctuations of demand in order to optimize the utilization of limited production capacities

- minimization of the total logistic costs of a supply chain or within a supply network

Examples of stores with regular replenishment are:

- *production supply stores* for raw and auxiliary material and parts
- *intermediate stores* of components, parts and modules
- *finished goods, delivery* and *spare part stores*
- *central, regional* and *local stores*
- *consumption stocks* of households

The stock level is determined by the *replenishment strategy*, which depends on demand, replenishment time and costs of replenishment and storing.

11.1.3 Keeping

Predetermined quantities of articles, shipments or goods can be kept in a store for defined time. The goal is to *bridge time* for different reasons. A *keeping store* typically contains the stock of a relatively small number of articles, products or shipments. As shown in *Fig. 11.1*, the stored quantity has either been delivered entirely or built up by consolidating part deliveries. It can be delivered entirely or in part shipments at planned dates.

Examples for *keeping stocks* are:

- *stocks of crops or raw material* that are piled up at harvest time and consumed during one or several years
- *stocks of sales promotion articles* that are procured in advance and shipped to the outlets at the promotion date
- *stocks of made- or procured-to-order material and parts* for a building or construction project that are collected on a storing area
- *shipments, freight* or *cargo* that is collected until departure date or until the capacity of the transport mean is full
- *temporary waiting queues* in front of a performance station or a transport node caused by cyclical operation (see *Sects. 13.3* and *13.5*)
- *batches of supplied goods* waiting for further distribution
- *sorting batches* of accumulated packages, pallets or load units to be sorted according to destinations or other criteria (see *Sect. 18.6.3*)

The scheduling of a keeping store follows the *push-principle*. Stocks and storing times are determined by an operating plan, a sales plan or production plan, a timetable or a sorting strategy. For example, the quantities for a *sales promotion* are collected in a central store. At the beginning of the sales campaign, part of the total quantity is shipped to outlets due to an *allocation plan*. The remaining quantity is a *demand reserve* and shipped later to the outlets with highest sales. The share and allocation plan for the initial distribution are *strategy parameters* of the sales campaign.

Special examples of keeping stores are stores for documents, books, film copies, data carriers and furniture, such as *libraries*, *archives* and *depots*. *Safes* for money and *safe-keeping stores* for valuable goods must fulfill specific access and safety requirements.

11.2 Criteria for Storekeeping

For each supply chain, it is necessary to decide, for which articles a storage station should be inserted between production, delivery, shipment, sales and consumption. A producing company must decide, which finished goods should be made to order and from which stage of the production process which material and parts should be made or procured anonymously to stock. Retailers must decide, which articles have to be kept on stock on which stage of the supply network and which articles should be procured to current customer orders.

Articles made or procured to customer orders are called *customer articles* or *order goods*. Articles that are anonymously supplied for stock replenishment are *storekeeping articles* (Chopra and Meindl 2007; Tempelmeier 1995). The decision whether a customer order should be *delivered from stock* or *delivered to order* depends on the required *service level*, the *costs* for direct delivery in relation to the costs for replenishment and storing and on the *risks* of storekeeping (see *Fig. 11.2*).

11.2.1 Service Effects

For *storekeeping articles* the *service level* can be optimized. By keeping a sufficient *stock*, a required availability can be ensured. The *delivery time* ex stock is solely determined by the administrative processing and the shipping time. It can therefore be extremely short.

In contrast to storekeeping articles, the service level for *order articles* is generally uncertain and fluctuates. The delivery time for order articles depends on the actual utilization of the production respectively the ex-stock availability of the supply station and on the delivery time of the *freight chain*.

11.2.2 Cost Effects

The production *costs per piece* depend on the product, the processing technique and the *setup costs* per order. For customer orders with small quantities these are higher than for stock replenishment orders with larger quantities.

The proportionate setup costs of order production can be decreased by collecting single customer orders until a sufficient quantity is reached, which is then processed in one pass. However, for low demand and small order quantities such a *serial production* of customer orders prolongs delivery times.

Also non-producing supply stations have administrative order processing costs, although these are generally lower than the setup costs of production. The propor-

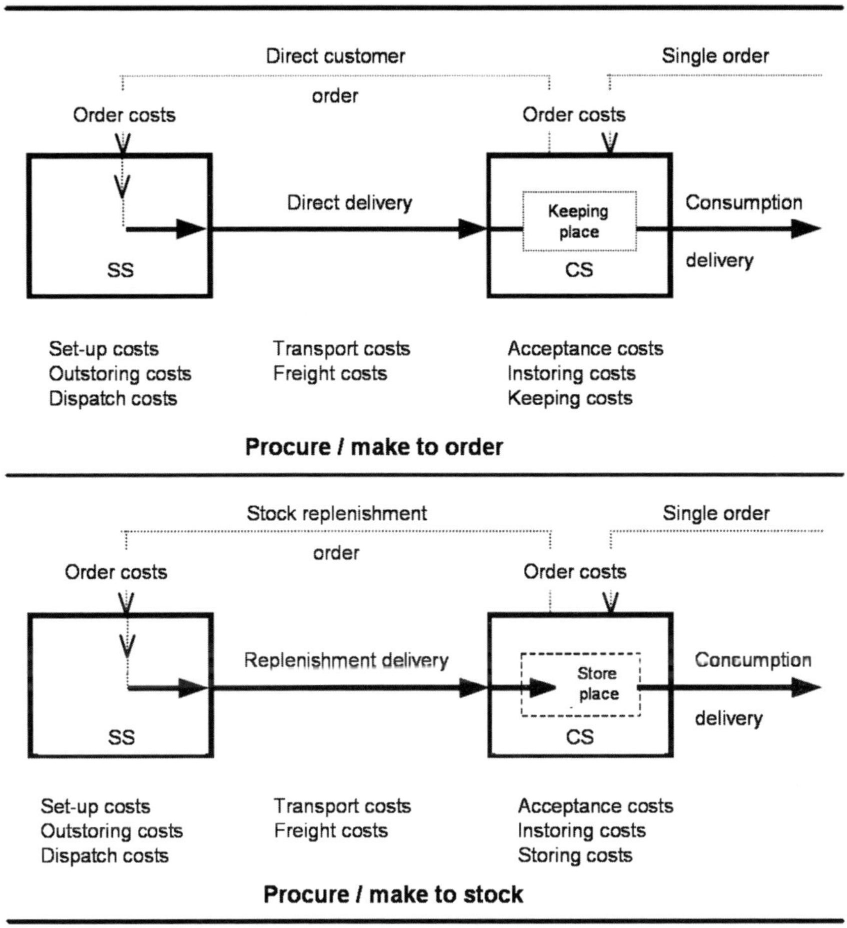

Fig. 11.2 Elementary process chains and relevant cost factors for procure- or make-to-order and procure- or make-to-stock

SS: supply, production or storage station
CS: consumption, sales or delivery station

tionate setup and order processing costs of external suppliers are often compensated by a *surcharge* for small orders or kept under control by requiring a *minimal order quantity*.

With order production, storing costs only occur, if the order is executed in advance and stored until delivery. Storekeeping articles always generate storing costs, which depend on the required ex-stock availability, the value, volume and weight of the articles and on the replenishment quantity. They can be minimized by *optimal replenishment quantities*.

11.2.3 Inventory Risks

Any inventory that is not backed up by binding customer orders holds the risk that some of the stored goods will not be required or sold.

An inventory risk for goods that are produced specifically for customers only exists if they are pre-fabricated without obligation of the customer to pay for them. *Noncommittal orders* are quite common in certain industries, e.g. in the *textile industry* and the *automotive industry*. Here, the suppliers often produce due to unsecured *block orders* or for non-obliging *frame orders*.

For anonymous storekeeping articles, the inventory risk is inevitable. It is the most important criterion for the storekeeping decision and depends on

- *innovation time*, which can be extremely short for *fashion articles* or *computer products*
- *deterioration, ageing and obsolescence risk* of merchandise
- *marketability* that is determined by the range of use and the number of customers or consumers for the article
- *inventory coverage*, i.e. the current relation of stock to consumption

In some cases, the inventory risk is compensated high profits. This holds for *spare parts* that are manufactured in advance in larger quantities and for *commodities* that are purchased speculatively at low price.

For goods with long lasting demand, the inventory risk is given by the *inventory risk interest*, which is calculated from the average inventory losses of the past. A *rule of experience* is:

- ▸ *Inventory risk interest* for articles with long lasting demand and broad application lies between 3 and 7% p.a, whereas for products with limited application and short demand it can reach 15% p.a. and more.

In order to reduce inventory risk, for each article or *inventory category* the *maximal acceptable inventory coverage* has to be defined. This threshold limits replenishment quantities and safety stock.

In *Table 11.2*, the consequences and relevant influence factors of the decision between make-or procure-to-order and make-or procure-to-stock are listed. From this, the *decision criteria* of *Table 11.3* and the following general *delimitation principles* for order articles and storekeeping articles can be derived:

- ▸ Order production and order procurement are opportune if the demand is temporary, the value of the article is high, the order quantity is large, the article units are big or the sales risk is high.
- ▸ Storekeeping is necessary if short delivery times and high ex-stock availability are preconditions for *marketability*.
- ▸ Storekeeping is opportune if the demand is predictable and the costs of replenishment and storing are lower than the costs of order procurement.

The cost comparison for the last decision will be performed in *Sect. 11.14*.

Table 11.2 Effects and influence factors of make-or produce-to-order and to-stock

Objectives	Make/Procure to Order	Make/Procure to Stock
SERVICE		
Availability	**insecure, varying**	**reliable high**
Influence factors	lot sizes, utilization	safety stock
Delivery Time	**insecure, varying**	**short and reliable**
Influence factors	lot sizes, utilization	administrative order execution
COSTS		
Storing	**minimal**	**high**
Influence factors	delivery date, scheduling	replenishment, availability
Production	**higher**	**minimal**
Influence factors	customer order quantities	replenishment quantities
RISKS	**minimal**	**higher**
Influence factors	unbinding orders	stock level, inventory coverage

Table 11.3 Criteria, properties and strategies of order articles and storekeeping articles

Criteria	Order articles	Storekeeping articles
Product	**specific**	**universal**
Application	limited	broad
Price	medium to high	low to medium
Volume	medium to large	small to medium
Service	**varying**	**high**
Delivery Times	varying	short
Availability	insecure	as scheduled
Demand	**temporary**	**longer lasting**
Predictability	bad	good
Order Quantities	larger $m_{Ropt}/2$	up to $m_{Ropt}/2$
Sales Risk	**high**	**calculable**
Innovation Time	short	long
Deterioration	high	low
OPTIMIZATION	**order bundling**	**supply bundling**
Strategy Variable	batch size	replenishment quantity

11.3 Scheduling of Storage Chains and Networks

A *single-stage storage station* is separated from other storage stations of a logistic chain by a preceding and a succeeding production or processing station without stocks or buffers. Examples for single-stage storage stations are production stores that directly supply machines and sales stores in outlets that directly serve customers.

Several storage stations in succession make up a *multi-stage storage chain*. Examples for two-stage storage chains are material buffers located at machines supplied by upstream *reserve stores*. Another example for an internal three-stage storage chain is a *commissioning system* with *access stock* on the picking place, *buffer stock* on places close to the picking place and *reserve stock* in a separate store. An example for an external three-stage storage chain is the finished goods store of a manufacturer that delivers to a central store of a retailer, which serves the sales buffers in the outlets.

For inventory scheduling, the present values of all parameters of the storage chain that influence the supply, replenishment and storing costs must be known. These are the *cost rates* for the replenishment and storekeeping processes of all related storage stations, the *article-logistic data* and the current *demand*.

Figure 11.3 shows the standard process of *dynamic inventory scheduling* for an individual article. The maximal stock is the sum of replenishment quantity and safety stock. The reorder point is the sum of the expected consumption during replenishment time and safety stock. Replenishment quantity and safety stock are the two basic *strategy variables of inventory management*:

- The *replenishment quantity* is the strategy variable of *replenishment scheduling* which can be adjusted to *minimize the storekeeping costs*.
- The *safety stock* is the strategy variable of *service scheduling* which is used to ensure the *required ex-stock availability*.

In a *multi-stage storage network* the demand of a storage station is the sum of the replenishment orders of all adjacent downstream storage stations. Therefore, the inventories within a multi-stage storage network can be scheduled due to the *pull-principle* by *retrograde inventory scheduling*:

- Within a *multi-stage storage network*, first the safety stocks, reorder points and replenishment quantities of the final storage stations are calculated from their demand. From the resulting replenishment demand, the corresponding parameters of the next upstream stores are calculated. Their replenishment demand is used as input for the further upstream stores and so on.

This is performed until the initial storage stations are reached, which are procured by production stations or external suppliers. Retrograde inventory scheduling can be executed centrally by an *order center* or a *central computer* or *locally* by the individual storage stations. Provided there are no *bottlenecks* at any stage of the supply network *local scheduling* is generally *self-regulating*. Under this provision, local scheduling leads to minimal costs for the total network while keeping the required availability, if each station schedules its own cost-optimal replenishment quantities and safety stock.

Any change in the transport means or load units automatically affects inventory, as this alters the cost rates for replenishment and storing. If a station switches to another supply chain, its safety stocks and replenishment quantities will change as well, since the replenishment times and cost rates of the other chain are generally different.

The mutual dependency of inventories in the delivery chains between industry and retailers have attracted a lot of attention over the last couple of years. Under the phrase *Efficient Consumer Response* (ECR), huge saving and improvement potentials by optimization of the supply networks are propagated (Simchi-Levi et al.

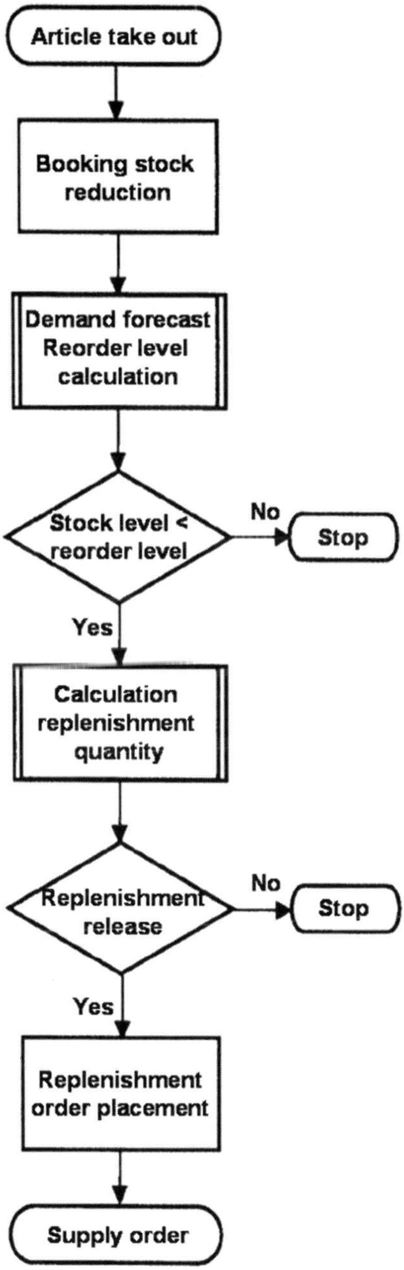

Fig. 11.3 **Process of dynamic inventory scheduling for a single article**

2008). This was initiated by the implementation of *electronic data interchange* (EDI) between companies and by improved *forecasting methods* and *scheduling software*. However, in spite of high expectations, many ECR-projects failed due to shortcomings of scheduling strategies and/or incorrect data and parameters used by the actors (Breiter 1996; Bucklin 1966; Christopher 2005; Cooper et al. 1997; Kotzab 1997, 1999; Laurent 1996; Ritter 1997; Toporowski 1996).

11.4 Scheduling Parameters

Optimal inventory scheduling is only possible if the scheduling parameters are completely known and correctly used. The central parameters for inventory scheduling are *article logistic data*, *order data* and *replenishment parameter*.

11.4.1 Article Logistic Data

For inventory scheduling, the following *article logistic data* are relevant:

- *measure units* [MU = piece, kg, *l*, m, m^2, m^3, ...] of the article
- *volume* v_{CU} [*l*/CU], weight w_{CU} [kg/CU] and *content* c_{CU} [MU/CU] of the *consumption unit* [CU], *sales unit* [SU] or *storekeeping unit* [SKU]
- *production or purchase price* P_{MU} [€/MU] per measure unit or P_{CU} [€/CU] per consumption unit
- *capacity* C_{LU} [CU/LU] of the *load units* [LU] for replenishment and storing
- required *ex-stock availability* η_S [%]
- *replenishment time* T_R [PE] and its *standard deviation* s_T [PE]

The replenishment time is equal to the *delivery time for repeated orders*. For the first order the replenishment time is generally longer. Length and standard deviation of the replenishment time for established suppliers can be derived by exponential smoothing from the delivery times of the past. New suppliers should guarantee reliable delivery times for replenishment orders.

In many companies, the article logistic data are not completely known or not registered in the computer database. This prevents the application of the most effective scheduling strategies and causes wrong replenishment proposals (Dittrich et al. 2000; Gudehus 2006). In addition to the logistic data, schedulers should know the relevant properties, sources and applications of the articles, they are responsible for.

11.4.2 Order Data

For any kind of scheduling, the *order data* must be known with sufficient accuracy. These are

- *order entry rate* λ_O [Ord/PE] and its *standard deviation* $s_{\lambda O}$
- *order quantity* m_O [CU/Ord] and its *standard deviation* s_{mO}

From these order data result the *demand flow* or *consumption rate*:

$$\lambda = m_O \cdot \lambda_O \qquad [\text{CU/PE}]. \tag{11.1}$$

If a longer lasting demand or consumption shows a regular pattern, the future demand respectively consumption can be forecasted by a program which operates with the mathematical forecasting methods of *Chap. 9*. However, even in these cases, the scheduler should critically asses the computer forecast and correct it, if deviating information about the future development is known.

11.4.3 Replenishment Parameter

The first strategy variable of inventory scheduling is the

- *replenishment quantity* m_R [CU/ROrd]

The quantity of a *replenishment order* [ROrd] placed with an external supplier is the *reorder quantity*. With an internal supplier or production station it is the *supply quantity* or *production lot*. From a *mean replenishment quantity* m_R and a stationary demand λ results the *replenishment frequency*:

$$f_R = \lambda / m_R \qquad \text{[ROrd/PE]} \tag{11.2}$$

and the *replenishment cycle time*:

$$T_{RC} = 1/f_R = m_R/\lambda_R \qquad \text{[PE]} \tag{11.3}$$

The replenishment cycle time is equal to the mean *range of the replenishment quantities*.

The minimal *number of load units* with a capacity C_{LU} [CU/LU] containing the replenishment quantity m_R is given by:

$$M_R = \{m_R/C_{LU}\} \qquad \text{[LU/ROrd]} \tag{11.4}$$

The curly brackets indicate the *rounding up* to the next integer. The dependency of the number of load units on the replenishment quantity is a *step function* as shown in *Fig. 12.10*. The *mean number of load units* for varying replenishment quantities is (see *Sect. 12.5*):

$$M_R = \text{MAX}(1; m_R/C_{LU} + (C_{LU} - 1)/2C_{LU}) \qquad \text{[LU]} \tag{11.5}$$

As long as $m_R \leq C_{LU}$ only one load unit is needed. If the mean quantity becomes $m_R > C_{LU}$, each replenishment delivery contains in addition to completely filled load units one partly filled unit with mean *filling loss* $(C_{LU}-1)/2C_{LU}$.

The additional costs caused by the filling losses can be eliminated by a *quantity adjustment strategy*, i.e. by rounding up or down of the replenishment quantity to the next multiple of the load unit capacity.

A *minimal order quantity* m_{Rmin}, which is required by a supplier or producer, leads to the *lower restriction*

$$m_R \geq m_{Rmin} \qquad \text{[CU/ROrd]} \tag{11.6}$$

for the replenishment quantity. Minimal quantities and lower restrictions should regularly be questioned, since it might be possible to reduce the total costs by allowing for smaller replenishment quantities.

11.5 Storekeeping Parameters

For *stationary consumption* and *closed dispatch* of the replenishment quantities, the time dependency of the *current stock* $m_S(t)$ has a *saw-tooth pattern* as shown in *Fig. 11.4*. Such a *cyclic stock* varies with equal probability between a minimal and a maximal stock (Chopra and Meindl 2007). Caused by stochastic fluctuations of demand and by the integer consumption quantities, the temporal course of the stock becomes a *step function*, as shown in *Fig. 11.1*, *11.4* and *11.20*, which fluctuates around a mean saw-tooth pattern.

The *minimal stock* varies around the

- mean *safety stock* m_{safe} [CU]

The safety stock is the second strategic variable of inventory scheduling, since it will be shown later:

► For stochastic demand and varying delivery times, the *safety stock* m_{safe} prevents delivery inability with a probability equal to the required *stock availability* η_S.

With *closed dispatch* of the replenishment quantities, the *maximal stock* is the sum of safety stock and replenishment quantity:

$$m_{Smax} = m_{safe} + m_R \qquad [CU]. \tag{11.7}$$

From the fact, that the *current stock* $m_S(t)$ varies with equal probability between safety stock and maximal stock, i.e. within the range:

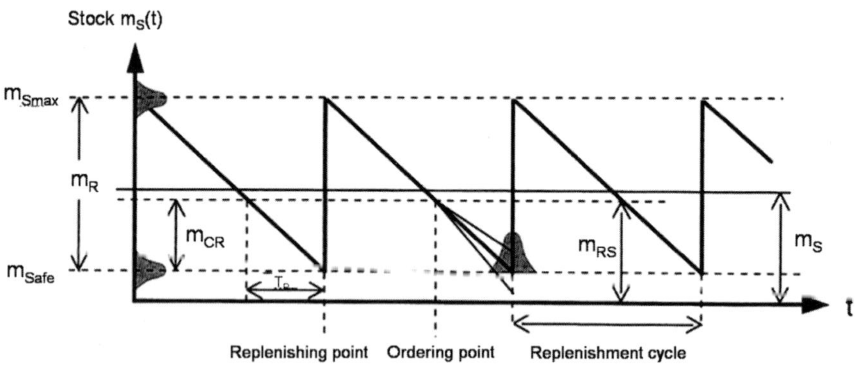

Fig. 11.4 Time dependency of stock for continuous demand and closed replenishment of complete quantities

$m_S(t)$	current stock
m_{safe}	safety stock
m_{Smax}	maximal stock
m_R	replenishment quantity
$m_S = m_{safe} + m_R/2$	mean stock level
$m_{CR} = \lambda \cdot T_R$	consumption during replenishment time T_R
$m_{RS} = m_{safe} + \lambda \cdot T_R$	reorder stock

$$m_{safe} \le m_S(t) \le m_{Smax} \qquad (11.8)$$

and from relation (11.7) follows:

- The *mean inventory* or *stock level* m_S of an article with stationary consumption and closed replenishment dispatch is the sum of the safety stock and the half replenishment quantity, i.e.:

$$m_S = m_{safe} + m_R/2 \qquad [CU]. \qquad (11.9)$$

Replenishment quantity and safety stock are the two central leverages, by which the stock level can be influenced. Whereas the consequences of the replenishment quantities are generally known, the importance of the safety stock is often neglected.

The minimal number of load units with capacity C_{LU} [CU/LU] containing the current stock $m_S(t)$ [CU] is $M_S(t) = \{m_S(t)/C_{LU}\}$. Without quantity adjustment, one load unit is only partly filled. From (11.9) follows, analogous to relation (11.5), the number of load units occupied by the *mean stock*:

$$M_S = MAX(1; (m_{safe}+m_R/2)/C_{LU}+(C_{LU}-1)/2C_{LU}) \qquad [LU] \quad (11.10)$$

By rounding up or down the replenishment and the delivery quantities to multiples of the capacity, the filling loss can be avoided. With such a *quantity adjustment strategy*, the term $(C_{LU}-1)/2C_{LU}$ in (11.10) disappears.

As shown later in *Sect. 16.4*, the number of storeplaces occupied by the number of load units (11.10) depends on the type of the storage system and on the *storing strategy*. For *single place stores*, the mean number of storeplaces occupied by the mean stock (11.9) is:

$$N_{SP} = \{ (m_{safe} + f_{PO} \cdot m_R)/C_{LU}\} \qquad [SP] \qquad (11.11)$$

Herein, f_{PO} is the *place order factor*

$$f_{PO} = \begin{cases} 1/2 & \textit{for free place order} \\ 1 & \textit{for fixed place order} \end{cases} \qquad (11.12)$$

Without quantity adjustment, the mean number of storeplaces required for the article stock is

$$M_{SP} = MAX(1; (m_{safe} + f_{PO} \cdot m_R)/C_{LU} + (C_{LU} - 1)/2C_{LU}) \qquad [SP] \qquad (11.13)$$

The required number of store places for *multi place stores*, e.g. for a *block place store*, can be calculated by analogous formulas (see *Chap. 16*).

From the stock level m_S [CU] and the mean consumption rate λ [CU/PE], several *inventory key indicators* are derived, e.g. the inventory turnover (11.16), the replenishment frequency (11.2) and the inventory range:

- The *inventory range* is the consumption time for the mean stock, assuming that the mean consumption of the past lasts on in future, i.e.:

$$IR = m_S/\lambda \qquad [PE]. \qquad (11.14)$$

Correspondingly, the *range of the replenishment quantity*, the *range of the safety stock* and the *range of the maximal stock* are defined.

The inventory range (11.14) differs from the mean storage time of the single article unit. The storage time depends on the *removal strategy*, such as FIFO (first-in-first-out) or LIFO (last-in-first-out), and on the demand. The storage time of a single unit can not be predicted but only noticed after the unit is taken from the storeplace.

In order to reduce the inventory risk, the maximal stock is often limited by a *maximal inventory range* IR_{max} [PE], resulting in the *risk limitation* for the sum of replenishment quantity and safety stock:

$$m_{safe} + m_R \leq IR_{max} \cdot \lambda. \qquad (11.15)$$

The reciprocal of the inventory range (11.14) is the *inventory turnover*:

$$IT = 1/IR = \lambda/m_S \qquad [1/PE]. \qquad (11.16)$$

The inventory turnover (11.16) should not be confused with the replenishment frequency (11.2). From (11.2) and (11.16) follows:

- The *inventory turnover* differs from the *replenishment frequency* by the factor $2m_R/(2m_{safe}+m_R)$.

For small safety stocks the inventory turnover is twice as high as the replenishment frequency. The inventory turnover decreases with increasing safety stock. From this follows the *inventory rule*:

▸ Inventory turnover below the replenishment frequency indicates too high safety stocks.

Inventory range and turnover are often reported for a whole store containing different articles. However, these indicators only make sense for a *homogenous assortment* that fulfils the requirements of the *mean value theorem of logistics* given in *Sect. 9.4.4*. That means:

▸ For a whole *heterogeneous assortment* of articles with different properties and demand, inventory range and turnover are misleading indicators.

With an expected demand λ [CU/PE], the *consumption during replenishment time* is

$$m_{CR} = \lambda \cdot T_R \qquad [CU] \qquad (11.17)$$

In order to prevent unavailability, a replenishment order must be placed as soon as the actual stock reaches the reorder stock. This leads to the rule:

- The *reorder stock* is the sum of the safety stock and the expected consumption during the replenishment time

$$m_{RS} = m_{safe} + m_{CR} = m_{safe} + \lambda \cdot T_R \qquad [CU]. \qquad (11.18)$$

The solution of the equation $m_S(t_{RO}) = m_{RS}$ is the *reorder time* t_{RO}, at which the current stock $m_S(t)$ reaches the reorder stock. For instationary consumption, safety stock and consumption during replenishment time can change. In this case, reorder stock and reorder point are *dynamic variables* which must be calculated for the actually expected mean demand (see *Fig. 11.3*).

11.6 Cost Rates for Replenishment and Storing

In order to evaluate the costs which are affected by inventory management, it is necessary to investigate carefully the whole storekeeping process with its main parts which are *replenishment* and *storing* (see *Fig. 11.2*). Correspondingly, the *storekeeping costs* are the sum of replenishment costs and storing costs:

- The *replenishment costs* depend on the *replenishment order rate*. They are caused by the replenishment process which starts with scheduling and ends with the deposition of the ordered quantity in the store.
- The *storing costs* depend on the stock level. They result from interest and risks of the inventory and from the place costs for the stored quantities.

To calculate the storekeeping costs, present values of the corresponding *cost rates* must be known. Values and benchmarks of other companies are generally not applicable for this purpose. Without correct cost rates, a minimization of the total costs is not attainable (Dittrich et al. 2000; Gudehus 2004).

11.6.1 Replenishment Order Costs

The replenishment order costs are the sum

$$k_{\text{ROrd}} = k_{\text{COrd}} + k_{\text{SOrd}} \qquad [€/\text{ROrd}] \tag{11.19}$$

of the costs which are caused per replenishment order in the consumption station and in the supply station:

- The *replenishment order costs in the consumption station* k_{COrd} [€/ROrd] are caused by scheduling and release of the replenishment order, communication with the supply station, acceptance of the shipment and registration of the order entry.
- The *replenishment order costs in the supply station* k_{SOrd} [€/ROrd] result from order acceptance, order processing, production scheduling, setup of production respectively of order picking or retrieval from store, dispatch and shipment, and from invoicing and communication.

For internal supply stations, the order costs are calculable. For external suppliers, they are generally included in the purchase price. In order to optimize *inter-company supply chains*, order costs and production setup costs of the external supplier must be taken into account and for this purpose separated from the purchase price.

11.6.2 Shipment and In-Storing Costs

If each replenishment is shipped separately, the shipment costs per replenishment order are the sum $k_{\text{ship}} = k_{\text{Oship}} + k_{\text{LUship}} \cdot M_{\text{R}}$ [€/ROrd] of the *shipment order price* k_{Oship} [€/ROrd] and the *load shipment price* k_{LUship} [€/LU/ROrd] times the *number of load units* M_R [LU/ROrd]. They are either prices of an external freight forwarder or internal cost rates (see *Sect. 21.15*).

The load shipment costs are the same for load units which are stored and load units which are dispatched directly without storing, i.e. they are not affected by inventory scheduling. Hence, the order costs of the shipments must be taken into account only, if the replenishments for the single articles are shipped separately. Otherwise, they do not concern inventory scheduling.

After arrival, the replenishment quantity is stored in. This causes per load the *specific in-storing cost* k_{LU} [€/LU]. For own stores, the in-storing costs can be calculated as explained in *Sect. 16.13*. For external stores, they are determined by the *in-storing price* of the *logistic service provider*. Some orientation values for in-storing cost rates for standard bins and pallets are given in *Table 11.4*.

11.6.3 Article Price

For inventory scheduling either the *article price per measure unit* P_{MU} [€/MU] or the *article price per consumption unit* P_{CU} [€/CU] has to be known. They are connected by the *content per consumption unit* C_{CU} [MU/CU] via the relation:

$$P_{CU} = C_{CU} \cdot P_{MU} \qquad [€/CU]. \tag{11.20}$$

For own products, the article prices for inventory management are the pure production costs without setup costs and overhead costs. For external supplies, the article price is the present net-purchasing price for the mean replenishment quantity minus all discounts and allowances.

11.6.4 Inventory Interest Rate

The *inventory interest rate* z_I is the sum

$$z_I = z_C + z_{risk} \qquad [\%/PE] \tag{11.21}$$

of the present *capital interest rate* z_C [%/PE] for the capital bound in the inventory and an *inventory risk rate* z_{risk} [%/PE] that takes into account the storekeeping risks, such as shrinkage, ageing, deterioration, obsolescence and loss (see *Sect. 11.2.3*).

The *terms of payment* of the supplier have no influence on the interest costs of the inventory, as capital no longer bound in stocks can be reinvested elsewhere.

As long as an article is not sold, neither its sales price nor its full costs should be used for the calculation of the interest costs of inventory.

11.6.5 Storeplace Costs

The costs for keeping a load unit in a store are the product of the *storing time* and the *storeplace costs* k_{SP} [€/LU-PE]. The storeplace costs depend on the dimensions and weight of the load unit and on the technique and dimension of the storage system.

For an internal store, the storeplace costs, as well as the in- and out-storing cost rates, can be calculated from investment and operating costs, as explained in *Sect. 16.13*. For an external store the storeplace costs are given by the *store place price* of the *logistic service provider*.

Table 11.4 Selected cost rates for in-storing and storeplace

		Performance Cost Rates				
		Small stores		**Large stores**		
PERFORMANCE	Performance unit	from	to	from	to	Price/cost unit
Order cost	Storage bin					
Scheduling	R-Order	3.00	4.00	2.50	3.50	€/R-Order
Order-entry control	R-Order	7.00	13.00	4.00	6.00	€/R-Order
Instoring	including internal transport					
Bins	Bin	0.30	0.40	0.20	0.30	€/Bin
Pallets	Pallet	2.50	3.50	1.50	2.00	€/Pal
Storing	single place storing					
Bins	Bin-Cday	0.04	0.06	0.02	0.03	€/Bin-Cday
Pallets	Pal-Cday	0.25	0.40	0.15	0.20	€/Pal-Cday
Outstoring	without order picking and internal transport					
Bins	Bin	0.20	0.30	0.15	0.20	€/Bin
Pallets	Pallet	1.50	2.50	0.80	1.40	€/Pal
Inventory interest	Rate	**9.0%**	**20.0%**	**7.0%**	**17.0%**	p.a.
Capital interest	Rate	4.0%	12.0%	4.0%	12.0%	p.a.
Storekeeping risk	Rate	5.0%	8.0%	3.0%	5.0%	p.a.
Load units	LU	**Volume**		**Dimensions**		
		Inner	Outer	Length	Breadth	Height
Bins	Bin	74	63	600	400	310 mm
Pallets	Pal	1,008	860	1,200	800	1,050 mm

cost rates without overhead; price basis 2006 (Germany)
1 year = 365 calendar days [CD] = 250 operating days [OD]
R-Order: replenishment order

The cost rates for the load units and for alternative storage systems can differ enormously (see *Table 11.4*). Therefore, scheduling must calculate with the specific cost rates of the individual store, which should be updated regularly.

Neither the storeplace costs nor the in- and out-storing cost rates depend on the stock value. *Storage cost drivers* are *movements* and *storing time* of the *load units*. In spite of this, many companies calculate storing costs – if at all – as a percentage of the inventory and cannot achieve minimal storekeeping costs.

11.7 Storekeeping Costs

The goal of inventory scheduling is to ensure the required stock availability at minimal costs. The *storekeeping costs* $K_{RS}(m_R)$, which are affected by inventory scheduling, are the sum of *replenishment costs* and *storing costs*:

$$K_{SK}(m_R) = K_R(m_R) + K_S(m_R) \qquad [€/PE]. \tag{11.22}$$

Both, replenishment costs and storing costs, depend on the *reorder quantity* m_R. For stationary consumption rate λ [CU/PE] and constant reorder quantity m_R, the reorder frequency is given by relation (11.2). Each replenishment order causes replenishment order costs (11.19) and in-storing costs for the number of load units given by (11.4). The product of the reorder frequency with the sum of order costs and in-storing costs gives the *replenishment costs* per period:

$$K_R(m_R) = (\lambda/m_R) \cdot (k_{ROrd} + k_{LU} \cdot \{m_R/C_{LU}\}) \qquad [€/PE]. \tag{11.23}$$

For a replenishment quantity m_R and a safety stock m_S, the mean inventory value is the product of the mean stock level (11.11) and the article price P. The inventory value multiplied by the interest rate (11.21) gives the interest costs per period. The storeplace costs are the product of the mean number of occupied storeplaces (11.13) and the storeplace costs k_{SP}. Therefore, with closed replenishment dispatch, the *storing costs* per period are:

$$K_S(m_R) = P \cdot z_S \cdot (m_{safe} + m_R/2) + k_{SP} \cdot \{(m_{safe} + f_{PO} \cdot m_R)/C_{LU}\} \qquad [€/PE]. \tag{11.24}$$

The resulting dependency of storekeeping costs (11.22) on the replenishment quantity m_R is shown in *Fig. 11.5* for an example with $C_{LU} = 1$.

Relations (11.22), (11.23), (11.24) and *Fig. 11.5* show the *dependencies of storekeeping costs*:

- Replenishment costs vary reciprocally to replenishment quantity.
- Storing costs are proportional to replenishment quantity and safety stock.
- For smaller replenishment quantities, the storekeeping costs decrease with increasing replenishment quantity until minimal costs are reached and increase with further increasing replenishment quantities.
- With the *optimal replenishment quantity* m_{Ropt} the storekeeping costs are minimal.
- Storekeeping costs change only slightly in the vicinity of the optimal replenishment quantity.

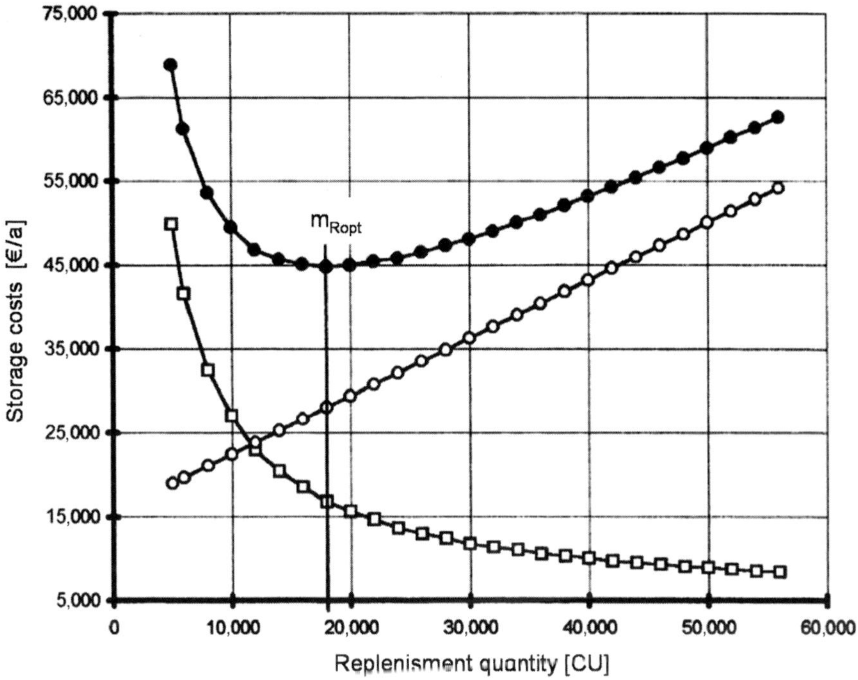

Fig. 11.5 Storekeeping costs as function of replenishment quantity

Example: Finished goods store for cigarettes
Squares: replenishment costs
Circles: storing costs
Black dots: replenishment costs + storing costs
CU: consumption unit (package unit with 10 cigarette packs)

The slight variation around the optimal value with minimal costs leads to the following *rules of replenishment scheduling*:

- Calculations of replenishment quantities must not be too precise.
- Calculated replenishment quantities have not to be adhered exactly, in particular if by rounding full load units are achieved.
- Inaccurate scheduling parameters and cost rates affect the optimal replenishment quantity and the minimal storekeeping costs only slightly.
- Inaccuracies of different parameters and cost rates are partly balanced out due to the law of error compensation (see *Sect. 9.4.3*).

However, the small influence of the inaccuracy of parameters and cost rates does not justify wrong storeplace costs or omitting single parameters.

For load units with a capacity $C_{LU} > 1$, the storekeeping costs (11.22) are a *step-function* of the replenishment quantity. In order to calculate the minimum, it is necessary to smooth out the steps by inserting the averaging functions (11.5) and (11.13) instead of the curly brackets (see *Fig. 12.10*). This results in the *mean replenishment costs*:

$$K_{Rm}(m_R) = k_{LU}\cdot\lambda/C_{LU}+(\lambda/m_R)\cdot\left(k_{ROrd}+k_{LU}\cdot(C_{LU}-1)/2C_{LU}\right) \quad (11.25)$$

and in the *mean storing costs* for closed replenishment dispatch:

$$K_{Sm}(m_R) = P\cdot z_S\cdot(m_{safe}+m_R/2)+k_{SP}\cdot \left((m_{safe}+f_{PO}\cdot m_R)/C_{LU}+(C_{LU}-1)/2C_{LU}\right) \quad (11.26)$$

The sum (11.22) of the expressions (11.25) and (11.26) are the *mean storekeeping costs* $K_{SKm}(m_R)$, which are a steady function that can be differentiated in respect to m_R. Differentiation of the mean storekeeping costs, setting of the result equal to 0 and solving this equation regarding m_R gives the

- *Optimal Replenishment Quantity* or *Economic Order Quantity* (EOQ):

$$\mathbf{m_{Ropt}} = \sqrt{\mathbf{2\cdot\lambda\cdot\left(k_{ROrd}+k_{LU}\cdot(C_{LU}-1)/2C_{LU}\right)/\left(P\cdot z_S+2f_{PO}\cdot k_{SP}/C_{LU}\right)}} \quad (11.27)$$

The replenishment quantity is restricted by the *minimal order quantity* (11.6) and by the *risk limitation* (11.15).

The *replenishment formula* (11.27) holds for *closed dispatch* with replenishment times independent of the replenishment quantity. For *continuous dispatch* of replenishment quantities, which exceed the daily *limit performance* μ [CU/d] of the supply station and are produced and dispatched successively in several days, the right hand side of Eq. (11.27) must be multiplied by the factor $1/(1-\lambda/\mu)$.

Resulting from relations (11.9) and (11.27) is the *optimal stock level*:

$$m_{Sopt} = m_{safe} + m_{Ropt}/2. \quad (11.28)$$

By inserting the optimal replenishment quantity (11.27) into (11.22), follow the *minimal storekeeping costs*:

$$K_{SKmin} = K_{SK}(m_{Ropt}) \qquad [€/PE]. \quad (11.29)$$

If the load unit equals the consumption unit, i.e. for $C_{LU} = 1$, and the storeplace costs are without influence, i.e. for $k_{SP} = 0$, the general formula (11.27) for the optimal replenishment quantity evolves in the

- *Harris-formula* for the *economic order quantity* (EOQ) (Harris 1913):

$$m_{Ropt} = \sqrt{2\cdot\lambda\cdot k_{ROrd}/P\cdot z_S}. \quad (11.30)$$

A consequence of the general replenishment formula (11.27), as well as of the special Harris-formula, is the *square root rule for the optimal order quantity*:

- ► The optimal replenishment quantity increases proportionally to the square root of the consumption rate (see *Fig. 11.6*).

Consequences of the general replenishment formula, which are disregarded by the Harris-formula, are:

- ► Storekeeping costs, economic order quantity and optimal stock level depend on the type and the capacity of the *load units* (see *Fig. 11.16*).

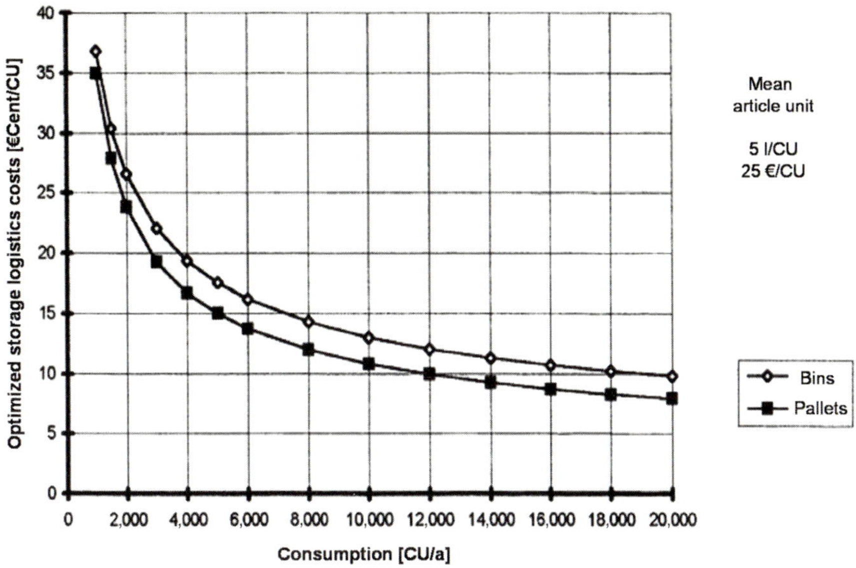

Fig. 11.6 Dependency of minimal storekeeping costs on consumption

Example: retailer store
Parameter: mean article unit = mean consumption unit [CU]
Load units: see *Table 11.4*

- Too big load units cause high filling losses and storeplace costs, whereas too small units generate too many movements and high in-storing costs.
- Storeplace costs, economic order quantity and optimal stock level depend on the *store-placing strategy*. For small safety stocks, the *fixed-place strategy* leads to higher costs than the *free-place strategy*.
- Value and size of storekeeping articles influence storekeeping costs, economic order quantity and optimal stock level (see *Fig. 11.7*).
- Without adjustment strategy, the additional costs for partially filled load units contribute considerably to the storekeeping costs if the replenishment quantities are small.
- Storing costs calculated as a percentage of inventory value cause too high stocks for articles with low value and big volume.
- Optimal replenishment quantity and stock level increase and decrease with order costs and in-storing cost rate.
- Inventory and optimal replenishment quantities decrease with the value of the articles, interest rate and store place costs.

In many cases, the performance costs depend on the utilization of production, storage system, transport means or other fixed resources. Correct cost rates for scheduling result from the following *costing rules* (see *Sect. 6.9*):

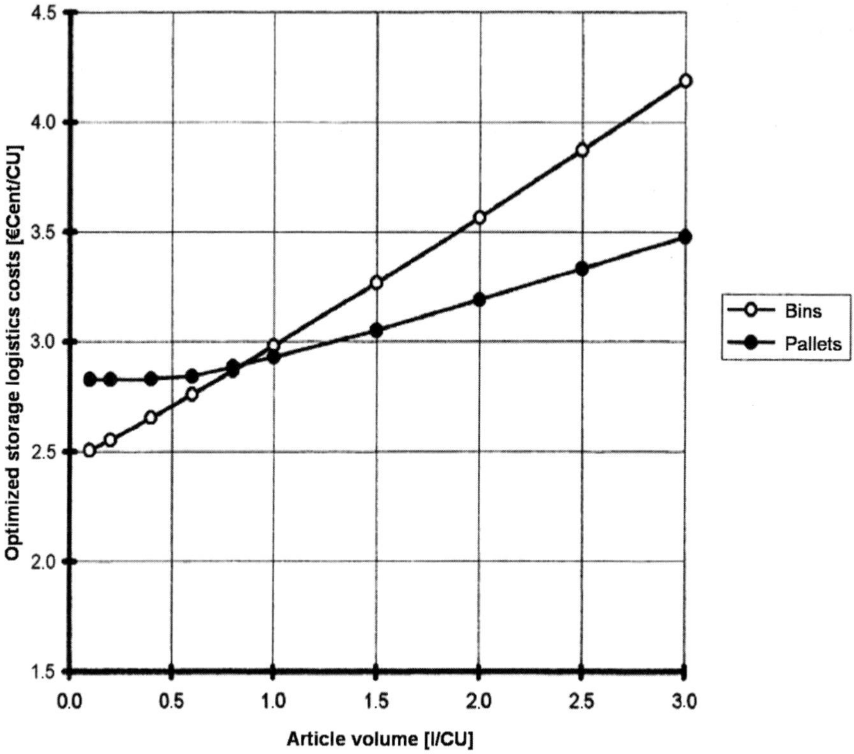

Fig. 11.7 Minimal storekeeping costs as function of article volume
Parameter: alternative load units

- For external services current *performance prices* have to be used, which result from offer and demand in the markets for logistic services.
- For internal services, *performance cost rates* are relevant, which result from full costs at maximal utilization of own resources.

If utilization-dependent cost rates are used, the cost rates increase with decreasing utilization and the optimal replenishment quantity (11.27) is reduced. This leads to a further increase of the cost rates and so on. Correspondingly, a higher utilization reduces the cost rates and causes an even higher utilization.

These negative effects can be avoided by *counter utilization-dependent cost rates*:

- ▸ In times of low utilization, the performance cost rates are reduced to marginal costs; in times of high utilization, they are raised to full cost.

This strategy achieves *self-regulating* the cost optimal utilization of internal and external resources.

11.8 Stock Availability and Safety Stock

Fixing the delivery ability for orders and the availability of storekeeping articles is a decision with far reaching consequences. The *stock availability* depends on the *safety level*, which is needed for the specific business or required by single customers. For this decision, it is necessary to distinguish between interruption reserve and safety stock:

- The *interruption reserve* is a *constant reserve stock* kept permanently for longer lasting interruptions of the supply by *critical events*, such as plant breakdown, urgent repair, transport damage, strikes or bottlenecks.
- The *safety stock* is a *randomly varying reserve stock* kept to ensure the stock availability during the replenishment time against random variations of demand and/or replenishment times.

The interruption reserve is blocked in normal times and used only in cases where the critical events happen. Its height is determined by the product of the mean consumption with the maximal expected interruption time.

The safety stock is a strategy parameter of inventory management. It can be used up completely at the end of a replenishment period, if the supply arrives delayed or if the demand during the replenishment time has been far higher than expected.

The safety stock, which is necessary to ensure a required stock availability, is with certain assumptions calculable by *probability theory* (Chopra and Meind 2007; Tempelmeier 1995; Simchi-Levi et al. 2008). However, the exact mathematical solution is quite complicated. The safety stock can also be calculated with sufficient accuracy by *approximation formulas*, which ensure a somewhat higher stock availability as required.

11.8.1 Delivery Ability and Stock Availability

The *total order delivery ability* is the probability that the actual stock is sufficient to execute a delivery order for storekeeping articles completely. A weaker type of delivery ability is the *partial order delivery ability* which is fulfilled if at least one unit of each ordered article is available. The delivery ability for multi position orders results by the product rule of probability theory from the *stock availability*[1] of the single articles:

► The *order delivery ability* η_O [%] for a multi-position order is the product of the *stock availabilities* η_{Si} [%] for the single articles A_i.

For single-position orders, the order delivery ability equals the stock availability, for many-position orders it is lower. For instance, the delivery ability of orders with an average of 5 positions of articles with a stock availability $\eta_S = 98\%$ is $\eta_D = \eta_S{}^5 = 98\%^5 = 0.90 = 90\%$.

[1] The commercial *stock availability*, i.e. the availability of an article stock, should not be mixed up with the technical *storage availability*, i.e. the availability of the storage system (see *Sect. 13.6*).

Order delivery ability and stock availability can be related either to orders or to operating days (see *Sect. 9.8*):

- The *order availability* is the relation between the number of orders executed in-time and completely from stock to the total number of orders within a considered period of time.
- The *daily availability* for a certain number of operating days is the relation of the number of days on which all delivery orders have been completely fulfilled to the total number of days.

Operating days, on which the stock is not sufficient to fulfill all incoming delivery orders, are days of *delivery inability*, even if some orders are executed. Therefore, a safety stock for a required daily availability ensures an order availability, which is equal or somewhat higher than the daily availability. Hence, the safety stock can approximately be calculated for a daily availability.

The stock availability of a single article is maximal immediately after arrival of a replenishment quantity and minimal at the ends of the replenishment cycles. It varies stochastically from cycle to cycle. Therefore, the *mean availability* of an article must be measured over several full replenishment cycles and smoothed out with formula (9.39) by weighting the measured availability of the actual cycle and the mean availability of the previous cycle (see *Sect. 9.8.4*).

Relevant for business practice is the mean *stock availability* η_S during several replenishments cycles. It is identical with the so called *β-availability* $\eta_\beta = \eta_S$ of OR. The *α-availability* η_α of OR is the ability to deliver during the replenishment time. For the calculation of the safety stock for the α-availability an explicit formula is known (Inderfurth 1994, 1999; Schneeweiß 1981; Tempelmeier 1995). This standard formula is often applied, although it generates safety stocks, which can be far higher than necessary.

11.8.2 Availability during Replenishment Time

Caused by the random variations of the demand λ [CU/d] and/or the replenishment time T_R [CU/d] the consumption during the replenishment time fluctuates around a *mean value* m_{CRm}, which is given by the product (11.1). Due to the *law of large numbers*, the demand within a longer replenishment time is approximately normal distributed. If s_λ [CU/d] is the standard deviation of the daily demand and s_T [d] the standard deviation of the replenishment time measured in days, due to the *law of error propagation* the *standard deviation of the demand in the replenishment time* is (see formula 9.32):

$$s_{CR} = \sqrt{T_R \cdot s_\lambda^2 + \lambda^2 \cdot s_T^2} \tag{11.31}$$

Due to the *general safety law* of *Sect. 9.4.1*, the *safety stock* which prevents inability to deliver at the end of the replenishment time T_R with probability η_α is:

$$m_{\alpha\,\text{safe}} = f_S(\eta_\alpha) \cdot s_{CR} = f_S(\eta_\alpha) \cdot \sqrt{T_R \cdot s_\lambda^2 + \lambda^2 \cdot s_T^2} \quad \text{for } \eta_\alpha > 50\% \tag{11.32}$$

Table 11.5 Safety factors for different safety levels

Safety Level $\eta_S(\%)$	Safety Factor $f_S(\eta)$
50.0	0.00
80.0	0.84
85.0	1.04
90.0	1.28
95.0	1.64
98.0	2.05
99.0	2.33
99.9	3.09

Safety level: delivery ability, ex-stock availability, overflow safety, etc.

The *safety factor* $f_S(\eta)$ in this formula is given by the *inverse standard normal distribution* (9.20) and calculable with the EXCEL-operation NORMSINV(η). A good approximation of this function is:

$$f_S(\eta) \approx (2\eta - 1)/(1-\eta)^{0,2} \qquad \text{for } \eta > 50\% \,. \tag{11.33}$$

The general dependency between availability and safety factor is shown in *Fig. 5.4*. The diagram also demonstrates the accuracy of the *approximation formula* (11.33) for values up to 99.5%. In *Table 11.5* the safety factors for common safety levels are given.

The safety factor, and consequently the safety stock, exceeds all limits if the required availability reaches 100%. That means:

- ▶ For randomly fluctuating demand, 100% availability is impossible.

For availabilities of up to 50%, the safety factor is 0 and no safety stock is necessary.

11.8.3 Necessary Safety Stock

In order to calculate the necessary safety stock for the practically relevant *β-availability*, one must take into account that during the time from the arrival of replenishment until the reorder point is reached, the ex-stock availability for most of the orders is 100%. During this time, the stock is higher than the reorder stock and sufficient to fulfill any order quantity smaller than the demand in the reorder time. As shown in *Fig. 11.4*, the time length with availability 100% is the *replenishment cycle time* $T_{RC} - m_R/\lambda$ minus the *replenishment time* T_R. With the *α-availability* η_α during replenishment time, the *mean stock availability* over the whole replenishment cycle time T_{RC} is given by the weighted mean value:

$$\eta_S = 1 \cdot (T_{RC} - T_R)/T_{RC} + \eta_\alpha \cdot T_R/T_{RC} \tag{11.34}$$

The solution of (11.34) with respect to η_α gives the dependence of the α-availability on the stock availability:

$$\eta_\alpha = 1 - (1 - \eta_S) \cdot m_R/(T_R \cdot \lambda) \qquad \text{if } m_R/\lambda > T_R \tag{11.35}$$

That means:

- If an *availability* η_S is required, the safety stock can be calculated with the standard formula (11.32) using the α-availability (11.35).

If the range of the replenishment quantity is shorter than the replenishment time, i.e. for $m_R/\lambda \leq T_R$, the α-availability equals the stock availability, i.e. $\eta_\alpha = \eta_S$.

11.8.4 Dynamic Safety Stock

If demand and/or replenishment times vary systematically in time, the safety stock has to be calculated dynamically with the current values for the mean demand $\lambda_m(t)$ and its standard deviation $s_\lambda(t)$, and for the mean replenishment time T_R and its standard deviation $s_T(t)$. These mean values and standard deviations can be derived from the values of the last periods by the method of *dynamic forecasting* with *exponential smoothing* as outlined in *Sect. 9.8*.

Inserting the current values for demand and replenishment time into formula (11.32) leads to the

► *formula for the dynamic safety stock*

$$\mathbf{m_{safe}}(t) = f_S(\eta_\alpha) \cdot \sqrt{\mathbf{T_R}(t) \cdot \mathbf{s_\lambda}(t)^2 + \lambda_m(t)^2 \cdot \mathbf{s_T}(t)^2} \tag{11.36}$$

with the *dynamic α-availability*:

$$\eta_\alpha(t) = \begin{cases} 1 - (1 - \eta_S) \cdot m_R(t)/(T_R(t) \cdot \lambda_m(t)) & \text{if } m_R(t) > T_R(t) \cdot \lambda_m(t) \\ \eta_S & \text{if } m_R(t) \leq T_R(t) \cdot \lambda_m(t) \end{cases} \tag{11.37}$$

The *cost optimal dynamic replenishment quantity* $m_R(t)$ results from inserting the current mean demand $\lambda_m(t)$ into formula (11.27).

The inability to deliver occurs with highest probability in the last days before the replenishment arrives. Therefore, the time of inability to deliver is generally shorter than the total replenishment time. Consequently:

► The stock availability achieved by a safety stock calculated with formulas (11.36) and (11.37) is slightly higher than required.

This consideration is confirmed by a comparison with the exact solution and by many simulations. As shown for an example in *Fig. 11.8*, the simulated mean stock availability turns out to be significantly higher than the required stock availability (Gudehus 2005). This also holds for instationary demand.

Figure 11.9 shows the dependency of the safety stock on the stock availability, if calculated with the conventional formula (11.32) and with the improved safety stock formulas (11.36) and (11.37). The comparison demonstrates:

► For the same stock availability, the conventionally calculated safety stock (11.35) is far higher than the safety stock calculated with the improved formula (11.36).

The improved safety stock formula (11.36) has been successfully applied and proven in practice. It can be implemented as add up to the standard scheduling software as offered by SAP, J.D. Edwards, Navision and others (Gudehus 2004/2007).

11.8.5 Influence Factors of Safety Stocks

The mean demand λ_m is the product (11.1) of the mean order quantity m_O with the mean order entry rate λ_O. Hence, if s_{mO} and $s_{\lambda O}$ are the standard deviations of

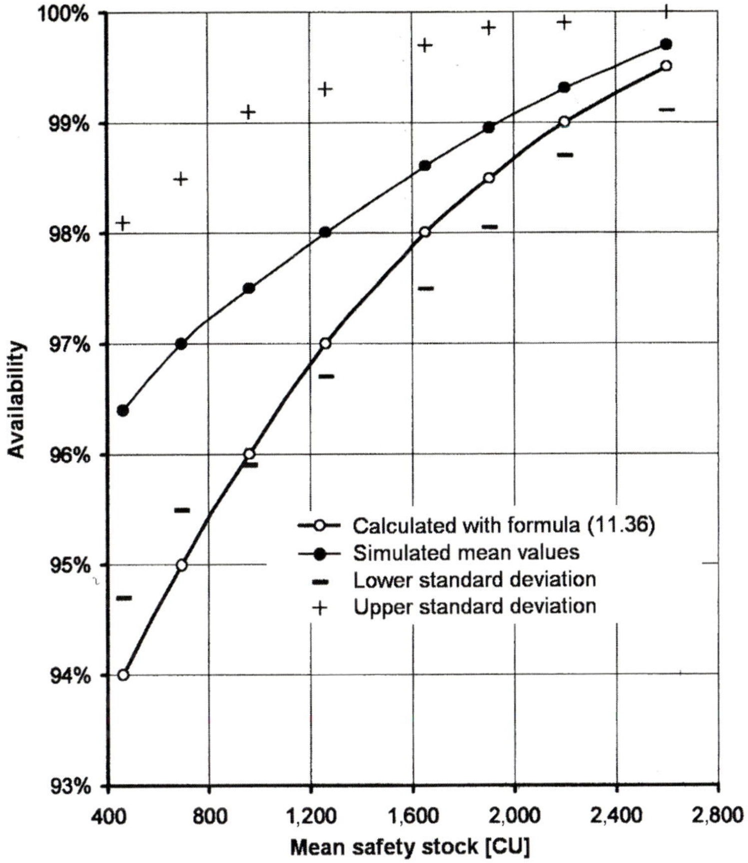

Fig. 11.8 Calculated and simulated stock availability as function of the safety stock

Replenishment time:	$T_R = 5 \pm 2$ d
Demand:	$\lambda(t) = 700 \pm 400$ CU/d
Replenishment quantity:	$m_R = 12.500$ CU/ROrd

stochastically varying order entry rate and order quantities, the standard deviation of the demand is

$$s_\lambda = \sqrt{m_O^2 \cdot s_{\lambda O}^2 + \lambda_O^2 \cdot s_{mO}^2} \tag{11.38}$$

From relation (11.38) and the fact, that the standard deviation for a random flow of single orders is $s_{mO} = \lambda_O$ (see *Sect.* 9.4.2), follows:

▸ The stochastic fluctuation of the demand is caused by the stochastic variation of order entry and order quantities.

► For constant order quantities, the standard deviation of the demand is proportional to the square root of the demand.

If the variations of the order quantities could be reduced, the stochastic variation of the demand and consequently the required safety stock can be diminished. This can be achieved by the *service strategy for big size orders*:

► *Big orders* are separated and executed completely or to a larger part not from stock, but by direct supply from the source.

As shown in *Sect. 11.12*, this strategy is also opportune for the costs.
From relations (11.36), (11.37) and (11.38), the most important influence factors on safety stock can be derived:

- The safety stock first increases slowly with the stock availability, but for higher availability it increases rapidly. For availability close to 100% the safety stock exceeds all limits (see *Figs. 11.9* and *11.10*).

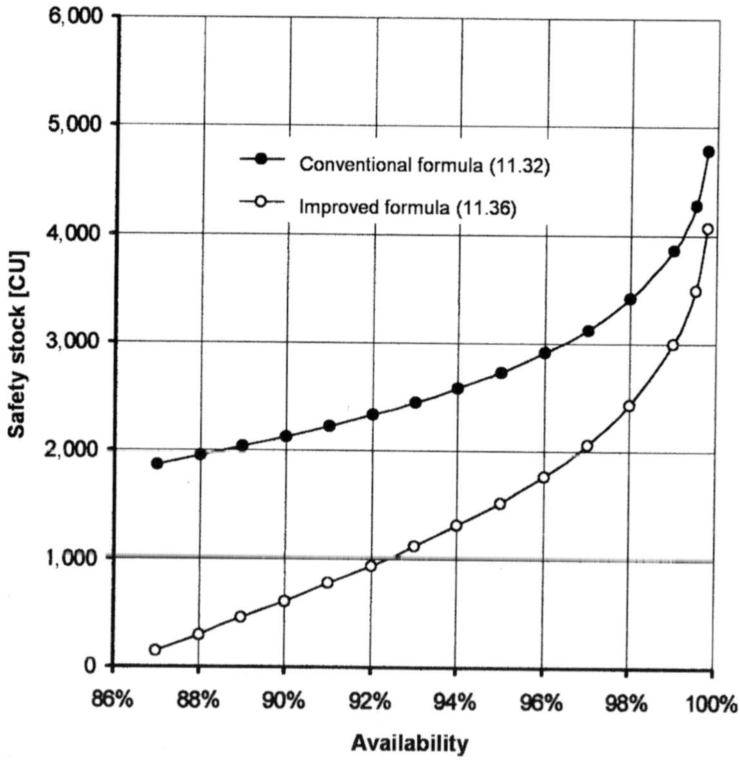

Fig. 11.9 Dependency of the safety stock on the required availability calculated with the conventional formula (11.32) and the improved formulas (11.36) and (11.37)

Parameters: see *Fig. 11.8*

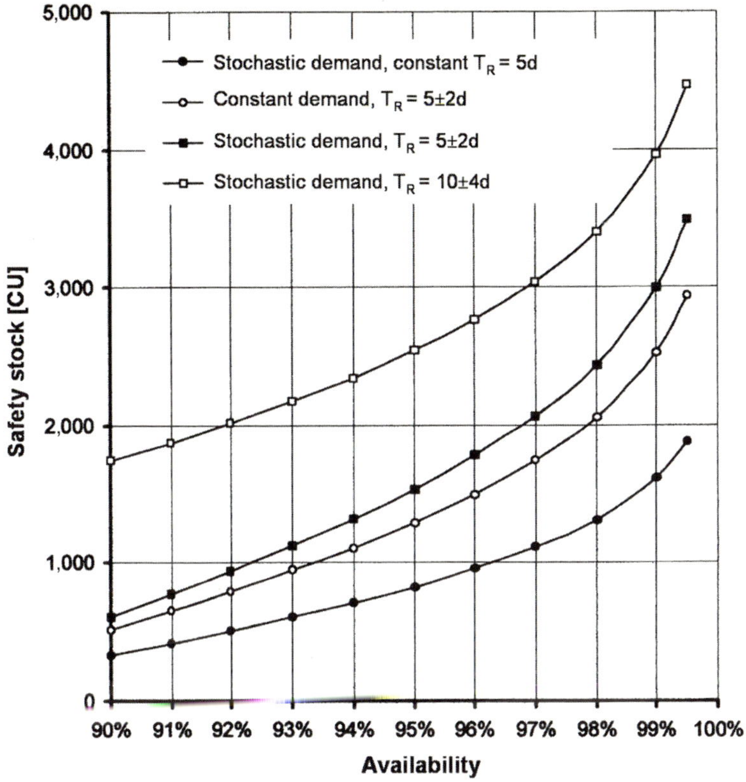

Fig. 11.10 Dependency of the safety stock on required availability for constant and stochastic demand and/or replenishment times

Other parameters: see *Fig. 11.8*

- The safety stock increases slowly with smaller variation and proportionally with higher variation of the demand (see *Fig. 11.11*).
- Above a certain minimal value, the safety stock increases with the square root of the replenishment time (see *Fig. 11.12*)
- For long replenishment times, the safety stock increases proportional to the standard deviation of the replenishment time (see *Fig. 11.13*).

The lower limit of the replenishment time in *Fig. 11.12* results from the fact, that an inability of delivery during the replenishment phase does not matter as long as the delivery times are short in relation to the replenishment cycle.

The dependencies of the safety stock on the different influence factors should be known and considered by schedulers, planners, purchasing managers, and sales people. The most important conclusion is:

► Long and even more unreliable replenishment times cause high safety stocks.

This proves the general principle: "punctuality before speediness".

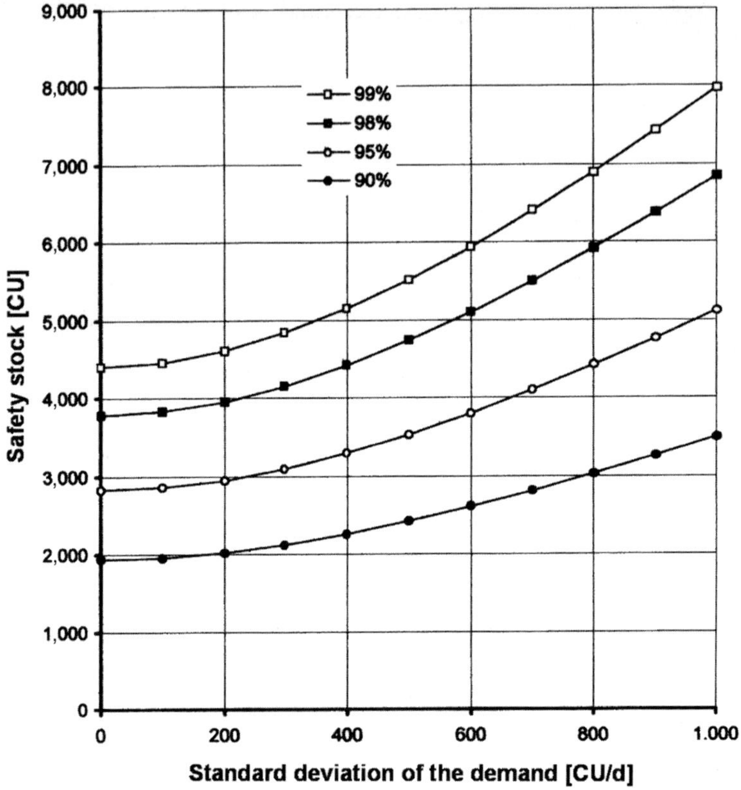

Fig. 11.11 Dependency of the necessary safety stock on the variation of demand for different required availabilities

Parameters: mean demand $\lambda = 700$ CU/d, others see *Fig. 11.8*

Production and suppliers should know the negative effects of long and varying delivery times. Customers and sales should be informed that 100% availability can never be achieved.

If the single orders arrive *randomly* with the same order quantity m, i.e. if $s_{mO} = 0$ and $s_{\lambda O} = \sqrt{\lambda_O}$, from relation (11.38) and the *theorem of large numbers* (9.23) follows:

$$s_\lambda = m \cdot \sqrt{\lambda} \qquad \text{for constant m and random } \lambda_O. \tag{11.39}$$

Inserting this into relation (11.32) gives the *square root law of safety stocks*:

► The optimal safety stock for random orders with constant quantities varies proportionally to the square root of the demand, if the variation of the replenishment time is small.

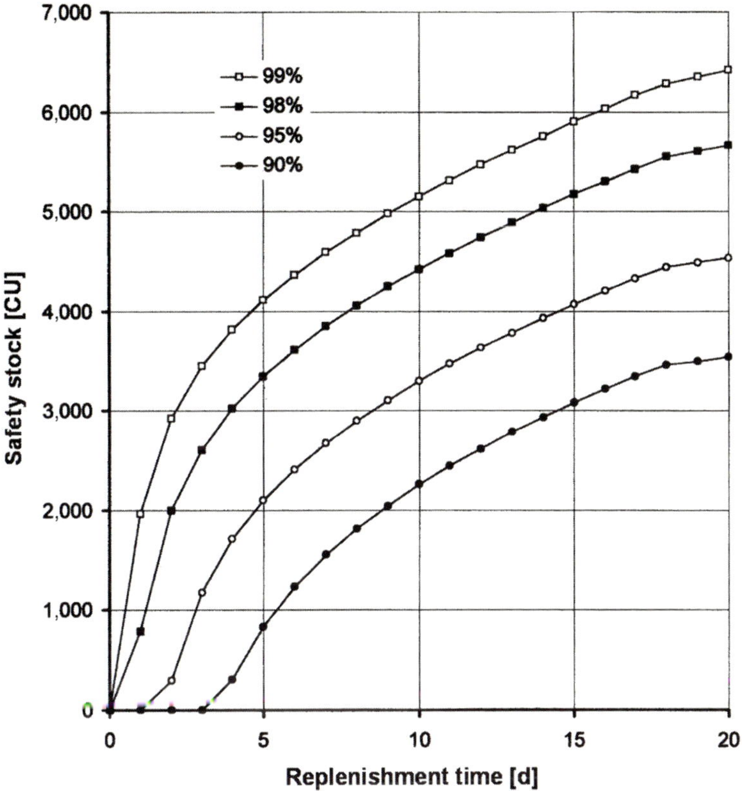

Fig. 11.12 Relation between safety stock and replenishment time for different stock availability

Other parameters: see *Fig. 11.8*

In combination with the square root law of the optimal replenishment quantity this leads to the *square root law of stock centralization*, as outlined in the next section.

11.8.6 Safety Costs and Stock Availability

Safety is expensive. This also holds true for the availability. The safety costs ensuring a required stock availability η are the costs for keeping a permanent safety stock $m_{safe}(\eta)$. Per consumption unit, these are the *specific safety costs*:

$$k_{safe}(\eta) = (P \cdot z_S + k_{SP}/C_{LU}) \cdot m_{safe}(\eta)/\lambda \qquad [€/CU]. \qquad (11.40)$$

Figure 11.15 shows that the specific safety costs increase enormously with the required availability η when approaching the 100% level. Safety costs decrease inversely proportional to the square root of the demand λ due to the square root law of safety stocks. They increase with the length and variation of the replenish-

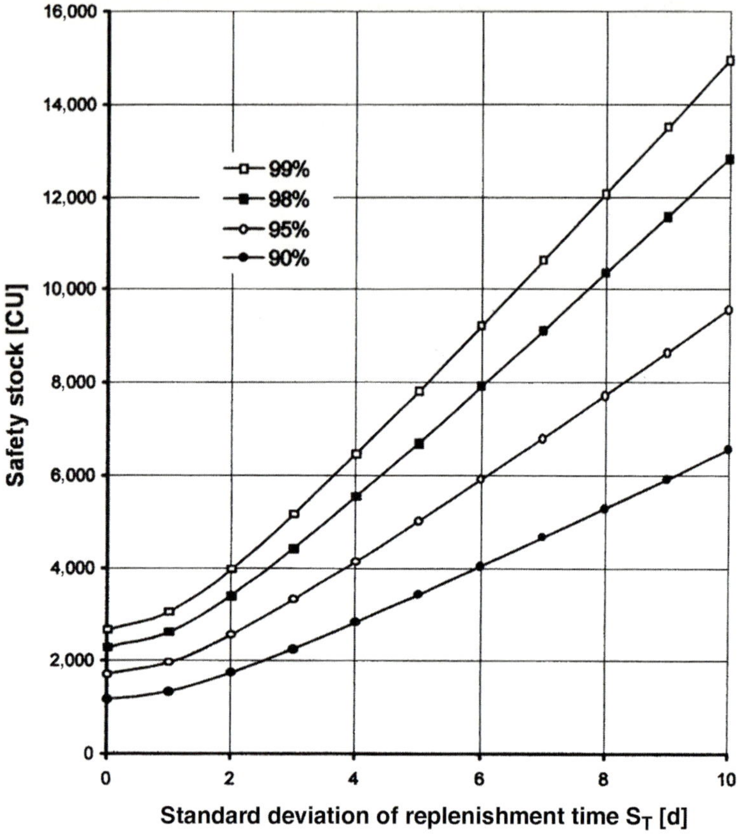

Fig. 11.13 Relation between safety stock and variation of the replenishment time for different availabilities

$T_R = 20 \pm S_T$ d *Other parameters*: see *Fig. 11.8*

ment time. For articles with high value and/or big size, the safety costs are far higher than for articles with low value and small sizes.

The counterpart to the safety costs are the *out-of-stock costs*. They increase proportionally to the *delivery inability* 1-η and decrease with increasing availability. Out-of-stock costs can be:

- *Loss of profit* or *loss of margin* for the sales shortfall of products or merchandise
- *Costs of production interruption* and *idle times* due to missing material or lack of supply
- *Deadlock costs* due to missing spare parts
- *Penalties for delayed delivery* or *fees for missing deadlines*

In some cases the *specific out-of-stock costs* k_{oost} [€/CU] per unit can be calculated, in other cases they can at least be estimated. Since out-of-stock costs occur

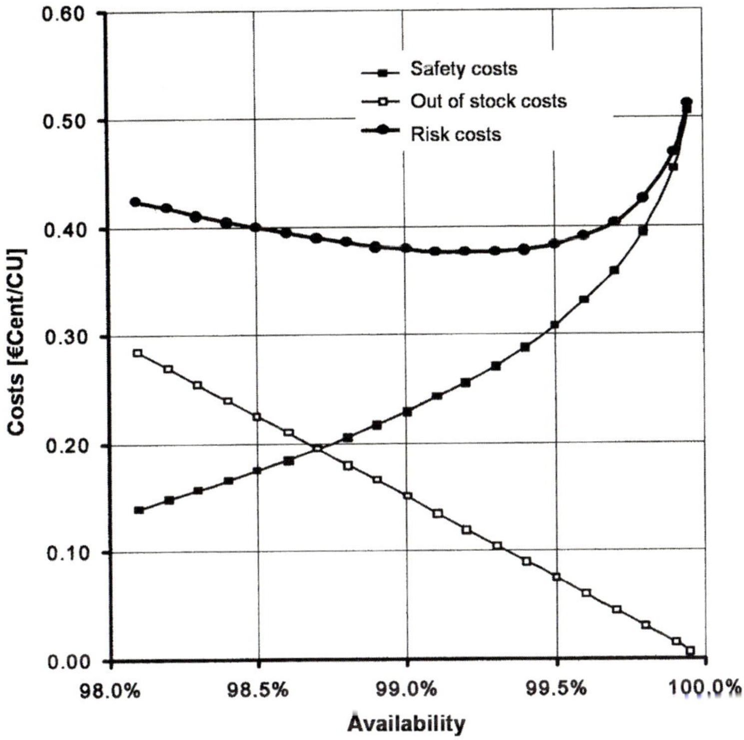

Fig. 11.14 Relation between risk costs and ex-stock availability

purchasing price: 2.50 €/CU *out of stock costs*: 0.15 €/CU
sales: 100 CU/d *other parameters*: see *Fig. 11.8*

with probability 1–η, the effective out-of-stock costs are $(1-\eta)\cdot k_{\text{oost}}$. The sum of safety costs and out-of-stock costs are the *specific risk costs*:

$$k_{\text{risk}}(\eta) = k_{\text{safe}}(\eta) + (1-\eta)\cdot k_{\text{out}} \qquad [€/\text{CU}]. \tag{11.41}$$

After inserting (11.40) and the formulas (11.36) and (11.37) for the safety stock, this relation gives the dependency of risk costs on the availability η. As illustrated in *Fig. 11.15*, the risk costs first decrease with increasing availability. After passing a minimum at an optimal value, they increase up to infinity. In the example, the optimal availability is $\eta_{\text{opt}} = 99.3\%$.

If the out-of-stock costs are known, the *optimal availability* η_{out} can be calculated by determining the point of minimal risk costs (11.41). Although this method might be tedious, it is the only objective way to determine the appropriate safety level.

Otherwise the availability has to be fixed by sales department or top management. This however, causes debates or is connected with arbitrariness. At least the availability for *product categories* should be determined with the help of model

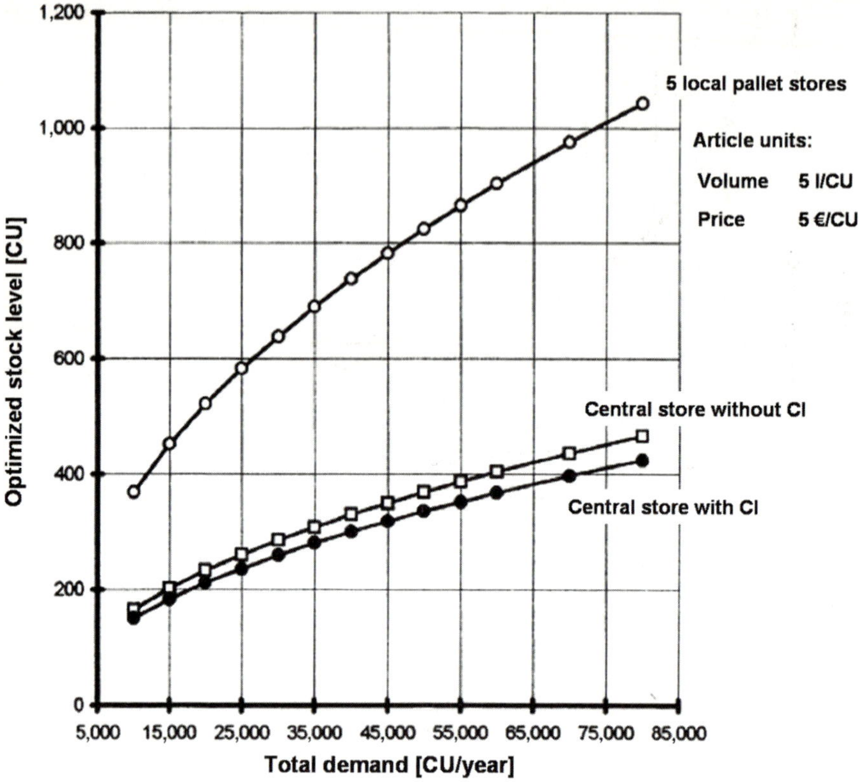

Fig. 11.15 Dependency of optimal stock on demand for a central store and for local stores
Local stores: sum of 5 local stores with equal demand
Central store without and with technical cost improvement (CI)

calculations based on relation (11.40) and (11.41). By this method, one can at least estimate whether a standard value of 95%, 98%, 99% or even 99.5% is appropriate and necessary.

Mathematical methods help to access the risk for events depending on measurable influence factors. As not all events are predictable, not all variations are measurable and not all costs are known, fixing the stock availability remains a top management decision with unavoidable risks.

11.9 Demand Dependency of Stock and Storekeeping Costs

The cost optimal stock level of a storekeeping article with stationary demand is the sum (11.28) of the safety stock and half of the optimal replenishment quantity. Due to relation (11.27), the optimal replenishment quantity is proportional to the square

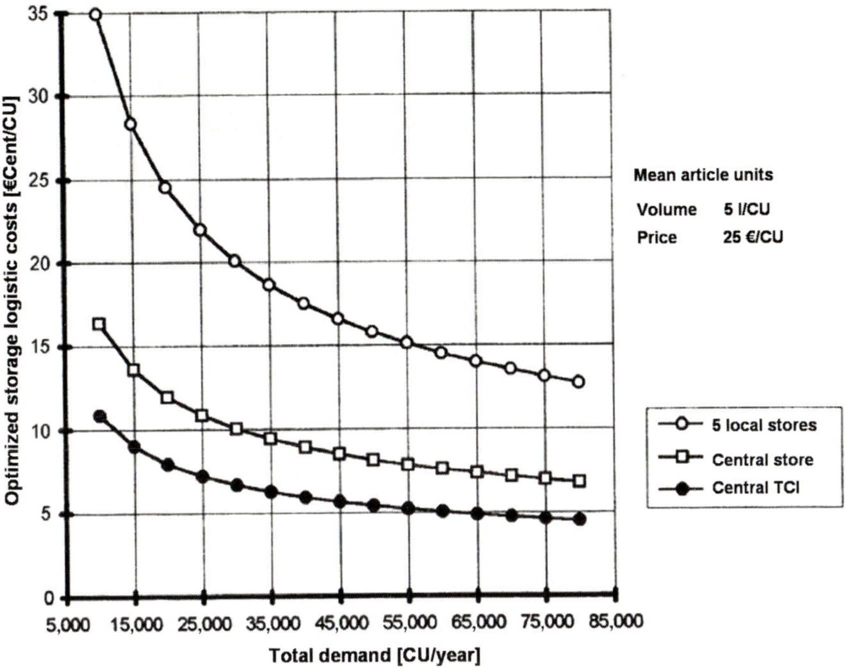

Fig. 11.16 **Dependency of specific storekeeping costs on demand for a central pallet-store with and without cost improvement compared with the sum of 5 local pallet stores**

Provision: optimal inventory management

root of demand. For sufficiently reliable replenishment times, also the safety stock depends on the square root of the demand. That leads to the *square root law of stocks*:

- The optimal stock level of articles with continuous demand is proportional to the square root of the demand λ

$$m_{\mathrm{Sopt}}(\lambda) = F_{\mathrm{S}} \cdot \sqrt{\lambda} \tag{11.42}$$

The *storage structure factor* F_S depends on the scheduling parameters, the required availability and the specific cost rates of the storage system. If these values are completely known, the structure factor can be calculated. From the square root law of stocks follow the *planning rules*:

- If the demand for an article is expected to change by a factor f_D, an optimally scheduled stock level changes only by the square root of this factor.
- By optimal inventory management, the *stock growth factor* f_S of the stock can be kept equal to the square root of the *demand growth factor* f_D:

$$f_{\mathrm{S}} = \sqrt{f_{\mathrm{D}}} \tag{11.43}$$

These planning rules can be applied to calculate the stock for increasing and decreasing sales, as well as for seasonal fluctuations. For example, if the demand doubles, the stock level increases only by a factor of 1.41 or 41% provided the inventory scheduling is performed optimally.

If the inventory scheduling is optimal and the effects of partially filled load units can be neglected, the relations (11.22), (11.27) and (11.29) lead to the *demand dependency of specific storekeeping costs*:

$$k_{\mathrm{Ropt}}(\lambda) = \mathrm{K}_{\mathrm{RSmin}}/\lambda = k_{\mathrm{o}} + k_1/\sqrt{\lambda} \qquad [€/\mathrm{CU}]. \tag{11.44}$$

The specific *fixed costs* k_o are determined by the transport- and in-storing-cost-rates which are independent from the throughput. The variable share of the specific storekeeping costs (11.44) with the *proportionality factor* k_1 is determined by the order costs and storing costs which both decrease with the throughput. The general dependency (11.44), which is shown for an example in *Fig. 11.17*, leads to the *square root law of storekeeping costs*:

▸ If replenishment and stocks are optimally scheduled, the specific storekeeping costs decrease inversely with the square root of the demand until asymptotically the specific transport and in-storing costs remain.

A further result is that the specific storekeeping costs are much lower if the total demand of an article is delivered from a central store instead of several local stores. This is due to the lower optimal stock of the central store and to the *economy of scales* and *lower cost rates* which can be achieved by modern storage technology for big storage and handling systems.

11.10 Centralization of Stocks

In order to optimize a supply network and to select the optimal delivery chains, it is necessary to estimate the potential savings of *stock centralization* and to quantify how far the consolidation of local stocks may reduce the total stock. If λ_{Ai} [CU/PE] are the *regional demands* of an article A which are served by N *local stores* $\mathrm{LS_i}$, i = 1, 2,....N, the *total demand* for that article is the sum:

$$\lambda_{\mathrm{A}} = \sum_{\mathrm{i}} \lambda_{\mathrm{Ai}} \qquad [\mathrm{CU/PE}]. \tag{11.45}$$

If optimally scheduled, due to relation (11.42), the single stocks in the local stores are:

$$\mathrm{m_{Ai}} = F_{\mathrm{LS}} \cdot \sqrt{\lambda_{\mathrm{Ai}}} \qquad [\mathrm{CU}]. \tag{11.46}$$

The *local storage structure factor* F_{LS} depends on the scheduling parameters, the cost rates and the required availability of the local stores. Correspondingly, the stock level of an optimally scheduled *central store* CS, which serves the total demand (11.45), is:

$$\mathrm{m_{AC}} = F_{\mathrm{CS}} \cdot \sqrt{\lambda_{\mathrm{A}}} \qquad [\mathrm{CU}]. \tag{11.47}$$

with the *central storage structure factor* F_{CS} of the central store. Solving Eq. (11.46) with respect to λ_{Ai} and inserting the result into (11.47) leads to the *law of article stock centralization*:

► By consolidation of optimally scheduled stocks of the same article from local stores in an optimally scheduled central store, the sum of the local stocks m_{Ai} can be reduced to the central stock:

$$m_{AC} = (F_{CS}/F_{LS}) \cdot \sqrt{\sum_i m_{Ai}^2} \qquad [CU]. \tag{11.48}$$

This relation holds for each article A separately. For example, with equal storage structure factors F_{CS} and F_{LS}, the consolidation of three local stocks $m_1 = 300$ CU, $m_2 = 400$ CU and $m_3 = 500$ CU of the same article, which add up to $m_L = 1{,}200$ CU, leads to the central stock $m_C = \sqrt{300^2 + 400^2 + 500^2} = 700$ CU. The stock reduction by centralization is 41% in this case. If the same sum stock $m_L = 1{,}200$ CU has been located more unequally, with $m_1 = 1{,}000$ CU in the first and $m_2 = m_3 = 100$ CU in the second and third store, the achievable central stock is $m_C = \sqrt{1,000^2 + 100^2 + 100^2} = 1,010$ CU. The central stock is 43% higher and the reduction of 14% far lower than in the other case.

By summation of (11.48) over all articles A which are stored in the local stores results the *law of stock centralization*:

► By consolidation of the stocks from several local stores with the same range of articles in a central store the sum of the local stocks for all articles can be reduced to the total central stock:

$$m_C = (F_{CS}/F_{LS}) \cdot \sum_A \sqrt{\sum_i m_{Ai}^2} \qquad [CU]. \tag{11.49}$$

If the regional demands λ_{Ai} of the single articles do not differ very much, the sums over A and i can be interchanged without affecting the result substantially. This leads to the *approximative law of stock centralization*

► If the regional demands of articles A kept on stock in N local stores LS_i, $i = 1,2,...N$, are of the same order of magnitude, the *sum of the total local stocks* m_{Li}

$$m_L = \sum_i m_{Li} \qquad [CU]. \tag{11.50}$$

can be reduced by consolidation in one central store with optimal scheduling to the *total central stock*

$$m_C = (F_{CS}/F_{LS}) \cdot \sqrt{\sum_i m_i^2} \qquad [CU]. \tag{11.51}$$

If the local and the central storage structure factors are equal, i.e. if $F_{CS} \approx F_{LS}$, relation (11.51) results in the *square root rule of stock centralization*:

► The optimally scheduled central stock is equal to the square root of the sum of the squares of the optimally scheduled local stocks.

For local stores with approximately the same assortment and demand, relation (11.51) leads to *Maisters Square Root Law* (Maister 1976):

► By centralization of the stocks of N local stores with similar demand the total stock can be reduced by a factor $\sqrt{N}$.

For example, if 4 approximately equal local stocks are centralized, the resulting total stock could be ½ of the sum of the local stocks. The centralization of 16 local stocks can reduce the total stock by a factor ¼.

Maisters simple square root law is often applied in business practice without considering the restrictive prerequisites, such as equal local stocks and equal assortments, optimal scheduling and equal structure factors. This can cause exaggerated expectations, incorrect storage planning and wrong decisions that cannot be corrected after a central store has been realized. Centralising non-overlapping assortments does lead to stock reductions, even if they are optimally scheduled.

A centralization of local stocks not only reduces the costs for interest and store-place occupation, but also the total storekeeping costs, since the optimal replenishment quantities increase and the replenishment frequency is reduced by the same factor as the stock level. That means:

► The centralization of stocks reduces the total storekeeping costs by the same factor as the total stock level.

Provided all premises are fulfilled, the square root laws indicate substantial potentials for stock reduction and cost savings by centralization. However, when applying the square root laws, one should always keep in mind that relation (11.51) is an *approximation* and that the rules derived from this relation are *rules of thumb*. If the regional demands differ by more than a factor 10 and the local article stocks by more than a factor 3, the approximation (11.51) can no longer be applied. In these cases, the central stock calculated with (11.51) is more than 15% lower than the correctly calculated central stock (11.49).

But even if calculated with the correct formula (11.51), the expected stock reduction and costs savings can only be reached by *optimal inventory management*. As in many companies scheduling programs and parameters are not optimal and remain unadjusted after centralization, the central store does not lead to the expected results.

Further deviations can be caused by different *storage structure factors*, which depend on the required availability and on the cost rates of the storage systems. If e.g. the centralization of stocks is used in order to improve the *availability*, the central storage structure factor will become higher than the local storage factors. This leads to a higher central stock and reduces savings.

Furthermore, the following cost improvements can lead to differences of storage structure factors for small local stores and large central stores:

- Due to higher throughput and capacity, the in- and out-storing costs and the place costs of a central store are lower than the respective cost rates of local stores, even if the same storage technique is used.

- The implementation of modern storage and conveyor techniques reduces the performance costs of a larger store significantly, when compared to the performance costs of smaller stores with conventional techniques.
- In a central store with higher stock, load units with larger capacity and higher filling degree can be used, which reduces the costs further.

For example, an automated high bay store is more cost efficient in comparison to a manually operated fork lift store, if the total stock exceeds 5,000 pallets (see *Sect. 16.13*). These and further effects impact the performance cost rates for central and local stores as shown in *Table 11.4*. This results in the *rule of thumb*:

- The structure factor for a large central store with the same availability is 10 to 20% lower than the structure factor for small local stores.

Due to this effect, even higher stock reductions and cost savings are possible by centralization. *Figure 11.15* shows the stock dependency on the total demand for 5 local stocks of the same size compared with one central store with and without cost improvements. The cost savings can be read from *Fig. 16.16.*

Even if the deliveries from a central store are faster than the direct deliveries from a manufacturer, in many cases after stock centralization, the local stations still have to keep *buffer stocks* or *sales stocks* in order to offer competitive delivery times. Albeit reduced by the centralization, the sum of the local stocks adds up to the central stock. Their replenishment from the central store reduces the total savings. To keep the local stocks as small as possible, they have to be scheduled with the optimal *replenishment strategy* as outlined in the following section. Another strategy, which combines the advantages of local stocks with the potentials of a central stock, schedules the sum of local stocks as a *virtual central store* (Gudehus 2006).

Besides stock reduction and improvement of availability, the main driving force for stock centralizing is reduction of total costs. For example, *Fig. 11.16* shows how centralization reduces storekeeping costs and how these costs can be lowered further by modern storage techniques.

However, when estimating the savings of a central store, one has to take into account the additional transports that result from the replenishment transports between the central store and the local stations. They reduce the total savings. In order to assess all the effects of a *logistic center* correctly, it is necessary to consider the total supply network including external transports as explained in *Chap. 21.*

11.11 Replenishment Strategies

Dependent on the trigger for a replenishment order, three different *replenishment methods* can be distinguished:

$$\begin{array}{l}\text{replacement method (r)}\\ \text{reorder-point method (s)}\\ \text{cycle-time method (t)}\end{array} \tag{11.52}$$

The *replenishment trigger* for the *replacement method* is emptying the content of an *access unit* or the consumption of an *access quantity*. For the reorder point method, it is reaching the *reorder stock* (11.18). For the cycle-time-method, the necessity of replenishment is checked cyclically at certain *scheduling dates*.

Options for the replenishment quantity are

fixed quantity (F)
cost optimal quantity (Q) (11.53)
fill-up quantity (S)

The fixed quantity is an externally required *minimal supply quantity* or an internally *fixed replenishment quantity*, such as a full *pallet*, *bin* or another *replenishment unit*. The cost optimal quantity can be calculated dynamically by the general *EOQ-formula* (11.27) or the simpler *Harris-formula* (11.30). The fill-up quantity is needed to regain a certain *target stock*, which is determined by the capacity of a fixed storeplace, sales shelf, buffer space or by other criteria.

The combination of the three replenishment methods (11.52) with the three options (11.53) results in the 9 different *replenishment strategies*:

(r;F) (r;Q) (r;S)
(s;F) **(s;Q)** (s;S) (11.54)
(t;F) (t;Q) (t;S)

The most important characteristics and application criteria for these replenishment strategies are listed in *Table 11.6*. As will be shown later, in most cases the *standard strategy* (s;Q), i.e. reorder point replenishment of cost optimal quantities, is opportune and applicable. It achieves minimal total costs and keeps the required availability.

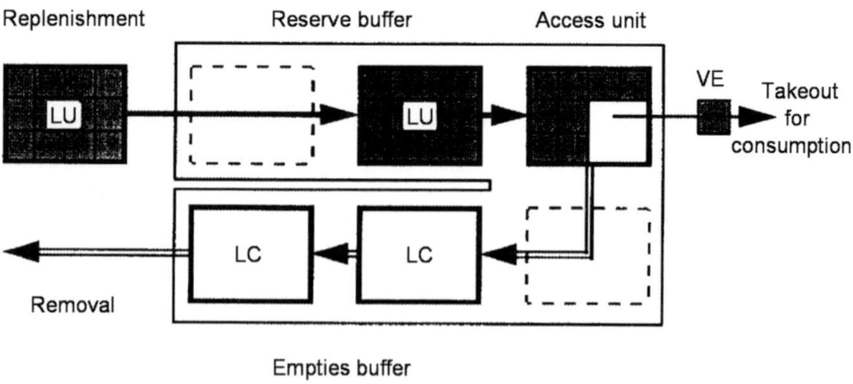

Fig. 11.17 Provision buffer and replenishment process

Buffer place capacity: $C_{BP} =$ 3 full load units (LU)
3 empty load carrier (LC)

Replacement unit capacity: $C_{RU} = 12$ consumption units

Table 11.6 Characteristics and criteria of replenishment strategies (11.54)

	Replenishment Methods		
	Replacement	**Reorder Point**	**Cycle Time**
Characteristics			
Dependency on	consumption	stock level	cycle
Checking Points	filling/take-out	take-out	time sched.date
Replen.Trigger	empty access unit	reorder quantity	reorder time
Replen.Quantity	full load unit	fix/opt/fill-up	fix/opt/fill-up
Criteria			
Advantages	minimal stock self-regulating	minimal costs self-regulating	low costs, opt. article bundling
Disadvantages	high repl.costs risky availability	limited article bundling	higher stock ext.control
Application	buffer stocks few expensive buffer places	reserve stocks sufficient cheap store places	sales stocks sufficient cheap store places
Conditions	short&reliable replen.times	longer&varying replen.times	planned replen.cycles

11.11.1 Replacement Method

The replacement method is applicable for the replenishment of *provision buffers*. The consumption station which must be supplied without interruption can be a machine, working station, assembly line, picking place, shelf in a retail outlet or a service station with continuous demand. The basic design of a *provision buffer* and the *replenishment process* are presented in *Fig. 11.17*. A *reserve unit* stands by in a *pre-buffer* behind an *access unit* on a buffer place close to consumption. It will be moved to the access place after the content of the access unit has been emptied.

The removal of the empty unit can be triggered by

1. the last item taken from the access unit
2. the last reserve unit moved to the access place
3. reaching a fixed reorder stock

The first two triggers differ only if the pre-buffer has capacity for more than one load unit. The load units can be bins, boxes or small containers with a removable *identification label*, *tag* or *card*. In Japanese such a card is called "Kanban". Hence, the replacement method with bins and cards is generally known as *Kanban*. Invented originally in Japan, nowadays Kanban is widely applied in the *automobile industry* and other *manufacturing plants*.

When the trigger event happens, the card is removed and attached to a board signaling a supply demand. Depending on the capacity of the provision buffer *One-Container-Kanban*, *Two-Container-Kanban* or *Many-Container-Kanban* are possible (Schonberger 1984, 1987 and 1988; Slack et al. 2004). The removable card of the Kanban-container can be replaced by a fixed *barcode label* or a *transponder* which is detectable by an electronic device. This advanced Kanban without cards is called *electronic Kanban* or *e-Kanban* (see *Sect. 12.7*).

If the buffer has a capacity for only two load units, e.g. for 2 containers or 2 pallets, the replacement method is called *Flip-Flop-Method*. The first unit is the access unit. The other one is either a full reserve unit or an empty load carrier waiting for replacement. When the last item has been picked, the two units exchange their role. The reserve unit becomes the access unit and the access unit the empty unit. The Flip-Flop-Method can be found in conventional *commissioning systems* (see *Chap. 17*) and in the *production*.

Replacement is also possible *without load carriers*. In this case, the single *consumption unit* itself is the access and reserve unit. The buffer can be a *rolling conveyer* in a flow shelf or simply a place in a rack. If the buffer has capacity for one consumption unit only, a replacement of single items without load is necessary. This so called *One-Piece-Flow-strategy* is applied for slow moving articles of high value or of big size, for which the demand during replenishment time is less than 1 CU. Examples are *spare part stores* and *jewlery shops*.

Due to the triggering of the replacement, the replenishment frequency (11.2) is determined in a *self-regulating* manner by the consumption and the content of the replacement unit (Gudehus 2006). This results in the main advantage:

► The replacement method requires neither a demand forecast nor a dynamic calculation of the replenishment quantity.

Free *strategy parameters* of the replacement method are:

- *capacity* C_R [CU/LU] of the replacement units, i.e. of the load unit LU.
- *number* n_R [LU/ROrd] of replacement units per replenishment

In order to ensure high availability and to avoid an inefficiently high replenishment frequency, the capacity must be fixed according to the *dimensioning principle*:

► The n_R replacement units should at least contain the safety stock plus the demand during the *maximal replenishment time*

$$n_R \cdot C_R \geq m_{safe} + T_{Rmax} \cdot \lambda \tag{11.55}$$

The replenishment quantity is restricted by the *buffer capacity* C_B for full units, i.e.:

$$n_R \cdot C_R \leq C_B - 1 \tag{11.56}$$

In order to minimize costs, the total replenishment quantity $n_R \cdot C_R$ should be approximately equal to the *economic order quantity* (11.27). Therefore, relations (11.27) and (11.56) determine also the capacity of the provision buffer.

In most cases, the strategy parameters of the replacement method are not used systematically to ensure a required availability and a cost optimal supply process. If the capacity of the load unit is smaller than (11.55), the risk of an empty access unit is high. If the product $n_R \cdot C_R$ of the number n_R of replacements with the capacity C_R of the replenishment unit deviates too much from the economic order quantity, the costs of the replacement method become far higher than the costs for the reorder-point method with optimal replenishment.

The risks and disadvantages of false strategy parameters increase, if the demand is not stationary and the parameters are not dynamically adapted to a changing demand. A *dynamic adaptation* of strategy parameters is possible with *e-Kanban*, as outlined in *Sect. 12.7*.

11.11.2 Reorder-Point Method

The reorder-point method is applicable for any type of storekeeping article. After each incoming order the scheduler or a program checks whether the reorder point is reached. If the stock falls below the *reorder stock* (11.18), there are two *replenishment options*:

- *Single article reorder point-strategy* (*Fig. 11.3*): After reaching the reorder stock, a replenishment order with fixed, cost optimal or fill-up quantity is released independently of the replenishment of other articles.
- *Consolidated reorder-point-strategy*: When the reorder stock is reached for one article, it is checked for all other articles from the same source, whether the *target stock difference*

$$\Delta_S(t) = m_{max}(t) - m_S(t) \tag{11.57}$$

  between the maximal stock (11.8) and the actual stock exceeds a minimal replenishment quantity m_{Rmin}. For a cost optimal share of these articles, a consolidated replenishment order is released.

The consolidation strategy anticipates the replenishment for articles that have not yet reached the reorder stock. Potential *advantages of a consolidated replenishment* are:

- ▶ By consolidated replenishment, the optimal *exploitation of discounts* for larger order sizes and order values can be achieved.
- ▶ If the demand from the same source is high, number and replenishment quantities of the consolidated articles can be adapted in such a way that *full load units* or – even better – *full truck loads* result in total.
- ▶ Different articles produced by the same production station without longer switch times can be produced in sequences with *minimal setup times*.

For example, stored liquids can be bottled into cans or bottles of different sizes with individual labels. Consolidated replenishment generates lower proportionate order and transport costs for the single article. The *strategy parameter* to maximize the savings is the maximal *pre-order time* for the anticipation of future replenishments.

The reorder-point method requires dynamic demand forecasts, calculation of the current reorder stock and inventory control after each incoming order. If performed manually, this is quite tedious and expensive. Nowadays, the scheduling can be supported and performed by *scheduling programs* using the above formulas and algorithms. For inventory scheduling by computer holds the 1st *replenishment scheduling rule*:

- ▶ Provided the replenishment quantity is calculated with formula (11.27) and the safety stock with (11.36) and (11.37) using correct parameters and cost rates, the *reorder-point method* (s;Q) *is the optimal replenishment strategy* for articles with regular demand.

With the standard strategy (s;Q), the required availability can be achieved at minimal costs, as long as there are no *bottlenecks* in the supply chain.

11.11.3 Cycle-Time Method

The cycle-time method is adequate for manual scheduling and for articles from sources with cyclic replenishment or cyclic supply tours.

The necessity for replenishments is checked at *scheduling dates* $t_{Sj} = t_o + j \cdot T_{SC}$, j = 1, 2, 3..., which are determined by the

- *scheduling cycle time* T_{SC} [PE] or *scheduling frequency* $f_{SC} = 1/T_{SC}$ [1/PE].

Customary scheduling cycles are *monthly scheduling* on certain calendar dates, *weekly scheduling* on a fixed day of the week or *daily scheduling*. *Replenishment options* are:

- *Single article cycle-time strategy*: For all articles whose current stock has already reached or will reach the reorder point until the next scheduling date, a replenishment order is released.
- *Consolidated cycle-time strategy*: For all articles which are supplied from the same source, it is checked whether the *target stock difference* (11.57) for a fixed *pre-order time* exceeds the minimal order quantity. For a cost optimal share of these articles, a consolidated replenishment order is released.

In both cases, the reorder quantity can be a fixed, fill-up or cost optimal quantity. The advantages of cyclically consolidated replenishment are the same as for the consolidated reorder-point method. However, with cyclic scheduling optimally filled load units and transport means are easier to achieve.

On the other hand, cyclic scheduling leads to higher stocks than the reorder-point method, as the replenishment orders are released earlier. The *anticipation time*, which is on average $T_{SC}/2$, increases the *mean stock of a cyclically scheduled article* to

$$m_{S\,cycle} = m_{S\,opt} + T_C \cdot \lambda/2 \qquad (11.58)$$

From this formula results the *2nd replenishment scheduling rule*:

► The stock level with cyclic scheduling exceeds the stock level with reorder-point scheduling by half of the consumption during the scheduling cycle.

For very short cycle times, i.e. for $T_{SC} \to 0$, cyclic scheduling becomes reorder-point scheduling. A consequence is the *3rd replenishment scheduling rule*:

► If replenishment scheduling is performed cyclically, the scheduling cycles should be as short as possible in order to avoid excessive inventory.

The mean range of inventory can be reduced by 8 working days, if scheduling is changed from a monthly to a weekly basis and by 2.5 working days when changing from weekly to daily scheduling. For example, the total inventory of a leading German *chemical company* was reduced by more than a third after switching from monthly to weekly scheduling.

11.12 Cost-Opportunity of Storekeeping

In addition to the general rules of *Sect. 11.2*, the *opportunity threshold of storekeeping* is a quantitative criterion to decide, whether it is more profitable to hold an article in stock or not. This threshold results from comparison of the cost dependency on demand for *delivery-to-order* and *delivery-from-stock*. The opportunity threshold also separates orders, which are cheaper to deliver from stock or to order.

11.12.1 Delivery-to-Order Costs

As long as the required delivery times are met, the incoming orders can be collected for a certain *order consolidation time* T_{OC}. Depending on competition and urgency of the orders, common consolidation times are several days up to a few weeks. If very short delivery times have to be met, the longest consolidation time is one working day. After consolidation, the collected orders are sent to the supplier as one single *direct order* [DOrd]. This reduces the proportionate order costs of procurement and production.

Analogous to the *replenishment order costs* k_{ROrd} [€/ROrd] for a storekeeping article of *Sect. 11.6.1*, the *direct order costs* k_{DOrd} [€/DOrd] consist of the administrative ordering costs of the delivering station and the order processing costs of the supply station, which for producing stations are mainly *setup costs* (see *Fig. 11.2*).

If the ordering process is the same, the direct order costs are equal to the replenishment order costs. In some cases direct orders are processed manually whereas replenishment orders are mainly processed by computer. In other cases, the delivery times for direct orders are much shorter than for replenishment orders. In these cases the order costs for direct delivery are higher.

For stationary demand λ [CU/d], an order consolidation time T_{OC} [d] leads to the mean *direct order quantity* $m_D = \lambda \cdot T_{OC}$ [CU/DOrd]. Therefore, the *specific costs for delivery-to-order*, i.e. the delivery costs per consumption unit, are:

$$k_{DL}(\lambda; T_{OC}) = k_{DOrd}/m_D = k_{DOrd}/(\lambda \cdot T_{OC}) \qquad [€/CU]. \qquad (11.59)$$

From relation (11.59) follows the general *rule for delivery to order*:

► For delivery to order the mean costs per consumption unit decrease inverse proportional with demand and consolidation time.

Figure 11.18 shows the dependency of the specific costs on the demand for two different consolidation times. Relation (11.59) holds not only for stationary but also for dynamic order entry rates with stochastically and systematically varying arrival times and quantities. This has been tested by many *simulations*.

11.12.2 Storekeeping Costs

The costs per consumption unit for deliveries from stock with optimal scheduling are the minimal storekeeping costs related to the demand λ [CU/PE]. Neglecting the costs for partly filled load units, the insertion of relation (11.27) for the optimal replenishment quantity into relations (11.22), (11.25) and (11.26) and the division by λ gives the *specific costs for delivery from stock*:

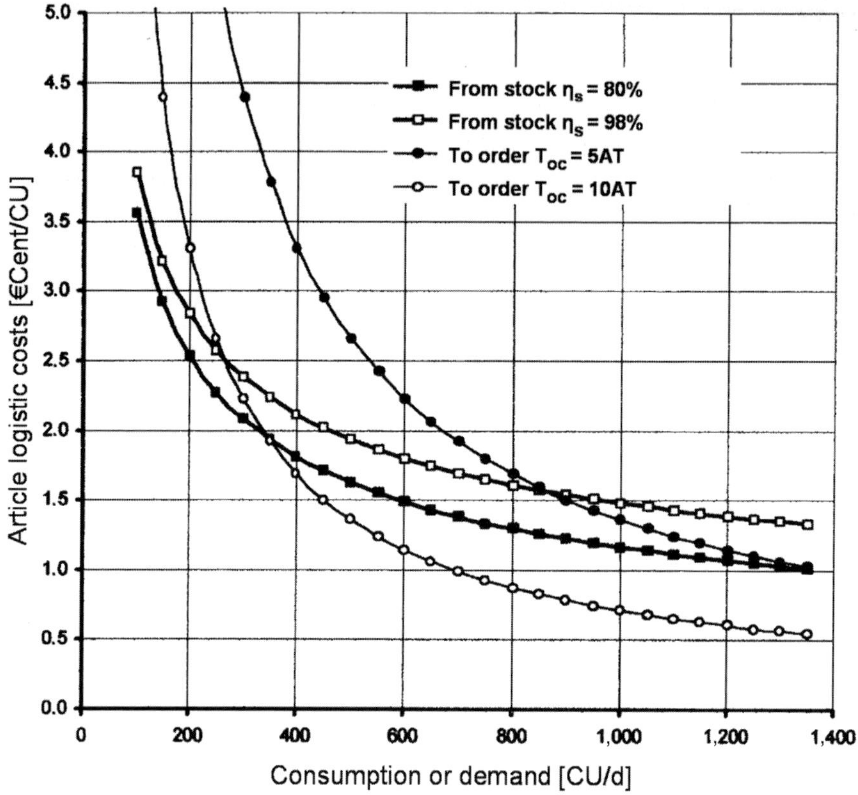

Fig. 11.18 Dependency of mean costs per consumption unit on the demand for delivery to order and delivery from stock

Deliver to order with order consolidation: relation (11.59)
Consolidation time: 5 and 10 d
Direct Order costs: 65.00 €/DOrd
Deliver from stock with optimal replenishment: relation (11.60)
Ex-stock availability: 80% and 98%
Replenish Order costs: 65.00 €/ROrd
Purchasing price: 2.50 €/CU
Inventory interest rate: 9% p.a.
Store place costs: 0.25 €/Pal-d
Pallet capacity: 3,200 CU/Pal

$$k_{SL}(\lambda; \eta_S) = K_{SK}(m_R)/\lambda =$$
$$= (P \cdot z_S + k_{SP}/C_{LU}) \cdot m_{safe}/\lambda + (2k_{Rord} \cdot (P \cdot z_S + 2f_{PO} \cdot k_{SP}/C_{LU})/\lambda)^{1/2} \quad [€/CU] \tag{11.60}$$

The first part are the safety costs and the second the optimal storage- and replenishment costs per consumption unit. The dependency of the specific costs for delivery from stock on the demand is shown in *Fig. 11.18* for the same example as for delivery to order. Parameter is the *availability*, here 80% and 98%, which determines the specific safety costs.

Taking into account that the safety stock increases with the square root of the demand, relation (11.60) leads to the general *rule for deliveries from stock*:

► For deliveries from stock, the mean costs per consumption unit decrease inverse proportional to the square root of the demand and increase over proportional with the required availability.

The specific delivery from stock costs (11.60) decrease slower with increasing demand than the delivery to order costs (11.59). As shown in *Fig. 11.18*, the two cost curves intersect at the *opportunity-threshold of storekeeping*.

11.12.3 Opportunity-Threshold of Storekeeping

The difference between the specific costs (11.59) for delivery to order and (11.60) fo delivery from stock is the specific *storekeeping profit*

$$p_{SK}(\lambda) = k_{DL}(\lambda; T_{OC}) - k_{SL}(\lambda; \eta_S) \qquad [€/CU] \tag{11.61}$$

The storekeeping profit depends on the demand λ, the tolerable consolidation time T_{OC} for direct delivery, and on the stock availability η_S. For lower demand, the storekeeping profit is positive and storekeeping is advantageous. For very high demand, it is negative and delivery to order is more profitable. At the *break even point*, the specific costs for delivery to order and for delivery from stock are equal and the storekeeping profit vanishes. After insertion of relations (11.59) and (11.60) into (11.61), the solution of the equation $p_{SK}(\lambda) = 0$ with respect to λ gives the

- *storekeeping opportunity-threshold*

$$\lambda_{SKopp} = \left(k_{DOrd}/T_{OC} + (P \cdot z_S + k_{SP}/C_{LU}) \cdot m_{safe}\right)^2 / \left(2k_{ROrd}(P \cdot z_S + 2\, f_{PO} \cdot k_{SP}/C_{LU})\right). \tag{11.62}$$

The opportunity threshold depends on the tolerable consolidation time for direct delivery and on the required stock availability. The dependency of the storekeeping opportunity-threshold on the consolidation time is shown in *Fig. 11.19*. The cost advantage leads to the general *storekeeping rule*:

► As long as the regular demand or consumption is lower than the storekeeping threshold (11.62), it is opportune to deliver articles from stock instead consolidating and sourcing them directly.

From relation (11.62) the following *decision rules* can be derived:

► The storekeeping opportunity-threshold decreases with increasing value and size of the article units.

► Higher interest rates and storeplace costs reduce the storekeeping opportunity threshold to lower demand.

► Storekeeping threshold and profit increase with higher direct order costs and lower replenishment order costs and by reducing the stock availability.

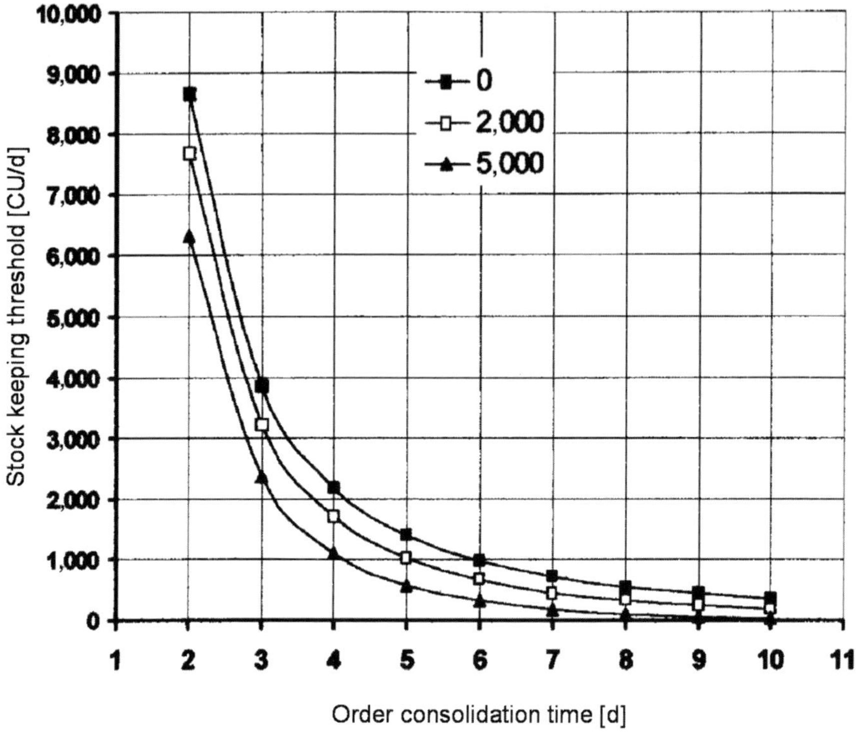

Fig. 11.19 Dependency of the storekeeping opportunity-threshold on order consolidation time

Parameters:	safety stock: 0/2,000/5,000 CU
	article price: 2.50 €/CU
Other parameters:	see *Fig. 11.18*

These rules correspond to the *experience of storekeeping*:

- ▸ Small and cheap articles with lower demand are generally kept on stock, valuable articles and articles with very high demand are made to order.

The formulas (11.61) and (11.62) enable the decision between make-to-order and make-to-stock for each single article. A *scheduling program* can calculate the profitability and the opportunity-threshold of storekeeping for each article, if the relevant *article logistic data* are available. When the demand of an order article has fallen below the opportunity threshold and the storekeeping opportunity profit becomes positive, the software suggests re-categorizing the article into a storekeeping article and vice versa.

11.12.4 Cost Optimal Delivery

In most cases, the safety costs are small compared to replenishment and storing costs and the additional costs for partly filled load units in formula (11.27) are negligible. Than the nominator in relation (11.62) is $k_{\mathrm{DOrd}}/T_{\mathrm{OC}}$ and the denominator equals $2\lambda k_{\mathrm{ROrd}}/m_{\mathrm{Ropt}}{}^2$. With these approximations, follows from relation (11.62) the dependency of the opportunity-threshold on the optimal replenishment quantity m_{Ropt}:

$$\lambda_{\mathrm{SKopp}}(m_{\mathrm{Ropt}}) \approx (k_{\mathrm{DOrd}}/k_{\mathrm{ROrd}}) \cdot m_{\mathrm{Ropt}}/\, 2T_{\mathrm{OC}}. \qquad (11.63)$$

The explanation for this result is quite simple: The direct order costs k_{DOrd} for make to order occur once within the order consolidation time T_{OC}, whereas the replenishments order costs k_{ROrd} occur twice per optimal replenishment quantity m_{Ropt}: the first time when the replenishment quantity is ordered and the second time during the storing time, since with cost optimal scheduling the storing costs are equal to the reorder order costs.

For minimal order consolidation time, i.e. for $T_{\mathrm{OC}} = 1$ d, and with equal order costs for direct sourcing and store replenishment, relation (11.63) leads to the *thumb rule for storekeeping*:

- ▶ Make or procure to stock is cheaper than make or procure to order as long as the daily demand is lower than half the optimal replenishment quantity.

This rule of thumb holds as well for the single orders of articles that can be either sourced directly or delivered from stock. That leads to the *opportunity principle for delivery quantities* of storekeeping articles:

- ▶ Orders with delivery quantities above half the optimal replenishment quantity are cheaper to be delivered directly from source than from stock.

Using this principle, *big orders* should be separated by program in order to source them directly. For urgent orders, a minor part is delivered from stock, whereas the major part is delivered to order. A further advantage of the separation of big orders for direct supply is that the stochastic variation of the store demand is reduced and, due to this, safety stock and costs are lower.

11.13 Dynamic Inventory Scheduling

Objectives of dynamic inventory scheduling are *minimal costs* and required *stock availability* for articles with stochastically fluctuating and systematically varying demand. These objectives are achievable by dynamic inventory scheduling with the following procedure.

Extensive simulations, as shown e.g. in *Figs. 11.20* and *11.21*, and many applications have proven that a suitable program for *dynamic inventory scheduling* generates optimal replenishment proposals for more than 95% of the *standard articles* (Gudehus 2003/2007). Only *critical articles* with irregular demand and *special orders* must still be judged and scheduled by a qualified person. For standard articles, the program selects delivery from stock or delivery to order, adapts scheduling parameters and method and proposes the cost optimal supply chain. Also the selection of *optimal load units* and *transport means*, which cause minimal costs and are technically appropriate, are possible by computer.

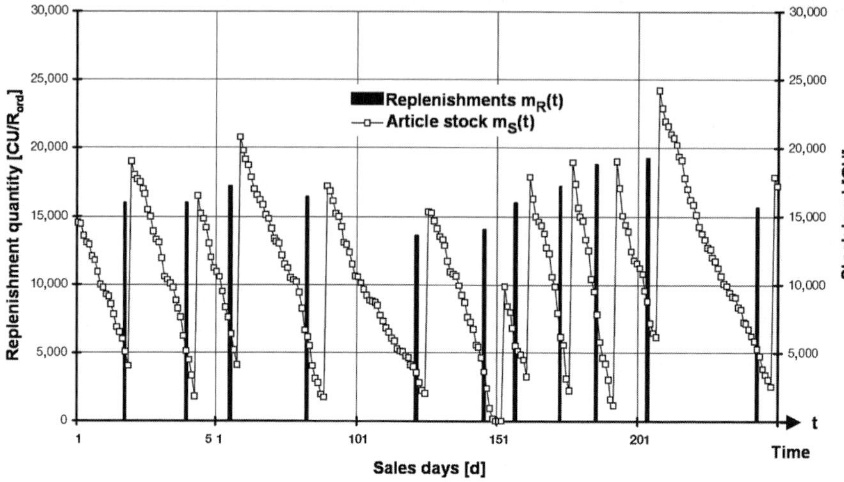

Fig. 11.20 Simulated stock and replenishment quantities for a store-keeping article with regular stochastic demand

Time dependency of the demand: see *Fig. 9.10*
Replenishment strategy: (s, Q)

11.13.1 Procedure of Dynamic Scheduling

Dynamic scheduling is performed in the following *steps*:

1. Periodical *forecast of the current demand* and its standard deviation based on the forecast values and the demand of the last period (see *Sect. 9.8*)
2. Periodical *updating of the replenishment times* and their standard deviation for all articles for which replenishments have arrived in the last period (see *Sect. 9.13.3*.)
3. Calculation of the *dynamic optimal replenishment quantity* (11.27) for all store-keeping articles from the current demand and the article logistic data.
4. Calculation of the *dynamic safety stocks* (11.36) for all storekeeping articles from the current demand and replenishment time for the required availability.
5. Calculation of the *dynamic reorder stocks* (11.18) for all articles from their current demand.
6. Determination of the *current replenishments orders* at the *check points* of the scheduling method, i.e. when *relevant events* have happened, after each *order entrance*, at the end of each day or at the fixed scheduling dates.
7. Display or print out of a *replenishment order list for uncritical articles* whose reorder stock has been reached or which should be ordered in advance in order to achieve *replenishment consolidation*.
8. Display or print out of a *warning list of critical articles* with abnormal demand behavior and of big or special orders, which should be scheduled personally.

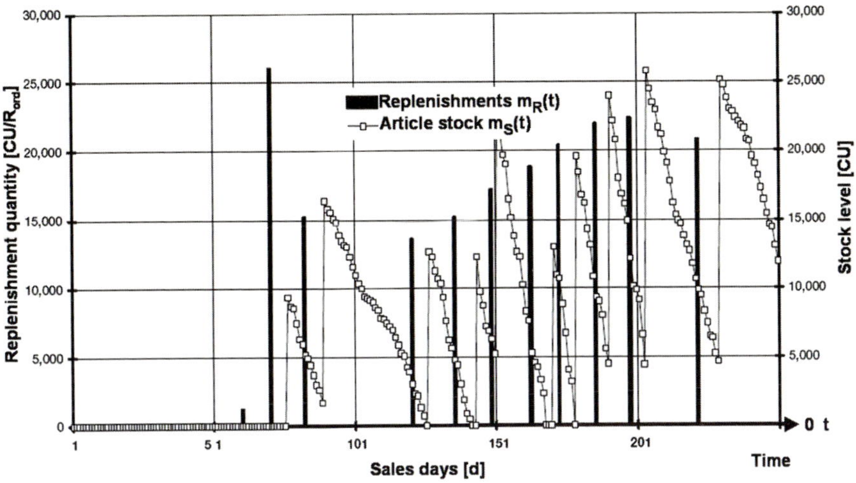

Fig. 11.21 Simulated stock and replenishment quantities of a store-keeping article with abruptly starting stochastic demand

Time dependency of the demand: see *Fig. 9.13*
Replenishment strategy: (s, Q)

The effort for scheduling and computing decreases with the length of the scheduling periods, whereas forecasting quality, storekeeping costs and adaptation decrease. For this reason, the scheduling period should be as short as feasible. Nowadays, modern computer and qualified programs enable *daily scheduling* without unreasonable effort.

The above scheduling steps must be followed exactly in the stated sequence. Otherwise, the dynamic scheduling is not *self-regulating* and may not achieve optimal results.

The replenishment proposals for *uncritical articles* are checked by a qualified person and normally released without changes. Together with the direct orders for larger deliveries and for non storekeeping articles they are sent immediately, e.g. by EDI, to the suppliers. *Critical articles* are

1. articles that do not fulfill the *predictability conditions* of *Sect. 9.5*
2. articles with *irregular increase or decrease of the demand*
3. articles with too long *inventory range*
4. article with a *zero period share* over 90%
5. articles with changing *storekeeping opportunity*

If an article, which is actually storekeeping, shows a negative storekeeping profit, the scheduler decides whether the article should be reclassified as *order article*. Correspondingly, the program calculates the storekeeping profit for all order articles. For positive storekeeping opportunity, an order article can in future be kept on stock.

11.13.2 Adaptation of Replenishment Method

The characteristics, conditions and criteria for the three replenishment methods (11.52) have been compiled in *Table 11.6*. From these criteria and further analysis follow the *selection rules* for *standard replenishment methods*:

- The *reorder-point method* is optimal if the production or supply station accepts orders at any time and executes them immediately.
- The *cycle-time method* is opportune if the production or supply station operates and/or delivers periodically.

If during the replenishment time, respectively during the cycle time, only a few replenishment orders for articles from the same source are generated, the *single article strategy* is sufficient. The *consolidation strategy* is opportune, if the savings generated by order bundling exceed the higher inventory costs.

For the simple *replacement method* neither a demand forecast nor a calculation of replenishment quantities is necessary. It does not need a scheduler or a computer and is within certain limits *self-regulating*. However, this method does not ensure reliable availability at lowest costs. From this results the *selection rule for the replacement strategy*:

- Self-regulating *replacement strategies*, such as *Kanban* and *Flip-Flop*, are opportune only for articles of low value with long lasting demand.

If the replacement units and their content are registered by computer, it is possible to examine dynamically whether the conditions for the replacement strategy are still fulfilled or not.

11.13.3 Selection of Optimal Load Unit

If for an article *load carriers* with different capacity, such as containers of different size or pallets with different dimensions or loading height, are technically possible and actually available, the corresponding *load units* LU_j with their *capacities* C_{LU} [CU/LU] can be filed within the *article master data*. The *cost rates* for storing and handling of the available load units are part of the *logistics master data* (see *Sect. 12.6*).

With these data, the number of load units (11.5), which is required for the cost optimal replenishment quantity (11.27), can be calculated for all technically suitable load carriers. By comparing the resulting *minimal storekeeping costs* (11.29), the optimal load unit is determined.

As long as the handling costs for in- and out-storing are higher than the storeplace costs, the following *selection rule for the optimal load unit* holds:

- Of all possible load units, the optimal load unit needs for the cost optimal replenishment quantity the smallest number of load units with highest filling degree.

With decreasing demand, replenishment quantity and safety stock become smaller and smaller. If the filling degree of the smallest available load carrier is far below 50%, *loose replenishment* without load carrier becomes optimal. For extreme low demand *one piece replenishment* without safety stock becomes opportune. At that point it is advisable to check, whether the article should remain storekeeping or whether it should be classified as order article.

11.13.4 Effects of Dynamic Scheduling

For an article with regular demand as shown in *Fig. 9.10*, which fluctuates stochastically around a systematically varying course, *Fig. 11.20* shows the time dependency of replenishments and stock resulting from dynamic scheduling. The calculated replenishment quantities, as well as the minimal, mean and maximal stock, follow the systematic variation of the demand corresponding to the square root laws of *Sect. 11.9*. The daily stochastic fluctuations of demand cause the step-wise stock reduction.

Figure 11.21 shows the time dependency of replenishments and stock for the same article and scheduling parameters, but with a different course of demand as shown in *Fig. 9.13*. For the first 60 days, the demand is 0. It starts suddenly on the 61st day. Right after the first order day, the dynamic demand forecast follows and generates a replenishment order. Until arrival of the first replenishment the stock availability is 0 and the delivery orders must wait. After arrival of the replenishment the waiting orders and further incoming orders can be executed from stock.

If an *initial stock* has been built up in advance for a new article, the store is able to deliver from the very first day. However, as the example *Fig. 11.21* demonstrates, the required stock availability is reached quickly even without initial stock. Only 10 days after the first demand, the pattern of the stock for the abruptly starting article becomes similar to the stock pattern of the same article with longer lasting demand.

The two examples and thousands of other simulations clearly demonstrate that dynamic scheduling leads to good results even for articles with irregular demand. The model calculations also indicate that the initial values of expected demand, safety stock and α-factor are rather uncritical. Only the *actual stock* must be inserted absolute correctly at the beginning and kept under control permanently.

11.14 Inventory Optimization

An optimization of inventories is only possible if the decisive cost rates, the required stock availability and all other relevant influence parameters are known. Without this knowledge, discussions on inventory optimization are pointless.

The dependency of optimal replenishment quantity, safety stock and total storekeeping costs on the scheduling parameters and cost rates can be calculated for the different strategies by the above formulas and algorithms. With these formulas optimal inventory management strategies can be developed, which lead to the above planning rules and dependencies.

The *inventory management strategies* can be differentiated into *stock reduction strategies* and *inventory optimization strategies*. Stock reduction strategies aim for reducing storing costs, regardless of the effects on stock availability and replenishment costs. Inventory optimization strategies aim to reduce the total storekeeping costs and take into account the consequences for stock availability. Inventory optimization does not necessarily result in a reduction of stocks, but may also increase stocks.

Stock reduction strategies with positive impact on costs are:

- ▶ Assessment and adjustment of the storekeeping assortment by examination of the opportunity and necessity of storekeeping for the single articles
- ▶ Direct delivery of larger order quantities which exceed half the optimal replenishment quantity direct from source and not from stock
- ▶ Limitation of replenishment quantities by maximal inventory ranges, depending on profitability and risks.
- ▶ Reduction of the scheduling frequency or conversion from cycle-time scheduling to reorder-point scheduling
- ▶ Reduction of replenishment times and their fluctuation by improving the reliability in the whole supply chain
- ▶ Selection of reliable and penalizing of unreliable suppliers and service providers

As outlined in *Sect. 11.5*, an inventory turnover below the replenishment frequency is an indicator of too high safety stocks. Other indicators for incorrect scheduling are:

- The *peak factors* of the seasonal stock development are higher than the square root of the peak factors of demand and consumption (see *relation* (11.43)).
- The *Lorenz-curve* of the stock for articles with regular demand runs above the Lorenz-curve of the demand (see *Sect. 5.8*).

Many attempts to reduce stocks by setting unexamined *benchmarks* for maximal range or minimal turnover for the whole company or the total assortment are not optimization strategies. They aim only for stock reduction, neglecting possible negative effects on total costs and stock availability. Real *inventory optimization strategies* are:

- ▶ Application and adaptation of suitable strategies
- ▶ Implementation of cost optimal instead of fixed replenishment quantities
- ▶ Scheduling and calculation with correct algorithms and formulas
- ▶ Permanent control and update of scheduling parameters and cost rates
- ▶ Conversion of storekeeping articles into order articles and vice versa due to the storekeeping opportunity
- ▶ Measuring the stock availability for the different articles, categories and customer groups, in order to adjust it to the necessities
- ▶ Permanent examination and adjustment of safety stocks
- ▶ Rounding up replenishment quantities to full load units and of shipments to full truck loads
- ▶ Replenishment consolidation of articles from the same source and harmonization of their supply with capacity and frequency of transport
- ▶ Stock consolidation for the supply of many consumption stations or sales outlets in a *real central store* (see *Sect. 11.10*)

- Implementation of a *virtual central store* for scheduling the local stocks of stations with the same range of storekeeping articles [Gudehus 2003/2007]
- *Dynamic forecasting* of the demand and permanent control of the forecast based on the actual *Point-of-Sales data* from the ends of the supply network
- *Pull-principle* and *retrograde scheduling* of inventories within a supply network for stations with regular demand (see *Sect. 11.3*)

The last four strategies require differentiated *cost accounting* for all articles and supply chains, from the demand stations upstream to the sources. They must also take into account the costs of the suppliers, as far as they depend on the replenishment strategies. In addition, one has to keep in mind, that a cost reduction by a real central store or a *logistic center* is only achievable, if the flow of goods and the consolidated stocks exceed a certain *critical mass* (see *Sect. 6.8*).

Chapter 12
Logistic Units and Master Data

Logistic units are the physical objects flowing through logistic networks. For planning, scheduling, controlling and optimization of these networks, the *logistic master data* of all relevant logistic units, performance stations, transport means and their relations must be known.

As shown in *Fig. 12.1*, loose goods, article units, parcels, cargo or other *filling units* can be bundled by *load carriers* into *load units*. To bridge shorter internal distances, the load units are handled and moved by fork lift trucks, conveyors, cranes or storing devices. To bridge longer distances between the dispatch ramp of one station and the receiving ramp of another station, the load units are carried either directly or via intermediate stations of an external logistic chain. Within the logistic stations, load units are transhipped, unpacked, sorted, repacked, buffered or stored.

In order to optimize the transports and processes in logistic networks, it is necessary to solve the *load unit task*:

- For the efficient flow through the logistic chains, smaller logistic units have to be bundled in load units by optimal selection, allocation, packing and labelling of suitable load carriers.

The costs of handling, transport and storing are determined by the size, weight and number of the load units. In order to calculate the necessary number of load units for

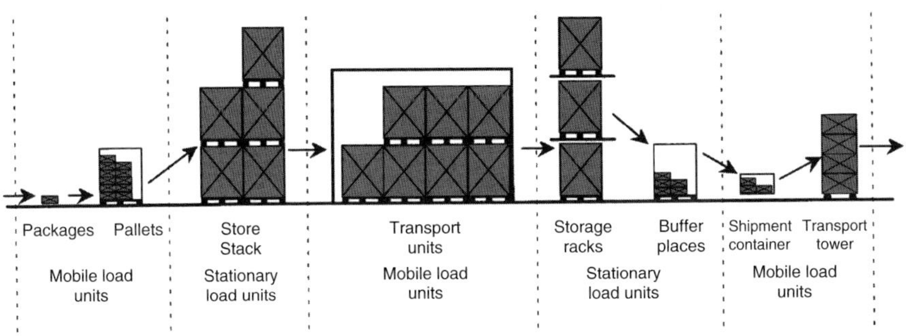

Fig. 12.1 Filling units and load units in the logistic chain

T. Gudehus, H. Kotzab, *Comprehensive Logistics*,
DOI 10.1007/978-3-540-68652-1_12, © Springer-Verlag Berlin Heidelberg 2009

a given filling quantity, the achievable *packing degree* and *load unit capacity*, which depend on the *packing strategy*, and the possible *filling degree*, which is determined by the *filling strategy*, must be known.

In this chapter, solutions of the load unit task are developed and its central parameters are analysed. The resulting formulas for calculating the capacity and the number of load units are crucial for dimensioning storage, handling and transport systems (see *Chaps. 16, 17* and *18*), and for the optimisation of supply networks (see *Chap. 21*). At the end of the chapter a *logistic database* is presented, which is needed for the efficient administration and processing of the logistic data. Finally, the advantages, which are achievable by combining the potentials of logistic units and logistic data, are demonstrated by the concept of *Electronic Kanban*.

12.1 Functions of Load Units

The repeated packing of *filling units* [FU] into *load units* [LU] within a logistic system, company or supply network results in a *packaging hierarchy* with several *packaging stages* (Kearney 1997). As an example, *Fig. 12.2* shows the packaging hierarchy for *consumer goods* in the supply chains from manufacturers to sales outlets.

Not all goods must be packed into load carriers in order to pass a logistic chain. The logistic unit of a lower packaging stage can also be kept as the load unit of the next higher packaging stage. However, bundling of filling units in load units offers several *advantages* and *optimization potentials*:

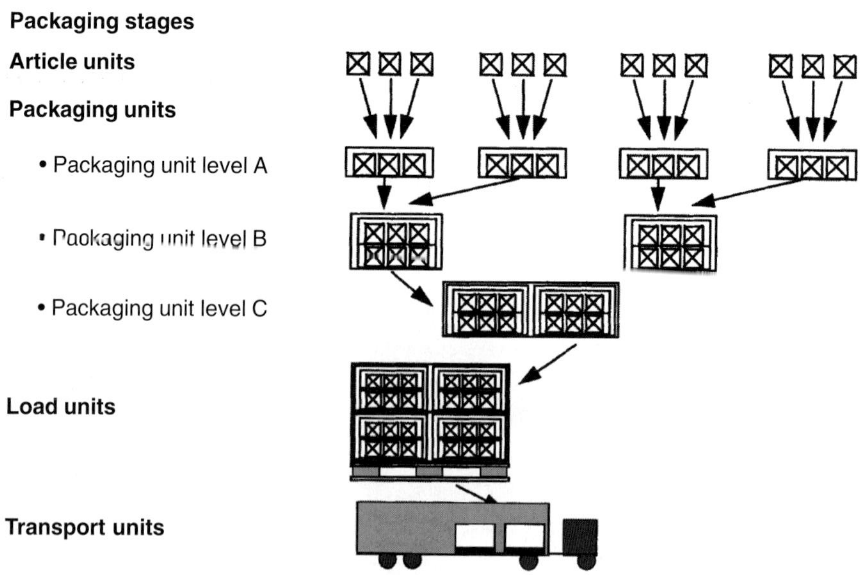

Fig. 12.2 Packaging hierarchy in the supply network for consumer goods

- *Packages* enable the efficient handling, conveying, shipment and stacking of loose items and bulk goods.
- *Standardized packing units* are prerequisites for automatic handling and order picking systems and the application of *robots*.
- Proper *load carriers* make goods storable, which due to their nature are difficult to store.
- *Standard storage units* are prerequisites for advanced storage systems, e.g. for automatic high bay stores.
- *Harmonized load units* help to rationalize and to speed up the transition between storage, handling and conveyor systems as well as the transfer from internal to external transport systems.
- The formation of *standard transport units* enables the application of transport techniques for cargo and shipments that are not suitable to be loaded loosely or difficult to transport.
- Standardized and harmonized filling units, load units and transport units allow packing and filling strategies with minimal packing loss and optimal capacity utilization.
- *Closed load carriers* and transport units protect goods against damage and secure the content against *theft* or *losses*.

These advantages lead to the *economy of scale for load units*:

► By bundling large numbers of filling units in small numbers of load units, substantial reductions of the proportionate handling, storage and transport costs are achievable.

Despite all the potential improvements and savings, load units and transport units are connected with several *disadvantages*:

- Filling, securing, unloading and labelling of load and transport units require manipulation and generate additional costs.
- *Load carriers* and *transport carriers* have to be bought or rented, maintained and cleaned.
- Empty load and transport carriers must be shipped back and provided in time at the filling stations.
- If the inner dimensions of the load units and the outer dimensions of the filling units do not harmonize, the volume utilization is reduced by *packing losses*.
- If the filling quantities are not whole-numbered multiples of the capacity of the load units, partly filled units are generated, which cause *filling losses*.
- The *dead weight* and *dead volume* of empty packages and load carriers reduce the effective capacity of transport means and storeplaces.
- The collection of a larger number of units to fill up a load unit or transport unit of larger size takes time and elongates transit times.

It is a challenge for logisticians to make optimal use of the potentials of load and transport units and to minimize their disadvantages.

12.2 Filling Units and Filling Orders

Filling units are either elementary or compounded logistic units:

- *Elementary logistic units* are the smallest handling units in the system and remain unchanged while flowing through the logistic chains.
- *Compounded logistic units* are created by bundling and packing smaller units and disappear after unbundling and unpacking.

Elementary logistic units of packaging stage zero are *loose goods*, such as bulk goods, solids, liquids or gases, or *unpacked goods*, such as cars, machines, *article units* [AU] and *consumption units* [CU]. Elementary units, which are packed e.g. in order to create *sales units* [SU], are filling units of the first *packaging stage*. In the second stage, smaller packages are further bundled by outer cartons into *packing units* [PU]. In the third stage, several packing units are stacked on pallets or filled in containers and become *shipping units* [SU] (s *Fig. 12.2*).

In a similar way, *storage units* [SU] are formed by bundling packages in pallets, containers or other load carriers. For external transport, load units are filled in *ISO-Containers*. *Transport units* [TU] are generated by loading transport means or transport tanks (see *Chap. 18*). Special "filling units", which are transported by cars, trains, airplanes or ships, are *animals* and *persons*.

12.2.1 Master Data of Filling Orders

Filling orders specify the packing, palletizing, storing, handling, shipping and loading of given filling units. The counter parts of filling orders are *unpacking orders*. Master data of *filling orders* [FO] are:

- *kind and purpose of filling*: packing, palletizing, storing, shipping or loading
- *specification of the filling units* [FU]: condition, content, outer dimensions and weight
- *number of order positions* n_{FO} [Pos/FO]: number of articles of a handling or packing order, number of deliveries of a shipping order or number of shipments of a loading order
- *quantities per order position* m_{FOi} [FU/Pos]: number of filling units of order position $i = 1, 2, ..., n_{FO}$.

From the number and quantities of the positions results the *filling order quantity*:

$$m_{FO} = \sum_i m_{FOi} \tag{12.1}$$

Depending on number and content of the order positions, filling orders are classified into *single item orders* and *multi item orders*, *single article orders* and *mixed article orders* or *single shipment orders* and *mixed shipment orders*.

Often a logistic chain or station has to be designed for a varying number of articles with different characteristics and for filling orders with changing quantities and numbers of positions. For this purpose, it is necessary to define proper *clusters* of similar orders and to calculate the *mean number of load units* from the mean order values of the different order clusters.

12.2.2 Master Data of Filling Units

The master data of filling units determine the packaging degree and the effective capacity of the load units. They are:

- *content* m_{FU} [QU/FU]: quantity of loose goods or number of logistic units contained in a filling unit
- *outer dimensions*: *length* l_{FU} [mm], *breadth* b_{FU} [mm], *height* h_{FU} [mm] of cubical units or *diameter* d_{FU} [mm] of cylindrical or spherical units
- *gross weight* w_{FU} [kg/FU]: total weight of the filling unit.

From the data of the single filling units, the minimal, mean and maximal values of a larger number or cluster of filling units can be calculated. For non-cubical filling units, which deviate from rectangular shape, many logistic calculations can be performed with the dimensions of the *cubical hull shape*.

The building of load units is limited by the following *packing and filling restrictions*:

- *maximal pressure resistance*: maximally tolerable surface pressure
- *stacking factor*: maximal number of units permitted to be stacked
- *stacking direction*: prescribed top, base or outward side
- *batch factor* and *batch dimensions*: maximal number of units that can be interleaved in a batch and outer dimensions of this batch.

Figure 12.3 shows how the design of a bin and of a product affects the volume of batch. Empty bins or products which are logistically designed can be stacked in a space-saving manner. The resulting batch of interleaved items is a *compounded logistic unit*. The tolerable maximal number of compounded items is the *stacking factor*.

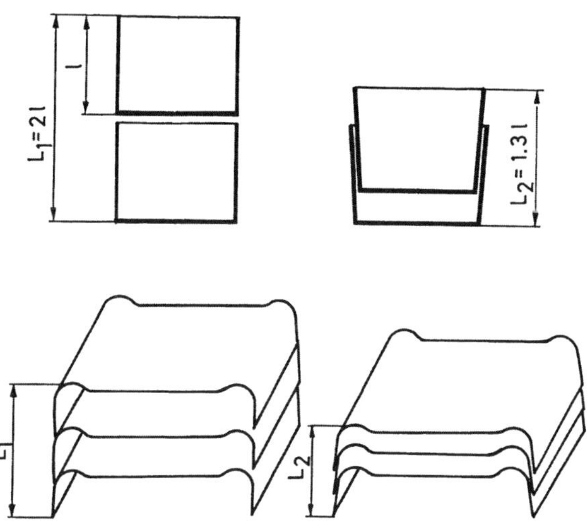

Fig. 12.3 Logistically designed bins and parts

If the vertical direction is not prescribed by the stacking restriction, the outer dimensions of a filling unit can be defined so that the relation holds:

$$l_{FU} \geq b_{FU} \geq h_{FU} \qquad (12.2)$$

The *volume of a filling unit* with cubical size is given by:

$$v_{FU} = l_{FU} \cdot b_{FU} \cdot h_{FU} \qquad (12.3)$$

The *specific weight of the filling unit* with volume v_{FU} and gross weight w_{FU} is:

$$\gamma_{FU} = w_{FU}/v_{FU} \qquad [g/cm^3] \qquad (12.4)$$

Filling units with very small volume compared to the volume of the load units, such as screws or granulates, are *bulk goods*. The extreme are homogenous solids, liquids or gas with units of molecular size.

12.2.3 Mean Dimensions of Filling Units

In order to calculate the mean capacity of the load units for a variety of filling units with different sizes, their mean dimensions and their mean weight must be known. These mean values can be calculated if the *logistic master data* of all filling units and their distribution frequencies are available. Otherwise, the dimensions of representative samples of filling units can be measured in a logistic station. However, this procedure is quite tedious and costly.

In many cases the mean volume of the units is known or can be calculated from the specific weight by relation (12.4). Then also the mean dimensions of the filling units can be determined, provided the dimensions are equally distributed. If the minimal height is small in comparison to the lengths and breadths and the dimensions are equally distributed within the boundaries (12.2) the mean dimensions of the units are approximately:

$$b_{FU} = (l_{FU} + h_{FU})/2 \quad \text{and} \quad h_{FU} = b_{FU}/2 \qquad (12.5)$$

The solution of these equations gives $b_{FU} = 2/3 \cdot l_{FU}$ and $h_{FU} = 1/3 \cdot l_{FU}$ and results in the general *relation of the mean dimensions*:

$$l_{FU} : b_{FU} : h_{FU} = 3 : 2 : 1 \qquad (12.6)$$

A simulation of 1,000 cubic units with random dimensions has confirmed the theoretical relation (12.6) of the mean side lengths. From the relations (12.6) and (12.3) between the mean volume and the mean dimensions follows the *approximation rule for mean filling units dimensions*:

► A large number of cubical filling units with mean volume v_{FU} and equally distributed dimensions has the mean dimensions

$$l_{FU} = 1.65 \cdot {v_{FU}}^{1/3} \quad b_{FU} = 1.10 \cdot {v_{FU}}^{1/3} \quad b_{FU} = 0.55 \cdot {v_{FU}}^{1/3} \qquad (12.7)$$

A measurement of the dimensions of more than 3,000 article units of a department store assortment resulted in a relation of the mean length to the mean breadth of 1.7 and a relation of the mean breadth to the mean height of 2.4. These experimental values are somewhat higher than the theoretical prediction (12.6). The differences are caused by the finite minimal height of the article units.

12.3 Load Units and Load Carriers

For the calculation of capacities and the development of packing and filling strategies, it is useful to define an *abstract load unit* as illustrated by *Fig. 12.4*:

- A load unit is a space with inner and outer dimensions which can be filled with units of smaller size.

This definition includes *mobile load units* that can be moved as well as *immobile load units* that have a fixed location. Concerning the calculation of capacity and filling degree, mobile load units do not differ from immobile load units.

The inner space of a mobile load unit is generally given by a *load carrier*, which is either standardized or adapted to a specific handling and transport technique. In some cases the load unit is simply formed by wrapping or strapping a stack or block of filling units. For big and heavy special goods a load carrier may not be necessary at all.

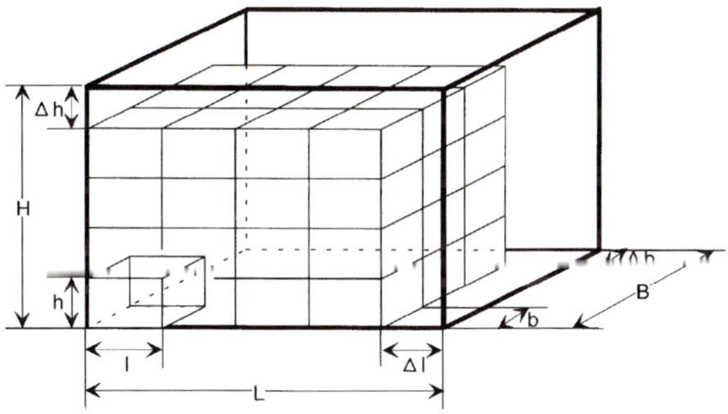

Fig. 12.4 Abstract load unit with filling units

l, *b*, *h*: outer dimensions of the filling unit
L, *B*, *H*: inner dimensions of the load unit
Δ*l*, Δ*b*, Δ*h*: lost measures for parallel packing

12.3.1 Immobile Load Units

Dependent on the function, an immobile load unit is a storeplace, buffer space or provision place. Examples of immobile load units are (see *Chap. 16*):

- floor places in the goods receiving and dispatch area
- provision places in front and buffer spaces behind working stations
- block places of an open-air ground store
- stationary rack-storeplaces
- store channels in a flow-rack-system
- parking lots and lanes for cars, busses, trucks or trailers
- buffer tracks in a vehicle transport network

12.3.2 Semi-Mobile Load Units

Semi-mobile load units can be moved only over short distances in a restricted area. Examples are

- *drawers* within *shelves*
- *shelf trays* of a *mini load storage system*
- places of a *rotating store* or *paternoster store* (see *Fig. 16.7*)

12.3.3 Mobile Load Units

Mobile load units can be moved, transported and dispatched without restrictions. For example,

- *Packing units* [PU] are bottles, bags, packages, cartons, trays and other packing means filled with liquid, gas, loose goods or discrete articles.
- *Storage units* [SU] are load units containing goods or packing units for storing purposes.
- *Dispatch units* [DU] are bins, boxes, pallets, freight containers and other load carriers filled with smaller logistic units that have to be dispatched.
- *Loading units* [LU] are load units containing goods, packing units or dispatch units for transport purposes.
- *Passive transport units* [TU] are transport means without own drive such as semi-road trailers, swap trailers or wagons.
- *Active transport units* [TU] are transport means with own drive (see *Sect. 18.1*).

Similar to the Russian babouschka puppet, a larger load unit can contain smaller load units that contain an even smaller load units, and so forth. This *self-similarity* is typical for logistic units in logistic chains as illustrated in *Fig. 12.2*.

The different types, constructions, measures and other features of standard load units and common load carriers, such as packages, containers, pallets and transport means, are defined and specified in various national and *European norms* and instructions (CCG 1985; DIN-, EN- and ISO-norms; Woxenius 1998).

In *Table 12.1* the key features of frequently used load units and load carriers are listed. The measures of *semi road trailers* [SRT] are shown in *Fig. 12.5* and of *swap body trailers* [SBT] in *Fig. 12.6*. These standardized transport units are increasingly used in European road transport as their internal measures are harmonized with the EURO-pallet dimensions and the standard heights of CCG1- and CCG2-pallets [CCG 1985; GS1].

The *circulation of empty load carriers* and the organisation of *load carrier pools* are subjects of *empties logistics*. In this field as well as in *reverse logistics*, ecological goals and requirements are of special importance (see *Sect. 3.4*). They are partly regulated by state law, e.g. by the German *packing regulations* (Verpackungsverordnung 1991). Many strategies and methods, which are described in this book, can also be applied for empties logistics and reverse logistics.

Table 12.1 Master data of standard load units and common load carriers

Load Unit		Abbr.	**Outside Measures**				**Inside Measures**				**Volume Efficiency (%)**
Load Carrier	Size	LU	length l [mm]	breadth b [mm]	height h [mm]	volume v[l]	length L[mm]	breadth B[mm]	height H[mm]	volume V[l]	η_{Veff}
Carton		**Cart**									
Standard	small		400	300	400	48	380	280	380	40	83.3
	medium		600	300	400	72	580	280	380	62	86.1
	large		650	450	450	132	630	430	430	116	87.9
Tray	standard	**Tray**	600	400	200	48	600	400	180	43	89.6
Box		**Box**									
Industry-Foldbox	small	IND-Box	400	300	235	28	365	280	215	22	78.6
	large	IND-Box	600	400	335	80	558	350	315	62	77.5
Euro-Foldbox	small	EU-Box	400	300	207	25	370	270	198	20	80.0
	large	EU-Box	600	400	307	74	570	370	298	63	85.1
Pallet		**Pal**									
Half-Pallet	standard	Half-Pal	800	600	1,050	504	800	600	900	432	85.7
Euro-Pallet	flat	Flat-Pal	1,200	800	600	576	1,200	800	450	432	75.0
	medium	CCG1	1,200	800	1,050	1,008	1,200	800	900	864	85.7
	high	CCG2	1,200	800	1,950	1,872	1,200	800	1,800	1,728	92.3
Industry-Pallet	high	IND-Pal	1,200	1,200	1,950	2,808	1,200	1,200	1,800	2,592	92.3
Container		**Cont**									
ISO-Container	short	20″-CONT	6,058	2,438	2,438	36,008	5,867	2,330	2,197	30,033	83.4
	long	40″-CONT	12,192	2,438	2,438	72,467	11,998	2,330	2,197	61,418	84.8
Transport-Units		**TU**									
Swap Body Trailer	short	SBT	7,150	2,550	2,600	47,400	7,000	2,435	2,400	41,623	83.3
Semi Road Trailer	long	SRT	14,150	2,550	2,600	93,810	14,000	2,435	2,450	83,520	89.0

Mean volume efficiency: 85%

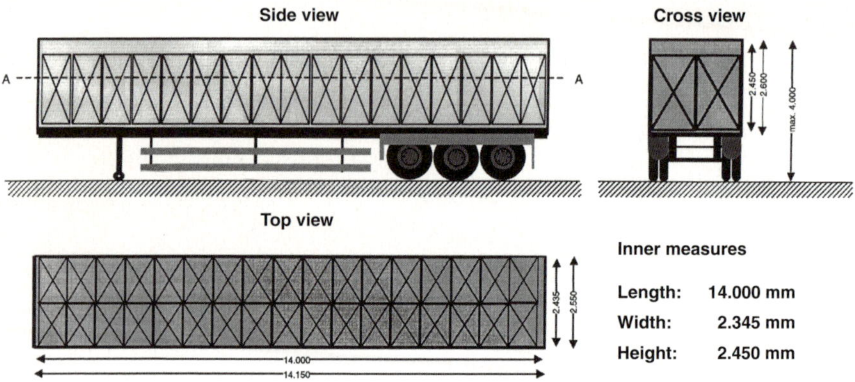

Fig. 12.5 Semi road trailer (STR) filled with CCG2 pallets

Transverse load capacity: $2 \cdot 17 = 34$ CCG2-Pallets/SRT
Longitudinal load capacity: $3 \cdot 11 + 2 = 35$ CCG2-Pallets/SRT

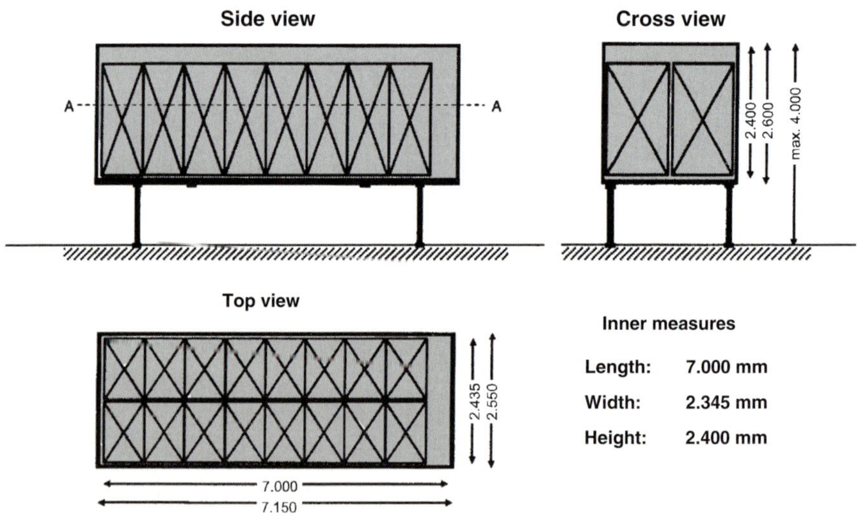

Fig. 12.6 Swap body trailer (SBT) filled with CCG2 pallets

Transverse load capacity: $2 \cdot 8 = 16$ CCG2-Pallets/SBT
Longitudinal load capacity: $3 \cdot 5 + 2 = 17$ CCG2-Pallets/SBT

12.3.4 Master Data of Load Units

The space required for storing and transporting a load unit depends on the *outside data of the load unit*:

- *outer dimensions*: outside length l_{LU} [mm], breadth b_{LU} [mm] and height h_{LU} [mm] of cubical load units and diameter d_{LU} [mm] for cylindrical units
- *gross weight* w_{LU} [kg/LU]: total weight of the filled load unit including load carrier

Capacity and packing degree are determined by the *inside data of the load unit*:

- *inner dimensions* or *load space measures*: inner length L_{LU} [mm], breadth B_{LU} [mm] and height H_{LU} [mm] of a cubical load space and further measures for an irregular load space
- *maximal filling weight, payload* or *net weight* W_{LU} [kg/LU]: maximally permitted weight of the content of the load unit

Further important features of a load unit are:

- *resilience*: maximally permitted weight burden on the top of a full load unit
- *stacking factor*: maximally tolerable number of stacked load units
- *filling directions*: sides from which a load unit can be filled and emptied

Depending on the possible filling directions, load units can be classified into:

- *one-side access units* with possible filling and depletion from one side, such as bins, boxes, containers, trailers and drive-in store places
- *two-sides access units* with possible filling and depletion from two sides, e.g. flow channels of a dynamic storage system
- *all-sides access units* with free access from all upper sides, such as flat pallets and other flat load carriers

Figure 12.7 shows a one side access unit with different longitudinal *packing strategies*. In *Fig. 12.8* a EURO-pallet is shown, that can be filled from all five upper sides with cartons due to a given *packing scheme*.

The *external volume* or *gross volume* of a cubic load unit can be calculated from the outer dimensions:

$$v_{LU} = l_{LU} \cdot b_{LU} \cdot h_{LU} \tag{12.8}$$

The *inner volume* or *net volume* of a cubic *loading space* is determined by the inner dimensions:

$$V_{LU} = L_{LU} \cdot L_{LU} \cdot L_{LU} \tag{12.9}$$

The relation of the net volume to the gross volume is the *volume efficiency*:

$$\eta_{Veff} = V_{LU}/v_{LU} \qquad [\%] \tag{12.10}$$

The volume efficiency is determined by the *dead volume* of the load carrier, which depends on the construction and causes the *technical volume loss*. *Table 12.1* shows that the volume efficiency of common load units is in a range between 75% and 93% with a mean value 85%. The volume efficiency is quite independent of the size of the load unit.

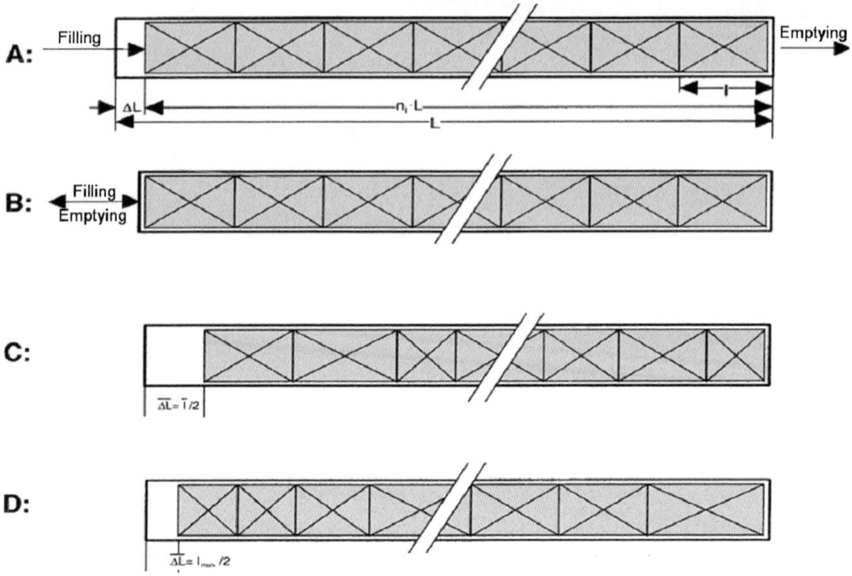

Fig. 12.7 One side access load units with different packing strategies

A: longitudinal packing of identical filling units without length adjustment *capacity*: $C = [L/l]$, *lost length* $\Delta L = L - [L/l] \cdot l$
B: longitudinal packing of identical filling units with length adjustment *capacity*: $C = [L/l] = N_l$, *lost length* $\Delta L = 0$
C: random longitudinal packing of different filling units *mean capacity*: $C = L/l + l/2, = N_l$, *mean lost length* $\Delta L = l/2$
D: optimal longitudinal packing of different filling units *mean capacity*: $C = L/l + l_{\min}/2, = N_l$, *mean lost length* $\Delta L = l_{\min}/2$

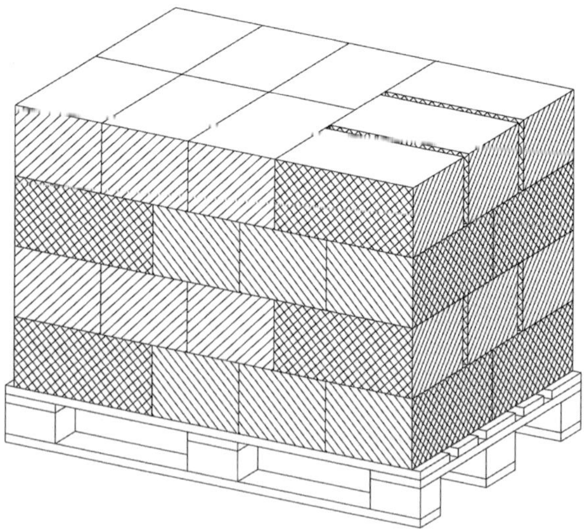

Fig. 12.8 Optimally packed CCG1-EURO-pallet (ISO master module)

The relation of the maximal net weight to the maximal gross weight of a load unit is the *weight efficiency*:

$$\eta_{Weff} = G_{LU}/g_{LU} \qquad [\%] \qquad (12.11)$$

The weight efficiency is determined by the *dead weight* of the load carrier, which causes the *technical weight loss*. The weight efficiency for passive load units ranges between 94% and 96%. For active transport means, it has values between 80% and 95%.

12.3.5 Capacity and Packing Degree

The capacity determines the number of load units necessary for a given load or filling quantity:

- The *load unit capacity* C_{LE} [FU/LU] is the number of filling units maximally achievable by a *packing strategy* under given *packing restrictions*.

In the following the indices FE and LE are omitted, if the meaning of the values without indices is clear from the context. Small letters *l*, *b*, *h*, v and *g* denote the outer dimensions, volume and weight of the *filling units*. In the case of unequal filling units, they indicate the respective *mean values*. Capital letters *L*, *B*, *H*, V and W denote the inner dimensions, net volume and filling weight of the *load units*.

For *weight determined load*, the down rounded quotient [W/w] of the maximal filling weight of the load unit to the weight of the filling unit is smaller than the down rounded quotient [V/v] of the net volume of the load unit to the volume of the filling unit: [W/w] < [V/v]. In this case the maximal capacity is given by *rounded weight quotient* [W/w]. For *volume determined load* the rounded volume quotient is smaller than the *rounded weight quotient*, i.e. [V/v] < [W/w] and the maximal capacity is given by rounded volume quotient [V/v]. Hence, the *maximal capacity of a load unit* is:

$$C_{max} = \mathrm{MIN}([W/w]; [V/v]) \qquad [FU/LU] \qquad (12.12)$$

Here, and in the following formulas, *squared brackets* [...] indicate a rounding down of the bracket content to the next smaller integer.

If the load capacity is weight determined, no optimization of the space utilization by a packing strategy is necessary. For a *mixed load*, the capacity is weight determined for one share and volume determined for the other share. In this case, the number of load units can be minimized by the *load allocation strategy* described in *Sect. 12.5.5*.

Volume utilization and *volume capacity* of a load unit depend on the *packing strategy*. Measure of the strategy efficiency is the *packing degree*:

$$\eta_{pack} = C \cdot v/V \qquad [\%] \qquad (12.13)$$

The packing degree is the share of the inner volume of a load unit occupied by filling units, if the capacity C is fully used. Correspondingly, the *packing loss* at full utilization is the unoccupied share of the inner volume:

$$\eta_{ploss} = 1 - \eta_{pack} = (V - C \cdot v)/V \qquad [\%] \qquad (12.14)$$

The solution of relation (12.13) with respect to C results in the *effective capacity* of a load unit with *mean packing degree* η_{pack}:

$$C_{eff} = \eta_{pack} \cdot V/v \qquad [FU/LU] \qquad (12.15)$$

The main goal of packing strategies is to maximise the effective capacity. As complicated strategies are more difficult to implement than simpler strategies, a new packing strategy must achieve a larger capacity than any simpler strategy. Due to the fact, that the capacity for discrete filling units is an integer, which may not be reduced by small improvements of the packing degree, this is not always the case.

12.3.6 Load Unit Demand and Filling Degree

The *load unit demand* is the minimal number of load units necessary for a given quantity m_{FO} of filling units. For load units with capacity C this number is:

$$M_{FO} = \{m_{FO}/C\} \qquad [LU/FO] \qquad (12.16)$$

The *curly brackets* $\{\ldots\}$ in this and the following formulas denote a rounding up of the bracket content to the next higher integer.

The load unit demand depends on the *filling strategy*. The efficiency of a filling strategy is measured by the achievable filling degree:

- The *filling degree* is the relation of the content m_{FO} to the maximally possible content $M_{FO} \cdot C$ of the M_{FO} load units

$$\eta_{fill} = m_{FO}/(M_{FO} \cdot C) \qquad [\%] \qquad (12.17)$$

Only if the filling order quantity m_{FO} is a whole numbered multiple of the capacity the filling degree can reach 100%. Otherwise each filling order generates at least one partly filled load unit and causes a filling degree below 100%.

If a filling strategy achieves the mean filling degree η_{fill}, the *mean load unit demand* for filling orders with mean quantity m_{FO} is

$$M_{FO} = m_{FO}/(\eta_{fill} \cdot C) \qquad [LU/FO] \qquad (12.18)$$

The main goal of filling strategies is to minimize the load unit demand by maximizing the filling degree.

12.4 Packing Strategies

Filling units with dimensions significantly smaller than the inner dimensions of the load units are generally filled into a load unit randomly, i.e. without any order. The gaps between the single filling units and the inner surface of the load unit cause a packing loss given by (12.14). The *random packing loss* increases with the size of the filling units in relation to the inner volume of the load units.

Packing losses can be minimized by packing strategies:

- A *packing strategy* or *loading strategy* is a method to arrange discrete units in a limited space in a way that the available space is maximally used and the packing restrictions are kept.

Packing strategies are *sequencing strategies* (see *Sect. 5.2.2*). Their *strategy parameters* are the orientations of the filling units in relation to the three dimensions of the loading space (Isermann 1987).

A packing strategy can be specified by *general packing rules* for the sequence, orientation and stacking of the filling units. It can also be determined by a *packing scheme* that prescribes in detail the spatial arrangement of the single filling units in the load unit, as shown e.g. in *Fig. 12.8* for a pallet. For a packing scheme, the capacity of each load unit has to be calculated separately. For a packing strategy with general rules, it is possible to derive *calculation formulas* for the *mean capacity* and filling degree, which can be used for the design of logistic systems and the calculation of logistic costs.

12.4.1 Packing Restrictions

When discrete filling units are bundled in a load unit, *general packing restrictions* must be taken into account. These are:

- *Indivisibility*: The content is a whole number of the discrete filling units.
- *Weight restriction*: The content cannot be heavier than the net weight.
- *Inner dimension limitation*: The largest dimension of the filling units must be smaller than the largest inner dimension of the load unit, the second largest dimension smaller than the second largest inner dimension and the smallest dimension smaller than the smallest inner dimension of the load unit.
- *Outer dimension limitation*: The outside dimensions of the completely filled load unit must not exceed certain maximal measures.

The maximal height of a *closed load carrier*, e.g. of a bin, box or container, is determined by the construction of the load carrier. The *loading height* of flat load carriers such as pallets is a free *design parameter* and can be used as an *optimization variable*. It is limited by the *maximal loading height* H_{max} or by the *stacking factor* of the filling units. For pallets with tolerable *oversize*, the length and breadth of the load unit are additional optimization parameters.

If load carriers with different capacities are available, further optimization is possible by allocating filling orders with different quantities and goods to the best suited load carriers. For this purpose, proper *selection rules* and *allocation strategies* are needed (see *Sect. 11.15.13, Chap. 16* and *Chap. 19*).

In addition to the general restrictions, in many cases *special packing restrictions* have to be respected, such as:

- *Stacking restrictions*: To prevent crushing and tilting, the filling units should only be stacked up to a certain height.
- *Upside prescriptions*: A certain side of a filling unit has to be kept upside.
- *Orientation prescriptions*: One side of a filling unit must be located in a specific direction or accessible from outside, as e.g. a label must be readable.
- *Safety restrictions*: The filling units have to be equally balanced within the load unit to prevent overbalance, slipping or tilting. The load safety can be improved by stacking the different layers into each other or contorted.

For oversized and flat products or bulk goods, additional *technical restrictions* and special stacking prescriptions exist.

12.4.2 Packing Optimization

Packing optimization is a *multi-dimensional cuttings problem* (Gilmore and Gomory 1965; Isermann 1987). Today powerful *packing optimization programmes* based on OR-algorithms are available to solve the tasks of packing and stowing.

On a computer all possible packing schemes can be created by systematic permutation of the filling units and their orientation in the load space. The *optimal packing scheme* is selected by comparing the packing degrees of all tolerable solutions. The processing time for this *optimization by total enumeration* increases quickly with the capacity C of the load unit, because the number N_{PS} of three-dimensional packing schemes is in the range $3! \leq N_{PS} \leq 3! \cdot C!$.

Packing optimization software is available for *equal filling units* and for *unequal filling units*. The resulting packing scheme holds for the given scenario. Any change of parameters requires a new calculation. General formulas for the calculation of the *mean capacities* for different filling units and quantities do not result from optimization programmes or by simulation.

For the analytical design and optimization of delivery chains and logistic networks, the dependency between capacity, packing strategies and filling quantity must be known. In the following sections, general packing strategies with packing restrictions are developed and formulas are derived for the mean capacity and packing degree.

12.4.3 Packing Strategies for Equal Units

The simplest strategy for packing equal cubical units into a cubical space is the elementary *packing strategy 1* shown in *Fig. 12.4*:

- *Parallel packing with fixed orientation*: Starting in a lower corner of the load unit, the filling units are packed side by side and layer by layer with the same orientation parallel to the inner sides of the load unit.

From allgnment of the filling units with *l* parallel to *L*, *b* parallel to *B* and *h* parallel to *H* results the load unit capacity for this basic packaging strategy:

$$C(l;b;h) = [L/l] \cdot [B/b] \cdot [H/h] \tag{12.19}$$

Again the squared brackets [...] denote rounding down. The packing degree results by inserting (12.19) into relation (12.13). By permutation of *l*, *b* and *h* keeping *L*, *B* and *H* fixed, the capacity and the packing degree for all 6 possible orientations of the filling units in relation to the sides of the load unit can be derived from (12.19).

If only the upward direction of the filling units is prescribed, packing strategy 1 can be improved by side permutation. The result is *packing strategy 2A*:

- *Parallel packing with height-restricted side permutation*: From the two possible packing schemes resulting from packing strategy 1 by permutation of *l* and *b* with fixed *h*, the one with the larger capacity is taken.

With C($l;b;h$) and C($b;l;h$) given by (12.19) follows the load unit capacity for packing strategy 2A:

$$C_{2A}(l;b;h) = \mathrm{MAX}(C(l;b;h); C(b;l;h)) \tag{12.20}$$

If no orientations of the filling units are prescribed, the next improvement is *packing strategy 2B*:

- *Parallel packing with complete side permutation*: Of the 6 possible packing schemes resulting from packing strategy 1 by permutation of l, b and h, the one with the highest capacity is selected.

With the permutations

$$\begin{aligned}&\mathrm{PERM}(C(l;b;h))\\&\quad = (C(l;b;h); C(b;l;h); C(l;h;b); C(b;h;l); C(h;b;l); C(h;l;b))\end{aligned}$$

of the dimensions l, b, h in formula (12.19) results the load unit capacity for packing strategy 2B:

$$C_{2B}(l;b;h) = \mathrm{MAX}(\mathrm{PERM}(C(l;b;h))) \tag{12.21}$$

As shown in *Fig. 12.7* for longitudinal packing, the parallel packing causes the *loss lengths* Δl, Δb and Δh in each direction, which cannot be filled by discrete units without turning them. For filling units with dimensions, which vary randomly around the mean values l, b and h, the *mean loss lengths* are $\Delta l = l/2$, $\Delta b = b/2$ and $\Delta h = h/2$. The packing loss caused by the loss lengths can be calculated with relation (12.14).

If the dimensions of the load units or of the filling units are not fixed, packing losses can be eliminated by the

- *Dimension adjustment strategy*: The inner dimensions of the load unit and the outer dimensions of the filling units are related by whole numbers:

$$L = n_l \cdot l \quad B = n_b \cdot b \quad H = n_h \cdot h \tag{12.22}$$

The capacity of adjusted load units is the product of the integers n_l, n_b, n_h:

$$C_{DA} = n_l \cdot n_b \cdot n_h \tag{12.23}$$

The dimension adjustment leads to a packing degree η_{pack} of 100%. It can be achieved in two different ways: The filling unit dimensions are adjusted to given load unit measures, or the load unit measures to given filling unit dimensions. A result of the adjustment of load units and filling units are the standardized logistic units listed in *Table 12.2*.

If the dimensions of load units and filling units cannot be adjusted, the unavoidable loss measures can partly be used by turning the last filling units in different directions. This leads for filling units with the prescribed height direction to *packing strategy 3A*:

- *Parallel packing with fixed height direction and loss space utilization*: If after the parallel packing strategy, b is smaller than the rest length $L/l - l \cdot [L/l]$ another pile with interchanged length l and breadth b is inserted in the gap.

Table 12.2 Standard Load Units

Load Unit		Abbr.	Outside Measures				Inside Measures				Volume Efficiency (%)
Load Carrier	Size	LU	length l [mm]	breadth b [mm]	height h [mm]	volume v[l]	length L[mm]	breadth B[mm]	height H[mm]	volume V[l]	η_{Veff}
Carton		**Cart**									
Standard	small		400	300	400	48	380	280	380	40	83.3
	medium		600	300	400	72	580	280	380	62	86.1
	large		650	450	450	132	630	430	430	116	87.9
Tray	**standard**	**Tray**	600	400	200	48	600	400	180	43	89.6
Box		**Box**									
Industry-Foldbox	small	IND-Box	400	300	235	28	365	280	215	22	78.6
	large	IND-Box	600	400	335	80	558	350	315	62	77.5
Euro-Foldbox	small	EU-Box	400	300	207	25	370	270	198	20	80.0
	large	EU-Box	600	400	307	74	570	370	298	63	85.1
Pallet		**Pal**									
Halfpallet	standard	Half-Pal	800	600	1050	504	800	600	900	432	85.7
Euro-Pallet	flat	Flat-Pal	1.200	800	600	576	1.200	800	450	432	75.0
GS1-Pallet	medium	CCG1	1.200	800	1.050	1.008	1.200	800	900	864	85.7
	high	CCG2	1.200	800	1.950	1.872	1.200	800	1.800	1.728	92.3
Industry-Pallet	high	IND-Pal	1.200	1.200	1.950	2.808	1.200	1.200	1.800	2.592	92.3
Container		**Cont**									
ISO-Container	short	20″-CONT	6.058	2.438	2.438	36.008	5.867	2.330	2.197	30.033	83.4
	long	40″-CONT	12.192	2.438	2.438	72.467	11.998	2.330	2.197	61.418	84.8
Transport-Units		**TU**									
Swap Body Trailer	short	SBT	7.150	2.550	2.600	47.400	7.000	2.435	2.400	41.623	83.3
Semi Road Trailer	long	SRT	14.150	2.550	2.600	93.810	14.000	2.435	2.450	83.520	89.0

With $b < l$ the resulting load unit capacity for packing strategy 3A is:

$$C_{3A}(l;b;h) = [L/l]\cdot[B/b]\cdot[H/h]+[(L/l-l\cdot[L/l])/b]\cdot[B/l]\cdot[H/h] \tag{12.24}$$

If no directions for the filling units are prescribed, the packing degree can be further improved by *packing strategy 3B*:

- *Parallel packing with free orientation and loss space utilization*: After the parallel packing strategy the loss lengths in all three directions are filled by side permuted filling units.

With $l \geq b \geq h$ the load unit capacity resulting for packing strategy 3B is:

$$\begin{aligned} C_{3B}(l;b;h) = [L/l]\cdot[B/b]\cdot[H/h] + [(L/l - l\cdot[L/l])/b]\cdot[B/l]\cdot \\ [H/h] + [(L/l - l\cdot[L/l])/b]\cdot[B/l]\cdot[H/h] \end{aligned} \tag{12.25}$$

The parallel packing strategies with loss space utilization can be combined with side permutation strategies. This leads to *packing strategy 4A*:

- Parallel packing with loss space utilization and height-limited permutation,

respectively, to *packing strategy 4B*:

- Parallel packing with loss space utilization and unlimited side permutation.

The load unit capacities for these combination strategies result from relation (12.21) by inserting for $C(l,b,h)$ the expression (12.24) and (12.25) respectively.

The packing optimization can be systematically continued by successive 90°-turns of the horizontal layers. Other possibilities are to turn single vertical piles and to combine the results with the strategies of side permutation and loss space utilization. The capacity formulas for the combined packing strategies become more and more complex with increasing numbers of turns and permutations. Finally the general packing strategy results in a singular *packing scheme*.

Table 12.3 contains the mean packing degrees which are achievable with different packing strategies. The theoretical values have been calculated with the following *master formula* (12.29) and the listed cutting loss factors. The table also contains the results of a *simulation*, by which sets of equal packages have been generated randomly and stacked on pallets following the different packing strategies. The volume relation V/v has been kept constant at 100 for these simulations.

The results of the simulation agree quite well with the analytical values and confirm the master formula (12.29). Total enumeration gives the optimal packing scheme with mean packing degree of 87.7%. This sets the *benchmark* for the other strategies.

From the results of *Table 12.3* and from further simulations, where all relevant parameters such as volume relation V/v, orientation of the filling units and relation $L : B : H$ of the inner load unit dimensions have been varied systematically, the following *packing rules* can be concluded:

- ▶ The differences of the packing degrees for the different packing strategies decrease with increasing volume relation V/v of load units and filling units.
- ▶ Packing strategy 4B, with loss space utilization and side permutation, leads to mean packing degrees that differ from the mean packing degrees of optimal

Table 12.3 Mean packing degrees and cutting-loss factors for different packing strategies

Abbr.	Packing Strategy	Restriction	Packing Degree Theory (%)	Packing Degree Simulation (%)	Cutting-Loss Factor f_{CL}
OPT	optimal packing scheme	none	87.6	87.7	0.20
4B	side permutation + loss space utilization	none	87.6	87.3	0.20
4A	side permutation + loss space utilization	height	84.7	84.3	0.25
3B	parallel packing + loss space utilization	none	81.8	81.6	0.30
2B	parallel packing + side permutation	none	81.8	82.2	0.30
3A	parallel packing + loss space utilization	height	76.3	75.9	0.40
2A	parallel packing + side permutation	height	76.3	77.2	0.40
2A	parallel packing	height	71.0	72.8	0.50

Relative load space: V/v = 100
Mean side relations: $l : b = 1.6$ $b : h = 2.2$
Theory: mean packing degrees calculated with master formula (12.29)
Simulation: mean packing degrees for 50 different cubical packages

packing less than 0.5% for filling units with dimensions at least 5 times smaller than the inner dimensions of the load unit, i.e. if the volume relation is V/v > 100.

- For larger filling units with volume relation V/v < 10, i.e. for filling unit dimensions up to 1/3 of the inner load unit dimensions, the optimal packing strategy compared to packing strategy 4B results in up to 10% higher packing degrees.
- A height restrictions reduces the achievable packing degree for smaller filling units with V/v > 100 by less than 5% and for larger units with V/v < 10 on average between 5 and 10%.
- Packing strategy 2B of parallel packing with side permutation results on average in the same packing degrees as the more complex parallel packing strategy 3B of parallel packing with loss space utilization
- For volume relations of V/v > 100, by the packing strategies 2B and 3B packing degrees are achievable that are on average up to 5% lower than the packing degrees of the optimal packing scheme.
- For volume relations V/v < 10, the packing degrees for packing strategies 2B and 3B are on average 10% lower than the packing degree with optimal packing scheme.
- For volume relations V/v up to 1,000, the packing degrees with packing strategy 1 are 10 to 20% lower than with optimal packing scheme.

The last statement means, that the basic packing strategy 1 is sufficient only for very small filling units with volume relations V/v above 1,000.

The volume relation or *relative load unit size* V/v is a dimensionless measure for the size of the load unit in relation to the size of the filling units. As the dependency

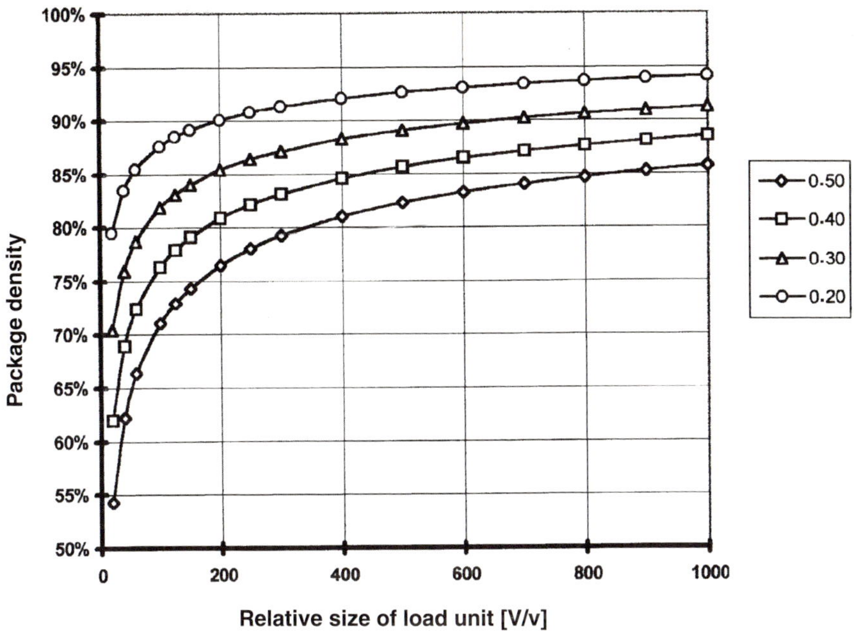

Fig. 12.9 Packing degree as function of the relative load unit size

Parameter: cutting-loss-factors of the different packing strategies of *Table 12.3*

of the packing degree on the relative load unit size in *Fig. 12.9* shows, it is the most important influence factor on the space utilization.

12.4.4 Mean Capacity and Packing Degree

In order to select, dimension and optimize load units for logistic systems with filling units of different sizes, calculation formulas for the *mean capacity* and the *mean packing degree* of the load units are needed. These formulas are derived in a first step for a large number of orders with equal filling units, which differ from order to order, and in a second step for orders with filling units of different size.

If the dimensions of the single filling units vary randomly by at least ± 25% around the mean values, the whole number function in the above formulas can be replaced by the approximations:

$$[L/l] \approx L/l - {}^1\!/_2 \quad [B/b] \approx B/b - {}^1\!/_2 \quad [H/h] \approx H/h - {}^1\!/_2 \quad (12.26)$$

These approximations reflect the simple fact, that down rounded figures are on average by ½ smaller than the exact figures. As shown in *Figs. 12.4* and *12.7* this explains the mean cutting losses *l/2*, *b/2* and *h/2* for filling units with randomly varying dimensions.

Insertion of the approximations (12.26) into relation (12.19) results in the formula for the mean capacity:

$$C(l, b, h) \approx (1 - \tfrac{1}{2} \cdot L/l) \cdot (1 - \tfrac{1}{2} \cdot B/b) \cdot (1 - \tfrac{1}{2} \cdot H/h) \cdot V/v \quad (12.27)$$

of cubical load units with inner dimensions *L*, *B*, *H* filled with cubical units of mean dimensions *l*, *b* and *h* due to the basic packing strategy. For all values of $L \geq B \geq H$ and $l \geq b \geq h$ it can be proven mathematically that the expression (12.27) calculated for all permutations (12.21) of the filling unit dimensions yields smaller values than for the side aligned order (*l*, *b*, *h*). This leads to the *side alignment strategy*:

- Packaging the filling units with the longest edge parallel to the longest inner side and with the second longest edge parallel to the second longest inner side of the load unit leads on average to the highest packing degree.

The mean dimensions of a great variety of filling units are related to the mean volume by the Eq. (12.7). The same holds for the mean dimensions of load units and their volume. Inserting these relations into Eq. (12.27) leads to the *master formula for load unit capacity*:

- The mean capacity of cubical load units with inner volume V containing aligned cubical filling units with mean volume v is

$$C \approx (1 - f_{CL} \cdot v/V)^{1/3})^3 \cdot V/v \qquad [FU/LU]. \qquad (12.28)$$

The *cutting-loss-factor* f_{CL} depends on the packing strategy. It is a *strategy parameter* with values in the range $0.5 < f_{CL} < 1.0$. For the basic packing strategy 1 follows from relation (12.27) the value $f_{CL1} = \tfrac{1}{2}$. The cutting-loss-factors for other packing strategies are listed in *Table 12.2*. They have been derived by analytical considerations and confirmed by simulations. The simulations confirm the applicability of the master formula (12.28) for volume relations V/v > 20 with accuracies better than ± 2%.

Inserting Eq. (12.28) for the capacity into relation (12.13) gives the *master formula for mean packing degree*:

$$\eta_{pack} \approx (1 - f_{CL} \cdot v/V)^{1/3})^3 \qquad [\%] \qquad (12.29)$$

The dependency of the mean packing degree on the relative load unit size V/v, as calculated by (12.29) for different packaging strategies, is shown in *Fig. 12.9*. The master formula (12.29) enables to calculate the possible improvements by the different packing strategies. By *packing restrictions*, the cutting-loss-factors are increased and the packing degree is reduced.

12.4.5 Loading Strategy for Unequal Filling Units

If the filling units have different dimensions, the mean loss length in the three spatial directions can be reduced by loading the units according to descending size. For packaging in one direction, the effect can be seen from *Fig. 12.7C* and *Fig. 12.7D*: By loading according to descending length the mean loss length is reduced from half of the mean filling unit length *l*/2 for random filling to half of the minimal length $l_{min}/2$ of the filling units.

This leads to the *loading strategy for unequal filling units*:

1. The filling units of an order are sorted in descending size.

2. The units with the largest volume are filled into a necessary number of load units following the parallel packing strategy with side permutation.
3. The next smaller units which just fit into the loss lengths are packed into the remaining space of these load units following the parallel packing strategy.
4. With the next largest units step 3 is repeated until no units are left which fit into the empty space of the load units.
5. For the remaining filling units, steps 1 to 4 will be executed until all filling units are packed into load units.

By this packing strategy for unequal units the load space is used on average at least as efficiently as by the parallel packing strategy with side permutation for equal units. Due to the shorter measures of the smallest units, the residual lengths are on average shorter compared to filling with equal units. This leads to the *capacity rule for unequal filling units*:

▶ Mean capacity and packing degree of load units filled optimally with unequal units are given by the relations (12.28) and (12.29) with loss-factor $f_{CL} \approx 0.15$

If the dimensions of the filling units differ from each other by more than a factor 10, it is necessary to cluster the units into groups of similar sizes. For these clusters the optimal packing and the capacity calculations are performed separately. The *clustering* of filling units and their *allocation* to suitable load units are further strategies for planning and optimisation of logistic systems.

12.5 Filling Strategies and Load Unit Demand

Whereas the goal of a packing strategy is to maximize the volume utilization, it is the goal of a filling strategy to minimize the load unit demand. That means:

- A filling strategy is a method to allocate the units of a given filling order to a minimal number of load units keeping the filling restrictions.

Filling strategies are *sequencing strategies* for optimal capacity utilization of given load units. The *strategy parameters* are the *partitions* of the filling units and their *allocation* to load units.

12.5.1 Filling Restrictions

Filling restrictions do not only affect the utilization of the load capacity, but can also reduce packing degree and capacity. Typical *filling restrictions* are:

- *Separate position filling*: A load unit should contain only filling units of the same order position.
- *Prescribed filling sequence*: The filling units have to be loaded in a certain sequence.

Separate position filling is typical e.g. for *shipping orders* with goods for different customers or destinations, which must be loaded separately.

Specific filling sequences are necessary in order to enable an *optimal transport tour*, to ensure a required *replenishment order* for a sales outlet or to keep the *FIFO principle* in a store.

12.5.2 Weight Determined Filling Strategy

If the load is weight determined, i.e. if $W/w > V/v$, the net weight, the indivisibility of the filling units and their limited pressure resistance can be kept by the following *weight determined filling strategy*:

- Starting with the heaviest and most resistible filling units the lighter and less resistible units are stacked on top of the heavier and more resistible units until the net weight of the load unit is reached.

The weight determined filling strategy can be performed in analogous steps as the filling strategy for unequal units by sorting the filling units due to descending weight instead of volume.

From relations (12.12) and (12.16) follows the *load unit demand for a weight determined filling quantity* m_{FO}:

$$M_{LU} = \{m_{FO}/[W/w]\} \quad [LU/FO] \tag{12.30}$$

Herein W is the net weight of the load units and *w* the mean weight of the filling units. Again, the curly brackets denote rounding up and the squared brackets rounding down.

12.5.3 Quantity Adjustment and Capacity Adjustment

If it is tolerable to adapt the order quantity, 100% filling degree is achievable by the *quantity adjustment strategy*:

- The quantities of the filling orders are rounded up or down to a whole numbered multiple of the load unit capacity.

Quantity adjustment is quite common in practice: lot sizes of a production are adapted to the capacity of the pallets; replenishment quantities are rounded up or down due to the storage unit capacity; shipments are collected until a transport unit is completely filled.

If the filling quantities cannot be changed but do not vary too much, the filling degree can be optimized by the *capacity adjustment strategy*:

- The load units are dimensioned or selected in such a way, that the whole numbered multiple of their capacity is slightly smaller than the order quantity.

For packing orders, cartons with appropriate size are selected. For loading orders, the dimensions of the pallets are adjusted. The capacities of transport units are selected corresponding to the size of the shipments.

12.5.4 Mean Load Unit Demand

If no quantity or capacity adjustment is possible, the load unit demand for orders without filling restrictions is a number in the range

$$m_{FO}/C \leq M_{FO} \leq (m_{FO} - 1)/C + 1 \quad \text{for} \quad m_{FO} > C. \tag{12.31}$$

For larger order quantities, which vary randomly around a *mean value* m_{FO}, the mean load unit demand equals the mean value of the boundaries (12.31), i.e.

$m_{FO}/C+(C-1)/2C$, and for small filling orders the load unit demand is 1. This leads to the *master formula for mean load unit demand*:

▸ For filling orders without weight or filling restrictions, the mean number of load units with capacity C necessary to hold the mean filling quantity m_{FO} without quantity or capacity adjustment is

$$M_{FO} = MAX(1; m_{FO}/C + (C-1)/2C) \qquad [LU/FO] \qquad (12.32)$$

This means: For small filling quantities with $m_{FO} < C$ only 1 load carrier of capacity C is needed. For larger quantities with $m_{FO} > C$ the mean number of load units is the sum of in average m_{FO}/C *full load units* plus one *partly filled load unit* with the mean *filling loss* (C–1)/2C. For C = 1 FU/LU, the filling loss is 0 and the number of load units equals the number of filling units. For large capacities, i.e. for $C \to \infty$, the mean filling loss is 50% or ½ LU per order.

The master formula (12.32) for the mean load unit demand of a large number of filling orders with varying quantities is the steady interpolation of the step function (12.16) for the load unit demand for orders with equal quantities. This is shown by *Fig. 12.10* for the case C = 5 FU/LU.

Other than the step function (12.16), the steady function (12.32) can be differentiated. Therefore the master formula (12.32) is applicable for analytical optimizations (see *Sect. 11.7*). If a separate position filling is required for orders with n > 1 order lines, on average per position the quantity m_{FO}/n has to be filled into separate load units. For small position quantities at least n load units are necessary and for larger quantities n partly filled units are generated. This leads to the *mean load unit demand for separate position filling*:

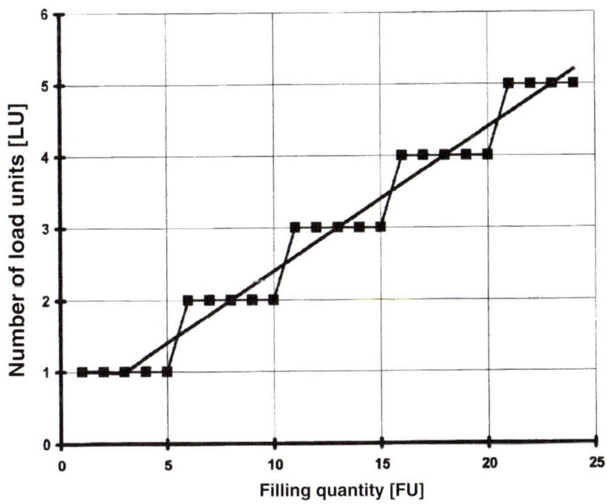

Fig. 12.10 Dependency of load unit demand on the filling quantity

Capacity: C = 5 FU/LU
Step function: formula (12.16) for filling orders with equal quantities
Steady function: formula (12.32) for filling orders with different quantities

► For separate position filling, the mean number of load units necessary to hold filling orders with n positions and mean order quantity m_{FO} is

$$M_{FO} = MAX(n; m_{FO}/C + n \cdot (C-1)/2C) \qquad [LU/FO] \qquad (12.33)$$

Insertion of formula (12.33) into relation (12.17) results in the *master formula for mean filling degree*:

► With separate position filling of orders with n positions and mean order quantity m_{FO} the mean filling degree of the load units is

$$\eta_{fill} = m_{FO}/MAX(n \cdot C; m_{FO} + n \cdot (C-1)/2) \qquad [\%] \qquad (12.34)$$

Without separate position filling the mean filling degree is given by formula (12.34) with n = 1. For this case, the resulting dependency of the mean filling degree on the load unit capacity is shown in *Fig. 12.11* and on the mean filling quantity in *Fig. 12.12*. From the master formula (12.34) result the *general filling rules*:

- The filling degree decreases with increasing *load capacity* due to the higher filling loss per order.
- The filling degree increases with the *filling quantity* because the filling loss per order becomes smaller and smaller.

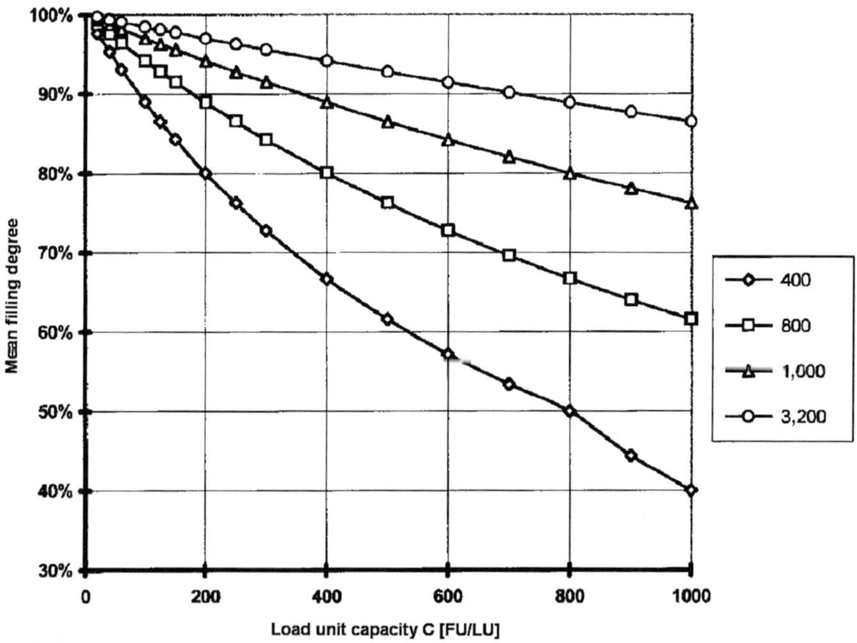

Fig. 12.11 Dependency of the mean filling degree on load capacity
Mixed position filling: n = 1
Parameter: mean filling quantity m_{FO} = 400/800/1,600/3,200 FU/FO

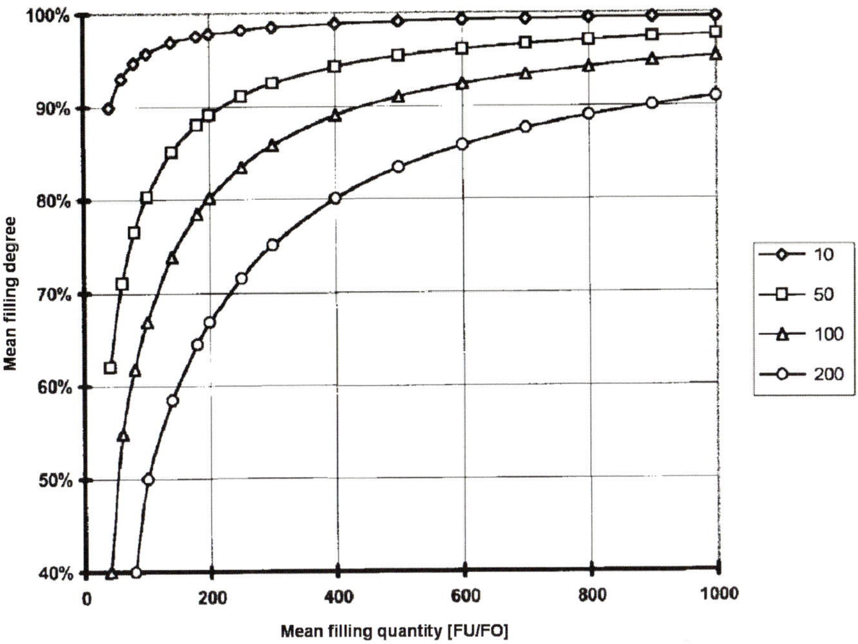

Fig. 12.12 Dependency of the mean filling degree on mean filling quantity
Mixed position filling: n = 1
Parameter: load unit capacity C = 10/50/100/200 FU/LU

These filling rules, which are illustrated by the examples of *Figs. 12.11* and *12.12*, are very import for the selection and dimensioning of load units.

12.5.5 Optimal Allocation of Mixed Freight

In some cases, the freight for the same destination consists of a quantity m_A of goods A, for which the capacity of the transport units is weight-determined, and of a quantity m_B of goods B, for which the capacity is volume-determined. In this case the number of transport units and the transport costs can be minimized by optimal allocation of the partial quantities to transport units.

The *specific net weight* of a transport unit with net weight W_{TU} and net volume V_{TU} is $\gamma_{TU} = W_{TU}/V_{LU}$. The *weight capacity* for heavy filling units with specific weight $\gamma_A = w_A/v_A \geq \gamma_{TU}$ is:

$$C_A = [W_{TU}/w_A]. \tag{12.35}$$

The *volume capacity* for light filling units with specific weight $\gamma_B = w_B/v_B < \gamma_{TU}$ and a packing degree η_{pack} is:

$$C_B = \eta_{pack} \cdot V_{TU}/v_A. \tag{12.36}$$

With these capacities, the *transport unit demand for separate loading* of the corresponding freight quantities m_A and m_B is:

$$M_{sep} = \{m_A/C_A\} + \{m_B/C_B\}. \qquad (12.37)$$

With separate loading, the volume of the $\{m_A/C_A\}$ weight determined transport units is only partly filled. For the $\{m_B/C_B\}$ volume determined transport units, the net weight is only partly used.

For combined loading of heavy freight A and light freight B, volume and net weight of the transport units are utilized maximally if the *partial contents* fulfil the conditions:

$$\begin{aligned} C_{Aopt} \cdot w_A + C_{Bopt} \cdot w_B &= W_{TU} \\ C_{Aopt} \cdot v_A + C_{Bopt} \cdot v_B &= \eta_{pack} \cdot V_{TU} \end{aligned} \qquad (12.38)$$

The solution of these equations results in the *optimal partial capacities* of heavy and light freight:

$$\begin{aligned} C_{Aopt} &= (\gamma_{TU} - \gamma_B)/(\gamma_A - \gamma_B) \cdot \eta_{pack} \cdot V_{TU}/v_A \\ C_{Bopt} &= (\gamma_A - \gamma_{TU})/(\gamma_A - \gamma_B) \cdot \eta_{pack} \cdot V_{TU}/v_B \end{aligned} \qquad (12.39)$$

If the relation $m_A{:}m_B$ of the partial quantities of the freight order differs from the relation $C_{Aopt}{:}C_{Bopt}$ of the optimal partial capacities (12.39), the condition (12.38) can not be kept for all transport units. However, even in this case maximal utilization of the transport units can be achieved by the *strategy of optimal load allocation of mixed freight*:

- The heavy and light units of a freight order with partial quantities m_A and m_B are filled together in the relation $C_{Aopt}{:}C_{Bopt}$ of the optimal partial capacities (12.39) into a minimal number of *mixed transport units*

  $$M_{AB} = \text{MIN}(\{m_A/C_{Aopt}\}; \{m_B/C_{Bopt}\}). \qquad (12.40)$$

  The remaining quantity of heavy freight $m_A{-}M_{AB}{\cdot}C_{Aopt}$ is filled into the minimal number of *heavily loaded transport units*

  $$M_A = \text{MAX}(0; \{(m_A - M_{AB} \cdot C_{Aopt})/C_A\}). \qquad (12.41)$$

  The remaining quantity of light freight $m_B{-}M_{AB}{\cdot}C_{Bopt}$ is filled into the minimal number of *lightly loaded transport units*

  $$M_B = \text{MAX}(0; \{(m_B - M_{AB} \cdot C_{Bopt})/C_B\}). \qquad (12.41)$$

Resulting from the optimal allocation strategy is the *optimal demand of transport units for mixed freight*:

$$M_{opt} = M_{AB} + M_A + M_B. \qquad (12.42)$$

Figure 12.13 shows the dependency of the transport unit demand on the quantity of heavy freight for separate loading (12.37) and for optimal allocation (12.42). The comparison results in the *allocation rule*:

- With *optimal quantity relation* of the heavy to the light freight

  $$m_A : m_B = C_{Aopt} : C_{Bopt} = (v_B/v_A) \cdot (\gamma_{TU} - \gamma_B)/(\gamma_A - \gamma_{TU}) \qquad (12.43)$$

  the reduction of the transport unit demand by optimal allocation is maximal.

In the example *Fig. 12.13* the number of transport units can be reduced by optimal allocation of heavy and light freight up to 20% compared with separate loading.

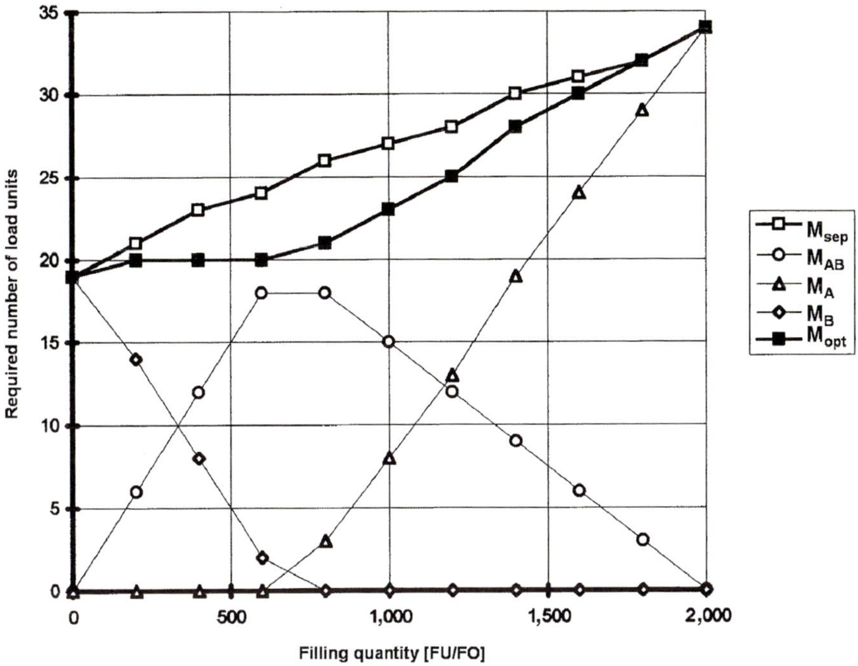

Fig. 12.13 Dependency of the transport unit demand on the quantity of heavy freight for different loading strategies

Total freight order quantity: $m_{FO} = m_A + m_B = 2{,}000$ FU
M_{sep}: transport unit demand for separate loading
M_{opt}: transport unit demand for optimal load allocation
M_{AB}: mixed filled transport units
M_A: heavy loaded transport units M_A: light loaded transport units

Optimal load allocation is well known in the freight business. Experienced *freight forwarders* try to acquire additional orders for *heavy freight* as long as the *volume freight* dominates the filling of the transport units and to get additional orders for volume freight if they have too many heavy freight orders. The formulas (12.39), (12.40), (12.41), (12.42), (12.43) enable a quantified *load optimization*.

The optimal load allocation strategy can also be applied for the joint distribution of goods with very different weights and volumes. In a practical case, the distribution of heavy bottles for a distillery has been combined with the distribution of packages containing light consumer goods (Behrentzen and Reinhardt 2002).

12.6 Logistic Master Data

Although the logistic master data are indispensable for optimizing logistic networks and processes, many companies still do not register these data completely and update them regularly. This explains why in many companies inadequate packages or load carriers are used, wrong storage systems are built and overly expensive freight or supply chains are selected. Correct and complete logistic master data are required for the calculation of

- performance cost rates and cost-based prices
- article-, category- and order-logistic costs
- procurement costs of the company and of single suppliers
- distribution costs of the company and for different customers

Logistic master data are also prerequisites for

- optimal selection and allocation of load carriers and transport means
- packing and filling strategies for load units and transport units
- cost optimal order scheduling and inventory management
- selection of optimal sourcing, delivery and transport chains
- calculation of the needed number of load carriers and transport means
- calculation of order picking and packing performances
- determination of the personnel for internal logistics
- performance dependent remuneration of external logistic services
- optimal selection of storage, commissioning and handling systems
- design, dimensioning and optimization of logistic systems and networks

In order to apply the logistic data for these purposes, a relational *logistic data base* is necessary consisting of *reference tables*, *directories* and *datasets* with logistic data of the same category. Each group of logistic data is stored only in one reference table which is used by all programs.

The logistic data base is part of the *data warehouse* of the company. It has to be designed in an open architecture as logistic datasets can change during time. The *dynamic logistic data* must be adapted regularly to a varying demand.

The logistic data base of a company generally contains the following *directories* and *datasets*:

- *order and article directory*

order logistic data
article logistic data (12.44)

- *logistic stations directory*

supplier logistic data
operational logistic data
sales outlet logistic data
customer logistic data (12.45)

- *logistic units directory*

$$\begin{array}{l}\text{package data}\\ \text{load carrier and load unit data}\\ \text{transport means and transport unit data}\\ \text{storage system and storeplace data}\\ \text{data of sales shelves}\end{array} \tag{12.46}$$

- *logistic cost directory*

$$\begin{array}{l}\text{handling cost rates and prices}\\ \text{storage cost rates and prices}\\ \text{transport cost rates and prices}\\ \text{freight costs and freight rates}\\ \text{administrative cost rates and prices}\end{array} \tag{12.47}$$

The content of the data sets (12.44) to (12.47) will be explained in the following. They refer partly to other commercial and technical data directories such as *price lists*, *material specifications*, *service and performance catalogues* and *labelling directories*. The general logistic data have to be completed by industry- and company-specific data [CCG 1993; GS1].

For *planning* logistic systems, further technical, commercial and other data are needed which can be found in *Sect. 3.7*. The logistic data which are needed for *dynamic scheduling* of orders and inventories will be presented at the end of this section.

12.6.1 Order Logistic Data

Order logistic data specify the logistic performances and requirements connected with the order execution. They initiate the logistic processes from the *source* to the *sink* (see *Sect. 2.1*).

For *supply orders*, the order logistic dataset contains:

- *addresses* of the delivery, production and receiving stations
- *numbers and names* of the articles to be procured, produced and delivered
- *required quantities* per article
- *dates* and *time requirements* for collection, dispatch and delivery

An *order line* or *position* of a supply order contains the name, number, specification and quantity of a single article. Correspondingly, there are *single* and *multi position orders*, and *one piece* and *multi piece orders*.

The position of a *performance order* contains the identification number, the name and the quantity of a performance or service. The identification number refers to the standards and specifications of a *performance and service catalogue*. The required quantities are measured in *performance units* (see *Sect. 6.5*).

12.6.2 Article Logistic Data

Article logistic data contain all information and data of an article which are needed to perform the logistic processes, to schedule orders and inventories, to manage logistic stations and to optimise the company logistic network. For these purposes the article logistic database contains:

- *article number*, such as *European Article Number (EAN)*, *Global Trade Item Number* (GTIN) or internal identification number
- *article name*: short description or specification
- *producers*: name and location of manufacturers and their production sites
- *suppliers*: name and addresses of suppliers and their delivery stations
- *article properties*: material type, hazardousness, fire class ...
- *article unit*: measuring unit for loose goods; quantity unit of discrete articles
- *supply units*: standard load units for the supply and their specific capacity
- *delivery units*: standard load units for the delivery and their capacity
- *packing and filling restrictions* for supply and delivery
- *regulations* for handling, storing and transport
- *delivery ability* for storekeeping articles
- *replenishment times*, such as production, procurement and delivery time
- *article value*: production costs or purchasing price per unit

The general article logistic data have to be completed by specific article data, which depend on the individual supplier or customer.

The *article specific capacity* of the load units can either be measured or calculated from the technical data and dimensions of the article units and the load carriers with the help of the above formulas.

12.6.3 Logistic Station Data

The logistic data of a station contain all information and data that affect the execution of logistic services and performances.

The *supplier logistic data* comprises of:

- *delivery addresses* of one or several dispatch points
- *dispatch area data*: number of docks, buffer capacity, control facilities
- *supply chain data*: transport connections, transhipment points, freight chains
- *supply conditions*: pricing for the different supply chains
- *supply unit data* of load carriers and transport means used for supply
- *time conditions*: operating times, pickup times and procurement times

The *operational logistic data* specify the logistic conditions of the production sites, stores and logistic stations of the company and the logistic service providers. They comprise:

- *receiving area data*: address, number of docks, buffer capacity and control
- *store data*: locations, type, capacity and limit performances of stores
- *commissioning data*: locations, capacities and performances of order picking
- *dispatch area data*: location, number of docks, buffer capacity and control
- *delivery chain data*: transport chains, transhipment points, freight chains

- *delivery conditions*: pricing for the different delivery chains
- *delivery unit data* of load carriers and transport means used for delivery
- *time conditions*: operating times, dispatch times and delivery times

The *sales outlets* of a retailer, such as shops, department stores, branches or markets, are the *final sinks* of all logistic networks and the *point of sales*. These special logistic stations are designed to fulfil customer needs and to support sales activities. The *sales outlet logistic data* comprises:

- *addresses* of the sales outlets
- *receiving area data*: number of docks, buffer capacity and control facilities
- *buffer stores*: type, capacity and limit performances of the buffer stores
- *sales area data*: places, racks, isles, shelves and capacities in the sales area
- *load unit data*: types and dimensions of acceptable load units
- *time conditions*: operating times and sales times

Many logistic station data, such as the load unit data, can be restricted to name and identification number, which refers to the general logistic unit directory.

12.6.4 Logistic Unit Data

Corresponding to the different functions of logistic units, their data base can be organized in the following directories and datasets:

- *elementary logistic units*
  article units
  storekeeping units (SKU)
  packing units
- *compounded logistic units*
  sales units or consumption units
  supply and replenishment units
  storage units
  handling and picking units
  packing and shipping units
  freight units, load units and transport units
- *store places*
  reserve storeplaces
  access places
  sales shelf places

For each of these logistic units the dataset comprises:

- *name and identification number* of the logistic unit
- *dimensions, weight and further technical data* (see *Sect. 12.3.4*)
- *name and number of the load carrier* referring to a *load carrier directory*
- *general packing restrictions* (see *Sect. 12.4.1*)
- *coding prescriptions*, such as type, size and position of the coding

A *load carrier directory* contains all necessary data about the supplier, purchasing prices and technical data of the load carriers, which are accepted to be used within the company. The coding prescription may also refer to a *coding directory*, which contains *coding standards* for the load units.

As the capacities of load units depend on the kind and size of the filling units, the load unit capacity is either part of the article logistic data or is calculated from the data of the filling order.

12.6.5 Logistic Cost Data

The performance cost rates and prices that are required to calculate the article- and order-logistic costs and that are needed for inventory management are registered in the logistic cost directory. The logistic cost rates and prices comprised in the different *cost datasets* (12.47) are defined and discussed in *Chaps. 7* and *8*. Their application in inventory management has been explained in *Chap. 11*.

12.6.6 Data for Dynamic Scheduling

For the computer-based dynamic scheduling of orders and inventories, complete, correct and up-to-date article and logistic master data are required. To ensure this, the origin of the data has to be defined and the responsibility for correctness, data input and data administration must be clearly regulated.

The data for dynamic scheduling include the above specified *static logistic data*. From the static data and the current demand or sales figures the *dynamic logistic data* can be calculated as described in *Chap. 11*. Also the *scheduling parameters* and scheduling strategies have to be adapted to changing conditions.

Most scheduling errors are caused by insufficient article data. This is quite often recognized too late after a new programme has been installed. That means:

- ▸ Precise knowledge, proper collection and efficient organisation of the necessary data are prerequisites for the success of dynamic scheduling.

Ignoring this experience rule can cause enormous trouble. In some cases companies went almost *bankrupt* by insufficient scheduling programmes, incorrect logistic data and wrong scheduling parameters (Dittrich et al. 2000).

12.7 Electronic Kanban

Electronic kanban or *e-kanban* is an example for the improvements which are achievable by combining the methods of logistics and the possibilities of information technology. As described in *Sect. 11.11.1* and shown in *Fig. 11.18*, *kanban* is a method for *self-regulated replenishment* of a consumption station with full bins by identification *cards*, which are called in Japan *kanban*.

Widely known is the conventional *two-bin-kanban* (Geiger 2003). The replenishment is initiated after the *access bin* has been completely emptied. The supply is triggered by setting out the empty bin or by fixing the card on a board near the station. During the replenishment of a newly filled bin the article units are taken from a *reserve bin* which has replaced the access bin.

An alternative is the *one-bin kanban* where the replenishment is triggered by fixing the card on a board after the content of the bin has reached the *replenishment stock* noted on the card. If the replenishment stock is sufficient, the newly filled bin arrives *just in time* before the access bin is emptied.

Kanban is very popular, especially in the automotive industry, because it is quite simple and easy to operate. However, the *disadvantages of conventional kanban* are often overlooked:

out-of-stock risks
uneconomic bin sizes
high handling and process costs
efforts and time losses by the card handling
(12.48)

These disadvantages can be reduced or completely avoided by *electronic kanban* with *coded bins* instead of cards. The code carrier can be a conventional *barcode label* or a programmable *smart label*, i.e. a *transponder* readable by RFID (Finkenzeller 2002; Shephard 2004).

The replenishment is initiated by scanning the bin code after the access unit has been emptied completely or after the replenishment stock is reached. The bin-information is sent by EDI directly to the supply station where it triggers the refilling and delivery of a supply bin. The content of the full bin, the article number and the quantity is stated on a renewed label or programmed on the smart label.

Empty bins, flat pallets or flap boxes are collected in the consumption station and sent as consolidated batches directly or via a *container pool* to the supply station. Larger bins, empty containers, box pallets or other unfoldable load carriers can be returned by the same vehicle that delivers the supply unit.

As electronic kanban operates without cards, it avoids the paperwork and card handling of conventional kanban. The return of the empties can be separated from the flow of full bins and organized more efficiently. The replenishment times are reduced considerably by the immediate transfer of the information from the consumption station to the supply station. With the help of a suitable scheduling programme, replenishment stock, supply quantity and bin size can be adjusted dynamically to a changing demand. By this procedure the supply strategies described in *Chap. 11* can be applied, the stock out risk is reduced, cost optimal bins are used and *minimal logistic costs* can be achieved.

Chapter 13
Limit Performances and Queuing Effects

Logistic, production and performance stations, systems and networks are passed by *discrete objects* [DO]. These objects are *logistic units*, like article units, load units or vehicles, and *information units*, like orders, documents, data and other digital information. Within the stations the objects are consumed, processed, generated, assembled or dispatched. The *limit performances* of the stations and transport elements determine the performance capability of the total system. *Waiting queues* extend the *transfer times* from the entries and sources to the exits and sinks. Hence, for the design of new systems and the optimisation of existing systems, the limit performances and the queuing effects must be known.

In this chapter, the flows of discrete objects through networks are studied. Formulas for the calculation of the *technical limit performances* are derived, *operating strategies* are described and the *limit performance laws* for the different types of stations and transport elements are developed (Arnold 2003; Gudehus 1976). If the incoming flows approache or exceed the limit performance, *waiting queues*, *congestions* and *backward blocking* arise. These *queuing effects* are calculable with the help of formulas derived from *queuing theory* (Arnold 2003; Gross 1998; Gudehus 1976; Kleinrock 1975; Mirasol 1964; Saaty 1957). The limit performance and queuing laws will be explained by several examples and used later for dimensioning and optimisation of storage, commissioning, transport and production systems (see *Chaps. 16, 17, 18, 20*).

Short interruptions and longer breakdowns reduce the limit performance of a system. In order to calculate the *effective limit performances*, the technical *reliability* and *availability* of elements, performance chains and systems are analysed. The definitions, methods and formulas presented here are essential for *capability analysis* and *acceptance tests* of *discontinuously operating systems* with *discrete flows* (Gudehus 1976/79; VDI 3581/3580/3694; FEM 9,222).

13.1 Throughput and Performance Rates

As shown in *Fig. 1.3*, N_I *inflows* λ_{Ii} [DO_i/PE], i = 1,2, ...N_I of discrete objects DO_i enter the system through *entrances* or *incoming stations* IS_i. Through N_O *exits*

T. Gudehus, H. Kotzab, *Comprehensive Logistics*,
DOI 10.1007/978-3-540-68652-1_13, © Springer-Verlag Berlin Heidelberg 2009

or *outgoing stations* OS_j, j = 1,2...N_O, *outflows* λ_{Oj} [DO_j/PE] leave the system. Within the system, the inflows are distributed to performance and logistic stations where they are merged, branched, terminated or transformed into internal flows or outflows.

Production, transport and logistic systems are operative subsystems of companies and of the whole economy. Their characteristics are:

- A *production system* consumes physical objects during a technical process and converts them into other objects (see *Chap. 20*).
- A *transport system* moves physical objects over longer distances to destinations where they leave the system physically unchanged (see *Chap. 18*).
- A *logistic system* transfers, stores, configures and/or rearranges incoming physical objects, which leave the system after certain time physically unchanged.

The performance rates and throughput of these systems are restricted by the limit performance of one or a few bottlenecks:

- *Bottlenecks* are the highest occupied stations and transport elements of a performance chain or network.

Flows, throughput and performance rates refer to a certain *period of time* [PE]. As explained in *Sect. 8.1*, the suitable time unit is determined by the required *punctuality* and by the relevant *systematic variations*. In most cases, the design and dimensioning of a production, transport or logistic system has to cope with the demand of the *peak hour of the peak day* of a year or a longer planning period. This leads to the *time scaling rule*:

► The measurement, planning and calculation of flows, throughput and performance rates should normally be based on the time unit of an *hour* [h].

In addition, it is necessary to differ between stationary and instationary flows with constant or random rates of single units or batches (*see Sect. 9.1*).

13.2 Limit Performances of Elementary Stations

In order to calculate the limit performances and queuing effects, it is necessary to differ between elementary and compound stations (see *Figs. 1.3* and *1.5*):

- An *elementary station* consists of a *single process zone*, which is passed by all flows, and cannot be separated into simpler stations.
- A *compound station* consists of *several process zones* and can be decomposed into parallel or serially connected elementary stations.

Performance and throughput capability of a compound station result from the limit performances of the constituting elementary stations. The further analysis can therefore be restricted to the elementary stations of different types:

- An *elementary station of type* (n,m) and *order* o = n + m converts within its *process zone* n *inflows* λ_{Ii} entering through entrances I_i into m *outflows* λ_{Oj} leaving the element through exits O_j.

The simplest elementary stations are sources, sinks and service stations. Special elementary stations are the *transport elements*:

- A *transport element of type* (n,m) and *order* o = n + m moves, joins or diverts n inflows λ_{Ii} of transport units or load units via a *transfer device* or in a *transfer zone* into m outflows λ_{Oj} of the same units.

The simplest examples are transport connections:

- A *transport connection* is an element of the type (1;1) and order 2 that carries or moves a flow λ of load or transport units from an incoming point I along a *route of length* s [m] to an outgoing point O.

The *structures of some transport elements* with ascending order are shown in *Fig. 13.1*. *Figure* 13.9 shows the structure of a general *irreducible transport node* of the type (n,m) with the order n + m. In the following, several examples for the technical realisation of different stations and transport elements are given.

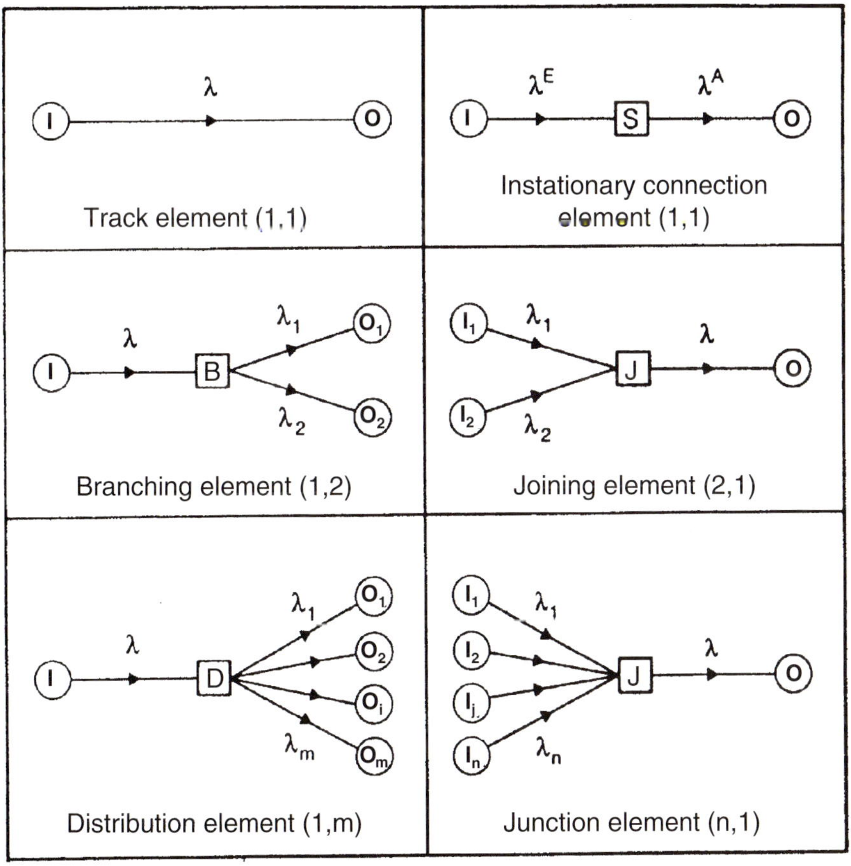

Fig. 13.1 Simple transport elements in ascending order

I_i: incoming points (entrances)
O_j: outgoing points (exits)

13.2.1 Sources

Sources are elementary stations of type (0,m) with m outflows. Examples of *simple source stations* are:

$$\begin{array}{l} \text{natural sources of raw material} \\ \text{incoming stations of a system} \\ \text{elementary production stations} \\ \text{assembly stations} \\ \text{filling and packing stations} \\ \text{storage modules} \\ \text{unloading and break down stations} \end{array} \tag{13.1}$$

If they have only one exit, these are *source stations* of order $o = 1$. If a source has two or more exits for the same or different objects, it is a station of higher order $o > 1$.

The outflows of a source are determined by the *cycle time* τ [s/cycle] of the process or production and by the *batch length* c [PU/cycle] of the operation, i.e. the number of process units [PU] generated at once. From this *technical key data* of a source, the intensity of the outflow can be calculated:

- The outflow of a source with *cycle time* τ and *batch length* c is

$$\lambda(c) = 3600 \cdot c/\tau(c) \qquad [\text{PU/h}]. \tag{13.2}$$ [1]

For the calculation of system performances and the assessment of queuing effects, all incoming and internally generated flows have to be known, since all flows entering internal stations and leaving the system are determined by stationary or time dependent *source flows*. The time dependency can be caused either by a time dependent cycle time, by time dependent batch lengths or by both, time dependent cycle times and batch lengths (see *Fig. 9.2* in *Sect. 9.1*).

The *minimal cycle time* τ_{min} and the *maximal batch length* c_{max} determine the maximal possible outflow, which is the

- *limit performance* of the source:

$$\mu = 3600 \cdot c_{max}/\tau_{min} \qquad [\text{PU/h}]. \tag{13.3}$$

Many production systems and most transport systems are operated by the *push principle*. Then, as explained in *Sect. 8.8*, also all processes within the system are triggered by incoming flows. From this follow the *dimensioning rules for push-operated systems*:

- ▸ *Design rule for push-operated systems*: The system with all its elementary stations and transport elements must be designed and dimensioned *down-stream* from the sources to the sinks starting with the entrance stations.
- ▸ *Performance rule for push-operated systems*: The maximal requirements to the system elements are determined by the limit performances of the external and internal *sources*.

[1] The scale factor 3600 s/h is only necessary when the measurement of flows, throughputs and performances is based on hours [h] and the cycle times are measured in seconds [s].

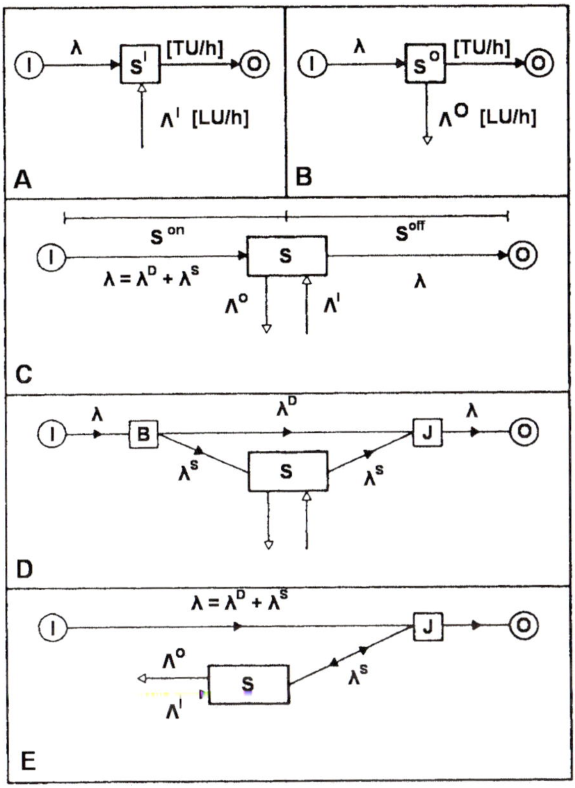

Fig. 13.2 Transfer stations between different transport systems

A: online loading station on a track system
B: online unloading station on a track system
C: online loading/unloading station on a track system
D: offline sidewise loading/unloading station besides a track
E: offline backwards loading/unloading station besides a track
Λ: flow of load units [LU/h]   λ: flow of transport units [TU/h]
B: branching element   J: junction element

With exception of natural sources of raw material, all sources have one or several entries for material, load units or orders. Examples are the different unloading stations for transport units shown in *Fig. 13.2*. By the unloading process, these stations become a source of leaving load units. It depends on the aspect and the specific problem as to whether the inflows of a source station are taken into account or not.

If n incoming flows are taken into account, a source with m exits becomes an elementary station of type (n,m). For example, a bottling station for beverages with an outflow of cases containing 24 bottles, is a source station of type (3,1). The inflows are the liquid beverage, empty bottles and empty cases. The output are the cases containing the bottled beverage.

13.2.2 Sinks

Sinks are elementary stations of type (n,0) with n inflows. Examples of *simple sink stations* are:

exits stations of a system
storage stations
consumption stations
elementary production stations
packing stations (13.4)
loading and build-up stations
depalletizer
disposal sites

Sinks absorb incoming objects, either single or in batches with $1 < c < c_{max}$, with a cycle time $\tau(t)$ which is longer than or equal to the minimal *consumption cycle time* τ_{min}. Consumption flows and limit consumption rates of sinks can be calculated from the cycle time and batch lengths by the formulas (13.2) and (13.3) which hold for sinks and sources as well.

Commissioning systems, retrieval systems and logistic centers are generally operated by the *pull principle*. Then, as explained in *Sect. 8.8*, all processes within the system are triggered by the required outflows. This leads to the *dimensioning rules*:

- *Design rule for pull-operated systems*: The whole system with its stations and transport elements must be designed and dimensioned *upstream* from the sinks to the sources starting with the exit stations
- *Performance rule for-pull operated systems*: The maximal performance requirements to the system elements result from the limit performances of the external and internal *sinks*.

That means, for dimensioning pull-operated systems, the maximally required consumption flows of all external sinks and the limit performances of all internal sinks must be known.

All sinks, except final disposal sites, have one or more exits from which generated products, waste, empties or load units leave the station after a certain process, storing or waiting time. Again, it depends on the aspect and the specific problem as to whether the outflows of a sink station are taken into account or not. If m different outflows are included, a sink with n entrances becomes an elementary station of type (n,m).

13.2.3 Service Stations

Service stations are elementary stations of the type (1,1) and order 2 which are entered and left by a single flow. Examples are:

dispatch stations
toll collection stations
work stations
labelling stations
control points (13.5)
operating places of call centres
compilation stations
gauging and verification stations
reading and identification stations

Fig. 13.3 shows the operation and parameters of a *general service station.* After a possible delay in a *waiting queue,* the incoming objects are transformed *single* or in *groups* or *batches* $1 < c \leq c_{max}$ with a *service cycle time* $\tau(t) < \tau_{min}$. The minimal service cycle time depends on the specific service process.

With respect to their performance, two principally different types of service stations can be distinguished:

- *Steady service stations*: As long as no queuing happens, the objects move continuously through the service station while the service is performed.
- *Unsteady service stations*: Each object is brought to a halt within the station in order to perform the service.

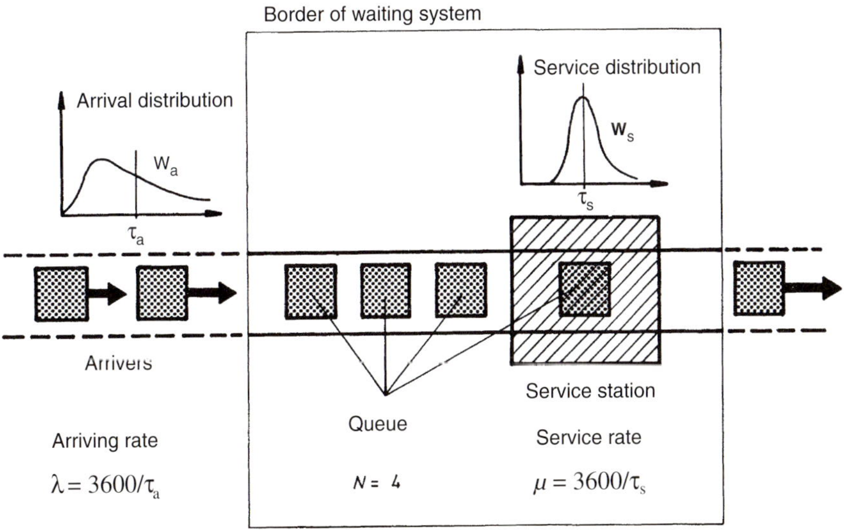

Fig. 13.3 Service station or waiting system of type $W_a/W_s/1$

W_a arrival time distribution
τ_a mean arrival cycle time
λ arrival rate or inflow
W_s service time distribution
τ_s mean service cycle time
μ maximal service rate or limit performance

Whereas the flow $\lambda(t)$ determines the *actual performance rate*, the *maximal flow* is given by the limit performance of the station, which is calculable with the *limit performance formula*:

- The *limit performance* of a station operating with *cycle time* $\tau_S(c)$ [s/cycle] and *batch length c* [PU/cycle] is

$$\mu_s(c) = 3600 \cdot c/\tau_S(c) \qquad \text{[PU/h]}. \tag{13.6}$$

The cycle time can depend on the batch length. With *clock-wise operation*, the cycle time is constant, with *stochastic operation* the cycle times vary randomly around a mean value. For *single-unit operation c* equals 1 PU and for *pair-wise operation c* equals 2 PU per cycle. With *constant batch operation* the batch length is constant, with *stochastic-batch operation* it varies around a mean value c_m. The stochastic fluctuations of the arrivals and/or service rates and/or batch lengths cause *queuing effects*.

13.2.4 Steady Connections

In a *steady connection* or *track element*, incoming load units or transport units move to an exit without halting as long as the exit is not blocked. The objects stop only for other objects with right of way, when a technical interruption occurs or if the following transport element or station causes a *backlog*.

Steady connection elements of conveyor systems are:

gravity conveyors and chutes
powered conveyors
roller and belt conveyors (13.7)
chain conveyors
S-conveyors

For example, *Fig. 13.4* shows an S-conveyor which transfers parcels or pallets continuously with high performance between horizontal conveyors.

Track elements are the elementary links in the network of *vehicle systems*. On the tracks run active transport units or vehicles, such as trucks, automatic guided vehicles (AGV), cars or trains. Examples of track elements are:

driving lanes,
traffic routes
road ways (13.8)
running tracks
railway tracks

The limit performance of a steady connection is the maximal possible passing flow. If $a_{min}(c)$ [m] is the *minimal distance* of the endpoints of two succeeding batches or groups of transport units [TU] and v_S [m/s] is the speed of the units, the minimal cycle time of a steady operating element is

$$\tau_S(c) = a_{min}(c)/v_S \qquad \text{[s]}. \tag{13.9}$$

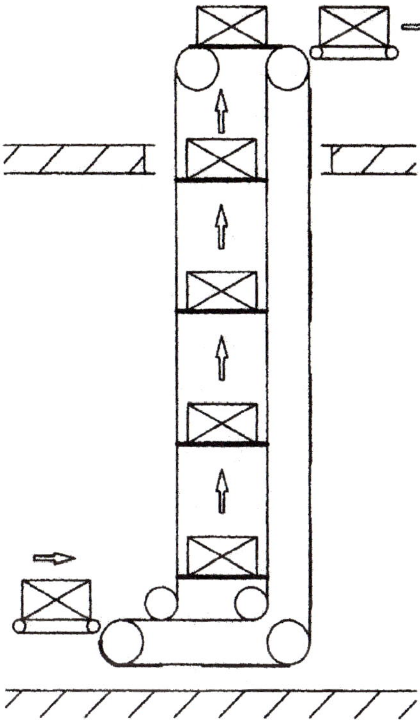

Fig. 13.4 Example of a steady vertical connection element
S-conveyer for parcels, containers or pallets

Inserting this into (13.6) results in the

- *Limit performance of a steady connection* with *speed* v_S [m/s] and *minimal distance* a_{min} [m] of the endpoints of batches of c transport units

$$\mu_S(c) = 3600 \cdot c \cdot v_S / a_{min}(c) \qquad [TU/h]. \qquad (13.10)$$

For a *conveyor system* the minimal distance of a batch of c units equals the *length of the load units* l_{LU} [m] times the batch length c plus a *construction distance* l_{con} [m]:

$$a_{min}(c) = c \cdot l_{LU} + l_{con} \qquad [m]. \qquad (13.11)$$

The construction distance is determined by the geometry of the objects, technical conditions and safety requirements. In the most favourable case it is 0.

Table 13.1 shows the limit performances of steady connection elements of conveyor systems for bins and pallets which are calculated from the technical data with the help of the formulas (13.10) and (13.11).

In a *vehicle system* the minimal distance of the single transport units depends on the *distance control*, the *length of the transport units* l_{TE} [m] and the necessary *safety clearance* l_{safe} [m] between the units. In order to avoid a collision, if a vehicle

Table 13.1 Limit performances of steady connection elements within conveyor systems

Conveyer element	Load unit LU	End point distance [m]	Speed [m/s]	Limit performance [LU/h]
Roller conveyer	Standard pallet	1.4	0.30	**771**
	Standard bin	0.7	0.50	**2,571**
Chain conveyer	Standard pallet	1.5	0.20	**480**
	Standard bin	1.5	0.50	**1,200**
Belt conveyer	Standard pallet	1.5	0.30	**720**
	Standard bin	0.7	0.80	**4,114**
Endless conveyer	Standard pallet	2.5	0.20	**288**
	Standard bin	1.0	0.40	**1,440**
S-Conveyar	Standard pallet	2.0	0.20	**360**
	Standard bin	0.6	0.40	**2,400**

Spreadsheet calculation using formulas (13.10) and (13.11)
EURO-pallet: $l \times b \times h = 1{,}200 \times 800 \times 1{,}800$ mm
Standard bin: $l \times b \times h = 600 \times 400 \times 300$ mm

driving ahead stops, the safety clearance has to be at least as long as the length of the way passed during the reaction time, plus the length of the de-acceleration. This length is the so called *stopping shadow*, which runs virtually ahead of each vehicle and should never touch an obstacle. From mechanics follows:

- The length of the *stopping shadow* of a vehicle with *speed* v [m/s], a *reaction time* t_R [s] and *emergency-deacceleration* b^- [m/s^2] is

$$l_{stop} = v \cdot t_R + v^2/2b_e^- \qquad [m]. \tag{13.12}$$

For vehicles with *active distance control*, the driver or a moving control system permanently checks the distance to the ahead driving vehicle and ensures that it always exceeds the stopping shadow. Hence, with active distance control, the minimal distance of transport units with length l_{TU} [m] is:

$$a_{min} = l_{TU} + v \cdot t_R + v^2/2b_e^- \qquad [m]. \tag{13.13}$$

For *passive distance control*, the track is divided into *block sections*. A stationary *block section control system* prevents the entrance of a transport unit with its stopping shadow into the next block section, as long as this is occupied by another unit. With *block section length d* [m], the minimal distance of succeeding transport units is:

$$a_{min} = d \cdot \text{ROUNDUP}(\ (l_{TU} + v \cdot t_R + v^2/2b_e^-)/d\ ) \qquad [m]. \tag{13.14}$$

A comparison of the relations (13.13) and (13.14) shows that the minimal distance is generally longer for block section control then for distance control. Only if the length of the block sections equals the minimal distance (13.13), the limit performances are the same. This leads to the *track control rule*:

► The limit performance of a track with block section control is lower then the limit performance of the same track with distance control.

Inserting (13.13) into (13.10) results for single unit operation, i.e. for c = 1, in the *limit performance formula for track elements*:

- The *limit performance of a track element* passed by single transport units of *length* l_{TU} [m] with *driving speed* v [m/s], *emergency-deacceleration* b_e^- [m/s²] and *reaction time* t_R [s] is

$$\mu(v) = 3600/(t_R + l_{TU}/v + v/2b_e^-) \qquad [TU/h] \qquad (13.15)$$

Table 13.2 presents the technical data and the resulting limit performances of several track elements for different vehicle systems. The dependency (13.15) of the limit performance of a road lane on the speed is shown in *Fig. 13.5* for vehicles of different length.

A result of the general limit performance formula (3.15) which holds for all kinds of track elements is the *maximal throughput law*:

► The limit performance of a track depends on vehicle speed and emergency-deacceleration and reaches the *maximal throughput*

$$\mu_{max} = 3600 \Big/ \left(t_R + \sqrt{2 \cdot l_{TU}/b_e^-} \right) \quad [TU/h] \qquad (13.16)$$

at the *throughput-optimal speed*

$$v_{opt} = \sqrt{2 \cdot l_{TU} \cdot b_e^-} \quad [m/s] \qquad (13.17)$$

As shown in *Fig. 13.5*, the throughput-optimal speed of a road lane is in the range 30 to 50 km/h. The speed dependency of the limit performance (13.15) and the formulas (3.16) and (3.17) are basis of the *flow-dependent speed control* in *traffic networks*. However, the throughput-optimal speed is generally not the *cost optimal speed*, which differs also from the *maximal speed*.

Table 13.2 Technical data and limit performances of track elements in vehicle systems

Track element	Vehicle type Transport unit TU	Vehicle length [m]	Dead time [s]	Optimal speed [m/s]	Emergency deaccelaration [m/s²]	Limit performance [TE/h]
Overhead track	Electric vehicle	1.5	0.5	1.2	0.5	**1,221**
ASV-track	AGV vehicle	2.5	0.7	1.9	0.7	**1,067**
Transport track	Truck	8.0	1.0	4.0	1.0	**720**
Road track	Mini car	2.5	1.0	6.3	8.0	**2,011**
	Standard car	5.0	1.0	8.4	7.0	**1,640**
	Long car	6.5	1.0	8.4	6.0	**1,456**
	Load truck	18.0	1.0	13.4	5.0	**977**

Spreadsheet calculation using formulas (13.15) and (13.17)
1 m/s = 3.6 km/h; 1 km/h = 0.28 m/s

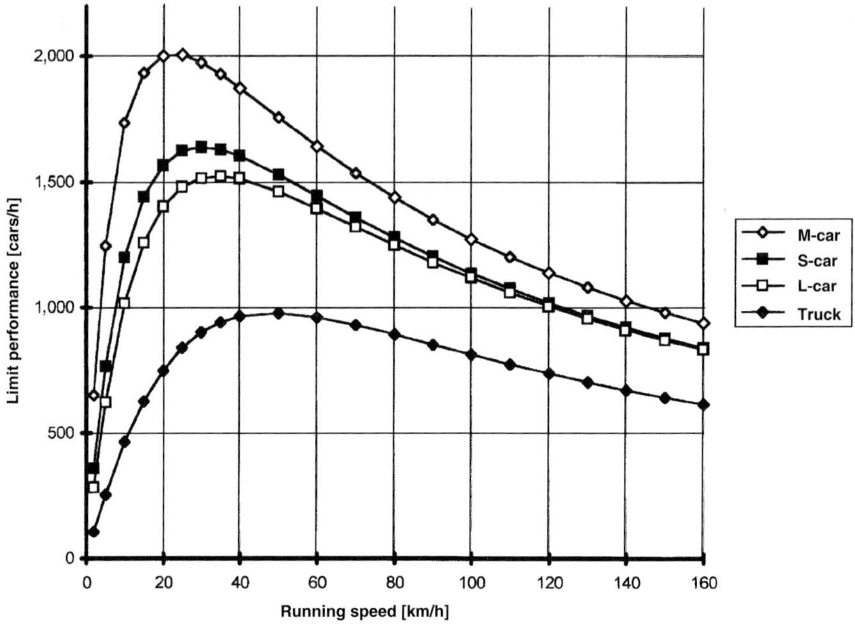

Fig. 13.5 Speed dependency of limit performances of a traffic lane

M-car mini passenger cars (e.g. Smart-car)
S-car standard passenger car
L-car large passenger car
Truck tractor trailer unit
Technical data: see *Table 13.2*

13.2.5 Unsteady Connections

An unsteady connection element carries $c \leq c_T$ load units by an intermittent operating *transfer element* [TE] with *capacity* c_T [LU/TU] forth and back along a *transfer path of length s* [m]. Examples for transfer elements are:

transfer carriers
turning platforms
swing-aside tables
lifting stations
elevators (13.18)
cranes
storage and retrieval units (S/R-units)
shuttles
transport vehicles

Also vehicles starting from a station, reaching another station after a certain travelling time and returning to the first station after being unloaded and/or reloaded, are transfer elements. A transfer element can be used in two different *operating modes*:

- *Single-transfer cycles*: The transfer element or vehicle is loaded only in one direction and empty in the other direction.
- *Combined-transfer cycles*: The transfer element or vehicle is loaded in both directions and carries load units forth and back in the same cycle.

The *single-transfer-cycle time* of a connection element or vehicle carrying c load units in one direction is the sum of the *on-loading time* $t_{on}(c)$, twice the *travelling time* $t_{trav}(s)$ for the *path length* s, and the *off-loading time* $t_{off}(c)$:

$$\tau_1(c;s) = t_{on}(c) + 2\,t_{trav}(s) + t_{off}(c) \quad [s] \tag{13.19}$$

The *combined-transfer-cycle time* or *double cycle time* for a forth and back transport of c load units is:

$$\tau_2(c;s) = 2 \cdot (t_{on}(c) + t_{trav}(s) + t_{off}(c)) \quad [s] \tag{13.20}$$

The on-loading and off-loading times are determined by the *loading technique* and by the organisation of the *loading process*. Loading can be performed by *loading device* mounted on the vehicle or by a *stationary loading device* such as a crane or another transfer element. For *single unit loading*, where the single units are loaded separately, the loading times are longer than for *batch loading*, where a batch or group of units is loaded at once.

The dependency of the travelling time of a transfer element with *acceleration* b^+ [m/s²], *maximal speed* v_m [m/s] and *deacceleration* b^- [m/s²] on the *path length* s [m] is given by the general *travelling time formula* (see Sect. *16.10*):

$$t_{trav}(s) = \mathrm{IF}(s < {v_m}^2/b_m;\ 2\sqrt{s/b_m};\ s/v_m + v_m/b_m) \quad [s] \tag{13.21}$$

Herein, b_m is the *harmonic mean of acceleration and deacceleration*:

$$b_m = 2 \cdot b^+ \cdot b^- / (b^+ + b^-). \tag{13.22}$$

Inserting (13.19) and (13.20) into (13.6) leads to the general *performance formulas for unsteady connections*:

- If operating in *single-transfer cycles* with average load $c \le c_T$, the *limit performance of an unsteady connection, transfer element* or *vehicle* with *load capacity* c_T, *on-loading time* $t_{on}(c)$, *off-loading time* $t_{off}(c)$, and *travelling time* $t_{trav}(s)$ over *path length* s is

$$\mu_1 = 3600 \cdot c/(t_{on}(c) + 2t_{trav}(s) + t_{off}(c)) \quad [\mathrm{LU/h}] \tag{13.23}$$

If operating in *combined-transfer cycles*, the limit performance is

$$\mu_2 = 3600 \cdot c/2(t_{on}(c) + t_{trav}(s) + t_{off}(c)) \quad [\mathrm{LU/h}] \tag{13.24}$$

The dependency of the limit performance on the path length is shown in *Fig. 13.6* for the example of a *pallet transfer shuttle* with different load capacities.

Table 13.3 contains the limit performances of unsteady connections of conveyor systems for bins and pallets that have been calculated with formula (13.23) based on the indicated technical data (see also *Fig. 16.20*). *Figure 13.6* and *16.20* and the limit

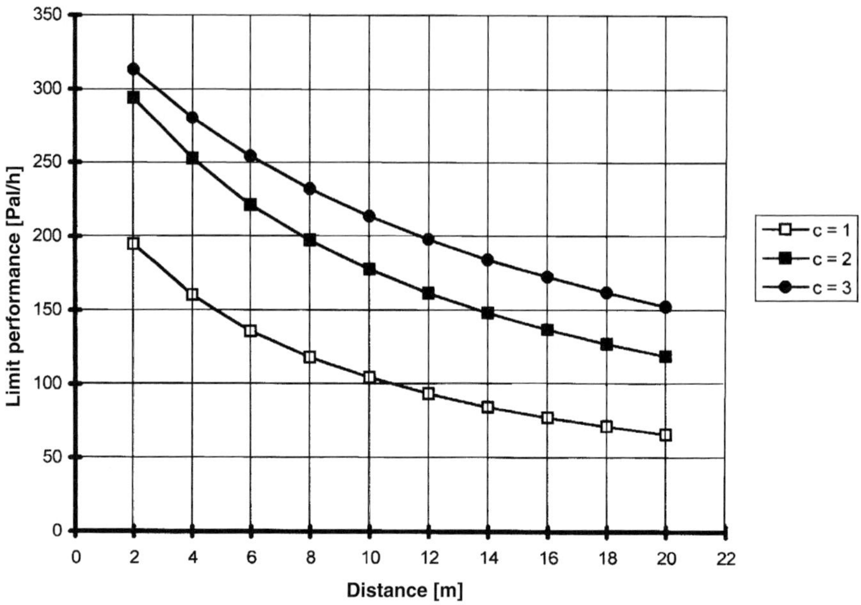

Fig. 13.6 Dependency of the limit performance of a transfer shuttle for pallets on the transfer path length

Parameter c [Pal]: load capacity of the transfer shuttle
Technical data: see *Table 13.3*

performances listed in *Table* 13.3 demonstrate the general *transport performance rule*:

- Limit performances of transfer elements and vehicles depend extremely on the load capacity and on the transport distance.

Hence, the limit performance of a transfer element or a vehicle can be improved most effectively by increasing the load capacity and by shortening the transport distance. For short distances, further improvements of the limit performance are achievable by a reduction of the loading times and by higher accelerations. The travelling speed is an effective lever to improve the transport performance only for longer distances.

13.2.6 Junction Elements and Branching Elements

As shown in *Fig. 13.7* for a lift station, *connection elements* can have $n > 1$ entries and/or $m > 1$ exits. With $n > 1$ and $m = 1$ they are *junction elements* and with $n = 1$ and $m > 1$ *branching elements*. A transport element with $n > 1$ and $m > 1$ can operate as a combined *junction&branching element*. This is an *irreducible transport knot* with a general structure as shown in *Fig. 13.9* .

Some examples for *unsteady junctions* and *branching elements* are listed in (13.18). *Steady junctions and branching elements* are

Table 13.3 Limit performances of unsteady connection elements in conveyor systems

Connection element Load transfer	Load unit LU	**Capacity** [LU/Trip]	**Enter and leave time** [s]	**Speed** [m/s]	**De/acceleration** [m/s^2]	**Drive way** [m]	**Limit performance [LU/h]**
Transfer	Pal.	1	12.0	0.20	0.3	4.0	**68**
carriage	Pal.	2	20.0	0.20	0.3	4.0	**117**
roller or chain conveyor	Bin	2	6.0	0.50	0.5	4.0	**300**
telescope fork	Pal.	1	22.0	0.50	0.5	4.0	**90**
	Pal.	1	22.0	0.50	0.5	8.0	**64**
Distribution	Pal.	1	12.0	1.00	0.8	4.0	**160**
carriage	Pal.	1	12.0	1.00	0.8	8.0	**118**
roller or chain conweyor	Pal.	2	20.0	1.00	0.8	4.0	**236**
Lift station	Pal.	1	12.0	0.20	0.3	3.0	**83**
roller or chain	Pal.	1	12.0	0.20	0.3	6.0	**49**
conveyor	Pal.	2	20.0	0.20	0.3	3.0	**140**
	Bin	2	6.0	0.50	0.5	3.0	**360**

Spreadsheet calculation using formula (13.23)
EURO-pallet: $l \times b \times h = 1{,}200 \times 800 \times 1{,}800$ mm
Standard bin: $l \times b \times h = 600 \times 400 \times 300$ mm

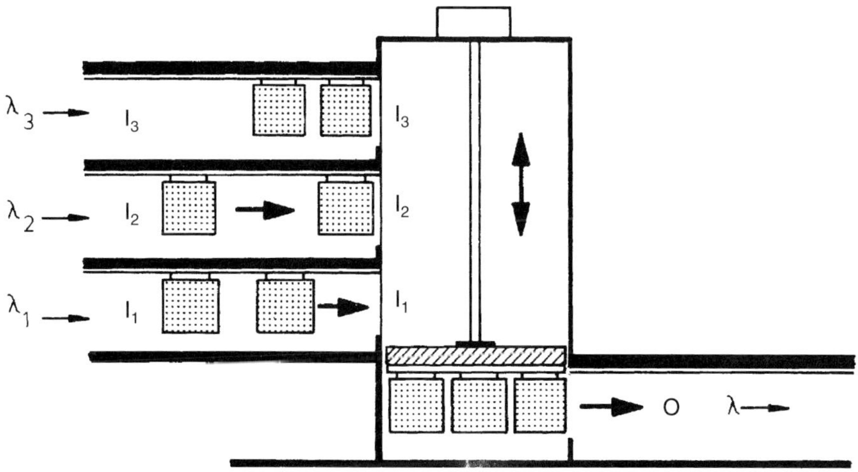

Fig. 13.7 Lift station with load capacity for c = 3 vehicles in a monorail transport system

railway switches
crossings
road junctions (13.25)
road forks
intersections

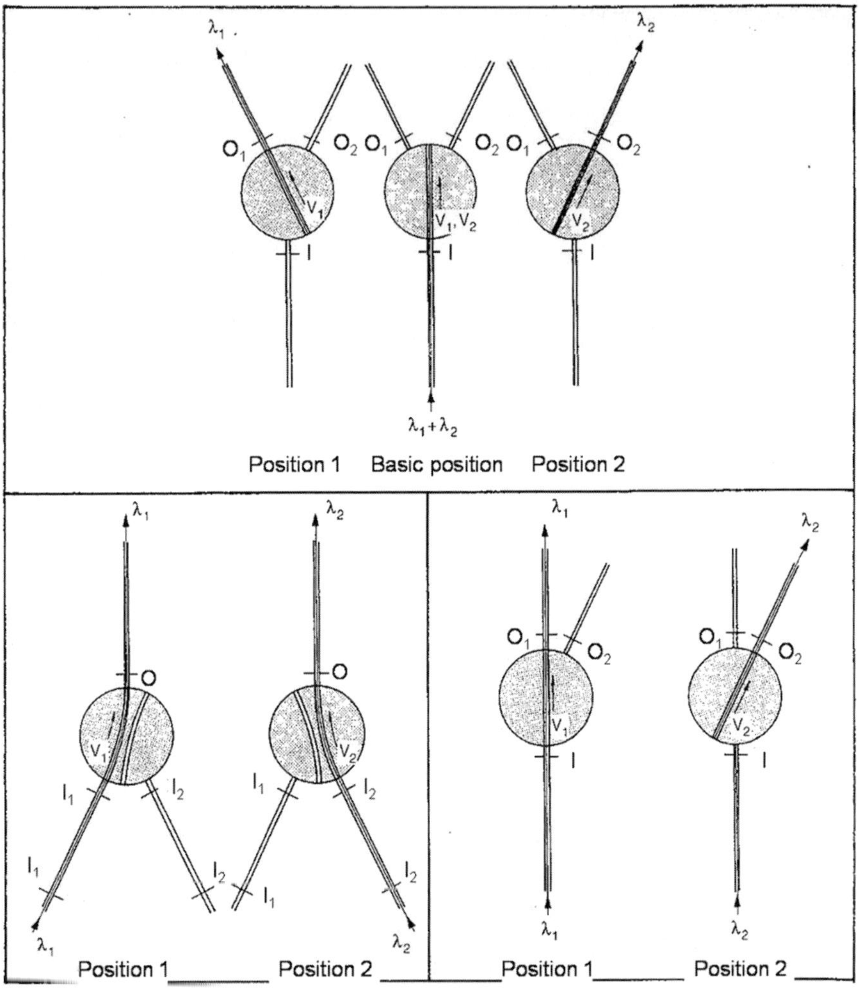

Fig. 13.8 Unsteady, half-steady and steady junction and branching elements of a monorail system

Top figure: unsteady branching element (turning device)
Left below: steady junction element (rail way switch)
Right below: half-steady branching element (passing & turning table)

In *Fig. 13.8* three different junction and branching elements of a monorail system are shown. The examples illustrate a general feature of all junction and branching elements: The connection between an entrance and an exit is either steady or unsteady. This feature is important for the calculation of the *partial limit performances* for the different connections of junction elements, branching elements, and irreducible transport knots:

Table 13.4 Limit performances and switch times of connection elements, junction elements and branching elements in conveyor systems for bins and pallets

Distribution element Junction element	**Load unit** LU	**Capacity** [LU]	**Transfer distance** [m]	**Swith time** [s]	**Limit performances** Passing [LU/h]	Branching [LU/h]
Rollers – turntable – rollers	EU-Pal.	1	2.0	0.0	**600**	**140**
	EU-Pal.	2	2.0	0.0	**600**	**210**
	Bin	1	1.0	0.0	**3,000**	**300**
	Bin	2	1.0	0.0	**3,000**	**450**
Rollers – lifter – chain conveyer	EU-Pal.	1	0.0	0.0	**650**	**180**
	Bin	1	0.0	0.0	**3,000**	**450**
Roller conveyer – transfer carriage	EU-Pal.	1	4.0	0.0	**700**	**70**
	Bin	1	1.5	0.0	**3,000**	**220**
	Bin	2	1.5	0.0	**3,000**	**400**
Rollers – shunt – rollers	Bin	1	0.0	9.0	**1,800**	**1,800**
Rollers – belt transfer – rollers	Bin	1	45°	1.7	**3,000**	**2,700**
Rollers – chain transfer	Bin	1	90°	0.0	**3,000**	**1,360**
	Bin	1	45°	0.0	**3,000**	**2,700**
Rollers – pusher – rollers	Bin	1	90°	0.0	**2,400**	**600**

EURO-pallet: $l \times b \times h = 1{,}200 \times 800 \times 1{,}800$ mm
Standard bin: $l \times b \times h = 600 \times 400 \times 300$ mm

- The *partial limit performances for steady connections* of an irreducible transport knot can be calculated with the universal formula (13.10).
- The *partial limit performances for unsteady connections* of an irreducible transport knot are calculable with the universal formulas (13.23) respectively (13.24).

In *Table 13.4*, the technical data and resulting limit performances are listed for different junction elements, branching elements, and higher order connections in transport systems for bins and pallets.

13.2.7 Irreducible Transport Nodes

The steady and unsteady connections, junctions and branching elements are special examples of order 1, 2 and 3 of a general *irreducible transport node* of type (n,m) and order o = n + m which is shown in *Fig. 13.9* .

Unsteady transport elements of higher order are the unsteady connections (13.18) with more than two entrances and/or exits. Examples for steady transport elements of higher order are:

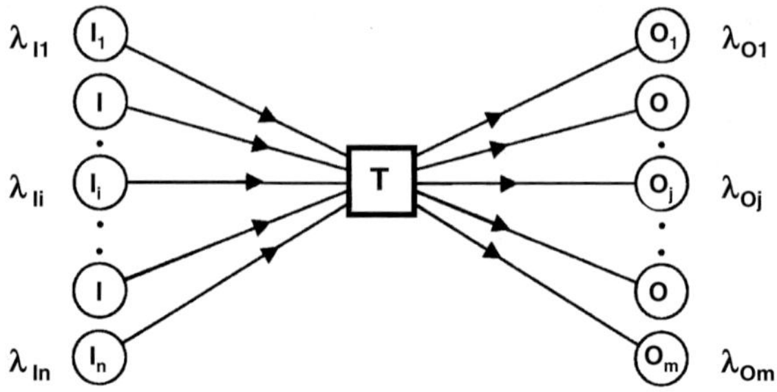

Fig. 13.9 Irreducible transport node or transport element of type (n,m) and order o = n + m

multiple switches
track crossings
crossing shunts
crossing nodes (13.26)

As shown in *Fig. 13.9*, from the n entrances I_i up to n·m *partial flows* λ_{ij} run to the m exits Oj of a transport node. The partial flows λ_{ij} are components of the n *inflows*:

$$\lambda_{Ii} = \sum_j \lambda_{ij} \tag{13.27}$$

Within the *transfer zone* of the transport node, the inflows are converted into m *outflows*:

$$\lambda_{Oi} = \sum_i \lambda_{ij} \tag{13.28}$$

The *total flow* passing the transport node is:

$$\lambda = \sum_i \lambda_{Ii} = \sum_j \lambda_{Oj} = \sum_i \sum_j \lambda_{ij} \tag{13.29}$$

The maximal possible throughput for connection $I_i \rightarrow O_j$ is the *partial limit performance* μ_{ij}, which can be calculated by formula (13.10) for steady connections and by formula (13.23) respectively (13.24) for unsteady connections. This leads to the *law of partial limit performance*:

- ► Each partial flow λ_{ij} of a transport node is restricted by the partial limit performance μ_{ij} of the respective connection

$$\lambda_{ij} \leq \mu_{ij} \tag{13.30}$$

The *partial flow utilisation* cannot exceed 100%, i.e. for all i and j holds:

$$\rho_{ij} = \lambda_{ij}/\mu_{ij} \leq 100\% = 1 \tag{13.31}$$

These conditions are necessary but not sufficient, since the transfer element is not only occupied by the passing flows but also by idle *switching moves*. The time losses for switching are determined by the *interruption time* or *switch time*:

- The *switch time* $\tau_{ij\,kl}$ [s] of a transport element is the minimal time difference between the time, when the last unit of partial flow λ_{ij} for the connection $I_i \rightarrow O_j$ has entered, and the time, when the first unit of flow λ_{kl} for another connection $I_k \rightarrow O_l$ can enter the transfer zone.

The switch times of transport elements correspond to the *setup times* of production stations (see *Sect. 20.5*). In *Table 13.4* the switch times of some transport elements for bins and pallets are listed.

The *switch time between steady connections* is the sum of the clearance time for the last batch, which has entered the transfer zone, and the turning time for the transfer element into the new direction. For example, the switch time of a one-lane road operated in two directions, which happens at construction sites, is the time between the last vehicle, which has entered into one direction, and the first vehicle, which can enter the lane into the opposite direction.

The idle *switch times for unsteady connections* are 0 as long as the turning times of the transfer element in both directions are shorter than the productive cycle times (13.19) for the transport moves of the transfer unit. Otherwise they are equal to the sum of the forth and back turning time minus the cycle times for the two connections. The switch times are lost for the productive use of a transport element, i.e:

▶ The productive utilisation of a transport element which is operated with *switch frequencies* $\nu_{ij\,kl}$ [1/h] and *switch times* $\tau_{ij\,kl}$ [s] between the directions $I_i \rightarrow O_j$ and $I_k \rightarrow O_l$ is reduced by the *switching occupation*

$$\rho_S = \sum_{i,j} \sum_{k,l} \nu_{ij\,kl} \cdot \tau_{ij\,kl}/3600 \quad [\%] \tag{13.32}$$

The switch frequencies are determined by the *operating strategy*.

13.3 Operating Strategies

Limit performances depend primarily on technique and cannot be changed after installation of the transfer element. However, the unproductive occupation by switching losses can be influenced by the operating strategy:

- An *operating strategy* regulates service, batches and priorities of the arriving units.

Like many other logistic strategies, most operating strategies are applications of the basic strategies of *Sect. 5.2*: *clustering*, *sequencing*, *securing*, and their counter strategies.

By an operating strategy different *goals* and *objectives* can be achieved such as:

- *Utilisation*

$$\text{maximal utilisation} \tag{13.33}$$

- *Performance*

$$\begin{gathered}\text{maximal throughput in all directions}\\ \text{maximal throughput in specific connections}\end{gathered} \tag{13.34}$$

- *Time Saving*

minimal transfer times for all directions
minimal transfer times for specific connections (13.35)

- *Reduction of Queuing*

minimal waiting queues and waiting times
no blocking of preceding stations (13.36)

- *Safety*

minimal breakdown probability
maximal traffic safety (13.37)
minimal risk of accidents

Some of these objectives are incompatible and cannot be achieved at once. Hence, before a strategy is applied, the objectives should be defined, accessed and prioritised (see *Sect. 5.1*).

13.3.1 Clustering Strategies

If the load capacity c_T of a station or a transport element exceeds 1, different numbers of units can be served and transferred collectively. The number of units to be served at once is regulated by one of the following *clustering strategies*:

- *Single-unit dispatch*: The single units are served and transferred to the different directions separately.
- *Constant-batch dispatch*: The arriving units enter the service station or transfer zone in batches of *constant length* $c \leq c_T$ and are collectively served and transferred to the required direction.
- *Variable batch dispatch*: The arriving units enter the service station or transfer zone in batches of *variable length* $c(t) < c_T$ and are collectively served and transferred to the required direction.

Single-unit dispatch is unavoidable if $c_T = 1$. If $c_T > 1$, single-unit dispatch is the fastest service strategy, but the resulting limit performance is lower than with *batch-wise dispatch*. Maximal utilisation can be achieved with constant batches of maximal length $c = c_T$. However, for lower flow intensities this strategy leads to longer transfer times as the first arriving units have to wait until c_T units have arrived.[2] On the other hand, dispatch in maximal batches causes lowest cost.

In order to avoid longer waiting times at low flow intensities, a variable batch dispatch can be applied. Here, $c(t) \leq c_T$ units, which arrived while the last batch has been served, are served together. This strategy is *self-regulating* since the occupation of the capacity and the partial limit performances improve with increasing flows. With variable batches, the service times at low flow intensities are far lower than with maximal batches but longer than with single unit dispatch, where the on- and off-loading times are shortest.

[2] This dilemma is known to all travellers who have been waiting for ferries or other shuttles that only start, if a sufficient number of passengers shows up.

13.3.2 Sequencing Strategies

If a station or transport element has more than one entry, not only the batch lengths but also the sequence, in which the incoming units of different entrances are served must be determined. This can be done by the *sequencing strategies*:

- *Random dispatch* (*equal right regulation* or *First-Come-First-Go*): The incoming units are dispatched in the random sequence of their arrival.
- *Weak priority dispatch* (*main road regulation*): Units that arrive from a second rank direction can enter the dispatch zone if the *limit time gap* between two passing units with priority is sufficient.
- *Absolute priority dispatch* (*stop road regulation*): Units that arrive from a second rank direction have to stop at an entry point and to wait until the *limit time gap* between two passing units with priority is sufficient.

For both priority regulations, it is necessary to sequence the incoming directions in a *priority sequence*:

$$\lambda_{I1} \text{ before } \lambda_{I2} \text{ before } \lambda_{I3} \ldots\ldots \lambda_{In-1} \text{ before } \lambda_{In}. \tag{13.38}$$

For a station with n inflows n! different priority sequences are possible. If only two flows are considered, the prioritised flow is called *main flow* λ_M and the other flow *secondary flow* or *next flow* λ_N. This does not mean that the main flow is always stronger than the next flow.

To ensure that a second rank unit can enter a main flow without disturbing any of its units, the time gap between two units of the main flow must be longer than the sum of the minimal cycle times τ_M and τ_N of the main flow and the secondary flow plus the forth and the back switch times τ_{MN} and τ_{NM}, i.e. longer than the *limit time gap*:

$$\lambda_L = \tau_M + \tau_N + \tau_{MN} + \tau_{NM} \quad [\text{s}]. \tag{13.39}$$

With absolute priority regulation, all units have to stop. This causes the additional *de-acceleration time* $v/2b^-$. Therefore the back switch time τ_{NM} is longer for absolute priority than for weak priority.

Priority regulation, compared with random dispatch, leads to minimal transfer times for the prioritised flows and to longer transfer times for the secondary flows. The maximal throughput is generally smaller with a priority dispatch. Absolute priority dispatch has the additional advantages of higher safety and functional reliability, but reduces the maximal throughput of the secondary flows considerably.

Random dispatch is quite easy to implement. However, at high partial utilisation it is difficult to detect the arrival times of the waiting units at the different entrances, which is necessary to ensure first-come-first-go. Due to the measurement and control of the limit time gaps, the costs for implementing a priority strategy are generally higher than for random dispatch.

13.3.3 Cyclic Dispatch

The highest safety can be achieved by cyclic dispatch with the two possibilities:

- *Constant-cycle dispatch*: The station or transport element changes operation from entrance I_i to another entrance after a *constant dispatch cycle time* T_{Ci}.
- *Adaptive-cycle dispatch*: The station or transport element changes the operation from entrance I_i to another entrance after an *adaptive cycle time* $T_{Ci}(t)$ which depends on the actual flow utilisations (13.31) or other criteria.

Strategy parameters of cyclic dispatch are the *dispatch cycle times* T_{Ci} [s], the *switch frequencies* ν_{Ci} [1/h] and the *priority sequence* (11.38) for the arriving flows. If each incoming direction I_i is served n_i times per sequence, the *total cycle time* for a complete sequence is:

$$T_C = \sum_i n_i \cdot T_{Ci} \quad [s] \tag{13.40}$$

The *change frequency of the sequences* is

$$\nu_C = 3600/T_C \quad [1/h] \tag{13.41}$$

and the *partial switch frequency* for the entrance I_i

$$\nu_{Ci} = n_i \cdot \nu_C \quad [1/h]. \tag{13.42}$$

The formulas (13.32), (13.40) and (13.41) lead to the *rules of cyclic dispatch*:

- ▸ Short dispatch cycle times lead to high operation frequencies and short waiting times, but cause lower performances due to the frequent switching.
- ▸ Long dispatch cycle times lead to higher limit performances but increase the waiting times due to the lower service frequencies.

Major disadvantage of cyclic dispatch are *systematic waiting queues* that build up in front of the entrances while other directions are served. If the cycle times of neighboured stations are not synchronized, the waiting queues can grow so much that they block preceding stations.

Cyclic dispatch achieves higher safety against disturbances and accidents as compared to the other strategies. However, the implementation of cyclic dispatch is more expensive. Due to the measurement of the partial flow utilisation, *adaptive cycle dispatch* is even more complex than constant cycle dispatch.

13.3.4 System Strategies

High limit performances, short transfer times and low operating costs are not only achievable by operating strategies for the single elements, but also by suitable system strategies (see *Sect. 5.3*):

- *System strategies* regulate the flow of objects through a chain or a network of several stations.

For the different systems, like production, handling, transport, storage and order picking systems, many system strategies are known. The special strategies for the different systems will be explained in the corresponding chapters. Some of them are combinations of the basic operating strategies which have been described before.

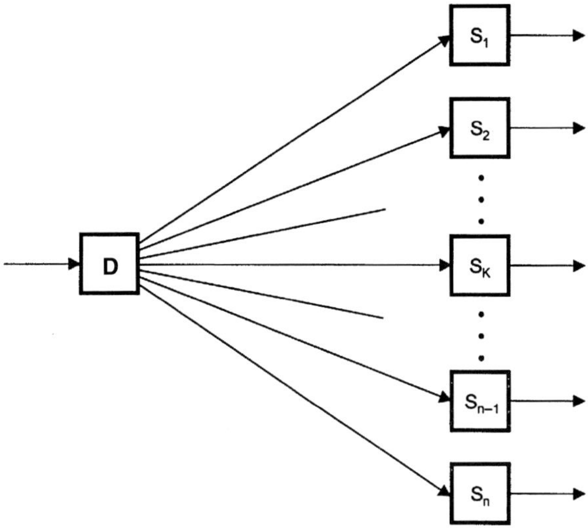

Fig. 13.10 Parallel stations or performance chains

DS: dispatch or allocation station

S_k: entrances of performance chains

If an input station, as shown in *Fig. 13.10*, has to allocate arriving units to parallel stations, which offer the same service or are entrances of parallel performance chains with the same function, in addition to clustering strategies the following *parallel operating strategies* are possible:

- *Cyclic single unit allocation* (*one piece flow*): The single arriving units are allocated cyclically to the parallel stations or performance chains.
- *Utilisation dependent single unit allocation*: The single units are allocated to the station or performance chain with lowest utilisation and shortest waiting queue.
- *Adaptive cycle batch allocation*: The units are allocated in adapted batches to the station or performance chain with lowest utilisation and shortest waiting queue.
- *Dynamic allocation*: The units are allocated to the station with the shortest waiting queue. If the utilisation of all operating stations exceeds n/(n+1) or causes an overflow of the queuing capacities, a further station is opened. If the utilisation drops below n/(n+2) one station is closed.

With dynamic allocation a larger number of parallel stations or performance chains can be adapted in a *self-regulating* manner to a varying intensity of the inflow (Gudehus 2005).

For the operation of a performance, logistic or transport chain, as shown in *Fig. 13.11*, the following *serial operating strategies* are possible:

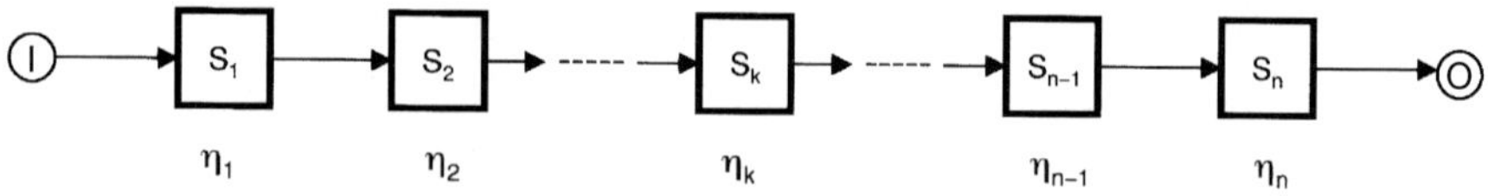

Fig. 13.11 Performance chain, logistic chain or transport chain
S_k: performance stations or transport elements
η_k: operating probabilities (reliability or availability)

- *Independent flow*: The arriving units flow independently from one station to the next, where they are served with suitable operating strategies.
- *Single unit throttled flow*: Single units that arrive with time distances shorter than the cycle time of the bottleneck station of the chain are throttled into a *tact rate* equal to the limit performance of the bottleneck.
- *Adaptive batch throttled flow* (*bottleneck allocation*): A dispatch station throttles and consolidates the arriving units into batches with length equal to the maximal capacity of the bottleneck station.
- *Synchronised flow* (*green wave regulation*): The cycle times of the subsequent stations with steady operation are synchronised in a way that a longer batch of units flows through the total chain without halting.

The parallel and serial operation strategies affect the partial performances, transfer times, reliability, availability, and process costs differently. The choice for either one or the other strategy depends on the objectives and priorities of the individual case. It is therefore necessary to analyse the strategy effects on system performance and queuing. This is possible with the help of the following *limit performance laws* and *queuing laws*.

13.4 Limit Performance Laws

In order to avoid increasing waiting queues in front of the entrances, the *total utilisation* of an elementary station, which is the sum of the *partial flow utilisations* ρ_α and the *switch utilisations* $\rho_{\alpha\beta}$, must be smaller 100% during the whole operating time, i.e.:

$$\rho = \sum_\alpha \rho_\alpha + \sum_\alpha \sum_\beta \rho_{\alpha\beta} < 100\% = 1 \tag{13.43}$$

Herein, α and β denote the *partial functions* F_α and F_β of the elementary station. By inserting $\rho_\alpha = \lambda_\alpha/\mu_\alpha$ and $\rho_{\alpha\beta} = \nu_{\alpha\beta}\cdot\tau_{\alpha\beta}/3600$ into (13.43) results the *universal limit performance law* (Gudehus 1975/76):

► *Necessary condition* for an elementary station with *limit performances* μ_α [PU/h], *switch times* $\tau_{\alpha\beta}$ [s] and *switch frequencies* $\nu_{\alpha\beta}$ [1/h] to cope with *partial flows* λ_α [PU/h] is

$$\sum_\alpha \lambda_\alpha/\mu_\alpha + \sum_\alpha \sum_\beta \nu_{\alpha\beta} \cdot \tau_{\alpha\beta}/3600 < 1. \tag{13.44}$$

Applications of the limit performance law (13.44) in logistics are *irreducible transport elements* of any order, where the sums in (13.43) and (13.44) have to be taken for all connections $\alpha = (i,j)$ and $\beta = (k,l)$ which cannot be passed at the same time. Other applications are *production stations*, which will be considered in *Chap. 20*. Here the functions F_α and F_β are the *execution processes* for different products or services and the switch times $\tau_{\alpha\beta}$ are the *change-over times* or *setup times* between the processes.

The fulfilment of the limit performance condition (13.44) is *necessary* for the functioning of stations and transport elements irrespective of the operating strategy. However, the condition is not always *sufficient*. Only for random dispatch of single units, for adaptive batches and for adaptive cyclic dispatch, condition (13.44) is necessary and sufficient. For priority regulation, constant batches and constant cycle dispatch, additional conditions must be fulfilled since these strategies cause additional time gaps.

If the process zone has a capacity $c_T > 1$, the partial limit performances $\mu_\alpha(c)$, which can be calculated by the above formulas, may differ for different batch lengths c. If this is the case, the partial flows λ_α have to be separated into a sum of flows $\lambda_\alpha(c)$, $c = 1, 2, \ldots c_T\text{-}1, c_T$ with equal batch length c. The partial utilizations $\lambda_\alpha/\mu_\alpha$ in (13.44) are replaced by:

$$\lambda_\alpha/\mu_\alpha \rightarrow \sum_c \lambda_\alpha(c)/\mu_\alpha(c) \tag{13.45}$$

If the units of an inflow λ, which is diverted by a *transport element* into different directions j, follow each other randomly, the probability that c units for the same direction j arrive in sequence is $(\lambda_j/\lambda)^c$. The probability that the next incoming unit does not belong to direction j is $(\lambda-\lambda_j)/\lambda$. The product of these probabilities is the *sequence probability* for the arrival of a batch of exactly c units destined for the same direction:

$$w_j(c) = (\lambda_j/\lambda)^c \cdot (\lambda-\lambda_j)/\lambda \tag{13.46}$$

Multiplying the sequence probability with the inflow λ, which consists of randomly mixed sequences of different length c for the same direction j, gives the partial flows:

$$\lambda_j(c) = \begin{cases} (\lambda_j/\lambda)^c \cdot (\lambda-\lambda_j) & \text{for } c < c_T \\ (\lambda_j/\lambda)^c \cdot \lambda & \text{for } c < c_T \end{cases} \tag{13.47}$$

By inserting this into formula (13.45) result the limit performance laws for transport elements with capacity $c_T > 1$ and randomly mixed input flows.

In the following, the universal limit performance law (13.44) is applied to junction and branching elements with different operating strategies. Most of the resulting rules hold also for transport elements of higher order.

The analytically calculated limit performances have been tested for several transport elements by *digital simulation*. The results of the simulations are shown in the following diagrams. They confirm the correctness of the limit performance laws.

13.4.1 Limit Performance Laws for Junction and Branching

For a junction element with two entries and one exit, the partial inflows λ_1 and λ_2 are joined to the *outflow*

$$\lambda_O = \lambda = \lambda_1 + \lambda_2 \tag{13.48}$$

For a branching element with one entry and two exits, λ_1 and λ_2 are the outgoing shares of the *inflow*

$$\lambda_I = \lambda = \lambda_1 + \lambda_2 \tag{13.49}$$

For both elements, the forth and back *switch frequencies* between the two directions 1 → 2 and 2 → 1 are equal

$$\nu = \nu_{12} = \nu_{21} \tag{13.50}$$

and the sum of the switch times is

$$\tau = \tau_{12} + \tau_{21} \tag{13.51}$$

With this notation, the condition (13.44) is reduced to the *limit performance law for branching and junction elements*:

$$\lambda_1/\mu_1 + \lambda_2/\mu_2 + \nu \cdot \tau/3600 < 1 \tag{13.52}$$

For elements with capacity $c_T > 1$, the inflows have to be separated into shares with the same batch lengths using the formulas (13.45) and (13.47).

13.4.2 Limit Performance Laws for Cyclic Dispatch

For cyclic dispatch of the inflows of a junction element with switch time sum τ and partial cycle times T_1 and T_2, the *total cycle time* is $T_C = T_1 + T_2$ [s] and the switch frequency is $\nu = 3600/T_C$. Inserting this into (13.52) results in the *limit performance law for junctions with cyclic dispatch*:

$$\lambda_1/\mu_1 + \lambda_2/\mu_2 + \tau/T_C < 1 \tag{13.53}$$

As shown in diagram *Fig. 13.12* for the rotary switch of *Fig. 13.8* with steady connections, the *operating point* $(\lambda_1;\lambda_2)$ of the partial flows is restricted by condition (13.53) to the area below the straight line $\lambda_1/\mu_1 + \lambda_2/\mu_2 + \tau/T_C = 1$. With diminishing cycle time, the straight line shifts towards the zero point due to increasing time losses for switching.

The fulfilment of condition (13.53), however, is only sufficient, if the partial cycle times T_1 and T_2 are adapted to the partial utilisations, i.e. if

$$T_1 = (\lambda_1/\mu_1) \cdot T_C \text{ and } T_2 = (\lambda_1/\mu_1) \cdot T_C. \tag{13.54}$$

If the partial cycle times T_1 and T_2 are kept constant, the necessary and sufficient limit performance laws for cyclic dispatch are:

$$\lambda_1/\mu_1 < T_1/(T_1 + T_2 + 3600 \cdot \tau) \text{ and } \lambda_2/\mu_2 < T_2/(T_1 + T_2 + 3600 \cdot \tau) \tag{13.55}$$

These conditions take into account not only the time loss by switching, but also for the *idle time* in the direction, which cannot not be used up to 100%. Only if T_1 and

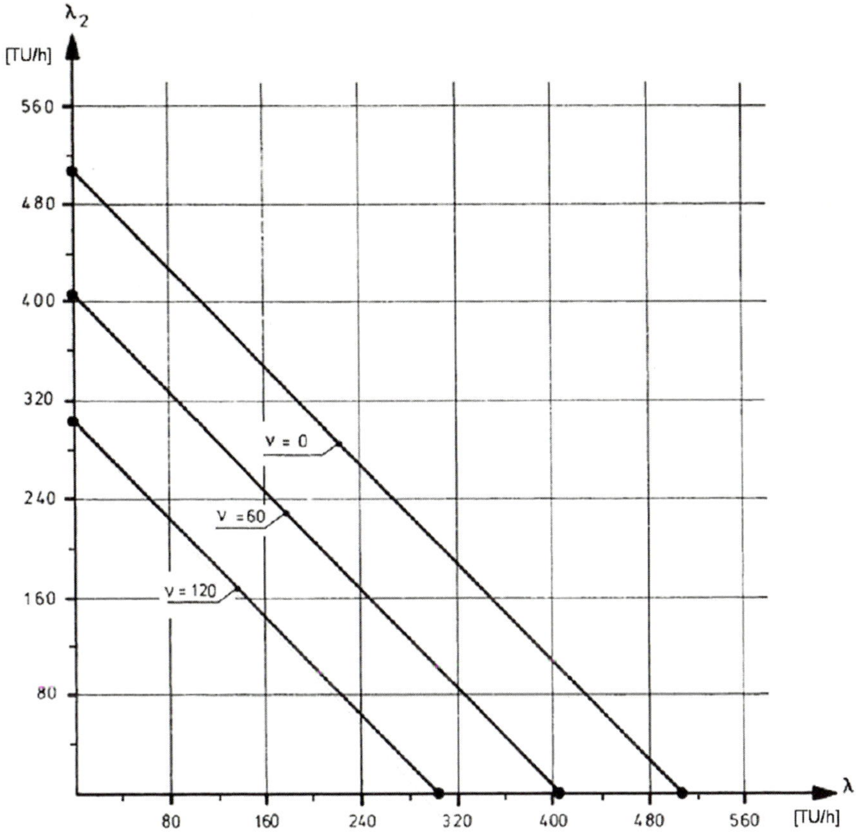

Fig. 13.12 Limit performance curves of a steady junction or branching element with cyclic dispatch

Parameter: switch frequency $\nu_C = 0/60/120$ 1/h
Steady connections: $\mu_1 = \mu_2 = 507$ LU/h and $\tau = 12$ s

T_2 fulfil exactly the equations (13.54), the sum of the two conditions (13.55) results in the necessary condition (13.53).

The formulas (13.53), (13.54), (13.55) can be applied for *traffic light control* in the network of *vehicle systems* (see *Sect. 18.7*).

13.4.3 Limit Performance Law for Random Dispatch

For a random dispatch of single units by a transport element, the probability of switching from one partial flow to the other is equal to the probability that a unit of the partial flow λ_j follows a unit of the partial flow λ_i. This *switching probability* is the product of the probability $\lambda_i/(\lambda_i + \lambda_j)$ that the last unit has been for direction i and the probability $\lambda_j/(\lambda_i + \lambda_j)$ that the next unit is for direction j, i.e

$$w_{ij} = (\lambda_i/(\lambda_i + \lambda_j)) \cdot (\lambda_j/(\lambda_i + \lambda_j)) \tag{13.56}$$

The *switch frequency* is the switching probability times the sum inflow $\lambda_i+\lambda_j$:

$$\nu_{ij} = w_{ij} \cdot (\lambda_i + \lambda_j) = \lambda_i \cdot \lambda_j / (\lambda_i + \lambda_i). \tag{13.57}$$

Insertion of (13.57) into (13.52), taking into account relation (13.51), leads to the *limit performance law of junction and branching elements with random dispatch*:

$$\lambda_1/\mu_1 + \lambda_2/\mu_2 + (\lambda_i \cdot \lambda_j/(\lambda_i + \lambda_i)) \cdot \tau/3600 < 1 \tag{13.58}$$

In *Fig. 13.13* the resulting *limit performance curves* for the junction and branching elements of *Fig. 13.8* are shown. The curves are given by formula (13.58) when replacing < by =.

For *full-steady elements* with two steady connections, the limit performance curve is a 45° rotated hyperbola running from point $(0;\mu_2)$ downwards to point $(\mu_1;0)$. The deviation of the hyperbola from the straight line connection of these two points is caused by the switch time.

The limit performance curves for the *half-steady element* and the *unsteady element* are straight lines running from $(0;\mu_2)$ downwards to point $(\mu_1;0)$ since the switch time is zero. As the limit performance for the unsteady direction is significantly lower than the limit performance for the steady connection direction, the limit performance curves for the half-steady and the unsteady element run far below the limit performance curve of the full-steady element. In all three cases, the operating

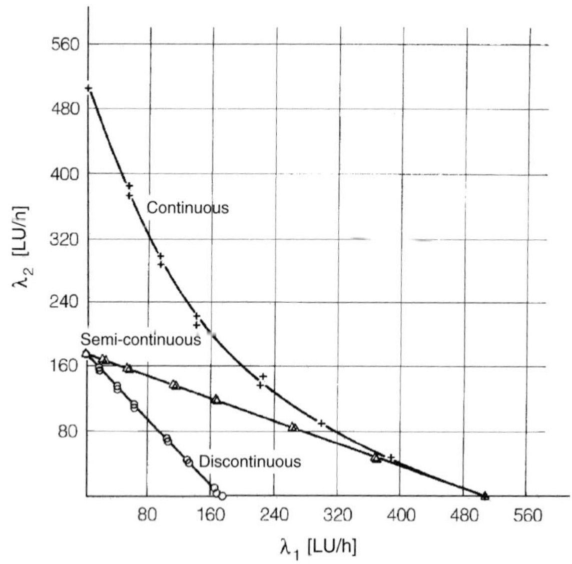

Fig. 13.13 Limit performance curves of junction or branching elements with random dispatch ($c_T = 1$)

Full-steady element: $\mu_1 = \mu_2 = 507$ LU/h , $\tau = 12$ s

Half-steady element: $\mu_1 = 507$ LU/h, $\mu_2 = 173$ LU/h, $\tau = 0$ s

Unsteady element: $\mu_1 = \mu_2 = 173$ LU/h , $\tau = 0$ s

crosses, circles, triangles: simulation results

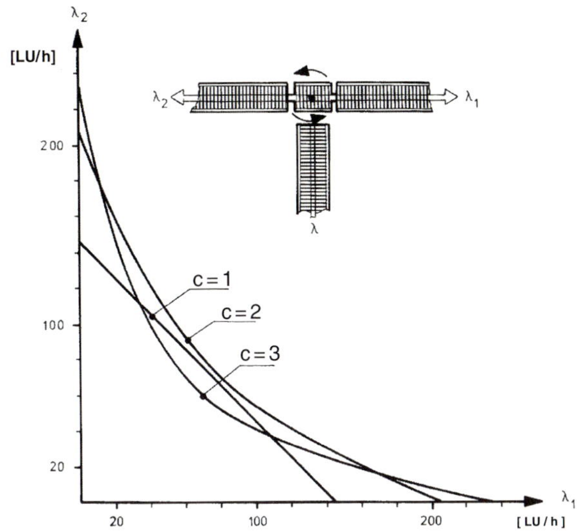

Fig. 13.14 Limit performance curves of a half-steady rotary switch with variable dispatch of randomly arriving batches

Parameter: load capacity of the turn table $c_T = 1/2/3$ LU (Pal)
Partial limit performances for
$c_T = 1$: $\mu_1(1) = \mu_2(1) = 144$ LU/h
$c_T = 2$: $\mu_1(1) = \mu_2(1) = 118$ LU/h, $\mu_1(2) = \mu_2(2) = 207$ LU/h
$c_T = 3$: $\mu_1(1) = \mu_2(1) = 100$ LU/h, $\mu_1(2) = \mu_2(2) = 178$ LU/h
$\mu_1(3) = \mu_2(3) = 231$ LU/h

point $(\lambda_1;\lambda_2)$ of the actual partial performance rates λ_1 and λ_2 lays always below the corresponding limit performance curve at all times.

Figure 13.14 shows the example of a half steady rotary switch of a roller conveyor with different capacities for $c_T = 1$, 2 and 3 pallets which can be moved at once. In this case the partial flows of different sequence length $c \le c_T$ must be separated using relation (13.47) and inserted after the replacement (13.45) into the limit performance law (13.58). Results are the limit performance curves of *Fig. 13.14* . The examples show that a larger capacity of the rotary table does not necessarily lead to a better performance limit for all operating points $(\lambda_1;\lambda_2)$ although the partial limit performances $\mu_1(c)$ are higher for $c > 1$. The reason is, that in the area where the partial flows λ_1 and λ_2 are nearly equal, the sequence probability (13.46) for longer batches destined for the same direction is low and the higher capacity is not fully used.

13.4.4 Limit Performance Law for Priority Dispatch

Not all time gaps between the prioritized units of a main flow are sufficient to allow a merging of secondary units. All time gaps of the main flow that are smaller than the required time limit gap (13.39) lead to a *performance loss*. Therefore, the limit performance law (13.57), although being necessary, is not sufficient to ensure a required throughput for the secondary flow.

The additional limit performance law for the secondary flow λ_N can be derived by analysis of the time gap distribution of the main flow λ_M. In many cases, e.g. for single cars on a traffic lane, the distribution of the time gaps is approximately a *modified exponential distribution* (9.10), as shown in *Fig. 9.3*. With such distribution of the time gaps between the end of the prioritised units holds the *limit performance law for priority dispatch* (Gudehus 1978/3):

$$\lambda_N < \lambda_M \cdot \mathrm{EXP}\big(-(\lambda_M \cdot \mu_M \cdot \tau/3600)/(\lambda_M - \mu_M)\big)/ \\ \big(\mathrm{EXP}((\lambda_M \cdot \mu_M/\mu_N)/(\lambda_M - \mu_M)) - 1\big) \quad (13.59)$$

Herein, λ_M is the prioritised *main flow*, which is restricted by the *main limit performance* μ_M, λ_N is the *secondary flow*, which is restricted by the *secondary limit performance* μ_N, and τ [s] is the *limit time gap* (13.39).

For the full-steady (continuous), half-steady (semi-continuous) and unsteady (discontinuous) junction elements of *Fig. 13.8* the resulting performance curves (13.59) for priority dispatch are shown in *Fig. 13.15*. The theoretical curves are confirmed by simulation as well as the curves of *Fig 13.13* for the same elements with random dispatch.

From the limit performance law (13.59) and by comparison of the diagrams follow the *rules for priority dispatch*:

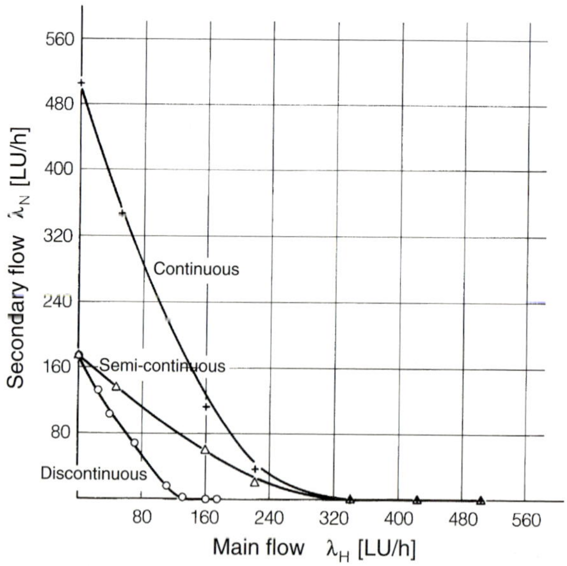

Fig. 13.15 Limit performance curves of junction elements with priority dispatch
Parameter: see *Fig. 13.13*
Crosses, circles, triangles: simulation results

- ▶ The possible throughput of the secondary flow is significantly lower for priority dispatch than for random dispatch.
- ▶ The possible throughput of the secondary flow decreases rapidly with increasing main flow and declines almost to zero long before the main flow has reached the limit performance.

This is known to any motorist who wants to enter a highly frequented main road and waits for a sufficient gap. Another example can be found in offices, where secretaries prioritize the jobs of the bosses and by this behaviour reduce the chance to fit in other jobs.

A consequence of the second rule is the *prioritization principle*:

- ▶ If a priority regulation is applied, the stronger flow should be prioritized.

The so called *Harders-formula*, which is widely used in *traffic planning*, is incorrect (Dorfwirth 1961; Harders 1968; Leutzbach 1956). It neglects the minimal distances between the vehicles caused by their own length and predicts for strong main flows far too high capacity for the secondary flow.

The more equally the gaps of the main flow are distributed, the more does the limit performance curve for priory dispatch deviate from the curve (13.59), which holds only for random gap distribution. As shown in *curve 6* of *Fig. 13.16* for a half-steady branching element, the limit performance curve becomes unsteady if the units of the main flow arrive in batches of equal length with constant gaps. This diagram demonstrates also how far the limit performance curves for the same transport element deviate for different operating strategies.

13.5 Waiting Queues and Queuing Laws

If the flows in a network reach or exceed the limit performances, *waiting queues* and *waiting times* are generated in front of the affected stations and *blockades* of upstream stations occur. These *queuing effects* are caused either by stochastic or systematic queues:

- *Stochastic queues* occur for stochastically varying arrival and/or service times when the total utilisation of a station (13.43) approaches 1.
- *Systematic queues* grow for any distribution of arrival and services times, when the total utilisation of a station exceeds 1 for longer time.

Analysis of the influence factors and calculation of the effects of stochastic queues are the subjects of *queuing theory* (Ferschl 1974; Gross 1998; Kleinrock 1975; Saaty 1957). However, many results and formulas of queuing theory are quite complex. Often, their prerequisites cannot be examined in practice or do not comply with reality. For logistics, *approximation formulas* are sufficient that can be derived from the exact formulas of queuing theory under relatively general conditions (Arnold 2002; Gudehus 1976). Their application is justified by the inaccurate values of the flows and the not exactly known distributions of the arrival and service times.

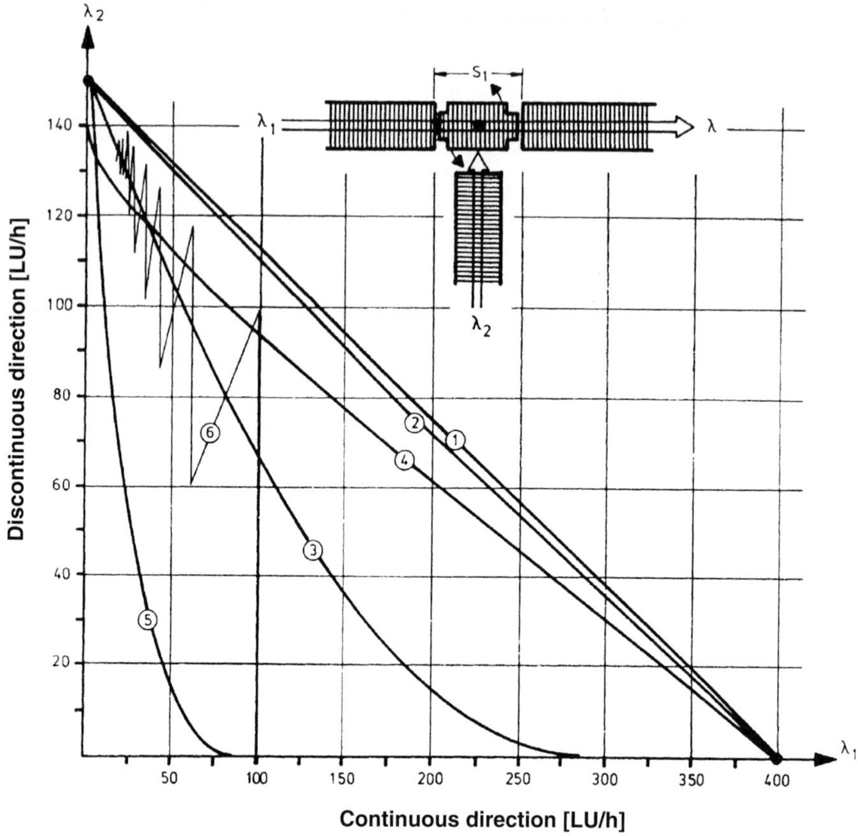

Fig. 13.16 Limit performance curves for different operating strategies

Branching element: roller conveyor/rotary table/roller conveyor for pallets
Parameters: $\mu_1 = 400$ Pal/h; $\mu_2 = 150$ Pal/h; $\tau = 1.25$ s
Curve 1: cyclic dispatch with a switching frequency $\nu << 60$ 1/h
Curve 2: random single dispatch
Curve 3: absolute priority of stochastic flow λ_1 with $\tau = 0$ s
Curve 4: absolute priority of stochastic flow λ_2, with $\tau = 0$ s
Curve 5: absolute priority of stochastic flow λ_1, with $\tau = 20$ s
Curve 6: absolute priority of flow λ_1 with equal time gaps and batches

If a logistic system is correctly dimensioned and operated, stochastic queues are not critical, whereas systematic queues can be far more unimportant.

13.5.1 Waiting Systems

In queuing theory, a general waiting system consisting of n parallel service stations with *arrival time distribution* W_a and *service time distribution* W_s is indicated by the *Kendall-notation* (Ferschl 1964):

$$W_a/W_s/n \tag{13.60}$$

For example, *Fig. 13.10* shows a waiting system $W_a/W_s/1$ with a single service station.

Unspecified *general time distributions* are indicated by G and a *general waiting system* with n parallel service stations by G/G/n. A random process with exponentially distributed service times is called a *Markov-process* and denoted by M. Correspondingly, M/M/1 is a single service station with exponentially distributed arrival and service times. This is the most familiar waiting system since it can be solved explicitly. If arrival or service times are constant, the distribution is called *Dirac-distribution* and denoted by D (see *Fig. 9.4*). Correspondingly, M/D/1 indicates a single service station with exponentially distributed arrival times and constant service times and D/M/1 a station with constant arrival times and exponentially distributed service times.

As explained in *Sect. 9.2.4*, the *Poisson-distribution* of a Markov-process and the Dirac-distribution are the special cases $k = 1$ and $k = \infty$ of the *k-Erlang distribution* E_k. A time distribution, for which only the *mean value* τ and the *standard deviation* s_τ are known, can be approximated by an Erlang-distribution with $k = (\tau/s_\tau)^2$. The corresponding waiting systems are indicated by $E_k/E_l/n$.

For Erlang-systems $E_k/E_l/1$, waiting theory has derived explicit calculation formulas for stationary queuing effects (Ferschl 1964). These formulas show that the queuing effects in front of a single service station depend, in first approximation, on the *total utilisation* ρ and on the *system variability* V. The system variability is the mean value of the *variability of the arrival times* $(S_a/\tau_a)^2$ and of the *variability of the service times* $(s_s/\tau_s)^2$:

$$V = (V_a + V_s)/2 = ((s_a/\tau_a)^2 + (s_s/\tau_s)^2)/2 \qquad (13.61)$$

This leads to the basic *rule of stochastic queuing*:

► Stochastic queuing effects only arise, if the system variability is larger 0.

The following *approximation formulas* can be used to calculate or estimate the queuing effects for general waiting systems. They lead to design and dimensioning rules for logistic and transport systems with stochastic flows and service rates. The queuing effects of a general waiting system G/G/n with more than one parallel station depend not only on the system variability but also on the *operating strategy*. The approximation formulas can also be used to assess the queuing effects for different operating strategies.
With the help of the queuing laws and by the right operating strategies, the stations of a logistic system can be decoupled. By this method, the functioning of the system is ensured as long as the flows are *stationary*. For a complex stochastic network with fast varying flows, however, a *stochastic simulation* is recommendable.

13.5.2 Determination of System Variability

For an existing system, the *arrival variability* $(s_a/\tau_a)^2$ of a stationary flow can be determined by measuring the single arrival times τ between the endpoints of the arriving units. However, in practice such measurements are quite difficult to perform and almost impossible for unsteady flows. For a planned system, the future arrival time distribution is unknown and only predictable with assumptions.

Measurements of the arrival times of passing vehicles on road lanes have shown that the distribution is a *modified exponential distribution* (Leutzbach 1956). Also in many other cases, the modified exponential distribution represents a random distribution of the arrival times quite well (see *Fig. 9.2*). As shown in *Sect. 9.3*, the variability of a modified exponential distribution with *mean arrival time* τ_a [s], i.e. with actual *inflow* $\lambda_a = 3660/\tau_a$ [PU/h], and *minimal arrival time* τ_{ao} [s], i.e. with *maximal flow* $\mu_a = 3660/\tau_{ao}$ [PU/h], is given by:

$$V_a = (s_a/\tau_a)^2 = ((\tau_a - \tau_{so})/\tau_a)^2 = (1 - \lambda_a/\mu_a)^2 \tag{13.62}$$

In the limit $\lambda_a \to 0$ of a weak flow, the arrival variability becomes $V_a = 1$ and the flow a *Poisson-flow*. In the other limit $\lambda_a \to \mu_a$, when the inflow reaches the maximal flow, the variability becomes $V_a = 0$ and the flow a *Dirac-flow* with constant arrival times τ_{ao}.

The *arrival variability* of a station with n entrances for the partial inflows λ_{ai}, i = 1, 2 ...n, which can reach maximally the flows μ_{ai}, is given by the weighted mean value of the variability (13.62) of the partial flows:

$$V_a = \sum_i (\lambda_{ai}/\lambda_a) \cdot (1 - \lambda_{ai}/\mu_{ai})^2 \tag{13.63}$$

If the service time distribution is known, the *service variability* V_s can be calculated with relations (9.7) and (9.8) of *Sect. 9.2*. If the distribution is rectangular, the service times vary randomly with equal probability between a *minimal service time* τ_{min} [s/PU] and a *maximal service time* τ_{min} [s/PU]. Then, the actual service rate is $\lambda_s = 2\cdot 3600/(\tau_{max} - \tau_{min})$ [PU/h] and the limit performance is $\mu_s = 3600/\tau_{min}$ [PU/h]. For a *rectangular distribution* of the service times, as shown in *Fig. 9.2*, the *service variability* is:

$$V_s = (s_s/\tau_s)^2 = (1/3) \cdot ((\tau_{max} - \tau_{min})/(\tau_{max} + \tau_{min}))^2 = (1/3) \cdot (1 - \lambda_s/\mu_s)^2 \tag{13.64}$$

In the limit $\lambda_s \to 0$ the service variability becomes $V_a = 1/3$. In the limit $\lambda_a \to \mu_a$, when the required service rate reaches the limit performance, the variability becomes $V_s = 0$ and the service times are equal to the minimal time τ_{min}.

A service station or transport element of higher order with the constant service times τ_α for the partial functions F_α, i.e. with the partial limit performances $\mu_\alpha = 3600/\tau_\alpha$ [PU/h], has a discrete service time distribution. In this case, from formula (9.12) results for partial service rates λ_α the *service variability of a multifunctional station*:

$$V_s = \sum_\alpha (\lambda_\alpha/\lambda) \cdot (1 - \mu/\mu_\alpha)^2 \tag{13.65}$$

Herein λ is the sum of the partial performance rates λ_α and μ is the weighted harmonic mean value of the partial limit performances μ_α. If the partial limit performances are the same for all functions, the service variability is 0 and the service times are constant.

If it is not possible to measure the system variability directly or to calculate it with the formulas (13.61) to (13.65), it is in many cases sufficient to calculate with estimated values which are derived from the following *thumb rules*:

- In the simplest case, arrival times and service times are constant and the system variability is 0.
- In the worst case, arrival and service times are exponentially distributed and the system variability is 1.
- In the intermediate case of stochastic arrival and constant service times or of constant arrival and stochastic service times the system variability is 0.5.

If the variability of the arrival times and of the service times are unknown, it is reasonable to assume for the system variability the *mean value* $V_s = 0.5$. With this assumption, the results are on the safe side, since at high utilisation, when the stochastic queuing effects become critical, either the arrival variability or the service variability tends to be 0.

13.5.3 Queuing Laws for Stochastic Queues

The number $N(t)$ of units waiting for service at time t, is the current *queue length* or *waiting queue*. The *current queue* $N(t)$ is a random number which cannot be calculated or predicted for a future time t. If at all, only the mean value, standard variation and probabilities of stochastic queues and their effects can be calculated.

For a single service station with n entrances, operated with the simplest operating strategy of *random dispatch of single units*, the following *queuing laws* can be derived from queuing theory (Arnold 1995; Ferschl 1964; Gudehus 1976):

- The *mean partial waiting queues* in front of the n entrances I_i, $i = 1, 2 \ldots n$, of the station are

$$N_{wi} = (\lambda_i/\lambda) \cdot (1-\rho + V \cdot \rho) \cdot \rho^2/(1-\rho) \quad = (\lambda_i/\lambda) \cdot N_w \qquad (13.66)$$

- The *mean total waiting queue in front* of all n entrances is the sum of the mean partial queues, i.e.

$$N_w = (1 - \rho + V \cdot \rho) \cdot \rho^2/(1 - \rho) \qquad (13.67)$$

- The *mean total waiting queue including the unit in service* is

$$N = (1 - \rho + V \cdot \rho) \cdot \rho/(1 - \rho) \qquad (13.68)$$

- The *standard deviation* of the actual total waiting queue around the mean queue, including the unit in service, is

$$s_N = \sqrt{V \cdot \rho \cdot N} \qquad (13.69)$$

- The *mean waiting time* of the units in front of the entrances is given by *Little's Law*

$$T_w = 3600 \cdot N_w/\lambda \quad [s] \qquad (13.70)^{©}$$

Herein, λ_i are the partial inflows, λ is the total inflow (13.29), ρ the total utilisation (13.43) and V the system variability (13.61). Although the formulas (13.66) to (13.70) only hold approximately, it has been proven by many simulations that their accuracy is sufficient for most applications as long as the flows and service rates are *stationary* and the single units are served *first come first go*. The solutions for batch-wise arrival or service and for priority dispatch are far more difficult. Explicit formulas are known only for special cases, e.g. for the priority dispatch of junctions with 2 entrances.

With priority dispatch, waiting queues occur only in front of the non-prioritised entrances for the secondary flows (Dorfwirth 1961). To calculate the mean waiting queue of the secondary flow of a junction with 2 entrances, the utilisation $\rho = \lambda_N/\lambda_{Nmax}$ has to be calculated with the maximal secondary flow λ_{Nmax} which is given by the right side of formula (13.59) (Gudehus 1976).

Figure 13.17 shows the dependency of the mean queue on the utilization for the single service station of *Fig. 13.3* with a system variability of $V = 0.6$. The curve has been calculated with formula (13.68). The points and circles are results of a *digital simulation* for two different arrival and service time distributions. The confirmation of the theoretical curve by the simulation is excellent.

In *Fig. 13.18* , the dependency (13.68) of the mean waiting queue on the system variability is shown for three different utilisations.

Many simulations of service stations with different arrival and service time distributions have been performed. They all confirm the theoretical calculations and the general *queuing rules*:

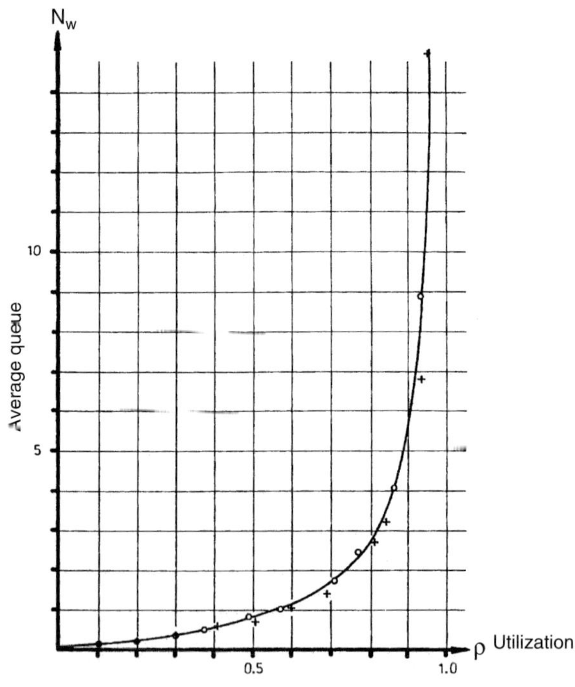

Fig. 13.17 Dependency of the mean waiting queue on the utilisation

Curve: analytical calculation for $V = (V_a + V_s)/2 = 0.6$
Crosses: simulation for $V_a = 0.2$ and $V_s = 1.0$
Circles: simulation for $V_a = 1.0$ and $V_s = 0.2$

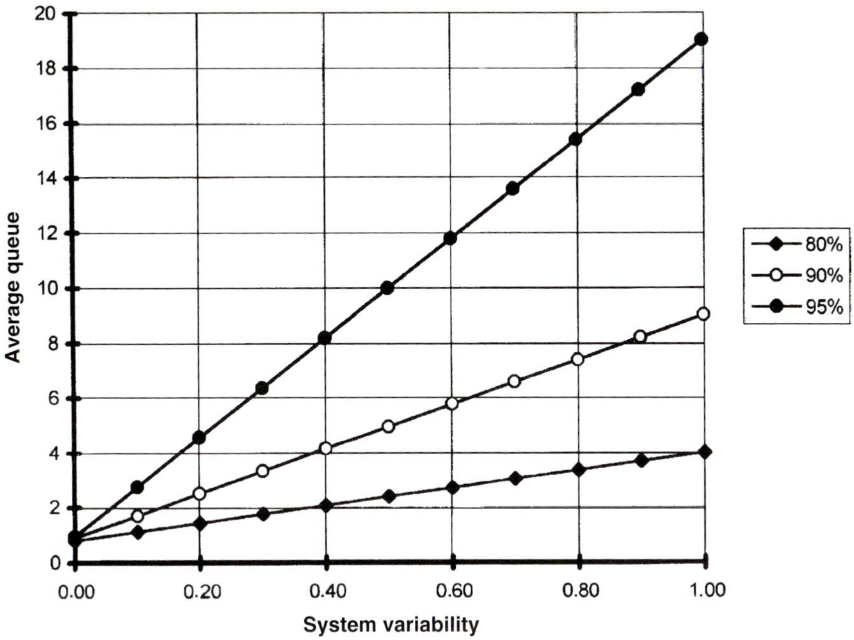

Fig. 13.18 Dependency of the mean waiting queue on the system variability
Parameter: utilization 80%/90%/95%

- As long as utilisation is below 50%, the mean stochastic queues are smaller than 1 and the queuing effects can be neglected even at maximal variability.
- Stochastic queues and related queuing effects increase over proportionally with increasing utilisation and are proportional to the system variability.
- At maximal system variability, the queuing effects grow fast with utilizations over 85%, and at mean system variability, for utilizations over 90%.
- The variation of the actual queues around the mean number increases with utilization and variability. At some times, the queue can grow very long, at other times it can be 0.
- Not the shape of the arrival and service time distributions, only the system variability has significant effect on queuing.

Figure 13.19 illustrates the queuing effects for a half-steady branching element. Here, the formula (13.66) has been used to calculate the mean queues for different utilizations in the transit direction. The points and circles are the results of a digital simulation.

13.5.4 Backlog Probability and Blocking Probability

As shown in *Fig. 13.20*, only a limited number of units can be accumulated on the connection element between the exit of a station S_1 and the entry point of a succeeding station S_0. As long as the current queue exceeds the *buffering capacity* B,

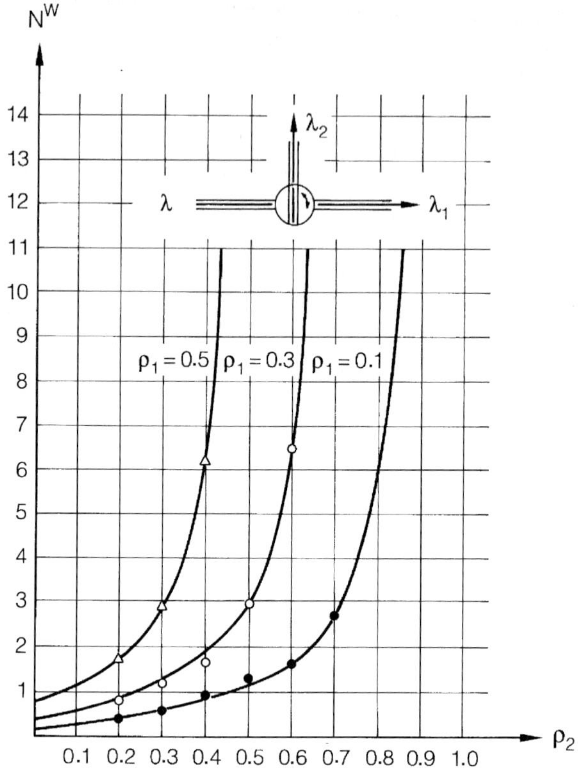

Fig. 13.19 Mean queue length in front of a half-steady branching element as function of the partial utilization in the branching direction

Parameters: partial utilization in passing direction
Points and circles: simulation results

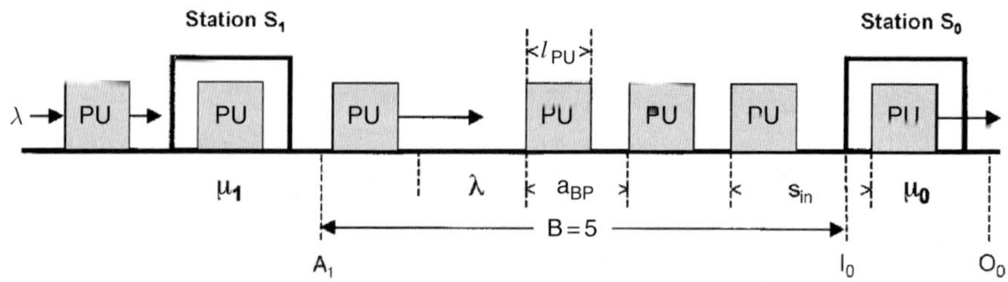

Fig. 13.20 2 service stations in sequence separated by a buffer for 5 units

B buffer capacity
PU process units
l_{PU} length of a single process unit
a_{BP} length of a buffer place
s_{in} entrance length
μ unit performances

the preceding station S_1 is blocked. The blocking reduces the *availability* of this station. The reduction is equal to the *blocking probability* P_B that a queue in front of a station is longer than the buffer capacity B on the connection up to the preceding station.

The blocking probability is determined by the *backlog probability* P_N that exactly N units are waiting before and within the system. For *single unit dispatch* by a service station with utilization ρ and system variability *V*, the backlog probability is (Ferschl 1964; Gudehus 1976):

$$P_N = ((1-\rho)/V) \cdot \left(V\rho/(1-\rho+V\rho)\right)^N \quad \text{for} \quad N > 0 \tag{13.71A}$$

As the probability of an occupied station is equal to the total utilization ρ, the probability to find the station unoccupied and a waiting queue of length N = 0 is

$$P_0 = (1-\rho) \tag{13.71B}$$

Figure 13.21 shows the dependency of the backlog probability calculated with the formulas (13.71A) and (13.71B) on the length of the waiting queue for a system with variability 0.75.

The blocking probability of the preceding station is equal to the sum of the backlog probabilities for waiting queues longer than the buffer capacity. This leads to the *blocking law*:

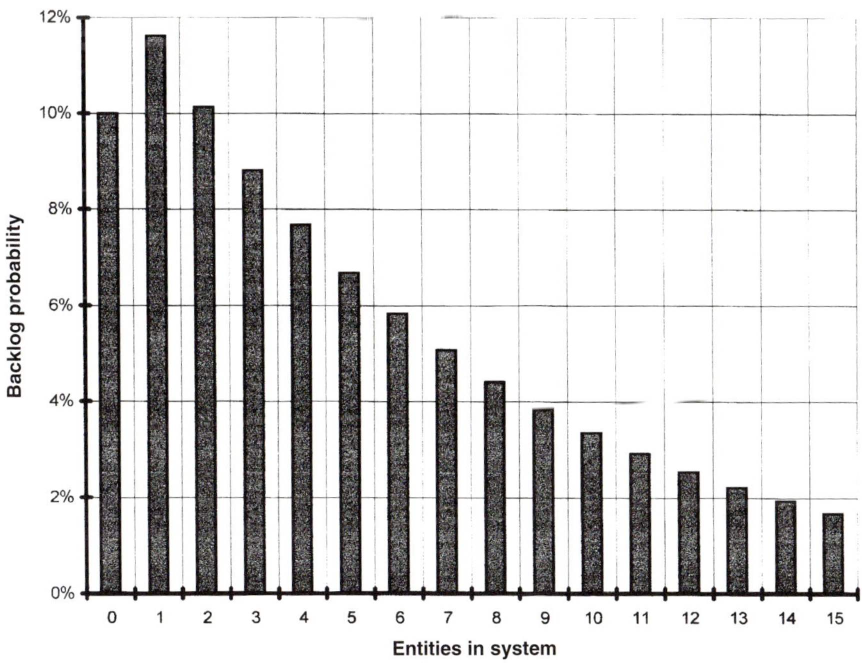

Fig. 13.21 Backlog probability for waiting queues
utilisation: ρ = 90%; system variability: *V* = 0.75
mean queue: *N* = 7 PU

► The *blocking probability* for a preceding station which is separated by a buffer capacity B from a following station with single unit dispatch, utilization ρ and system variability *V* is

$$P_B = \sum_{N>B} P_N = \rho \cdot \left(V \cdot \rho/(1-\rho + V \cdot \rho)\right)^B \qquad (13.72)$$

The approximate formulas (13.71) for the backlog probability and (13.71) for the blocking probability have also been confirmed by simulations.

Figure 13.22 shows the blocking probability as function of the buffer capacity for a system variability of $V = 0.75$. The probability that the waiting queue exceeds 7 units is in this case 30%, and that it exceeds 10 units is 20%.

From relation (13.72) follows the *blocking rule*:

► The blocking probability decreases exponentially with the buffer capacity between the consecutive stations.

Another important result is the *decoupling rule*:

► Highly utilized successive stations of performance networks and process chains which operate with great variability should be decoupled by a buffer with capacity significantly larger than the mean queuing length.

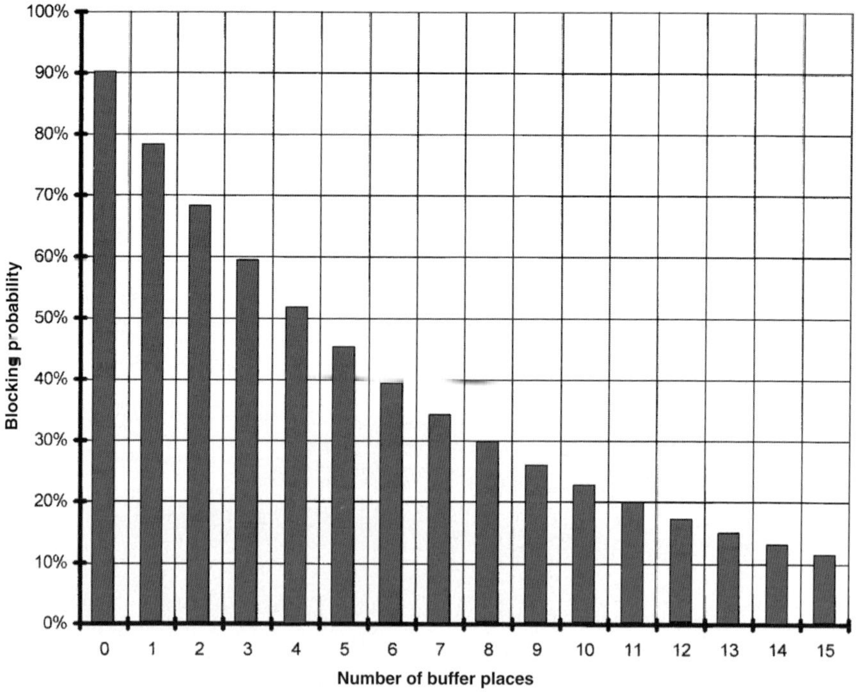

Fig. 13.22 Blocking probability for different buffer capacities
utilisation: $\rho = 90\%$; system variability: $V = 0.75$
mean queue: $N = 7$ PU

Of course, the decoupling rule should be applied only as far as technically feasible and economically reasonable.

Another consequence of the formulas (13.71) and (13.72) is the *throttling strategy*:

- Queuing effects and blocking probability in a chain of stations can be reduced by a throttling station at the entrance which reduces the variability of the arrival times.

A reduction of the arrival time variability results automatically by *cyclic dispatch* of an incoming stochastic flow to *parallel stations* as shown in *Fig. 13.10* .

13.5.5 Queuing with Parallel Stations

Waiting systems of the most' general type G/G/n with n parallel stations can be found in practice quite often, e.g.:

call centers
counters at banks and post offices
sales counters for tickets
toll stations on motor-highways (13.73)
control stations
incoming docks
vehicle pools

The cyclic dispatch of a stochastic flow λ_a with variability V_a to n parallel stations creates n equal flows $\lambda_{a\,n} = \lambda_a/n$ with variability $V_{a\,n} = V_a/n$ (Ferschl 1964). Due to relation (13.67), the mean length of the waiting queues in front of the n parallel station is:

$$N_{wn} = (1-\rho_n + V_n \cdot \rho_n) \cdot \rho_n^2/(1-\rho_n) \quad (13.74)$$

Herein, $\rho_n = \lambda_a/(n \cdot \mu_s)$ is the utilisation and $V_n = (V_a/n + V_s)/2$ the reduced system variability of the single stations with limit performance μ_s. That means:

► Cyclic dispatch of a stochastic flow to n parallel stations reduces the system variability and hence the queuing effects.

Further reductions of the queuing effects are possible by the parallel operation strategies of *Sect. 13.3.4*. However, it is difficult to quantity their effects.

The examples demonstrate the importance and potentials of operating strategies for systems with parallel and serial stations. The possible strategies for the many possible structures and conditions however require a deeper analysis that goes beyond the topic of this book.

13.5.6 Systematic Queuing

When the current inflow $\lambda(t)$ exceeds the limit performance $\mu(t)$, i.e. when the current utilisation $\rho(t) = \lambda(t)/\mu(t) > 1$, a waiting queue grows up in front of the service station. If N_0 is the waiting queue at time 0, the queue after a time T is:

$$N_w(T) = N_0 + \int_0^T (\lambda(t) - \mu(t))dt. \qquad (13.75)$$

During times with utilization below 100%, randomly varying arrival and service times cause the *stochastic waiting queues* which have been discussed before. When the *mean input flow* λ exceeds the *mean service performance* μ for a longer time T, the mean length of the systematic queue can be calculated by the integral (13.75), if the mean inflow λ and the service rate μ are known. Resulting are the general *laws of systematic queuing*:

- For a *stationary inflow* λ and a *stationary service* rate $\mu < \lambda$, the mean waiting queue at the end of an *overload time* T is

$$N_w(T) = (\lambda - \mu) \cdot T. \qquad (13.76)$$

- If the service or dispatch is interrupted for a time T, a stationary inflow λ piles up to a waiting queue

$$N_w(T) = \lambda \cdot T. \qquad (13.77)$$

- For a stationary stochastic inflow with variability of the arriving times V_a, the systematic waiting queue varies randomly around the mean value (13.77) with the standard deviation

$$s_N(T) = \sqrt{V_a \cdot \lambda \cdot T} = \sqrt{V_a \cdot N_w(T)}. \qquad (13.78)$$

Due to the *law of large numbers*, the queue for a Poisson-flow with maximal variability $V = 1$ varies randomly with a standard deviation equal to the square root $\sqrt{N_w}$ of the mean queue length N_w. For a Dirac-flow with $V = 1$, the deviation of the growing queue from the mean value (13.77) is 0.

The laws for systematic queuing can be applied in many areas. Formulas (13.76) and (13.78) can be used to calculate the mean queuing length and its variations during the stop phases of cyclic dispatch. If a station is supplied with batches of constant lengths which are served in constant times, the resulting *backlog* can be calculated by applying formula (13.76).

For illustration, *Fig. 13.23* shows the growing queue of cars in front of a traffic light during the red light cycle time T. If the queuing vehicles must keep a minimal endpoint distance a_{min} [m], the *congestion* in front of a *traffic throttle* with limit μ less than the inflow λ, grows up in the *throttling time* T [h] to the *congestion length*:

$$L_w(T) = a_{min} \cdot N_w(T) = a_{min} \cdot (\lambda - \mu) \cdot T \quad [m] \qquad (13.79)$$

From this relation results the *backlog speed* of the tailback:

$$v_w = \partial L_w(T)/\partial t = a_{min} \cdot (\lambda - \mu) \quad [m/h] \qquad (13.80)$$

If e.g. a flow of $\lambda = 1{,}200$ vehicles/h is reduced by a traffic control to $\mu = 100$ vehicles/h, the congestion grows from the control point, opposite to the driving direction, with the speed $v_w = 6.6$ km/h, if the endpoint distance of the waiting vehicles is $a_{min} = 6.0$ m. The fast backwards move of a *traffic tailback* can cause heavy accidents.

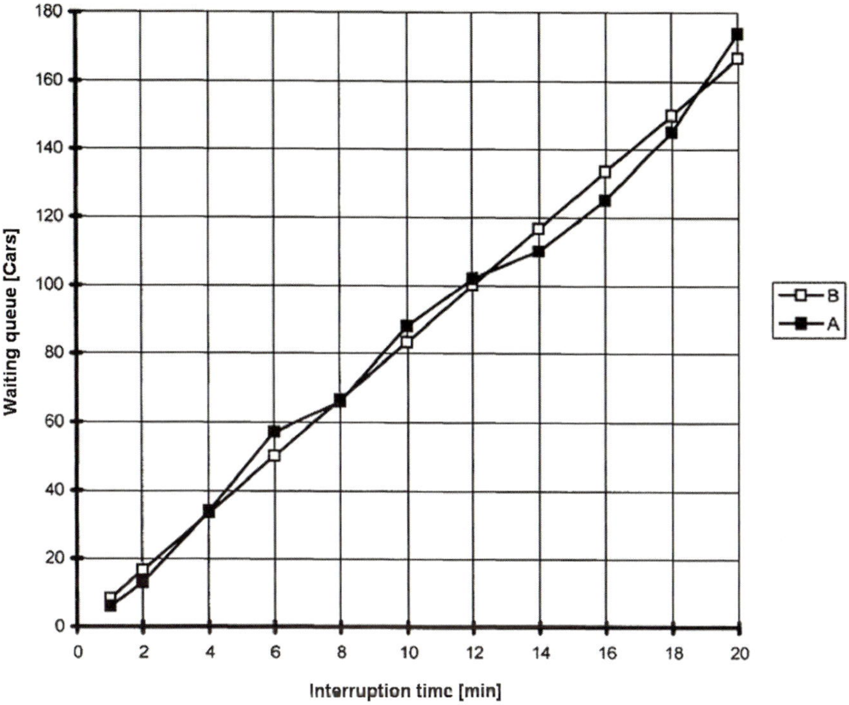

Fig. 13.23 Length of a systematic queue as function of the interruption time for a constant flow and a stochastic flow

Curve A: poisson-arrival time distribution $V_a = 1$
Curve B: constant arrival times $V_a = 1$
Input flow: 500 vehicles/h

13.5.7 Maximal Utilisation

Limited buffer capacities and restricted transit times can only be met if the stations of the system are not used up to their limit performances, since queuing effects reduce the maximal utilisation of the stations. From the previous analysis results the *dimension rule for buffers*:

- ▸ The buffer capacity between two consecutive stations has to be at least as large as the mean queue that can arise in front of the second station for the expected utilization.

The maximal utilisation of a station with buffer capacity B can be derived by solving relation (13.68) with respect to the utilisation after inserting $N = \mathrm{B}$. For the M/M/1 system with $V = 1$ this leads to the *rule of maximal utilisation*:

- ▸ The maximal utilisation of a station with buffer capacity B and system variability $V = 1$ is

$$\rho_{max} = \mathrm{B}/(1+\mathrm{B}) \quad [\%]. \tag{13.81}$$

If for instance the buffer capacity is B = 5 PU, a maximal utilisation $\rho_{max} = 5/6 = 83\%$ can be achieved for the worst system variability $V = 1$. If the system variability is smaller, a higher utilisation as (13.81) is possible. This example demonstrates that queuing effects can quite significantly reduce the *effective limit performance* of a station or total system.

The utilisation of preceding station is reduced by the blocking probability (13.72) to

$$\rho_{max} = 1 - P_B \quad [\%]. \tag{13.82}$$

For example, if a buffer capacity B is exceeded with a probability of 5%, the maximal utilisation of the preceding station is reduced to $\rho_{max} = 1 - 0.05 = 95\%$. A result is the *function securing rule*:

▸ Functioning of succeeding stations can be ensured either by sufficient buffer capacity or by higher limit performances.

If a certain *punctuality* η_T is required for a *transport time* or *lead time*, the single waiting times in a performance chain with n stations must be kept with probability $\eta_{Tn} = \eta_T{}^{1/n}$ (see *Sect. 8.8*). With this value the maximal tolerable queue can be derived from the backlog probability (13.71). The tolerable queue determines by formula (13.82) the maximal utilisation of the stations. The calculation of the maximal utilisation and the needed buffer capacities are essential for design and operation of *high performance systems* (see e.g. *Sect. 17.11* and *Fig. 17. 32*).

13.6 Reliability and Availability

Reliability and availability are key indicators of the *functional safety* of a system element, performance chain or system. They differ for continuously and for discontinuously operating systems (Gross 1998; Gudehus 1976/79). The *reliability* is defined as follows:

- The *reliability* $\eta_{rel\,\alpha}$ [%] of a system element, performance chain or system for a *partial function* F_α is the probability that the respective function is executed correctly without failure during the regular operating time.

The counterpart of the reliability is the *unreliability* or *error liability*:

$$\eta_{rel\,\alpha} = 1 - \eta_{rel\,\alpha} \quad [\%] \tag{13.83}$$

The definition of *availability* is:

- The *availability* $\eta_{ava\,\alpha}$ [%] for a *partial function* F_α is the probability to find the element, chain or system during the regular operating time in a state that allows a correct execution of the respective function without failure.

The counterpart of the availability is the *unavailability*:

$$\eta_{unav} = 1 - \eta_{ava\,\alpha} \quad [\%] \tag{13.84}$$

These definitions hold for elements, chains and systems as well. In practice, the reliability and availability of a performance chain or a complex system are difficult to measure. It is almost impossible to keep all technical, operational and conditions constant for a statistically sufficient period of time. On the other hand, reliability and availability of the *irreducible elements* of a system can be measured much easier during the regular operating time. From the results for the elements, reliability and availability of the process chains and for the total system with performance rates λ_α can be calculated.

The following definitions and formulas, which can be used for this purpose, are essential for *function and acceptance tests* of all kinds of systems processing discrete objects. They are basis of the corresponding German *VDI-regulations* [VDI 3581/3580/3639] and European *FEM-regulations* [FEM 9.222].

13.6.1 Measurement of Reliabilities

In order to measure the partial reliability $\eta_{rel\,\alpha}$ for a partial function F_α, the *irreducible element* has to be operated with the planned *partial performance rates* λ_α for a *test time* T_{tot}. In this time, the total *number of function tests* or *trials* of the function F_α is:

$$n_a = \lambda_\alpha \cdot T_{tot}. \tag{13.85}$$

During the test, the *numbers of failures* $n_{\alpha\,fail}$ with the functions F_α are counted. From the total number of trials (13.85), and the counted number of failures results the *number of correct performances* of function F_α:

$$n_{\alpha\,corr} = n_\alpha - n_{\alpha\,fail}. \tag{13.86}$$

With these numbers, the reliabilities can be calculated:

- ► The current *partial reliability* is the relation of the number of correct performances to the total number of trials of the respective function

$$\eta_{rel\,\alpha} = n_{\alpha\,corr}/n_\alpha = n_{\alpha\,corr}/(n_{\alpha\,corr} + n_{\alpha\,fail}). \tag{13.87}$$

The summation of the numbers of trials, the number of failures and the number of correct performances for *all* partial functions F_α leads to the total reliability:

- ► The current *total reliability* of the irreducible element is the mean value of the partial reliabilities (13.87) weighted with the *utilisation probabilities* λ_α/λ

$$\eta_{rel} = n_{corr}/(n_{corr} + n_{fail}) = \sum_\alpha (\lambda_\alpha/\lambda) \cdot \eta_{rel\,\alpha}. \tag{13.88}$$

From this relation follows the *structure dependency of total reliability*:

- ► For differing partial reliabilities, the total reliability depends on the current *performance structure*, i.e. on the *performance weights* $g_\alpha = \lambda_\alpha/\lambda$.

The application of these definitions, rules and formulas can be demonstrated by the automatic storing- and retrieval-unit of *Sect. 16.11* with the 3 partial functions: *in-storing* F_I, *out-storing* F_O and combined *in- and out-storing* F_{IO}. The respective technical *limit performances* are $\mu_I = \mu_O = 35$ Pal/h and $\mu_{IO} = 25$ Pal/h. During a test time of $T_{tot} = 60$ h, the R/S-unit performed $n_I = 360$ in-storing cycles, $n_O = 480$ out-storing cycles and $n_{IO} = 780$ in-out-storing cycles, i.e. in total n = 1,620 cycles. The numbers of failures are $n_{I\,fail} = 5$ in-cycles, $n_{O\,fail} = 6$ out-cycles, $n_{IO\,fail} = 17$ in-out-cycles and in total $n_{fail} = 28$ cycles with failures. From these data result the *partial performance rates* $\lambda_I = 360/60 = 6$ In-Pal/h, $\lambda_O = 480/60 = 8$ Out-Pal/h and $\lambda_{IO} = 780/60 = 13$ In-Out-Pal/h. With the total performance rate of $\lambda = 27$ cycles/h 19 Pal/h are stored in and 21 Pal/h have been stored out.

That means, that the *partial utilisations* have been $\rho_I = 6/35 = 17\%$, $\rho_O = 8/35 = 23\%$ and $\rho_{IO} = 13/25 = 52\%$, resulting in a *total utilisation* of $\rho = 17\% + 23\% + 52\% = 92\%$. The measured *partial reliabilities* are $\eta_{rel\,I} = (360–5)/360 = 98.61\%$, $\eta_{rel\,O} = (480–6)/480 = 98.75\%$ and $\eta_{rel\,IO} = (780–17)/780 = 97.82\%$. The partial reliability for in-out-storing is smaller than the reliabilities for separate in- or out-storing since the in/out-double-cycles last longer than the in- or out-single-cycles. The *total reliability* is $\eta_{rel} = (1{,}620-28)/1{,}620 = 98.27\%$.

The *statistical error* of a measured reliability results from the law of *large numbers* and the *law of error propagation* (see *Sect. 9.4.3*). It is

$$s_\eta = \sqrt{(1-\eta_{rel})/n}. \tag{13.89}$$

For the given example, the total reliability has been measured with the statistical error $s_\eta = 0.33$, i.e. $\eta_{rel} = 98.27 \pm 0{,}33\%$. The determination of the accuracy of the measurement is important if a *penalty* has to be paid for not reaching a guaranteed value (see *Sect. 13.8*). If for example, the manufacturer of the S/R-unit has guaranteed a total reliability of 99.0%, he would have failed in the test.

From formula (13.89) follows the *rule for reliability measurements*:

► In order to measure an expected reliability η_{rel} with a required statistical error s_η, the number of trials has to be

$$n = (1-\eta_{rel})/{s_\eta}^2. \tag{13.90}$$

If e.g. a guaranteed total reliability of 99.0% should be measured with an accuracy of ±0.25%, the total number of cycles to be performed is 1,600.

The relations (13.85) and (13.90) determine also the *necessary test time*:

► In order to achieve a required accuracy s_η for a reliability value η_{rel} with performance rate λ , the test time must be

$$T_{tot} > n/\lambda = (1-\eta_{rel})/({s_\eta}^2 \cdot \lambda) \quad [h]. \tag{13.91}$$

In the given example for a planned performance of 25 cycles/h the test must last at least $1{,}600/25 = 64$ h.

The described measuring procedure and the formulas (13.85) to (13.91) can be used also for the measurement of the reliability of performance chains and complete systems. However, with increasing number of functions and elements, the direct measurement of the process and system reliabilities becomes more and more difficult.

13.6.2 Measurement of Availabilities

Failures of any function during the operation time reduce the partial limit performances μ_α of an *irreducible system element* with functions F_α to the *available partial limit performances*:

$$\mu_{\alpha\,ava} = \eta_{ava} \cdot \mu_\alpha \tag{13.92}$$

The down time caused by a failure of one function interrupts the operation of the irreducible element for all other functions as the partial functions are executed in the same process zone. Therefore, the reduction factor in (13.92) is the *total availability* η_{ava} of the irreducible element.

In order to measure the total availability during a test time T_{tot}, all *down times* $\tau_{i\,down}$, $i = 1, 2, \ldots n_{fail}$ are measured, which last from failure until successful restart. The down times, which have been needed to restore the capability to perform, and the causes of the failures are registered in a *fault report*. The sum of the measured down times gives the *total down time*:

$$T_{down} = \sum_i \tau_{i\,down}. \tag{13.93}$$

The *mean down time* is:

$$\tau_{down} = T_{down}/n_{fail}. \tag{13.94}$$

The *available time* to perform any of the functions is the total test time minus the total down time:

$$T_{aval} = T_{tot} - T_{down}. \tag{13.95}$$

From the available time (13.95) and the total operation time, the availability can be calculated:

- The *measured availability* of an irreducible element is the relation of the available time to the total operating time

$$\eta_{ava} = T_{aval}/T_{tot} = T_{aval}/(T_{down} + T_{aval}). \tag{13.96}$$

The total down time divided by the total number of failures n_{fail} is the mean down time per failure or *mean time to restore*:

$$\mathrm{MTTR} = T_{down}/n_{fail} = \tau_{down}. \tag{13.97}$$

The available time divided by the total number of trails n is the *mean time between failures*:

$$\mathrm{MTBF} = T_{aval}/n. \tag{13.98}$$

With these definitions the availability of the element can be written as

$$\eta_{ava} = \mathrm{MTBF}/(\mathrm{MTTR} + \mathrm{MTBF}). \tag{13.99}$$

The total down time T_{down} is the mean down time τ_{down} multiplied by number of failures. For an irreducible element with total reliability η_{rel}, the number of failures n_{fail} is the product of the total performance rate λ and the unreliability $(1-\eta_{rel})$ multiplied by the total test time T_{tot}. The product $T_{down} = \tau_{dow}\cdot(1-\eta_{rel})\cdot\lambda\cdot T_{tot}$ can be inserted into relation (13.95) and this into (13.96). Resulting is the *connection between availability and reliability*:

- The total availability of an irreducible element depends on the actual performance rate λ, the mean reliability η_{rel} and the mean down time τ_{down}

$$\eta_{ava} = \mathrm{MAX}(0;\ 1-\lambda\cdot\tau_{down}\cdot(1-\eta_{rel})). \tag{13.100}$$

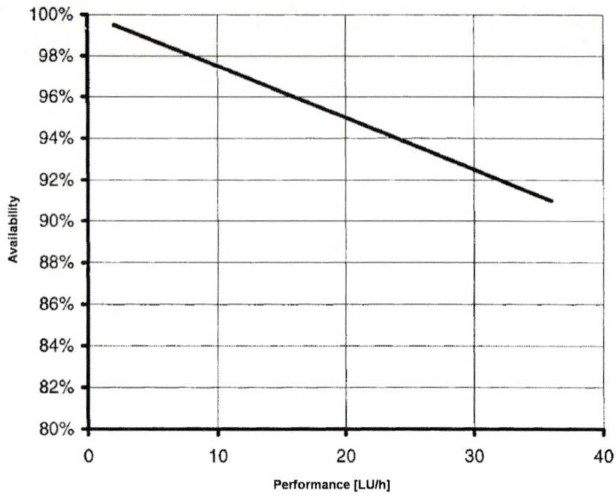

Fig. 13.24 Dependency of the availability on the performance rate

R/S-unit: storage and retrieval unit
limit performance: $\mu_S = \mu_O = 36$ Pal/h
availability: $\eta_{rel} = 99.00\%$
MTTR: $\tau_{down} = 15$ min

For the above example of a R/S-unit, the dependency of the availability on the performance rate is shown in *Fig. 13.24* and on the reliability in *Fig. 13.25*.

The example and formula (13.100) show:

- ▸ The availability is reduced by decreasing reliability, increasing down times and higher performance rate.

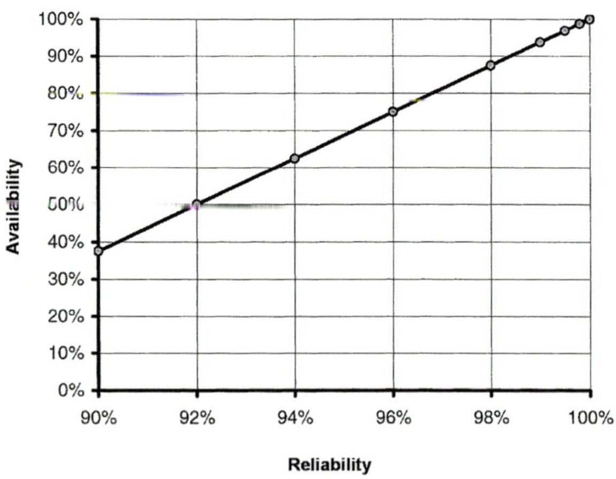

Fig. 13.25 Dependency of the availability on the reliability

R/S-unit: storage and retrieval unit
Performance rate: $\lambda = 36$ Pal/h
MTTR: $\tau_{down} = 15$ min

In the example of the S/R-unit with reliability 99.0% and a MTTR of 15 min, the availability goes down from 100% at low utilisation to 91.0% for utilisation close to 100% with the limit performance of 36 LU/h. Even more critical is the dependency on the reliability as shown in *Fig. 13.25*, where the availability is only 70% if the reliability falls to 95%.

Consequences of the dependency (13.100) between availability, down times and reliability are the *availability rules*:

- The availability has to be provided and guaranteed by the manufacturer for the performance required by the user, i.e. for the planned utilisation.
- The availability can be improved by reducing the down times as well as by improving the reliability.

The impact of the current performance on the availability has to be taken into account in particular when *high-performance systems* are dimensioned. The current performance depends on the *operator* and not on the *manufacturer* of the system. Many users and some manufacturers are not aware of these dependencies.

13.6.3 Influences on Reliability and Availability

Reliability theory generally deals with *continuously running systems* (Gross 1998). There, the *operation reliability* is defined as the probability of the running system *to operate for a time t without failure.* In order to distinguish it, the reliability of *discontinuously operating systems*, like logistic systems and performance systems, is called *function reliability*.

The function reliability of an irreducible element which processes discrete objects in a certain process time is determined by its operation reliability. If the failures are independent of each other and their occurrence is random, the operation reliability decreases exponentially with the *operation time* T_O:

$$R(t) = R_0 \cdot \exp(-T_O/t_m) \quad [\%]. \tag{13.101}$$

Herein, t_m is the *mean time between failures for continuous operation* and R_0 the *start-reliability*, i.e. the probability of a successful process start. For elements, which are running only if they are used, generally $R_0 < 1$. For permanently running elements, like a conveyor or a process control, $R_0 = 1$.

The mean time between failures for continuous operation is shorter than the mean time between failures for discontinuous operation due to the idle times between the single arrivals of discrete objects. The operating time T_O for a discontinuously operating element equals the *process time* τ_P for the relevant function. This can be the service time of a service station, the cycle time of a production station or logistic device, the passing time through a transport element or the transport time of a vehicle. From formula (13.101) follows the *reliability law of discontinuous elements*:

- The partial function reliability $\eta_{rel\,\alpha}$ of a discontinuously operating element decreases exponentially with the process time $\tau_{P\,\alpha}$ for the function F_α.

This law explains e.g., that the partial reliability $\eta_{rel\,IO}$ of the S/R-unit for longer lasting double-cycles is smaller than the reliability $\eta_{rel\,I}$ for shorter single cycles.

The reliability of the elements and the whole system are influenced by the *manufacturer* and the *user*. *Manufacturer-dependent influences on reliability* are:

design and construction
quality of materials
serviceability
accuracy and control of assembly
quality of implementation (13.102)
experience of the manufacturer
probation of the elements
technical running life time

User-dependent influences on reliability are:

qualification of operating personal
quality and regularity of maintenance
protection against damages
attention to manuals (13.103)
intensity and duration of utilisation
condition of process units
actual running time

The manufacturers of the elements of logistic systems ought to offer figures for the reliability of all critical elements, which are guaranteed under realistic operating conditions. This enables system supplier and user to calculate the reliability and availability of the specific system by the formulas of this section.

The *down time* is the sum of *failure-recognition time*, *show-up time* of maintenance personnel, *error-search time*, *spare-part replenishment time*, *repair time*, *test time* and *restart time*. Similar to the reliability, the down times are influenced by the manufacturer and the user. *Manufacturer-dependent influences on down times* are:

construction and control
easiness to repair and to maintain
completeness of the spare part recommendation (13.104)
quality and completeness of documentation
instruction and training of operation personnel

User-dependent influences on downtimes are:

qualification and alertness of personnel
qualification and availability of maintenance staff
compliance to maintenance instructions
availability of spare parts (13.105)
emergency organization
location of the system

For *complex systems*, the high number of influence factors on reliability and availability often causes conflicts between *manufacturers, general contractor* and *user*. Therefore it is advisable to specify the influence factors and to clarify the respective responsibility in advance [FEM 9.222].

From the influence factors (13.102) to (13.105) follow the *experience rules*:

- Due to *start problems* and *lack of experience*, reliability and availability of a system are low at the beginning of the technical life time, but improve soon within a *learning phase* after solving the start problems.
- Due to *wear and tear* of parts and material, reliability and availability go down at the end of the *technical life time*.

When the costs for repair and maintenance reach the earning value of the system, a new investment is necessary.

13.6.4 Function Safety of Serial and Parallel Stations

Both key figures of the functional safety of serial and parallel stations, *reliability* and *availability*, can be calculated from the respective values of the stations by the same rules of *probability theory*. In order to abbreviate the following explanations, the term *function safety* η stands for reliability η_{rel} as well as for availability η_{ava}. Both are probabilities.

For a process chain as shown in *Fig. 13.11*, consisting of n successive stations, from the *multiplication rule of probability* follows:

- The *partial function safety* η_α for the function F_α of a process chain is the product of the partial function safeties $\eta_{\alpha\,k}$ for the involved partial functions $F_{\alpha\,k}$ of the single stations S_k, k = 1,2. . .n:

$$\eta_\alpha = \eta_{\alpha\,1} \cdot \eta_{\alpha\,2} \cdots \eta_{\alpha\,n-1} \cdot \eta_{\alpha\,n-1} = \prod_k \eta_{\alpha\,k} \tag{13.106}$$

For example, if the single reliability of the 10 elements of a performance chain is 99.5%, the process reliability of the whole chain is only $0.995^{10} = 95.1\%$.

Apart from the performance stations and transport elements, the overall *process control* contributes to the successful execution of the different functions of a performance chain. Therefore, the *function safety of process control* has to be taken into account as a factor in the product (13.106).

A failure of a vehicle in a transport system affects the transport process as much as a failure of the system control. Therefore, also the *function safety of the vehicles* is a separate factor in the probability product (13.106) for a *vehicle system*. The same holds for *function safety of load units* in a *conveyor system*.

If the reliability is calculated by formula (13.106), the measured partial reliabilities (13.87) have to be inserted. They are independent of the actual utilisation by other performances but can deviate for the different functions. The product rule (13.106) leads to the general *complexity law of process reliability*:

- *Process reliability* decreases with increasing length of the performance chain.

For the availability calculation, the measured total availability (13.96) is relevant since the partial availabilities are equal for all functions. The current availability of each single element depends on the utilisation of all its functions. This leads to the *system interaction principle*:

- The *process availability* depends on the total performance flow through the performance chain and on the competing flows through the same stations by all crossing performance chains.

Due to this principle, it is not sufficient to examine only single chains. The examination also has to include the interactions between the different performance chains of the system. From the *interaction principle*, in combination with the product rule (13.106), follows the *complexity law of process availability*:

- The availability of a process decreases with increasing length of the performance chain and with the number of other processes using the same chain elements.

The availability of a supply chain is the *availability to supply*. The process reliability of a freight chain is called *mission probability*. It is the probability that a shipment arrives at the right destination on time, complete and undamaged. The mission probability of a freight chain is the *punctuality* if a correct and damage-free dispatch is taken for granted (see *Sect. 8.7*).

According to the complexity principle, performance chains should be as short as possible. This can be achieved by the following *design rules for performance chains*:

- *Decoupling* of a pre-process from a time critical order process by inserting a decoupling station with free stock, e.g. a store, or a production station with free capacity (see *Sect. 8.5.2*).
- *Separation* of performance chains by intermediate buffer places which allow ongoing of a part process when the preceding or following process is disturbed.
- *Redundancy chains* parallel to a high performance chain or to a functionally critical station (see *Fig. 13.26*).

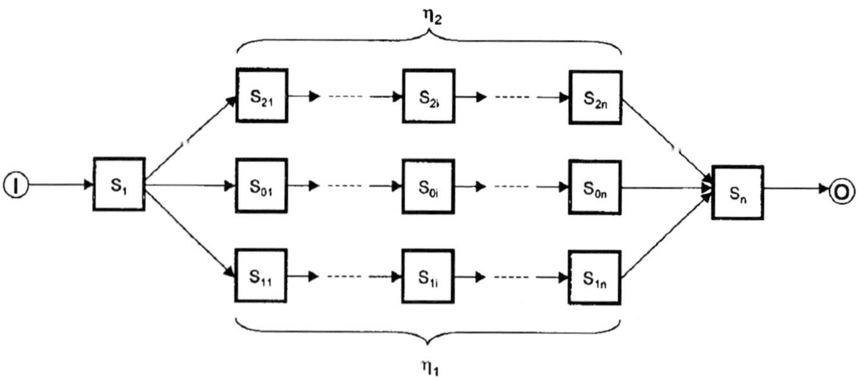

Fig. 13.26 Parallel process chains with twofold redundancy

PC_0: main chain PC_1, PC_2: redundancy chains
S_{rk}: k^{th} station of process chain PC_r
η_r: reliability or availability of process chain PC_r

Separation of a process chain by *intermediate buffers* is common within production engineering for linking machines. This avoids a production interruption in case of a short-term breakdown of a pre-ceding or successive machine. For this purpose, the capacity of the buffer has to be sufficient for the waiting queues caused by stochastically varying service or down times of the succeeding stations. On the other hand, the buffer capacity must ensure the uninterrupted operation of the station during the down time of a proceeding station.

By installation of n additional parallel chains with the same functional capability, an *n-fold redundancy* is achieved. For example, *Fig. 13.26* shows 3 parallel chains with redundancy n = 2. The effect of redundancy chains on function safety is calculable with the probability rules. A number of n parallel process chains PC_r, r = 1,2,. . .n that can take over a share p_r [%] of the performance of the main process chain PR_O has a *part redundancy* p_r if $p_r < 1$, and a *full-redundancy* if $p_r = 1$.

The probability, that a parallel chain with part redundancy p_r and function safety η_r can take over in case of failure of another chain a share p_r of the required performance rate, is the product $p_r \cdot \eta_r$. The probability that it fails to take over is the failure probability $(1-p_r \cdot \eta_r)$. The probability that none of the n parallel redundancy chains can take over a share of the required performance rate is the product of all failure probabilities.

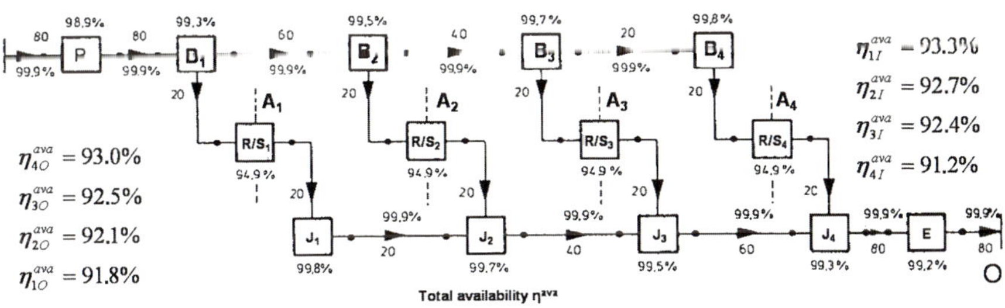

Fig. 13.27 Structure diagram with the availability analysis of the separated inward-conveyor system and outward-conveyor system of an automatic high bay store

P: profile control	B_i: branching elements
A_i: aisles of the store	E_i elevator station
J_i junction elements	S/R_i storage and retrieval units

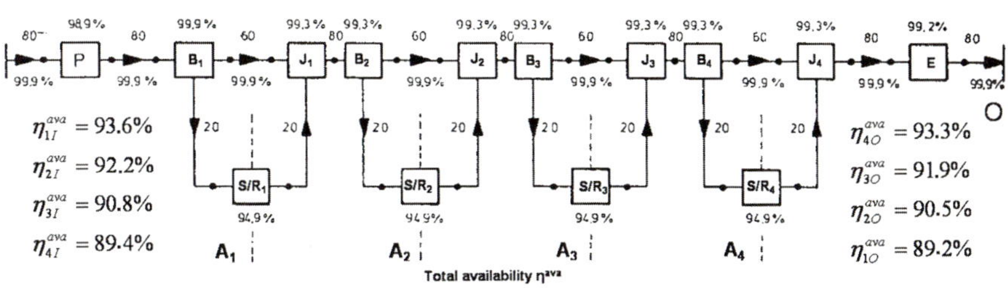

Fig. 13.28 Structure diagram with the availability analysis of the combined inward- and outward-conveyor system of an automatic high bay store

Notation: see *Fig. 13.27*

Following is the *redundancy law*:

► The function safety of a *redundancy chain* with n parallel chains offering the part redundancies $p_{\alpha r}$ with function safeties $\eta_{\alpha r}$ is

$$\begin{aligned}\eta_\alpha &= 1 - (1 - p_{\alpha 1} \cdot \eta_{\alpha 1}) \cdot (1 - p_{\alpha 2} \cdot \eta_{\alpha 2}) \cdots (1 - p_{\alpha n} \cdot \eta_{\alpha n}) \\ &= 1 - \prod_{r=1}^{n} (1 - p_{\alpha r} \cdot \eta_{\alpha r})\end{aligned} \tag{13.107}$$

For the example of *Fig. 13.26* with two parallel process chains, which each can take over the full performance rate with reliability $\eta_{\alpha r} = 90\%$, the resulting process reliability is $\eta_\alpha = 1-(1-0.90)^3 = 0.999 = 99.9\%$. If the two parallel chains offer only 50%-redundancy, i.e. if $p_1 = p_2 = 0.5$, the resulting process reliability is $\eta_\alpha = 1-(1-0.90) \cdot (1-0.5 \cdot 0.90)^2 = 0.967 = 96.7\%$. That means, the process reliability with twofold *part redundancy* is lower than with twofold *full redundancy*, but still significantly higher than the process availability of 90.0% without redundancy.

Each function F_α of a system uses a specific performance chain PC_α. If λ_α are the partial performances for the functions F_α, the total performance rate of the system is

$$\lambda = \sum_\alpha \lambda_\alpha \tag{13.108}$$

The probability that the function F_α is used by the next coming object is the *utilisation probability* λ_α/λ. The total function safety of the system is the sum of the partial function safeties η_α, that result with the relations (13.87), (13.96), (13.106) and (13.107), multiplied with the utilisation probabilities λ_α/λ. This leads to the *system rules*:

► The *system reliability* of a system with the process chains PC_α, partial performances λ_α and process reliabilities $\eta_{\alpha\,\mathrm{rel}}$ is

$$\eta_{\mathrm{rel}} = \sum_\alpha (\lambda_\alpha/\lambda) \cdot \eta_{\alpha\,\mathrm{rel}} \tag{13.109}$$

► The *system availability* of a system with the process chains PC_α, partial performances λ_α and process availabilities $\eta_{\alpha\,\mathrm{ava}}$ is

$$\eta_{\mathrm{ava}} = \sum_\alpha (\lambda_\alpha/\lambda) \cdot \eta_{\alpha\,\mathrm{ava}} \tag{13.110}$$

Figure 13.27 and 13.28 show the results of an *availability analysis* for two different inward-and outward-conveyor systems of a *high bay store*. This demonstrates the calculation of the partial and the total availability of a system based on the measured availabilities of the elements by applying the previous formulas. The results show that the total system availability for the same performance is 92.5% for the separated inward-and outward-conveyor system, and 91.4% for the combined inward-and outward-conveyor system. The lower availability of the combined system results from the higher utilisation of the elements along the conveyer line used together by the inward and by the outward flow (Gudehus 1976).

For systems without redundancies, insertion of relation (13.106) into (13.109) and some side-calculations lead to the *law of system reliability*:

▸ The *measured system reliability* of a non-redundant system is the relation of the observed number of correct performances to the total number of trials, which is the sum of the number of failures and successes

$$\eta_{rel} = n_{corr}/(n_{fail} + n_{corr}). \tag{13.111}$$

From formula (13.110) follows after insertion of relation (13.106) and some arithmetic the *law of system availability* (Gudehus 1976/III; VDI 3581):

▸ The *measured system availability* of a non-redundant system with partial performances λ_α, sum performance (13.108) and *partial downtimes* $T_{\alpha\,down}$ during an *operating time* T is

$$\eta_{ava} = 1 - \sum_\alpha (\lambda_\alpha/\lambda) \cdot (T_{\alpha\,down}/T) \tag{13.112}$$

Formulas (13.111) and (13.112) allow the direct calculation of the current system reliability and system availability from the counted failures and measured down times. The resulting values depend on the current performance rates λ_α. They do not indicate the function safety of the partial process chains and the critical elements of the system. Therefore, they are not sufficient for a *contractual agreement* about the function safety of a system. In addition, contractor and customer have to agree upon guaranteed values for the reliability and the availability of all *critical system elements* and important performance chains (see *Sect. 13.8*).

13.7 Capability Analysis

In order to avoid disappointing surprises and expensive changes after realisation of a logistic, production or transport system, it is necessary to perform a capability analysis in advance. For an existing system, a capability analysis identifies *bottlenecks, weak points* and necessary *extensions*. Steps of a *capability analysis* are:

1. Elaboration of the *structure diagram* of the system with all elementary stations, transport elements and their connections.
2. Establishing the *capacity diagram* by insertion of the calculated or measured *capacities* of stores and buffers, and of the installed *limit performances* of all stations and transport elements into the structure diagram.
3. Working out the *flow diagram* by insertion of the required performance rates and flows through the stations and transport elements.
4. Analysis of the used *operating strategies* for the stations and transport elements, and of the *system strategies* for the subsystems and total system.
5. Calculation of the *utilization diagram* with the total utilization of all stations and transport elements which result from the required performance rates, installed limit performances, switch times and operating strategies.
6. Establishing the *queuing diagram* with the calculated mean queues, waiting times and blocking probabilities.

7. Development of the *availability diagram* by inserting the expected or measured availabilities of the stations and transport elements.
8. *Assessment of the system capability* by analysing the utilization diagram, queuing diagram and availability diagram.
9. *Evaluation of the system availability* after calculation of the availability for the stations and transport elements and for the order and logistic chains.

A logistic, production or transport system is only capable of performing as required if the following *function rules* are observed:

- The *utilization* of any station or transport element should not exceed 85% during normal operation time and 95% during peak times.
- In *peak times* the mean waiting queues in front of the stations and nodes must be shorter than the buffer capacity on the connections.

If a mean waiting queue exceeds the available *buffer capacity*, it is necessary to calculate the *blocking probability* for the preceding station and to examine whether it can still cope with the blocking. The utilisation diagram and the queuing diagram immediately show the bottlenecks:

- *Bottlenecks* are all stations and transport elements of a system that are utilized during peak times over 95% and cause waiting queues which intolerably reduce the effective performance of preceding stations.

In many cases, the capability analysis of *high-performance systems* shows that their performance rate is limited by only one or a few bottlenecks. The elimination of these bottlenecks is the first step for improving the capability of the total system. *Measures to eliminate bottlenecks* are:

- *Drilling*: Improvement of the limit performance by increasing the dispatch capacity, lowering the switch times, reduction of idle times, higher acceleration and speed and shorter in- and out-flow times.
- *Strategy Change*: Random instead of priority dispatch or interchanging the priorities between main and secondary flows.
- *Bypassing*: Use of not fully utilized parallel performance or transport chains with the same function.
- *Doubling*: Installation of a parallel station or a redundant chain with the same function.

To illustrate the procedure of a capability analysis, *Fig. 13.29* shows the structure diagram of a bin conveyer system serving an *order picking system* with local picking stations and static replenishment (see *Chap. 17*). The picked items are placed into the bins which are provided and removed by the conveyor system to and from the picking stations in the rack aisles. After the last item for an order has been put into a bin, it is conveyed to the packaging zone. Bottleneck elements of this system are the junction elements J at the front side of the racks. Other bottlenecks are the picking stations P, as their backlogging can cause blockings of the preceding stations.

Critical for the function of a system are all stations and transport elements whose breakdown leads either to an interruption of important functions or to a performance

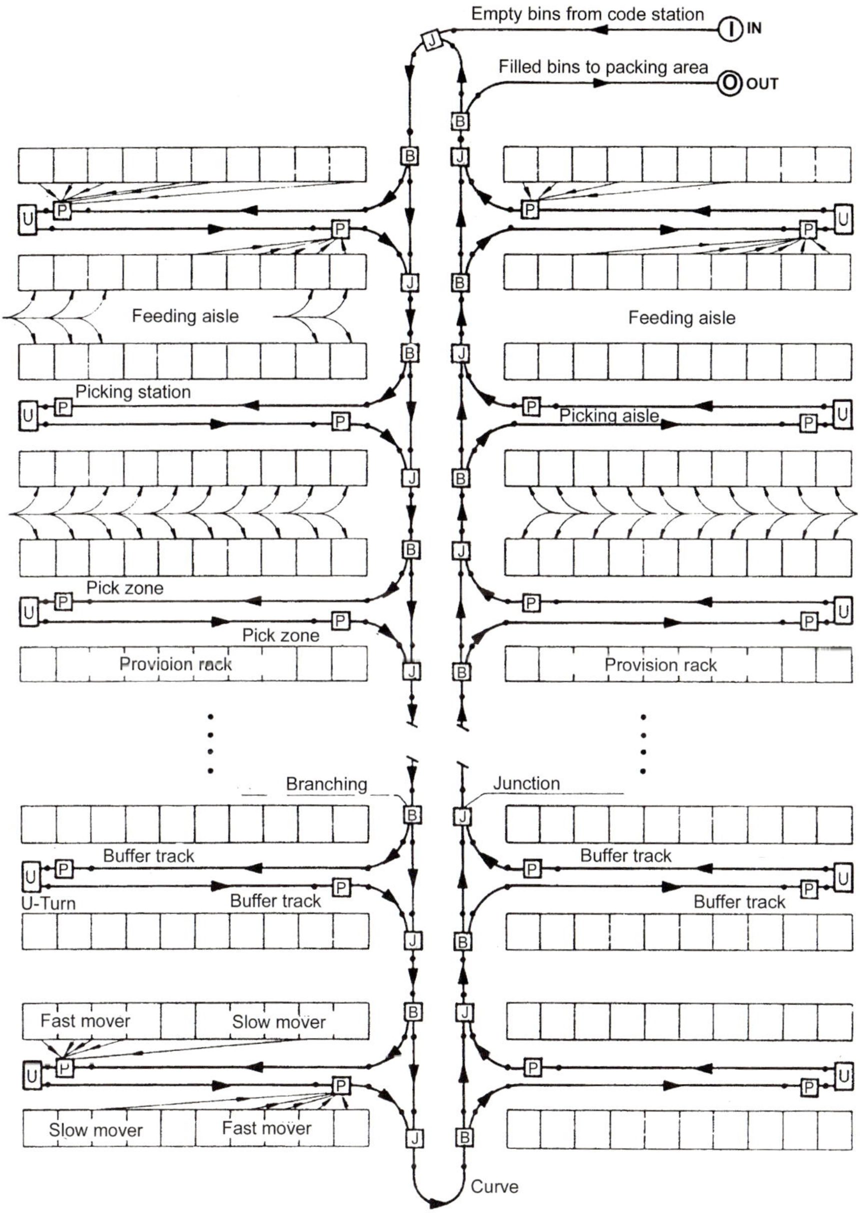

Fig. 13.29 Structure diagram of a conveyor system for order picking

B: branching elements U: u-turning disks
J: junction elements P: picking stations

reduction of more than 20%. The *function-critical elements* can be identified from the performance and availability diagram by calculating the performance reduction after setting to 0 the availability of the single elements.

Breakdown stations are stations with availability below 90%. They block preceding stations by frequent or longer lasting interruptions and cause an underutilization of subsequent stations. In many cases, only a few breakdown stations cause delivery delays and missed deadlines.

The bottleneck elements of the separated conveyor system of *Fig. 13.27* are the first branching element of the inward system and the last junction element of the outward system. The bottlenecks of the combined inward-outward conveyor system of *Fig. 13.28* are all branching and junction elements along the front side of the high bay store. In both cases the function critical elements are the S/R-units, as in this example their availability is only 95% at maximal utilization. The availability of a S/R-unit, however, should be above 98% at maximal utilisation if it is designed professionally and maintained regularly.

If *lead times* or *transport times* between the entries and sources and the sinks and exits of a system shall not exceed certain *benchmark times*, it is necessary to work out the *process chain diagrams* in addition. These diagrams show all possible performance, transport and order chains of the system separately. The transport times between the stations, the waiting times in front and the process times within the stations and transport elements of all these process chains have to be calculated and summed up (see *Sects. 21.8* and *21.9*). Then the resulting *transfer times* T_α for the different process chains PC_α starting from the entries and sources, and ending in the sinks and exits of the system, are compared with the benchmark times $T_{\text{bench}\,\alpha}$ (see *Sect. 21.4*).

A further result of the limit performance and queuing laws is the general procedure of *analytical system design*:

1. In the first step, the required performance rates and flows are considered as stationary, the processes times are assumed to be constant and all stations and elements are taken as 100% reliable. The limit performances and capacities of the stations and transport elements, the process chains and the structure of the system are designed and dimensioned for the requirements of the peak hour with 10% reserve.
2. In the second step, for the resulting system a capability analysis is performed, using empirical values for the reliabilities and down times of the elements. Based on the results of this analysis the identified bottlenecks, weak points and critical elements are eliminated or corrected.
3. In the third step, the limit performances of the bottleneck elements are adjusted to the stochastic fluctuations. The queuing effects are released by adequate buffer capacities. The function safety of the processes is improved by increasing the availability of critical elements or by redundancies.

If a system is used in different times for different performance rates and flows, it is essential to perform the capability analysis for all critical *scenarios*.

If the performance rates and/or structure are expected to change fast, e.g. in less than an hour, and if the expected stochastic fluctuations are high, in addition to the capability analysis, a simulation of the planned system should be performed. A *stochastic simulation* of an analytically planned system reveals the interdependency and consequences of fast systematic and stochastic changes. To cope with the dynamic effects identified by the simulation, the limit performances and buffer capacities are adjusted or the operation strategies are adapted (Pflug 1996; Kuhn 1993; Lanzendörfer 1975).

13.8 Acceptance of Plants and Systems

Before a system which has been delivered by a *contractor* is accepted by the *customer* it should be tested. The acceptance test of a plant, conveyor or vehicle system, an automatic high bay store or any other logistic system can be performed analogous to the above capability analysis [FEM 9.222].

After the contractor reports the completed installation and when the operating personnel has been instructed and trained sufficiently, the user has to prepare the necessary equipment, forklifts, pallets and workers to carry out the *acceptance tests*, which involve a function test and a performance test:

- During the *function test*, the guaranteed properties and functions of the system elements, the required processes and the whole system are checked and tested.
- During the *performance test* the capacities, cycle times, process times, failures and down times are measured for a *test time* of sufficient length with the contractually agreed performance rates and flows.

Function test and performance test are successful completed if all required functions are performed correctly, the contractual limit performances are fulfilled and the system availability is at least 80%. Then the plant or logistic system is handed over to the client, and the contract is fulfilled with the provision that the guaranteed reliability and availability will be reached within a *trial phase* of sufficient length, at least until the end of the *warranty time*.

If the required performance rates have not been achieved with minimal availability of 80%, the manufacturer is obliged to touch up the plant within adequate time, which under normal conditions should not be longer than a couple of weeks. After the deficiencies are corrected and the causes of failures are eliminated, the acceptance tests are repeated.

After successful acceptance tests and handing over, the system is used productively by the customer at his own risk. In the *trial phase* the user records all failures, measures the down times, analyses the causes of failures and notices the responsibility. From the observed data, the reliability and availability of critical elements, partial functions and the whole system are calculated. The trial phase ends after the contractually agreed values for the reliabilities and availabilities have been reached for a statistically sufficient time. Alternatively, on demand of the contractor, an availability test can be conducted:

- During the *availability test*, the system is operated by the customer in presence of the contractor with at least 70% of the planned performance rate and throughput for a statistically adequate period of time, which lasts normally between 16 and 80 operating hours.

Personnel from the contractor and the customer report all failures and measure the down times. Together they analyse the failures and determine the responsibility for the causes of failures and the shares of the down time. Taking into account only failures and down times, for which the contractor is responsible, the actual reliabilities and availabilities are calculated. If they exceed the guaranteed figures, the plant or system is finally accepted. Otherwise, the contractor is obliged to mend the failures. If this is not possible or done within the regular warranty time, which can lasts up to 12 months, the customer has the right for *price reduction* or *return*.

The contractually agreed deadlines, capacities, limit performances and due dates can be enforced by *fines* like a *deadline penalty*, *availability penalty* or a *performance penalty*. Each required penalty, however, increases the price of the system since the contractor will incorporate the costs of his risks as allowances into the calculation (see *Sect. 7.2.2*). A pragmatic solution of the goal conflict between enforcing the contractual agreements and higher prices are the *penalty limitations*:

- ▸ Single penalties are limited to a reasonable level of about 5% of the relevant price.
- ▸ The sum of all penalties and fines is limited up to 10% of the system price.

13.8.1 Deadline Penalty

To enforce the contractor to keep important *milestones* and the agreed deadline for the whole plant or system, the following *deadline penalty regulation* is adequate:

- ▸ As far as caused by the contractor, the customer is entitled to a reduction of 0.1% of the price for the affected subsystem or total system per day of delay.

Instead of a daily penalty also a weekly deadline penalty of 0.5% per started week of delay can be agreed.

If the customer is extremely interested in an earlier start of the productive operation, the parties can also agree on a corresponding *due date bonus*.

13.8.2 Performance Penalty

To enforce the contractor to fulfil the contractual capacities and limit performances, the following *performance regulation* is possible:

- ▸ If a guaranteed capacity or limit performance is not achieved by a part system or the whole system, the customer is entitled per 1% deviation from the guaranteed value to a reduction of 0.5% on the relevant price.

Correspondingly, a customer can offer a *performance bonus* for exceeding the contractual values.

13.8.3 Reliability and Availability Penalty

A system or plant can only be used efficiently if the installed functions and limit performances are executed with sufficient reliability and availability. Therefore, it is advisable to agree, in addition to the performance penalty, about an *availability and reliability penalty regulation*, e.g:

- If the guaranteed reliability and/or availability are not achieved with in the warranty period, the customer is entitled to a price reduction for the affected sub- or total system of 0.5% per 1% deviation from the contract value.

Only failures and down times are taken into account for which the contractor is responsible. Because much money is involved, it is of utmost importance to specify the mutual responsibilities in advance.

The effect of the penalty is illustrated by an example: The guaranteed availability for a plant with a delivery value of 5.5 Mio € is 98.0%. Although the contractor mended the plant several times, the availability reached only 95.0%. In this case, results an availability penalty of 3·0.5%·5,500,000 € = 82,500 €.

When measuring reliability and availability in order to test the adherence to the guaranteed values, it is eminent to take into account the above quality measurement principles. Otherwise the measurement errors can cause unnecessary differences. After the test has been completed, not only the mean values but also the *statistical errors* of the measurement have to be calculated. In order to allow a fair judgement on the fulfilment of a guaranteed value, the error of availability measurements must be lower than ±0.5% and of reliability measurements lower than ±0.1%.

Chapter 14
Purchasing, Sales and Logistics

Logistics depends on purchasing and sales and vice versa. Purchasing and sales people negotiate the *prices* of products and services, determine the *terms of trade and delivery* and initiate the supply and delivery chains between suppliers, companies and customers.

Physical distribution of finished goods is traditionally part of sales and marketing (Pfohl 1990; Wöhe 2000). Sourcing and procurement of raw material, pre-products and merchandise are looked at as tasks of purchasing. However, *modern logistics* and *supply chain management* open new competitive advantages and saving potentials (Bowersox et al. 1969; Bucklin 1966; Coughlan et al. 2006; Cooper et.al. 1997; Christopher 1992; Dobler 1996; Kotzab 1997; Schönsleben 1998; van Weele 2004). This, however, requires special competences and separation of logistics from sales and purchasing.

Relieved from the responsibility for stocks and distribution of finished goods, the sales department can concentrate on marketing and selling. Without the responsibility for physical procurement and materials management, the purchasing department can focus on strategic procurement, search for suppliers and negotiation of prices, terms and contracts.

If purchasing, sales and logistics are separated without defining their special tasks and mutual obligations, conflicts, resistance and misunderstanding can arise, which prevent the aimed advantages. Condition for the common success of logistics, sales and purchasing to the advantage of customers and company is observance of the *collaboration rules*:

- knowing the goals and restrictions of the company
- observance of the specific options of action
- respect for the tasks and contributions of each other
- adequate freedom of action and decision competencies
- partnership without subordination
- critical directness internally, agreement externally
- mutual support to the benefit of customers and company

T. Gudehus, H. Kotzab, *Comprehensive Logistics*,
DOI 10.1007/978-3-540-68652-1_14,

Starting with the core competencies of sales and purchasing, this chapter describes the *relations* and *interfaces* between sales, purchasing and logistics. It outlines how logistics can effectively support purchasing and sales and where sales and purchasing should take into account the requirements and potentials of logistics.

14.1 Core Competencies of Sales and Marketing

Core competencies and central tasks of sales and marketing are (Coughlan et al. 2006; Kotler 2000):

- investigation of markets and demand
- analysis of the competitive situation
- determination of customer service expectations
- planning the range of products, merchandize and services
- organization and implementation of marketing channels
- management of the sales organization
- demand and sales planning
- advertising and promotion
- identification of and calling on potential customers
- investigation of customers willingness and ability to pay
- calculation of prices and offers
- processing of customer requests
- preparation of quotations
- negotiation and contracting
- customer relations and consulting

The overall goal of *marketing*, *sales force* and *back office* is to create demand and to sell the products and services of the company.

Tasks of *marketing* are to set up the processes for creating, communicating, and delivering value to customers and to manage *customer relationships* in ways that benefit the company and its stakeholders. The marketing department plans, initiates and controls all sales activities, but should not be responsible for order fulfillment and physical distribution. The letter activities would extend marketing to the entire company. To incorporate physical distribution as *marketing logistics* (Pfohl 1990) is neither useful nor appropriate, as it would interrupt the logistic chains from the suppliers to the customers at the end of production.

The company specific organization of the *sales force* depends on the type of the business. Generally, it consists of branch offices, outlets, sales offices, representatives, dealers and franchise partners. The *back office* is responsible for *order acceptance* and *order preparation*. This includes commercial order examination, order collection and entering of the order data into the computer. After services, prices and delivery conditions have been negotiated and confirmed, the customer order becomes a binding external order which is handed over to *order scheduling* or to an *order center*.

14.2 Core Competencies of Purchasing

Core competencies and central tasks of purchasing are (Dobler et al. 1990; Kraljic 1983; Monczka et al. 2002):

- exploration of sourcing markets
- identification of potential suppliers
- investigation of market prices and available quantities
- determination of general purchasing conditions
- examination of references, credit and service quality of suppliers
- search for suppliers of the needed quantities and qualities
- tendering for major demand and projects
- negotiation of purchase prices and delivery conditions
- conclusion of contracts and frame agreements for the regular supply
- control of the performance and quality of external suppliers

The goals of strategic and operative purchasing are cost-optimal sourcing and efficient fulfillment of the company's demand. Demand scheduling, material management and the release of supply for the current demand of production or for customer orders are not necessarily tasks of the purchasing department. As outlined in *Sect. 2.2*, these activities can be performed more professionally by an independent *order center* or *order scheduling*.

14.3 Order Scheduling and Supply Management

Order scheduling, inventory management and current supply for production and customer orders are not core competencies of sales and purchasing. Therefore, order scheduling should not be attached to sales or purchasing, as it is still customary in many companies.

As outlined in detail in *Sect. 2.2* and *Chap. 10*, *order scheduling* transforms *external orders* into *internal orders* by applying appropriate *scheduling strategies*. The internal orders are allocated to production facilities, suppliers and service providers according to *rules* and *priorities* that have been established by top management in accordance with sales, production and purchasing. These rules regulate:

make or buy
make to stock or make to order
source to stock or source to order
allocation of limited resources (14.1)
replenishment quantities
service levels
procurement and delivery chains

These decisions can cause conflicts between sales, production and purchasing. In order to avoid them, order scheduling should be part of logistics. However, like all other departments with direct customer or supplier contact, order scheduling must observe the following *contact principles*:

- *Purchasing* has to be informed immediately on all important supplier contacts and to be asked before a procurement agreement is changed.
- *Sales* has to be informed immediately on all important customer contacts and to be asked before sales agreements are changed.

If the second principle is not kept, the sales representatives are no longer respected by the customer and the chances for follow-up orders dwindle.

As long as all actors observe the contact principles, the schedulers and order centers of the companies, which take part in a supply chain, can cooperate directly without permanent contact to purchasing and sales. IT-connection of the order centers via *EDI* or *Internet* enables an efficient *inter-corporate supply chain management* (see also *Sect. 21.17*).

For the single companies, efficient supply and delivery require

1. *sourcing strategies* (see *Sect. 7.7.6*) for material, pre-products, parts, modules, merchandise and logistic services (see *Sect. 22*)
2. *marketing strategies* (see *Sect. 7.7.7*) for the products and services of the company

The development of marketing strategies for logistic services is a central task of *logistic marketing*, which – in contrast to *marketing logistics* – is useful and necessary.

14.4 Products, Merchandize and Services

Sales management is responsible for the range of products, merchandize and services of the company. This not only involves planning and expansion of the assortment, but also the decision which article should be eliminated or sold out in order to reduce excess inventory. An assortment that is not permanently controlled will get out of hand. The share of dead, non-sellable and *slow moving articles* and the total inventory will grow continuously.

However, sales management often hesitates to eliminate articles or to devaluate inventories. Therefore, company logistics should assist sales and marketing in the analysis and adjustment of the assortment. For this purpose, the logistic department periodically analyses where the assortment is getting out of hand, which articles are *bestsellers*, which inventories are too high in relation to sales and which articles are *slow-movers* or dying (see *Sects. 5.8* and *5.9*).

The *ABC-analysis* of the sold variants of a product indicates which variant is accepted by the market and which can be eliminated. *Variant management* that uses this information can reduce costs enormously (Schönsleben 1998). Logistics also advises sales, purchasing and production in the decision as to which products and orders should be *made to stock* or *made to order*. Important contributions of logistics to the development and market introduction of new products are the design and selection of *packages* and *load carriers*.

14.4.1 Product and Packaging Logistics

Product and packaging logistics ensures that,

- materials and packages of the product can be recycled, are reusable or can be disposed in a cost-efficient manner,
- size of the product and dimensions of the package harmonize with load carriers and transport means, in order to minimize volume losses, to enable efficient storing and transport and to minimize package return,
- the costs for handling, storing, loading and unloading and for filling of the shelves in the outlets are minimal

Mixed pallets, *partly filled load units* and *single item order picking* should be avoided. Distribution costs can be reduced by minimal handling and by load units which are adapted to the dimensions of the sales units and the size of the shipments (see *Chap. 12*).

Time consuming commissioning, sorting and handling of small sales units are avoided by presenting the articles at the *point of sales* in larger units. Within the whole supply chain only these larger logistic units are handled. The *fine picking* of single sales units is left to the customer. Optimal *trays*, *displays* and full *article pallets* are the secret of *retail logistics*, as can be observed at the most profitable retail chains, such as *Tesco* in England and *ALDI* in Germany (Kotzab and Bjerre 2005).

14.4.2 Trays and Displays

Trays and displays are special logistic units for the presentation of sales units in outlets, markets, and other sales locations.

Trays are packages of several sales units of the *same article* for the presentation in a sales outlet. They consist of flat saucers, cardboards or ground bottoms, on which the article units, bags or blister packages are placed side by side. In order to protect them during handling and transport, trays may have a cover that is removed after it has been placed in the shelf. Typical dimensions of trays are 200×200, 200×400, 400×400 and 400×600 mm, which fit optimal to the standard measures of *Euro-Pallets*.

Displays are load units containing *different articles* for *sales promotions*. Customary load carriers for displays are *half-size Euro-pallets* with the basic measures 600×800 mm, also called *Chep-Pallets*. The article units are protected and kept together by cardboard parts, which can be printed in an eye-catching manner. The advantage of displays is that they are sent out to all sales outlets of different retailers with the same design. Disadvantages are the costs for scheduling, packing material and build up of the displays.

14.4.3 Full Article Pallets

The biggest logistic units for the sales presentation of consumables are pallets completely filled with the same article units. They are placed directly on the floor in the retail store. The customers serve themselves by picking the article units from the pallet. Full article pallets are the most efficient way to transport goods from the output of production to the *point of sales* without handling single article units.

Full article pallets are common in trade for fast moving consumables and promotions. They can also be used in production for the procurement of parts and material (see *Chap. 12*).

14.5 Delivery Service and Logistic Quality

In some industries, service, availability and ability to deliver are the most important *competitive advantages* (Bowersox et.al. 2007; Pfohl 1990). The *delivery service* includes:

adequate availability of storekeeping articles
punctual delivery for customer ordered goods
competitive delivery times
keeping promised delivery dates. (14.2)

Usually, the sales department sets the standards for the *service level* and the required logistic quality. The standards should be based on customer requirements and the competitive situation. *Benchmarks* for the *logistic quality* of a company are the quality standards that logistics, purchasing and sales jointly agree upon. Any deviations from these standards are *quality defects*, which should be reported periodically to purchasing and sales by *logistic controlling*.

Furthermore, logistics can suggest cost neutral service improvements and indicate the additional costs for service improvements that have to be charged to the customer or to be incorporated in the prices. Logistics should prevent that sales offers a *logistic over-kill service*, such as a nation-wide 24-hour-delivery service, that is needed and appreciated only by a small number of customers.

14.6 Sales Channels and Distribution Structure

Possible sales channels for the products and services of a company to the different groups of customers are (Coughlan et al. 2006):

- wholesale to retail
- key account manager to retail chains
- direct sales to *original-equipment-manufacturers* (OEM)
- factory or company owned outlets
- regional branches to local re-sellers and end-users
- organization of independent retailers or franchise partners
- *mail order companies* and *internet-sellers*

Physical distribution includes storing, handling and transport of finished goods from own production or of an externally sourced retail assortment to the points where customers take them over. As outlined in detail in *Chap. 21*, the *structure of the distribution network*, which is made up by logistic centers, storage locations, transshipment stations and transport connections, is determined by the *terms of delivery*, such as *ex works* or *free to door*, the quantities to be delivered, the distances and the delivery times.

The *distribution network* of many companies is traditionally linked with the *sales network*. Distribution areas and delivery regions are congruent with historically grown sales territories and sales regions. Stores and depots are often located at the sales bases. Tasks and objectives of logistics, however, differ from that of sales and marketing. That leads to the *strategic rules*:

- optimal distribution areas are not congruent with optimal sales regions
- sales channels and delivery chains do not have to be parallel or connected
- optimal locations of logistic centers, transshipment points and regional depots are not identical with the optimal locations of the sales force

Just as the destination of a delivery must not be identical with the location of the purchase department of the customer, it is not necessary that sales locations, offices or outlets are located at the same place as stores, transshipment points and dispatch stations. An optimal European-wide sales organization is generally organized in countries or language areas, while an optimal *European distribution network* is made up of transnational logistic regions that are adjusted to the metropolitan areas and industrial clusters, and take into account the main traffic routes (see *Fig. 21.20*).

14.7 Price Calculation and Logistic Costs

For the calculation of list prices and projects, the manufacturing-, sourcing- and order-costs, and the article- and order-logistic costs must be known. For negotiations with suppliers and customers, purchasing and sales should know the delivery costs, and take into account the cost difference between delivery *ex works* and *free to door*.

The calculation of correct article-and order-costs is only possible, if all *logistic master data* are available (see *Chap. 7*). Today, still only a few companies have correct and complete logistic master data at their disposal and are able to calculate their logistic costs. In many companies, handling, storing and transport costs are still taken into account by a general surcharge, which is a certain percentage of the goods costs. This can cause order losses due to too high logistic costs as well as wrong sales decisions and too low prices that do not cover the total costs.

Qualified contributions of logistics to *pricing* require competent *logistic costing and controlling* that registers the logistic, setup and performance costs within the supply chain and allocates them to the products and services dependent on consumption and utilization (see *Chaps. 6 and 7*).

14.8 Internal Logistic Services

As outlined in *Sect. 2.9*, the *core competencies of company logistics* are

1. Planning and implementation of an integrated logistic network between suppliers, production, sales locations and customers
2. Organization, scheduling and control of optimal replenishment, production and distribution processes within this network

This does not mean that company logistics has to operate itself all stores, logistic centers, transport means and freight systems. Some logistic activities can be left to suppliers and customers, others can be outsourced to *logistic service providers* (see *Chap. 22*). In any case, company logistics remains responsible for reaching the specific logistic objectives (see *Sect. 3.4*).

Where company logistics is not directly involved, it has to assist, consult, help and advise other departments of the company in all questions of logistics. This includes distribution of information material, spare part logistics, point of sales logistics, and logistic consulting of sales, purchasing and customers.

14.8.1 Distribution of Information Material

Documentation on products and spare parts, samples, advertising material, brochures, catalogues, and pricelists are necessary to win customers and to inform them about products and services. Share holders and business partners are informed on the situation of the company by quarterly and annual reports.

Frequency, type, and required availability of the information material determine the storing and distribution, which are neither core competence of sales nor of company logistics. If the distribution of information material cannot be integrated in the distribution of finished goods, it should be outsourced to specialized service providers.

14.8.2 Spare Part Logistics and After-Sales Services

Companies that produce sophisticated technical products, such as vehicles, electronic devices or machines, need an efficient spare part logistics and service organization. In a growing number of industries, availability and high speed provision of spare parts are important arguments for purchasing and sales. Some companies, such as the manufacturers of *elevators* and of *printers*, generate their profits mainly by spare parts and after-sales services.

If after-sales service is integrative part of selling, it should be organized and performed in responsibility of the sales department. Otherwise, after-sales services, spare part distribution and the logistics of repair shops are tasks of company logistics. They must often meet special requirements, such has high availability and extremely short replenishment times (Ihde et.al. 1989).

14.8.3 Point of Sales Logistics

Category management, presentation and explanation of products and merchandize and the operative sales activities around the *point of sale* (*POS*) and in the markets, shops and outlets are core competencies of sales and marketing. But also company logistics starts at the POS. Hence, condition for the success of local sales efforts is professional *POS-logistics*. Its tasks are:

- inventory management of storekeeping articles
- organization and technology of replenishment
- organization of goods receiving and control

- logistically optimal placing and presentation of the articles
- space saving storing of reserve stocks
- replenishment of sales stocks by own staff or by external *shelf jobbers*
- organization of the flow of goods and information around the cash-desk
- *reverse logistics* of empties, load carriers and package material
- logistics of returned goods and complaints
- organization of *home delivery service* for customers

Several investigations of the activities in the department stores, markets and outlets of retailers revealed that the employees are occupied between 30 and 40% of their time by logistic tasks. Additional time is needed for administrative activities and payment processes. Selling and consulting of customers make up only about 35% of the time and seem to be of minor importance. These figures indicate improvement potentials for the POS-logistics, which triggers the total logistic system of the company (see *Sect. 21.12*).

14.8.4 Internal Logistic Consulting

Company logistics is a cross-sectional function. It has to organize and control all logistic processes from the supplier of the supplier up to the final customer. Although company logistics cannot act operationally in all sections of the supply and delivery chains, it should at least offer advice and assistance to other departments as well as to customers and suppliers. The internal logistic consulting includes:

- advising sales and customers in logistic issues, such as delivery times, frequencies and dates, and inventory management (ECR and CRP)
- assisting sales when negotiating *logistic conditions* and *logistic discounts*
- supporting sales with planning of traveling routes of salesmen
- consulting purchasing and suppliers when establishing and organizing supply chains, in particular for new products and relationships
- development of *logistic standard conditions* for purchasing and sales
- calculation of logistic costs and support of pricing
- logistic consulting of top management for co-operations and acquisitions

The employed logisticians of a company are not always available or not qualified for all these tasks. Hence, for a big project or for special problems, it is opportune to employ logistically competent *external consultants*.

Part II
Systems, Networks and Operations

Chapter 15
Logistic Networks and Systems

The logistic network of a company is part of the *global logistic network* that is made up of the networks of forwarders, railways, airlines and shipping companies and of industrial enterprises, trading companies and service providers. The global logistic network has many owners and users. It serves different purposes and several interests (see *Fig. 0.1* and *15.1*).

Central tasks of *network management* are to delimit the logistic network of the company and to organize its connections to the networks of suppliers, customers and service providers. For this purpose, management has to decide, which logistic tasks can be left to suppliers and customers, which should be performed by the company itself and which are better outsourced to *logistic service providers*. The boundaries of the company logistic network depend on the core competencies and on the importance of logistics for the business. In some cases, e.g. for a *car manufacturer* with a logistic network as shown in *Fig. 1.15*, the network extends from the customers' customer to the suppliers' supplier.

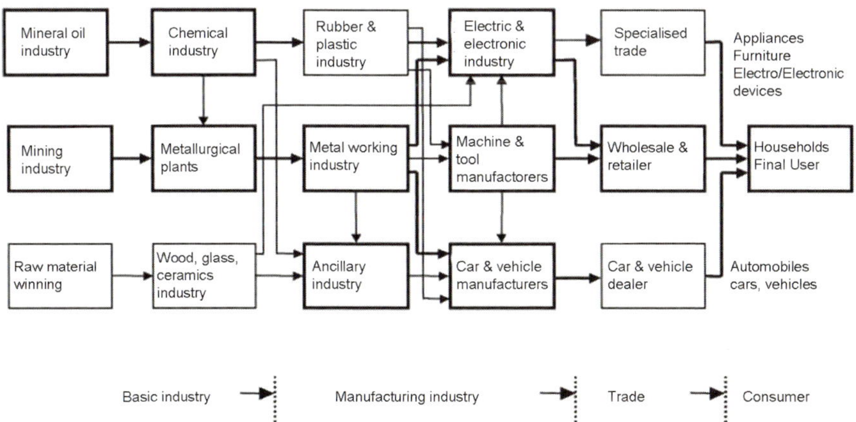

Fig. 15.1 Logistic network and supply chains for utility goods

DOI 10.1007/978-3-540-68652-1_15, © Springer-Verlag Berlin Heidelberg 2009

As explained in *Sect. 2.9*, the tasks of *logistic network management* (LNM), also called *supply chain management* (SCM), are (Chopra and Meindl 2007; Kuhn and Hellingrath 2002; Schönsleben 1998; Simchi-Levi et al. 2008):

- *Strategic Logistic Management*: In order to cope with *future demand*, systems are *planned*, *organized*, *set up* and *linked* to an optimal logistic network.
- *Operative Logistic Management*: In order to execute *current orders* at lowest costs, the available supply chains and resources are *scheduled* and *operated* efficiently.

In *Part I* of this book, the general *principles* and *strategies* of modern logistics and the organizational, technical, and commercial *options* of supply chain management have been described. In addition, the methods and strategies for scheduling orders and inventories were developed. In the following *Part II*, the general rules and strategies of *Part I* are applied to logistic technology and network management. *Logistic technology* comprises design, dimensioning and optimization of *storage*, *commissioning* and *transport systems*, layout principles for *logistic sites* and strategies for *production logistics*. *Network management* is concerned with the design of *dynamic networks* and the selection of *optimal supply chains*.

After the topics and methods of logistic technology and network management have been outlined, the characteristics of *logistic service providers* and the procedure for their employment will be described. The last chapter deals with the role and importance of *people in logistics*. It closes with an outlook on the tasks and challenges for logistics of the future.

15.1 Dynamic Networks

Logistic networks and the flows of physical objects within these networks are the *action fields of practical logistics* and the *research areas of theoretical logistics*. For a long time theoretical logistics has considered only stationary flows and stable networks, although the importance of *industrial dynamics* had already been emphasized by *Forrester* in 1961 (Forrester 1961).

The independent decisions of consumers and companies cause *stochastic fluctuations* of the order flows and material flows. Varying customer behavior, technical development and changing demand generate *systematic variations* of these flows. The dynamic of the flows determines and changes the logistic networks. In the short term, swelling, shifting and decreasing flows force the actors to adjust the static and dynamic capacities of the logistic stations and transport connections. In the long term, the changes of the flows compel the *redesign of the network* structure and the planning of new systems.

The research into dynamic networks and the investigation of its structures, processes, laws and principles are still in an initial stage. The strategies for planning, scheduling and operating dynamic networks have not yet been studied systematically. Possibilities and limitations of technique, organization and scheduling and the interactions between technology and economy of dynamic networks are still quite unknown.

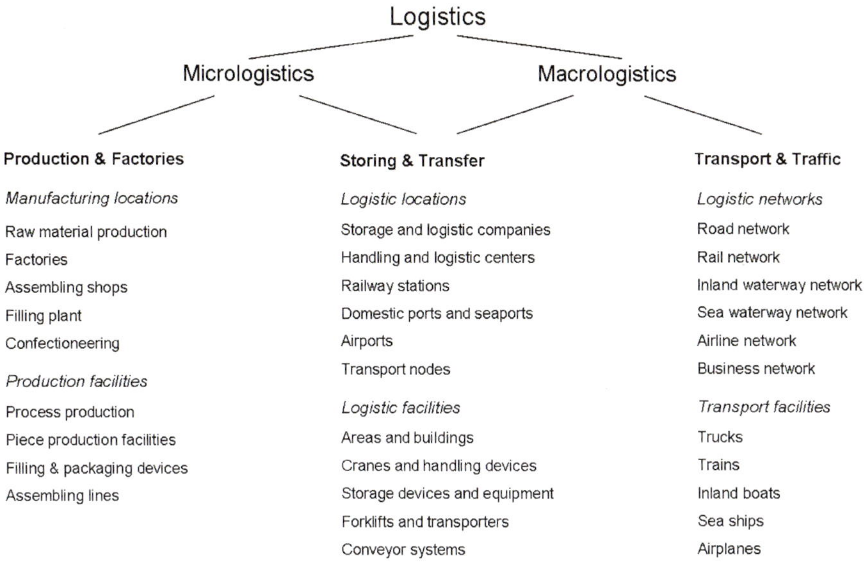

Fig. 15.2 Action fields and research areas of logistics

Similar to architecture or informatics, logistics is an applied science. It has to offer answers to questions of practical relevance and should develop methods to solve actual problems. Whoever wants to achieve certain goals in logistics must have a clear picture of the *action fields* shown in *Fig. 15.2*. Managers need plans of the network and systems of the company and records of the available resources, capacities, distances and interfaces of the stations and transport connections. They should know the logistic services, their costs and prices, and observe the *logistic markets*. In some field of logistics, the required information is available quite completely and accurately, in other fields it is still lacking.

15.2 Hierarchy of Logistic Systems

A *logistic network* is a number of sources, sinks and intermediate stations which are linked by transport connections and passed by physical objects. The *material flows* in the logistic network are initiated and controlled by *data flows*. Some data run together with the material flows, others are conveyed by separate *data networks*.

Analogously to *Internet*, *Extranet* and *Intranet* the logistic networks can be differentiated into *Intralog*, *Extralog* and *Interlog*:

- *Intralog* is the internal logistic network of a production site or logistic station.
- *Extralog* is the external logistic network spanned between the production sites and logistic stations of the company, its suppliers and its customers.
- *Interlog* is the connection of the logistic networks of all households, companies, service providers and other actors of an economy.

Generally, a logistic network is multi-functional and composed of sub- and part-systems with different and special functions.

Part-systems of the *Intralog* are machines, robots, stores, commissioning systems and handling stations, which are connected by cranes, conveyors and vehicles. Planning, building, connecting and operating these part-systems are tasks of *internal logistics* or *Intralogistics* (Ackerman et al. 1997; Arnold 1998; Apple 1972; Bode and Preuß 2004; Frazelle and Apple 1994; Günthner and Heptner 2007; Hung and Fisk 1984; Meller et al. 2004).

Subsystems of the Extralog are the *supply networks* for material and parts, the *distribution networks* for finished products, the *recycling networks* and the Intralogs, which are connected by external transport systems. These subsystems and their relations are subjects of *external logistics* or *Extralogistic.*

The Extralogs of all households, companies and logistic service providers of the world are subsystems of the global Interlog. The analysis and optimization of the Interlog are subjects of *general logistics* and *strategic logistic research.*

A closer look shows that each subsystem or element of a logistic network consists of subsystems and elements and so on. This leads to the *logistic systems hierarchy*:

1. The Interlog, i.e. the global logistic network, consists of
2. national and regional logistic networks composed of the
3. Extralogs of households, companies and service providers connecting the
4. Intralogs of consumption, production and logistic sites consisting of
5. handling, storing, commissioning, conveyor and vehicle systems made up by
6. machines and robots which consist of parts, components or modules.

It is not task of logistics to deal with the lowest level of this hierarchy, but of other disciplines, such as materials handling, conveyor engineering, vehicle construction, ship building, aircraft design and road construction. As outlined in *Sect. 3.11*, these technical disciplines focus on the construction and development of the specific equipment, robots and machine systems.

The logistic systems hierarchy is similar to the *packaging hierarchy* of logistic units (see *Sect. 12.1* and *Fig. 12.2*). It reflects the *self-similarity*, which is typical for *complex systems* (Simon 1962). To the hierarchy of the operative logistic system corresponds the *hierarchy of the control system*. For example, *Fig. 18.6* shows the levels of a hierarchical control system for a transport network. By the control system the different functions of the subsystems and elements are managed, coordinated and controlled in order to execute the *current orders* and requirements efficiently, correctly and reliable.

The analysis and design of the system hierarchy and its structures, of the subsystems and elements, and of the corresponding control systems are key activities of *strategic network management.* It has to differ between the *horizontal integration* of systems of the same hierarchy level and the *vertical integration* of systems of different hierarchical levels. The orders come from vertically integrated systems of higher levels or from horizontally integrated systems of the same level. The vertically integrated systems of lower levels determine the *limit performances* and *buffer capacities* of a logistic system. They also receive their orders from systems of a higher or the same level.

15.3 System Planning and System Optimization

Central task of *system planning* is to develop a system that is capable to fulfill the *performance demand* at minimal costs and keeps the given constraints. *System optimization* has either to improve the static and dynamic *capacities* or to reduce the *performance costs* of an existing system.

15.3.1 Demand and Capacities

The actual or expected demand results from the number and content of orders placed by operators, users or customers:

- The orders to a system specify the *kind*, *quantity* and *quality* of the required performances, products and services and prescribe the *date* and *location* of the delivery.

Orders to a production system are *processing orders*, *manufacturing orders* or *assembling orders*. Orders to logistic systems can be *pick up orders*, *transport orders*, *storing orders*, *handling orders*, *commissioning orders* or *delivery orders*.

A required or anticipated *order entry rate* λ_O [Ord/PE] with a mean *order content* m_O [LU/Ord], measured in *performance units* [PU] or *logistic units* [LU], leads to the *performance rate, throughput* or *material flow*:

$$\lambda_{LU} = m_O \cdot \lambda_O \quad [PE/PE \text{ or } LU/PE]. \tag{15.1}$$

A randomly fluctuating order entry causes a *stochastic flow*. If the order entry is time dependent, the throughput or material flow becomes *dynamic*. The order entry rate, throughput, performance rates and material flows constitute the *dynamic demand*.

Stochastically fluctuating dynamic flows can generate *order backlogs* and *material stocks* for which the system has to provide sufficient buffer capacity. Since *order scheduling* has the option to execute orders in advance, the *static demand* for storage and buffer capacities of the system results from the dynamic demand and the *scheduling strategies* for orders and inventory.

Static demand and dynamic demand determine what should be achieved by a system. The demand has to be compared with the *static* and *dynamic capacities*, which measure what a certain system can achieve under given conditions. As outlined in *Chap. 13*, the *dynamic capacity* of a system is determined by the *limit performances* μ_i [LU/PE] of the system elements SE_i and by the *operating strategies*. The *static capacity* of the system depends on the *buffers* and *storage capacities* C_i [LU] of the elements, and on the *scheduling strategies* for the orders and inventories.

In addition to the internal influence factors, the capacities of a system are affected by *external factors*, such as size and structure of the orders and the stochastic and dynamic of the order flow. All internal and external influence factors must be known in order to optimize and schedule an existing system or to plan and realize a new system.

15.3.2 Performance Costs

The *performance costs* k_{LU} [€/PU] are the *operating costs* $K(\lambda_{LU})$ [€/PE] related to the *performance rate* λ_{LU} [PU/PE]:

$$k_{LU} = K(\lambda_{LU})/\lambda_{LU} \quad [€/PU]. \tag{15.2}$$

If the system produces, processes, handles or transfers several *partial flows* λ_{LUr} [PU_r/PE] of logistic or performance units PU_r, the total operating costs can be split up into a sum $K = \sum K_r(\lambda_{LUr})$ of partial operating costs $K_r(\lambda_{LUr})$ due to the utilization of the system elements. With relations corresponding to (15.2), the *partial performance costs* k_i [PU_i/PE] can be calculated from the partial operating costs. As explained in *Chap. 6*, the operating costs are determined by the depreciations, interests and maintenance costs for buildings, machines and equipment, and by the costs for personnel, material and energy. The operating costs can be split into *fixed costs* $K_{fix}(\mu_{LU})$ and *variable costs* $K_{var}(\lambda_{LU})$, that depend on the performance rate λ_{LU} [PU/PE]. The fixed costs are independent of the current performance, but determined by the limit performances μ_{LU} [LU/PE].

15.3.3 Stepwise Improvement and Iterative Planning

With decreasing *utilization* $\rho = \lambda_{LU}/\mu_{LU}$ [%], the performance costs of a system increase due to the remaining fixed costs. Systems with longer lasting underutilization are oversized and generate high costs. In order to avoid this, the system should be planned and optimized in the following steps:

1. Different system solutions are designed, dimensioned and optimized for the expected *mean demand*. Out of these, the most efficient solutions with the lowest inventory and performance costs are selected by the methods described in *Sect. 3.9*.
2. The best of these solutions are adjusted to the expected *dynamic* of the demand. To cope with the demand of the *peak hour* of a regular *peak day*, the operating strategies, limit performances and storage capacities are adapted. The investment, performance costs and return on investment of the adapted systems are recalculated and compared again.
3. The most economic of these solutions is adjusted to the *stochastic* of the demand. By a *function and performance analysis* as described in *Sect. 13.7*, the limit performances, buffer capacities and operating strategies are adjusted to the stochastic demand. The result is the technically and economically *optimal system*.

This *stepwise improvement* in logistics is similar to *perturbation calculation* in physics, by which a problem that has no explicit mathematical solution can be solved to any required accuracy. In the first step, a solution is calculated for the main influence factors disregarding any higher order disturbances. In the next steps, further influence factors are taken into account in the sequence of their relative importance. This is done until a solution with the required accuracy has been reached.

The stepwise improvement is central part of the *iterative planning and optimization* of logistic networks, systems and subsystems which is shown in *Fig. 15.3*. After

Fig. 15.3 Iterative planning and optimization of logistic networks and systems

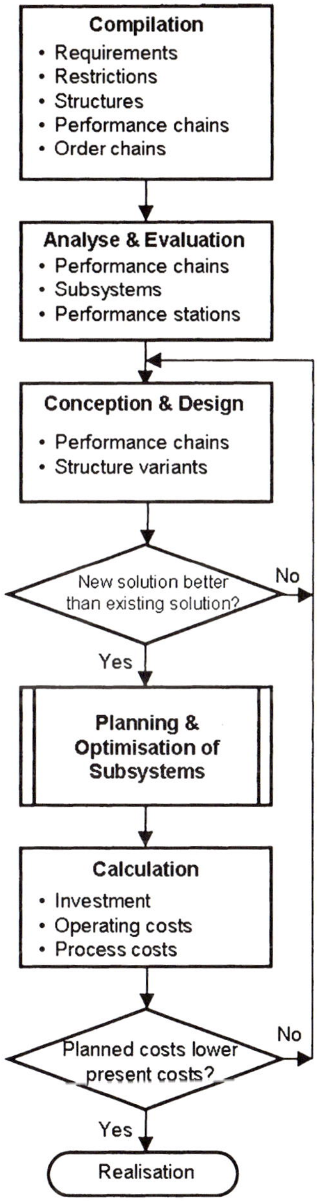

gathering the static and dynamic demand and the constraints, the next step is an analysis and evaluation of the available order and performance chains and of the existing systems. The further steps depend on the hierarchical position of the system.

If at the end of *network planning* no solution results, which fulfils the demand at significantly lower costs than the current state, the subsystems, which initially have been assumed as fixed, are analyzed, optimized and if necessary redesigned. For the planning and optimization of a subsystem, the same procedure as on the higher level can be applied.

15.3.4 System Planning and Design Principles

The following *system planning principles* help to reach with adequate effort in quite short time the optimal solution:

- *Structure follows processes*: Before the structure of the network is planned the performance processes and logistic chains have to be designed.
- *Data flow follows material flow and performance processes*: Needed material flow through the network and required performance processes determine the design of order processes and information flows.
- *Informatics follows logistics*: The logistic chains through the network and systems and the necessary operating and scheduling strategies determine the IT-systems, such as APS, ERP, PPS and MIS, and not vice versa.

The logistic functionalities of standard software are generally limited. In order to exhaust the potentials of logistics, the operation and scheduling strategies and other logistical IT-requirements must be specified in a manual before the architecture of the IT-system is designed and the software is selected or programmed.

Complex systems with many closely connected subsystems and elements are difficult to control and trouble prone. Even the most accurate calculations and highly sophisticated simulations cannot significantly improve a complex system. A reduction of the *complexity* of large networks and systems is achievable by the basic *principles of system design* (Gudehus 1975/II; Simon 1962):

- *Simplicity principle*: In many cases, the simplest solution with the shortest supply, delivery and performance chains, the smallest number of parallel elements and subsystems, and the lowest automation is the best solution. In any case it is *benchmark* for higher sophisticated solutions.
- *Decoupling principle*: If the total system is outlined and dimensioned in a way, that under normal conditions backlogs and feedbacks of the subsystems are improbable, the decoupled subsystems can then designed, optimized and scheduled separately.
- *Approximation principle*: The formulas and calculations for dimensioning, optimization and scheduling must not be more accurate than the planning data, input values and demand figures.

Two further design principles for logistic networks and systems result from the fact that transport, storing and handling of physical goods are far more expensive than transfer, storing and processing of data and information. These are:

- *Dominance of material flow*: In Intralogistics, the flows of physical goods and material, not the data flows, determine the optimal logistic system.
- *Dominance of logistic chains*: In Extralogistics, the operative supply, performance and delivery chains, not the order flows and administrative processes, determine the optimal logistic network of a company.

As shown in *Fig. 8.1*, an order flow ends after passing a chain of administrative stations in an operative station, where the order execution starts. Here the order chain is linked with the performance or delivery chain. The *decoupling station* is the interface between the anonymous section and the order specific section of the supply chain (see *Sect. 8.6*).

15.3.5 System Optimization and Greenfield Solution

Principally, there exist two extreme options to optimize and redesign the logistic network of a company:

- *System optimization*: Keeping the existing structures and systems as far as possible, the capacities are adapted to the changing demand and the performance costs are reduced primarily by organizational efforts, optimization of processes and better strategies. Investments are kept as low as possible.
- *Greenfield solution*: After thorough redesign of the processes and structures, optimal systems are planned and new operations are built on the green field at optimal locations using modern technology without restrictions.

The greenfield solution should always be compared with the optimized state and not with the existing state. Quite often, the optimization of the existing processes and systems reduces the costs to a satisfying level and can cope with an changing demand. If the resulting performance costs of the improved system are only slightly higher, the greenfield solution is not attractive.

Quite often, the difference of the fixed costs for the Greenfield solution and the existing system is higher than the achievable reduction of the variable costs. This happens, e.g., if the existing buildings, plants and machines are written off completely but can still be used. In such cases, the greenfield solution can be realized if at all only stepwise. Even if the greenfield solution is not realized, it is a reference for the development of the company logistics. The capacities and costs of the optimal solution are *analytical benchmarks* for the existing systems.

Chapter 16
Storage Systems

Storage systems and their operation are highly underestimated areas of logistics. Due to lack of knowledge, many stores are wrongly planned or not optimally operated. This causes underoccupation or *shortages* of storeplaces and underutilization or *bottlenecks* of storage devices. Comparatively few problems are originated by *storage technique*, which is well approved and reliable since many years. Another problem, caused by lacking knowledge of the *cost drivers of storing*, is the uneconomic use of the available stores.

The design, technique and operating strategies for storage systems can be derived from the basic *task of storing* which is:

- Keeping of stocks and provision of goods.

This task involves unavoidably in- and out-storing, since the complete *storing process* consists of the part-processes:

1. *in-storing* of the storage unit by a *storage device*
2. *keeping* and *provision* for access of the storage unit in a *storeplace*
3. *out-storing* or *retrieval* of the storage unit by the *storage device*

The *storage unit* can be a single article, a pallet, bin or container, or another load carrier filled with discrete articles or loose goods. In the simplest case, the *storage device* is a person. For many or heavy units it is a fork lift truck, stacker crane or another *storage- and retrieval unit*. These S/R-units can provide an ordered sequence of storage units. The ordered *provision of full load units* is the simplest realization of the *commissioning process*, which will be studied in the next chapter. In this chapter only *unit load stores* are considered. They can also provide and restore *access units* for *dynamic commissioning*.

The design, technique and strategies of storage systems are the topics of *storage technology*. The combinatorial possibilities for the layout and arrangement of storeplaces lead to the different *storage types*. The *storage technique* is determined by the storage units and by the technical realizations of places and racks, by the storage device and the in- and outward transport system. Decisive for the *storage control* are the *storage strategies* and the *inventory management*. For the optimal

T. Gudehus, H.Kotzab, *Comprehensive Logistics*,
DOI 10.1007/978-3-540-68652-1_16,

selection between stores with different cost rates, suitable *store-utilization strategies* are needed.

In this chapter the storage requirements are specified, the different methods and techniques of storing are described and strategies for the design, dimensioning, selection and operation of storage systems for discrete units are developed. The technical planning of bunkers, silos, blending beds and disposal sites for bulk material and of tank stores for liquids and gases are not dealt with here.

The investment and operating costs of a storage system depend on the prices and costs for the part systems and components. The operating costs can be separated into *static storage costs* for the storeplaces and *dynamic storage costs* for the in- and out-storing. Based on model calculations, the main influence factors on the static and dynamic storage costs are analyzed and compared for different standard pallet storage systems. In the last section of this chapter, the consequences for sourcing storage services are outlined.

16.1 Storage Requirements

For the planning of a new storage system, as well as for the outsourcing of storage services, the future storage requirements must be known. They consist of the *dynamic storage demand* or *throughput requirements* and the *static storage demand* or *storekeeping requirements*.

As outlined in *Chap. 11.3*, the static and dynamic storage demand results from the *order requirements* of inventory management. The dynamic storage demand can be calculated from the *supply and delivery orders* (see *Sect. 3.7*). The stock levels result from the replenishment quantities and safety stocks.

16.1.1 Order Requirements

The storage requirements are generated by the *storage orders* of internal or external users:

- An *in-storing order* [I-Ord] requires to pick up and to store a *stock quantity* M_I [SU/I-Ord] of *storage units* [SU].
- An *out-storing order* [O-Ord] requires to store out and to provide a *demand quantity* M_O [SU/O-Ord] of a *storage good*.

The storage good can be a *storekeeping article* of an assortment, which is stored permanently to serve a longer lasting demand, or a *promotion article*, which is stored for a sales promotion for limited time. Other storage goods are the load units of a shipment, which are buffered until delivery, or parts of a customer order, which have been produced, procured or picked in advance. From a logistical point of view, also transport units, such as waiting swap bodies and wagons, and vehicles, such as parking cars, are storage goods. Also waiting animals and people can be considered as storage goods.

The storage units are either single items or *load units*. If the storage units are not prescribed, the first step of storage planning is to select load carriers and to dimension storage units, which are optimal for the storing process. This can be achieved by the *packing- and filling strategies* of *Chap. 12*. With the capacity of the storage unit the number of storage units which are necessary to keep the in- and out-storing quantities is calculated from the order quantities with the formulas of *Sect. 12.5*.

The storage units of an in-storing order are collected from *pickup points* located in an input buffer area. The storage units of out-storing orders must be provided at certain *transfer points* in a commissioning zone, dispatch area or another output buffer area. The pickup points and transfer points are the *spatial interfaces* of the storage system with the adjacent logistic systems.

In addition to the *spatial requirements*, certain *time requirements* must be kept, such as

- the *retrieval time* T_{SU} [s/SU] for out-storing one single storage unit
- the *providing time* T_{Ord} [s/O-Ord] for out-storing *all* units of an order

The retrieval time is the sum of the cycle time of the storage device for the out-storing process and the transport time from the store to the transfer point. Due to the varying distances of the single storeplaces to the transfer points, the single times fluctuate randomly around a mean retrieval time.

If only one storage device is available, the providing time is the sum of the single retrieval times for the ordered units. The providing time can be considerably reduced by several storage devices which operate in parallel. For the calculation of the *mean retrieval time* and the *mean providing time*, also the *waiting times* caused by randomly incoming orders and stochastically varying cycle times have to be taken into account (see *Sect. 13.5*).

The time necessary to store out the whole store content is the *storage clearing time*. In some cases, this time is a critical requirement. For example, the total clearing time for commercially used *car parking systems* should be not much longer than 30 min.

16.1.2 Dynamic Storage Demand

The dynamic storage demand determines the number of storage devices, the limit performance of the in- and output transport system and the personnel. The in-storing and out-storing *order flows* λ_{IOrd} [I-Ord/PE] and λ_{OOrd}[O-Ord/PE] with the mean order quantities $M_I$and M_O cause the

- mean in-storing flow

$$\lambda_{Im} = M_I \cdot \lambda_{IOrd} \quad [SU/PE] \tag{16.1}$$

- mean out-storing flow

$$\lambda_{Om} = M_O \cdot \lambda_{OOrd} \qquad [SU/PE] \tag{16.2}$$

As long as the stock level does not change during a longer period of time, the mean out-storing flow and the mean in-storing flow equal the *storage throughput*:

$$\lambda_T = \lambda_{Im} = \lambda_{Om} \quad [SU/PE]. \tag{16.3}$$

During the hours of an operating day and for periods where the stock level changes, the in-storing flow and out-storing flow can be different and deviate from the mean throughput. The actual in-storing and out-storing flows are restricted by the limit performance of the storage system. Therefore, the flows of the *peak day* in the planning period, which must be served within the *peak hour*, determine the *dynamic storage demand*:

- The *in-storing demand* is the maximal hourly in-storing flow to be served

$$\lambda_{\mathrm{I}} \quad [\mathrm{SU/h}] \tag{16.4}$$

- The *out-storing demand* is the maximal hourly out-storing flow to be served

$$\lambda_{\mathrm{O}} \quad [\mathrm{SU/h}]. \tag{16.5}$$

The dynamic storage demand can be derived from the annual throughput by division through the number of *operating hours* of the year and multiplication with a *throughput peak factor* f_{Tpeak}, which takes into account the seasonal and daily variations of the input and output. From the dependence on the operating time results the importance of *time management* for storage planning, since:

► Dynamic storage demand and the required storage system are influenced by the length of the operating times.

By optimal time management as outlined in *Sect. 8.2*, e.g. by adaptive operating times, an extreme storaging demand can be smoothed out or avoided.

During peak hours, in- and out-storing orders are waiting in an *order buffer*. This buffer can be used to organize as many combined in-storing and out-storing cycles as possible in order to maximize the throughput performance of the storage devices. With *combined storage cycles*, the dynamic demand is reduced to the sum of:

► the *combined in-out-storing demand*

$$\lambda_{\mathrm{IO}} = \mathrm{MIN}(\lambda_{\mathrm{I}}; \lambda_{\mathrm{O}}) \qquad [\mathrm{SU/h}] \tag{16.6}$$

► and the *separate in-storing demand*

$$\lambda_{\mathrm{Isep}} = \lambda_{\mathrm{I}} - \mathrm{MIN}(\lambda_{\mathrm{I}}; \lambda_{\mathrm{O}}) \quad [\mathrm{SU/h}] \tag{16.7}$$

► or the *separate out-storing demand*

$$\lambda_{\mathrm{Osep}} = \lambda_{\mathrm{O}} - \mathrm{MIN}(\lambda_{\mathrm{I}}; \lambda_{\mathrm{O}}) \quad [\mathrm{SU/h}] \tag{16.8}$$

This leads to the *general storage-planning rule*:

► Operating strategies can alter or reduce the effective storage demand.

If a storage system is used for in- and out-storing of full storage units and at the same time for providing and restoring of *access units* for *dynamic commissioning*, the combined demand (16.6) is increased by the *provision demand* λ_{P} [SU/h]. Since provision and restoring can normally be executed in combined cycles, the *total in- and out-storing demand* is:

$$\lambda_{\mathrm{IOtot}} = \mathrm{MIN}(\lambda_{\mathrm{I}}; \lambda_{\mathrm{O}}) + \lambda_{\mathrm{P}} \qquad [\mathrm{SU/h}]. \tag{16.9}$$

By *postponement* of less urgent in- and out-storing orders or by separation of the operating times for commissioning and shipment, an augmentation of the dynamic demand can be avoided. With optimal *time management*, the dynamic storage demand is no longer the sum, but at best only the maximum of in- and out-storing demand and provision demand.

16.1.3 Static Storage Demand

The static storage demand determines the layout, dimensions and number of the storeplaces. It is given by:

- the *number of articles* N_A [Art] to be stored at the same time
- the *dimensions* l_{Su}, b_{SU} and h_{SU} [mm], *mean weight* w_{SU} [kg/SU] and *maximal weight* w_{SUmax} [kg/SU] of the *storage units* [SU]
- the *mean stock* M_S [SU/Art] and *maximal stock* M_{Smax} [SU/Art] of storage units per article.

The articles of a homogenous assortment can be consolidated in the same storage units and stored in a *homogenous store* with equal storeplaces. For dimensioning a homogenous store, only the mean values of the static and dynamic demand for all storekeeping articles must be known.

If the articles differ very much in size, weight, throughput or stock level, it is necessary or opportune to use different storage units and to separate the assortment into homogenous groups of logistically similar articles. For the partial assortments, suitable load carriers must be selected and optimally dimensioned. Resulting are *storage article groups*, such as *bin articles*, *pallet articles* and *container articles*. The stocks of a heterogeneous assortment can be stored either separately in different homogenous storage systems or combined in a *heterogeneous storage system*. The heterogeneous storage system consists of universally usable storeplaces or of places with different dimensions.

For the static dimensioning of a storage system, it is necessary to differ between temporary stock and permanent stock (see *Sect. 11.1*):

- A *temporary stock* of a good is kept only for a limited time which is generally known already when the good is stored in. Such a *push-stock* is stored and provided for a definite future demand.
- A *permanent stock* of a storekeeping article is kept over a longer period of time to serve a permanent demand. The replenishment of such a *pull-stock* depends on the current stock and demand.

An example of temporary stock is a quantity of article units, which has been purchased or manufactured for a *sales promotion*. Typical time dependencies of temporary stocks are shown in the lowest diagram of *Fig. 11.1*. If a promotion is prepared for an article which is regularly storekeeping, further storage capacity for the temporary push-stock is required.

The time dependency of the stock of an article, which is scheduled due to the inventory strategies of *Chap. 11* in order to serve a longer lasting demand, has a *saw-tooth pattern* as shown in the *Figs. 11.4* and *11.23*. For a larger assortment of storekeeping articles, the mean and the maximal *stock of storage units* with capacity

C_{SU} can be calculated by formula (11.10) and (11.13) from the *safety stock* m_{safe} [CU] and the *mean replenishment quantity* m_R [CU] of the *consumption units* [CU].

Neglecting partly filled units, the *maximal number of storage units* [SU] of a storekeeping article is approximately the sum of the safety stock M_{safe} [SU/Art] and of the replenishment quantity M_R [SU/Art], both filled into storage units:

$$M_{Smax} = M_{safe} + M_R \qquad [SU/Art]. \tag{16.10}$$

The *mean number of storage units* of the article is approximately the sum of the safety stock and half of the replenishment quantity in storage units:

$$M_S = M_{safe} + M_R/2 \qquad [SU/Art]. \tag{16.11}$$

The *mean stock* of N_A storekeeping articles with a mean safety stock M_{safe} and a mean replenishment quantity M_R during longer time is:

$$M_{S\ tot} = N_A \cdot (M_{safe} + M_R/2). \qquad [SU]. \tag{16.12}$$

If storeplaces are permanently reserved for the maximal stock of the single articles, the sum of the maximal stocks for all articles determines the required number of storeplaces. That means:

- ▶ With *fixed storage order*, the mean *capacity demand* for N_A articles with mean safety stock M_{safe} and mean replenishment quantity M_R is

$$M_{S\ fix} = N_A \cdot (M_{safe} + M_R) \quad [SU]. \tag{16.13}$$

If empty storeplaces can be used for incoming storage units of any article, only the places for the safety stocks are permanently blocked, whereas on average only half of the replenishment units occupy storeplaces. That means:

- With *free storage order*, the effective *capacity demand* for N_A articles with mean safety stock M_{safe} and mean replenishment quantity M_R is

$$M_{S\ free} = N_A \cdot (M_{safe} + M_R/2) \quad [SU]. \tag{16.14}$$

This is a further example for the above rule that the effective demand depends on the operating strategy.

The current total stock $M_{S\ tot}(t)$ of an assortment of many articles fluctuates stochastically around the mean value (16.14). In addition, it may vary seasonally during a year. This leads to the *capacity dimensioning rule*:

- ▶ In order to prevent shortages of storeplaces, the capacity of a store with free storage order must be designed for the *effective total stock*

$$M_{S\ eff} = f_{Speak} \cdot f_{Sover} \cdot M_{S\ free} \quad [SU]. \tag{16.15}$$

The *stock peak factor* f_{Speak} takes into account the seasonal and weekly stock changes. It results from an analysis of the annual stock pattern as shown for example in *Fig. 3.4*. Due to the *square root law of logistics* of *Sect. 11.9*, the stock peak factor for an assortment of storekeeping articles with optimal replenishment is the square root of the *throughput peak factor* f_{Tpeak}.

The *overflow factor* f_{Sover} can be calculated by the *theorem of large numbers* of *Sect. 9.4*, since the stochastic fluctuations of the total stock of many articles with uncorrelated behavior are normal distributed. The standard deviation of the total stock depends on the standard deviations of the stocks of the single articles. If the safety stocks are significantly lower than the maximal stocks, which is normally the case, the variance of a stock with saw-tooth pattern and a maximum M_S is $s_S{}^2 = M_S{}^2/12$. From the variance of the article stocks results with relation (9.23) the *variance of the total stock* (Gudehus and Kunder 1972):

$$s_{S\ tot}{}^2 = N_A \cdot s_S{}^2 = M_{S\ tot}{}^2/12\,N_A \qquad [\text{SU}]. \tag{16.16}$$

Hence, in order to prevent an overflow of the store capacity with probability η_S, the capacity demand (16.14) must be increased by the *overflow factor*:

$$f_{S\ over}(\eta_S) = 1 + f_S(\eta_S)/\sqrt{12 N_A} \tag{16.17}$$

The *safety factor* $f_S(\eta_S)$ in the overflow factor (16.17) has been defined in *Sect. 9.4* and can be taken from *Fig. 5.4* or *Table 11.5*. For example, to achieve a *capacity safety* of 98% for an assortment of 100 storekeeping articles, the store must be designed with the overflow factor $f_{\text{Sover}} = 1 + 2.05/\sqrt{12 \cdot 100} = 1.06$, i.e. the store capacity must include a *breathing reserve* of 6% of the total stock. If the store contains only 10 storekeeping articles, the breathing reserve is 19% of the total stock.

The relations (16.17), (16.13) and (16.14) and the influence of sales promotions lead to the *store capacity rule*:

- The reduction of the effective capacity demand by a free storage order is diminished by higher safety stocks, by the breathing reserve, which is larger for smaller numbers of storekeeping articles, and by the temporary stocks of promotion articles or other goods.

Since this rule is generally neglected or unknown, the positive effect of the free storage order strategy is often overestimated. It never reaches the often claimed capacity reduction by a factor of 2.

16.1.4 Store Time and Stock Range

Besides the static and dynamic storage demand, the *store time* T_S [PE] is crucial for the design of a storage system and for the selection of the most economic store. The store time for a push-stock is already known or foreseeable, when the stock is stored in. For a pull-stock it is unknown when the replenishment arrives.

For any stock with steady consumption and supply, the *mean store time* can be calculated. This so called *stock range* is the quotient of the mean total stock divided by the mean throughput for a time of sufficient length:

$$T_S = M_{S\ tot}/\lambda_T \qquad [\text{PE}]. \tag{16.18}$$

The inverse of the stock range is the *stock rotation* $R_S = 1/T_S = \lambda_T/M_{S\ tot}$ [1/PE], also called *inventory turnover*. Due to relation (16.18), stock range and stock rotation can vary even for a constant total stock if the throughput changes.

Stores can be differentiated corresponding to the mean store time (*Sect. 11.1*):

- In a *sorter buffer,* article units are collected, sorted and kept for a short time in order to provide them in a required sequence.
- In a *short time store* or *buffer store,* goods are collected and kept for several hours or a few days.
- In a *supply store* or *warehouse,* quantities for a range of articles are stored for indeterminate time in order to serve a longer lasting demand.
- In a *longtime store* or *depot,* goods are kept for many weeks, months or even years.

Special storage systems are *disposal sites* which keep waste or discarded goods forever.

16.2 Storeplaces and Storage Types

For the specification of the dimensions and the set up of a storage system, the *standard store coordinates*

x-coordinate: horizontal direction parallel to the storage aisles
y-coordinate: vertical direction orthogonal to the storage aisles
z-coordinate: horizontal direction orthogonal to storage aisles

are used in storage technology. They are shown in *Fig. 16.1* for a high-bay store.

Any store consists of a number *storeplaces* [SP] with *place length* l_{SP} [mm] in x-direction parallel to the aisles, *place breadth* b_{SP} [mm] in y-direction perpendicular to the aisle and *place height* h_{SU} [mm] in the upright z-direction.

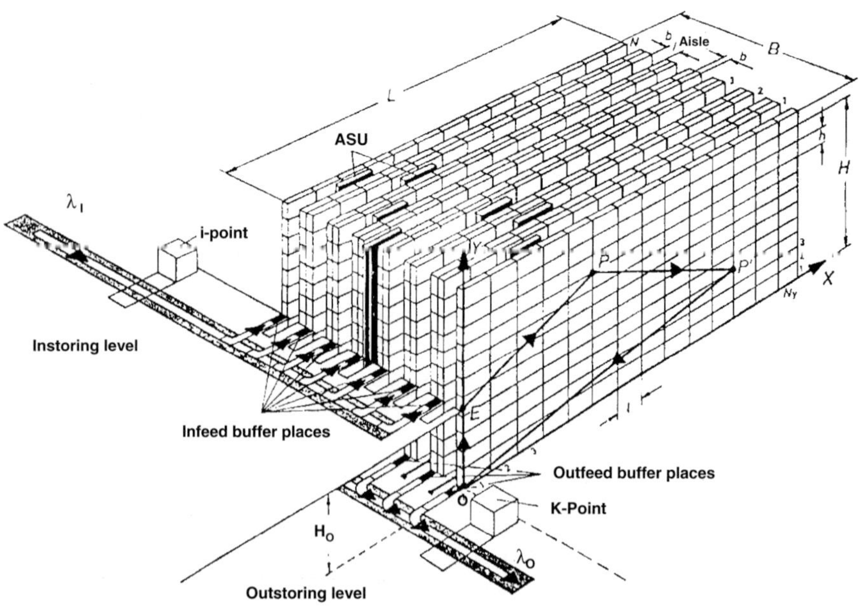

Fig. 16.1 Schematic diagram with store coordinates for a high-bay store

The *storeplace capacity* C_{SP} [SU/SP] is the maximal number of storage units which can be kept in one storeplace. It is the product

$$C_{SP} = C_x \cdot C_y \cdot C_z \qquad [SU/SP] \tag{16.19}$$

of pile length, stacking factor and pile depth:

- The *pile length* C_x is the maximal number of storage units that can be put side by side in x-direction of a storeplace.
- The *stacking factor* C_y is the maximal number of storage units which can be put on top of each other in a storeplace.
- The *pile depth* C_z is the maximal number of storage units that can be put behind each another in one place.

For a known storeplace capacity the number of storeplaces, which are needed to keep a number of storage units, can be calculated with the formulas of *Sect. 12.5*. A store with N_{SP} places and place capacity C_{SP} has the *total capacity*:

$$C_{S\ tot} = N_{SP} \cdot C_{SP} \qquad [SU] \tag{16.20}$$

The total store capacity is the upper limit for the static storage demand which can be fulfilled by an existing store. It also determines the *accessibility* of the storage units, which differs for:

- *Single-unit stores* with place capacity $C_{SP} = 1$ SU/SP for only one storage unit, where each unit is directly accessible.
- *Multi-unit stores* with place capacity $C_{SP} > 1$ SU/SP for several storage units, of which only the upper front units are directly accessible.

In a *homogenous store,* all places have the same dimensions and capacity, whereas in a *heterogeneous store* as shown in *Fig 16.4*, they have different dimensions and capacities.

A further property of a store with pile depth $C_z > 1$ is the *mobility of the storage units* within the storeplaces:

- In a *stationary storeplace,* the storage units remain in fixed position between in-storing and out-storing.
- In an *in-stationary storeplace,* the storage units move from an in-storing position to a spatially separated out-storing position.

The storeplaces can be arranged in two principally different ways:

- In *horizontal place arrangement* the storeplaces are located side by side in a horizontal plane.
- In *vertical place arrangement* the storeplaces are located side by side and on top of each other in a vertical plane.

Another property is the *mobility of the storeplaces*:

- The storeplaces of an *immobile store* cannot be moved.
- The storeplaces of a *mobile store* can be moved.

The aisles for in- and out-storing the storage units can be arranged in two different ways:

- *Spatially combined in- and out-storing aisles*: The storage units are stored in and out from the same side of the storeplaces, which for $C_z > 1$ are either *slide-in places* or *push-in-places*.
- *Spatially separated in- and out-storing aisles*: The storage units are stored in and out from opposite sides of the storeplaces, which for $C_z > 1$ are *flow-through places* or *push-through places*.

Stores with spatially combined in- and out-storing aisles can be either operated at the same time with single cycles for in-storing or out-storing only or at different times with combined in- and out-storing cycles. With spatial separation, the in-storing and out-storing can be executed independently in single cycles at the same time.

For storeplaces with pile depth $C_x > 1$, combined aisles generate automatically a *Last-In-First-Out* order and separate aisles a *First-In-First-Out* order of the storage units.

From combining the possible arrangements of the storeplaces with their different properties result 6 *basic types of storage systems*:

1. *Block Place Stores* with *block places* on a ground area in horizontal arrangement.
2. *Sorter Buffer Stores* with *buffer channels* in horizontal, sloping or vertical arrangement.
3. *Shelf Rack Stores* with stationary *shelf places* located in a fixed vertical rack.
4. *Flow Rack Stores* with *flow-through channels* mounted on a fixed vertical rack.
5. *Mobile Rack Stores* with shelf places located in vertical *racks movable in z-direction*.
6. *Rotary Stores* with shelf places mounted on a horizontally or vertically *movable carrousel*.

Most of these *storage types* can be designed with spatially combined or separate in- and out-storing aisles. They operate with different technical equipment more or less automatically. Resulting is a large variety of storage systems with specific advantages and disadvantages. In the following, *storage application rules* are derived from a comparison of properties and costs. These rules can be used to preselect a storage system for a given demand.

16.2.1 Block Place Stores

In a block place store, the places are arranged as shown in *Fig. 16.2* on the ground area on both sides of a combined in- and-out-storing aisle. Also spatially separated in- and-out-storing aisles are possible, but not common, since they need more space.

The block places can be served from the front or side by a *forklift truck*, which needs aisles between the blocks, or from the top by a *hall crane* or a *bridge crane*, which do not need aisles but free moving space above the storage units. Another possibility is a *stacker crane*, which takes the units from the side and lifts them vertically. Stacker cranes need only small aisles and free top space.

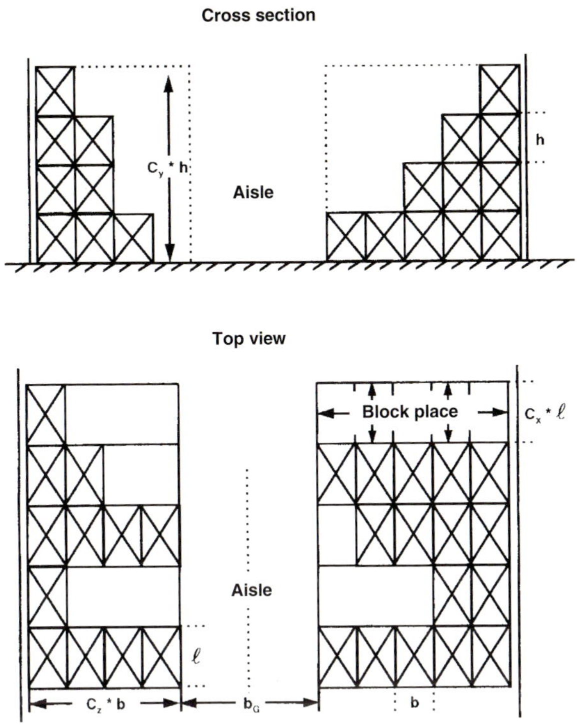

Fig. 16.2 Block place store with block places
C_x: pile length C_y: stacking factor C_z = pile depth

Examples for conventional *block place stores* (BPS) are:

- outdoor *container stores* served from one side by forklift trucks or from the top by van carriers or bridge cranes
- block place stores for stackable goods without load carrier, such as coils, drums or paper rolls
- indoor block place stores with forklift trucks or hall cranes for stackable pallets or box pallets
- *buffer stores* for bins, pallets, boxes or containers with accessibility from several sides by a forklift truck (see *Fig. 16.10*)
- *car-parks* and vehicle *parking areas* on a ground area or in a multi-story building

The *advantages* of conventional block place stores are:

- *no investment for racks*, only for painted markings on the ground
- *short cycle times* in block place stores with small total stock
- *short clearance times* if sufficient trucks and drivers are available
- *flexible adaptability* of place dimensions and layout

The *disadvantages* of block place stores are:

- *low filling degrees* for small stocks per article and single article placing
- *restricted accessibility* of the storage units
- no FIFO, but *LIFO* for block stores with combined in- and out-aisles
- *long cycle times* in extended block place stores

Due to their properties, the main *applications* of block place stores are:

- ► Intermediate buffer stores in the goods receiving, production or dispatch area for storage units with stacking factor higher 1.
- ► Supply stores for *pull-stocks* of articles with stacking factor higher 2 and replenishment quantities of 10 or more storage units per article
- ► Stores for *push-stocks* of smaller numbers of articles with more than 10 storage units and stacking factor higher 2.

Planning and operation of block place stores are generally underestimated. Although block place stores look quite simple, they are difficult to dimension and to operate with optimal utilization.

16.2.2 Sorter Buffer Stores

A sorter buffer store consists of parallel flow channels arranged in a horizontal, sloping or vertical plane as shown in *Fig. 16.3*. The channels can be tracks, rails, chutes, rollers or conveyors. If the storage units are not powered, such as vehicles, they are moved by gravity or on a powered conveyor from the in-storing side to the out-storing side of the channel.

The channels are served from separate filling and retrieval aisles by technical devices of the same or different kind. The in-storing and/or the out-storing can be executed by a forklift truck, by a shuttle with a transfer device or by a conveyor with branching and joining elements.

Examples and applications for *sorter buffer stores* (SBS) are:

- sorter buffers for packages, parcels, bins or pallets which arrive in batches and are sorted for different packing places, shipments or transports
- sorter buffers for packages, bins or pallets that arrive article mixed and have to be separated into article pure batches (see *Fig. 18.13*)
- vertical or sloping chutes for the fully automatic commissioning of small article units
- buffer and sorting tracks in a vehicle network
- switching harps for the formation of trains out of single wagons

The *advantages* of sorter buffer stores are:

- short retrieval, providing and clearance times
- high throughput performance
- low throughput costs if utilization is high
- minimal personnel by total automation
- FIFO with flow through channels
- good attainability of the storage units from the top

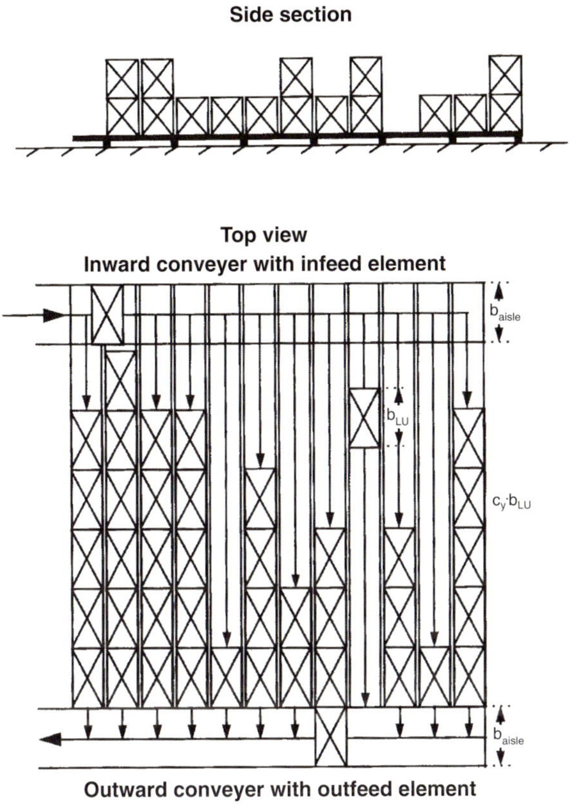

Fig. 16.3 Sorter buffer store with separate filling and retrieval aisles
Channel capacity $C_{CH} = C_x \cdot C_z = 6 \cdot 2 = 12$ LU

Their *disadvantages* are:

- *high investment* for construction, conveyors and automation
- *poor space utilization*
- *low filling degrees* of the channels
- *high place costs*
- *no single accessibility* of the storage units
- *low flexibility* for storage units with different dimensions

From the high throughput, the low throughput costs and the excessive place costs follow the *application rules for sorter buffer stores*:

- ▶ Sorter buffer stores are suitable for fully automatic sorting of equal units.
- ▶ Sorter buffer stores are useful for *short-term buffering*.

Such as the dynamic linear and rotating buffer sorters, which will be described in *Sect. 18.6.3*, the sorter buffer stores are *conveyor systems* for collecting and sorting storage units. Due to their high place costs, they are not suitable for keeping stocks for longer time.

16.2.3 Shelf Rack Stores

Shelf rack stores consist of *place modules* [PM] located in a rack construction. The place modules contain either only one storeplace, as shown in *Fig. 16.4*, or several places. The place modules are arranged in the racks similar to a honeycomb in vertical planes on both sides of the aisle. One aisle and two adjacent storage modules form an *aisle module* as sketched in *Fig. 16.5*. The total shelf rack store is made up of a number N_{AM} of parallel aisle modules. The storage devices can be forklift trucks, stacker cranes or automatic S/R-units.

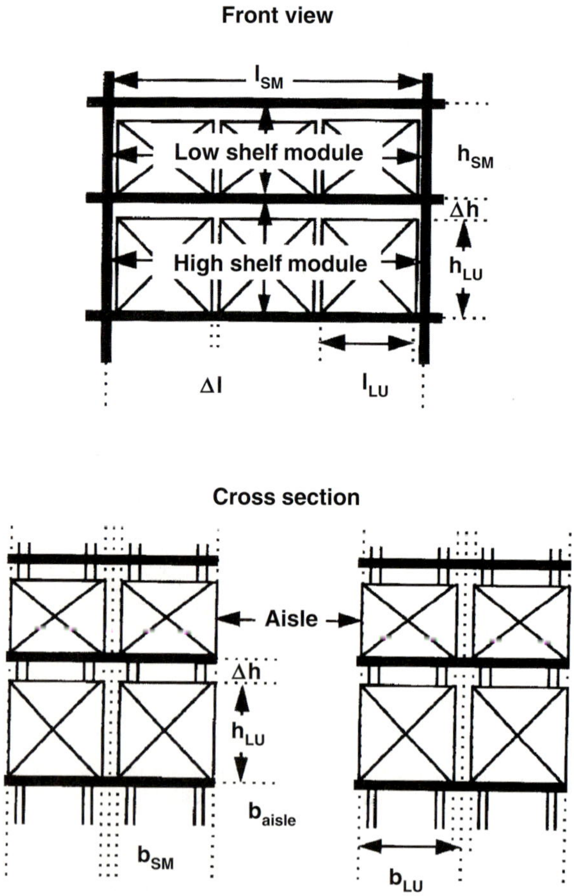

Fig. 16.4 Place modules in a rack with single-unit storeplaces

Storage unit dimensions: l_{SU}, b_{SU}, h_{SU}
Shelf clearances: Δl, Δb, Δh
Place module dimensions: l_{PM}, b_{PM}, h_{PM}

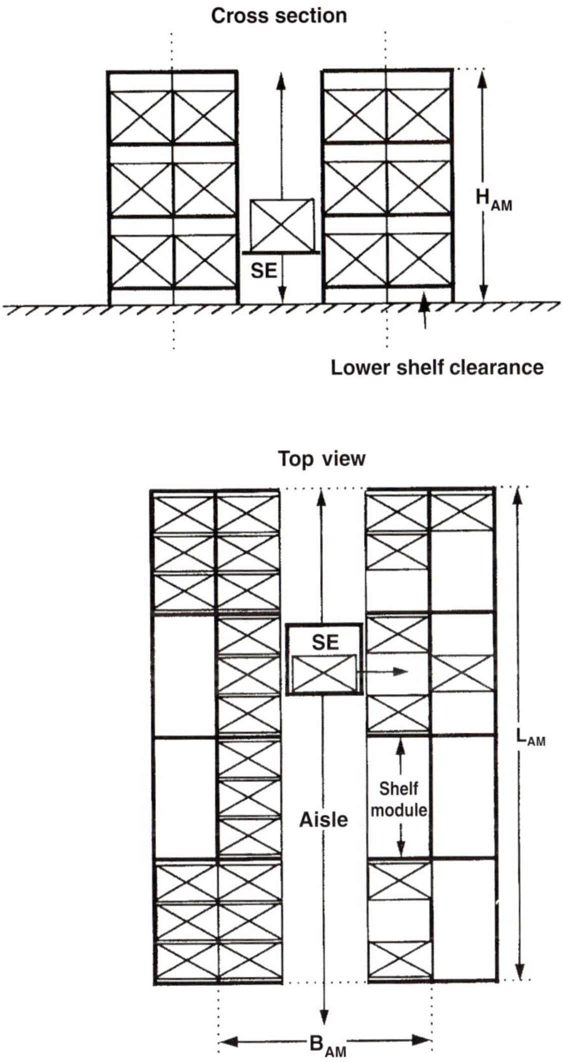

Fig. 16.5 Aisle module of a shelf rack store

Aisle module dimensions: l_{AM}, b_{AM}, h_{AM}
Storage element, device or S/R-Unit: SE

Typical examples for shelf rack stores are:

- manually operated *shelving stores* for small parts
- automatic *mini-load stores* (MLS; see *Fig. 17.34*)
- *conventional pallet rack stores* up to 8 m (CRS)
- *narrow aisles pallet stores* up to 15 m height (NAS)
- *automatic high bay stores* up to 50 m height (HBS)

The *advantages* of single-unit shelf rack stores are:

- *direct access* to all storage units
- up to *100% filling degree*
- *high space utilization* increasing with store height
- *short cycle times* and high in- and out-storing performance
- with conventional shelf racks suitable for decentral commissioning
- automatic MLS and HBS suitable for dynamic commissioning
- high flexibility for articles with varying stocks

Some of these advantages cause higher investment, e.g. for racks and S/R-units, which increase with the store height, for in- and out-feed transport systems and for automation. In spite of the higher investment, system comparisons in agreement with experience confirm the *application rule for shelf rack stores*:

► Shelf rack stores are generally the most economic solution for stocks with many articles and small stocks per article

If the stocks per article are high, a *multi-unit shelf rack* can be opportune, where 2 or 3 units are placed behind each other. However, some of the advantages get lost with this solution. The *disadvantages* of a multi-unit shelf rack store are reduced access and flexibility, lower filling degrees and broader aisles for some storage devices, e.g. with a deep-reaching telescope fork.

A combination of block storing and shelf storing are *drive-in rack stores*. Here, the storage units of a block are separated vertically by side-bars of a rack. Each *rack place* can keep several units behind each other with pile depth $C_z > 1$. The places are served by a high reaching fork lift truck, if the lower levels are not filled. A vertical set of rack places corresponds to a block place, where all front units are accessible. The drive-in rack store is suitable for keeping articles with high stock and small stacking factor.

16.2.4 Flow Rack Stores

As shown in *Fig. 16.6*, a flow rack store consists of several levels of parallel *flow channels* which are assembled in a rack construction. The single levels are similar to a sorter buffer. The channels are loaded from on side by a storing device and unloaded from the opposite side by a retrieval device. Conventional flow rack stores have spatially separated in- and out-storing aisles. Depositing and taking of the storage units are executed by a fixed fork or telescope fork or by a roller or chain conveyor mounted on a S/R-unit.

Examples for flow channel stores are:

- *live stores* with gravity rolls for packages
- *flow-through racks* with gravity rolls or powered conveyors for pallets
- *passive flow-through racks* with track bars for *roll-pallets*
- *compact stores* for pallets with satellites carried by automatic S/R-units

A special flow rack store is the *automatic satellite store* or *compact store*. Here, the S/R-units carry a *satellite trolley*. The empty trolley moves on side tracks below

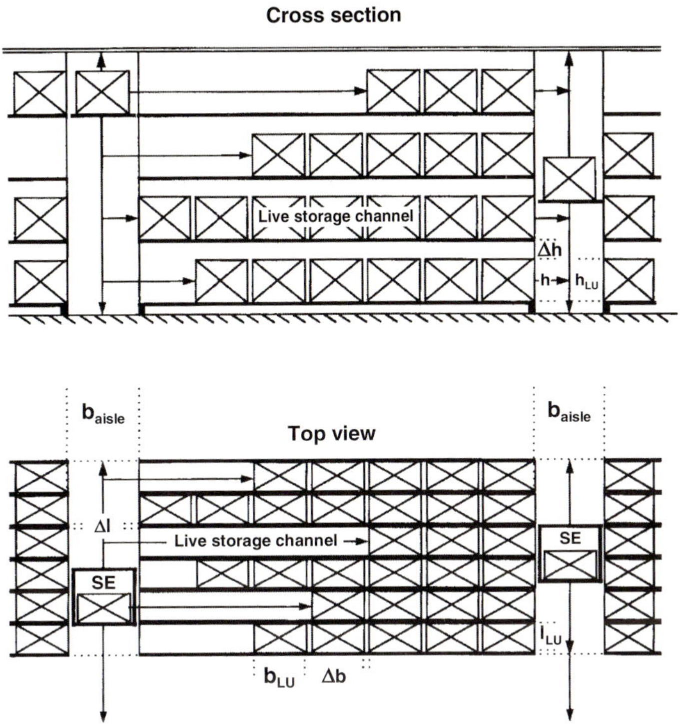

Fig. 16.6 Flow channel store with separated in- and out-storing
Clearances: Δl, Δb, Δh
Channel capacity: $C_{CH} = C_z = 7$ LU

a storage unit, takes it up, transports it into or out of the channel and puts it down. During the storage time, the unit stays on fixed side bars in the rack. Satellite stores are possible with spatially separated or with combined in- and out-storing aisles.

The *advantages* of flow channel stores are:

- good space utilization by compact construction
- First-In-First-Out for separate in- and out-aisles
- possibility of high automation
- integrated storing of reserve units and access units

Due to the last feature, life stores and flow-through racks are central elements of *commissioning systems*. The in-storing of the reserve units can be executed by hand or automatically by an S/R-unit, whereas the picking is done manually (see *Figs. 17.3, 17.8* and *17.9*).

The advantages are connected with higher investments for the racks and the equipment of the channels. For automatic systems, such as compact stores, additional investments are caused by the special S/R-units, in- and out-feed transport system and process control.

With exception of the better space utilization, flow rack stores have the same disadvantages as sorter buffer stores. Further disadvantages are the poor attainability of the backward storage units within the rack levels, and the relatively long cycle and clearing times. Hence, only for high stocks per article, a compact store can be a better solution than a shelf rack store.

16.2.5 Mobile Rack Stores

Similar to a shelf rack store, a mobile rack store consists of place modules located in single vertical rack planes or rack rows. The rack rows are mounted on wheels and are movable on trails in z-direction transversal to the aisles. The shifting is executed by hand or by a powered mechanism.

By this construction for more than 3 rack rows, the *proportionate aisles number* $n_{AS} = N_{AS}/N_{RR}$, i.e. the number of aisles N_{AS} in relation to the number of rack rows N_{RR}, can be smaller than for immobile rack stores, which have a *proportionate aisles number* $n_{AS} = 0.5$ for combined and $n_{AS} = 1$ for separate in- and out-aisles. Resulting are the main *advantages of mobile rack stores*:

- maximal space utilization and most compact storing

However, the smaller aisle number and the shifting imply several *disadvantages*:

- *no immediate acces* to the storage units in the closed aisles
- *long access times* for storage units in closed aisles
- *time losses* for rack shifting
- *low throughput performances*
- additional investment for the *shifting mechanism*

From these features results the *application rule for mobile rack stores*:

▸ Mobile rack stores can be opportune for long-term storing of many articles with low stocks.

Corresponding to this rule, mobile racks are common in *archives* for documents, files, films or electronic data carriers.

16.2.6 Rotary Stores

In a rotary store, the movable places are shelve modules, single bins or special devices, hanging below or mounted between circulating chains or cables. For in- and out-storing of storage units or for filling or picking of single items, all storeplaces are rotated until the place with the required article faces a stationary *basic station*. The basic station is manned or equipped with handling and control devices. As long as no access is required, the storeplaces are at rest.

As shown in *Fig. 16.7*, two basically different versions of *rotary storage systems* (RSS) are possible:

- *Paternoster Stores*: The storeplaces are mounted between parallel rotating chains or cables and are moved *vertically* to a basic station at the side.
- *Carrousel Stores*: The storeplaces are hanging below a rotating chain or cable and are moved *horizontally* to a basic station at the front.

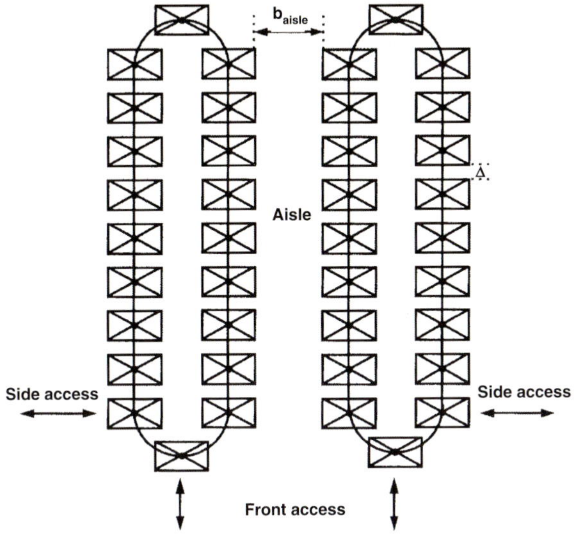

Fig. 16.7 Basic types of rotary stores
Paternoster Store: vertical rotation with side access
Carrousel Store: horizontal rotation with front access

Applications are paternoster stores for small parts, tools or spare parts in the production and for documents or files in the administration. A historical example is a parking paternoster for cars which was first built in the USA in 1920. Carrousel stores can be found in dry cleaning shops for buffering of hanging garments. Paternoster and carrousel stores are also used to buffer reserve units and to provide access units for the *dynamic commissioning*. Examples are the paternoster stores for carpet rolls in DIY-markets.

The most often claimed *advantage* of rotary stores is:

- *better space utilization* due to the avoided aisles

However, the clearances and dimensions of the movable storeplaces and the additional space for conveyor construction and basic station over-compensate in many cases the space savings by the avoided aisles. Additional *disadvantages of rotary stores* are:

- *no direct access* to the majority of storage units
- *long access times*
- *necessity to move the whole stock* for each access
- *limited performance* caused by only *one basic station*
- *high investment* for static construction, rotation mechanism and control

Due to these disadvantages, rotary stores are too expensive and too slow for storing containers or pallets and of limited use for commissioning. A general *application rule for rotary stores* is:

► Rotary stores are suitable as buffer and provision system for dynamic commissioning if the process times at the basic station are long and the throughput is low.

Very different from slow moving rotary stores are fast rotating buffer sorters. In such a *rotating dynamic buffer sorter*, article units or parcels circulate on a conveyor until they are released in the required sequence to packing places or other exits. Linear and rotating dynamic sorter buffers will be analyzed in *Sect. 17.13.3*. They are complex conveyor systems, which differ principally from all other storage systems.

16.3 Storage Technique

For the execution of the storing functions, a storage system consists of the *technical storage-subsystems*:

storage units with or without *load carrier*
storeplaces within or without *racks*
storage devices and *load handling devices*
in- and out-feed transport system (16.21)
storage process control and *administration system*
receiving and *dispatch area*
storage building with *technical installations*

Technical installations without specific storage function are fire detectors, sprinklers, smoke outlets, air conditioning and heating systems. Other storage-related installations are workshops for repair and maintenance, rooms for process control and computer and rooms for operating personnel.

The technical subsystems (16.21) can be realized in many different ways. The multitude of technical possibilities combined with the basic storage types lead to the vast number of different storage systems, which can be found in practice. Design, construction and realization of the technical subsystems are not tasks of logistics but of other disciplines, such as steel construction, mechanical engineering, material handling, process control and site building. The tasks of *logistic technology* and *storage planning* are selection, dimensioning and combination of the possible technical subsystems and to design a reliable and cost efficient storage system which fulfils the static and dynamic demand. For these tasks a logistician must know the options, free parameters and main features of storage technique (Ackerman et al. 1997; Arnold 1998; Bode and Preuß 2004; Günthner and Heptner 2007).

The *clearances*, *tolerances* and *positioning accuracy* are crucial for the required space, flexibility and reliability of a storage system. The storing performance is determined by the *mean transport length* between pickup points, storeplaces and transfer points, and by the *speed* and *acceleration* of the storage and handling devices. Other influence factors on the in- and out-storing performance are the *positioning times* and the *dead times* of the control system.

16.3.1 Storage Units and Load Carriers

Storage units are the goods to be stored with or without load carrier. Discrete goods of larger size with similar dimensions and sufficient stability, such as cartons, coils or paper rolls, can be handled and stored without load carriers. Small or heterogeneous units as well as loose goods, liquids and gases must be consolidated in suitable load carriers. Common load carriers for storing are standardized bins and trays, flat pallets, box pallets, movable storage racks, cassettes and containers (see *Chap. 12*).

The type of the load carrier and the dimensions of the storage units determine the storage type and restrict the possibilities of storage technique. The dimensions of the storeplaces and the width of the aisles are determined by the maximal tolerable external dimensions of the storage units including the allowances for excess load. The restrictions for the spatial orientation of the storage units limit the layout possibilities of shelf and aisle modules.

For example, many automatic pallet stores allow an excess load of $\pm$ 50 mm on both sides of the standard pallets. This allowance increases the effective dimensions of a $800 \times 1{,}200$ *EURO-pallet* to $900 \times 1{,}300$ mm and requires 22% more ground area per pallet. In spite of their effect on dimensions, utilization and flexibility of the whole storage system, allowances and clearances are often not sufficiently considered. The planner has to keep in mind:

- Too high allowances and too large *clearances* cause poor space utilization and higher investment. Too small *allowances* reduce the flexibility for deviating storage units and the *functional reliability*.

In most cases the vertical orientation of the storage units is prescribed, whereas the orientation of length and breadth in relation to the aisles is free. This leads to the *orientation option for storage units*:

- Storage units with different measures in two or three directions can be placed either *flat* with the longest measure *parallel* or *transversal* to the aisle or *upright* with the longest measure in vertical direction.

The orientation option can be used to minimize the required space and to maximize the performance. It is of particular importance for *long goods* with length much larger than diameter, such as pipes and bars, or for *flat goods* with height much smaller than length and breadth, such as panels and plates.

16.3.2 Storeplaces and Storage Racks

The most simple and flexible solution for storing are *block places*, which are marked on the ground. They can easily be adapted to changing requirements.

The fixed shelves in a storage rack for small parts can be wooden, metal or plastic plates. In mini load stores for standard bins or trays, the storeplaces are slide bars mounted laterally on the upright racking. The storeplaces in standard pallets racks are cross beams, which are fixed to the upright steel racking either in parallel or transversal to the aisle.

Instationary storeplaces for *passive storage units* can be chutes, slide bars, gravity or powered rollers or chain conveyors. Another possibility are fixed places on cross beams where the storage units are moved by a satellite trolley.

Storeplaces for *mobile storage units*, such as *load carriers with rolls*, and for *vehicles* are buffer tracks. Rolling units without power are moved from place to place by a shifting or pulling device. Vehicles can move on *parking lanes* with own power.

The dimensions of the place modules, i.e. *place module length*, *depth* and *height*, are free *design parameters* of shelf rack stores. A further option is the *partition of the place modules*, i.e. the number of storeplaces in x-direction between the racking uprights. The width of the racking uprights results from the weight and dimensions of the storage units, and from the static and material of the rack construction.

To avoid failures, in- and out-storing *clearances* are necessary between neighboring storage units and to the rack construction. Additional clearances, as shown in *Fig. 16.4*, are necessary for the free move of the handling device. They also take care of the *tolerances* of the storage units and the rack construction, and of the limited accuracy of the *positioning* of the S/R-unit:

► The necessary shelf clearances decrease with the positioning accuracy of the S/R-units and increase with the tolerances of storage units, rack construction and S/R-units.

For example, due to the European regulation FEM 9.831 for automatic high bay stores, the standard shelf clearance between storage units and racking uprights is 100 mm. This increases the space occupied by an 800×1,200 mm *Euro-pallet* by about 10%. Additional space is needed for the uprights and cross beams of the rack construction, and for the pipes and heads of a sprinkler system. That means:

- The shelf clearances and construction measures determine the *place differences* Δx, Δy and Δz between the effective *dimensions of the place modules*, l_{PM}, b_{PM} and h_{PM}, and the outer *dimensions of the storage unit*, l_{SU}, b_{SU} and h_{SU}, which include allowances for excess load.

A free design parameter for storages is the orientation of the storage units relative to the aisle. Without external restrictions, the positioning possibilities for storage units with the measures $l_{SU} > b_{SU} > h_{SU}$ lead to the *vertical orientations*:

- *flat positioning* with the longest measure l_{SU} horizontal
- *upright positioning* with the longest measure l_{SU} vertical

and to the *horizontal placing options*:

- *longitudinal placing* with the next longest measure parallel to the aisle
- *transversal placing* with next longest measure transversal to the aisle

In block place stores for pallets with less than 800 mm width, a longitudinal placing is necessary, in order to ensure enough space for the forklift truck between the stacks within the blocks. For other storage types, the orientation of the storage units is a free design parameter which can be used to optimize ground or space utilization, throughput performance, and costs.

16.3.3 S/R-Units and Load Handling Devices

The in- and out-storing can be executed by a person or by a mechanical *storage and retrieval unit* carrying a *load handling device*. The *S/R-unit* is either manually controlled or operates automatically (Bode and Preuß 2004; Günthner 2007).

Storage and retrieval can be executed by the S/R-unit in different *storing cycles*:

- In a *single in-storing cycle,* the S/R-unit takes up a storage unit from a pickup point, transports it to a storeplace, deposits it and comes back to the pickup point.
- In a *single out-storing cycle,* the S/R-unit drives from the transfer point to a storeplace, takes up a required storage unit, transports it to the transfer point and puts it down.
- In a *combined in-out-storing cycle,* an in-storing process, which ends at an *in-storing place*, is followed by an out-storing process, which starts after an intermediate empty drive at an *out-storing place* (see *Fig. 16.1*).

In addition to in- and out-storing a S/R-unit can *relocate* storage units from one storeplace to another in order change the *storage order*.

In-storing limit performance μ_I [SU/h], *retrieval* or *out-storing limit performance* μ_O [SU/h], *throughput* or *in- and out-storing limit performance* μ_{IO} [SU/h] and *transfer limit performance* μ_{IO} [SU/h] of the storing system are determined by the *number*, the *construction* and *moving behavior*, the *speed* and *acceleration* and by the *load capacity* C_{SD} [LU/SD] of the *S/R-units*. Further influence factors on the limit performances are the *mean driveway lengths*, the *aisle dependency* of the S/R-unit and the *operating strategies*.

Technical data and other features of customary S/R-units and storage devices are summarized in *Table 16.1*. The table contains also standard prices, which will be used in the following model calculations. The technical data of selected load handling devices are listed in *Table 16.2*.

Corresponding to their *moving behavior* S/R-units can be classified into:

- *One-dimensional moving storage devices*, which carry the storage unit in an additive drive- and lift-motion. Examples are storage trucks with fixed fork or swiveling fork.
- *Two-dimensional moving storage devices*, which carry the storage unit in a simultaneous drive- and lift-motion, such as narrow-aisle storage trucks, rail-bound rack feeders and S/R-machines and overhead storage devices.
- *Three-dimensional moving storage devices*, which can move the storage units on a spatial path in horizontal x- and y-direction, in vertical x- and z-direction and in all three x-y-z-directions simultaneously. Examples are overhead cranes, hall cranes and stacker cranes.
- *Additive moving storage devices*, which execute the in- and out-storing in additive part-moves in x-, y- and z-direction by different subsystems. These can be lifts for the vertical moves, shuttles for the horizontal moves in the aisles and a satellite trolley or telescope fork for the moves in z-direction within the storeplaces.

Table 16.1 Technical data and standard prices of storage devices

STORAGE EQUIPMENTS (SE) Load units	**Capacity** LU/SE	**Lifting height** up to app.	**Aisle width** EURO-Pal.	**Running** speed accell.	**Lifting** speed accell.	**Benchmark price** 2007 T €
High lift truck	1	4.5 m	3.0 m	2.0 m	0.2 m/s	10 to 30
Pallets			3.4 m	0.7 m/s^2	0.1 m/s^2	
Fork lift truck	1 to 4	6.5 m	3.0 m	3.0 m/s	0.3 m/s	20 to 40
Pallets			3.4 m	1.0 m/s^2	0.3 m/s^2	
Reach truck	1	7 m	2.2 m	2.5 m/s	0.5 m/s	30 to 40
Pallets			2.5 m	1.0 m/s^2	0.3 m/s^2	
Narrow aisle stacker	1 to 2	14 m	1.5 m	2.5 m/s	0.35 m/s	75 to 90
Pallets			1.8 m	1.0 m/s^2	0.5 m/s^2	
Storage & retrieval unit	1 to 2	40 m	1.5 m	5.0 m/s	2.0 m/s	150 to 250
Pallets				1.0 m/s^2	1.0 m/s^2	
Mini-loads	1 to 8	8 m	1.0 m	3.0 m/s^2	2.0 m/s^2	70 to 120
TransFaster	1 to 3	15 m	1.6 m	5.0 m/s	2.0 m/s	120 to 150
Pallets				1.0 m^2	1.0 m/s^2	
Shelf trolley	1 to 2	–	1.0 m	up to 6.0 m/s	–	30 to 40
Pallets, Bin			1.4 m	1.0 m/s^2	–	
Satellite shuttle	1	–	0	1.0 m/s	–	25 to 35
Pallets				0.5 m/s^2	–	
Stacker crane	–	8 m	0	1.0 m/s	0.3 m/s	100 to 180
Long goods, coils etc.				0.3 m/s^2	0.2 m/s^2	

Aisle widths: for longitudinal orientation (transversal second line) of 800·1,200 mm pallets
Standard prices: by 2007 inclusive drive unit and mobile control

Table 16.2 Technical data of load handling devices

LOAD LIFTING DEVICES LLD	**Load units** LU	**Capacity** LU/LLD	**Storage** equipment	**Depth** up to	**Moving** speed	accell.
Fixed fork	Pallets	up to 4	Lift trucks	2.0 m	0.3 m/s	0.2 m/s^2
Movable fork	Pallets	up to 2	Reach trucks	2.0 m	0.3 m/s	0.2 m/s^2
Side grips	Cartons	to 2	Lift trucks	1.5 m	0.5 m/s	0.2 m/s^2
	Drums	to 4	S/RU			
Telescope forks	Pallets	to 2	Lift trucks	2.0 m	0.5 m/s	0.3 m/s^2
Telescope table	Tablar	1	S/RU	1.5 m	1.0 m/s	0.8 m/s^2
	Bins	to 2	S/RU			
Push-and Pull device	Tablar	1	S/RU	1.5 m	1.5 m/s	1.0 m/s^2
Roller table	Pallets	1 to 2	S/RU	–	0.4 m/s	0.3 m/s^2
Chain conveyor	Bins	1 to 4	Transfer trolley			

Prerequisite for a reliable automation of a storage system is the accurate *track guidance* of the S/R-units within the aisles. Technical possibilities are *mechanical guidance* with running rail and guard rail, *inductive guidance* with guide wire, *optical guidance* with marker lines or guide marker, and *acoustical guidance* with control points for coupled navigation. Most flexible is *electronic guidance* with *transponders* in the ground or attached at the vehicle, whose signals are detected with *RFID* (Finkenzeller 2002; Shephard 2004).

Construction, moving behavior and track guidance together determine the *aisle dependency of storage devices* (Arnold 2003; Bode 2004):

- *Aisle-independent storage devices* such as forklift trucks and narrow-aisle trucks can change the aisles and leave the storage area without restrictions.
- *Aisle-changing storage devices* such as rack feeders in combination with a *aisle-transfer unit* or *curve-going rack feeders* can change the aisles with reduced speed, but do not leave the storage area.
- *Aisle-fixed storage devices* such as rail-guided rack feeders and automatic S/R-units without aisle transfer cannot leave the aisle and only serve the store-places within one aisle.

Changing aisles is a relatively time consuming process. Long *aisles changing times* τ_{AC} [s/AC] cause low *aisle changing limit performance* $\mu_{AC} = 3600/\tau_{AC}$ [AC/h]. The effective storage and retrieval performance is significantly reduced by aisle changing, if the *aisle changing frequency* λ_{AC} [1/h] is high.

Load handling or lifting devices take, hook or clamp storage units from above, below, front or side. Depending on the execution of the loading or unloading cycle, they can be differentiated into:

- *Load handling devices with empty move*: For each loading cycle an empty forward move and for each unloading cycle, an empty backward move is executed. Examples are fixed or swiveling forks, underpinning telescopic forks and telescopic tables, sideward clamping jaws, grippers or from above latching container-spreaders and satellite trolleys.
- *Load handling devices without empty move*: Loading and unloading cycles are executed in one move. Examples are pulling or pushing devices for bins and roller tables or chain conveyors for pallets.

Load handling without empty move achieves higher limit performances. The limit performance of a storage device depends also on the capacity and the number n_{HD} of load handling devices [HD] per storage device [SD]:

- The *capacity of the handling device* C_{HD} [SU/HD] is the number of storage units it can carry and move at the same time. The *capacity of the storage device* C_{SD} is the number of load handling devices n_{HD} [HD/SD] times the capacity per the device:

(16.22)

Stacker cranes and automatic S/R-units for pallets are built with capacities up to 4 SU/SD. S/R-units for bins, trays and packages are available with capacities up to 8 SU/SD. However, the costs for additional load handling devices are often higher than the savings by the higher limit performances, which increase less than proportional with the capacity.

16.3.4 In-Feed and Out-Feed Transport System

The *in-feed transport system* conveys in-storing units from the pickup points to the *in-feed places* of the storage device. At the *I-point,* the storage units are identified and registered by the storage administration system. In automatic stores, the units have to pass a *contour and weight control*. They are rejected, if their size or weight exceeds the technical limits of the storage system.

The *out-feed transport system* conveys out-storing units from the *out-feed- places* of the storage devices to the prescribed transfer points outside the storage area. After the *C-point*, where they are identified and released, the storage units leave the area of the storage control system.

Stores with aisle-dependent storage devices need an in- and out-feed transport system. Aisle-independent storage devices can transport the storage units from the pickup point into the aisles and from there to the transfer points. Stores with aisle-independent storage devices are equipped with a transport system only if the distances of the pickup and transfer points to the store area are long.

The in- and out-feed transport system can be a *vehicle system*, a *conveyor system* or a *combination* of these two basically different transport technologies, which are described in detail in *Chap. 18*. The in-feed and out-feed places of manually operated stores with aisle changing storage devices are normally cantilever shelves at the rack front. They can be served independent of the storage device by *storage distribution vehicles*, such as forklift trucks.

More sophisticated in- and out-feed vehicle systems are *electric overhead trolleys* (EOT) or *automatic guided vehicles* (AGV). To decouple the arrivals of the vehicles from the stochastically varying cycles of the S/R-units, the vehicle system is separated from the in- and out-feed places in front of the aisles and from the input and transition points by *buffer conveyor lanes* as shown in *Fig. 16.7* and *16.8*.

Another combined vehicle and conveyor system is the *distribution shuttle* shown in *Fig. 16.9*, which runs in front of the racks transversal to the aisle and connects the buffer lanes in front of the racks with the stationary conveyors of pickup, transfer or commissioning stations.

For an *in- and out-feed conveyor system,* many technical solutions are possible. Examples are chain or roller conveyors in combination with the branching and joining elements of *Sect. 18.6*. As shown in *Fig. 13.27* and *13.28,* the in- and out-feed conveyors can be arranged in two basically different ways:

- *Separated in-feed* conveyor system *and out-feed* conveyor system on two different levels or at two sides of the racks
- *Combined in- and out-feed* conveyor system on one level at the same side of the racks.

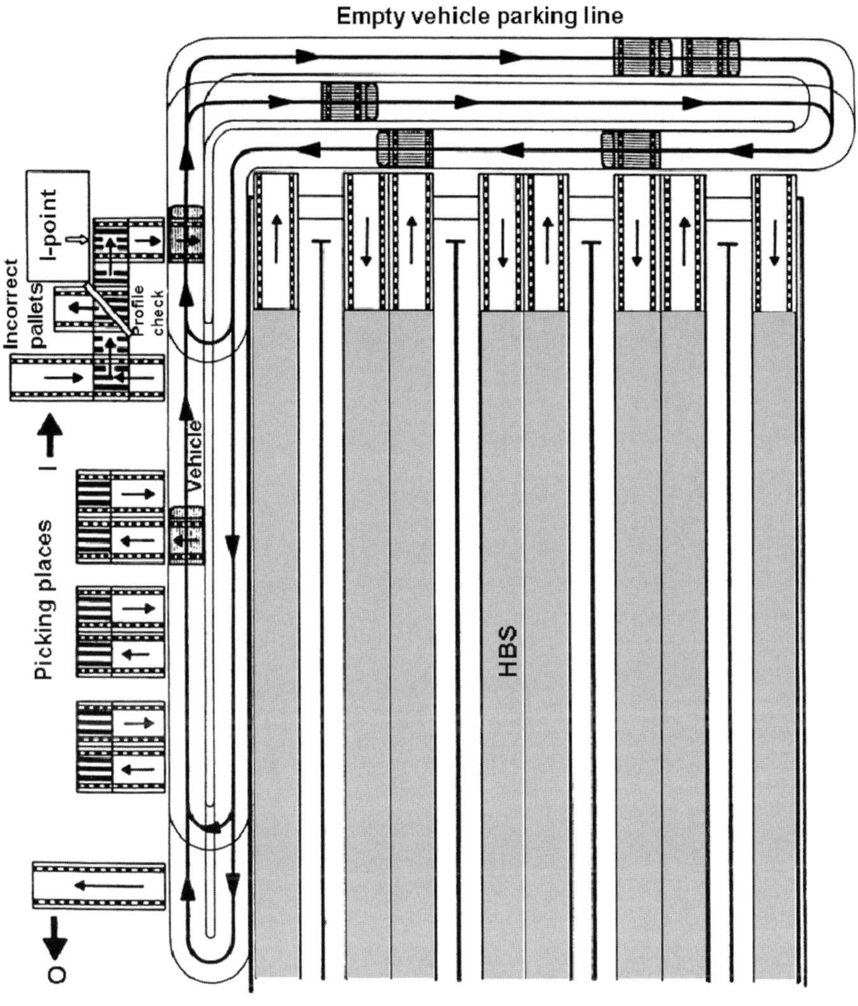

Fig. 16.8 Distribution shuttle and conveyors for the one level in- and out-feed transport system of an automatic high-bay store

Vehicles: electric overhead trolleys or automatic guided vehicles

Separated systems are more expensive, but have more buffer places and offer additional space, which can be used e.g. for picking stations of *dynamic commissioning*. However with separated systems, combined in-out-storing cycles take longer due to the empty moves between the in- and out feed places.

Combined transport systems are less expensive and enable combined in-out-storing cycles without intermediate empty moves. Their disadvantages are fewer buffer places on the combined parts of the conveyor system and reduced *availability* (see *Sect. 13.6.4*).

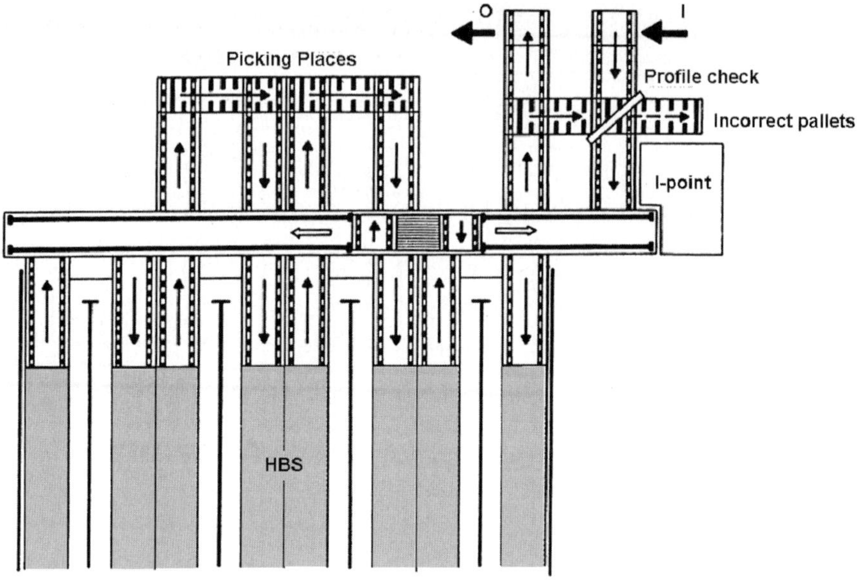

Fig. 16.9 One-level combined vehicle (transfer shuttle) and conveyor in- and out feed system of an automatic high-bay store

16.3.5 Storage Management System

A *storage management system* (SMS) consists of a central *storage administration system* (SAS) and subordinated *storage control systems* (SCS). It is generally part of a *warehouse management system* (WMS), which has additional functions, such as the management of commissioning, internal transports, packaging and dispatch. Both, SMS and WMS, are *hierarchically* organized as shown in *Fig. 2.1* and described in *Sect. 2.5*.

Tasks of the *storage control system*, which is made up by local *programmable control modules*, are *tracking* and *guiding* of the storage units on the in-feed and out-feed transport system and *steering* and *positioning* of the S/R-units.

For the control of aisle-independent storage devices, forklift trucks and other transport vehicles, special *vehicle guidance systems* (VGS) and *transport management systems* (TMS) are in use, which operate with remote data transmission based on radio frequency or infrared.

To position an automatic storage device in front of a destination place, two *different positioning techniques* are possible:

- *Absolute positioning*: The control of the storage device focuses on a positioning mark which is located directly at the destination place.
- *Relative positioning*: The control of the storage device is geared by positioning marks that are fixed on the mast of the device and along the aisle.

The positioning technique impacts the *reliability* and *availability* of the storing device significantly, whereas the positioning time affects the *limit performances*. The major advantages of absolute positioning are higher accuracy and independency from tolerances. However, it is more complex and slower. Relative positioning depends on the rack tolerances and on the adjustment of the device, but is less complex and faster. Most reliable and fastest, but more complex and expensive is a combination of both techniques.

The storage control modules are linked to the *storage administration system*, which has the tasks of

- *receiving*, *buffering* and *control* of the in-storing and out-storing orders
- *allocation* of the storage units to the storeplaces due to optimal *placing strategies*
- *track-*, *buffer-* and *stock-control* of all storage units moving or resting between I-point and C-point
- *directing* and *coordinating* in- and out-feeding transport systems and storage devices due to optimal *moving strategies*
- *generation of transport orders* for the in- and out-feed vehicles

Some of these *tasks* can be executed by a *foreman* or the *warehouse manager*. Others are supported or taken over by a *storage management system* (SMS) or a *warehouse management systems* (WMS). Nowadays, qualified software is capable of performing most of the storage management tasks (Warehouse Logistics 2008; Wolf et al. 2006). However, many *standard SMS* do not offer complete and optimal software to realize all needed storage strategies. On the other hand, some SMS have redundant or useless functions and strategies. It is therefore advisable to specify all necessary functions and required storage strategies in a *duties record briefing* before selecting, inquiring and programming a storage management system.

The storage management system can operate either *off line* or *on line* with a super-ordinate business administration and ERP system, such as SAP, and with the sub-ordinate control modules. In the off-line mode, no time critical links with other systems exist, while in the on-line mode, the storage management system is permanently linked with the super-ordinate and/or the sub-ordinate systems. A guideline for the design of a storage management system is the *decoupling rule of hierarchies* (see *Chap. 2*):

- ▸ High *reliability* and *availability*, and minimal *idle times* can be achieved by a *system hierarchy,* which decouples the storage management system from the super-ordinate and from the sub-ordinate systems.

Idle times can occur at any step of the movement of a storage unit from the I-point and to the storeplace and from the storeplace to the C-point. They are caused by data reports, data processing and requests for instructions. Idle times of several seconds or minutes significantly affect the performance of a storage system. Reasons are high user frequencies during rush hours, wrong allocation of functions in the systems hierarchy, weaknesses of the software and insufficient capacity of the computer hardware.

16.3.6 Goods Entry and Dispatch

The performance of a storage system, as well as of a logistic center or a production site, depends on the design and dimensioning of the goods entry and dispatch area. Bad design, wrong location and false organization can cause *bottlenecks* and high transport effort (see *Chap. 19*).

If the storage system is located in a separate building, the goods entry and dispatch is made up of *gate modules*, consisting of *ramps*, *gates* and *provision areas*, and of *functional areas* for entrance control and for the brake-down, build-up and securing of storage units. In the dispatch area, also *dynamic commissioning places* and *packaging stations* can be located.

Layout possibilities for the goods entry and dispatch area are (see *Chap. 19*):

- *Separated goods entry and goods dispatch areas* with in-gate modules located at one side and out-gate modules at another side of the building.
- *Partly combined goods entry and goods dispatch areas* with groups of in-gate modules and out-gate modules along one side and the option to alter the function of the gate modules located in the middle.
- *Combined goods entry and dispatch area* with bi-functional in-out-gate modules along the same side of the building that can be used alternately for loading or unloading.

A combined goods entry and dispatch area offers the greatest flexibility and enables combined in- and out-feed tours of the transport means.

Experience and theoretical considerations lead to the following *planning rules for goods entry and dispatch*:

- The goods entry and dispatch area should be designed in a *modular layout* with equal *gate modules* (GM) as shown in *Fig. 16.10.*
- The *number of gate modules* N_{GM} is determined by the frequency and the capacities of the incoming and outgoing transport means during peak hours, by the loading and unloading times and by the buffer capacity.
- The *number of buffer places* N_{BP} and the layout of the buffer area in a gate module depend on the throughput, the process times for the goods entry and dispatch, and on the needed decoupling between the internal and external logistic chains that are connected at the ramps.

Buffer places and handling can be avoided, if the storage units are transported immediately to and from the storage system directly out of and into the docking swap trailers or semi-trailers. This *just-in-time loading*, however, requires perfectly synchronized internal and external logistics processes.

16.3.7 Storage Buildings

Block place stores and car parks can be located on open ground space without protection. Smaller supply stores are placed in a multi-functional building or in a production hall. Larger stores need a separate *storage building*. This can be a *conventional hall*, a *multi-storey building* or a special construction, such as a *high-bay store silo* or a *parking garage*.

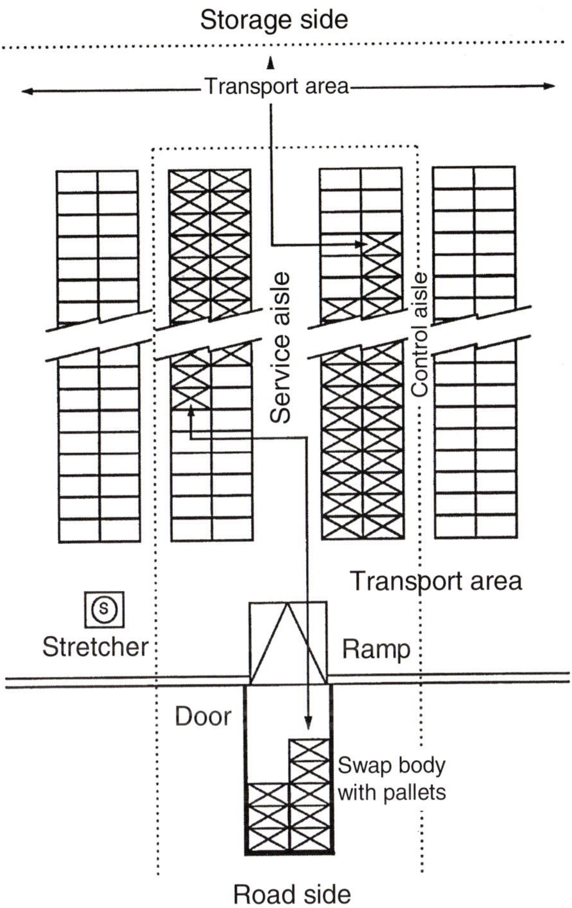

Fig. 16.10 Gate module for goods entry and goods dispatch
Parameter: number of buffer places per gate module

Due to elevators and long transport times to the upper floors, *multi-storey buildings* are in most cases uneconomic for stores with high throughput and capacity. For a *high-bay store silo* with heights over 15 m, the rack is erected on a concrete basement and the roof and wall claddings are mounted directly on the rack construction. Practical experience and system comparisons confirm the *application rule for high bay stores*:

▸ A *high bay store* is the most economic solution, if ground is expensive, capacities exceed 5,000 pallets and/or the stock rotation is higher 6 p.a.

The basic *planning rule for logistic buildings* is:

▸ *Stores and logistic buildings should be planned from the inside to the outside and not from the outside to the inside.*

Type, construction and technique of the storage system determine the storage building. A building erected on the *green field* should not hamper the layout and functional areas. An existing building with fixed dimension, gates and distances of the uprights generally restricts the layout of the storage system and enforces suboptimal solutions.

Safety regulations for goods and people cause general *safety restrictions of storage buildings*:

- Stores with large capacity and high *fire load* must be divided in several separate *fire sections* of limited size.
- The distance to the next exit within each fire compartment should not exceed the tolerable *escape way length.*

The size of a fire section depends on the *hazardous category* of the stored goods, on the *room height* and on the *insurance fees*. It ranges from 1,200 m^2 up to 6,000 m^2. German VDI-regulations prescribe a *maximal escape radius* of 50 m from each point to the next fire exit (VDI 3564/1999).

16.4 Storage Strategies

Performance and costs of a storage system depend on the *operating strategies*. If a new storage system is planned, investment and operating costs can be minimized by optimal storage strategies. Storage strategies can also improve the performance, utilization and operating costs of existing stores.

Corresponding to the static and dynamic demand, storage strategies can be differentiated into *static* or *placing strategies* and *dynamic* or *moving strategies*.

16.4.1 Placing Strategies

Placing strategies determine the allocation of the storage units to storage zones and storeplaces. Main goal of the static storage strategies is to fulfill the capacity demand with a minimal number of storeplaces. Further goals are safety, accessibility and availability, short storing cycles and minimal relocations.

The most important placing strategies are:

- *Fixed storage order*: Fixed storeplaces are reserved for the maximal expected stock of an article and blocked for other articles.
- *Free storage order*: Free storeplaces can be used for storage units of any article.

As explained before, the capacity demand can be reduced considerably by free storage order only if many different articles with deviating behavior are stored at the same time in the same store. The free storage order is also called *chaotic storage order*, since it leads to randomly changing locations of the articles. Nowadays, a free storage order can be realized without chaos by standard storage administration software.

Compromises between fixed and free storage order are the placing strategies:

- *Uniform aisles distribution*: The stock of one article is equally distributed over the aisles in order to achieve maximal accessibility and availability.
- *Zone-fixed storage order*: Certain storage zones are reserved for defined categories of articles or can only be used for specific storage units.

A nearly uniform aisle distribution results automatically by *cyclic allocation* of the incoming storage units to the aisles.

A zone-fixed storage order increases the capacity demand due to the higher *breathing reserve*, which is necessary for the smaller number of articles in the separated zones. This leads to the *planning rule*:

▶ The number of fixed storage zones should be as small as possible.

Possible *placing strategies* for *multi-unit stores* with place capacity $C_{SP} > 1$ are:

- *Place adjustment*: Small storeplaces are used for small storage units and articles with low stock, large places for big storage units and articles with high stock (see *Sects. 12.4* and *12.5*).
- *Article-pure placing*: One storeplace is used only for storage units of the same article or the same batch.
- *Article-mixed placing*: One storeplace can be used for storage units of N_{AP} different articles or batches.
- *Place clearing*: The storage units from partially filled places are stored out first.

With article-mixed placing, higher filling degrees for lower article stocks are achievable. However, this strategy causes *relocation moves* since:

- For a storeplace occupied by storage units of N_{AP} different articles on average $(N_{AP}-1)/2$ storage units have to be removed in order to access a storage unit of a required article.

Relocating a storage unit is about 30 times more expensive than the daily storeplace costs. That leads to the *placing rule*:

▶ Article-mixed placing is only opportune for goods which are stored for at least a month or longer or if relocations can be executed during unproductive waiting times.

Article-mixed placing is quite often introduced when additional capacity is required. However, in addition to the relocation moves, it needs special storeplace administration and can cause confusion und mistakes. If the storage system has been selected and dimensioned correctly, mixed placing is avoidable.

A static storage strategy for improving the effective performance, is

- *Fast mover concentration*: In order to reduce the mean drive ways of the storage devices and shorten the average storage cycle times, fast moving articles are located in storeplaces near the in-feed and out-feed places.

A fast mover concentration can be realized with *fixed A-B-C-zones* for A-B-C-articles with the above disadvantages of a zone fixed order, or by a *dynamic in-storing* of the arriving goods into free storeplaces at distances, which increase with the expected storage time. The effect of a fast mover concentration is often overrated. Simulations and analytical calculations lead to the *fast mover rule*:

- ▸ By fast mover concentration, the throughput performance of a storage system can be improved up to 15% only for very long and high storages and if the ABC-distribution of the assortment is very distinctive.

In most cases the *fast mover effect* is significantly below 10%.

16.4.2 Moving Strategies

Moving strategies determine the *execution sequence* for in-storing and out-storing orders and relocations. Goals of dynamic storage strategies are maximal throughput performance, fulfilling time requirements and keeping storing restrictions. Such *storing restrictions* are:

- *Strict First-In-First-Out* (FIFO): The storage unit, which has been stored in first, must be stored out first.
- *Moderate First-In-First-Out*: The storage units from the *batch*, which has been stored in first, must be stored out first.
- *Last-In-First-Out* (LIFO): The storage unit, which has been stored in last, must be stored out first.

The strict FIFO can be kept without relocations in single-unit stores and flow rack stores with spatially separated storage and retrieval. The moderate FIFO prohibits the article mixed utilization of multi-unit places. LIFO is unavoidable without relocation for block place stores, multi-unit stores and flow rack stores with spatially combined storage and retrieval.

The most important *moving strategies for storage devices* are:

- *Single-cycle strategy*: In- and out-storing orders are executed in separate in-storing cycles and out-storing cycles. If goods dispatch is urgent, out-storing orders are prioritized. In-storing orders are prioritized, if arriving storage units pile up.
- *Double-cycle strategies* (see *Fig. 16.1*): The in- and out-storing orders are executed in combined in-out-storing cycles as far as compatible with the time requirements.
- *Optimal cycle strategies* (see *Fig. 16.9*): For a storage device with load capacity C_{SD}, a number $n_O \leq C_{SD}$ of storage units for out-storing orders and a number $n_I \leq C_{SD}$ of empty storeplaces for in-storing orders are selected and sequenced in such a way, that the total travel time for the combined n_I/n_O-cycle is minimal.

Single cycles are necessary, if the in-storing or the out-storing has absolute priority. They can be executed without performance loss as long as there are only in-storing orders or only out-storing orders. The throughput performance is improved by double cycles, since they avoid the empty inward and outward moves of the single cycles. However, with double cycles the in-storing, retrieval and providing times are longer.

For double cycles of single-unit storage devices with capacity $C_{SD} = 1$, different optimal cycle path strategies are possible. However, they achieve only small effects (Gudehus 1973). For storage devices with load capacity $C_{SD} > 1$ up to C_{SD} in-storage units and up to C_{SD} out-storage units can be selected and ordered in many different ways (see *Sect. 5.2*). The task to identify the tour with the shortest cycle time leads to the *tour planning* and *travelling salesman problems* which will be analyzed in *Sect. 18.11*. A very efficient method to generate approximately optimal round tours is the *two-stripe strategy* shown in *Fig. 16.19*, which was developed by *Miebach 1971*. The effects achievable by optimal cycles will be calculated in *Sects.16.10* and *17.10*.

Moving strategies with the goal to minimize relocations and aisle changes are:

- *Relocation strategy*: Relocations in order to keep FIFO or to clear article-mixed places are executed in times without in- or-out-storing orders.
- *Aisle-change strategy*: If a storage device serves more than one aisle, the orders for units in the same aisle are collected for a fixed or adaptable *aisle collection time* T_{AC} and executed in *cyclic batches*. The aisle changing times are minimized by serving the neighboring aisles in sequence.

The product of the mean *aisle changing time* τ_{AC} with the *aisle changing frequency* $\lambda_{AC} = 1/T_{AC}$ gives the time loss by aisles changes.

Moving strategies for the in-feed and out-feed transport systems are:

- *Out-feed strategies*: Urgently required storage units are prioritized at the joining elements of the conveyer system.
- *In-feed strategies*: Incoming storage units are conveyed to the in-feed places in front of the aisles either in *batches* or single in *cyclic sequence*.

Not all placing and moving strategies are compatible. For example a fast mover strategy reduces the effects of combined cycles and optimal path strategies, since the travel length can not be reduced several times. This leads to the *strategy selection rule*:

- ▸ Only compatible and really effective strategies should be realized in order to avoid unnecessary programming and long waiting times for the result of a complex calculation by the process control system.

Before a strategy is realized, it is necessary to check the potential *strategy effect* and the *compatibility* with other, more effective strategies.

16.5 Place Demand and Filling Degree

The *place demand* N_{SP} [SP], i.e. the number of storeplaces necessary to store a given number of storage units, is determined by the achievable filling degree, which depends on the place capacity and on the placing strategies. The place demand is generally higher than the number of storage units to be stored, i.e. than the *capacity demand* M_S [SU]. This leads to the *storage dimensioning rule*:

- ▸ The first step when dimensioning a storage system is the calculation of the place demand for the intended placing strategy from the capacity demand and the place capacity.

The calculation of the place demand can be executed with the help of the formulas of *Sect. 12.5*. With free storage order, storeplaces are needed only for the *mean article stock* (16.11). That means:

- With *free storage order* and *article-pure placing,* the *mean number of storeplaces* with capacity C_{SP} [LU/SP] necessary to store a mean article stock of M_S storage units per article is

$$N_{SPfree} = MAX(1; M_S/C_{SP} + (C_{SP}-1)/2C_{SP}) \quad [SP/Art]. \tag{16.22}$$

As long as the article stock is not 0, at least one place is occupied per article. If the stock is higher than the place capacity, on average a share $(C_{SP}$-1)/2C_{SP} of one article place is empty. The mean *filling loss* $(C_{SP}$-1)/2C_{SP} increases the number of places per article. The filling loss for single-unit stores is 0. For multi-unit stores with very large place capacity, i.e. for $C_{SP} >> 1$ LU/SP, it is on average 0.5 storeplace per article.

The N_{SPfree} places are filled on average with M_S storage units and contain maximal $N_{SPfree} \cdot C_{SP}$ units. That means:

► The possible mean filling degree of a store with *free storage order* and article-pure placing is

$$\eta_{S\ free} = M_S/(N_{SPfree} \cdot C_{SP}) = M_S/MAX(C_{SP}; M_S + (C_{SP}-1)/2) \ [\%]. \tag{16.23}$$

As shown in *Fig. 16.11*, the possible filling degree of multi-unit stores decreases with increasing place capacity. A filling degree of 100% for single-unit stores is only possible with free storage order. The filling degree of multi-unit stores with $C_B > 1$ is generally less then 100%. However, as shown in *Fig. 16.12*, the possible filling degree increases with the mean stock per article.

With fixed storage order, storeplaces are needed for the maximal article stock (16.10). That means:

- With *fixed storage order* and *article pure placing* the *number of storeplaces* with capacity C_{SP} [LU/SP] necessary to store a maximal article stock of M_{Smax} storage units is

$$N_{SPfix} = MAX(1; M_{Smax}/C_{SP} + (C_{SP}-1)/2C_{SP}) \quad [SP/Art]. \tag{16.24}$$

- The maximal *possible mean filling degree* of a store with *fixed storage order* and article-pure placing is

$$\eta_{S\ fix} = M_{Smax}/MAX(C_{SP}; M_{Smax} + (C_{SP}-1)/2) \quad [\%]. \tag{16.25}$$

Due to the relations (16.10) and (16.11), the maximal stock and the mean stock per article depend on the *safety stock* M_{safe} and on the *replenishment quantity* M_R. *Fig. 16.13* shows the dependency of the possible store filling degree on the safety stock, which has been calculated with relation (16.23) and (16.25).

The filling degree for free storage order, as well as the filling degree for fixed storage order for multi-unit stores with $M_{safe} > 0$, are less then 100%, since a certain number of storeplaces is permanently blocked by the safety stock. In the example of *Fig. 16.13*, the possible filling degree exceeds 90% for free storage order, independent of the safety stock. For fixed storage order, it is far less than 90% as long as the mean stock is much higher than the safety stock.

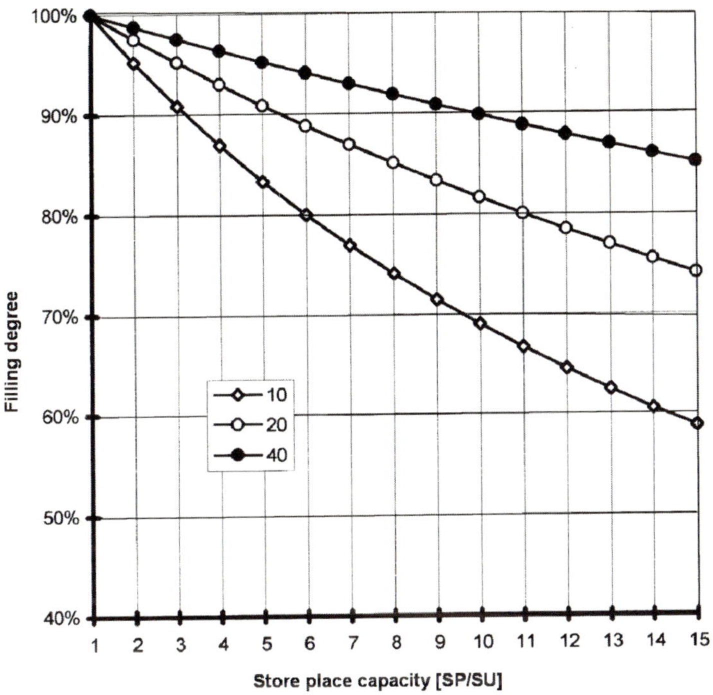

Fig. 16.11 Dependency of the possible mean store filling degree on the place capacity for different mean article stocks
Parameter: mean article stock $M_S = 10/20/40$ SU
free storage order, article pure placing

The difference between the filling degree for fixed storage order and free storage order is maximal when the safety stocks are small. The difference disappears when the safety stocks become high and the single article stocks do not change during the storage time. That leads to the *storage order rule*:

▸ A fixed storage order does not affect the store capacity, if the stocks per article or per batch remain constant during the whole storage time.

In order to calculate the effect of article mixed placing on the store capacity, it is useful to introduce the *storage order factor*:

$$f_{SO} = \begin{cases} 1 & \text{for fixed storage order} \\ 1/2 & \text{for free storage order} \end{cases} \quad . \tag{16.26}$$

If article mixed placing is allowed for up to N_{AP} articles per place, the filling loss per article is reduced by the factor $1/N_{AP}$. That leads to the *general store dimensioning formulas*:

- With *mixed placing* of N_{AP} different articles per place, the *number of storeplaces* with capacity C_{SP} [LU/SP] necessary to store an mean stock of M_S and a safety stock of M_{safe} storage units per article is

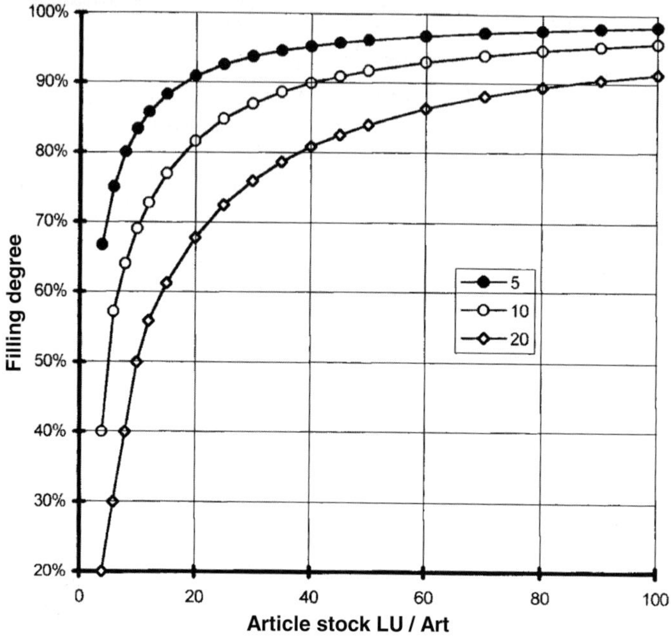

Fig. 16.12 Dependency of the possible mean store filling degree on the mean article stock for different place capacities

Parameter: $C_{SP} = 5/10/20$ SU/SP, free storage order, article pure placing

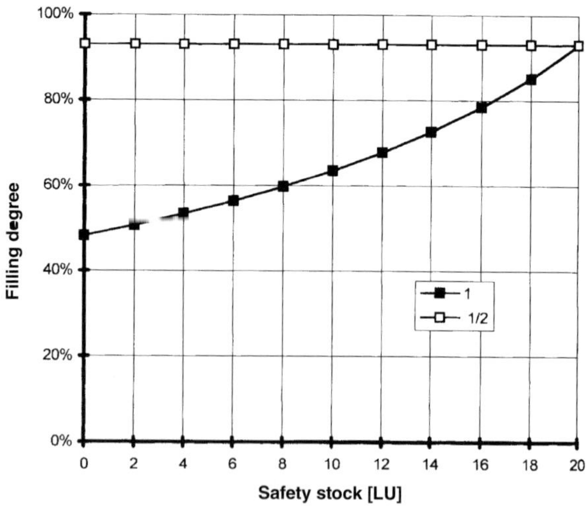

Fig. 16.13 Dependency of the mean store filling degree on the mean safety stock per article

$f_{SO} = 1$: fixed storage order $f_{SO} = 0.5$: free storage order
storeplace capacity: $C_{SP} = 5$ SU/SP
mean article stock: $M_S = 20$ SU

$$N_{SP} = MAX\big(1/N_{AP}; (M_S + 2f_{SO} \cdot (M_S - M_{safe}))/C_{SP} + (1/N_{AP}) \cdot (C_{SP}-1)/2C_{SP}\big) \quad [SP/Art]. \tag{16.27}$$

- The *possible filling degree* of a storage system with a mean stock of M_S and a safety stock of M_{safe} storage units per article and article mixed placing of N_{SP} articles per place is

$$\eta_S = M_S/MAX\big(C_{SP}/N_{AP}; (M_S + 2f_{SO} \cdot (M_S - M_{safe}) + (1/N_{AP})(C_{SP}-1)/2)\big) \quad [\%]. \tag{16.28}$$

By multiplication with the total number of articles N_A, relation (16.27) leads to the general *formula for total storeplace demand*:

- The *total storeplace demand* for a total mean stock $M_{S\,tot}$ and a total safety stock $M_{safe\,tot}$ of N_A articles, of which N_{AP} can be stored mixed in the same storeplace of capacity C_{SP} [LU/SP], is

$$N_{SP} = MAX\big(N_A/N_{AP}; (M_{S\,tot}/C_{SP} + 2f_{SO} \cdot (M_{S\,tot} - M_{safetot}))/C_{SP} + + (N_A/N_{AP}) \cdot (C_{SP}-1)/2C_{SP}\big) \quad [SP]. \tag{16.29}$$

For stochastically fluctuating and season-dependent stock levels, the total stock M_{Stot} in formula (16.29) has to be replaced by the *effective total stock* M_{Seff} given by relation (16.15).

The general dimensioning formulas (16.27), (16.28) and (16.29) can be universally applied for dimensioning and optimizing all types of stores. *Storage dimensioning software* that operates without these formulas leads to wrong results and is unsuitable for storage optimization.

Instead of the effective store capacity, planners calculate quite often with the 100%-store capacity. The *100%-store capacity* is the product of the number of storeplaces N_{SP} and the place capacity:

$$C_S = N_{SP} \cdot C_{SP} \quad [SU] \tag{16.30}$$

It is the maximal number of storage units which can be filled in the storeplaces irrespective of the placing strategy. The 100%-store capacity is reduced by the possible filling degree (16.28) to the *effective store capacity*:

$$C_{S\,eff} = \eta_S \cdot C_S \quad [SU] \tag{16.31}$$

From relations (16.27), (16.28), (16.29), (16.30) and (16.31), the following *storage rules* can be derived:

- ▸ The effective store capacity for multi-unit stores is lower than the 100%-store capacity.
- ▸ The deviation of the effective store capacity from 100% capacity is caused by filling losses, which depend on the placing strategies.
- ▸ The filling losses of a multi-unit store with place capacity $C_{SP} > 1$ increase with decreasing mean stock per article.

These storage rules are important for storage planning and for the calculation of storeplace costs. Due to the neglected effects of article stocks, place capacity and storage strategies, many block place and flow rack stores contain as much empty space as used space although no free places are available.

From relation (16.27) follows the *mixed-placing rule*:

► The filling degree of a multi-unit store can be improved by article mixed-placing with $N_{SP} > 1$ articles per place, with the effect, that to reach a certain article, on average $(N_{SP}-1)/2$ units of other articles must be removed.

If the additional handling is tolerable, article-mixed placing helps to improve the effective capacity of an existing multi-unit store. However, when designing a new store, article-mixed placing should, if at all, only be considered for long-time storing over 30 days.

16.6 Ground Area per Storage Unit

The *net ground area* of a storage system is the ground area occupied by all storeplaces, the rack construction and the aisles, without front and backside area for the transport system or for aisle changing of S/R-units. The *100%-ground area per storage unit* is the net ground area divided by the 100%-store capacity (16.30).

Decisive for the necessary ground area of a store for a certain total stock is the *effective ground area per storage unit*, i.e. the net ground area divided by the effective store capacity (16.31). It is equal to the 100%-ground area divided by the filling degree η_S and generally larger than the 100%-ground area per unit. The effective ground area per storage unit is the key for *storage optimization*:

► By minimization of the effective ground area per storage unit, the investment for the ground area and for the storage building can be reduced.

The 100%-ground area per storage unit is determined by the capacity and dimensions of the storeplace and by the number and width of the aisles per rack module.

As also illustrated in *Figs. 16.2, 16.3, 16.4, 16.5* and *16.6*, the base dimensions of a storeplace, i.e. the *length* l_{SP} parallel and the *depth* b_{SP} orthogonal to the aisles, result from the external measures of the storage units including allowances, from their orientation to the aisles, from the required shelf clearances and from the proportionate measures of the rack construction. *Table 16.3* contains typical place dimensions for different types of pallet stores which will be used for the following model calculations.

The *width of the aisles* b_{aisle} is also determined by the dimensions and the orientation of the storage units relative to the aisles. In addition it depends on the construction of the storage device, the load handling and the aisle clearances. For model calculations, the aisle widths of typical storage devices for pallets are listed in *Tables 16.1* and *16.3*.

The *number of store levels* N_y is the number of places on top of each other. For stores with horizontal place arrangement, such as block place stores and sorter buffers, the number of levels is $N_y = 1$. For stores with vertical place arrangement, such as rack stores, the number of store levels is a free *design parameter* $N_y > 1$ that can be used for storage optimization.

The *number of aisles* N_{aisle} is also a design parameter for storage planning which determines the *proportionate number of aisles* n_{aisle}. The *proportionate number of*

Table 16.3 Effective storeplace dimensions and ground areas per pallet for different pallet storage systems

Storage system	Pallet orientation to aisle	Storage place dimensions		Service aisles		Stacking factor pallets on top	Vertical levels up to	Storage height [m]	Effective floorspace optimized [m²/Pal]
		Length [mm]	Depth [mm]	Width [mm]	Prop. number				
Block place store	longit.	1,300	850	3,000	1/2	2 to 6	1	4 to 6	0.4 to 1.8
Drive-in rack store	longit.	1,400	850	3,000	1/2	1	6	6 to 8	0.3 to 0.6
Flowrack store	transv.	1,000	1,200	3,000	2	1 or 2	4	3 to 8	0.25 to 0.6
Convent. rack store	transv.	950	1,300	2,500	1/2	1	8	6 to 8	0.4 to 0.6
Narrow aisle store	transv.	950	1,300	1,800	1/2	1 or 2	10	8 to 14	0.2 to 0.4
High bay store	transv.	950	1,300	1,500	1/2	1 or 2	30	16 to 40	0.07 to 0.2

Storage units: CCG1 pallets
Dimensions: 800·1,200·1,050 mm
Aisle width: see also *Table 16.1*

aisles n_{aisle} is the number of aisles in relation to the number of storeplaces N_{SPz} orthogonal to the aisles:

$$n_{aisle} = N_{aisle}/N_{SPz}. \tag{16.32}$$

Stores with stationary places on both sides of a combined storage and retrieval aisle have the proportionate aisle number $n_{aisle} = 0.5$. If the places are located only on one side of a combined aisle, $n_{aisle} = 1$. For flow channel stores with separate storage and retrieval aisles, the proportionate aisle number is $n_{aisle} = 2$.

With N_z store levels, a proportionate aisle number n_{aisle}, pile length C_x, and pile depth C_z results for the *ground area per storeplace*:

$$a_{SP} = (C_x \cdot l_{SP}) \cdot (C_z \cdot b_{SP} + n_{aisle} \cdot b_{aisle})/N_z\cdot \tag{16.33}$$

The division of (16.33) by the capacity of the storeplace (16.19) gives the

- *100% ground area per storage unit* for a store with N_y store levels, place capacity $C_{SP} = C_x \cdot C_y \cdot C_z$ and a stacking factor of C_y

$$a_{SU}(C_{SP}) = a_{SP}/C_{SP} = l_{SP} \cdot (b_{SP} + n_{aisle} \cdot b_{aisle}/C_z)/(N_z \cdot C_y)\cdot \tag{16.34}$$

The 100%-ground area per storage unit decreases with the number of store levels and with the storeplace capacity. The division of (16.34) by the store filling degree leads to the

- *effective ground area per storage unit* for a store with N_y store levels, place capacity C_{SP}, stacking factor C_y and possible store filling degree η_S

$$a_{SUeff}(C_{SP}) = a_{SU}(C_{SP})/\eta_S(C_{SP})\cdot \tag{16.35}$$

The effective ground area per storage unit depends on the place capacity C_{SP} and through relation (16.8) also on the mean stock and the safety stock per article. Starting with $C_{SP} = 1$, it decreases due to the reduction of the 100%-ground area with increasing place capacity until a minimum is reached at the *optimal place capacity*. For higher place capacities it increases due to the reduction of the possible filling degree. This leads to the general *storage optimization rule*:

- The ground area, the ground investment and the costs for the storage building can be minimized by using the *optimal storeplace capacity*.

To determine the optimal place capacity, relations (16.28) and (16.34) have to be inserted into (16.35). The minimum of the resulting function $a_{SUeff}(C_{SP})$ can be found by variation of the place capacity C_{SP}. Due to relation (16.19), the place capacity can be varied in three different directions. That makes the general solution of the *storage optimization task* somewhat difficult.

An explicit solution of practical importance is possible for multi-unit stores with free storage order and article-pure placing. Inserting the relations (16.28), (16.23) and (16.34) into (16.35) gives:

- The *effective ground area-per storage unit* for a store with N_y levels, free storage order, article-pure placing, pile length C_y, pile depth C_z, stacking factor C_y and an mean article stock $M_S > C_{SP} = C_x \cdot C_y \cdot C_z$ is

$$a_{SUeff}(C_x, C_y, C_z) = (M_S + (C_{SP}-1)/2) \cdot l_{SP} \cdot (C_z \cdot b_{SP} + n_{aisle} \cdot b_{aisle})/(C_x \cdot C_z \cdot N_y \cdot M_S) \tag{16.36}$$

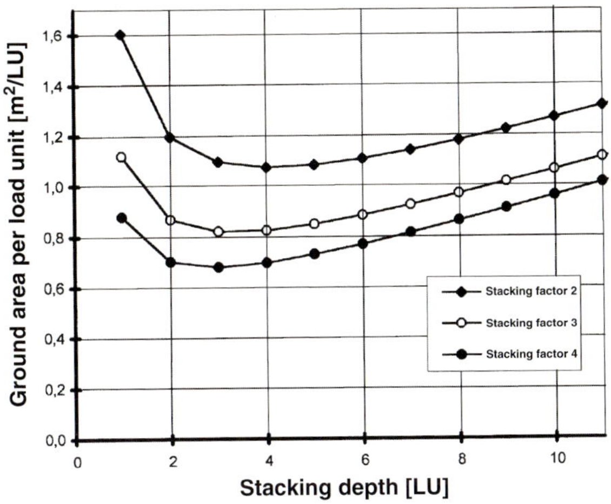

Fig. 16.14 Dependency of the effective ground area per storage unit on the pile depth

Storage type: block place store for Euro pallets
Parameter: stacking factor $C_y = 2 / 3 / 4$ SU/stack
Place dimensions: $l_{SP} = 850$ mm, $b_{SP} = 1{,}250$ mm
Aisle width: $b_{aisle} = 3{,}000$ mm
Proportionate aisle number: $n_{aisle} = 0.5$
Mean article stock: $M_S = 10$ pallets/article
Placing strategies: free storage order, article pure placing

An example is given in *Fig. 16.14*, where the dependency of the effective ground area per pallet on the pile depth C_z has been calculated with relation (16.36) for a block place store with $N_y = 1$ and combined storage and retrieval aisles, i.e. with the proportionate aisle number $n_{aisles} = 0.5$. The curves of *Fig. 16.14* illustrate the following *dependencies* and *effects*, which can be derived from the general relations (16.35) and (16.36):

1. The effective ground area per storage unit first decreases with the *pile depth* C_z until the *optimal pile depth* $C_{z\,opt}$ is reached and increases with higher pile depth.
2. The effective ground area per storage unit decreases with increasing *stacking factor* C_y whereas the optimal pile depth shifts to smaller values.
3. The effective ground area for constant place capacity decreases with increasing *article stock.*
4. The effective ground area increases linear with the pile length C_x. It is minimal at $C_x = 1$, i.e. for storeplaces containing only one unit in aisle direction.
5. Wrong dimensioning of the storeplaces can lead to deviations from the minimal ground area and optimal storage volume of 25 to 40%.
6. For *multi-unit stores,* wrong placing strategies cause deviations up to 25% from the optimum.
7. For *single-unit stores,* no place capacity optimization or placing strategies are necessary, as $C_{SP} = 1$ SU/SP

Since the storeplace capacity considerably affects the effective ground area, the total volume, the filling degree and the place costs, its optimization is the first step of the *storage planning* (Gudehus and Kunder 1972).

16.7 Storeplace Optimization

The *optimal place capacity* C_{SPopt} for multi-unit stores is the product of optimal pile length $C_{x\,opt}$, optimal stacking factor $C_{y\,opt}$ and optimal pile depth $C_{z\,opt}$:

$$C_{SPopt} = C_{x\,opt} \cdot C_{y\,opt} \cdot C_{z\,opt} \tag{16.37}$$

The optimal place capacity can be determined with the help of the formulas (16.35) and (16.36). From the linear increase of the effective ground area with the pile length C_x results the *1st stacking rule*:

- The *optimal pile length*, i.e. the optimal number of storage units positioned side by side in aisle direction of the same place is

$$C_{x\,opt} = 1\ \mathrm{LU/SP}. \tag{16.38}$$

The stacking of the storage units is generally limited by a *technical stacking factor* $C_{y\,tech}$, i.e. by the number of storage units that can maximally be stacked on top of each other. The effective area per storage unit decreases as long as the stacking factor is larger than the mean article stock, i.e. for $C_y > M_B$. These facts lead to the *2nd stacking rule*:

- The *optimal stacking factor* for a given technical stacking factor $C_{y\,tech}$ and a mean article stock M_S is:

$$C_{y\,opt} = \mathrm{MIN}(C_{y\,tech}; M_S). \tag{16.39}$$

By equating the differentiation of function (16.36) with respect to the pile depth C_z to 0, and solution of this equation after C_z follows the *3rd stacking rule*:

- The *optimal pile depth* for mean article stocks M_S with stacking factor C_y in a store with free storage order, article-pure placing, storeplace depth b_{SP}, proportionate number of aisles n_{aisle} and aisle width b_{aisle} is

$$C_{z\,opt} = \sqrt{(2\,M_S - 1) \cdot n_{aisle} \cdot b_{aisle}/(C_y \cdot b_{SP})}. \tag{16.40}$$

Corresponding formulas for the calculation of the optimal pile depth for fixed storage placing and article-mixed placing can be derived from the general function (16.35) (Gudehus 1974).

Function (16.40) shows the following *dependencies* and *effects*:

1. The optimal pile depth does not depend on the number of store levels, as the space in each level is simultaneously optimized.
2. The optimal pile depth increases by the square root of the article stock, since the storeplaces for higher article stocks can be made deeper to compensate the space loss by the aisles.
3. The optimal pile depth increases with increasing aisle width and with the proportionate aisles number, as a greater aisle space loss has to be compensated by deeper places.

4. For larger stacking factors, the optimal pile depth is lower and the ground area utilization is higher than for smaller stacking factors.

With the help of relations (16.37), (16.38), (16.39) and (16.40), the optimal place capacity can be calculated for any article group with similar article stocks. The resulting pile depth is minimal 1 and must be rounded to the next integer.

For example, if articles with a mean article stock of $M_S = 10$ pallets with dimensions 800·1,200 mm and stacking factor $C_y = 4$ are stored in a block place store with place depth of $b_{SP} = 800+50 = 850$ mm, aisle width $b_{aisle} = 3{,}000$ mm and combined storage and retrieval aisles, i.e. $n_{aisle} = 0.5$, the optimal storeplace depth is $C_{z\,opt} = \sqrt{}(2{\cdot}10\text{-}1){\cdot}0.5{\cdot}3{,}000)/(4{\cdot}850) = 2.9$ pallets. Hence, for these article stocks, the optimal block place has a depth of 3 pallets and a capacity $C_{SPopt} = 3{\cdot}4 = 12$ pallets. *Figure 16.14* shows that in this case the optimized effective floor place per pallet is 0.7 m^2.

The *optimal article stock* is the quantity that can be stored with the lowest effective ground area in places with pile depth C_z. It results from solving equation (16.39) with respect to the mean article stock M_S.

$$M_{opt}(C_z) = C_y \cdot b_{SP} \cdot (C_z + 1/2)^2/(2n_{aisle} \cdot b_{aisle}) \qquad (16.41)$$

This formula leads to the *general placing rule*:

- ▸ Storeplaces with pile depth C_z and article-pure placing are optimally used to store articles or batches with mean storage quantities M_S in the range

$$M_{opt}(C_z - 1) < M_S \le M_{opt}(C_z) \qquad (16.42)$$

In an existing store with storeplaces of different pile depths C_z, an in-storing order with the expected mean stock quantity M_S during the storing time is optimally stored in a place with capacity in the range (16.42). For example, merchandise articles with regular demand, arriving with a replenishment quantity M_R, have a mean stock $M_S = M_R/2$. If the in-storing quantity for a *sales action* is M_I and will leave the store unchanged, the mean stock is $M_S = M_I$.

As an example in *Table 16.4*, the optimal pile depths and the quantity borders (16.42) have been calculated for a pallet block place store. Similar tables can be calculated for other multi-unit stores with places of different dimensions, such as flow channel stores or satellite stores, and used in practice for the optimal placing of in-storing order quantities. The formula (16.41) can also be programmed and applied by a storage management system to determine the space optimal storeplace for any in-storing order.

From relation (16.42) follows the *storage design rule for heterogeneous stocks*:

- ▸ The total stock is divided into groups with mean article stocks in the range (16.42) for $C_z = 1, 2, 3\ldots$ and for each group, the place demand is calculated for the optimal pile depth with the help of formula (16.29).

For the stock of an assortment with pronounced ABC-structure, a heterogeneous block place store with different pile depths that has been designed by this rule requires between 10 and 20% less ground area than a homogeneous block place store with equal pile depths.

Table 16.4 Optimal pile depth of pallet block storeplaces for pallets with different mean storage quantities

Stacking factor 2			Stacking factor 3			Stacking factor 4		
Stock quantity Pal/article		Pile	Stock quantity Pal/article		Pile	Stock quantity Pal/article		Pile
from	to	depth Pal	from	To	depth Pal	from	to	depth Pal
1	1	1	2	2	1	3	3	1
2	4	2	3	5	2	4	7	2
5	7	3	6	10	3	8	14	3
8	11	4	11	17	4	15	23	4
12	17	5	18	28	5	24	34	5
18	24	6	27	36	6	35	48	6
25	32	7	37	48	7	49	64	7
33	41	8	49	61	8	65	82	8

Storage strategies: article-pure placing and free storage order
Aisle width: $b_{aisle} = 3{,}000$ mm transversal orientation
Proportionate aisle number: $n_{aisle} = 0.5$
Place dimensions: $l_{SP} = 850$ mm $b_{SP} = 1{,}300$ mm
Pallet dimensions: $l_{LU} = 1{,}200$ mm $b_{LU} = 800$ mm

The optimized effective ground area per storage unit can be calculated by inserting the optimal pile depth (16.40) into function (16.36):

$$a_{SUopt} = a_{SUeff}(C_{zopt}) \qquad [m^2/SU]. \tag{16.43}$$

Figure 16.15 shows the relationship between the optimized effective ground area per storage unit and the mean stock per article for the example of the pallet block place store and different stacking factors. This leads to the rule:

- The optimized effective ground area per storage unit decreases with increasing article stocks and reaches asymptotically the 100%-ground area (16.34).

The optimization of the effective ground area per storage unit with the help of the above rules and formulas is an important step when planning and dimensioning a multi-unit store. The space optimization not only minimizes the ground area, but also the total storage volume, the place costs and the path lengths for in- and out storing. After the ground area has been optimized, the other free design parameters can be used to achieve a cost optimal multi-unit store.

16.8 Storage Planning and Dimensioning

Before planning a new storage system or executing a store extension, a *target planning* is necessary to determine the expected static and dynamic storage demand and the restrictions. This includes the critical examination of which articles should be kept in stock at all, and the assessment of safety stocks and replenishment quantities. In many cases, an extension of the store can be avoided, and the size of a new store reduced by optimal *inventory management* (see *Chap. 11*).

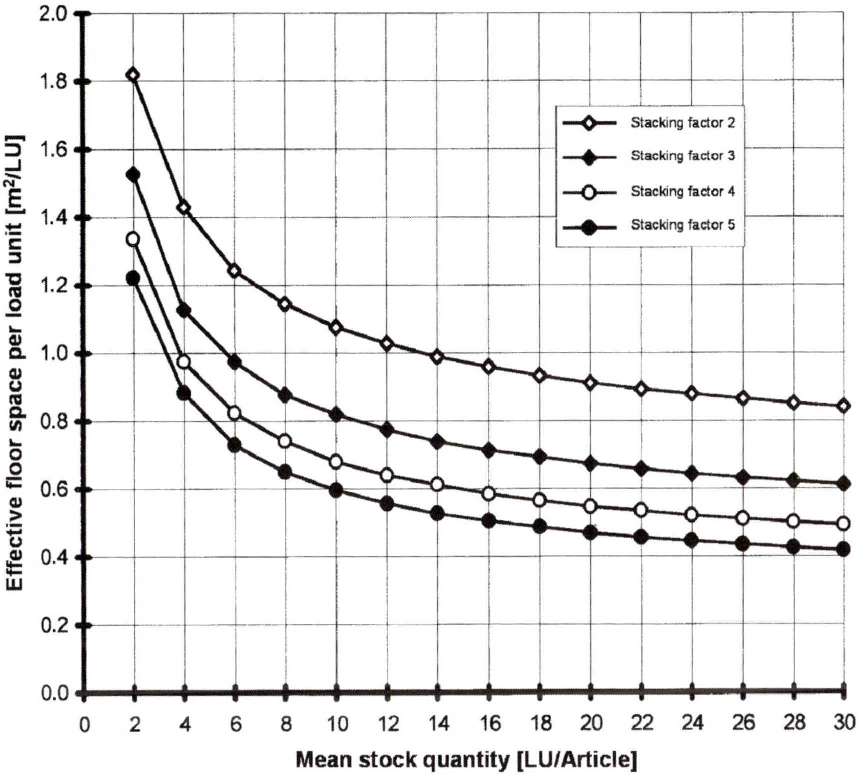

Fig. 16.15 Dependence of the optimized effective ground area per storage unit on the mean stock quantity

Parameter: stacking factor $C_y = 2/3/4/5$ other parameters: see *Fig. 16.14*

According to Sect. 3.2, the target planning is followed by *system planning*, *layout planning* and *detail planning*. *Steps of storage system planning* are:

1. *Synopsis* of the complete static and dynamic storage demand and of all relevant restrictions.
2. *Segmentation* of the storekeeping articles in homogenous groups with similar storage demand, mean stocks and technical properties.
3. Selection of load carriers, dimensioning of the storage units, and determination of the stacking factors for the different homogenous article groups.
4. *Pre-selection* of suitable storage types for the different article groups and *elimination* of insuitable storage technologies based on the previous rules and criteria.
5. *Determination* of the optimal storeplace *capacity* and calculation of the *place demand* for the different article groups.
6. *Technical conception* of the storeplaces, shelves and rack modules, storage devices and in- and out-feed transport system.

7. *Static storage dimensioning* of the technically designed systems in order to fulfill the static storage demand using the static design parameters and optimal placing strategies.
8. *Dynamic storage dimensioning* of the single modular systems in order to cope with the dynamic storage demand using the dynamic design parameters and optimal moving strategies.
9. Design of the *storage management system* (SMS) with storage administration and local control units.
10. *Calculation of investment* and *operating costs* based on budget prices and target cost rates.
11. *Overall optimization* of the resulting modular storage system by variation of the remaining free parameters.
12. *Selection and combination* of the most economic storage systems for the different article groups.

In the following *layout planning* phase, the modular storage systems are combined with the separately planned commissioning systems, goods entry and dispatch area and other functional areas, and connected by suitable transport systems. The result is a space and cost optimal warehouse, logistic centre or production plant with a *modular set-up* that enables stepwise extension and flexible adaptation to cope with future demand (see *Chap. 19*).

System planning and layout planning are *iterative processes* (see *Sect. 3.2*). Depending on the results of the following steps some of the previous steps are repeated several times. Up to a certain extent, storage planning can be executed and supported by suitable *storage planning software*. The program calculates, with the help of the formulas and algorithms presented in this book, the dimensions, capacity, limit performances, investment and operating costs for the subsystems and the total storage system for a given static and dynamic storage demand. Further input values are restrictions, technical data and prices of the storage device, building parts and technical installations, available ground area and personnel. Storage planning software can be used to

- simulate the system behavior for time dependent demand
- calculate different scenarios
- compare alternative system solutions
- determine the application fields for different storage types
- calculate and minimize place costs and throughput costs
- sensitivity analysis for different requirements

Conditions for a qualified storage planning software are correct formulas and algorithms, which take into account all relevant design parameters. The *analytical simulation* of a storage system for dynamic demand with the help of such software can replace a time consuming and costly *stochastic simulation* (see *Sect. 5.4*).

Target figures of storage planning are the total operating costs, the place cost rates and the throughput cost rates. They can be optimized by varying the free parameters. Independent *design parameters* for planning and optimization of storage systems are:

- *Aisle Orientation*
  - longitudinal placing $l_{SU} \parallel L_{AM}$
  - transversal placing $l_{SU} \perp L_{AM}$
- *Place parameters*
  - storeplace capacity C_{SP}
  - place module capacity C_{PM}
- *Static storage parameters*
  - number of store levels N_y (16.44)
  - number of aisle modules N_{AM}
  - front-side buffer places N_{BP}
- *Dynamic storage parameters*
  - number of storage devices N_{SD}
  - capacity of storage devices C_{SD}
  - speeds of storage devices v_x, v_y, v_z
  - acceleration values b_x, b_y, b_z

By systematic variation of these parameters, a given storage demand can be fulfilled within the restrictions at minimal operating costs and/or investment. Other parameters and key performance indicators of a storage system can be derived from the design parameters (16.44). In some cases it is opportune to replace one of the parameters (16.44) by another parameter. For example, instead of the vertical number of store levels N_y the horizontal number of place modules N_x can be used.

Storage planning is much more sophisticated than generally assumed. With increasing degree of detail, more and more features have to be considered. For example, a higher store capacity can be achieved within the same total space by using the front area of the racks for additional storeplaces or by bridging transport ways with overhead racks. The limit performance can be increased by using multi-unit storing devices and by dynamic storage strategies.

In order to assess the *feasibility* of a theoretically possible solution, *technical expertise* and *practical experience* are indispensable. On the other hand, engineers and planners should not be blinded by technique and lose sight of the potential optimization of the total system.

16.9 Static Storage Dimensioning

The task of the *static storage dimensioning* is to fulfill the static storage demand at lowest investment and place costs.

A storage system [S] consists of *storage modules* [SM], which are – as shown in *Figs. 16.16* and *16.17*– the combination of N_{AM} parallel *aisles modules* [AM]. Each aisle model, such as example *Fig. 16.5*, consists of one S/R-aisle and a number of *place modules* [PM], which contain the single *storeplaces* [SP] and are located horizontally and vertically along one or both sides of the aisle.

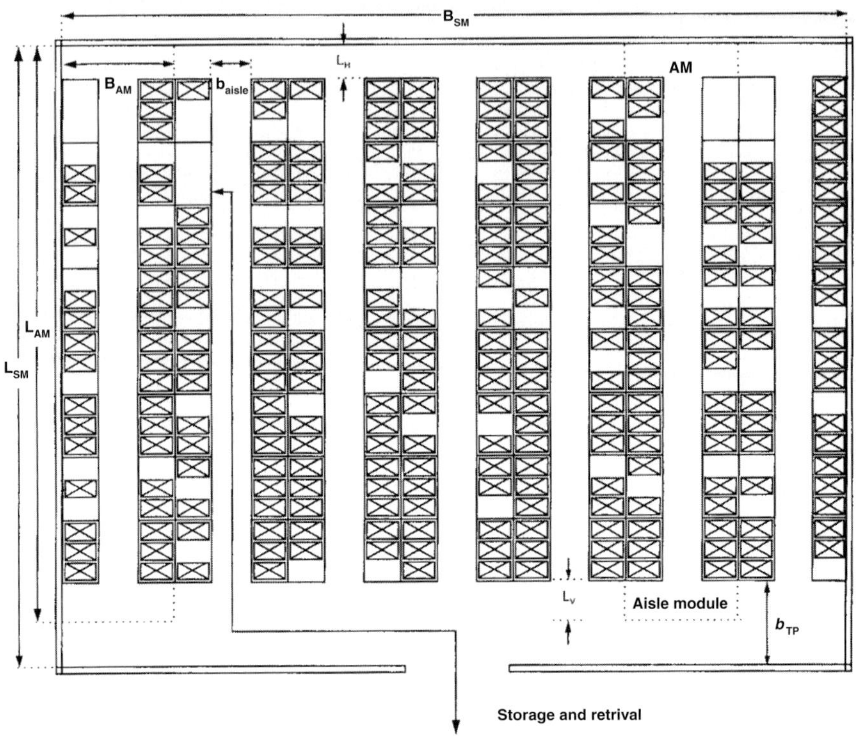

Fig. 16.16 Storage module with parallel aisles arrangement
AM: aisle module
b_{TP}: transport path width

Several storage modules can be arranged in a *fire section*, which is separated from other sections by *fire walls*. Storage modules, which are served by the same in- and out-feed transport system, as shown in *Figs. 16.8* and *16.9*, form a *conveyor section*. In layout planning, the storage modules are combined and connected with other *function modules* (see *Chap. 19*).

The static storage dimensioning must take into account the following *restrictions* and *constraints*:

- Total *storage length* L_S, *width* B_S and *ground area* $A_S = L_S \cdot B_S$ are limited by the dimensions L_{Smax} and B_{Smax} of the available ground:

$$L_S \le L_{Smax} \qquad B_S \le B_{Smax} \qquad A_S \le L_{Smax} \cdot B_{Smax} \tag{16.45}$$

- The *storage height* H_S is restricted by the inner height of an existing building or by a technically feasible or tolerable *construction height* H_{Smax}:

$$H_S \le H_{Smax}. \tag{16.46}$$

- For a store with manually operated storage devices, the floor dimensions of a storage module which constitutes a fire section are limited by the maximal *escape way length* s_{Exit} to the next exit:

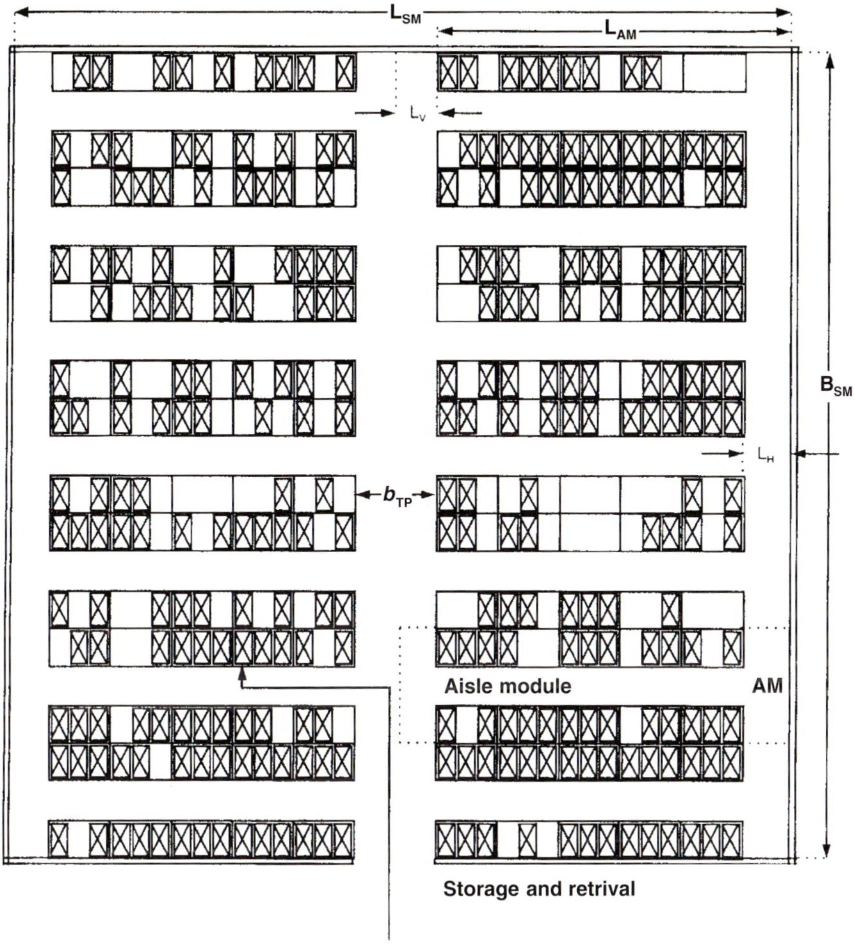

Fig. 16.17 Storage module with opposite aisles arrangement
AM: aisle module
b_{TP}: transport path width

$$\sqrt{(L_{SM}/2)^2 + (H_{SM}/2)^2} \le s_{Exit} \tag{16.47}$$

- To ensure that the storage devices can run freely within an aisle on shortest paths the number of aisles modules should be larger or at least equal to the number storage devices:

$$N_{AM} \ge N_{SD} \tag{16.48}$$

- To avoid expensive aisle transfer units and performance reduction by aisle changes, for automatic high bay stores the number of aisles module should be equal to the number of S/R-units: $N_{AM} = N_{SD}$.

- The number of aisle modules in a conveyor section is limited by the maximal number of storage devices that can be served by the same in- and out-feed transport system:

$$N_{AM} \leq N_{AMmax} \tag{16.49}$$

For example the distribution shuttle of *Fig. 16.8* can serve maximally 6 automatic S/R-units. If high capacity is required, a restricted ground area can be *KO-criterium* for storage types with limited technical height.

Considering the given restrictions, the static storage dimensioning can be performed in the following steps:

16.9.1 Place Module Design

In a place module, several storeplaces of equal or different size are combined in one construction unit with smallest possible ground area and volume. Their design must enable a flexible arrangement of many place modules side by side or on top of each other.

The simplest solution is a place module identical with the storeplace. For example, the place module of a block place store is – as shown in *Fig. 16.2* – the block place. A place module of a satellite store consists – depending on the shelf construction – of one or more parallel rack channels.

Place modules with two or more places and with different dimensions, which are arranged in a rack, are shown in *Figs. 16.4* and *17.20*. They are used alternatively for 3 *EURO pallets* 800·1,200 mm or 2 *Industry pallets* 1,000·1,200 mm. Within place modules of different height, CCG1-pallets and CCG2-pallets can be stored together. The possibility to vary the placing of storage units with different dimensions in the same module is a central *advantage of place modules*:

- A storage system for storage units of different size can be build up from equal place modules, which are designed for different storage units.

Design and number n_{PM} [SP/PM] of storeplaces in a place module together with the dimensions of the storage units determine the measures l_{PM}, b_{PM} and h_{PM} of the place module.

The capacity of a place module with n_{PM} [SP/PM] storeplaces of capacity C_{SP} [SU/SP] is $C_{PM} = n_{PM} \cdot C_{SP}$ [SP/PM]. Hence, the required *number of place modules* for a demand of N_{SP} storeplaces is:

$$N_{PM} = \{N_{SP}/C_{PM}\} = \{N_{SP}/(n_{PM} \cdot C_{SP})\} \qquad [PM]. \tag{16.50}$$

The curly brackets {...} denote the rounding up of the bracket content to the next integer.

16.9.2 Aisle Module Design

A *standard aisle module* is made up by N_x place modules, which are arranged horizontally side by side in x-direction on both sides along the S/R-aisle, and N_y place modules arranged vertically on top of each other in z-direction. In an *irregular aisle module*, the place modules are located only on one side. The following formulas, which demonstrate the principle procedure of static storage dimensioning, hold for standard aisle modules.

For block place stores the number of vertical levels $N_y = 1$. For stores with vertical arrangement of the place modules, such as rack stores, it is a variable $N_y > 1$. For these systems, the *vertical number of place modules* N_y is, within the height restriction (16.46), a free design parameter.

If a total of N_{PM} place modules is required, which are arranged on top of each other in N_{AM} aisle modules with the *vertical number of place modules* N_y, the necessary *horizontal number of place modules* is

$$N_x = \{N_{PM}/2\,N_y \cdot N_{PM}\} \quad (16.51)$$

For place modules with the measures l_{PM}, b_{PM} and h_{PM} result the *measures of the aisle module* (see e.g. *Fig. 16.5*):

$$L_{AM} = N_x \cdot l_{PM} + L_{pos} \quad (16.52)$$

$$B_{AM} = 2 \cdot b_{PM} + b_{aisle} \quad (16.53)$$

$$H_{AM} = N_y \cdot h_{PM} + H_{pos} \quad (16.54)$$

Herein, L_{pos} is the sum of the front- and backside *positioning measures*, which are necessary for the S/R-unit to reach the storeplaces in the first and last horizontal position. H_{pos} is the sum of the lower and upper positioning measures necessary to reach the lowest and the highest places. The positioning measures and the *aisle width* b_{aisle} are determined by the construction of the S/R-unit and of the handling device.

16.9.3 Arrangement of Storage Modules

As shown in *Figs. 16.16* and *16.17*, the aisle modules can be combined to a compact storage module in two *standard aisle arrangements*:

- *Parallel aisles arrangement* of all N_{AM} aisle modules with a *transport path* along the front
- *Opposite aisles arrangement* of two equal cluster of $N_{AM}/2$ aisle modules with a transport path in the middle

From *Fig. 16.16*, with a width of the *front transport path* b_{TP}, result the *ground measures* of the storage module with *parallel aisles arrangement*:

$$L_{SMpar} = L_{AM} - L_{pos} + b_{TP} \quad (16.55)$$
$$B_{SMpar} = N_{AM} \cdot B_{AM}.$$

The *ground area* of the storage module with parallel arrangement is

$$A_{SMpar} = N_{AM} \cdot B_{AM} \cdot (L_{AM} - L_{pos} + b_{TP}). \quad (16.56)$$

From *Fig. 16.17*, with a width of the *middle transport path* b_{TP}, result the *ground measures* of the storage module with *opposite aisles arrangement*:

$$L_{SMopp} = 2 \cdot (L_{AM} - L_{pos}) + b_{TP} \quad (16.57)$$
$$B_{SMopp} = N_{AM} \cdot B_{AM}/2.$$

The *ground area* of the storage module with opposite arrangement is

$$A_{SMopp} = N_{AM} \cdot B_{AM} \cdot (L_{AM} - L_{pos} + b_{TP}/2). \tag{16.58}$$

The storage height is for both arrangements equal to the height (16.54) of the aisle module. Therefore the total *volume of the storage module* is:

$$V_{SM} = L_{SM} \cdot B_{SM} \cdot H_{AM}. \tag{16.59}$$

The difference of the ground areas (16.56) and (16.58) for the standard arrangements is

$$\Delta A_{SM} = N_{AM} \cdot B_{AM} \cdot (b_{TP} - L_{pos})/2. \tag{16.60}$$

This leads to the *aisle module arrangement rule*:

- The opposite aisles arrangement requires less ground area then the parallel aisles arrangement if the transport path is broader than the horizontal positioning measure for the S/R-units.

This rule, however, is of minor importance in practice as the arrangement of the aisles is in many cases not only determined by the required ground area but more by the dimensions and the construction of an existing hall, by the combination with other function modules, such as the commissioning systems, and by shortest travel paths for S/R-units and transport vehicles. Due to these influences for block place stores and conventional rack stores, the opposite aisle arrangement is opportune. For high bay stores with aisles longer than 20 m, the parallel aisle arrangement is the better solution.

If the dimensions of the sum of the required storage modules exceed the fire and security restrictions, it is necessary to divide the total number of aisles into 2, 4 or more equal *storage clusters* of smaller dimensions, which are separated by fire walls and automatic security doors in different *fire sections* (see e.g. *Fig. 17.21*).

16.10 Travel Time Formulas

The necessary number of storage devices is determined by the cycle times for in- and out-storing. They depend on the travel times for the partial moves of the load unit in the three dimensions of the storage (see *Fig. 16.1*).

If a transport device takes a load unit from a start point in a one-dimensional movement along a *travel distance* s [m] to a stop point, the speed depends on time as shown in *Fig. 16.18*. The load unit is moved with a *mean acceleration* b^+ [m/s²] until the *maximal speed* v_m [m/s²] is reached. In order to stop at the target point the speed is reduced just in time with a *mean de-acceleration* b^- [m/s²]. The time for this *elementary transport move* is determined by the travel distance, the maximal speed and by the *harmonic mean value* of acceleration b^+ and de-acceleration b^-, i.e. the *acceleration value*:

$$b_m = 2 \cdot b^+ \cdot b^- / (b^+ + b^-) \quad [m/s^2]. \tag{16.61}$$

Maximal speed and mean acceleration values are listed for some storage devices in *Table 16.1* and for selected handling devices in *Table 16.2*. Corresponding data

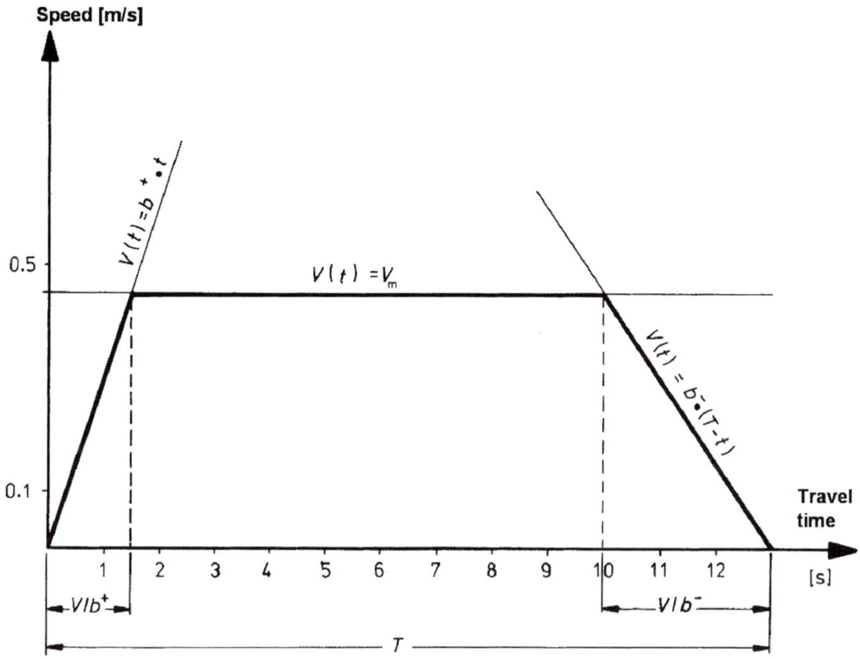

Fig. 16.18 Idealized dependency of the travel time on the speed

b^+: mean acceleration
b^-: mean deacceleration
v_m: maximal speed

for people and order picking devices are given in *Table 17.2*. These standard values are suitable for model calculations and planning. Real data have to be specified and guaranteed by the manufacturers.

From the time dependency of the speed follows (Gudehus 1975):

- The *travel time for an elementary move* over a travel distance *s* [m] with an *acceleration value* b_m and *maximal speed* v_m [m/s²] is

$$t_m(s) = \mathrm{IF}\left(s < v_m^2\ ;\ 2\sqrt{s/b_m}\ ;\ s/v_m + v_m/b_m\right) \quad [s] \qquad (16.62)$$

The universal *one-dimensional travel time formula* (16.62) is valid for any elementary move with the time dependency of *Fig. 16.18*. If the transport load is heavy, the travel times must be calculated with the mean values of acceleration and speed with and without load.

For horizontal moves of a storage device or a crane only in x-direction, m = x and the corresponding technical data have to be inserted. For vertical moves of a storage device m = y, and for horizontal moves of a crane or a telescope fork in z-direction, m = z.

The travel time for an *additive transport* along a path $s = \Sigma s_i$ with the *partial distances* s_i between starts and stops is the sum of the *partial travel times* $t_m(s_i)$ for the elementary moves:

$$t_{tr}(s) = \sum t_m(s_i). \tag{16.63}$$

Some transport means can execute moves in two or three directions simultaneously (see *Fig. 16.1*). A *two-dimensional moving storage device* transfers the load simultaneously in x-direction over a transport length *l* and simultaneously in y-direction over a lifting height *h*. For this case holds the *two-dimensional travel time formula*:

- The *travel time for a two-dimensional move* over a length *l* with acceleration value b_x and speed v_x and simultaneously over a height *h* with acceleration value b_y and speed v_y is

$$t_{xy}(l;h) = \mathrm{MAX}(t_x(l); t_y(h)). \tag{16.64}$$

This formula is valid for any kind of transfer device, which moves simultaneously in two directions. The one-dimensional travel times $t_x(l)$ and $t_y(h)$ are given by formula (16.62) with m = x and m = y. With z instead of y and *b* instead of *h*, formula (16.64) holds for the horizontal moves of overhead cranes.

In the most general case of a simultaneous three dimensional move, e.g. by a *stacker crane* or of the *hand of an order picker*, the traveling time is the maximum of three elementary travel times:

$$t_{xyz}(l;h;b) = \mathrm{MAX}(t_x(l); t_y(h); t_z(b)). \tag{16.65}$$

Travel times can be reduced by increasing the maximal speed only if the elementary moves reach the maximal speed. Therefore, too high speeds are ineffective and only cause overpowered and expensive transport and storage devices. A consequence of this fact is the *1st speed selection rule*:

- The *maximal effective speed* of a transport device with acceleration value b_m for moves with path lengths s in the range $0 < s < S$ is

$$v_{eff} = 1/2 \cdot \sqrt{S \cdot b_m} \tag{16.66}$$

For example, the maximally effective horizontal speed for a storage device with acceleration value $b_x = 0.5$ m/s serving storage modules with the length L = 60 m is $v_{xeff} = \frac{1}{2}\sqrt{60.05} = 2.7$ m/min = 160 m/min.

The load carrier of a storage device, which moves simultaneously in x- and y-direction, follows a line parallel to the *speed line* $y = (v_y/v_x)\cdot x$. The mean travel time is minimal, if the number of storeplaces reached with maximal running speed equals the number of places reached with maximal lifting speed (Gudehus 1973). This leads to the *2nd speed selection rule*:

- The *optimal speed relation* between running speed v_x and lifting speed v_y of a storage device, which serves a rack with length L and height H in simultaneously two dimensional moves, is

$$v_y/v_x = H/L. \tag{16.67}$$

If the partial speeds are selected corresponding to this rule, the speed line is the diagonal of the rack plane (see *Fig. 16.1*).

The limit performance of transport means, vehicles and storage devices, which serve a large number of places along a path, on a plane or within a space, is determined by the weighted mean value of the single travel times. For one-dimensional moves, the mean value of the travel times and the travel time for the mean distance are approximately equal, as long as the maximal speed is reached for the majority of the moves. A consequence is the *one-dimensional travel time rule*:

- The *mean one-dimensional travel time* $t_m(n)$ between the neighbors of n randomly chosen points on a path with total length S equals the travel time $t_m(s_n)$ for the *mean distance* s_n between the neighbored points, which is

$$s_n = S/(n+1). \tag{16.68}$$

Special cases of this general rule are:

- The mean distance between the endpoint and the single points on a path of length S is S/2, if all points are visited with same frequency.
- The mean distance between any of two points on a path of length S is S/3, if all points are visited with same frequency.

An application of the first rule is:

- The mean drive-in path for storeplaces with depth b and pile depth C_z is

$$B_m = b \cdot (C_z + 1)/2. \tag{16.69}$$

For two-dimensional moves, the mean travel time to the single points is not equal to the travel time to the mean point. For the optimal speed relation (16.67) and equal visiting frequency for all points of a rack plane with length L and height H, the mean value of the single travel times (16.64) can be calculated explicitly. Results are the *two-dimensional travel time rules* (Gudehus 1973):

- The *mean two-dimensional travel time of single cycles* between the corner point E = (0;0) and the points of a rectangular plane with length L and height H is the sum

$$t_1(L, H) = t_{xy}(E, P_1) + t_{xy}(E, P_1) \tag{16.70}$$

  of the two-dimensional travel times $t_{xy}(E,P_1)$ and $t_{xy}(E,P_1)$ from the entrance to the two *target points*:

$$P_1 = (2L/3; H/5) \quad \text{and} \quad P2 = (L/5; 2H/3). \tag{16.71}$$

- The *mean two-dimensional travel time of double cycles* is the sum of the two-dimensional travel times from the corner E to the first target point P_1, from there to the second target point P_2 and back to the corner:

$$t_2(L; H) = t_{xy}(E, P_1) + t_{xy}(P_1, P_2) + t_{xy}(E, P_1). \tag{16.72}$$

The two dimensional travel-times $t_{xy}(E,P_1)$, $t_{xy}(E,P_1)$ and $t_{xy}(P_1,P_2)$ are given by formula (16.64). The accuracy of the travel times calculated with these formulas is better than 2%, also if the optimal speed rule (16.67) is not exactly kept.

Storing and commissioning devices with load capacity for C_{DC} storage units can serve $n \le 2C_{DC}$ storeplaces in a combined in-out-storing cycle. These *n-point-cycles* can be executed effectively with the *2-stripe strategy* of *Miebach* as shown in

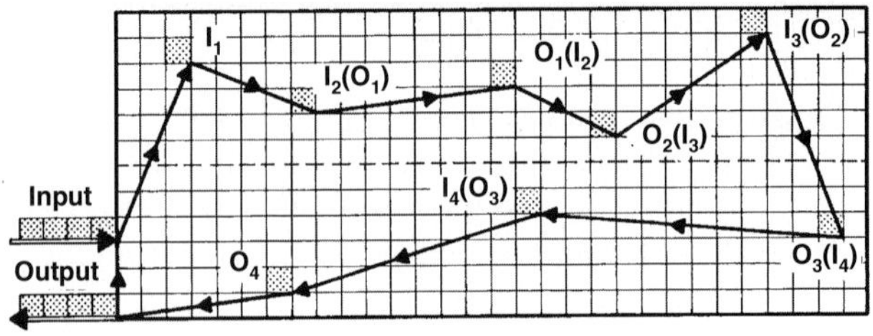

Fig. 16.19 Combined storage and retrieval cycle with 2-stripe strategy
in-storing places: I_i out-storing places: O_j capacity of storage device: $C_{SD} = 4$ LU

Fig. 16.19 (Miebach 1971). For this strategy holds the approximate *n-point travel time formula* (Gudehus 1973):

- The *mean n-point travel time* of a storage and retrieval tour starting at the lower corner point E and visiting n randomly distributed places of a plane with length L and height H following the 2-stripe strategy and performing two-dimensional moves between the neighbor points is:

$$t_n(L;H) = \begin{cases} t_y(3H/4) + t_y(H/2) + (n-1)\cdot t_x(2L/(n+2) & \text{if } 4 \le n < 6 \\ t_y(3H/4) + t_y(H/2) + t_y(H/4) + (n-2)\cdot t_x(2L/(n+2)) & \text{if } 6 \le n \le 10 \\ t_y(3H/4) + t_y(H/2) + t_y(H/4) + (n-2)\cdot t_x(H/6) & \text{if } n > 10 \end{cases} \quad (16.73)$$

Herein, $t_x(..)$ and $t_y(..)$ are the one-dimensional travel times given by the universal formula (16.62) with m = x and m = y. The travel time for the special case n = 3 can be calculated by interpolation between the result of the formula (16.73) for n = 4 and the double cycle travel time (16.72).

The accuracy of the given travel time formulas is in most practical cases sufficient for the calculation of the travel and cycle times, since the inaccuracy of the speed and acceleration values is higher than of the approximations.

16.11 Dynamic Storage Dimensioning

Tasks of dynamic storage dimensioning are to determine the number of storage devices and to design the in- and out-feed transport systems in order to fulfill the dynamic storage demand at lowest costs. The throughput performance can be optimized by proper moving strategies. The number of storage devices and the total costs can be minimized by the layout of the storage and by the dynamic storage parameters (16.44).

Restrictions for the dynamic storage dimensioning are:

- acceptable retrieval times for single storage units
- tolerable providing times for complete storing orders
- prescribed storage strategies such as FIFO and uniform aisles distribution

Taking these restrictions into account and using the above travel time formulas, dynamic storage dimensioning can be executed in the following steps:

16.11.1 Calculation of Cycle Times

The *in-storing cycle time* is the time a storage device needs to take up to C_{SD} storage units from a pick up point, to transfer them to empty storeplaces, to put them down and to return empty to the takeover up point. The *out-storing cycle time* is the time a storage device needs to move to one or more storeplaces, to collect up to C_{SD} storage units and to transfer them to a transfer point, where they are deposited.

The *combined in-out-storing cycle time* is the total time necessary to take over and to transfer up to C_{SD} storage units to empty storeplaces and to collect, during the same run, up to C_{SD} required storage units, to transfer them to the transfer point and to deposit the out-stored units. If takeover point and transfer point are separated, the combined cycle ends with the *empty move* of the storage device from the exit to the entrance.

The *relocation cycle time* is the time needed to move a storage unit from one storeplace to another place including the empty travel time between two succeeding relocations.

In any of these four cases, the cycle time is the sum of *load handling cycle times* for taking and/or depositing storage units, *travel times*, and *idle times between* the successive moves. The length of the load handling cycle depends on the loading technique, which can be taken into account by the *load handling factor*:

$$f_{LH} = \begin{cases} 1 \text{ for load handling with empty move} \\ 2 \text{ for load handling without empty move} \end{cases} \tag{16.74}$$

The *idle times* between the different steps of a storage cycle are caused by the delayed reaction and response of the control system and by the positioning of the storage device. State of the art for *mean idle times* t_o of storage devices is:

- The mean idle time between the moves of *manually operated devices* depends on the skill and practice of the operator and ranges from 1 to 2 s.
- The mean idle time between the moves of a well adjusted *automatic storage device* is between 0.5 to 1 s.

Insufficient adjustment, slow control electronics and bad storage management can increase the idle times significantly and reduce the performance of the storage devices considerably.

For stores without fast-mover zones and with one-dimensionally moving storage devices with one load handling unit, the cycle times can be calculated from the length L and the height H of the storage module and from the mean drive-in-depth B with the *cycle time formulas for additive moves*:

- The *mean single cycle time* for separate in-storing or out-storing cycles of storage devices with additive moves is

$$\begin{aligned} \tau_I(L;H) &= \tau_O(L;H) \\ &= 4 \cdot t_o + 2 \cdot t_x(L/2) + 2 \cdot t_y(H/2) + 2 \cdot f_{LH} \cdot t_z(B) \end{aligned} \tag{16.75}$$

- The *mean double cycle time* for combined in-out-storing cycles of storage devices with additive moves is

$$\begin{aligned} \tau_{IO}(L;H) = 6 \cdot t_o + 2 \cdot t_x(L/2) + 2 \cdot t_y(H/2) \\ + t_x(L/3) + t_y(H/3) + 4 \cdot f_{LH} \cdot t_z(B) \end{aligned} \tag{16.76}$$

- The *mean relocation cycle time* for relocations in the same aisle with storage devices with additive moves is

$$\tau_R(L;H) = 4 \cdot t_o + 2 \cdot t_x(L/3) + 2 \cdot t_y(H/3) + 2 \cdot f_{LH} \cdot t_z(B) \quad (16.77)$$

Herein, the single travel times $t_m(..)$ with $m = x, y, z$ are given by the one-dimensional travel time formula (16.62).

For two-dimensional moving storage devices with optimal speed relation (16.67) and with entrances and exits at the same corner of the storage module, the cycle times are given by the *cycle time formulas for simultaneous moves* (Gudehus 1973; FEM 9.851 1973; VDI 4446 1999):

- The *mean single cycle time* for separated in-storing or out-storing cycles of storage devices with simultaneous moves is

$$\begin{aligned}\tau_I(L;H) &= \tau_O(L;H)\\ &= 2 \cdot t_o + t_{xy}(2L/3;H/5) + t_{xy}(L/5;2H/3) + 2 \cdot f_{LH} \cdot t_z(B)\end{aligned} \quad (16.78)$$

- The *mean double cycle time* for combined in-out-storing cycles of storage devices with simultaneous moves is

$$\begin{aligned}\tau_{IO}(L;H) = 3 \cdot t_o &+ t_{xy}(2L/3;H/5) + t_{xy}(L/5;2H/3)\\ &+ t_{xy}(14L/30;14H/30) + 4 \cdot f_{LH} \cdot t_z(B)\end{aligned} \quad (16.79)$$

- The *mean relocation cycle time* for relocations in the same aisle with storage devices with additive moves is

$$\tau_R(L;H) = 2 \cdot t_o + 2 \cdot t_{xy}(L/3;H/3) + 2 \cdot f_{LH} \cdot t_z(B) \quad (16.80)$$

Herein, the single travel times $t_{xy}(..)$ are given by the two-dimensional travel time formula (16.64). If entrance and exit are separated, single and combined cycle times increase by the empty travel time between exit and entrance.

By locating the entrance and exit not at a corner but on half height or on half length of the storage module, the mean driving times can be reduced. However, even for very high and long stores, the improvements of the limit performances by entrance and exit location on half of the height are less than 5%, and for a location on half the length are maximal 10%. These small effects generally do not justify a special construction.

By *optimal cycle strategies*, e.g. by selection of an in-storeplace within the out-traveling area or of an out-storeplace within the in-traveling area, the mean double cycle time can be reduced by the mean travel time $t_{xy}(14L/30;14H/30)$. For large stores the time savings by these strategies reaches up to 10% of the double cycle time. That means:

- By optimized double cycles the effective in-out-storing performance can be improved for stores with large dimensions by up to 10%.

Much higher improvements of the performances can be achieved by storage devices with several handling units. For example, *Fig. 16.20* shows the effects of higher load capacities on the limit performance for a pallet high bay store.

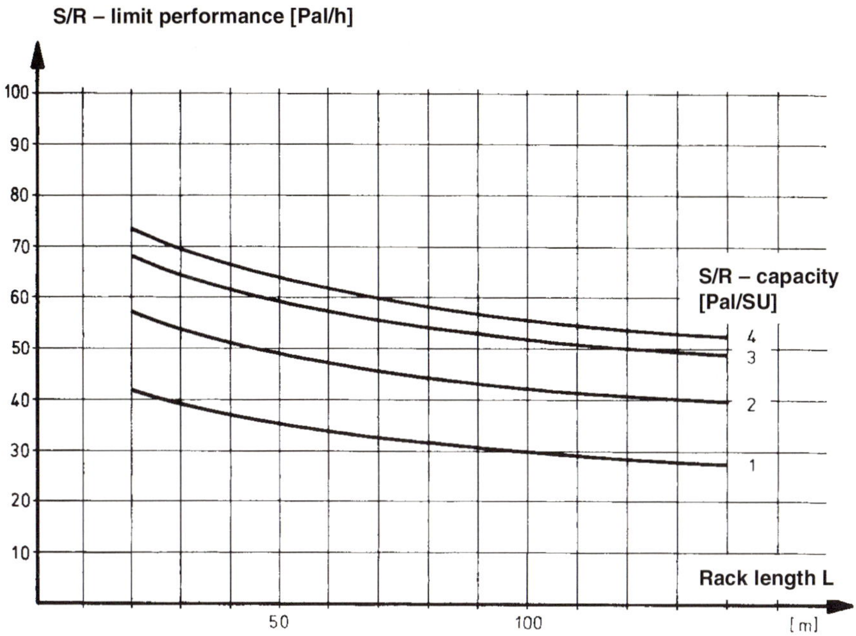

Fig. 16.20 Dependency of the in-storing and out-storing limit performance of a S/R-unit on the length of rack module and on the load capacity (Gudehus 1972)

Storage system: automatic high bay store for pallets
Parameter: load capacity of storage device $C_{SD} = 1/2/3/4$ Pal
Moving strategy: 2-stripe strategy of Fig. 16.19

The limit performance of S/R-units with $n > 2$ load handling devices operated with the 2-stripe strategy can be calculated with the *cycle time formulas for multi-unit storage devices*:

- The achievable *mean single cycle time* for separated in-storing or out-storing cycles of storage devices with $n > 2$ load handling devices is

$$\begin{aligned}\tau_{In}(L; H) &= t_{On}(L; H)\\ &= (n+1) \cdot t_o + t_n(L; H) + (n+1) \cdot f_{LH} \cdot t_z(B)\end{aligned} \tag{16.81}$$

- The *mean double cycle time* for combined in-out-storing cycles of storage devices with $n > 2$ load handling devices is

$$\tau_{IOn}(L; H) = (2n+1) \cdot t_o + t_{2n}(L; H) + (2n+1) \cdot f_{LH} \cdot t_z(B). \tag{16.82}$$

Herein $t_n(L;H)$ and $t_{2n}(L;H)$ are the n-point, respectively the 2n-point travel time given by formula (16.73).

Fig. 16.20 shows that the achievable improvement of the limit performance by two telescope forks on the S/R-pallet storage unit instead of one is about 40%. The further improvements by additional telescope forks decrease with the number of forks. As the investment increases at least linear with the additional load handling units, it is necessary to assess in any case whether it is profitable to use storage devices with multi-load handling.

For storage devices, which can change the aisle module, in addition to the in- and out-storing cycle times the aisle transfer times must be taken into account:

- The *mean aisle transfer cycle time* of an aisle changing storage device for stores with length L and a mean *transfer path* B_u is

$$\tau_{AC}(L;BT) = 3 \cdot t_o + 2 \cdot t_x(L/2) + t_u(B_u). \tag{16.83}$$

The travel times $t_i(..)$ can be calculated for $i = x$, u by the universal one-dimensional travel time formula (16.62) with the corresponding technical values for speed and acceleration of the storing device and the transfer unit.

The mean transfer path for aisle changes without strategy is one third of the width of the number N_{AM} of aisle modules served by one storage device, i.e. $B_u = N_{AM} \cdot B_{AM}/3$. The mean transfer path for *cyclic aisle change* between neighboring aisles is the breadth of the aisle module, i.e. $B_u = B_{AM}$. The comparison leads to the aisle transfer rule:

- The effective performance of stores with aisle changing storage devices can be improved and the number of storage devices can be reduced by cyclic instead of random aisle changes.

In many cases, the times for relocations and aisle changes and the possible moving strategies are not appropriately considered while dimensioning a store. Their effects on the performance are often neglected or incorrectly calculated.

16.11.2 Calculation of Number of Storage Devices

At the same time a storage device can execute only one of the *partial storage functions*:

$$\begin{array}{ll} \text{I}: & \text{in-storing in single cycles} \\ \text{O}: & \text{out-storing in single cycles} \\ \text{IO}: & \text{in-out-storing in double cycles} \\ \text{R}: & \text{relocations in separate cycles} \\ \text{AC}: & \text{aisle changes} \end{array} \tag{16.84}$$

The *partial utilization* $\rho_i = \lambda_i/\mu_i$ of the functions i = I, O, IO, R, AC is determined by the *partial demands* λ_i and the *partial limit performances* μ_i of the storage devices. From the general limit performance laws of *Sect. 13.4* follows the

- *Limit performance law for storage devices*:

$$\rho = \lambda_I/\mu_I + \lambda_O/\mu_O + \lambda_{IO}/\mu_{IO} + \lambda_R/\mu_R + \lambda_{AC}/\mu_{AC} < 1. \tag{16.85}$$

λ_I and λ_O [SU/h] are the in-storing demand (16.7) and the out-storing demand (16.8) respectively, which are executed in single cycles. λ_{IO} [SU/h] is the combined in-out-storing demand (16.6) respectively (16.9) executed in double cycles. λ_R [SU/h] are the relocations within the same aisle and λ_{AC} [AC/h] the frequency of aisle changes.

Due to relations (13.6) and (13.92) the *effective limit performances* of a storage device with *load capacity* C_{SD} can be calculated from the corresponding cycle times by the formula:

$$\mu_{i\,eff} = \eta_{ave} \cdot \eta_{uti} \cdot C_{SD} \cdot 3600/\tau_i \quad [\text{SU/h}] \quad \text{for i} = \text{I, O, IO, R, AC} \tag{16.86}$$

This formula takes into account, that the maximal possible partial limit performances μ_i are reduced by the technical *availability* η_{ave} [%] and by the stochastic *utilization* η_{uti} [%] to the *effective limit performances* $\mu_{i\ eff}$.

The technical availability of a storage device depends on the construction, the load, the maintenance and other factors, which have been analyzed in *Sect. 13.6.* Condition for an efficient and reliable storage system is:

- The *technical availability* for automatic storage devices as well as for manually operated devices must be at least 98%.

As explained in *Sect. 13.5,* a stochastic utilization results from the randomly arriving in- and out-storing orders and from the varying cycle times of the storage devices. The stochastic utilization depends on the number of *buffer places* on the in-feed and out-feed conveyors before the entrances and behind the exits of the storage aisles. Analytical calculations based on queuing theory and stochastic simulations of the queuing system lead to the *buffer dimensioning rule*:

- It is necessary to install at least 3 inward buffer places and 3 outward buffer places per aisle when a *stochastic utilization* of the storage devices of more than 97% is required.

If the in- and out-storing orders are executed by N_{SD} storage devices, the performance demand per storage device is $1/N_{SD}$ of the total dynamic storage demand (16.6), (16.7) and (16.8). To meet the limit performance condition (16.85) sufficient storage devices are necessary. Their number depends on the operating strategy and can be calculated by the *formulas for the number of storage devices,* which follow from the limit performance law (16.85):

- With only *single cycles,* the required number of aisle-fixed storage devices for a simultaneous in-storing demand λ_I and out-storing demand λ_O is

$$N_{SD} = \{\lambda_I/\mu_{Ieff} + \lambda_O/\mu_{Oeff}\} \tag{16.88}$$

- With *optimal double cycles,* the required number of aisle-fixed storage devices for simultaneous in-storing demand λ_I and out-storing demand λ_O is

$$N_{SD} = \left\{MIN(\lambda_I;\lambda_O)/\mu_{IOeff} + (\lambda_I - \mathrm{MIN}(\lambda_I;\lambda_O))/\mu_{Ieff} + (\lambda_O - \mathrm{MIN}(\lambda_I;\lambda_O))/\mu_{Oeff}\right\} \tag{16.89}$$

The curly brackets indicate a rounding up to the next integer since the number of storage must be a whole number. This causes a stepwise increase of the number of storage devices with steadily increasing performance demand.

If for example, a demand of $\lambda_I = 120$ in-pallets/h and $\lambda_O = 160$ out-pallets/h must be executed in single cycles with a limit performance $\mu_1 = \mu_I = \mu_O = 35$ Pal/h, the required number of storage devices is $N_{SD} = \{120/35+160/35\} = \{8.0\} = 8$. With optimally combined in- and out-cycles and a double cycle limit performance $\mu_2 = \mu_{IO} = 25$ Pal/h, the number of storage devices is reduced to $N_{SD} = \{120/25+40/35\} = \{5.9\} = 6$. The difference of 2 storage devices demonstrates the important *effect of double cycles.*

If in addition to the productive in- and out-storing orders, during the same hour λ_R [SU/h] unproductive relocations are executed with relocation limit performance

μ_R, the time loss must be taken into account by adding the partial occupation λ_R/μ_R on the right side of (16.88) and (16.89). By e.g. $\lambda_R = 30$ SU/h, which are executed with a limit performance is $\mu_R = 65$ SU/h, the number of storage devices increases to $N_{SD} = \{120/35+160/35+30/65\} = \{8.5\} = 9$ for only single cycles and to $N_{SD} = \{120/25+40/35+30/65\} = \{6.4\} = 7$ for optimal double cycles. A consequence is the *relocation advice*:

► Unproductive *relocations* can affect the productive performance of the storage devices considerably and should not be executed during peak hours.

For aisle changing storage devices, the available time for productive in- and out-storing cycles is reduced by aisle changes. By an *aisle change frequency* λ_{AC} [AC/h] and an *aisle change limit performance* μ_{AC} [AC/h], an *aisle change occupation* $\rho_{AC} = \lambda_{AC}/\mu_{AC}$ [%] of the storage devices is caused. This has to be taken into account by addition of the unproductive occupation ρ_{AC} on the right sides of relations (16.88) and (16.89). In the above example, a frequency of $\lambda_{AC} = 8$ AC/h executed with $\mu_{AC} = 15$ AC/h increases the number of necessary storage devices from 8 to 9 for single cycles and from 6 to 7 for optimal double cycles. This leads to the *aisles-changing advice*:

► *Aisles changes* should be avoided and not executed during peak hours.

The influence of operating hours and peak factors on the number of required storage devices is often neglected. If it is possible to increase the operating hours flexibly, from e.g. 8 to 16 hours per day or more, and to distribute the order execution equally over the operating time for an existing store, the performance can be increased by a factor of 2 or more, and for a planned store, the number of storage devices can be reduced.

By extension of the operating hours, depreciation, energy consumption and maintenance costs remain unchanged. Personnel for control even increase for highly automatic stores. This has to be considered for the cost calculation and shows:

► Storage planning is not a technical but rather a sophisticated organizational, managerial and technical task, i.e. an essential logistic task.

16.11.3 Layout of In- and Out-feed Conveyor System

For stores with *aisle-independent storage devices*, the cycle times (16.75) to (16.82) have to be increased by the mean travel times for the transports from the pickup places to the storage modules, and from the storage modules to the transfer places. The additional travel times decrease the limit performances and increase the number of storage devices.

Stores with *aisle-dependent storage devices* can either be served by free moving vehicles, such as forklift trucks or AGV, or by a fixed installed conveyor system. The number of vehicles can be calculated similar to the number of storage devices by the formulas (16.86) and (16.88) with the corresponding cycle times.

If the in- and out-feed transport is executed by a conveyor system, the maximal throughput is limited by its *bottleneck elements*. As shown in *Figs. 13.27* and *13.28*, these are the first branching element on the inward-conveyor line, and the last junction element on the outward-conveyor line. From the limit performance laws for the bottleneck elements result the dimensioning rules for the in-feed and out-feed conveyor system:

- A total in-storing demand λ_I served by N_{SD} parallel storage devices is limited by the partial limit performances μ_- and μ_L of the bottle-neck element of the in-feed conveyor system

$$\lambda_I \cdot (1 - 1/N_{SD})/\mu_- + (\lambda_I/N_{SD})/\mu_L \leq 1. \quad (16.90)$$

- A total out-storing demand λ_I served by N_{SD} parallel storage devices is limited by the partial limit performances μ and μ_L of the bottle-neck element of the out-feed conveyor system

$$\lambda_O \cdot (1 - 1/N_{SD})/\mu_- + (\lambda_O/N_{SD})/\mu_L \leq 1. \quad (16.91)$$

The conditions (16.90) and (16.91) determine the maximal number of storage devices N_{SDmax}, which can be served by a conveyor system with given partial performances, and limit the size of the *conveyor sections*.

16.12 Storage Investments

For system finding and layout planning it is sufficient to calculate the total storage investment based on *standard prices* for the site, the technical subsystems (16.21) and their single parts. For detail planning and investment decisions, the standard prices must be replaced by the *quoted prices* of manufacturers and suppliers.

For systems with high numbers of the same kind of technical devices, it is furthermore necessary to take into account the *economies of scales* in manufacturing and assembly. In the planning phase, the decrease of the storage investment with size and number can only be estimated. A precise quantification is possible only by tendering.

A storage system and its parts can be divided into static and dynamic parts:

- ▸ *Static storage parts* are necessary for keeping and provision of the storage units. They determine the *storeplace costs*.
- ▸ *Dynamic storage parts* are required for in- and out-storing. They cause the *throughput costs*.

Certain subsystems, such as the *storage management system* (SMS), are used for dynamic and for static functions. Their investment and operating costs must be divided corresponding to the relative utilization into a static and dynamic share. However, it is quite difficult to evaluate the investment P_{SMS} [€] for the hard- and software of a SMS in advance, since it depends on the requirements and on the storage size. The SMS-investment ranges between 30 and 200 T€ and can be quantified more precisely only by tendering.

16.12.1 Static Storage Investment

The most important parts of the static storage investment with its *cost drivers* and *cost factors* are:

- *Site*: The investment for the storage site depends on the real estate price and the preparation costs. The site investment increases proportional to the storage ground area A_S by a *site and preparation factor* P_{SP} [€/m^2].
- *Foundation*: The investment for foundation and base plate is determined by the foundation conditions, the kind of the storage building, the rack construction and the total floor load. The foundation investment increases proportional to the ground area A_S by a *foundation and base factor* P_{FB} [€/m^2].
- *Building*: The building investment depends on the construction and quality of the storage building and on the technical installations for heating, cooling, air conditioning etc. For halls, the investment increases approximately by a *hall building factor* P_{HB} [€/m^2] proportional to the ground area A_S, with a *wall cladding factor* P_{WC} [€/m^2] proportional to the total wall area F_W and by a *roof construction factor* P_{RC} [€/m^2] proportional to the roof area, which is roughly equal to the ground area of the store.
- *Racks*: The investment in racks depends on the storage type, the storeplaces or place modules, and on the rack construction. For moveable storeplaces, the rack investment also depends on the conveyor technique in the storage channels. The rack investment varies proportional to the number of storeplaces N_{SP} by a *rack place factor* P_{RP} [€/SP] or to the number of place modules N_{PM} by a *place module factor* P_{PM} [€/PM].
- *Sprinkler*: Sprinkler and other fire protection installations, such as fume and heat outlets, can be taken into account by a *sprinkler and protection factor* P_{SP} [€/SP]. This investment depends on the fire class of the storekeeping articles and increases proportional to the number of storeplaces N_{SP}.

For pallet stores, some benchmark values for the cost factors of the static storage investment are given in *Table 16.5*. These standard values can be used for model calculations, system comparisons and budgeting, but not for the calculation of the precise storage investment. If the investment for the SMS is taken into account with 50%, the dependency of the total static storage investment on the *storage investment cost drivers* L_S, B_S, H_S and N_{SP} is:

$$\begin{aligned} I_{Sstat} = {} & B_S \cdot L_S \cdot (P_{SP} + P_{FB} + P_{HB} + P_{RC}) + 2H_S \cdot (L_S + B_S) \cdot P_{WC} \\ & + N_{SP} \cdot (P_{RP} + P_{SP}) + P_{SMS}/2. \end{aligned} \tag{16.92}$$

The dependency (16.92) leads to the *storage capacity rule*:

▸ The static storage investment does not directly depend on the installed throughput performance and increases by the *cost drivers* L_S, B_S, H_S and N_{SP} linear with the *effective storage capacity*.

The required throughput, however, influences the static investment indirectly by the selected storage type and the number of storage aisles.

The static investment (16.92) related to the effective storage capacity (16.31) is the static storage investment per storage unit or effective *storeplace investment*. For this static key indicator of a store holds the *technical economy of capacity*:

Table 16.5 Investments parts and standard cost factors for palette stores

Storage Investment Parts	Standard cost factors range	model price	price units
Storage site			
Real estate and preparation (without traffic installation)	40 to 100	70	€/m^2
Silo Building			
Concrete bare plate	250 to 300	275	€/m^2
Wall clothing	80 to 100	90	€/m^2
Internal concrete walls	100 to 150	125	€/m^2
Roof construction incl. FHS	130 to 180	150	€/m^2
Mezzanine levels	90 to 120	100	€/m^2
Storage Hall (H: 6 to 12 m)			
Foundation, base plate, construction, hall facilities etc.	150 to 200	175	€/m^2
External walls	250 to 300	275	€/m^2
Internal fire walls	100 to 150	125	€/m^2
Roof construction incl. FHS	50 to 80	65	€/m^2
Gate module incl. installations and proport. traffic space	120 to 160	140	T€/GM
Rack Construction			
Free standing pallet racks	40 to 60	50	€/PalPlace
Silo racks	100 to 140	120	€/PalPlace
Block place marking	10 to 20	75	€/BlockPlace
Sprinkler Installation			
Nozzles. tube network, central control (without reservoir)	30 to 50	40	€/PalPlace
In- and Outfeed system			
Conveyer installation incl. buffer places per aisle	40 to 80	60	T€/aisle
Pallet conveyors	6 to 10	8	T€/m
Storage Management and Control System			
Hard- and software	100 to 400	250	T€

Costs and prices for model calculations (base: 2007)
FHS: fume and heat outlets
gate module: see *Fig. 16.10*

► The effective *storeplace investment* decreases with the effective storage capacity and approaches an asymptotic value for very high capacities, which differs for the different storage types.

This *technical economy of scale* holds for all storage types. It is caused by the decreasing space losses of the front-, back-, up- and downward *positioning measures* and by the diminishing contribution of the SMS. The proportionate storeplace investment decreases until the maximal possible height of a storage system is reached or other technical limits enforce additional storage modules. A further effect results for multi-unit storages from the increase of the filling degree (16.23) with the number of storage units per article.

Via depreciation and interest, the storeplace investment influences the *storeplace costs*. They decrease with increasing capacity, if the storeplaces are optimally filled.

16.12.2 Dynamic Storage Investment

The most important dynamic storage investments and their *cost drivers* and *cost factors* are:

- *Storage devices*: The investment for storage devices and load handling units includes the moving construction, machine, drives and control units and the correlated stationary power supply, guidance and aisle equipment. It varies with the number of storage devices N_{SD} proportional to the *storage device price* P_{SD} [€/SD].
- *Transfer vehicles*: The transfer vehicle investment for forklift trucks or AGV and their control system is proportional to the vehicle number N_{TV} and the *transfer vehicles price* P_{TV} [€/TV].
- *Conveyor system*: The in- and out-feed conveyor investment depends on the technique, the transport lengths, the number N_{SM} of storage modules to be served and on the required throughput performance. This investment can only roughly be calculated by standard price factors, such as price per meter or price per conveyor element. Storage optimization can be performed with a *conveyor system price* P_{CS} [€/SD] per storage device.

Table 16.1 contains standard prices of selected storage devices and transfer forklift trucks. Standard price factors for automatic pallet conveyors are listed in *Table 16.5*. These values can be used for system comparisons, optimizations and for rough cost budgeting.

If the proportionate investment for the SMS is 50%, the dependency of the total dynamic storage investment on the cost drivers N_{SD} and N_{TS} is:

$$I_{Sdyn} = N_{SD} \cdot (P_{SD} + P_{CS}) + N_{TV} \cdot P_{TV} + P_{SMS}/2. \qquad (16.93)$$

The number of storage devices and transfer vehicles increases stepwise less than proportional with the throughput performance since the proportionate transport distances decrease. On the other hand the transport distances for higher storage capacities are longer.

This leads to the *storage throughput rule*:

► The total dynamic storage investment increases less than proportional with the required throughput and store capacity.

The proportionate storage throughput investment is the relation of the dynamic storage investment (16.93) to the maximal possible throughput of storage units. For this key indicator holds the *economy of throughput*:

► The proportionate *storage throughput investment* decreases with the throughput and approaches an asymptotic value at high performances.

The storage throughput investment, however, is no practical benchmark and unsuitable for storage planning.

16.12.3 Investment Comparison of Storage Systems

For the comparison of different storage types and alternative systems, sometimes the total storage investment related to the total capacity K_S is used. However, the specific *storeplace investment*

$$I_{SP} = (I_{Sstat} + I_{Sdyn})/K_S \qquad [€/SP] \tag{16.94}$$

depends on many factors, which are difficult to separate. Therefore, the indicator place investment is quite misleading and must be applied carefully. If used at all, the place investment for multi-unit stores, such as block place stores or satellite stores, should be related to the *effective storage capacity* (16.31) and not to the 100%-storage capacity (16.30). Related to the 100%-capacity, multi-unit stores seem more attractive than they really are.

The following *Figs. 16.21* to *16.29* show the dependencies of the total storeplace investment and of the throughput costs on storage capacity and on limit performance for four different pallet stores: a *block place store* (BPS) served by forklift trucks; a *conventional rack store* (CRS) with forklift trucks; a *narrow aisle store* (NAS)

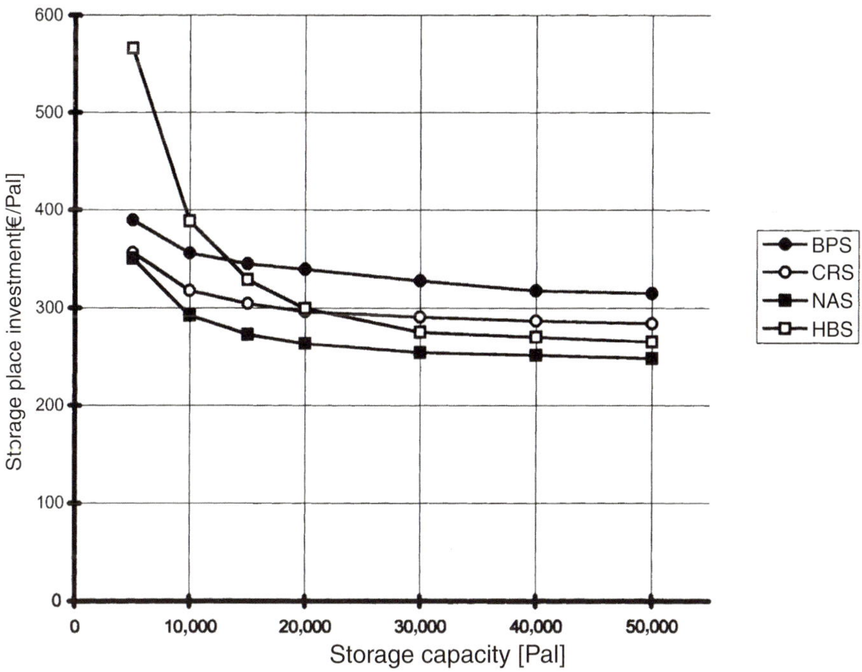

Fig. 16.21 Dependency of the total storage investment per pallet on the effective storage capacity

Parameter: turnover rate 12 p.a., other see *Table 16.6*

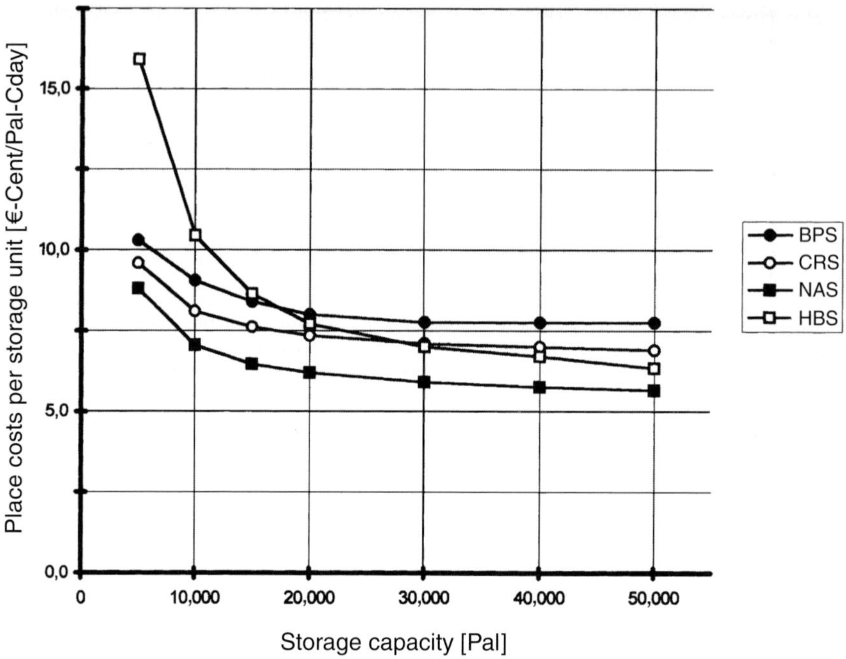

Fig. 16.22 Dependency of the storeplace costs on the storage capacity at 100% filling degree
turnover rate: 12 p.a. other parameters and terms see *Table 16.6*

operated by high reaching trucks; an automatic *high bay store* (HBS). Each system has been designed and optimized for the static and dynamic demand listed in *Table 16.6* with the help of a storage planning software using the above formulas and dimensioning rules. *Table 16.6* also contains the planning data, the resulting storage dimensions and key data, and the calculated costs of these systems. The goods entry and dispatch area are not included in these calculations, since they are project dependent and do not influence the storage system decision.

Figure 16.21 illustrates the dependency between the total investment per pallet and the storage capacity at a constant annual turnover of 12. This dependency and the above relations lead to the general *storage rules*:

- For all storage types, the investment per pallet decreases fast with increasing capacity at small capacities, and slower and slower at higher capacities until it reaches an asymptotic value, which depends on the storage type.
- The asymptotic value of the investment per pallet is reached at a capacity of about 20,000 pallets for block place stores and narrow aisle stores and at about 30,000 pallets for automatic high bay stores.
- For a narrow aisle store and the assumed turnover 12, results the lowest storeplace investment at all capacities.

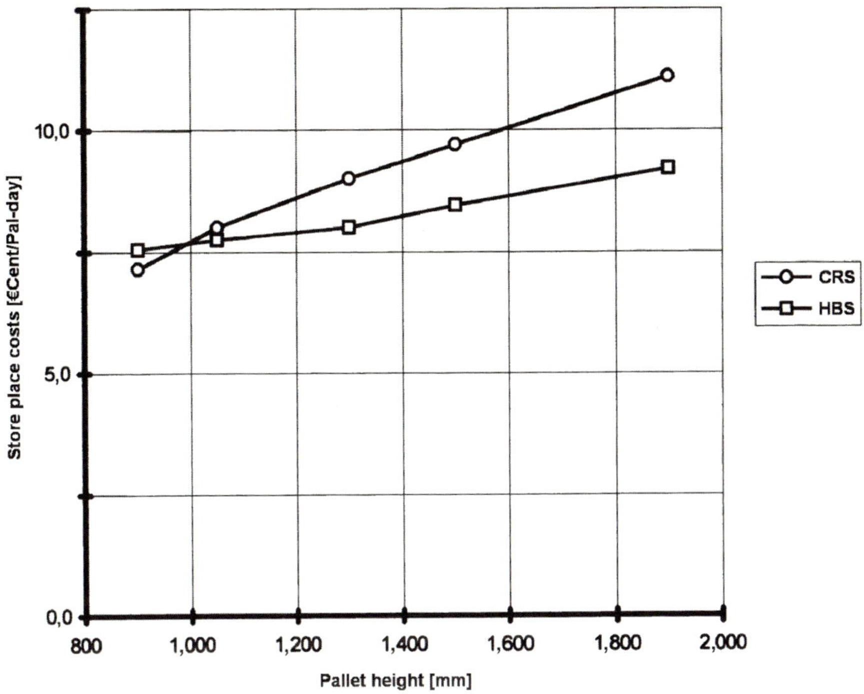

Fig. 16.23 Dependency of the storeplace costs on pallet height
Capacity: 20,000 pallets, *turnover rate*: 12 p.a.
other parameters and terms see *Table 16.6*

- Up to a capacity of 15,000 pallets, the block place store requires a lower investment per pallet than the automatic high bay store, but for all capacities a higher investment than the conventional and the narrow aisle store.
- Up to 15,000 pallets the storeplace investment for a high bay store exceeds the place investment for all other storage types. From 20,000 pallets upwards the investment per pallet undercuts the place investment for a conventional store and approaches for more than 30,000 pallets the place investment for a narrow aisle store.

These general rules for the dependencies and the relative magnitude of the investments for the compared storage systems are in a wide range independent of the throughput. They also hold true for other turnover rates. However, this is not the case for the absolute values. From the influence of the dynamic investment (16.91) on the total storage investment results the *storage investment rule*:

- The total investment per pallet critically depends on the installed limit performances and increases significantly with the required throughput.

It is common practice to estimate the total investment for a planned store by multiplying a benchmark price for the place investment of another project with

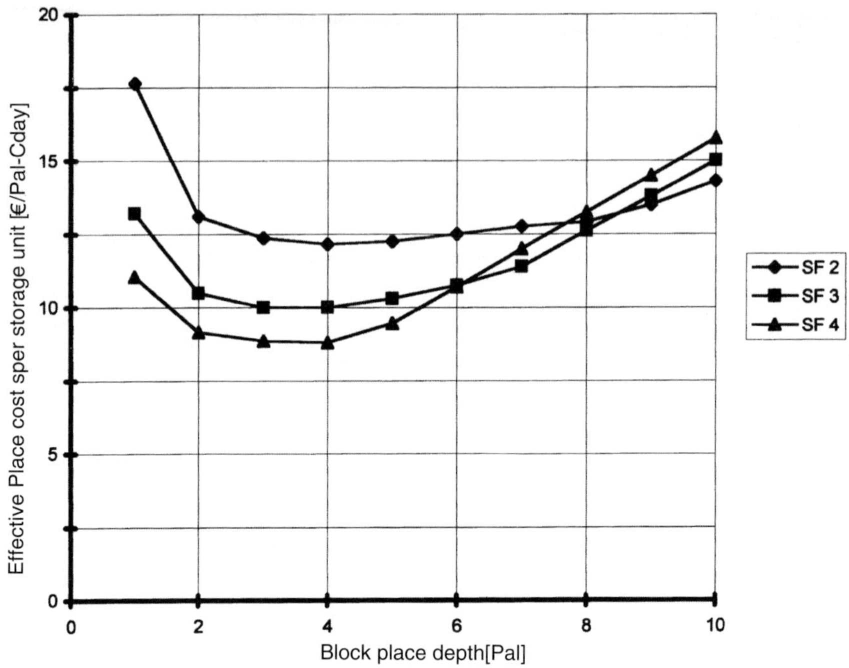

Fig. 16.24 Dependency of the effective storeplace costs for a block place pallet store on the storeplace depth

Stacking factors SF: $C_y = 2 / 3 / 4$
*Capacity:*20,000 pallets, *turnover rate*: 12 p.a.
other parameters and terms see *Table 16.6*

the required storage capacity. However, this practice can lead to extremely incorrect results, as the place investment depends critically on storage type, capacity and throughput. The estimation of the investment by *benchmarks* is tolerable only if the requirements are comparable and the storage type is the same.

16.13 Storage Operating and Performance Costs

A solution with higher investment, which enables by cost reduction a return on investment of less than 5 years, is in most cases economically preferable. The operating and performance costs are therefore far more important for the comparison and selection of storage systems than the investment.

Analog to the separation of the investment it is possible to separate the total *storage operating costs* K_S [€/PE] into *static storage costs* K_{Sstat} and *dynamic storage costs* K_{Sdyn}:

$$K_S = K_{Sstat} + K_{Sdyn} \qquad [€/PE]. \tag{16.95}$$

The static storage costs are caused by the operation of the static storage parts. The dynamic storage costs are generated by the dynamic storage parts and can be

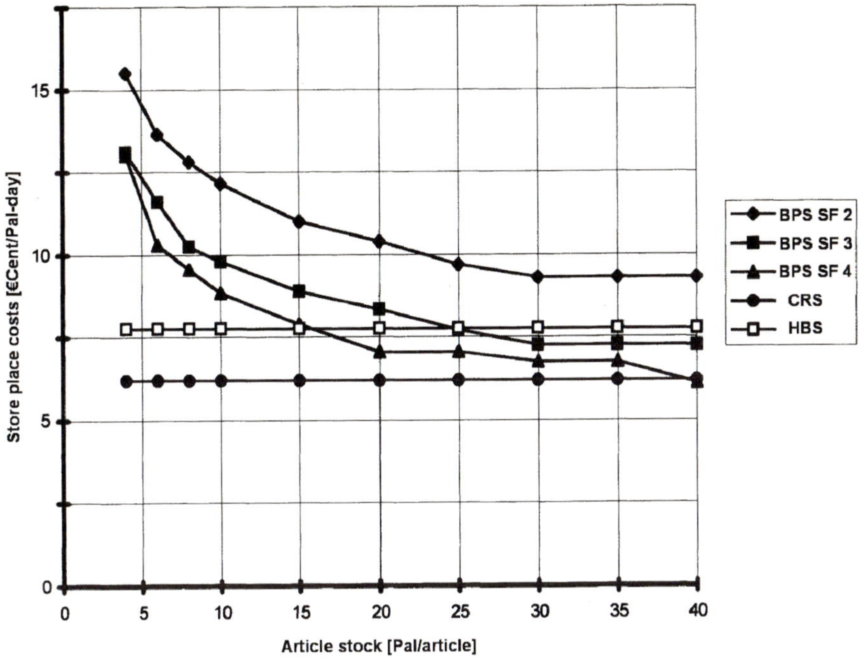

Fig. 16.25 Dependency of the effective storeplace costs on the stock per article for different optimally dimensioned pallet stores

Stacking factor SF: $C_y = 2/3/4$
Capacity: 20,000 pallets, *turnover rate*: 12 p.a.
other parameters and terms see *Table 16.6*

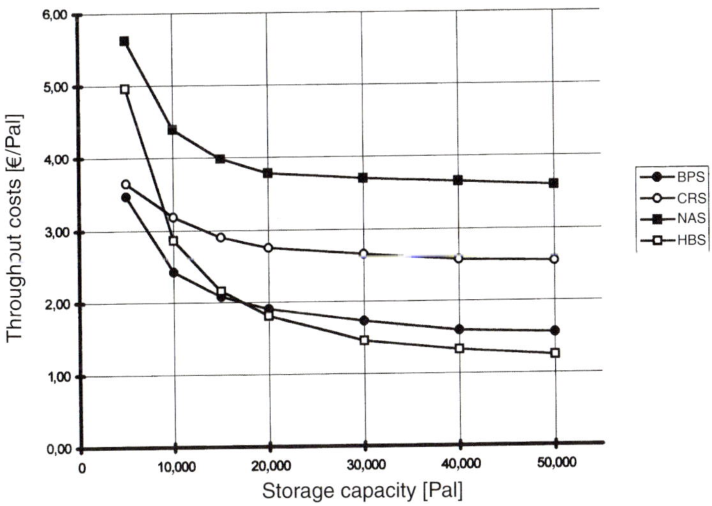

Fig. 16.26 Dependency of the storage throughput costs on the effective store capacity

turnover rate: 12 p.a. *stacking factor*: $C_y = 3$
other parameters and terms see *Table 16.6*

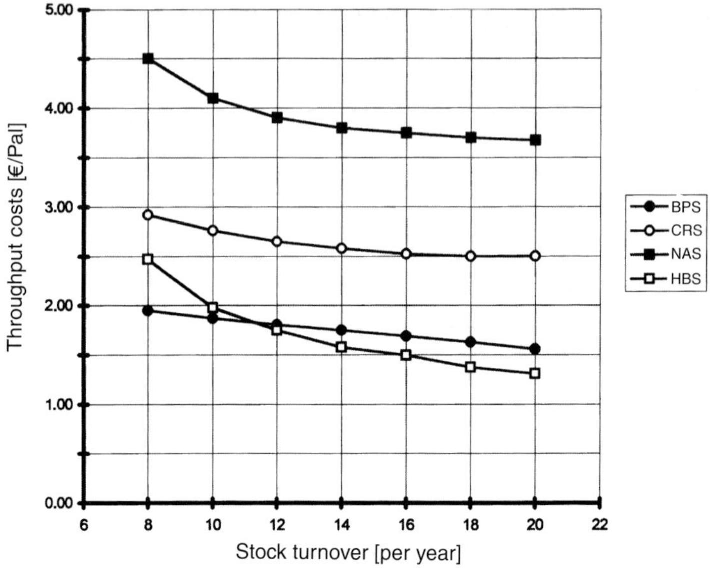

Fig. 16.27 Dependency of the storage throughput costs on the stock turnover
capacity: 20,000 pallets *stacking factor*: $C_y = 3$
other parameters and terms see *Table 16.6*

separated further into *in-feed costs* K_{Sin} and *out-feed costs* K_{Sout}, which are caused alone by in-storing and out-storing respectively.

The total storage operating costs K_S for period PE related to the storage throughput λ_T [SU/PE] during this period are the *storage turnover costs*:

$$k_{ST} = K_S/\lambda_T \qquad [€/SU]. \tag{16.96}$$

Correspondingly, the *storage capacity costs* $k_{SC} = K_S/(C_S \cdot T_{PE})$ [€/SP-day] are the total operating costs divided by the capacity and the length T_{PE} of the accounting period PE measured in *operating days*. However, k_{ST} and k_{SC} depend on both *storage cost drivers*, on *storeplace-days* and on *throughput*.

The separate relations of the partial operating costs K_{Sstat} and K_{Sdyn} to the static storage cost driver, i.e. the *used place-days per period* λ_P [SU-days/PE], and to the dynamic cost driver, i.e. the *throughput* λ_T [SU/PE], respectively are the *partial storage performance costs*, i.e. the *place costs* $k_P = K_{Sstat}/\lambda_P$ [€/SU-day] and the *throughput costs* $k_T = K_{Sdyn}/\lambda_T$ [€/SU]. They reflect the static and the dynamic aspect of a store separately and are basis of the *storage cost rule*:

► For comparison and selection of storage systems, for remuneration of storage services and for scheduling inventories, only the separate static place costs and the dynamic throughput costs are relevant.

When calculating operating costs and performance costs, it is necessary to differentiate between *standard costs* for a planned static and dynamic demand and *actual costs* for a present or past period with given static and dynamic performance (see

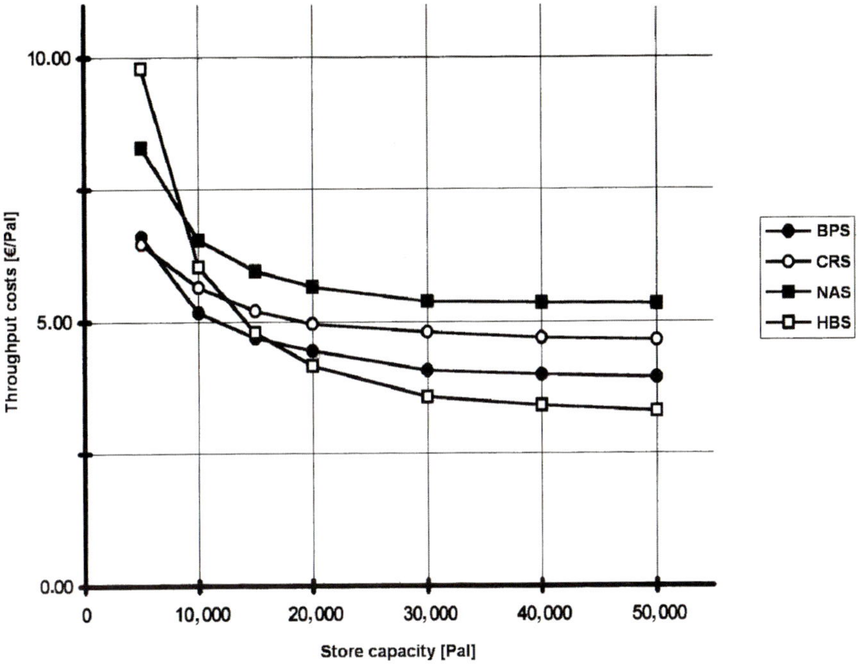

Fig. 16.28 Dependency of the storage turnover costs on the store capacity
store capacity: 20,000 pallets *stacking factor*: $C_y = 3$
other parameters and terms see *Table 16.6*

Sect. 6.2). The partial performance costs critically depend on the *utilization* of the storeplaces and the limit performances. This causes the *fixed costs dilemma of stores* with consequences, which have been described already in *Sect. 6.8*.

Figures 16.22 to *16.28* show the standard performance costs that have been calculated for the 4 compared pallet storage systems of *Table 16.6* assuming 100% utilization. An underutilization of the installed capacities causes an increase of the partial performance costs. This is shown in *Fig. 16.29* for the underoccupation of the storage capacity.

16.13.1 Storeplace Costs

The static storage operating costs result from deprecitiation, interests, energy, repair and maintenance, proportionate costs for the storage management system and other running costs for the static storage parts.

Cost drivers of the static storage operating costs are the installed storage capacity and the mean number of storeplaces occupied during the accounting period of lenghth T_{PE} [days/PE]. The mean total stock of storage units M_{Stot} is given for a pull-stock by relation (16.12). At maximal utilization, it reaches the *effective storage capacity* C_{Seff}, which is given by formula (16.31).

Table 16.6 Storage demand, key storage data and results of a model planning of 4 different pallet storage systems

	Demand	STORAGE SYSTEMS				Unit
		BRS	**CRS**	**NAS**	**HBS**	
Performance requirements						
Stock assortment	Peak	1,000	1,000	1,000	1,000	Article
Throughput	Peak	960	960	960	960	Pal/ODays
Stock	Peak	20,000	20,000	20,000	20,000	Pal
Turnover rate	Peak	12	12	12	12	per year
Load units	CCG-1 Pallets incl. tolerances			Stacking factor: 3		
Length	max	1,300	1,300	1,300	1,300	mm
Breadth	max	900	900	900	900	mm
Height	max	1,050	1,050	1,050	1,050	mm
Volume	max	1.23	1.23	1.23	1.23	m^3/LU
operation						
Calendar days		365	365	365	365	CDays/year
Operating days		250	250	250	250	ODays/year
Daily operating time		12	12	12	12	h/ODay
Storage design						
Storage devices		5 FST	7 SMS	5 SGS + 4 VTS	4 RBG	R/S-units
Length		63	42	44	112	m
Breadth		300	254	146	17	m
Height		4.4	8.4	13.6	35.0	m
Ground area		18,900	10,634	6,424	1,926	m^2
Total volume		82,215	89,325	87,366	67,424	m^3
Investment		**6,780**	**5,925**	**5,270**	**5,992**	**T €**
Per pallet place		339	296	263	300	€/Pal
Operating costs	100%	**1,068**	**1,174**	**1,359**	**998**	**T€/year**
for utilization	80%	973	1,046	1,178	924	T€/year
Fixed costs		593	532	454	630	T€/year
Variable costs		475	643	906	369	T€/year
Life time		15	15	15	15	Years
Performance costs						
Throughput costs	100%	1.90	2.66	3.78	1.81	€/Pal
for utilization	80%	1.94	2.70	3.82	1.91	€/Pal
Place costs	100%	8.40	7.40	6.22	7.80	€C/Pal-CDay
for utilization	80%	10.30	9.00	7.60	9.60	€C/Pal-CDay
Turnover costs	100%	4.45	4.89	5.66	4.16	€/Pal
for utilization	80%	5.07	5.45	6.13	4.81	€/Pal
ROI for utilization	80%	7.4	5.0	0.0	2.8	Years

BPS: block place store (stacking factor 3) CRS: conventional rack store (aisle width 2.5 m)
NAS: narrow aisle store (aisle width 1.75 m) HBS: high bay store
ROI: return on investment from savings by additional investment

The mean number of occupied storeplaces M_{Stot} times the number of days per period T_{PE} is the number of used place-days per period, i.e. $\lambda_P = M_{Stot} \cdot T_{PE}$. The static storage operating costs divided by the *static cost driver* λ_P gives the

- static *storeplace costs* per storage unit and operation day

$$k_P = K_{Sstat}/\lambda_P = K_{Sstat}/(M_{Stot} \cdot T_{PE}) \geq K_{Sstat}/(C_{Seff} \cdot T_{PE}) \qquad [€/SU\text{-day}] \tag{16.97}$$

For pallet stores the place costs are also called *pallet overnight costs*, since they correspond to the overnight price of a hotel.

The dependencies of the place costs on the storage capacity at constant turnover, which result from the model calculations at 100% filling degree, are presented in *Fig. 16.22* for the 4 pallet stores specified in *Table 16.6*. These and further results of the model calculations, which are presented in *Fig. 16.23*, *16.24* and *Fig. 16.25*, lead to the *static storage rules*:

- ▶ At small capacities the storeplace costs decrease rapidly and at higher capacities slower and slower with increasing storage capacity until they reach an asymptotic value, which depends on the storage type.
- ▶ In a wide range, the storeplace costs are independent of the installed throughput performance.
- ▶ The storeplace costs depend on the mean filling degree of the storeplaces but not on the utilization of the installed limit performances.
- ▶ The storeplace costs increase less than proportional with the volume and the weight of the storage units (see *Fig. 16.23*).
- ▶ The storeplace costs of multi-unit stores such as block place stores depend critically on the storeplace depth, the stock per article and the stacking factor (see *Fig. 16.24*). Only for single-unit stores, the storeplace costs are independent of these parameters.

The dependencies of *Fig. 16.24* reflect the consequences of the storeplace depth and the stacking factors on the ground area per storage unit, which are shown in *Fig. 16.14*. Further model calculations lead to the *multi unit storage rule*:

- ▶ *Space-optimal storeplace capacity* equals *cost-optimal place capacity*.

This means that the above storeplace optimization for multi-unit stores leads to minimal place costs. Without optimization, the effective place costs of a multi-place storage can be significantly higher due to insufficient space utilization.

16.13.2 Storage Throughput Costs

The dynamic storage operating costs result from deprecitiation, interest, energy, repair and maintenance, operating personnel, proportionate costs for the storage management system and other running costs for the dynamic storage parts.

The *operating personnel* includes the drivers of vehicles, forklift trucks and storage devices and the involved foremen, maintenance and control center staff. In manually operated stores, personnel is a major cost factor. The total costs for personnell are calculated from the *number of operating people* N_{OP} [Pers] by multiplication with the *personnel cost rate* P_{OP} [€/Pers-PE] for full-time employees. The qualification is taken into account by different cost rates.

The number of operating people with daily *working time* T_W [h/d], which are necessary to operate N_D manually operated devices during a daily *operating time* T_{OP} [h/d], can be calculated with the universal *personnel demand formula*:

$$N_{OP} = \{N_D \cdot (T_{OP}/T_W)/\eta_{ave}\} \qquad (16.98)$$

The personnel demand formula (16.98) holds universally for any kind of manually operated vehicles and devices in intra- and extra logistics. The curly brackets denoting the up-rounding integer operation are necessary, if only *full-time personnel* is employed, and can be omitted for *part-time personnel.*

The *staff availability* η_{ave} [%] in (16.98) is the product of an *allowance factor* and an *absentee factor*. Under good working conditions, personal allowances between 80 and 90% can be achieved. The absentee factor for vacation and illness, which reduces the paid annual working time, is e.g. for German industrial workers about 80%.

The number of devices N_D, which are necessary to meet a dynamic demand, can be reduced by extension of the daily operating time T_{OP}. This organizational measure reduces the dynamic investment. However, due to relation (16.98) it does not alter the required personnel. The depreciation increases by extension of the operating time due to higher wear and tear. This leads to the general *operating rule for mechanical devices*:

- ▸ By an extension of the daily operating time for mechanical devices, the total investment but not the operating costs for personnel, depreciations, repair and maintenance can be reduced.

Since this rule is neglected, the cost savings by extended operating times are often overestimated.

Cost drivers of the dynamic storage costs are the installed limit performances and the throughput of storage units during the accounting period. The dynamic storage costs K_{Sdyn} [€/PE] for a period PE divided by the throughput λ_T [SU/PE] are the *storage throughput costs:*

$$k_T = K_{Sdyn}/\lambda_T \qquad [€/SU] \qquad (16.99)$$

Certain tasks, such as inventory management, require a further differentiation of the dynamic operating costs $K_{Sdyn} = K_{SI}+K_{SO}$ into *in-feed costs* K_{SI} and *out-feed costs* K_{SO}. The separate relations to the mean in-storing and out-storing performance (16.4) and (16.5) are the specific *in-storing costs* $k_{SI} = K_{SI}/\lambda_I$ [€/SU] and *out-storing costs* $k_{SO} = K_{SO}/\lambda_O$ [€/SU].

The dependencies of the throughput costs on the storage capacity at constant turnover, which result for the above four pallet stores at 100% utilization of the installed limit performance, are shown in *Fig. 16.26. Figure 16.27* shows the dependency on the stock turnover rate at constant storage capacity.

Not included in the model calculations are the throughput costs for the goods entry and dispatch area. These costs are between € 1.70 and € 2.50 per pallet, depending on the specific requirements and organization (based on costs per 2007).

The presented dependencies and further model calculations show:

- Similar to the static placecosts, the storage throughput costs decrease with the store capacity.
- The storage throughput costs depend on the utilization of the installed limit performance, but not on the filling degree of the storeplaces.
- For stores of the same type and capacity, the throughput costs at 100% utilization do not differ much for different turnovers.

The degression of the throughput costs with the store capacity in *Fig. 16.26* is the result of the decreasing contribution of the proportionate costs for storage management system, staff in the control center and for the in- and out-feed conveyors. These *technical economies of scale* are more effective for stores with high automation than for those with low automation.

16.13.3 Cost Comparison of Storage Systems

From the dependencies of the performance costs shown in *Fig. 16.22* to *16.28* and from further model calculations, result the following characteristics for the compared storage systems:

- Place costs and throughput costs for *conventional rack stores* do not change very much with capacity and turnover. For over 5,000 pallets, their performance costs are significantly higher than the costs of the other storage systems with exception of the block place store.
- For all capacities, the storeplace costs of a *narrow aisle store* with high reaching forklift trucks are lower than the place costs of other storage systems, whereas the throughput costs are significantly higher in the whole capacity range.
- In the whole capacity range, the place costs of *block place stores* are higher than the place costs of the narrow aisle store and, above 20,000 pallets, also higher than the costs of the high bay store.
- The throughput costs of block place stores are considerably lower than the throughput costs of narrow aisle stores, and up to a capacity of 20,000 pallets also lower than the throughput costs of high bay stores.
- For more than 20,000 pallets, the place costs of an automatic *high bay store* are lower than the place costs of a block place store and above 30,000 pallets, lower than the place costs of conventional rack stores. They approach at very high capacities asymptotically the place costs of narrow aisle stores.
- The throughput costs of a high bay store decrease strongest with capacity and throughput. For more than 20,000 pallet places, the throughput costs of a high bay store are lower than for all the other storage systems.

Figure 16.28 shows the calculated dependency of the *turnover costs* (16.96), i.e. of the total operating costs related to the throughput, on the store capacity for the four pallet storage systems. This leads to the *advantage rule for high bay stores*:

- At capacities above 10,000 to 15,000 pallets, automatic high bay stores have the lowest turnover costs in comparison to all other storage types.

The model calculations and the general comparison of the storage performance costs and their dependencies cannot replace the planning for a specific project. The results only show the possible options and general tendencies. Without qualified storage planning and careful calculation of the performance costs, essential optimization possibilities would be missed. The result could be a wrongly designed system, which does not meet the requirements or causes unnecessary high costs.

16.13.4 Utilization Dependency of Performance Costs

Due to the *fixed cost dilemma* of *Sect. 6.8*, the real performance costs of a store can become far higher than the planned costs rates, if the installed capacity and limit performances are not fully used. The effect of an *underutilization* of store capacity and limit performance on the turnover cost rate is shown in *Fig. 16.29* for the four different storage types. This diagram and further model calculations lead to the dependencies:

- With decreasing filling degree of the installed capacity, the place costs increase initially relatively slowly. The same holds true for the dependency of the throughput costs on the utilization of the limit performance.
- With further decreasing utilization, the increase of the cost rates becomes more and more critical and affects the profitability of the whole storage operation.
- For stores with high fixed costs share, the dependency of the performance costs on utilization is stronger than for stores with lower fixed costs share.

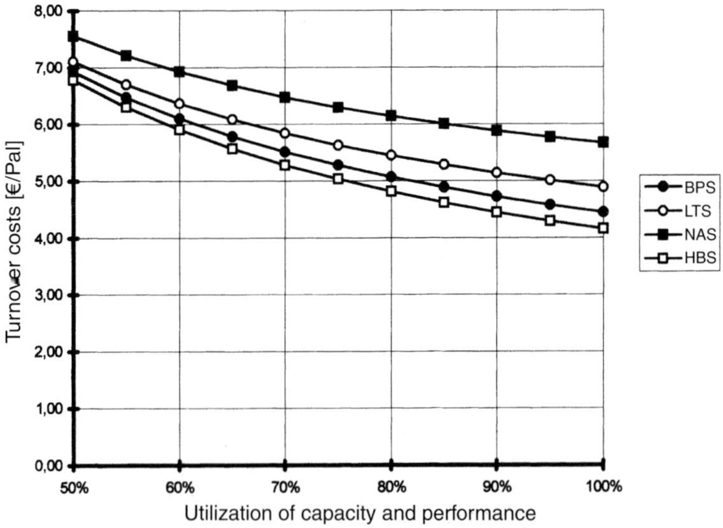

Fig. 16.29 Dependency of the storage turnover costs on the utilisation of capacity and limit performance

capacity: 20,000 pallets *turnover*: 12 other parameters and terms see *Table 16.6*

- For all storage types, the fixed costs share of the static operating costs exceeds 85%. Therefore, the dependency of the place costs on the filling degree is in all cases very sensible.
- The fixed costs share of the dynamic operating costs increases with the degree of automation. It ranges from 5% for manually operated stores, such as the narrow aisle store, to 25% for fully automatic systems, such as the automatic high bay store.
- An underutilization of the installed limit performance is less critical than an underutilization of the installed capacity, if operating times and staff can be flexibly adapted to a lower demand.

Conclusions from these dependencies are the *storage utilization and selection rules*:

- The installed storage capacity should be used with highest possible filling degree even if the utilization of the storage devices is reduced.
- Automatic stores with high fixed costs can only be operated economically if the mean filling degree of the capacity is higher than 80%, and if the mean utilization of the storage devices is better than 70% for an operating time of at least 8 h per day.
- Automatic stores with high fixed costs are unsuitable for a strongly varying static and dynamic demand. Manually operated stores with lower fixed costs are more flexible and better suited for varying demand.

These utilization and selection rules are of special importance for a *logistic service provider*, who plans and builds a storage system in order to offer the installed storage services on the market.

16.14 Procurement of Storage Services

Any company which needs storage services has to answer the question, whether and to which extent these services should be executed by the company itself, or should be sourced from the market and executed by a third party. As outlined in more detail in *Chap. 22*, outsourcing of storage functions has the following advantages:

- Planning, setting up and operation of stores are *core competencies* of specialized logistic service providers, who are more effective in these tasks than companies with other core competencies.
- A storage service provider can achieve economic and technical economies of scale and lower operating costs by building larger stores which are used for many clients.
- The personnel costs in the storage and logistic business are generally lower than in other industries.
- By intelligent pricing and sales management, a storage service provider may gain other clients with opposite seasonal demand and by this means achieve a higher and more even utilization.

A service provider wants to cover the whole operating costs with the sales revenues and to achieve appropriate profit. Therefore, the prices for storage performances include surcharges for administration, sales, risks and profit. Depending on the market situation, on the size of the operation and on the duration of the contract, *storage service providers* calculate with a *SOP-surcharge* p_{SOP} for sales, overhead and profit between 15 and 20%.

A service provider is willing to build and operate a customized store only if the contract will run for at least five years. For larger and automatic stores the contract should run for about 10 years. In addition the service provider needs a guarantee for a sufficient utilization during the contract period.

The *utilization risk* is the most difficult problem when calculating storage performance prices. The risk of higher operating costs due to lower utilizations is generally taken into account by a *utilization risk surcharge* p_{UT} [%] on the prices. This surcharge can exceed 20% if a company does not guarantee a minimal utilization of the provided capacity and limit performances.

The surcharges for sales, overhead, profit and risks can over-compensate the cost advantages of an external service provider and cause prices for external storage services which are higher than the internal cost rates for self-executed storage services.

Finally, the *make-or-buy* problem can only be solved on the market by tendering. For this purpose, the required services, expected stock levels and throughput, frame conditions, guaranteed minimal utilization, intended contract duration and other important conditions have to be carefully specified in a tender document, which is sent out to a selected number of qualified storage service providers.

The quality of the *tender documents* is crucial for the bidding success (see *Chap. 22*). It is important to specify the required performance packages and to predetermine the price structure of the remuneration prices, which should be offered in a *price blanket*. This refers not only to the specific storage services but also to handling, loading, unloading and additional services in the goods entry and dispatch. However, it is advisable not to prescribe the storage type and technology in order to stimulate the service provider to develop an optimal solution that takes into account his strengths, resources and other users.

As storage services and influence factors on prices differ from case to case, the *price transparency* on the *storage service market* is low. Even if performance requirements and constraints are comparable, the prices are not only determined by costs. They depend crucially on regional market conditions and on the current offer and demand. In order to evaluate the offered prices of storage service providers it is necessary to calculate, parallel to the bidding, own storage operating costs.

If a new storage has to be built, either by a logistic service provider or by the company itself, a qualified storage system selection and planning are unavoidable. For planning, dimensioning, optimization and cost calculations, the above formulas, rules and planning tools can be applied. If it turns out, that an own store operated by the company and not outsourcing is the best solution, the results of system planning can be used to find a *general contractor* for the realization. Even with a general contractor, independent planning is opportune in order to evaluate the offered solutions and prices.

16.15 Store Allocation and Selection

If several stores are available within a certain radius, it is necessary to decide for each in-storing order to which one of the stores it should be sent. This decision depends on the different costs rates of the available internal stores and on the performance prices and guaranteed utilization of the external stores of service providers.

If the *mean stock quantity* M_S and the *mean storage time* T_S [d] of an arriving in-storing order are known, the storing costs for the different stores S_i can be calculated from the place costs k_{Pi} [€/SU-d] and the throughput costs k_{Ti} [€/SU]. Including the transport costs, which are caused by the moves along the *transport path* s_i [km] between arrival point, store location and destination point with *transport cost rate* k_{Tr} [€/SU-km], the expected storing costs per storage unit for store S_i are:

$$k_{Si}(T_S) = k_{Tr} \cdot s_i + k_{Ti} + k_{Pi} \cdot T_S \qquad [€/SU]. \tag{16.100}$$

Since the fixed costs for own stores are independent of utilization, only the variable part of the performance cost rates count for the store selection. As long as their utilization is above the guaranteed level, the cost rates for the external stores are the place and throughput prices, which are paid to the service provider. Otherwise the external cost rates are 0, since they have to be paid independent of the occupation as long as the guaranteed utilization is not reached.

With the individual values for s_i, k_{Ti} and k_{Pi}, the storing costs (16.100) for all available stores S_i, $i = 1,2,\ldots N_S$, can be calculated and used for the *store allocation strategy*:

► An in-storing order is allocated to the available store which causes the lowest storing costs (16.100).

Available are all stores that are technically suitable for the storage units and offer sufficient free storeplaces for the input quantity.

The store allocation strategy leads in a self-adjusting manner to minimal total storage costs. Following this strategy, the in-storing orders are automatically allocated to the stores due to the following prioritization:

1. If no suitable external store is underutilized, the in-storing order is allocated to the available own store with lowest storing costs (16.100).
2. If all own stores are totally filled, the in-storing quantities are stored in the external store with lowest storing costs (16.100).
3. If one or more external stores have not reached the guaranteed utilization, the in-storing order is allocated to the underutized external store with the lowest transport costs, provided that these are lower than the sum of the transport and throughput costs for the available own stores.

The dependency of the storage costs (16.100) on the storage time is the *characteristic storage cost curve*. As shown for three examples in *Fig. 16.30*, these curves increase linear with the storage time. The cost curves for stores with lower place costs are less steep than the curves for stores with higher place costs. At the cutting points of the cost curves the *storing opportunity* of the compared stores changes.

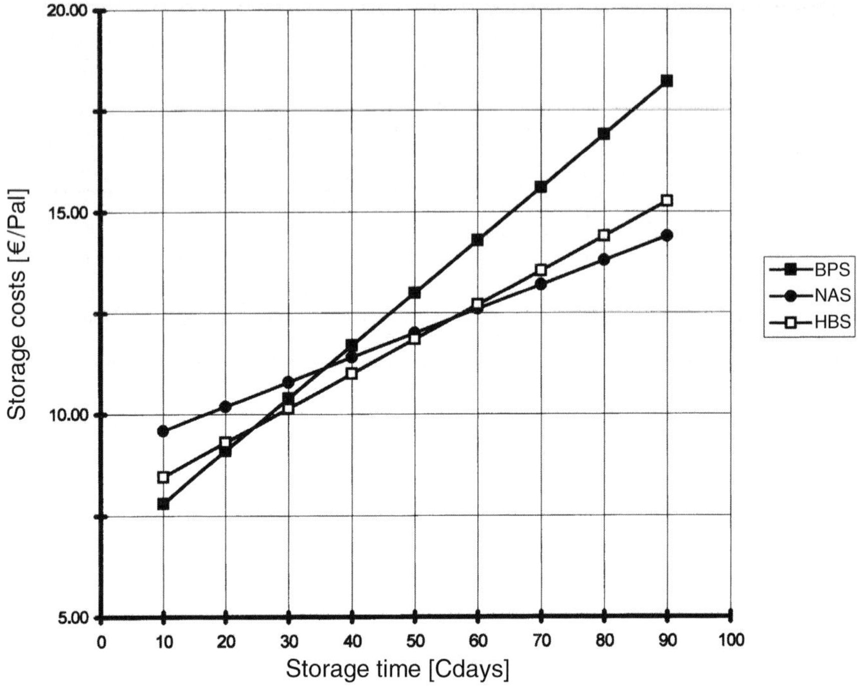

Fig. 16.30 Characteristic storage cost curves for three different pallet stores

Cost rates:	transport costs	throughput costs	place costs
BPS block place store	1.25 €/Pal	5.25 €/Pal	0.13 €/Pal-d
NAS narrow aisle store	1.75 €/Pal	7.25 €/Pal	0.06 €/Pal-d
HBS high bay store	4.00 €/Pal	3.75 €/Pal	0.08 €/Pal-d

If the store allocation strategy is applied to the three stores of *Fig. 16.30*, in-storing orders with mean storage times up to 25 calendar days are stored at minimal costs in a block place store. In-storing orders with mean storage times between 25 and 58 calendar days are stored in a high bay store, and orders with mean storage times over 58 calendar days in a narrow aisle store. If the free capacity of the cost-optimal store is not sufficient, the store with the next highest costs is taken. The decision depends on the storage filling degree and changes if the storage prices depend on the utilization.

The store allocation strategy was developed for and implemented first by a German chemical company in 1998. Simulations for the existing number of internal and external stores have proved that by this strategy the total storage costs can be reduced by more than 20% compared with the former store selection. Later the company replaced the small stores by a new central high bay store which cut the total costs dramatically, primarily due to the eliminated external transports.

The algorithm of the store allocation strategy can be easily programmed. The program controls the utilization of the stores, calculates the storage costs for the in-storing orders and proposes the optimal store to the scheduler.

The cost curves are not only useful for the allocation of storages, but can be applied also for the selection of the cost optimal storage system in the planning phase. They lead to the general *storage selection rules*:

- Storage systems with low throughput costs are adequate for stocks with long storing times and high turnover even if the store placecosts are high.
- Storage systems with low place costs are suitable for stocks with long storing times and low turnover even if the throughput costs are relatively high.

Narrow-aisle stores have low place costs and high throughput costs. They are therefore adequate for stocks with long storing time. Automatic *high bay stores* have lower throughput costs – in particular at high capacity – and slightly higher place costs. They are better suited for storing stocks with high turnover.

Optimally designed *block place stores* with stacking factor above 2 and with more then 10 pallets per article have low throughput costs and low place costs. If these conditions are fulfilled, block place stores are very well suited for fast turning long-term stocks, for low turning short-term stocks and for stocks with medium turnover. Their main advantage is high flexibility.

Also *conventional rack stores* with normal aisle widths and fork lift trucks have relatively low throughput costs, but slightly higher place costs as compared to narrow aisle stores. Conventional rack storages are therefore well suited for stocks with medium turnover and for fluctuating dynamic demand.

Buffer stocks with very short storage times up to one or two days can also be stored in *satellite stores* or *flow rack stores* with extremely high place costs and relatively low throughput costs.

In most cases with one of these four storage types the practical demand for pallet storing can be matched at minimal costs. Other storage types for pallet storage, such as compact stores or mobile-rack stores should only be used under special conditions.

The above application areas for the different storage types for pallet storing, the allocation strategy and the selection rules can be applied for the storage pre-selection in the system finding phase. However, the final decision and the determination of the precise economical borders between the possible storage systems for pallets and other storage units require knowledge of the specific demand, competent system planning and correct logistic cost accounting.

The presented applications and results demonstrate the practical use of differentiated storage performance cost rates, throughput costs and place costs, and of the separate consideration of static and dynamic operating costs. The total turnover costs, which depend on static and dynamic storage costs together, are less useful and misleading.

The lengths of the in- and out-storing cycles and of the internal transport distances increase with the store capacity and cause a limit for the *economy of scale* for storage systems (Nowotny 2003). This *counter economy of scale*, i.e. the increase of the performance costs for very high capacities and throughput demand will be analyzed in *Sect. 20.11*.

Chapter 17
Commissioning Systems

Commissioning of orders is the most difficult and underestimated task of *intra logistics*. It is far more than just picking articles (Gudehus 1973b; VDI 1976):

- ▸ *Commissioning* is the collection and consolidation of required quantities from an assortment of articles due to given orders.

The general underestimation of this task is reflected by the customary use of the words *order picking* instead of *commissioning*. The difficulties are caused by the large number of technical solutions, organizational possibilities and operating strategies, but also by the many *cost drivers* of commissioning systems.

Commissioning is reduced to simple out-storing, if only *full load units* of the same article are requested. If load units of different articles are required by the same order, they must be consolidated after out-storing. From the last chapter follows:

- ▸ *Commissioning of full load units* can be executed alone by *unit load storage systems*.

Commissioning of quantities less than the content of an access unit requires a separation of the order quantity from the provided quantity. That means:

- ▸ Gripping and collecting *part quantities* are the central tasks of commissioning systems.

The *gripping* or *picking* is executed either *manually* by people with or without mechanical support or *automatically* by mechanical devices. Both, picking person and picking device, are called *picker*. For non-uniform article units, it is difficult to mechanize and automate the gripping process. Gripping of single pieces requires in many cases the longest time and causes the highest costs of the whole commissioning process.

The *part-processes of commissioning* are:

1. *provision* of the article assortment on access places
2. *moving* of the picker to these places
3. *taking* of required quantities from the access places
4. *deposition* of the quantities in a bin, on a conveyor or a transport mean

T. Gudehus, H. Kotzab, *Comprehensive Logistics*,
DOI 10.1007/978-3-540-68652-1_17, © Springer-Verlag Berlin Heidelberg 2009

5. *consolidation* of the picked order quantities
6. *replenishment* of the access places

The spatial and temporal combinations of the possible solutions for these part-processes lead to a large variety of *commissioning methods*. They are multiplied by the different technical solutions, which are possible to realize the part-processes, i.e. by the *commissioning technique* (Arnold 2002; Bode and Preuß 2004; Borries 1975; Gudehus 1973; Miebach 1971; VDI 1976, 1977, 1994, 1998; Vogt 1993).

By combining single storage, collection and replenishment systems, commissioning systems with increasing *stages* and *complexity* can be built up:

- *Elementary commissioning systems* are combinations of a single *collection system*, which executes the commissioning, with a single for the supply of the access places.
- *Compounded commissioning systems* are combinations of several elementary storage and commissioning systems in *parallel* or in *series*.

For the organization and operation of elementary and compounded commissioning systems, different strategies are possible. Similar to the storage strategies, the *commissioning strategies* can be differentiated into placing, execution, moving and taking strategies. The supply for the access places is guided by *replenishment strategies*. For the removal of empty load carriers and for the supply of empty bins, *empties-strategies* are applied. If different commissioning systems are available, in addition *selection strategies* and *allocation strategies* are needed.

After an analysis of the requirements for commissioning systems, the different commissioning methods are developed and the customary techniques are described in this chapter. The set up of elementary and compounded commissioning systems is presented in the following sections. This includes an analysis of the advantages, disadvantages and main applications for the different systems. Several possibilities to improve the performance and to reduce the *commissioning costs* are presented. The effects which can be achieved by sophisticated strategies will be demonstrated.

In order to dimension commissioning systems and to improve and minimize their costs, it is necessary to determine their *limit performances* and to calculate their *performance costs*. For this purpose, formulas for path time, picking time and total commissioning time are derived. They allow the systematic planning, selection and optimization of commissioning systems which meet the requirements and constraints at minimal costs.

17.1 Commissioning Requirements

The insufficient knowledge of the requirements is the most common cause for commissioning problems. The qualitative and quantitative requirements for commissioning systems can be derived from the orders, functions and services, which have to be executed in a certain period of time.

The *qualitative functions* and *services* of a commissioning system are:

- *Basic functions*

$$\begin{array}{l}\text{picking of article quantities}\\ \text{filling of order bins or dispatch units}\\ \text{consolidating the picked quantities}\end{array} \tag{17.1}$$

- *Preparation functions*

$$\begin{array}{l}\text{order preparation}\\ \text{provision of the assortment}\\ \text{filling the access places}\\ \text{replenishment of reserve units}\\ \text{storing of reserve units}\\ \text{scheduling of inventories and replenishment}\end{array} \tag{17.2}$$

- *Additional services*

$$\begin{array}{l}\text{price marking, coding and labeling of article units}\\ \text{packing the order quantities}\\ \text{building up dispatch units}\\ \text{labeling and identification of the dispatch units}\end{array} \tag{17.3}$$

The preparation functions are initiated by the basic functions. They ensure an uninterrupted and error-free commissioning process. Additional services – also called *added values* – are not in all cases necessary for commissioning. They can be executed before or after the completion of the orders. Before comparing the performance and costs of commissioning systems, the additional services have to be specified and taken into account separately.

The *quantitative requirements* consist of *primary requirements*, which are the expected article assortment and the orders to be executed, and *secondary requirements*, which can be derived from the primary requirements, such as the picking performance, throughput and stocks. The performance demand fluctuates stochastically and can vary systematically during a longer period of time (see *Chap. 9*). The systematic variations during a year, a week or a day can be taken into account by specific *peak factors* or different *operating scenarios*. As far as orders cannot be postponed, the system must be dimensioned for the maximal hourly demand of the peak day.

The collection, preparation and structuring of the quantitative performance requirements imply decisions which may affect the planning results. Therefore, it is advisable to start the planning with *preliminary data*. The preliminary data have to be completed or rearranged, if it turns out that further data are needed.

Table 17.1 presents the commissioning requirements for two different *business cases*. In addition to these values, the seasonal variations and the stochastic fluctuations have to be taken into account. These data are used in the following for model calculations of the dependencies of performances and costs.

Table 17.1 Commissioning requirements for two business cases

	Central Store Industry	**Distribution Center** Retailer	
ARTICLES	Nonfood	Merchandize	
Number	800	30,000	
Share A-Article	10%	10%	
B-Article	40%	40%	
C-Article	50%^	50%	
PICK UNITS (Picks)	Packages	Article units	PK
Throughput	18,750	252,000	PK/Day
Volume	3.8	5.0	l/PK
Weight	1.9	2.5	kg/PK
ORDERS	Customer orders	Outlet orders	Ord
Lead time	4	8	Hours
Order throughput	250	1,500	Ord/day
Positions per order	15.0	12.0	Pos/Ord
Order quantity	5.0	14.0	PK/Pos
PICKING DEMAND	3,750	18,000	Pos/day
Share A-Article	50%	60%	
B-Article	30%	35%	
C-Article	20%	5%	
PROVISION UNITS	Paletts	Paletts and Bins	PU
Capacity	209	Free parameter	PK/PU
Throughput	90	System dependent	PU/day
DISPATCH UNITS	Paletts	Paletts and Boxes	DU
Capacity	185	Free parameter	PK/DU
Throughput	250	System dependent	DU/day

17.1.1 Assortment Requirements

The assortment requirements specify the *variety of articles*, which have to be provided in order to meet the expected orders. They comprise the:

- *number of articles* N_A to be provided for order picking
- *article properties*, such as value, content, shelf-life, hazard category, fire class, size and bulkiness
- *article units* [AU] with their dimensions l_{AU}, b_{AU}, h_{AU} [mm], volume v_{AU} [l/VU] and weight w_{AU} [kg/AU].
- *provision units* [PU] with capacity C_{PU} [AU/PU], dimensions l_{PU}, b_{PU}, h_{PU} [mm], volume v_{PU} [l/PU] and weight w_{PU} [kg/PU].

The number of articles determines the number of *access places* as well as the required *access length* of commissioning systems with static provision. The *article units* can be single articles, bottles, barrels or packages containing liquids, powder or loose goods, or cartons containing different articles. Depending on their destination, the article units are also called *sales units* [SU] or *consumption units* [CU].

The *provision units* are generally the *access units* out of which the article units are picked. They can be pallets, containers or other load units, but also the article units themselves or even single parts, which are provided without load carrier. In many cases, the load carriers and provision units are determined by the supplier. If they are not prescribed, the dimensions of the provision units and the allocation to the article groups are free *design parameters* which can be used to optimize the commissioning system.

Handling units [HU] are the smallest units which are handled in a logistic system. For commissioning systems, this is the *picking unit*, also called *pick unit* or simply *pick* [PK]. For *fine picking*, the picks are the article units: PK = AU. For *package picking*, they are packages, such as *cartons*, *shrink-wrapped quantities, trays* or *displays*, which contain several article units.

For a homogenous assortment with similar article units, it is sufficient to know the mean dimensions and weight of the picking units. If the article units differ much, it is essential to divide the assortment into sufficiently *homogenous article groups*, such as small, medium, large, heavy and bulky parts, which are stored and provided in different load units.

If the *volume throughput* [m^3/PE] of the articles differs significantly, it is advisable to segment the articles further into *throughput groups*, e.g. into A-articles, B-articles and C-articles, which result from an *ABC-analysis* of the volume flow (see *Sect. 5.8*). The different throughput groups can be provided in load units of different size, such as bins for small volume and pallets for high volume throughput.

17.1.2 Order Requirements

Commissioning orders can be *external orders*, such as delivery orders, dispatch orders or spare part requirements, or *internal orders*, such as supply orders for the production or assembly line. Other internal orders are *part-orders* for parallel commissioning zones and *series-orders* for the first step of a two-step commissioning system.

The order requirements include:

- type of the *commissioning orders* [Ord]
- *order entrance flow* λ_{Ord} [Ord/PE] per period [PE = year, day, hour]
- *order positions* or *order lines*, i.e. *articles per order* n_{pos} [Pos/Ord]
- *position quantity* m_{Pos} [AU/Pos] or *grips per position* m_{grip} [Grips/Pos]
- *dispatch unit* [DU] with capacity C_{DU} [AU/DU], dimensions l_{DU}, b_{DU}, h_{DU} [mm], volume v_{DU} [l/DU] and weight w_{DU} [kg/DU].
- *order lead time* T_{Ord} [h]

If the order entry fluctuates stochastically and varies systematically, the mean value and the variance of the order throughput must be known for the peak day.

If the urgency of the orders differs, they have to be grouped into *urgency classes*. Resulting are *rush orders*, *express orders*, *standard orders* and *due date orders*. For fixed operation times and limited lead times, also the hourly order entry for the peak day must be known.

For orders that do not differ much, it is sufficient to calculate with the mean number of positions and the mean picking quantity. If the orders differ significantly, it is necessary to group them into homogenous *order clusters*, such as *single-position orders* and *multi-position orders*, or small, medium and big orders.

The *mean order quantity* results from the mean number of positions and the mean quantity per position:

$$m_{Ord} = n_{Pos} \cdot m_{Pos} \qquad [PK/Ord]. \tag{17.4}$$

The *mean order volume* v_{Ord} and the *mean order weight* w_{Ord} result from the order quantity and the mean volume v_{PK} [m^3/PK] and mean weight w_{PK} [kg/PK] of the picking units:

$$\begin{aligned} v_{Ord} &= n_{Pos} \cdot m_{Pos} \cdot v_{PK} \qquad [m/Ord] \\ w_{Ord} &= n_{Pos} \cdot m_{Pos} \cdot w_{K} \qquad [kg/Ord] \end{aligned} \tag{17.5}$$

The dispatch units can be bins, cartons, boxes, pallets, containers, folding boxes or other kinds of load carrier. If the load carriers required for the dispatch are different, the commissioning orders must be classified into bin-, pallet- and container-orders and taken into account separately.

If the dispatch units are not predetermined, the design, selection, dimensions and allocation criteria of the dispatch units are *free design parameters*. The capacity of a dispatch unit C_{DU} [PK/DU] with given dimensions results from the dimensions of the picking units with the formulas of *Chap. 12*. In a *pick&pack system*, the picks are deposited directly into the dispatch units. In these systems, the throughput of dispatch units determines the design and limit performances of the out-feed conveyor of the picking area.

17.1.3 Throughput Requirements

The throughput requirements are determined by the assortment and the order requirements. For system design and dimensioning, the following throughput data are required:

- *Volume flow*

$$\lambda_V = v_{Ord} \cdot \lambda_{Ord} \qquad [m/PE] \tag{17.6}$$

- *Weight flow*

$$\lambda_W = w_{Ord} \cdot \lambda_{Ord} \qquad [kg/PE] \tag{17.7}$$

- *Position pick demand*

$$\lambda_{Pos} = n_{Pos} \cdot \lambda_{Ord} \qquad [Pos/PE] \tag{17.8}$$

- *Article pick demand*

$$\lambda_{AU} = n_{Pos} \cdot m_{Pos} \cdot \lambda_{Ord} \qquad [AU/PE]. \tag{17.9}$$

The article pick demand induces the *throughput of provision units* with *capacity* C_{PU} [AU/PU] which is:

$$\lambda_{PU} = \lambda_{AU}/C_{PU} \qquad [PU/PE]. \tag{17.10}$$

The throughput of provision units determines the limit performance of the replenishment system. Replenishment frequency and number of feeding devices decrease with increasing capacity and size of the provision units, whereas the access length and the picking paths increase. The optimum of these opposite effects is the *cost optimal provision unit*. This leads to the general *provision unit rule*:

► For each project there are provision units with optimal size and capacity.

The *throughput of dispatch units* with capacity C_{DU} [AU/DU] is given by:

$$\lambda_{DU} = \lambda_{AU}/C_{DU} + \lambda_{Ord} \cdot (C_{DU} - 1)/2C_{DU} \qquad [DU/PE] \qquad (17.11)$$

The additional term $\lambda_{Ord} \cdot (C_{DU} - 1)/2C_{DU}$ is caused by the one partially filled dispatch unit per order, which is reduced by a factor $1/N_{Ord}$, if N_{Ord} different orders are consolidated and delivered in the same dispatch units (see *Sect. 12.5*).

The dispatch unit throughput is increased by partially filled units in particular for small order quantities. However, a larger capacity and size of the dispatch or collection units causes a lower output flow rate and reduces the limit performances of the output transport system. The optimization of these opposite effects leads to the general *dispatch unit rule*:

► For each project there are collection units or dispatch units with optimal capacity and size.

The optimal size and capacity of the provision, collection and dispatch units are difficult to determine. They depend on many influence factors and differ from project to project. In practice, a limited variety of load carriers has already been standardized and agreed with suppliers and customers. The load unit optimization reduces then to the order dependent selection of the optimal load carriers for supply and dispatch.

17.1.4 Stock Requirements

The article stocks on the access places in the picking area enable the uninterrupted commissioning with shortest paths at minimal costs. From these goals, the following *stock dimension principles* can be derived (see *Chap. 11*):

► The *minimal stock* or *pull-stock* for an article in the commissioning system is achieved by optimal *inventory and replenishment scheduling* for the access places.

► If the article stock exceeds the pull-stock, it should be kept in the picking area only as far as free places are available and the storing does not affect the performance of the commissioning system.

A push-stock and reserve quantities beyond this level must be stored in a separate *reserve store*. Due to these principles, for commissioning systems with static provision the total stock of an article is contained in

- an *access unit* on the access place for picking
- a *provision unit* placed near the access place
- *reserve units* stored in the commissioning zone or a separate reserve store

The content of the access unit and the provision unit is the *pull-stock* (see *Fig. 17.8*). The total space required for the pull-stock is determined by the number of articles and by the placement strategy for access units. The store space for the reserve units depends on the storage order as explained in *Sect. 16.5*.

17.2 Commissioning Methods

In order to execute the picking process, three *function elements of commissioning* have to be joined at one place as shown in *Fig. 17.1*. These are:

- *access units* A_i, $i = 1,2,\ldots N_A$, which contain sufficient quantities of N_A articles,
- *order collection units* O_j, $j = 1,2,\ldots N_B$, on which the picking quantity m_{ij} from article A_i for order O_j are deposited.
- *pickers* P_k, $k = 1,2,\ldots N_K$, who execute the picking.

The possible methods of commissioning are determined by the decision as to where the picking is executed, which elements are permanently located at the picking place, and which are moved only temporarily to this place.

This results in six different *commissioning methods*, where either two elements come to a third stationary element, or one element comes to two stationary elements:

1. pickers come with order collection units to stationary access units
2. collection units are moved to stationary pickers and access units
3. access units are conveyed to stationary pickers and collection units

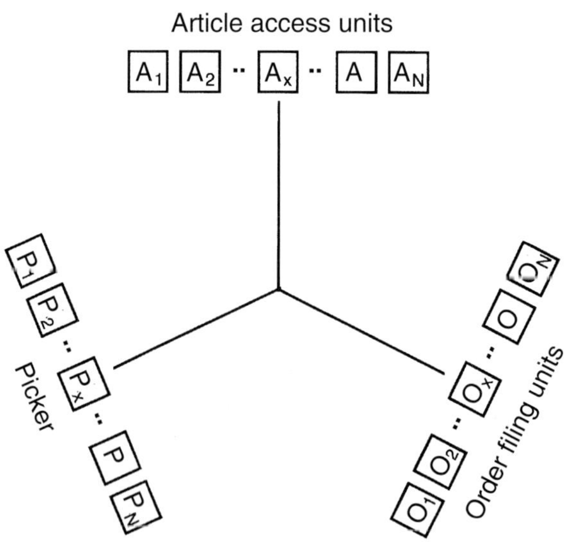

Fig. 17.1 Central elements of a commissioning system

Access units: A_i, $i = 1,2,\ldots N_A$
Order collection units: O_j, $j = 1,2,\ldots N_O$
Pickers: P_k, $k = 1,2,\ldots N_P$

4. pickers with access units come to stationary collection units
5. pickers with collection units move to stationary access units
6. access units and collection units are moved to stationary pickers

Figures 17.2, *17.3*, *17.4*, *17.5* and *17.6* illustrate the realization of five of the six basic commissioning methods.

A seventh method is *total stationary commissioning* with or without collection unit. Here, one access unit and a picker are located at a fixed place. The picker is a mechanical device, which pulls a requested quantity of article units from the access place and drops it on a conveyor or into a collection unit. The access place can be e.g. the end of a flow rack place (see e.g. *Fig 17.14E*).

With *single-stage commissioning,* the picked units are finally conveyed to a consolidation place or a packing station. With *two-stage order picking,* the picked units pass a *sorter system*, which accumulates the goods from different sources and finally distributes them to the packing stations or other destinations.

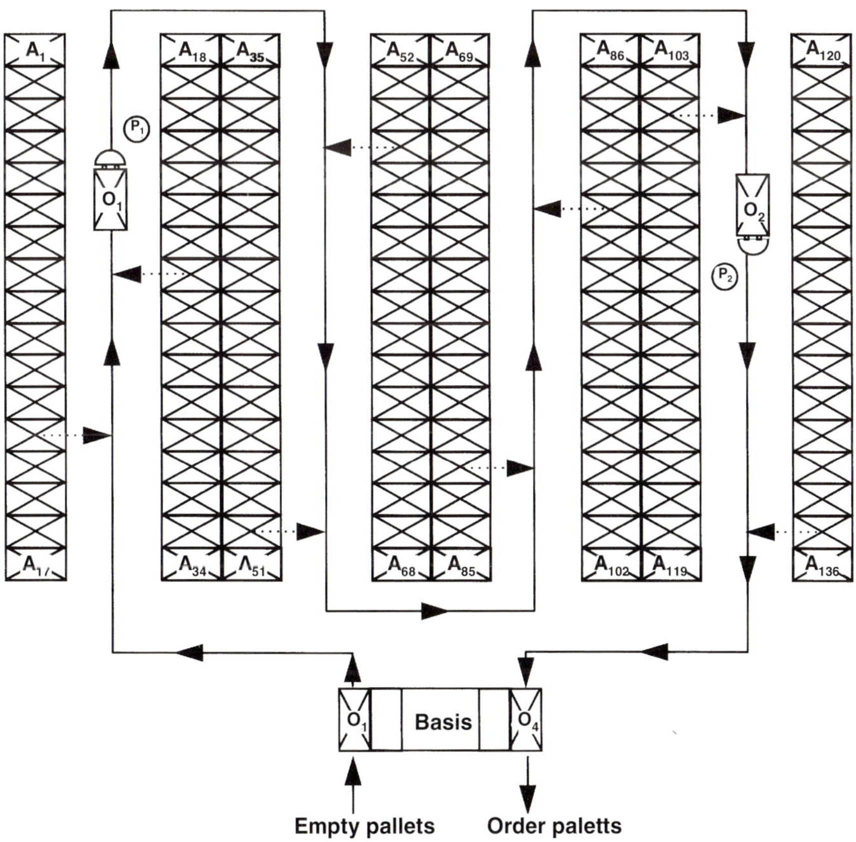

Fig. 17.2 Conventional commissioning with static article provision and spatially combined feeding and picking aisles

17.2.1 Conventional Commissioning with Static Article Provision

Conventional commissioning - shortly called *person-to-article* (P2A) or *man to the good* (M2G) - is well known from *self-service shops*. As shown in *Fig. 17.2*, the access units are located on fixed places next to each other on the floor or in racks. By optimal place layout short paths, good availability and optimal space utilization are achieved.

The picker moves with a collection unit to the article places. The articles to be picked and their location are indicated by a *pick list* (*pick-by-paper*), by a mobile *electronic display*, by stationary *lights displays* (*pick-by-light*) or acoustically through a *head phone* (*pick-by-voice*). The picker starts at a *basic station*, takes over the empty collection units, moves to the first picking place, picks the requested quantities, puts them into the collection unit, quits the number and reads the next picking place. This is repeated until the last article for the order has been picked. The picker returns to the basic station, where the bins with the order quantities are deposited and new orders are taken over.

In most conventional commissioning systems, the picking is executed manually. With article units or packages of cubic size, which are delivered on pallets with a fixed *packing scheme*, the picking can also be executed automatically by a movable *robot* or a *depalletizer*.

The *advantages* of conventional order picking are:

- minimal technical efforts
- simple organization with or without computer support
- short order lead times
- rush orders, single orders, series order, partial orders as well as complete orders can be executed simultaneously
- high flexibility with respect to a varying picking demand and changes of assortment
- applicable for all kinds of goods ranging from small and light to large and heavy articles.

Due to these advantages, conventional commissioning is the most common method in business practice. However, its *disadvantages* are sometimes ignored and often under-estimated:

- The long distances for a wide assortment of many articles with access units of larger dimensions require more pickers than other solutions.
- Additional ground areas are needed for access places and aisles, if picking aisles and feeding aisles are spatially separated.
- Problems with replenishment after the *depletive grip*, when the last unit has been picked and further units are required by the same order.
- Effort and interruptions by the disposal of the empty load carrier after the depletive grip.

Many of these disadvantages can be avoided or minimized by optimal design of the picking places and layout of the system, by suitable operating strategies and by technical support. In many cases, an optimally designed and operated conventional commissioning system is the most efficient commissioning method. It is especially suited for

- Commissioning from pallets to pallets (*pick-to-pallet*) of a small assortment of up to 100 articles with stocks of up to 3 pallets per article.
- Commissioning of small article units from a broad assortment of many articles that are provided in rack shelves or in flow channels.

The first situation can be found in warehouses which serve outlets or retailers with consumer goods. The second situation is given e.g. in *mail-order companies* and wholesalers of *pharmaceuticals*.

17.2.2 Local Commissioning with Static Article Provision

For local commissioning, the access units are located as shown in *Fig. 17.3* on fixed places in the *working area* of a picker who serves a limited number of article places. The commissioning orders are indicated locally by display or moved to the picking areas with the collection unit on a conveyor. The bins or cartons are diverted to the picking area and wait there on a buffer lane until the requested quantity is picked and deposited. Afterwards, the bin or carton with the picked quantities is conveyed to the subsequent picking area.

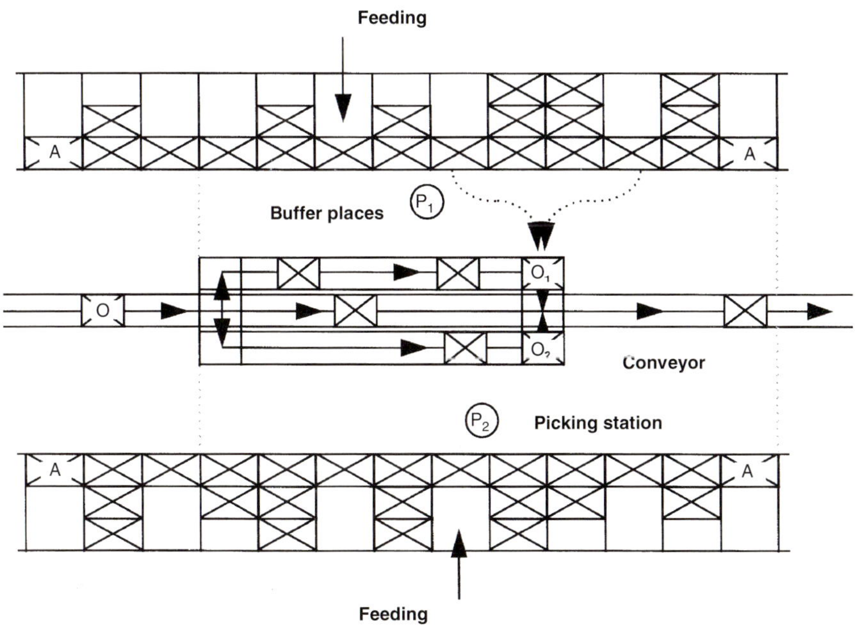

Fig. 17.3 Local commissioning with static article provision and spatially separated picking and replenishment

Finally, the collection units are transported by a consolidation and sorting system to the packing places. In a *pick&pack system*, the order cartons are finally checked, closed, sealed and send directly to the dispatch area (see *Fig. 13.29*). If the picks are laid down on a conveyor belt without a bin, the local commissioning is called *pick-to-belt* (P2B). If they are collected in a *bin* or a *tote*, it is named *pick-to-bin* or *pick-to-tote* respectively.

The *advantages* of local commissioning are:

- short paths for the pickers
- continuous working
- no setup and waiting times at a central basis
- high picking performances

But there are also several *disadvantages* involved:

- The work of the pickers in the subsequent commissioning zones is interdependent.
- Low flexibility when performance requirements change or fluctuate.
- Large space demand for the conveyor system, buffer lanes, picking aisles and spatially separated feeding aisles.
- A separate reserve store is needed, if the article stocks are high.
- Necessity of *two-stage commissioning* or *batch processing* of *series orders*.
- Long order lead times due to batch processing or two-stage order picking.
- Bad utilization and long waiting times due to stochastically varying picking times in particular if many small orders occur.
- Problems after the *depletive grip* and with disposal of empty load carriers.

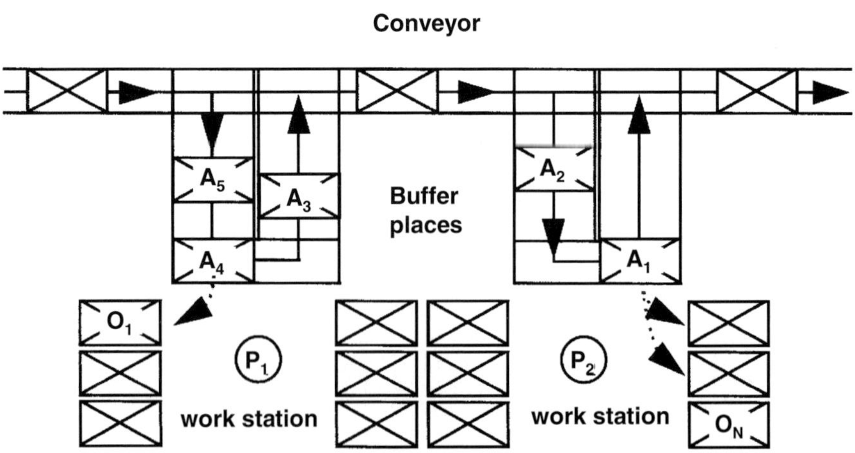

Fig. 17.4 Static commissioning with dynamic article provision

These disadvantages of a local commissioning system can be partly reduced by optimal design, allocation and dimensioning of the access places, working zone and conveyor system.

Local commissioning can be more efficient than other commissioning methods for extreme high picking demand with more than 100,000 orders per day of less than 5 order positions and for a wide assortment of more than 10,000 articles of small size and low weight. Applications of this kind of *high-performance order picking* can be found in *mail-order companies* and in the *pharmaceutical industry*.

17.2.3 Static Commissioning with Dynamic Article Provision

Static commissioning with dynamic article provision, also called *dynamic order picking*, *article-to-person* (A2P) or *goods to the man* (G2M)- is well known from shops with *sales-counter service*. Here, a sales person takes a box with the requested article from a shelf, brings it to the counter, takes out the required quantity and returns the box with the remaining quantity to the shelf.

As shown in *Fig. 17.4*, in static commissioning systems with dynamic article provision the picking is executed at a fixed *workstation*, where the collection units are waiting. The *access units* with the required articles come from an integrated *provision- and reserve-store*. They are provided by a *conveyor system* or a *vehicle system*. The access units wait in the workstation and are moved to the picker on demand. Because the units stay only during the picking process this kind of article provision is called *dynamic*.

If single orders of sufficient volume are processed, for each order, one or more collection units, which can be pallets, boxes, bins or small containers, are placed at the picking station. In a *pick&pack system*, they are the dispatch cartons. With *single-stage serial processing*, several waiting orders requiring the same article are served at once from the same access unit. After an order is completed, the collection unit is transported to a packing place or directly to the dispatch.

Two-stage series order processing is opportune, if many small orders have to be served. The picked quantities are laid down on a *conveyer* and transported to the second stage where they are separated and sorted. In both cases, access units with rest quantities are either conveyed to the next picking station or back to the provision- and reserve-store.

The *advantages* of dynamic order picking are:

- no or minimal walking for the picker
- ergonomic workplaces
- support by handling devices for heavy and bulky articles
- high picking performances due to eliminated path times
- no problems after the depletive grip
- simple removal and supply of empty load carriers
- high flexibility in case of changing assortment and order structure
- space-saving automatic provision- and reserve-store

- optimal protection of stocks against unauthorised access
- low space demand due to the elimination of picking aisles
- simple realization of pick&pack
- picking stations can be located close to the packing or dispatch area

However, dynamic order picking is connected with the *disadvantages*:

- high investment for the automatic provision- and reserve-store and for the conveyor or transport system
- high provision costs for articles on pallets and in big containers
- long order lead times in peak hours for two-stage order picking
- limited flexibility against fluctuating picking demand with high peaks due to limited performance of the provision system
- load securing before restoring access units may be necessary

These disadvantages can be partially eliminated or reduced by a high performance provision store, an advanced process control system, intelligent operating strategies, and by multi-shift operation with flexible working hours (see *Sect. 17.15*).

The consolidation of orders, which require the same articles in batch or series orders, is another possibility to minimize the required provision performance. The *batch size* is an important optimization parameter of dynamic order picking systems, as the number of provisions per single order decreases with increasing number of simultaneously processed orders. However, the order lead times increase with the batch size.

Performance and cost comparisons for many projects result in the *application rule for dynamic order picking*:

► Dynamic order picking systems are suitable for a high steady picking demand from a varying assortment of many articles either if the position quantities are large, or if many orders require the same articles.

Dynamic order picking can also be the best solution for picking heavy, bulky or *odd sized articles* which need support by handling devices or if connected with additional services (17.3) and/or with special devices. Another application is the commissioning of *high-value articles* which must be protected against unauthorized access.

Dynamic order picking is especially suited for article-based commissioning of serial orders in the first stage of a two-stage order picking system. However, as explained later, two-stage order picking requires a complex organization and process control, and additional investment for the second commissioning stage.

Technical requirements of dynamic provision are conveyable load units with compatible dimensions, such as standardized bins or trays and norm pallets. Article units which are stacked on flat load carriers must be sufficiently secured.

Most common applications of stationary commissioning with dynamic provision are the *mini load systems* for *standard bins* and *shelf trays* as shown in *Fig. 17.34*. Nowadays, a mini load module can reach extremely high performance of more than 120 in- and out-store cycles per hour. With this, the provision costs are reduced to an acceptable level. Dynamic order picking from *pallet to pallet* is only efficient in connection with an automatic high bay store of high performance at low operating costs. Hence, for dynamic pallet picking, further innovations are necessary in order to make the performance costs competitive.

Special technical solutions for dynamic provision are vertical and horizontal *rotary stores* as shown in *Fig. 16.7*. They keep the access units on mobile store places and provide them to the stationary picking station on demand. However, as explained in *Sect. 16.2.6*, rotary stores have many disadvantages, such as limited provision performance, high space costs and inefficient replenishment. In addition, the waiting times between single provisions can reach 20 to 60 s per cycle. This offsets the saving of path times for the pickers. Therefore, paternoster and carousel stores are suitable only for storing and for *low performance picking* of small parts, such as tools and spare parts, of documents, film rolls and archive material, and for the commissioning of long goods, such as pipes and rug rolls.

17.2.4 Inverse Commissioning

Inverse commissioning is the reversal of conventional commissioning. Here the roles of the collection units and the access units are interchanged. As shown in *Fig. 17.5*, the order bins or dispatch units are located at fixed places on the ground or in shelves. They are filled by the picker who visits these places one after another together with the article access unit.

As with dynamic order picking, the article units are provided dynamically by a conveyer or transport system from a *provision- and reserve-store*. In stock-free *transshipment stations,* the access units for the inverse order picking are provided directly from the goods receiving area. The remaining quantities in the access units are provided for the next order series or are restored.

Inverse commissioning offers the following *advantages*:

- short paths for small numbers of simultaneously processed orders
- high picking performance for large order quantities
- integrated automatic provision and reserve storage
- high flexibility for changing assortments
- no problems with empty load carriers
- reduced space demand due to elimination of long picking aisles
- order consolidation places can be arranged close to the dispatch area
- direct picking into dispatch units (*pick&pack*)
- simple organization of the picking area

The *disadvantages* of inverse commissioning are:

- high investment for the automatic provision and reserve store and for the conveyor or transport system, if not supplied by transshipment units
- complex organization and process control for batch processing of order series
- long order lead times due to batch processing
- limited flexibility, when the access units are provided by an automatic store and the picking demand fluctuates heavily during peak hours

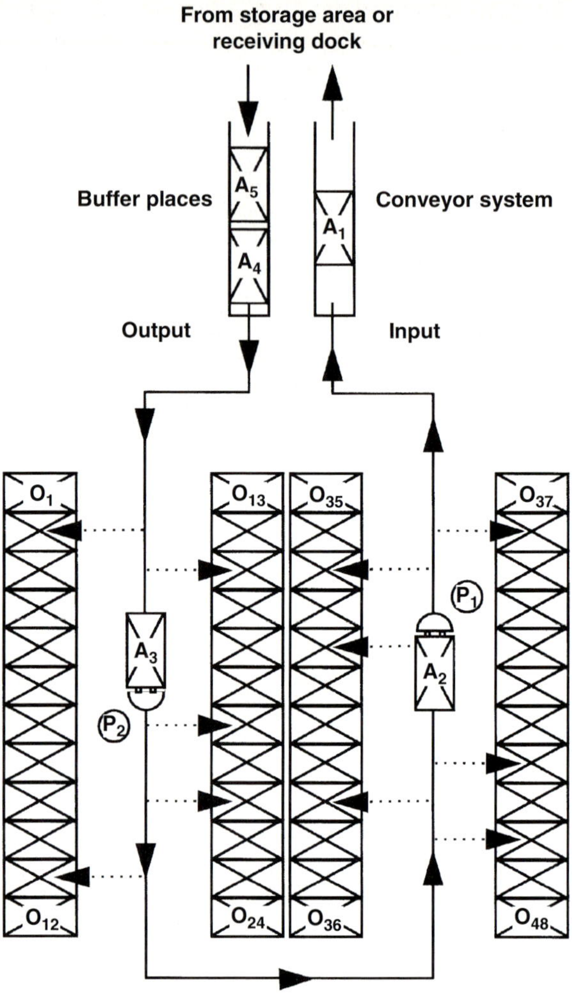

Fig. 17.5 Inverse commissioning of palletized articles

These disadvantages are less important in a *logistic center* with powerful provision systems operating in multiple shifts.

Inverse commissioning is especially suited for a limited number of orders with few positions and large quantities of the same articles. It can also be used for a relatively wide assortment of significantly more than 1,000 articles with distinctive ABC-distribution. Just as with dynamic order picking, the optimization parameter of inverse commissioning is the *batch size* of the simultaneously processed orders, by which the number of provisions can be reduced whereas the order lead times increase.

Inverse commissioning to pallets or to roller towers with dispatch containers is applied in the logistic centers of *retail chains* for the replenishment of sales outlets with *promotion articles*. Another application is *two-stage crossdocking* of daily delivered single-article pallets into dispatch pallets in stock-free *transshipment stations* (see *Sect. 20.1*).

In both cases the number of daily requested articles can be limited by adequate scheduling strategies for the outlet supply. For example, the different articles groups are delivered cyclically on fixed week days.

17.2.5 Mobile Commissioning with Static Provision of Articles and Orders

For mobile commissioning, the access units with the articles and the collection units for the orders are arranged as shown in *Fig. 17.6* on fixed places along two sides of an aisle. Between these places moves a picking device within the aisle and fills the collection units with the articles which have been taken from the access units.

Filled collection units are removed from the backside and transported to the packing places or the dispatch area. After this, another empty collection unit is set up. Emptied access units are replaced by full units in the same way.

Mobile commissioning is typical for fully *automatic order picking* by a movable *picking robot* or *depalletizer*. The application of *mechanical picking devices* is however limited to dimensionally stable, cubic or cylindrical standard article units where dimensions do not differ much. The access units must be equal to the load units, such as flat norm pallets, which are always loaded in the same *stacking scheme* per article.

Mobile commissioning by robot or depalletizer requires high investments. It is therefore only efficient for high throughput volumes of many units per position in *multi-shift operation*. As these conditions are seldom fulfilled, mobile commissioning is quite uncommon. Only a few practical applications can be found for *fast moving consumer goods*, such as *beverages*.

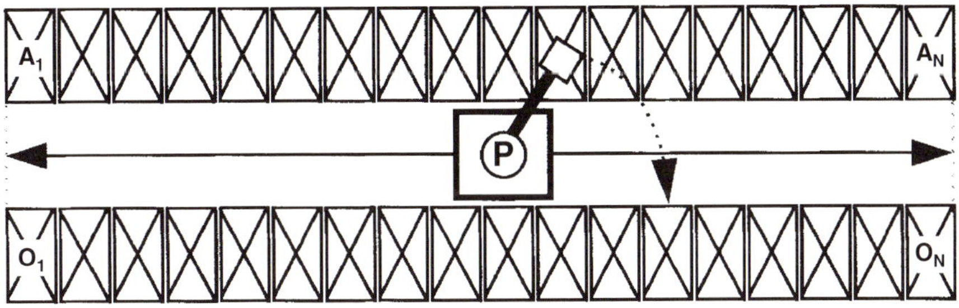

Fig. 17.6 Mobile commissioning with static provision of articles and orders

17.2.6 Static Commissioning with Dynamic Provision of Articles and Orders

In this last of the sixth commissioning methods, the access units and the order collection units are moved dynamically to a stationary picking station. Here, the required quantities are picked manually and deposited into one of several waiting order collection units. The access units are provided from an automatic buffer store by a high performance conveyer system and removed to it after the picking. Empty collection units are supplied, and filled order units are removed by another transport system.

This method allows extreme picking performances of up to 1,000 picks per hour and person. It is suitable for small articles and has been realized mainly in the *pharmaceutical industry*. However, stationary commissioning with dynamic provision involves complex material-handling and control systems.

17.3 Commissioning Technique

The single components of a commissioning system can be technically realized in many different ways (Arnold and Furmans 2003; Bode and Preuß 2004; Frazelle and Apple 1994; Günthner and Heptner 2007). As shown in *Fig. 17.7*, a classification of 16 different elementary commissioning systems results from the combination of the possible technical solutions for the four part processes

- *provision* of the access units: *static* or *dynamic*
- *moving* of the picker: *one-dimensional* or *two-dimensional* (17.13)
- *picking* of the articles: *manual* or *mechanical*
- *deposition* of the order quantities: *central* or *local*

This classification is standardized in the German guidelines VDI 1976, 1977, 1994 due to the proposals of one of the authors (Gudehus 1973). However, this

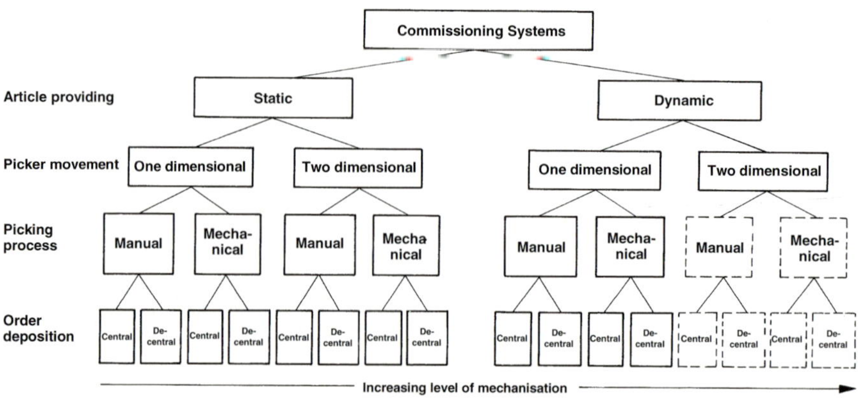

Fig. 17.7 Classification of elementary commissioning systems

scheme does not include inverse commissioning and lacks many technical variants of replenishment, provision, deposition and of information display.

The possible designs and technical variants are manifold. The combination with the above commissioning methods results in far more than 1,000 different commissioning systems. However, less than 50 of these theoretically possible systems are of practical value. In the specific case, only a few solutions are economical (Bode and Preuß 2004; Borries and Fürwentsches 1975; Gudehus 1973; Miebach 1971; VDI 1976, 1977, 1994).

17.3.1 Article Provision

The access places and the provision of the access units can be designed as follows:

- The access place is *stationary* on a fixed place, as in *Figs. 17.2*, *17.3* and *17.8*, or changes *dynamically* during the commissioning process, as illustrated in *Figs. 17.4* and *17.5*.
- The access places are arranged horizontally in *one dimension* or horizontally and vertically in *two dimensions*.
- The access units can be positioned with their longitudinal side either *parallel* or *transversal* to the picking aisle.
- The access places can be supplied from the front side *spatially combined* with the picking, as in *Fig. 17.2*, or from the back side *spatially separated* from the picking, as shown in *Figs. 17.3*, *17.8*, *17.9* and *17.10*.

For stationary commissioning with dynamic article provision, the design of the stationary picking stations and the information display are decisive for the picking performance. The performance is critically affected by the number of buffer places on the conveyor upstream and downstream of the picking places.

The access place of a commissioning system with static article provision is a place on the floor, in a rack or on a shelf. In systems with spatially separated picking and supply, they are the exits of roller or sliding *flow channels* which are fed from the back side, as shown in *Figs. 17.8*, *17.9* and *17.10*. Behind each access unit one or more provision units are located as reserve. They move forward to the picking front after the depletive grip.

The disadvantage of spatially separated feeding and picking is the space for additional aisles. The aisle breadth is determined by the size of the provision units and by the technology of the replenishment devices.

No additional aisles are necessary in systems with spatially combined feeding and picking. Here, it has to be decided whether feeding and picking are executed by the same technical device or by different devices which are specialized only on picking or on feeding. If different picking and feeding devices are used, appropriate operating strategies are necessary to avoid conflicts and congestions. Although spatially combined feeding and picking has the advantage of low space demand, it is connected with several *disadvantages*:

- reduced picking performance
- limited replenishment performance without automation
- no immediate replenishment after the depletive grip

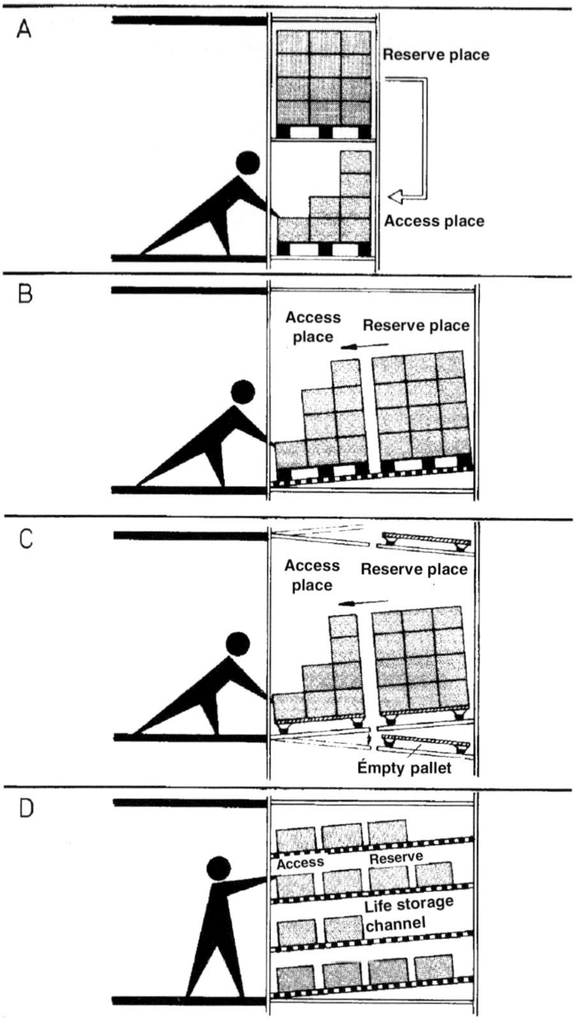

Fig. 17.8 Possible solutions for access unit and reserve units for systems with spatially separated feeding and picking

The disadvantages of combined aisles can be either minimized or eliminated by applying adequate operating strategies, such as *free place order* in combination with the *flip-flop-method*. From the advantages and disadvantages result the application rules:

- *Spatial separation* of feeding and picking is better if high throughput requires several replenishments per day and more than one picker per aisle.
- *Spatial combination* of feeding and picking is sufficient, if throughput and pick demand are low, or is necessary, if the available space is limited.

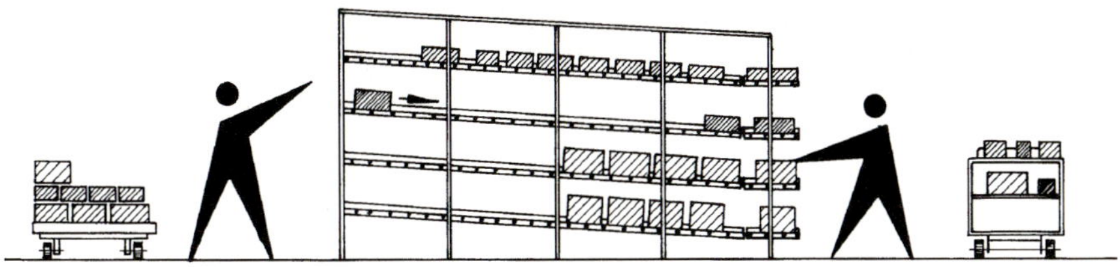

Fig. 17.9 Static provision of single article units in a live store sack with separate feeding- and picking-aisles

Fig. 17.10 Static pallet provision with feeding by an automatic S/R-unit and manual picking

The precise decision between these solutions depends on the specific circumstances of the project.

17.3.2 Movement of Pickers

In commissioning systems with static article provision, the picker moves to the access units. This is possible in the following ways:

- The picker *walks* with a hand driven trolley from place to place and picks the articles (see *Fig. 17.11*)
- The picker drives with a *horizontal picker truck* or with a special *pick-mobile* on the floor to the access places on the ground.
- The picker drives on a *high-picker truck* in an *additive* horizontal and vertical *move* to the access places in a rack.
- The picker moves *simultaneously* in vertical and horizontal direction on a special *rack-feeder unit* (see *Fig. 17.12*).

The first three kinds of movements are *one-dimensional*, the last kind is *two-dimensional*. Simultaneous movement in two dimensions reduces the driving times as compared to one-dimensional movement, if only a small number of places out

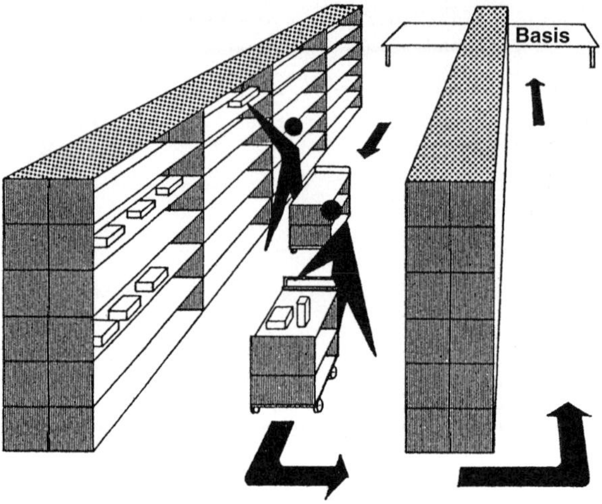

Fig. 17.11 Commissioning system with static article provision, one-dimensional movement, manual picking and central deposition

of many access places have to be visited. The compact construction of the racks by narrow aisles is a further advantage of two-dimensional picking.

The commissioning performance depends on the kind of movement, the capacity of the collection units, and on the speed and acceleration of the *commissioning devices* [CD]. The aisle breadths and the technically possible height determine the

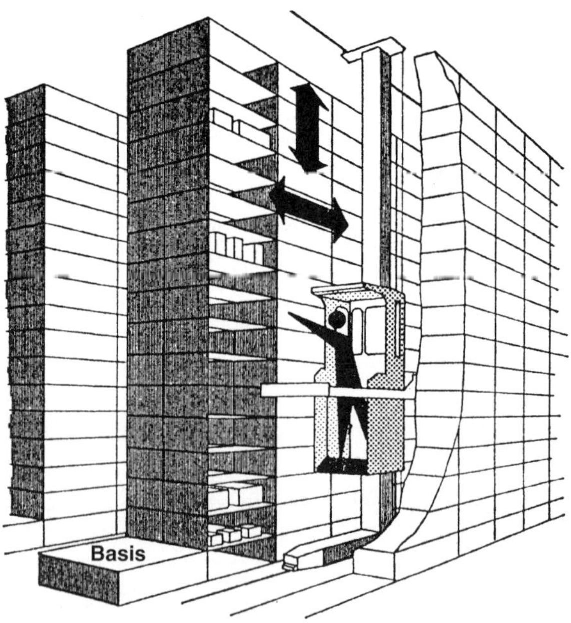

Fig. 17.12 Commissioning system with static article provision, two-dimensional movement, manual picking and retrieval and central deposition

Table 17.2 Technical key data and target prices for commissioning devices

COMMISSIONING DEVICE Picking units	**Load** up to	**Lifting** height up to	**Aisle** width appr.	**Driving** speed accell.	**Lifting** speed accell.	**Invest prices** 2007 T€
Walker without handcart	1 kg	–	1.0 m	2.1 m/s	–	–
Small items			1.5 m	1.4 m/s^2	–	
Walker with handcart	40 kg	–	1.5 m	1.3 m/s	–	1 to 2
Bins, cartons, article units			2.5 m	1.0 m/s^2	–	
Electro-handlift truck	1,200 kg	1.0 m	1.5 m	1.7 m/s	0.05 m/s	3 to 4
Bins, cartons, pallets	1 EURO		2.5 m	0.5 m/s^2	0.2 m/s	
Horizontal pick car	2,000 kg	0.5 m	1.5 m	2.5 m/s	0.1 m/s	15 to 20
Bins, cartons, pallets	2 EURO		2.5 m	0.7 m/s^2	0.3 m/s^2	
Vertical pick car	1,000 kg	5.5 m	3.2 m	2.2 m/s	0.2 m/s	30 to 40
Bins, cartons, pallets	1 EURO		–	0.7 m/s^2	0.5 m/s^2	
Rack-picking unit distance	1,000 kg	10 m	1.4 m	2.0 m/s	0.5 m/s	70 to 100
Bins, cartons, pallets	1 EURO		–	0.5 m/s^2	0.5 m/s^2	

Aisle breadth: with and without overtaking possibility, deposition on Euro-pallets
Prices: with electric power, manual operation, without display

space and ground area of the commissioning system. *Table 17.2* presents these technical key data as well as target prices for selected commissioning devices.

Special commissioning devices, such as the pick trolleys of *Fig. 17.14B*, can carry collection units for several orders. By this means, *small series order picking* can be executed in one stage, which reduces the path time per position and avoids the double handling of two-stage order picking.

17.3.3 Picking

The picking process is the most difficult part of the total commissioning process due to the necessary separation of the single article units. Picking can be executed as follows:

- *Manual picking* by a person without technical support as shown in *Fig. 17.11*.
- *Mechanical picking* by a person with support by a technical handling device such as a vacuum lifting pad hanging on a rotary or slewing crane.
- *Automatic picking* by a picking robot or depalletizer as shown in *Fig. 17.6*.
- *Automatic discharging* by an extracting device as shown in *Fig. 17.4E* which separates single article units from a flow channel or a chute and drops them into a collection unit or on a conveyor belt.

The manual picking performance depends critically on the *ergonomic design* of the picking place, the *dimensions* of the access and the deposition places and on the *distance* and *angle* between pick and drop location. *Fig. 17.13* shows these *design parameters* which can be used to optimize manual picking.

Prerequisites for fully automatic picking by robots or by mechanical devices are uniform picking units with suitable surface. The picking units have to be provided

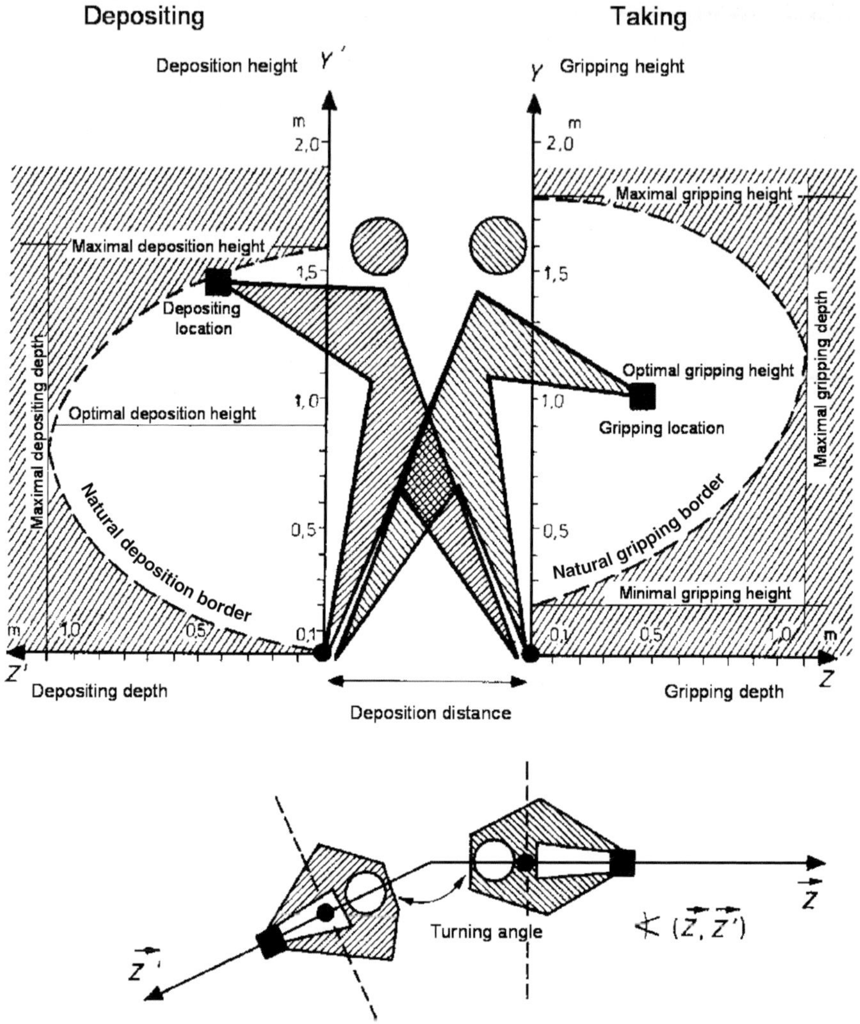

Fig. 17.13 Spatial influence factors of manual gripping

measures refer to people with body height 1.70–1.80 m
disadvantageous picking areas are hatched

either single or on pallets in the same stacking scheme per article. If the article units are not delivered in adequate form and stacking, they must be separated for the automatic picking. In many cases the separation effort compensates the rationalization of automatic picking. Examples are manually filled chutes with an extractor device for small packages, that operates similar to a *cigarette automate*.

The reliable and fast automatic *grip into a box*, which is filled with mixed irregular units, is technically still unfeasible. Even if all technical requirements are fulfilled, automatic picking is in most cases too expensive. Only with continuous high utilization during the whole year, where at least two shifts can be ensured, automatic commissioning may reduce the costs as compared to manual commissioning.

17.3.4 Deposition

Figure 17.14 presents several technical solutions for the article deposition after picking. It also shows different connections to the outward conveyor system:

- The picking units can be deposited *loose* on a conveyor, into internal collection units or into an external dispatch unit.
- The filled collection or dispatch unit is brought to a *central basis*, as in *Figs. 17.2*, *17.4* and *17.11*, where the picking tour ends, or is left back at a *local basis* as in *Figs. 17.3* and *17.15*.

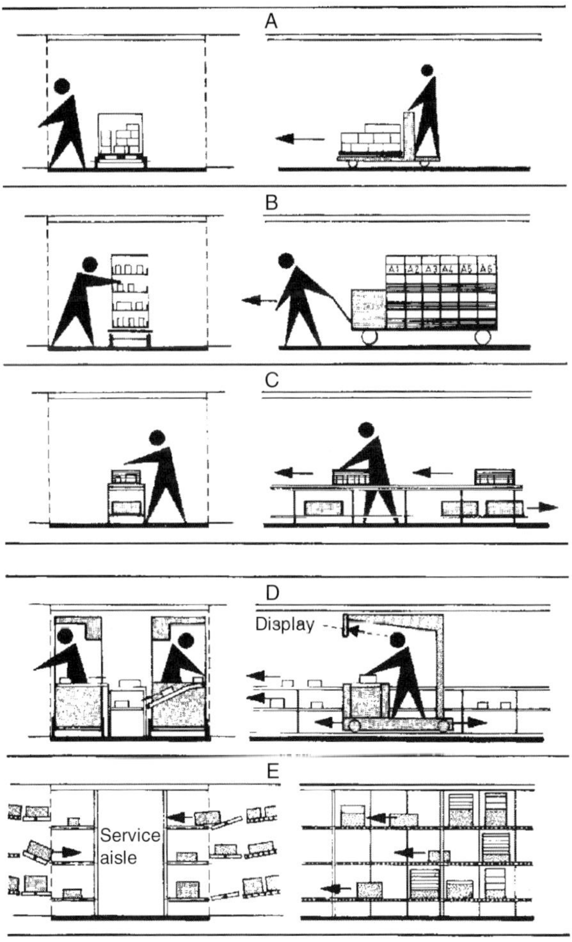

Fig. 17.14 Possible solutions for the deposition after picking

A: Conventional commissioning to pallets on a truck
B: One-stage serial commissioning to shelves on trolley
C: Local deposition into order bins on a conveyor belt
D: Commissioning from pick-mobile with local deposition
E: Automatic picking by an extractor device

Fig. 17.15 Commissioning system with static article provision, one-dimensional movement, manual picking and local deposition

- The further transport of the collection or dispatch unit is executed by the picking trolley, a commissioning machine or by a separate conveyor.

The main advantage of *loose deposition* on a conveyor is that the quantities are not limited by the capacity of a load carrier. This enables continuous picking as long as orders are available. Loose deposition is used in the first stage of a two-stage order picking process.

If the picking units are laid down in a bin or on a pallet several order bins or pallets can be simultaneously transported on a movable *honeycomb rack* or in a *cabinet pallet*. This allows for *single-stage commissioning* of small *order series*.

The use of internal collections units causes double handling. The first handling occurs at the picking place and the second at the packing station or in the dispatch area. Double handling can be avoided by pick&pack:

► In a *pick&pack system*, the picked units are directly laid into the external dispatch unit, which can be a container, pallet or carton.

Pick&pack has been proven in many applications to be the most effective method to minimize picking efforts and commissioning costs.

17.3.5 Conveyor and Sorter Systems

Conveyor and sorter systems can supply provision units and empties. Other systems transport, distribute and sort the collection units or picked quantities as far as these tasks are not executed by the picker or the picking devices.

Pallets and heavy load units can be transported by a *heavy conveyor system* with roller tracks, chain conveyors and transfer units or by a *vehicle system*, such as an overhead *monorail system* or floor based *automatic guided vehicles* (AGV). In multi-storey buildings, horizontal transport systems are vertically connected by lifts and elevators. The limit performances of pallet conveyor systems range from 50 to 200 pallets per hour (see *Tables 13.3* and *13.4*).

The transport and sorting of smaller units, such as cartons and bins, can be executed by a *light conveyor system* with belts, gravity or powered rollers, junction and diversion elements and elevators or lift stations. Their limit performances range from 2,000 to 3,000 units per hour (see *Table 13.4*).

For *two-stage order picking*, special *buffer sorter systems* are necessary to collect, buffer and distribute the picked units from the first stage to the second stage. This can be a *rotary sorter* or high-performance *linear sorters* with sorting performance of 10,000 units per hour and more (see *Figs. 18.13, 18.14* and *21.2*).

The selection, dimensioning and layout of conveyor, vehicle and sorter systems and their connection with commissioning systems and other functional areas depend on the individual project. Picking performance and throughput of the total system are determined by the limit performances of the *bottleneck elements*. These have to be identified, carefully planned and dimensioned as explained in *Chap. 18*. The overall performance of a commissioning system can be improved by optimal *operating strategies*.

17.3.6 Packing and Consolidation

The commissioning process ends with the provision of consolidated order quantities in load units ready for shipment. Therefore, the articles have to be packed either continuously during a pick&pack process or finally in special packing stations. Afterwards, the parcels, pallets and load units are moved to consolidation places in the dispatch area.

Packing is performed in parallel stations. The necessary number of stations can be calculated from throughput, order structure and packing time. The packing time depends on the ergonomic design of the workplace, the provision of the packing material, the conveying of the goods to be packed and on the removal of the ready parcels.

The goods can either be provided in a chute, on an accumulating roller conveyor, in collection units or on a transport vehicle. To ensure efficient packing without interruptions, a buffer lane with a capacity for at least two orders per packing place is necessary. Alternatively, two parallel lanes per place operating with the *flip-flop-principle* are possible.

The commissioning zone and the packing stations can be directly connected by a transport system or decoupled by a stationary or dynamic *buffer sorter* (see *Figs. 18.13* and *18.14*). If directly connected, the number of active packing stations and their buffer capacity limits the *batch size* of the series order in the picking area. If packing and stations are decoupled, the batch size is limited by the capacity of the intermediate buffer sorter system and the number of exits to the packing stations (see *Figs. 18.13* and *18.14*).

The number and the buffer capacity of the packing stations, the limit performance of the transport systems, and the performance and capacity of the sorter system are important design and dimensioning parameters for the complete commissioning system including packing and dispatch.

If different load carriers such as pallets or containers are used for dispatch, the parcels, packages and article units must be filled, packed and stacked into these *load units* with maximal *filling degree*. This is achieved most effectively by the *packing strategies* and *filling strategies* explained in *Sect. 12.4* and *12.5*. A further concentration is possible by *sandwich pallets*, which are built up from stacks of several flat loaded pallets. Finally the dispatch units must be labeled and secured for the transport by strapping, plastic or shrinking foil.

A *shipment* or *transport load* is normally made up of many picking orders. Hence, the last steps in a logistics station are consolidation of the picked quantities with load units coming *directly* from a store or from *crossdocking* into deliverable *shipments* and checking of correctness and completeness. The consolidation and final control can take place either on buffer areas in front of the dispatch ramps, or directly in the waiting transport means (see *Fig. 16.10*).

17.3.7 Commissioning Control

Tasks of commissioning control are the *initiation*, *steering* and *control* and the optimal *coordination* of the part processes of commissioning in order to execute the internal and external orders correctly and most efficiently. This includes the scheduling of stocks and replenishments of storage, buffer and access places.

These tasks are executed by the operative personnel, a foreman or a supervisor supported by a *warehouse management system* (WMS) or a special *commissioning control system* (CCS) and by the *local control units* of trucks, vehicles and other devices (Warehouse Logistics 2008; Wolf et al. 2006). WMS and CCS get orders from the superior ERP and receive information from the subordinated part systems. The more tasks and functions are executed by WMS and ERP, the more the warehouse manager can concentrate on the supervision of *personnel* and the control of *performance* and *quality*.

Besides the relief of the personnel, the most important advantage of *computer-aided commissioning* are the quick data registration and display, and the possibility to execute optimal operating strategies. In addition, low error rates, on-line operation, shorter delivery times and paperless work are achievable by computer support

In order to work on, the picker must know the next picking place, the requested articles, the picking quantities and the deposition place. The necessary data and information for this purpose can be announced in two different ways:

- *Documented information* on a hard copy such as *pick-lists* or *order forms*.
- *Paperless information*, either visually via screen or light displays (*pick-by-light*) or acoustically via headsets (*pick-by-voice*).

For conventional and inverse commissioning, the pick-list is taken over at the basis and moves with the picker. For local commissioning, the order forms travel with the collection units to the picking stations.

The information announcement in paperless-picking systems can be *stationary* in the picking zone or *mobile*, traveling with the picker. The so-called *pick-by-light* with light displays at the access places is optimal in commissioning zones of limited

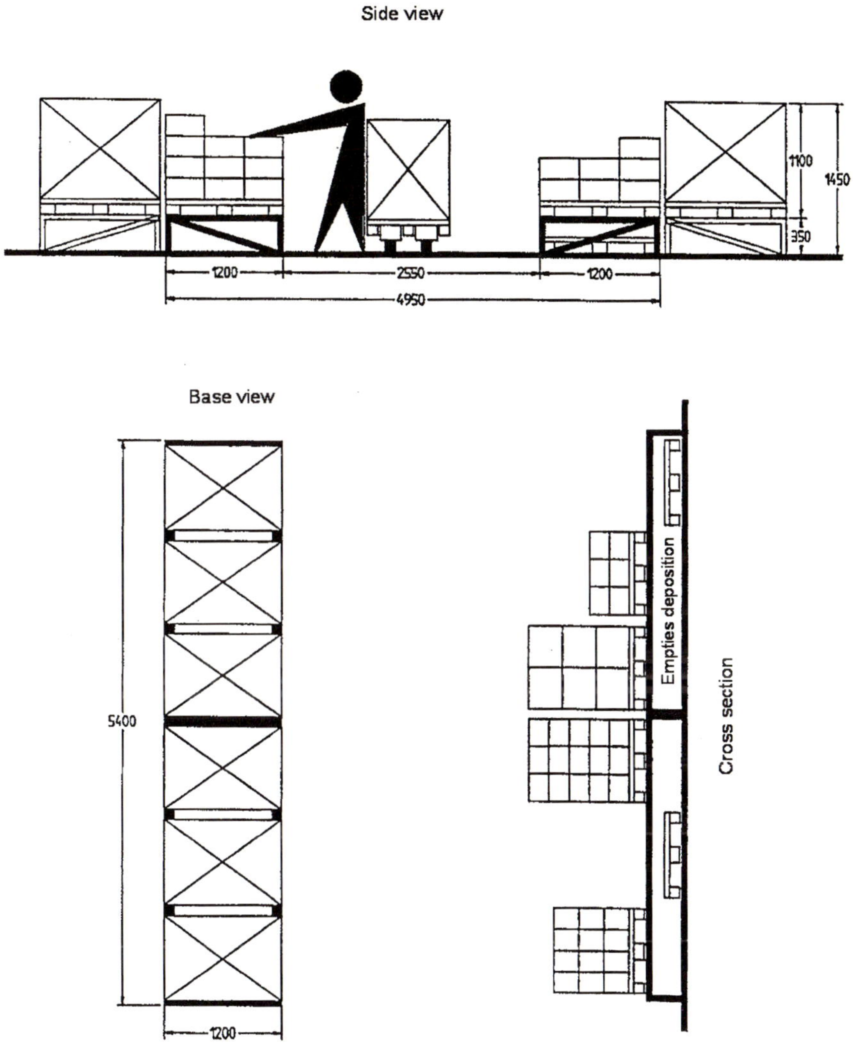

Fig. 17.16 Optimal provision for manual picking from pallet to pallet
provision of CCG1-pallets in optimal picking level
empty pallet buffer places below the access places
flexible flip-flop operation with changing access places
measures in mm

length, as in local and stationary commissioning. A mobile display is preferable, if many access places are located in aisles longer than 20 m.

The picker needs time to find, read and consider the information and to confirm the order execution and control data. This unproductive *dead time* can exceed the productive time for picking, especially if the display is inadequate, the input technique is inappropriate and if too much information has to be punched in. Too long *waiting times* for information can be a big problem for on-line controlled commissioning systems. In some installations, the waiting times in peak hours exceed 5 s. This is avoided by the *control system design rule*:

► For the on-line control of a high performance commissioning system, a separate software unit is necessary, which operates independently below the central ERP system.

The completeness and presentation of information for the picker and the processing speed of the computer determine dead times, waiting times and error rates. They also affect performance, quality and costs of commissioning.

17.4 Commissioning Quality

Commissioning quality is measured by the position picking quality and by the order picking quality:

- *Position picking quality* is the relation of the correctly picked order positions to the total number of positions for a certain measuring period.
- *Order picking quality* relates the correctly and punctually executed commissioning orders to the total number of orders for a certain period.

The commissioning quality is affected by the *availability* of the articles in the access places. If an article is *out of stock*, orders requiring this article cannot be executed. The task of replenishment scheduling for the commissioning zone is to ensure article availability in the access places. This is guaranteed by suitable *replenishment strategies*. The task of *inventory and replenishment scheduling*, which has been described in *Chap. 11*, is to ensure the overall article availability including reserve stocks.

The order picking quality also depends on the *order structure*, i.e. on the mean number of order lines. It is the key figure for customers, who expect correct and complete orders. The position picking quality is a key figure for the performance of the pickers, as they are not responsible for the order structure.

Both quality figures are determined by the errors of the order picking process. Typical *picking errors* are:

picking from the wrong access unit
mix-up of articles
picking the wrong quantity
deposition in the wrong collection unit (17.14)
skipping of positions
leaving out picking orders
too late order completion

The probability of picking errors η_{Perr}, i.e. the position error rate, determines the position picking quality $\eta_{PQ} = 1 - \eta_{Perr}$, i.e. the probability that a position is correctly picked (see *Sect. 3.4.4*):

- The *position error rate* η_{Perr} [%] is the relation of the incorrectly executed picking positions to the total number of executed picking positions.

As long as all articles are in stock, the position error rate should be considerably below 1% for all commissioning systems. Nowadays, position error rates below 0.1% are possible with special precautions.

The *order picking quality* $\eta_{OQ} = 1\text{-}\eta_{Oerr}$ is determined by the order error rate:

- The *order error rate* η_{Oerr} [%] is the relation of the number of completely and correctly executed orders to the total number of executed orders.

The probability that an order with n positions is correctly and completely executed is the product of the position picking probability for the single articles. This leads to the *order picking quality rule*:

▶ The order picking quality for orders with on average n positions is $\eta_{OQ} = \eta_{PQ}{}^{n} = (1\text{-}\ \eta_{Perr})^{n}$ and the order error rate is $\eta_{Oerr} = 1\text{-}(1\text{-}\eta_{Perr})^{n}$, if the position error rate is η_{Perr} and the position quality is $\eta_{PQ} = 1 - \eta_{Perr}$.

For example, the picking order error rate for orders with on average 5 positions and a position error rate of 1.0% is $1\text{-}(1\text{-}0.01)^5 = 4.99\%$. This confirms the general experience:

▶ It is far more difficult to achieve high order picking quality for orders with many positions than for orders with few positions.

Picking errors can be measured and registered by *quality control* in the packing station, before dispatch or by the customer. However, more effective than measuring is to avoid errors at the origin. Picking errors are avoided, or at least reduced, by high *quality awareness* of the personnel. This can be achieved by *control information*, which has to be scanned or punched in by the picker before or after each pick. Other solutions are *control weightings* and *automatic counting* of the picked quantities after deposition.

The goal of all these efforts is *zero-defect commissioning*, although this is principally impossible. Therefore, the organization of any commissioning system has to care for error avoidance, but must also be prepared for the occurrence of errors. Errors are possible in the best system and should never lead to a breakdown of the operation.

17.5 Combined Storage and Commissioning Systems

Within an elementary commissioning system, only small numbers of orders can be executed from a limited number of homogenous articles with low stocks. High throughput and picking demand, heterogeneous assortments and high article stocks require the combination of several elementary systems.

Elementary systems of the same or different kind can be arranged and operate in parallel or in series. *Fig. 17.17* shows the structure of a complex network of storage systems and commissioning systems which are connected by transport and information systems.

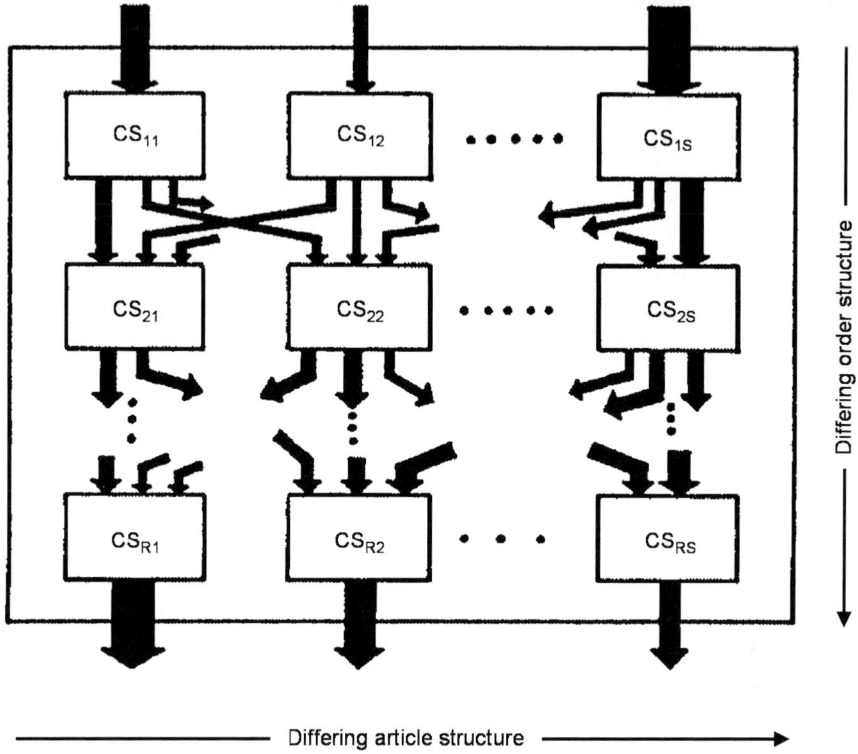

Fig. 17.17 Network of parallel and serial connected elementary commissioning systems CS_{ij}

The general system structure is determined by the *network rules*:

- The higher the picking demand and the larger the number of articles and their differences in size, throughput and stock, the more *parallel* storage and commissioning systems are necessary.
- The higher the throughput, stocks and picking demand, the smaller the size of the picking units and the more external orders differ, the more necessary are systems in *series* with *multi-stage order picking*.

It is quite difficult to plan a commissioning system when articles and orders differ very much. The adequate differentiation and the right combination of the optimal systems are central tasks of *system planning*. In the operating phase, the *optimal article allocation* and the best use of existing systems must be found. For this purpose, selection, operation and allocation strategies are needed.

17.5.1 Parallel Commissioning Systems

For a assortment with many similar articles, several equal elementary commissioning systems are arranged in parallel. Each elementary system is a separate *picking zone*. *Figs. 13.29*, *16.16*, *16.17*, *17.18* and *17.19* demonstrate how 2, 4, 6 or more

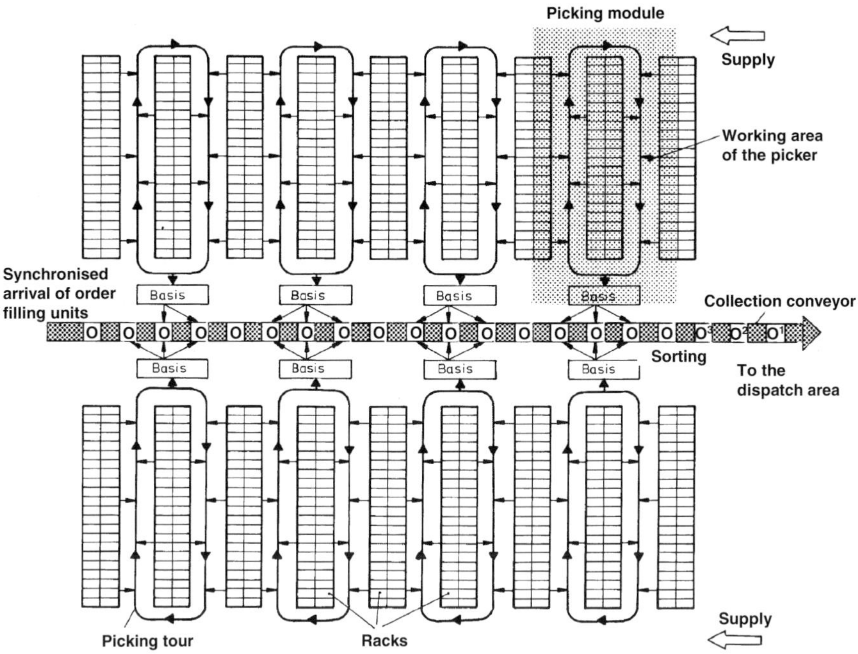

Fig. 17.18 Parallel commissioning Zones in the first stage of a two-stage order picking system

conventional commissioning systems can be combined in *commissioning modules* for the parts of an assortment.

If the articles of an assortment differ much in their physical properties and/or throughput, it is necessary to set up article-group specific commissioning systems. Common solutions are:

- small articles with low volume throughput in *shelve-* or *bin-systems*
- big articles and high volume throughput in *standard pallet-systems*
- long, heavy, bulky and special goods in *specialized systems*

The more different the assortment, the more different commissioning systems are possible. However, each specialized system must be dimensioned for the peak demand. Further disadvantages of specialization are partial underutilization, splitting of external orders into internal part orders and the necessity to consolidate the part order quantities. This leads to the *system planning rule*:

► The number of different systems should be kept as low as possible. They should be as universal as possible and only as specialized as necessary.

Articles that can be served by several commissioning systems must be allocated to the most efficient system. The most important *allocation strategy* depends on the *volume throughput*. It locates articles with small volume flow in bin- or shelve-systems, and articles with high volume flow in pallet-systems. If the volume flow is extremely high or if the order quantities for the same article are very different, it is opportune to keep this article at the same time in two or more picking zones or in different systems.

17.5.2 Separation of Storing and Commissioning

If the total stock of an assortment fills more load units than can be kept in the commissioning system without affecting the picking path length, it is better to separate the excess stock and to keep it in a *reserve store*. The excess stock is the part of the total stock that exceeds the content of the access place and of one additional provision unit per article.

The place costs in a separate reserve store with high capacity, which can be fully automatic, are significantly lower than the place costs in a commissioning system, which is designed for efficient order picking. However, a separate reserve store has the disadvantage that the load units are handled more often and have to be transferred in good time by a transport system or forklift truck to the commissioning system. This leads to the store *planning rule*:

► A separate reserve store is more economic than an integrated store only if the total stock exceeds considerably the place capacity within the commissioning system.

The replenishment of the reserve and access places in the commissioning system is triggered by the *pull-principle*. It should be executed as far as possible in full load units.

17.5.3 Two-Stage Order Picking

A two-stage order picking system consists of two commissioning systems in series, or of a commissioning system followed by a sorter system:

- In the *first stage*, the quantities for several external orders are picked *article oriented* due to internal *batch* or *series orders*.
- In the *second stage,* the collectively picked quantities of the first stage are *order oriented* separated or sorted due to the external orders.

The systems of the first stage can be conventional commissioning systems with static article provision as shown in *Fig. 17.18*, or static commissioning systems with dynamic article provision as shown in *Fig. 17.4*. The system in the second stage can be an *inverse commissioning* system as shown in *Fig. 17.5*. The second step can also be a sorting process, which is executed by a conveyor system or by *high performance sorters* as shown in *Fig. 18.14*. They collect the articles quantities of the first stage and distribute them to consolidation stations where they are packed to parcels or stacked on dispatch pallets.

Two-stage order picking with static provision in the first stage reduces the proportionate path, dead and basis times. The time savings increase with the number of orders per tour and the quantities per article. With dynamic order picking in the first stage, the number of out- and in-storing moves and the proportionate dead times are considerably reduced by batch orders for many external orders which require the same articles.

However, two-stage order picking involves *twofold handling* of the picking units. In the first stage they are picked from the access places, in the second stage out of the collection unit or from the exit station of the conveyor or sorter. Further disadvantages of two stage order picking are long order lead times. These disadvantages partly compensate or even cancel the savings of batch picking in the first stage. Most affected by the disadvantages are *single-position orders* and *rush orders*. The advantages and disadvantages lead to the following *conditions for two-stage order picking*:

- many orders (>1,000 per day) with few positions (2–5 Pos/Ord) and small picking quantities (up to 10 pieces) of a wide assortment (>10,000 articles)
- orders with different structure, many positions and large picking quantities
- continuous or bundled order entry (once or twice a day) and batch wise dispatch (up to four dispatches per day)
- small, light and conveyable picking units (< 7 kg per piece)
- no rush orders and no special services
- high utilization during the whole year for at least 8 hours per day

These conditions are generally fulfilled in *mail-order companies* and in the wholesale of *pharmaceuticals*. But even here, it remains open whether two-stage order picking is more efficient than one-stage order picking. The answer can only be given for the single project by comparing the effective commissioning costs for the optimal planned one-stage and two-stage system.

17.5.4 Channel-Store Commissioning

A channel-store commissioning system is a space saving combination of conventional or local commissioning systems with a narrow-aisle storage system. It is made up by a number of parallel *picking- and storing-aisle modules* as illustrated in *Fig. 17.19*. Each aisle module consists of two half *replenishment aisles*, two sideward *storage racks* and several *picking channels* on top of each other between the racks and the replenishment aisles. The tunnel-like picking channels are up to 60 m long and between 2.5 and 3.0 m high. The picking channels can also be placed within or along the outside of an automatic high-bay store.

Narrow-aisle trucks or automatic S/R-units operate in the replenishment aisles and feed the access and reserve places. The access places of the articles are arranged as shown in *Fig. 17.20*, side by side or on top of each other in a flexible *access place module*. Reserve units are located above the access places. The access place module can also be equipped with *flow channels* as illustrated in *Fig. 17.8*.

The pickers operate in the spatially separated picking aisles. They drive with a hand-cart, a pick-mobile or a horizontal picking device starting from a central basis to the access places on both sides of the picking channel. The article places are either indicated on a picking list or a mobile display. The picker takes the requested quantities from the indicated place, deposits them on a conveyor, or takes them back to the basis on the picking device. Full collection units and pallets are dropped there and conveyed by trucks or by a transport system to consolidation stations in the

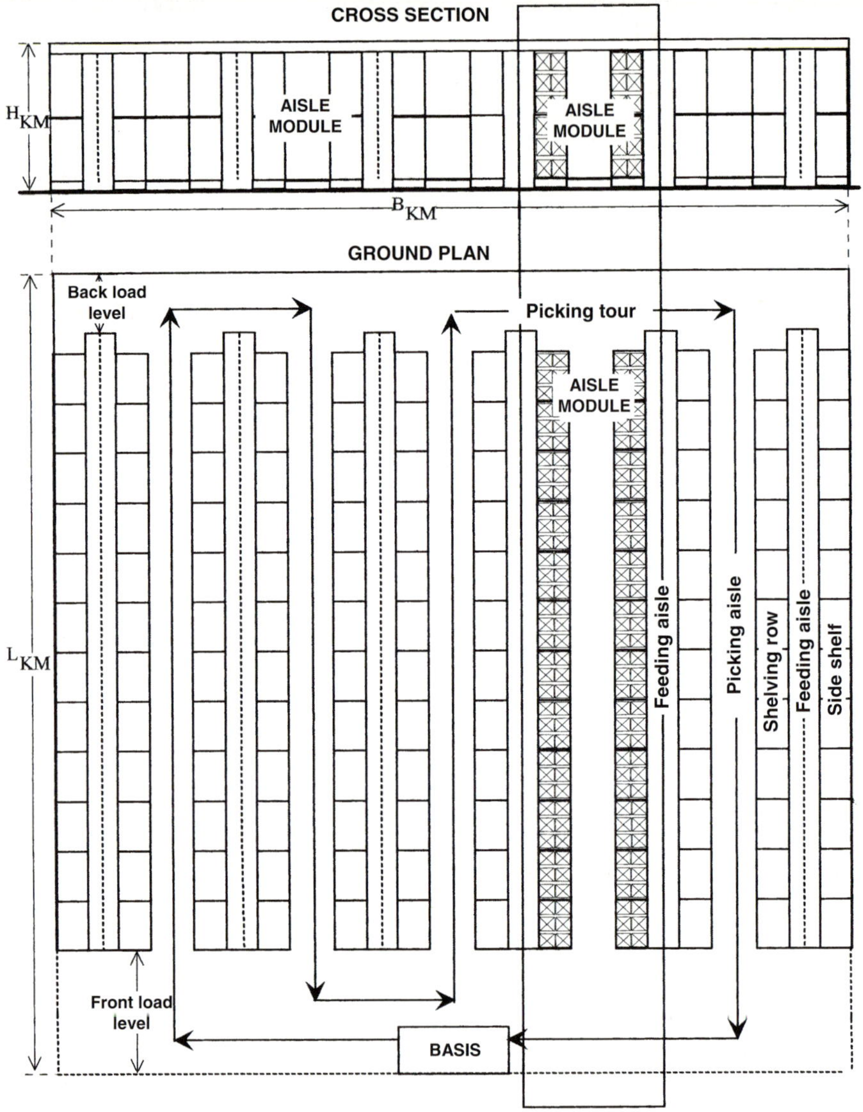

Fig. 17.19 Commissioning module of a two level channel-store commissioning system

dispatch area. If a conveyor system is installed in the channels, the picked quantities are moved loose or in collection units to the packing stations or the dispatch area.

Dimensions and number of aisle modules which constitute one commissioning module are limited by the maximal *escape length* (< 50 m) and by the maximal allowed size of a *fire section*. As shown in *Fig. 17.21*, several commissioning

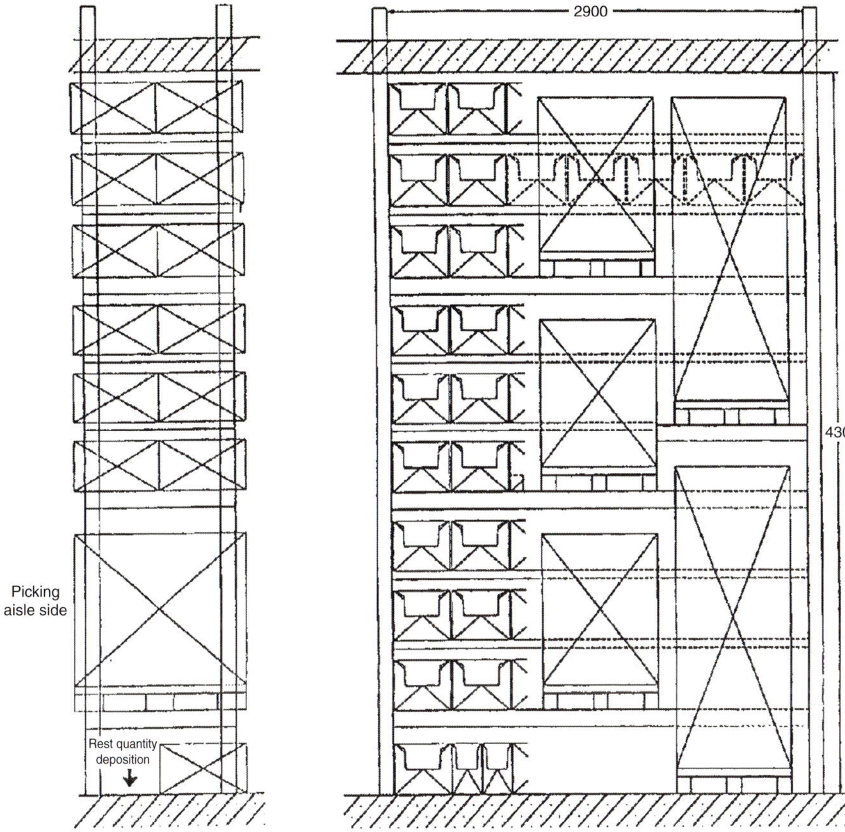

Fig. 17.20 Flexible access place module for bins and pallets

modules can be arranged in opposite and in parallel. Large channel store commissioning systems can consist of 8 to 16 modules with 6 to 8 picking channels per module on two or three levels. Such systems have been build and operated in Germany by large department store retailers in the last 40 years. They replenish retail outlets with an assortment of between 30,000 and 50,000 articles. Nowadays, also industrial companies operate such systems.

The main *advantages* of channel-store commissioning systems are:

- high picking performances for assortments with more than 1,000 articles
- high flexibility towards fluctuating demand and assortment changes
- effective replenishment by narrow aisle stackers or automatic S/R-units
- modular set up, compact construction and stepwise extension

However, after several-channel store commissioning systems have been built the following problems and *disadvantages* became obvious:

- protection of the pickers from falling into replenishment aisles

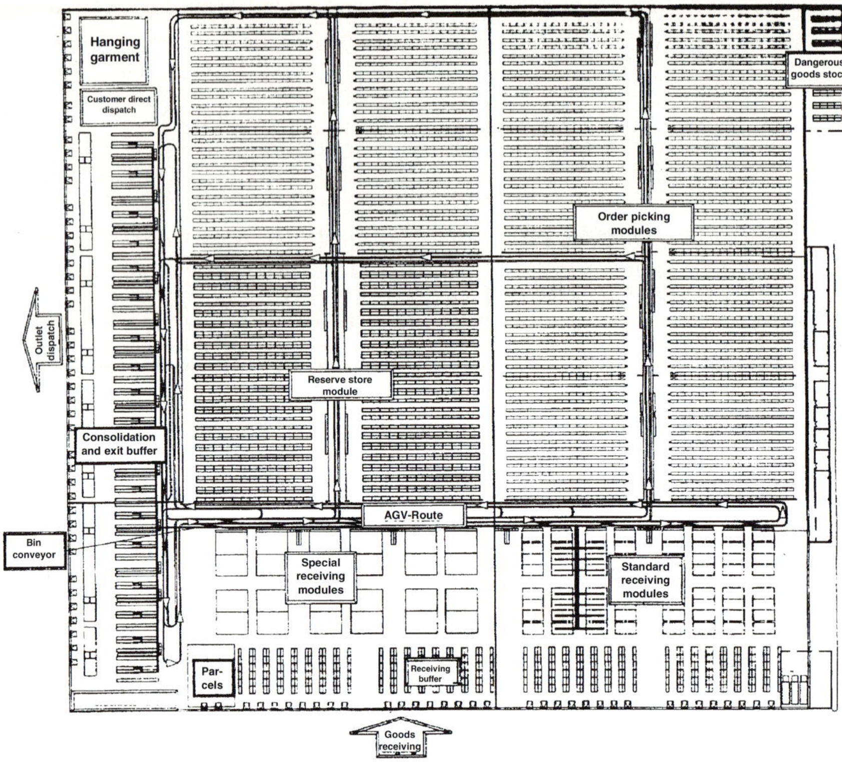

Fig. 17.21 Logistic center of a retail group for consumer goods

12 commissioning modules with 8 aisle modules for articles in bins and pallets with regular demand
4 storage modules for promotion articles with narrow aisle forklift trucks

- limited availability of the upper access places
- difficult escape from the long channels
- limited replenishment performance
- dead times by congestion effects
- high place costs
- limited reserve capacity
- difficult supply of empty bins and pallets
- tedious disposal of emptied load carriers
- removing of filled collection bins and pallets from the upper levels.

These disadvantages and their consequences can partly be avoided or minimized by qualified planning. Practical experience has shown:

► High picking performances and the economic utilization of a channel-store commissioning system are only achievable by optimal operating strategies and by an appropriate warehouse management system.

Unqualified planning, wrong operating strategies and insufficient software can cause enormous problems for channel-store commissioning systems.

17.6 Commissioning Strategies

The strategies for commissioning systems can be classified into:

placing strategies
preparation strategies
execution strategies
path strategies
take-out strategies
replenishment strategies
empties strategies

By these strategies different goals can be achieved. The prioritization of the goals determines the selection and combination of the strategies. But not all strategies are compatible. Incompatible strategies can partly or completely compensate against the effects of other strategies. The same goal can be achieved by different means. It is therefore advisable to assess effect and compatibility of the strategies to ensure that the effort for their realization is justified.

17.6.1 Placing Strategies

Placing strategies determine, on which places and in which zones which articles are stored and provided. The goals of placing strategies are efficient space utilization, short paths and minimal replenishment efforts. The most important placing strategies are:

- *Static pick-place order*: For each article a fixed access place is reserved as long as the article belongs to the active assortment.
- *Dynamic pick-place order*: Free access places are used for the next article that requires an access space. The article remains in its place only until the depletive grip.
- *Fixed reserve-place order*: For each article, fixed reserve places are reserved.
- *Free reserve-place order*: Free reserve places can be used for any article.
- *Zone-wise free-place order*: The places within certain *provision zones* can be used freely for a defined group of articles
- *Fast mover concentration*: If the article provision is static, the access units of fast moving articles are placed near the basis. For dynamic article provision, the fast moving articles are placed near the exits of the provision store. In both cases, path times are reduced.

- *Packing-optimal placing*: In order to avoid crushing and to achieve a high packing degree, the articles are located along the commissioning aisles in the order of decreasing volume and weight and increasing sensitivity (see *Sect. 12.4*).
- *Grip-optimal placing*: Fast moving articles and articles which are difficult to handle are provided in the optimal picking height. Slow moving and easy to handle articles are provided on lower and higher places (see *Fig. 17.13*).
- *Separation of reserves*: Smaller numbers of reserve units are stored above the access places, larger numbers in a separate reserve store.
- *Static flip-flop*: Each articled has two neighbored access places, one for the actual access unit, an other for the access reserve unit. The picker switches the places after the depletive grip.
- *Dynamic flip-flop*: The number of access places N_{AP} is made higher than the number of articles N_A of the assortment and occupied by the N_A articles in free place order. The N_{AP}-N_A unoccupied places are used flexibly for the replenishment of the access units which have reached the reorder level. After the *depletive grip*, the picker switches to the new place.
- *Single-article place utilization*: One access place and one access unit contains only one article.
- *Multi-article place utilization*: One access place or one access unit is occupied by several articles.
- *Throughput-dependent allocation*: An article is allocated to the most efficient commissioning system for the expected volume throughput.

The advantages of a *static pick-place order* are that it is quite easy to organize and allows for the arrangement of articles along the picking aisles in a required sequence, e.g. in the order of descending weight or volume or due to a customer-requested sequence. The article order can also correspond to the sequence of shelves in the sales room of the outlets. However, the static place order requires re-organization and replacing of the picking area whenever the assortment changes, e.g. at the beginning of a new season.

A *dynamic pick-place order* enables variable place utilization in combination with flip-flop. A free reserve place order reduces the place demand for the reserve units considerably. Free-place order, however, requires reliable place control and dynamic order scheduling.

The *separation of the reserve units* from the access units in commissioning systems with static provision leads to smaller *access lengths* and shorter distances. Insufficient replenishment, however, can cause an interruption of picking after the depletive grip and generate incomplete orders.

This can be avoided by the *flip-flop method*. Here, the picking process can be continued after the *depletive grip* at the second access place, where a reserve unit has been replenished in time. For *static flip-flop*, the length of the picking front is doubled and the path lengths are longer. *Dynamic flip-flop* requires only 10 to 20% more access places than articles, if the replenishment is optimally scheduled.

In conventional commissioning systems, by *fast mover concentration* reductions of the path lengths of up to 30% are possible, whereas the picking performance can be improved only up to 10%. The improvement depends on the ABC-distribution, the number of articles, the allocation of the picking places and on the path strategy. The main disadvantage of fast mover concentration is the increased probability of mutual *blocking* of pickers in front of the fast mover places, which are visited most frequently.

Another difficulty is, that fast movers may become slow movers and vice versa. A solution for articles with changing frequency is shown in *Fig. 17.22*. Here, the access units of the articles are replaced if in a self-regulating manner (Reinhardt 1993). Due to its disadvantages, the potentials of the fast mover strategy in conventional commissioning systems are smaller than generally expected. Its application is limited.

By *throughput-dependent allocation,* articles with low volume flow are stored and provided in small bins, single cartons or loose in flow channels whereas articles with larger volume flow are stored and provided in large quantities on pallets. Strategy parameter of throughput-dependent allocation is the *critical volume throughput*. The optimal value of this threshold depends on the individual project and must be calculated for the specific case.

The described placing strategies for conventional commissioning systems can also be applied for other systems with static article provision and for inverse commissioning.

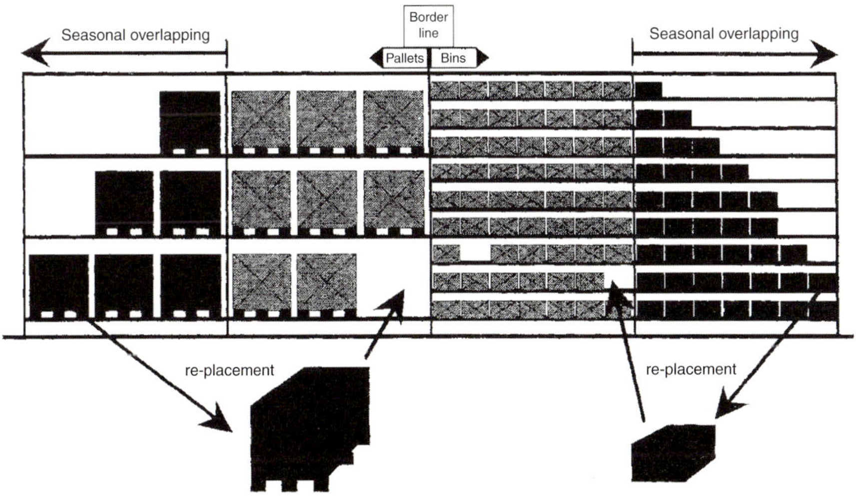

Fig. 17.22 Flexible placing of access units in bins and on pallets

17.6.2 Preparation Strategies

External orders are checked, sorted and prepared before they are executed. This is done by *schedulers* supported by computer programs. It can be executed due to different *order preparation strategies*:

- *Real-time-processing*: Each order is immediately processed and passed completely or as part-orders to the operative commissioning zones.
- *Time-batch-processing*: The incoming orders are collected for a certain *cycle period* T_C [h], jointly processed and passed as part-, batch- or series-order to the commissioning zones.
- *Quantity-batch-processing*: Incoming orders are collected up to a prede-fined *batch quantity* M_B, processed together, and passed to the respective picking areas as part- or series-order.

Real-time-processing is common for conventional commissioning systems with small picking demand and for express orders. However, with real-time-processing, many effective execution strategies cannot be realized. It also causes heavy fluctuations of the picking demand. This can generate *waiting queues* in front of the commissioning zones or at the access places of fast movers in peak hours. If the order entry is low, the pickers have to wait longer for the next job.

Batch-processing allows for the application of execution strategies that improve picking performance and enable a balanced utilization of the different picking zones. A disadvantage of batch-processing is a longer order lead time, which increases with the length of the cycle time and the batch quantity.

With time-batch-processing, the batch orders contain a varying number of external orders. This leads to batch orders with different quantities and deviating process times. The cycle time is a free *strategy parameter*, which can be adjusted to keep the maximal tolerable lead time. In the limit $T_C \to 0$, time-batch-processing becomes real-time-processing.

With quantity-batch-processing, the order batches always contain the same quantity and have nearly the same process time. In this case, the strategy parameter is the batch quantity. It can be adjusted to achieve optimal picking efficiency and to keep the maximal tolerable lead time.

17.6.3 Execution Strategies

Single orders as well as batch orders can be executed in the picking zones due to different *order execution strategies*:

- *Single-order execution*: All orders or selected orders such as rush orders are executed separately as single orders.
- *Fixed-batch execution*: A fixed batch of s external orders is executed as a *series order*. A new series order is started after the previous series order has been completely finished.
- *Floating-batch processing*: A floating batch of s varying external orders is executed in a progressing picking process. A new external order is started after an external order has been finished and left all picking zones.

- *Sequential order execution*: Single or series orders with articles from several picking zones pass these zones one after another.
- *Parallel order execution*: Single or series orders with articles from several picking zones are divided in part-orders and executed in parallel.
- *Prioritization of rush orders*: Rush orders are executed in all picking zones with first priority.

The advantages of single-order execution are short order lead times and a simple organization. Therefore, it is applied when short delivery times are essential. Disadvantages of single-order execution are unbalanced utilization of picking zones, lower picking performance and the neglect of consolidation effects.

These effects can be achieved by batch execution, however, with the disadvantage of longer lead times and a more complex organization. The *batch size* or *series length*, i.e. the number of consolidated external orders, is a *free strategy* parameter by which any compromise between lead time and consolidation effect can be achieved. With series length s = 1, the batch execution is reduced to single-order execution. The series length of order execution can be different from the batch size of order scheduling. It is limited either by the number and the capacity of the packing places and consolidation stations or by the capacity of the intermediate transport and sorter system.

With fixed-batch execution, a *switch time* of sufficient length between two consecutive order series is necessary in order to clear picking zones and sorter. Then, the next series can start. For dynamic-batch execution, no switch times are necessary.

The sequential execution of undivided orders has the advantage that the order is completed after it has passed the picking zones. If the orders are picked into the dispatch units, the completed order units are immediately available for shipping. Only one partially filled unit is generated per series order. If the series order equals the shipping order, no consolidation is necessary in the dispatch area.

Further advantages of sequential execution are the simple organization and the undivided responsibility of one picker for one order, unless partly executed orders are not passed to other pickers in the following zones. The disadvantages are long lead times for orders with many positions from different picking zones.

Parallel order execution shortens lead times, improves the utilization of the pickers and increases the picking performance. With parallel execution of smaller orders, one partly filled unit per picking zone is generated. These have to be compacted in the dispatch area. For big orders with several dispatch units, part-orders are generated automatically.

Single and batch order execution can be combined with sequential and parallel execution by computer-aided picking control. The different strategy parameters can be used to exploit the advantages optimally and to avoid the disadvantages of *combined strategies*.

17.6.4 Path Strategies

Path strategies, also called *traveling, moving* or *tour strategies*, determine the picking path and the sequence of the picking locations. To find the path of shortest length leads to the well-known *travelling-salesman-problem*. It can be either solved exactly by *full-enumeration* and comparison of all possible tours, or approximately by *OR-heuristics* (Churchman et al. 1961; Müller-Merbach 1970).

The full enumeration for n positions per tour results in n! = 1·2·3...(n-1)·n possible tours. This leads e.g. for n = 12 to about 480 million tours. Most of them are practically irrelevant. Even powerful computers calculate long time to find the shortest or approximately shortest tour for more than 10 positions per order with the help of OR-heuristics. This leads to unacceptable long waiting times for the picker in a real-time operation.

In addition, even the shortest tour may not be feasible in a system where several pickers are using the same aisle. A further disadvantage of OR-methods is, that they do not offer an explicit formula to calculate the *mean path time*. This time, however, is needed, for the analytical system optimization, for the calculation of picking performances and picking costs and for the assessment of the different influence factors. The experimental assessment of the influence factors and their effects on performance and costs is insufficient.

Therefore, it is necessary to develop *pragmatic path strategies* that result in feasible tours with lengths that do not deviate from the theoretical minimum by more than 10%. The strategy must be easy to understand, take care for regulated traffic, and should require only short preparation time. In addition, it must be possible to calculate the mean tour lengths for different order clusters.

For *one-dimensional move* of a picker in a rack module as shown in *Fig. 17.23,* three basic *aisle path strategies* are possible (Miebach 1971; Kunder and Gudehus 1975):

- *Aisle passing strategy* or *wriggling strategy*: The picker starts at the basis station, runs or drives in a wriggling path through the picking aisles, visits the requested articles one after the other and returns to the basis station. Aisles without requested articles are skipped.
- *Aisle entering strategy without repetition*: The picker starts at the basis station, moves along the front of the racks to the most remote aisle with requested articles, enters, picks all requested quantities and returns to the front. The picker repeats this procedure until all articles are collected and returns to the basis.
- *Aisle entering strategy with repetition*: The picker starts at the basis station, moves along the front of the racks to the most remote aisle with requested articles. For each article the picker enters the aisle and returns to the front separately. After all articles are collected the picker returns to the basis.

These path strategies can be further differentiated by *one-way aisles* or *two-way aisles* and by *one-side picking* or *two-side picking*.

The aisle passing strategy results in a simple and well controlled picking process. Blockages or encounters with other pickers can be avoided in small aisles by one-way traffic with *two-side picking* or in broader aisles with two-way traffic by *one-side picking*. The aisle passing strategy is applied in most conventional commissioning systems.

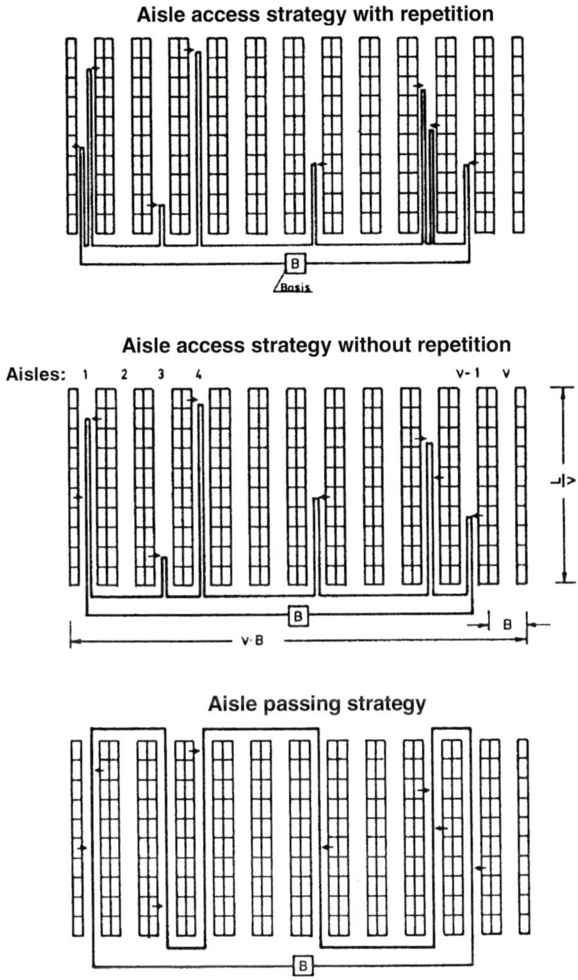

Fig. 17.23 Aisle passing and aisle entering strategies for one-dimensional movement in a picking aisle module

Strategy parameter: number of picking aisles $N_{PA} = 12$

Aisle entering strategies are necessary, if the aisle can only be reached from one front side of the rack. This is the case if the replenishment devices change the spatially separated aisles at the other front side. Also if the picker can pass on both sides, the entering strategy can be opportune, when fast moving articles are located close to basic-side rack front.

For aisle entering with repetition the picking aisle can be narrow, if the picker visits the article places without trolley and carries small quantities by hand to the front. The aisles must be broader for aisle entering without repetition where the picker moves into the aisle with a trolley in order to carry larger quantities. As a consequence, aisle entering with repetition can be opportune for small articles and quantities whereas aisle entering without repetition is necessary for larger articles and quantities.

The final decision between the three aisle path strategies depends on the mean path times which are determined by order structure, access length and article placing. They will be calculated later.

For the *two-dimensional movement* of the picker on a rack feeder or a high reaching order picking truck in a vertical rack module (see e.g. *Fig. 17.12*) *Miebach* 1971 invented the (see *Fig. 17.24*):

- *N-stripe strategy*: The vertical rack plane is divided into an even number of horizontal stripes N = 2, 4 or 6. The picker starts at the basis, moves to the closest requested article in the highest or lowest stripe, and visits the next articles by up and down moving along an S-shaped tour until the order is finished and the picker returns to the basis.

The strategy parameter N is used to minimize the mean travel time, which is given by the *N-point travel time formula* (16.71) of the previous chapter. Analytical calculations in agreement with simulations proved that the mean travel time achieved by the N-stripe strategy deviates less than 5% from the travel times for optimal tours. The optimal number of stripes for up to 25 order positions is N = 2 and for more than 25 order positions is N = 4.

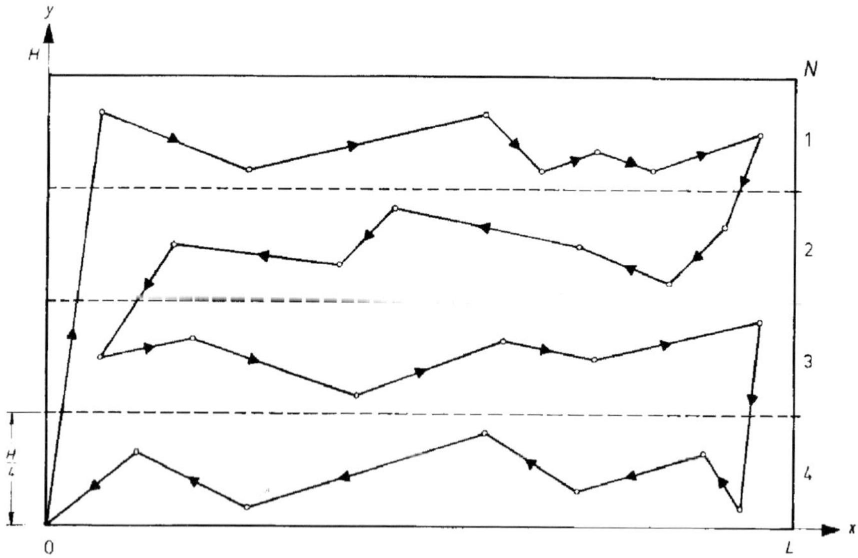

Fig. 17.24 N-stripe strategy for two-dimensional movement of the picker in front of a vertical rack plane

Strategy parameter: stripe number N = 4

Commissioning systems with dynamic article provision require no path strategies. Only strategies for the sequence of the articles and orders to be picked at a central station are needed. Path strategies are required in the preceding storage and provision system (see *Sect. 16.4*).

17.6.5 Take-out Strategies

Taking strategies determine from which access unit which quantity should be taken if the same article is attainable on more than one place. The most important *take-out strategies* are:

- *First-In-First-Out* (*FIFO*): Article units and load units that have been stored in first will be taken out first.
- *Priority of partial units*: The access unit with the lowest content is emptied first, even if this leads to position splitting and additional moves.
- *Quantity adjustment*: An article is taken from an access unit with content equal to or exceeding the requested quantity, even if this generates additional partially filled units
- *Taking the access unit*: If the content of the access unit is equal to or exceeds the requested quantity, the complete unit is taken after the excess quantity has been put aside.

Quantity adjustment is especially favorable for commissioning systems with dynamic article provision. Here, it avoids twofold provision for large position quantities. For commissioning systems with static article provision, quantity adjustment is not opportune as it complicates control, blocks several places per article and elongates path lengths.

Taking the access units can be combined with *fixed flip-flop* on two access places per article. It is advantageous for dynamic article provision if the removed excess quantity can be restored.

The advantages of taking the access unit are shorter picking times and use of the access load carrier. These advantages are especially important for large picking quantities. Instead of a complete pallet also *full layers of cartons* can be taken out.

17.6.6 Replenishment Strategies

Replenishment strategies are scheduling strategies for the supply of the access places in the picking zones. They should ensure high availability of articles with minimal effort. The replenishment for the picking zone implies the availability of reserve stocks. This is task of *inventory management* for the total stock (see *Chap. 11*).

Commissioning systems with dynamic article provision are supplied from an integrated storage- and provision system and do not require separate replenishment for picking. The replenishment strategies for static article provision are based on the *pull-principle*. These are:

- *Static flip-flop* or *two-bin Kanban*: Each article has two access units either side by side or behind each other. One is access unit and the other reserve unit. Replenishment of a full unit is initiated when the access place has been emptied. Picking is continued without interruption from the reserve unit.
- *Dynamic flip-flop* or *one-bin Kanban*: In addition to one access place per article, one or more reserve places are available in the picking zone. If the stock on the access place reaches the *reorder stock*, replenishment of a full unit is initiated. The arriving load unit is placed on a free place close to the last article place. After the depletive grip, the picking is continued at the new access place without interruption.
- *Refilling*: If the stock on the access place falls below the reorder stock, a sufficient replenishment quantity is requested to fill up the access place. After its arrival, the refilling quantity is added to the rest quantity in the access place.

The static flip-flop or two-bin Kanban is the simplest replenishment strategy for a picking zone (see *Sect. 12.7*). Its advantages are uninterrupted picking and avoidance of stock-outs. Additional advantages are that the picked quantities can be adjusted and that the access unit can be taken, if it is located close to the reserve unit. It also leaves enough time to remove empty bins and load carriers.

Static flip-flop with reserve units behind the access unit, e.g. in a *flow channel*, is optimal for picking from an assortment of many articles with high throughput as the path lengths are minimal. Static flip-flop with two parallel places doubles the length of the picking zone. For palletized goods it is therefore limited to a small assortment of up to 200 articles.

Dynamic flip-flop minimizes the number of additional places for reserve units, provides reliable scheduling and fast supply. It needs computer-assisted place management and dynamic scheduling.

As explained in *Chap. 11,* the *reorder stock* m_{RO} [AU] is the sum of a *safety stock* m_{safe} [AU] and the expected consumption until the replenishment arrives. For a *replenishing time* T_R [h] and a picking demand λ_{PS} [AU/h] it is:

$$m_{RO} = m_{safe} + \lambda_{PS} \cdot T_R \qquad [AU] \tag{17.15}$$

The reorder stock depends on the duration of the replenishment process and on the picking demand. Therefore, it must be permanently adapted to the current demand.

When the supply arrives, the stock has reached a remaining quantity which is on average equal to the safety stock. After the rest quantity has been picked the access place is available for other articles. This leads to the *picking zone dimensioning rule*:

▸ For dynamic flip-flop, the necessary *number of access places* N_{AP} in a picking zone for N_A articles with a mean safety stock m_{safe} [AU/Art] and load unit capacity C_{LU} [AU/LU] is

$$N_{AP} = (1 + m_{safe}/C_{LU}) \cdot N_A. \tag{17.16}$$

The replenishment for *refilling* is also triggered by the reorder stock. The refill quantity for an article with continuous demand should be the cost optimal replenishment quantity (see *Chap. 11*). If its value is unknown, the refill quantity is determined by the capacity of the access place. The advantage of refilling is the lower space requirement in the picking area. The disadvantage is the handling of single article units. Applications are picking of small pieces from flow channels or shelves and the supply of sales counters.

17.6.7 Empties Strategies

Empties strategies regulate the efficient removal of the empty load carrier after the *depletive grip*, and the supply of empty order collection bins or pallets in due time for picking.

Empties removal strategies are:

- *Empties removal by picker*: Empties are taken and carried away after the depletive grip by the picker.
- *Empties removal by picking device*: Empties are removed and carried away by the picking device.
- *Empties removal by conveyor*: Empties are removed by the picker, laid down on a special empties conveyor and conveyed to an *empties stacker*, an *empties collection centre* or to the next point of demand (see *Fig. 17.14c*).
- *Empties removal by replenishment device*: Empties are removed before replenishment and carried away by a replenishment device (see *Fig. 17.8c*).
- *Separate empties collection*: Empties are put aside or below the access place by the picker and removed later by a special *empties collector* (see *Fig. 17.16*).

Planning and organization of the empties are often neglected. The efforts for their removal are underestimated. This causes problems and costs during the operation of the commissioning system, in particular with heavy pallets of deadweights above 15 kg. The mechanization of pallet removal relieves the personnel and increases the picking performance in pallet commissioning systems with high volume flow.

Not only the removal of empties, but also the replenishment of empty collection units, such as bins or pallets, must be organized. *Empties replenishment strategies* are:

- *Stack-wise empties provision*: The required empties are provided in stacks, placed at the central basis or in the local picking zones and taken away by the pickers. The replenishment is initiated by the picker when the stack falls below the reorder stock.
- *Order-wise empties provision*: All empties for the order quantity are coded at an *empties preparation station*, laid on an *empties supply conveyor* and conveyed to the picking zones.

Stack-wise provision is the simplest and safest method. It requires the lowest technical and organizational efforts. The empties are "married" with the order by adding the order document.

Order-wise provision is more complex and less reliable. It is suited for commissioning systems with local deposition where a conveyor system runs along the picking zones or to the access places. Order-wise provision implies the danger that the picking process is interrupted by a lack of collection units. Another risk is mixing up collection units for different orders.

The provision of empties is of special importance for *pick&pack*. The dispatch cartons or packing means are either transported by the picker on a picking trolley, or provided by a conveyor at the pick places in an optimal position. *Single-position orders* where the picker can estimate the required packing means cause no problems. For multiple-position orders, the optimal size and number of cartons must be calculated by a computer program from the size and number of the article units of the total order.

17.7 Planning of Commissioning Systems

Commissioning is the most critical part process of *intralogistics*. It has to be harmonized and synchronized with the adjacent processes of goods receiving, storing, packing and dispatch. Hence, planning of the commissioning systems is closely interrelated with planning of the storage systems, transport systems and other subsystems of a *logistic center* (see *Sect. 3.2* and *16.8*) (Brynzér et al. 1994; Bode and Preuß 2004; De Koster and Van der Poort 1998).

In many projects, order picking requires most of the operative personnel and offers the best opportunities for saving costs and improving performances. It is therefore advisable to plan the commissioning system first, and to develop the storage and transport systems afterwards. *Planning steps for commissioning systems* are:

1. *Specification of the commissioning demand* and *synopsis of the restrictions and interfaces* to adjacent systems.
2. *Segmentation of the assortment* into article clusters of similar size, demand and volume flow.
3. *Clustering of the external orders* into order groups, such as single- and multi-position orders, small and large orders, or standard and express orders.
4. *Selection and dimensioning* of *provision units* and *collection units* for the different article clusters and order groups.
5. *Selection* of adequate methods, elementary systems, combinations and operating strategies for the different article clusters and order groups based on the above rules and features.
6. *System design* and *technical conception* of replenishment and provision, of place modules and picking modules, of the picking and replenishment devices, of conveyor technique and IT.

7. *Static design and optimization* of the provision zones and picking modules using the free design parameters and placing strategies.
8. *Dynamic design and optimization* of the commissioning systems including calculation of the pickers, the commissioning and replenishment devices, and of the conveyor systems using the free parameters and strategy variables of execution, path and replenishment strategies.
9. *Conception of the control system* with data flows, information displays and communication processes.
10. *Budgeting* of investment, operating and performance costs based on target prices and cost rates.
11. *Selection of the cost optimal commissioning systems* for the different article clusters and order groups, which are compatible with adjacent the systems.

After the commissioning systems have been planned, the goods receiving, storage systems, packing places and dispatch area are designed, dimensioned and optimized. The modular solutions for the different functional areas are combined by the rules and procedures of *Chap. 19* in a *layout plan* and connected by transport systems. The result is a cost optimal total solution for the *intralogistic* of a *company site*, a *logistic center* or a *factory*.

This iterative planning process can be supported by computer programs. *Planning software* for commissioning systems calculates the required access places, and the dimensions of the picking aisles and rack modules for a given demand. It uses the following formulas and algorithms for traveling times, cycle times and picking times to calculate the number of pickers and devices.

Standard *program modules* for elementary commissioning systems can be adjusted to the special characteristics of a project. Commissioning systems with static and dynamic article provision require different programs as their set up and operating strategies differ basically. The single program modules are linked in a program for the whole commissioning system. The analytical computer-model can be used to assess the influence factors and strategy variables, to simulate *operating scenarios* and to perform *sensitivity analysis*.

Some planners select the commissioning system based on qualitative comparisons, *utility analysis* or *benchmarks*. However, in many cases after systematic optimization, an un-favored initial solution turns out to be better than the originally favored solution. Planning requires more than computer programs and planning software. Technical expertise, experience and judgment are essential to ensure that the final solution fulfils the requirements, is reliable and flexible and can be operated at minimal costs.

17.8 Design Parameters and Strategy Variables

The optimal use of the design parameters and strategy variables is decisive for the successful planning of a commissioning system. They determine dimensions, path lengths, picking times, performances and operating costs.

Design parameters and strategy variables for all kinds of commissioning systems, as far as they are not externally predetermined, are:

dimension and assignment of access units
dimension and assignment of collections units (17.17)
batch sizes of the serial orders

Design *parameters for A2P-commissioning systems* with dynamic article provision are (see *Fig. 17.4*):

location of the stationary picking stations
number of pickers per station
number of buffer places before the access place (17.18)
number of buffer places behind the access place
number and position of order collection places per station

For a given picking demand (17.9), the parameters (17.17) and (17.18) determine the *number of picking stations* N_{PS} and the *provision performance* λ_P [AU/h]. The storage and provision system for the required performance and capacity can be designed as explained in *Chap. 16*.

Design *parameters for conventional P2A-commissioning systems* with static article provision are:

orientation of access units
capacity of access units C_{AV}
capacity of access modules C_{AM}
capacity of reserve modules C_{RM}
picking aisles per commissioning module N_{PA}
number of picking levels N_{PL}
replenishment aisles per commissioning module N_{RA}
number of replenishment levels N_{RL}
number of commissioning modules N_{CM}
orientation of commissioning modules (17.19)
speed and acceleration of replenishment devices
capacity of the replenishment devices C_{RD}
speed and acceleration of commissioning devices
capacity of commissioning devices C_{CM}
number of basis stations N_{BS}
location of basis stations
filled container capacity of the basis
empties capacity of the basis

Design parameters for *inverse commissioning* are the static order collection places and the number and length of the picking aisles (see *Fig. 17.5*).

17.9 Static Design of Commissioning Systems

The tasks of the static design of a commissioning system are to determine the number, locations and dimensions of the basis stations, of the access and reserve places and of the picking and replenishment aisles. Goals are short picking tours, optimal performance, small area demand and lowest costs. In addition, project specific restrictions, such as available areas, buildings and installation height, and legal restraints for safety and work places must be fulfilled.

Further restrictions are *fire sections* and the maximal tolerable *escape length*, i.e. the shortest distance between the working areas and the next exit of the fire section. The German *VDI-regulations* allow escape lengths of up to 50 m (VDI 3564, 1999).

The static design of A2P-commissioning systems with dynamic article provision is reduced to the modular design and optimal location of the stationary picking stations using the design parameters (17.18). *Figure 17.25* shows a modular work station for dynamic picking from pallets into dispatch bins standing on wheel-platforms.

A conventional P2A commissioning system with static article provision is made up by *access place modules* (AM) and *reserve place modules* (RM) located side by side or on top of each other along a *picking aisle*. The access and reserve place modules in a picking aisle and the attached *replenishment aisles* form a *picking module* (PM). Several picking modules make up a *commissioning module* (CM) as shown in *Figs. 17.19* and *17.23*. The steps of static design of conventional P2A-commissioning systems are:

Fig. 17.25 Stationary picking module for dynamic provision of palletized articles and picking into dispatch bins

17.9.1 Design of Access Place Modules

Design and dimensions of the access place modules determine the total space demand, the picking tour lengths and the picking performance of the whole system. The access place modules have to fulfill the following requirements:

- as many access places as possible along the shortest front
- sufficient places for the access reserve
- ergonomic provision of the picking units
- efficient replenishment for the access places
- precautions for the pickers from the replenishment aisles
- arrangements for empties return
- adaptability to changing access units

Figure 17.20 shows an example for a flexible access place module for bins and pallets that fulfills all these requirements. It has been designed for the channel-store and commissioning system of *Figs. 17.19* and *17.21*.

The length l_{AM}, the depth or breadth b_{PM} and the height h_{PM} of the access place module are determined by its design and construction. The capacity of the access place module is the sum of the capacity C_{AMA} of all access places and the capacity C_{AMR} of all reserve places within the module:

$$C_{AM} = C_{AMA} + C_{AMR} \qquad [LU/PM] \tag{17.20}$$

The total demand for access places depends on the number of articles and on the placing strategy. With the *access order factor*

$$f_{AO} = \begin{matrix} 1 & \text{for fixed access place order} \\ 2 & \text{for static flip-flop} \\ 1 + m_{safe}/C_{AP} & \text{for dynamic flip-flop} \end{matrix} \tag{17.21}$$

results the

► necessary *number of access place modules* with total access place capacity C_{AMA} [LU/ AM] for picking from an assortment of N_A articles:

$$N_{AM} = \{f_{AO} \cdot N_A/C_{AMA}\}. \tag{17.22}$$

In this and the next formulas the curly brackets $\{\ldots\}$ denote rounding up to the next integer.

The access place modules are located side by side on the two sides of the picking aisle. With access module length l_{AM} and N_{PA} picking aisles, the *total access length* of the P2A-commissioning system is:

$$L_{AL} = N_{AM} \cdot l_{AM}/2. \tag{17.23}$$

The total access length is divided into N_{PA} picking aisles. The *number of picking aisles* is a free design parameter. In addition to the access places, the picking aisle module should contain as many reserve modules as possible without elongating the picking paths. For the *flip-flop strategy*, it must have places for at least two replenishment units per article. One of them is the access place.

If the reserve capacity of the access place modules is not sufficient, additional reserve place modules can be located above the access place modules or at the opposite side of the replenishment aisles. The reserve place modules are designed similarly to the store place modules (see *Sect. 16.3.2*).

17.9.2 Design of Commissioning Modules

A picking aisle module consists of N_x access place modules which are located in x-direction on both sides of the picking aisles with aisle breadth b_{PA}. The commissioning module is made up by N_{PM} parallel picking aisle modules. If picking and replenishment are spatially separated, the picking modules are located between the replenishment aisles.

For a commissioning system with several picking levels as shown *Fig. 17.19*, the vertical distance between the floors can be used for reserve places. With N_{PL} picking levels, level distance h_{PL} and a *vertical positioning measure* H_{pos} of the picking and replenishment devices, the total height of the commissioning module becomes:

$$H_{PM} = H_{CM} = N_{PL} \cdot h_{PL} + H_{pos}. \qquad (17.25)$$

The number of access place modules in x-direction within the N_{CM} commissioning modules, which each consists of N_{PM} picking aisle modules in N_{PL} picking levels, is determined by the number of access modules (17.22) and given by:

$$N_x = \{N_{AM}/(2 \cdot N_{PM} \cdot N_{PL} \cdot N_{CM})\} \qquad [AM/CM]. \qquad (17.26)$$

With this number, the *length of the picking aisle module*, which equals the length of a commissioning module, can be calculated from the length of the access place module l_{AM} and the horizontal position measure L_{pos} of the picking and replenishment devices:

$$L_{PM} = L_{CM} = N_x \cdot l_{AM} + L_{pos} \qquad (17.27)$$

If picking and replenishment are spatially separated, a replenishment aisle is located on both sides of the picking module. The inner replenishment aisles of a commissioning module provide two picking aisle modules. The two outer replenishment aisles serve only one picking aisle module. For combined feeding and retrieval, the replenishment aisles vanish (see *Fig. 17.23*).

With the replenishment aisle factor

$$f_{AO} = \begin{cases} 0 \text{ for spatially combined replenishment} \\ 1 \text{ for spatially separated replenishment,} \end{cases} \qquad (17.28)$$

the depth of the access place modules b_{AM} and the breadth of the picking aisle b_{PA}, the *breadth of the picking aisle module*, including the proportionate replenishment aisle of breadth b_{RA}, becomes:

$$B_{PM} = 2 \cdot b_{AM} + b_{PA} + f_{AO} \cdot b_{RA}. \qquad (17.29)$$

Including the additional two half replenishments aisles at the outsides the *breadth of the commissioning module* is

$$B_{CM} = B_{PM} \cdot (2b_{AM} + b_{PA} + f_{AO} \cdot b_{RA}) + b_{RA}. \qquad (17.30)$$

With length (17.27) and breadth (17.30), the *ground area of the commissioning module* becomes:

$$A_{CM} = L_{PM} \cdot B_{PM}. \tag{17.31}$$

The number of commissioning modules, the number of aisle modules per commissioning module and the number of commissioning levels are free *design parameters*. The application of these parameters results in a goal conflict between optimal space and area utilization on one side, and minimal picking tour lengths on the other side. In addition, the project specific constraints and the safety restrictions must be kept.

With a small number of long aisles, the ground space becomes minimal as the losses for the horizontal position measures are more and more negligible. However, the overall length and breadth are limited by the *escape length* of 50 m. This leads to maximal dimensions of a commissioning module of $L_{PM} \cdot B_{PM} = 100 \cdot 100$ m and a maximal aisle length of about 80 m. However, the maximal aisle length does not lead to minimal tours lengths for conventional P2A-commissioning, as the probability to skip aisles without requested articles is lower for a small number of long aisles than for a high number of short aisles. The tour optimal aisle number will be determined in the next section.

17.9.3 Number and Location of Basis Stations

In a P2A-commissioning system with central deposition, the basis stations are located in front of the picking aisles. They are start- and end-point of the picking tours. The number and location of the basis stations are additional *design parameters* that can be used for system optimization.

A result of tour optimization is the *basis location rule*:

► In order to minimize the mean tour lengths, the basis stations must be located equally distributed in front of the picking aisles.

If there is only one basis station, it has to be located in the middle of the front. With more than one station, two different *basis visiting strategies* are possible:

- *Fixed basis stations*: Each picker has a fixed basis where orders and empties are received and the picking tour starts and ends.
- *Flexible basis stations*: The picker deposits the quantities of a completed order at the next basis and starts the next picking tour from there.

Flexible basis stations minimize the paths from and to the basis. They also reduce *dead times* if several pickers queue up at the basis stations. However, the flexible basis strategy must be well organized and computer controlled in order to ensure the supply of the pickers with orders and empties.

The probability of simultaneous arrival of two or more pickers at the same basis can be avoided by shifting the start time of the single pickers. This also helps to reduce dead times at the basis (see *Sect. 17.11*).

At each basis station, a terminal is installed and sufficient capacity for providing the pickers with empties must be available. Also the deposition and removal of filled collection units have to be organized. The basis stations of a larger logistic center are connected with transport systems to supply the empties and to remove the filled load units. In order to decouple the picking from the inbound and outbound transport, sufficient buffer places are necessary at the basis stations.

17.10 Minimal Tour Length and Optimal Aisle Number

The *travel time* for the tour of a picker from the basis to the places of the requested articles and back to the basis depends on the tour length, the speed and acceleration and on the number of stops and aisle changes. The tour length is determined by the path strategy, the number of positions per order, the placing of the articles, the number, length and breadth of the picking aisles and by the aisle changes.

In order to calculate the picking performance, only the *mean tour length* for the *mean number of visited aisles* must be known, which is given by the *theorem of aisle visits* (Kunder and Gudehus 1975):

- For an assortment randomly placed in N aisles, the mean number of aisles, where articles for orders with on average n positions are located, is

$$x = (1 - (1 - 1/N)^n) \cdot N. \qquad (17.32)$$

The proof of this theorem is: 1/N is the probability that a requested article is located in one specific aisle of the N aisles. Hence, (1-1/N) is the probability that the article is not located in this specific aisle. Consequently, $(1\text{-}1/N)^n$ is the probability that none of the n required articles and $(1\text{-}(1\text{-}1/N)^n)$ the probability that at least one of these articles is located in this aisle. As each single of the N aisles has the same probability to contain one or more of the n requested articles, the mean number of aisles is this probability times the number of aisles N.

Formula (17.32) has been approved by thousands of stochastic simulations. For far more than 10 aisles holds the *approximation formula*:

$$x \simeq (1 - e^{-n/N}) \cdot N \qquad \text{for} \qquad N >> 10 \qquad (17.33)$$

This formula is easier to differentiate than the exact formula (17.32). The dependency (17.32) of the mean number of visited aisles on the mean number of order positions n is shown for three different aisle numbers in *Fig. 17.26*.

This dependency illustrates the general *aisle visiting rules*:

- If the number of order positions is considerably smaller than the number of picking aisles, the mean number of aisles equals the number of positions, i.e. $x = n$, as they are located with highest probability in different aisles.
- With increasing number of order positions the mean number of visited aisles becomes smaller than the mean number of order positions, as more and more articles are located in the same aisle. For $n \approx N > 10$ the mean aisle number is $0.63 \cdot n$.
- With further increasing number of order positions, the mean number of visited aisles approaches the total number of aisles, i.e. $x \to N$.
- The mean number of visited aisles is independent of the total number of articles.

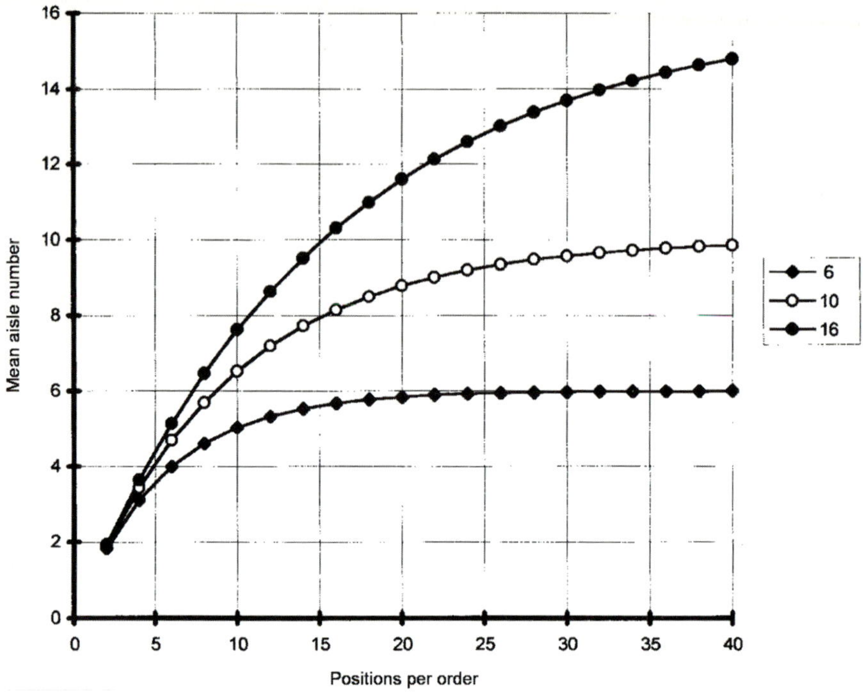

Fig. 17.26 Dependency of the mean aisle number on the mean number of positions per order
Parameter: number of picking aisles $N_{PA} = 6/10/16$

With the help of formula (17.32), the mean tour lengths and the mean travel times can be explicitly calculated for any number and arrangement of aisles and for the different path strategies.

Logic and qualitative comparisons already show that most of the theoretically possible aisle arrangements result in longer paths or are practically unsuitable. It has been proved by stochastic simulations and analytical calculations that the *parallel aisle arrangement* with a centered basis in front, as illustrated in *Figs. 16.6, 17.19* and *17.23* is the most practicable solution. With this arrangement, by *optimal aisle number* and *optimal path strategy,* the shortest travel times can be achieved (Kunder and Gudehus1975; Miebach 1971).

For the *opposite aisle arrangement* as shown in *Fig. 16.7*, the optimally achievable travel times are slightly longer than for the parallel arrangement. However, as demonstrated in *Sect. 16.9.3,* the area of this arrangement is somewhat smaller.

17.10.1 Path Length and Time with Aisle Passing Strategy

A picking tour, which starts and ends at a central basis and follows the aisle passing strategy as shown in *Figs. 17.2*, *17.19* and *17.23*, always passes an even number of aisles. For picking orders with many positions $n >> N$ the probability to pass aisles without requested articles is reduced by the *aisle number rule*:

► The total number of aisles for conventional commissioning should be even.

If the articles are equally distributed and the mean number of aisles with requested articles x is smaller than the total number of aisles N, i.e. if $x < N$, the number of visited aisles is x, if x is even, and x+1, if x is odd. Hence, for many orders with *mean position number* n, the mean tour length through the aisles of length L_{PM} is $(x+1/2)\cdot L_{PM}$, if $x < N$, and $N\cdot L_{PM}$ for $x = N$. The mean path along the two front sides of N picking aisles modules with breadth B_{PM} for x visited aisles is $2\cdot x/(x+1)$ times the maximal path $(N-1)\cdot B_{PM}$ along the front side for $x = N$. This leads to the *order tour length rule*:

- The *mean tour length* following the aisle passing strategy is for picking orders with mean position number n and N parallel aisles modules with length of L_{PM} and breadth B_{PM} given by

$$L(N) = \text{MIN}(N; x + 1/2) \cdot L_{PM} + 2 \cdot (x/(x+1)) \cdot (N-1) \cdot B_{PM} \qquad [\text{m}]. \tag{17.34}$$

If v is the maximal *speed* and b the *acceleration value* (16.61) of the picker and his vehicle, the travel time per tour, including the n accelerations and de-accelerations after and before each visited article, is $L/v+n\cdot v/b$ due to relation (16.62). In addition, 2x+1 de-accelerations and accelerations with time loss v/b happen at the front sides when the aisles are changed and the basis station is visited. This results in the *travel time rule*:

► The *mean order travel time* following the *aisle passing strategy* is for picking orders with mean position number n and N parallel aisles modules with length L_{PM} and breadth B_{PM} given by

$$t_n(N) = L(N)/v + (n + 2x + 1) \cdot v/b \qquad [\text{s}]. \tag{17.35}$$

The *mean path time per position* for the aisle passing strategy is:

$$\tau_{trar}(n) = t_n(N)/n \qquad [\text{s/Pos}]. \tag{17.36}$$

The mean number of aisles x with requested articles is given by formula (17.32) and the mean tour length L(N) by relation (17.34). Formulas (17.32) to (17.36) can be used to optimize the number of aisles and to calculate the picking performance for a given demand.

In *Fig. 17.27*, the dependency of the mean path time per position on the mean number of positions is shown for an example with the requirements of *Table 17.1*. The mean path time per position decreases with increasing number of order positions as the distances between the stops get shorter.

In order to provide the *total access length* L_{AL} given by relation (17.23), each picking aisle module must have the length:

$$L_{PM} = L_{AL}/N. \tag{17.37}$$

A functional dependency of the mean tour length L(N) on the number of aisles N results by inserting relation (17.37) in relation (17.34). For the above example, the calculated dependency is shown in *Fig. 17.28*. Following the aisle passing strategy,

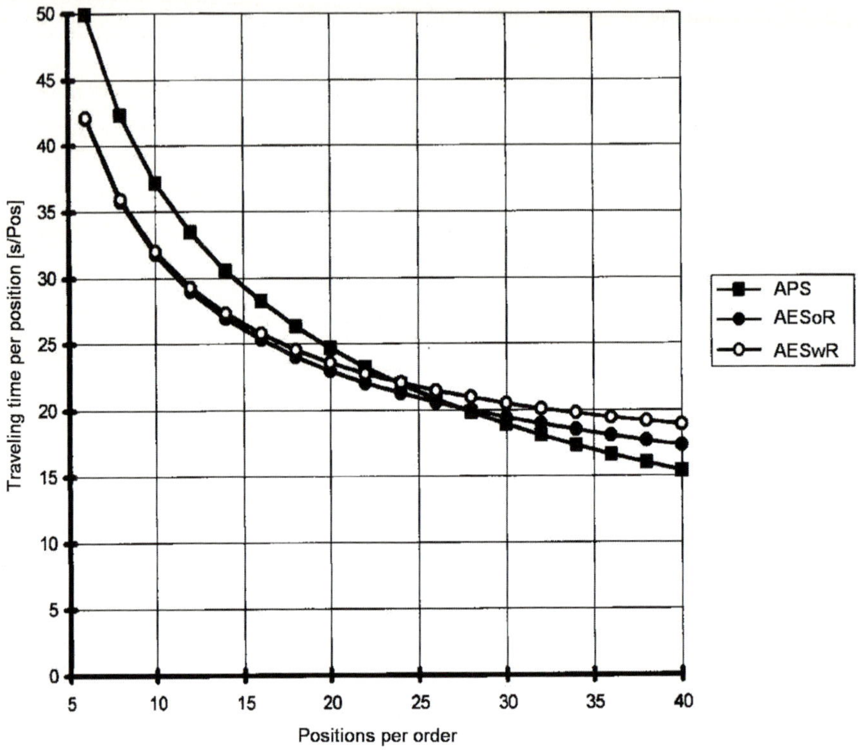

Fig. 17.27 Dependency between the mean path time per position and the mean number of order positions for different aisle path strategies

APS:	aisle passing strategy
AESoR:	aisles entering strategy without repetition
AESwR:	aisles entering strategy with repetition
Demand:	see *Table 17.1*, column 1
System:	see *Figs. 17.2* and *17.16* Placing : with fast mover concentration
Aisles:	N = 12, L = 36 m
Access units:	pallets 800 × 1,200 × 1,050 mm
Dispatch units:	pallets 800 × 1,200 × 1,050 mm
Picking device:	forklift trucks for 2 pallets

the mean tour length initially increases slightly with the number of aisles due to the additional front paths but decreases with higher aisles numbers due to the increasing probability of skipping aisles. However, from an optimal aisle number upwards the mean tour length increases again due to the increasing front paths. That means:

- Following the aisle passing strategy, the mean tour length does not change much with the aisle number.
- There exist two *optimal aisle numbers* resulting for on average minimal tour length, the first at N = 2 and the second approximately at

$$N_{opt} \approx \sqrt{0,3 \cdot n \cdot L_{PM}/B_{PM}} \quad \text{if} \quad n < N_{opt} \tag{17.38}$$

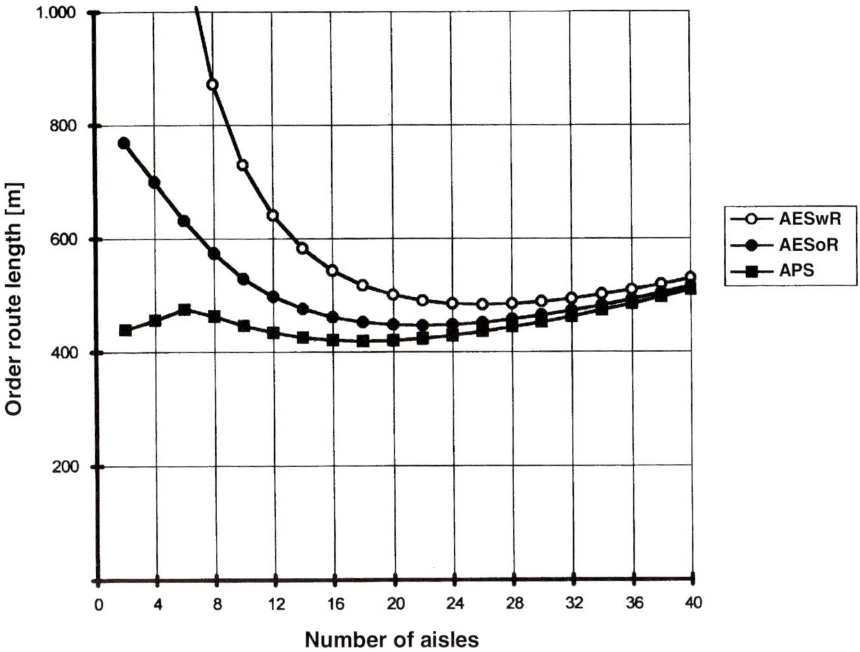

Fig. 17.28 Dependency of the order tour length on the number of picking aisles for different path strategies and equally distributed articles

Parameter: n = 15 Pos/order, others see legend *Fig. 17.27*

Formula (17.38) for the second optimal aisle number results by insertion of the approximate formula x ≈ 0,63·n into (17.34), differentiation with respect to N, setting the result equal 0 and solving it with respect to N. A further investigation shows that the mean tour length for the second optimum is only shorter than for two aisles, if the mean position number n is smaller than the value (17.38). Otherwise, the optimal number of aisles with shortest tour lengths is $N_{opt} = 2$.

For the considered example with a total access length of 432 m, an aisle module breadth of 5.1 m and on average 15 positions per order, the calculated second optimum for the aisle number is $N_{opt} = 20$ picking aisles. This value agrees quite well with the position of the exact minimum in *Fig. 17.28*.

Since with the aisle passing strategy, each visited aisle is completely passed, a concentration of fast moving A-articles near the basis front does not result in shorter travel times. On the contrary, the concentration of fast movers increases the probability that pickers block each other in front of these articles and cause longer waiting times. The same effect occurs if the A-articles are concentrated in the two or four middle aisles near the basis. The small savings of the front paths are generally overcompensated by the longer blocking times.

17.10.2 Path Lengths and Times for Aisle Entering with Repetition

Following the aisle entering strategy with aisle repetition of *Fig. 17.23*, the picker enters an aisle n times and travels a mean path length $2 \cdot L_{PM}/2 = L_{PM}$ within the aisles, if the articles are equally distributed. If fast moving A-articles with a share of the assortment p_A and a share of the order positions p_O are concentrated in the front sides of the aisle near the basis, the mean path length within the aisle is reduced with the *fast-mover factor*:

$$f_A = 1 + p_A - p_O \tag{17.39}$$

to $f_A \cdot L_{PM}$. If for example, the 20% A-articles make up 65% of the order positions, i.e. if $p_A = 0.20$ and $p_O = 0.65$, the fast mover factor is 0.55 and the mean path in the aisles is reduced by 45% with a fast mover concentration near the aisle fronts. The path length along the front sides is for the aisle entering strategies equal to the front path length for the aisles passing strategy. This leads to the *tour length rule*:

► The *mean tour length* following the *aisle entering strategy with aisle repetition* is for picking orders with mean position number n and N parallel aisles modules with length of L_{PM} and breadth B_{PM} given by

$$L(N) = f_A \cdot n \cdot L_{PM} + 2 \cdot (x/(x+1)) \cdot (N-1) \cdot B_{PM} \qquad [m]. \tag{17.40}$$

The functional dependency of the mean tour length L(N) on the aisle number N results after inserting relation (17.37) into (17.40). It is shown for the above example in *Fig. 17.28* for equally distributed articles and in *Fig. 17.29* for a concentration of 20% A-articles requested by 65% of the order positions.

Within the x visited aisles happen 2n and at their fronts x+1 accelerations and de-accelerations. This results in the *travel time rule*:

► The *mean order travel time* following the *aisle entering strategy with aisle repetition* is for picking orders with mean position number n and N parallel aisles modules with length of L_{PM} and breadth B_{PM} given by

$$t_n(N) = L(N)/v + (2n + x + 1) \cdot v/b \qquad [s]. \tag{17.41}$$

The mean number of visited aisles x is given by formula (17.32). The mean tour length L(N) can be calculated by formula (17.40). The path time per position results from the mean order travel time (17.41) with formula (17.36). Its dependency on the position number is shown in *Fig. 17.27*. The general formula and the examples show:

► For the aisle entering strategy with aisle repetition, the mean order tour length depends strongly on the number of aisles.

► With this strategy, the order tour length decreases with increasing number of aisles until the *optimal aisles number* is reached, as the paths within the aisles get shorter and more aisles are skipped. The mean order travel time increases slowly for higher aisle numbers.

► With an aisle entering strategy, by concentration of fast moving A-articles near the aisles entrances, the mean tour length can be reduced considerably, if the share of A-positions is much higher than the share of A-articles, i.e. for $p_O >> p_A$.

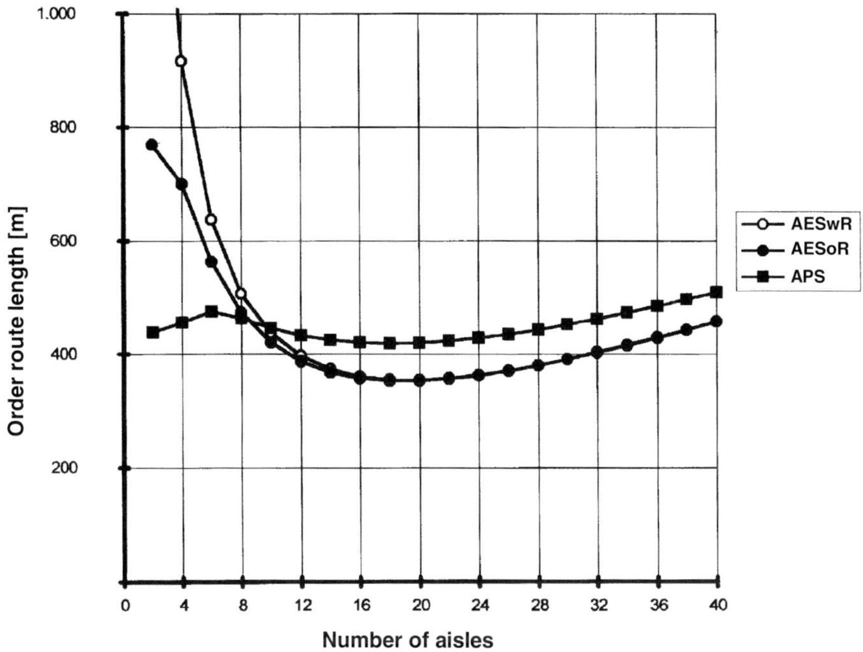

Fig. 17.29 Dependency of the order tour length on the number of picking aisles for different path strategies with fast mover concentration near the aisle entrances

Parameter: n = 15 Pos/order, $p_A = 20\%$, $p_O = 65\%$, others see *Fig. 17.27*

However, the savings of the path times at high picking demand are compensated by the longer waiting times due to the higher *blocking probability* at the concentrated A-article places. In the same way as before results:

- The *optimal aisles number* for the aisle entering strategy with aisle repetition

$$N_{opt} \approx \sqrt{0,5 \cdot f_A \cdot n \cdot L_{PM}/B_{PM}} \quad \text{if} \quad n < N_{opt} \tag{17.42}$$

For the above examples result, in agreement with the minimum positions of *Fig. 17.28* and *Fig. 17.29* respectively, the optimal aisle number $N_{opt} = 25$ without fast mover concentration, and $N_{opt} = 19$ with fast mover concentration.

17.10.3 Path Lengths and Times for Aisle Entering without Repetition

If all articles are equally distributed, the aisle entering strategy without aisle repetition leads to a mean number of requested articles per aisle n/x and a mean path length in the aisles $2x\cdot((n/x)/(1+n/x))\cdot L_{PM}$. If fast moving A-articles are concentrated near the aisle entrances, the mean path lengths in the aisles are reduced by the *fast-mover factor*

$$f_A = \text{MIN}(1\ ;\ 1 + p_A - p_O^{n/x}). \tag{17.43}$$

The path length along the front sides is equal to the front path length for the aisles passing strategy. This leads to the *tour length rule*:

▸ The *mean tour length* following the *aisle entering strategy without aisle repetition* is for picking orders with mean position number n and N parallel aisles modules with length of L_{PM} and breadth B_{PM} given by

$$L(N) = f_A \cdot (2xn/(n + x)) \cdot L_{PM} + 2 \cdot (x/(x + 1)) \cdot (N - 1) \cdot B_{PM}. \qquad (17.44)$$

Again, the functional dependency of the mean tour length L(N) on the aisle number N results after inserting relation (17.37) into (17.44). It is shown for the example in *Fig. 17.28* for equally distributed articles and in *Fig. 17.29* for fast mover concentration.

Within the x visited aisles, n+x and at the fronts x+1 accelerations and deaccelerations happen. This results in the *travel time rule*:

▸ The *mean order travel time* following the *aisle entering strategy without aisle repetition* is for picking orders with mean position number n and N parallel aisles modules with length of L_{PM} and breadth B_{PM} given by

$$t_n(N) = L(N)/v + (n + 2x + 1) \cdot v/b \qquad [s]. \qquad (17.45)$$

Again, the mean number of aisles x with requested articles is (17.32). The mean tour length L(N) is given by formula (17.44). The path time per position results from the mean order travel time (17.45) with formula (17.36). Its dependency on the mean position number is shown in *Fig. 17.27*.

In the same way as before results:

▸ The *optimal aisle number* for the aisle entering strategy without aisle repetition

$$N_{opt} \approx \sqrt{0,4 \cdot f_A \cdot n \cdot L_{PM}/B_{PM}} \quad \text{if} \quad n < N_{opt}. \qquad (17.46)$$

For the above example, in agreement with *Figs. 17.28* and 17.*29* respectively, the optimal aisle number without fast mover concentration is $N_{opt} = 22$ and with fast mover concentration $N_{opt} = 19$. The other rules are quite similar for aisle entering with and without aisle repetition.

17.10.4 Strategy Comparison and Optimal Aisle Arrangement

The comparisons of the mean travel times for the three examined path strategies and the examples *Figs. 17.27*, *17.28* and *17.29* show:

▸ Without fast mover concentration, the aisle passing strategy is more effective than the aisle entering strategies with and without aisle repetition.

▸ Fast mover concentration reduces the path times only for aisle entering strategies but not for the aisle passing strategy.

▸ Fast mover concentration with optimal number of aisles significantly reduces the travel times for aisle entering strategies. However, at high picking demand, the savings of travel times are compensated by waiting times due to blocking effects.

▸ Only if the number of fast moving articles is small and their share of the order lines is high, aisle entering strategies are opportune compared to the aisle passing strategy.

The above rules and conclusions together lead to the *planning rule for P2A-commissioning systems*:

▸ The optimal solution for P2A-commissioning systems is an even number of parallel or opposite picking aisle modules in the range between 2 and the value given by formula (17.38).

If not planning a *green field solution*, the initial solution can be adjusted to the existing building by stepwise changing the number and length of the aisle modules without affecting the path times very much.

The *standard strategy for P2A-commissioning* is aisle passing without fast mover concentration. If on average not more than one picker is working in an aisle, it can be advantageous to concentrate fast movers near the entrances and to pick orders with small number of positions, i.e. with $n < n_{opt}$, due to the aisle passing strategy and other orders with $n \geq n_{opt}$ due to the aisle passing strategy. The *opportune position number* n_{opt} can be determined by comparing the travel times (17.35), (17.41) and (17.45).

17.11 Pick Performance and Commissioning Times

The performance of a single picker can be measured in different ways:

- The *order-pick performance* μ_{Ord} [Ord/h] is the number of orders executed by a picker per hour.
- The *position-pick performance* μ_{Pos} [Pos/h] is the number of positions or order lines picked by a picker per hour.
- The *article-pick performance* μ_{AU} [AU/h] is the number of article units [AU] a picker picks and delivers per hour.
- The *collection-unit performance* μ_{CU} [CU/h] is the number of collection units [CU], such as bins or pallets, a picker fills per hour.

As the different performance measures depend on each other, only one measure is necessary. The selection of a suitable measure is a matter of convention and depends on company and project. If not explicitly stated differently, in this book commissioning performance means *position-pick performance*. When comparing the performances of different commissioning systems one has to be aware of the *influence factors*:

▸ Commissioning performance strongly depends on the order structure, the type of the commissioning system, the capacity of the provision and collection units, and on the number of articles.

If t_{Ord} is the mean *order picking time* for execution of orders with mean position number n_{Pos}, the mean *position picking time* is

$$\tau_{pos} = t_{Ord}/n_{Pos} \qquad [s/Pos] \tag{17.47}$$

With this time results the effective *limit performance* of a picker:

$$\mu_{Pos} = \eta_{ava} \cdot \rho_{max} \cdot 3600/\tau_{pos} \quad [Pos/h] \tag{17.48}$$

Herein η_{ava} is the *availability* of picker and technical devices and ρ_{max} the *maximally achievable utilization* of the commissioning system.

From the position picking performance and the mean quantity per position m_{AU} [AU/Pos], the *article-pick performance* can be derived:

$$\mu_{AU} = m_{AU} \cdot \mu_{Pos} \qquad [AU/h] \qquad (17.49)$$

Due to formula (12.32) of *Sect. 12.5*, the *mean number of collection units* with capacity C_{CU} for a mean order quantity $m_{Ord} = m_{AU} \cdot n_{Pos}$ is:

$$n_{CU} = MAX(1; m_{AU} \cdot n_{Pos}/C_{CU} + (C_{CU} - 1)/C_{CU}) \quad [CU/Ord] \quad (17.50)$$

This formula takes into account one partly filled collection unit per order. With (17.50) the *collection unit performance* can be calculated by the formula:

$$\mu_{CU} = n_{CU} \cdot \mu_{Pos}/n_{Pos} \qquad [CU/h] \qquad (17.51)$$

Due to the partly filled units the collection unit performance depends on the order quantity and on the capacity of the dispatch units. Therefore, the performance measurement by collection units is generally misleading.

If the provision and access units have the capacity C_{PU} [AU/PU], the number of emptied access units per order is:

$$n_{PU} = m_{AU} \cdot n_{Pos}/C_{PU} \qquad [PU/Ord]. \qquad (17.52)$$

The number of emptied access units (17.52) determines the interruption frequency of order completion by depletive grips.

In order to calculate the commissioning performance (17.48), it is necessary to quantify the picking time per position, the availability of the picker and the maximal utilization. The position picking time is the sum of *path time* τ_{trav}, *setup time* τ_{sup}, *grip time* τ_{grip} and *basis time* τ_{bas}:

$$\tau_{pos} = \tau_{trav} + \tau_{sup} + \tau_{grip} + \tau_{bas} \qquad [s/Pos] \qquad (17.53)$$

The relative contributions of the four part times to the total picking time differ very much for the different commissioning methods. For example, the path time for commissioning with dynamic provision is 0.

For conventional and inverse commissioning with one-dimensional movement, the path time per position is given by relation (17.36). For commissioning with two-dimensional movement it is equal to the mean travel time $t_n(L,H)$ for an n-point tour given by relation (16.71) divided by the number of positions n_{pos}.

For the calculation of picking times and performances, only the *mean values* for a large number of homogenous orders are of interest. If the assortment or the orders are heterogeneous, it is necessary to calculate the mean values for the different article classes and order groups separately.

In order to calculate the picking performance, all process times, which contribute to the picking time (17.53), must be related to the mean position number n_{Pos}. For example, if $t_{X\,PU}$ [s/PU] is the time necessary for a process X related to the provision unit PU, the contribution to the position time is:

$$\tau_{X\,Pos} = n_{PU} \cdot t_{X\,PU}/n_{Pos}. \qquad [s/Pos]. \qquad (17.54)$$

Herein n_{PU} is the number of provision units (17.52). If $t_{X\,CU}$ [s/CU] is the time necessary for process X connected with the collection unit CU, the contribution to the position time is:

$$\tau_{X\,Pos} = n_{CU} \cdot t_{X\,CU}/n_{Pos}. \qquad [s/Pos]. \tag{17.55}$$

Here n_{CU} is the number of collection units (17.50).

All activities related to the article unit AU have the strongest influence on picking time and picking performance. Next important are the activities related to the order position. Hence, these times must be calculated or measured most accurately. Processes connected with the provision and collection units are not as important, although they should not to be neglected.

17.11.1 Position Setup Time

The position setup time is the mean time per order line at the picking place before and after the grip process. It is caused by *information*, *positioning*, additional *handling activities* and regular *waiting*. Setup times can reach the order of magnitude of grip times.

Data and *information* are necessary to instruct the picker and to control the activities at the picking place. Such *information activities* are

reading of the next access place
search and identification of the access place
input of control information (17.56)
processing of pick-list and documents
scanning of access, article and collection units

Positioning is necessary before the grip to bring the picker and/or the access unit in the right position and after the grip to close the picking process. Depending on commissioning method and technique, typical *positioning activities* are

alignment of the picking device and/or the access unit
getting of and on the picking device
provision and removal of the access unit (17.57)
moving to and from the access place

Handling activities beyond the grip also depend on commissioning method and technique. Additional *handling activities* at the picking place are

setup of empty packages or dispatch units
closing of packages or dispatch units
coding and labeling of article units
removal of emptied load carriers (17.58)
pull-out of access reserve units
opening of provision units and packages

Regular *waiting* or *dead times* are

$$\begin{array}{l}\text{waiting times for replenishment}\\ \text{waiting times for information}\\ \text{blocking times caused by other pickers}\end{array} \qquad (17.59)$$

The waiting times for replenishment critically depend on the replenishment strategy and on the availability of the articles on the access place. The stochastically varying blocking times are either added to the setup time or taken into account as follows by the utilization.

Setup times at the picking place depend on many influence factors, such as working conditions, spatial situation, position, kind and content of the information display, and on attention and experience of the picker. These factors have to be well considered when designing access places and picking process.

The processing of *pick-lists* or documents is reduced by adequate displays either at the access place or on the commissioning device. The times for information input can be considerably reduced by *scanning*. Scanning also decreases failure rates. Waiting for supply units after the depletive grip is avoided by flip-flop-replenishment or by flow channels.

It is quite difficult to determine or to measure all parts of the setup time. The best method is to build up a pilot working place and to measure all activities with a stop watch. Well proven analytical methods are *MTM* or *work factor*. Some setup activities can be executed parallel to other activities, e.g. an empty pallet can be removed from the access place while waiting for the replenishment pallet. In these cases, only the longest lasting of the parallel activities contributes to the setup time.

17.11.2 Grip Time

The grip time is the time a picker needs to take out the article units from the access place and to deposit them in the collection unit or on a conveyor. The manual gripping process executed in the spatial picking situation of *Fig. 17.13* is shown in the flow diagram *Fig. 17.30*.

The main time for the gripping process is determined by the *elementary gripping steps*:

$$\begin{array}{l}\text{reaching forward}\\ \text{grabbing the good}\\ \text{moving the good}\\ \text{depositing the good}\\ \text{returning to start position}\end{array} \qquad (17.60)$$

In addition to the elementary steps, the gripping process can include *additional work-steps* such as:

$$\begin{array}{l}\text{clipping or cutting}\\ \text{weighing or cradling}\\ \text{admeasuring or measuring}\end{array} \qquad (17.61)$$

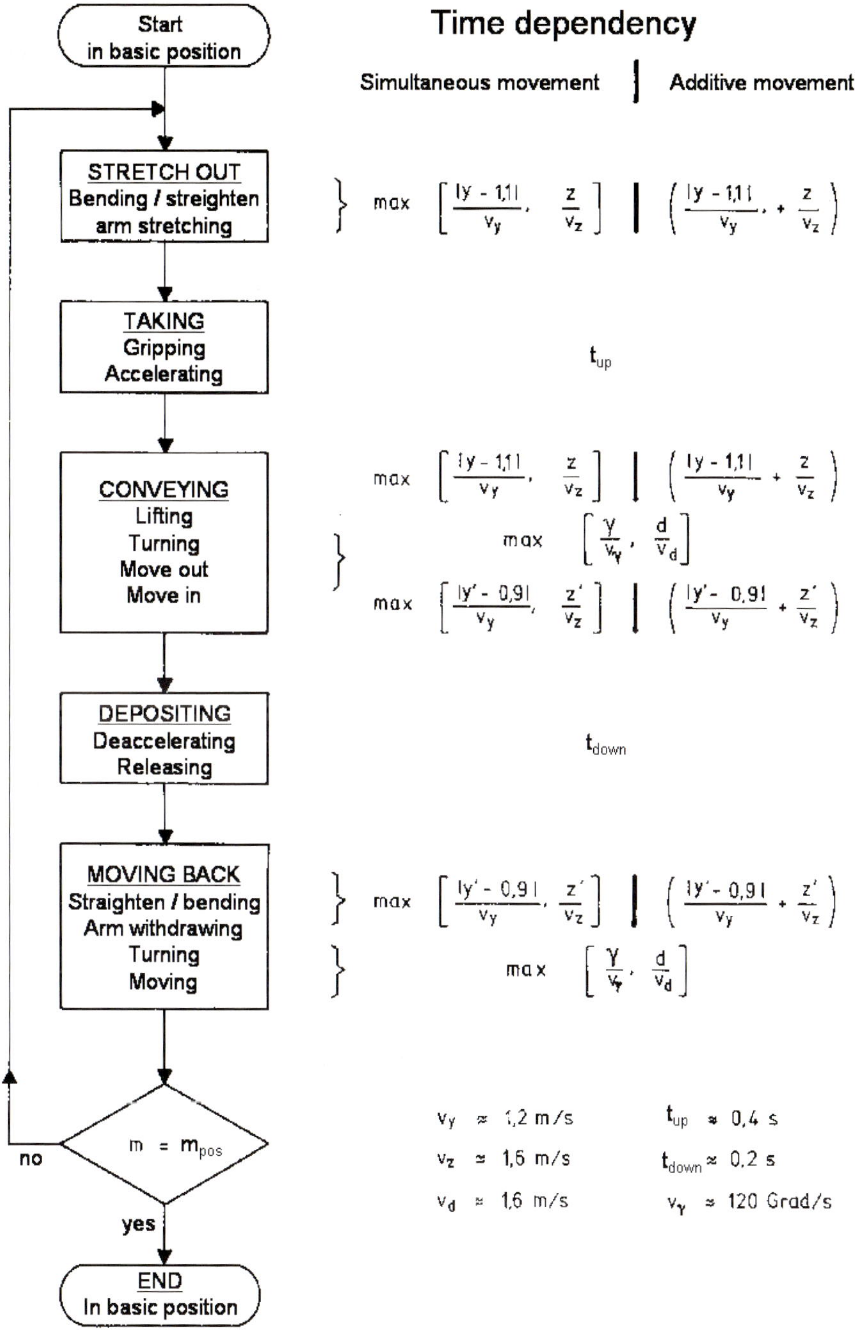

Fig. 17.30 Process flow and part times of the gripping process

For this purpose, often larger quantities are taken out and rest quantities are put back. The *spatial influence factors* of manual gripping are:

$$\begin{aligned}
&\text{minimal grip height } h_{gr\ min}[\text{m}] \\
&\text{maximal grip height } h_{gr\ max}[\text{m}] \\
&\text{mean grip depths } b_{gr}[\text{m}] \\
&\text{mean deposition angle } \gamma_{dep}[\text{m}] \\
&\text{mean deposition distance } d_{dep}[\text{m}] \\
&\text{minimal deposition height } h_{dep\ min}[\text{m}] \\
&\text{maximal deposition height } h_{dep\ max}[\text{m}] \\
&\text{mean deposition depth } b_{dep}[\text{m}]
\end{aligned} \tag{17.62}$$

As shown in *Fig. 17.13*, the spatial influence factors are determined by the design of the picking place and by the size and position of the access unit.

Additional influence factors are:

$$\begin{aligned}
&\text{mean picking quantity per position } m_{gr}[\text{AU/grip}] \\
&\text{mean volume per grip } v_{gr}[\text{l/grip}] \\
&\text{mean weight per grip } w_{gr}[\text{kg/grip}].
\end{aligned} \tag{17.63}$$

For *single-piece grips* with $m_{Pos} = 1$ AU/Pos, $m_{gr} = 1$ AU/grip, $v_{gr} = v_{AU}$ and $w_{gr} = w_{AU}$. For multi-piece positions, a larger quantity x can be taken at once, if the *article units* AU are small and light. For *multi-piece grips* of x pieces per grip, $m_{gr} = x$ AU/grip, $v_{gr} = x{\cdot}v_{AU}$ and $w_{gr} = x{\cdot}w_{AU}$. As multi-piece grips can cause counting errors, they should be excluded if high commissioning quality is required.

The different dependencies of the grip time on the most important influence factors (17.62) and (17.63) are illustrated by the diagrams of *Fig. 17.31*. These dependencies have been measured in the logistic center of a retailer.

The grip time per position is the product of the position quantity m_{Pos} [AU/Pos] and the *piece-grip time* or *grip time per article unit* τ_{AU} :

$$\tau_{grip} = m_{Pos} \cdot \tau_{AU} \quad [\text{s/Pos}] \tag{17.64}$$

The single steps (17.60) of the gripping process can be executed in *simultaneous moves* and/or in *additive moves*. The gripping move from a large shelf space or from a pallet and the deposition on a free place or on a pallet is simultaneous in two or three space directions. The gripping move is additive in two or three directions for grips from a small and deep rack-shelf-place and for the deposition into a honeycomb rack.

For mechanical gripping, the piece-grip time is the cycle time for the *gripping cycle* of the picking robot or picking device, which depends on speed, acceleration, movement and distances in the three space directions.

The time for a single manual grip can be measured either in a pilot station or a real pick place or determined analytically by *MTM* or *work factor*. For picking performance calculations only the *mean grip time* is required, which is given by the well approved

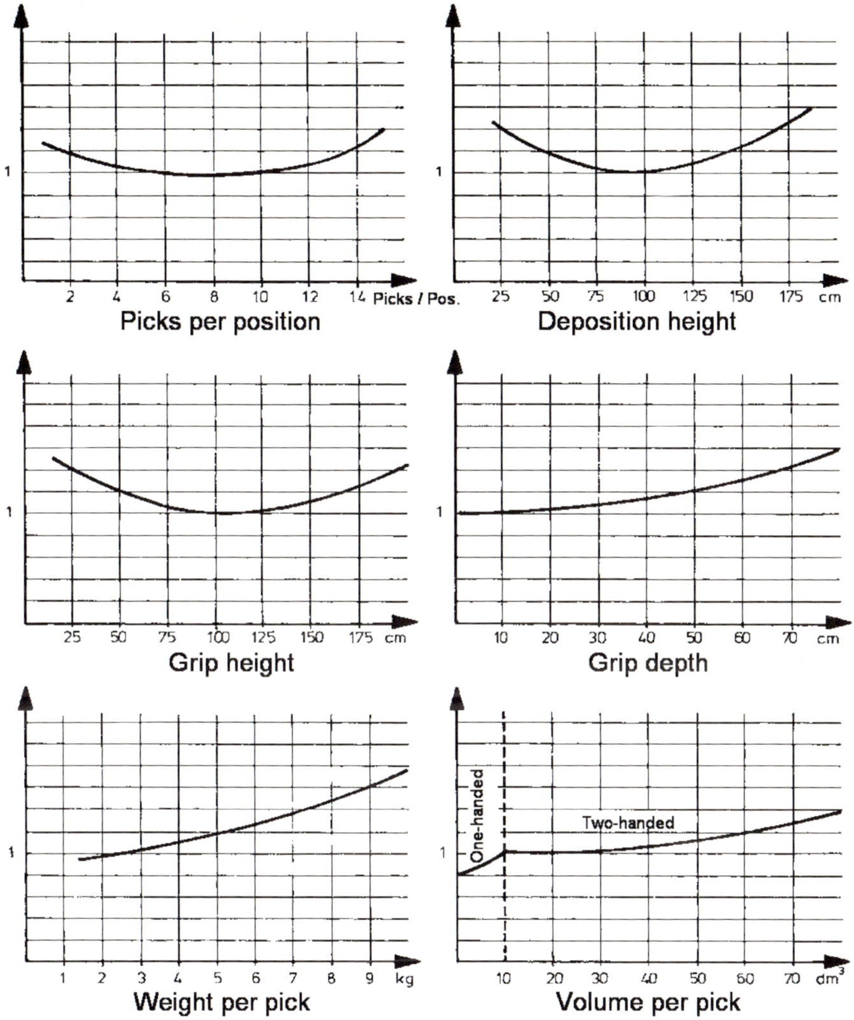

Fig. 17. 31 Measured dependency of the mean grip time per position on different influence factors

► *semi-empirical grip-time formula*

$$\tau_{AU} = 2 \cdot (1 + w_{AU}^2/110) \cdot (1 + v_{AU}^2/18000) \cdot (0.3 + \mathrm{MAX}(f(h_{gr})/1{,}2; b_{gr}/1.6) + \\ + \mathrm{MAX}(f(h_{dep})/1{,}2; b_{dep}/1.6) + \mathrm{MAX}(\gamma_{dep}/120; d_{dep}/1.6)) \quad [s/AU] \tag{17.65}$$

with the height-factor

$$f(h) = \begin{cases} (h_{max} - h_{min})/2 & \text{if } h_{min} < 1\text{ m} < h_{max} \\ (h_{max} - h_{min})/2 - 1 & \text{if } h_{min} > 1\text{ m or } h_{max} < 1\text{ m.} \end{cases} \tag{17.66}$$

This formula has been derived analytically (Gudehus 1973). The parameters are fitted to the results of many grip time measurements. The formula holds for *simultaneous moves* and *single-piece grips* within the *limitations*:

grip and deposition heights up to 1.8 m
grip and deposition depths up to 1.2 m
piece volumes up to 50 l
piece weights up to 10 kg
(17.67)

If some of the elementary steps (17.60) are executed *additive*, the respective term MAX(a;b) in formula (17.65) must be replaced by the sum (a+b). For *multi-piece grips* of an average of x article units, in formula (17.65) v_{AU} must be replaced by $x \cdot v_{AU}$ and w_{AU} by $x \cdot w_{AU}$ and the result has to be divided by x.

The mean grip times resulting with the grip time formula (17.65) are in the range from 2 to 10 s per piece. They depend strongly on the mean volume and weight of the pieces. The *mean grip time per position* can be calculated by inserting the result of formula (17.65) into (17.64).

17.11.3 Basis Time

The *basis time* or *order-setup time* is the time the picker spends at the basis station before starting and after returning from the picking tour. Only systems with central basis have basis times different from zero. Commissioning systems with dynamic article provision or with local deposition do not have a basis. Their basis time is zero.

The basis time is caused by the following tour-preparation and order-closing activities:

acceptance and deposition of order documents or pick-lists
preparation and sorting of pick-lists due to the tour strategy
positioning for take-over and delivery
take-over of empty collection units
deposition of filled collection units
coding and labeling of load units
(17.68)

The occurrence of these activities and the necessary times are determined by the external requirements, by commissioning method and technique and by the design and dimensions of the basis station. Paperless picking does not need time for preparation and deposition of pick-lists. The takeover of empties can be executed simultaneously with the deposition of full units.

Basis times are determined for a single project either by measuring the mean times for each single activity, or analytically by *MTM* or *work factor*. Waiting times caused by mutual blocking of several pickers arriving in a basis station at the same time are not part of the basis time. They are taken into account in the general utilization of the system.

The times for basis activities generally arise per commissioning order or per collection unit. To calculate the position picking performance, they must be converted in proportion to the position. If process X requires the time $t_{X\,Ord}$ [s/Ord] per order, the corresponding time per position is given by

$$\tau_{X\,Pos} = t_{X\,Ord}/n_{Pos}. \qquad [s/Pos] \qquad (17.69)$$

The required time for an action per dispatch unit can be converted with formula (17.55).

17.11.4 Availability

A picker is not always available. This holds for people, robots and machines as well. The working time of a picker is the sum of productive times and unproductive times:

- During the *productive time* T_{prod} within the working time, the picker is available to execute picking orders.
- The *unproductive time* T_{unpr} is caused by technical breakdowns, personal allowances and by other activities not directly related to commissioning.

Holidays, times for illness and maintenance times are not included in the unproductive time share as they occur outside the operating hours, although they affect the operating costs.

The loss of efficiency by the unproductive times of the picker and the picking devices is taken into account in formula (17.48) by the availability:

- The availability of a picker is the long term relation of the productive time to the working time as long as sufficient orders are available

$$\eta_{ave} = T_{prod}/(T_{prod} + T_{unpr}) \qquad (17.70)$$

The availability of a human picker operating with a mechanical device is the product of the *personal availability* and the *technical availability*:

$$\eta_{ava} = \eta_{P\,ava} \cdot \eta_{T\,ava} \qquad (17.71)$$

As explained in *Sect. 13.6*, the technical availability of a picking device or a picking robot is affected by the frequency and the duration of failures and interruptions. By adequate construction and preventive maintenance *technical availabilities* above 98% are state of the art.

Approved *experience values for personal availability* of pickers, which can be used to calculate picking performances, are:

$$\begin{array}{ll} \text{excellent working conditions and low stress} & \eta_{P\,ava} \approx 90\% \\ \text{decent working conditions and normal stress} & \eta_{P\,ava} \approx 85\% \\ \text{poor working conditions and high stress} & \eta_{P\,ava} \approx 80\% \end{array} \qquad (17.72)$$

These experience values are target values. If performance control is lacking or the management is bad, the availability can be far below these values. The losses by low availability are often ignored. Improvements of working conditions, management style and control can increase performance and reduce commissioning costs more than many technical measures.

The personal allowances for people depend on working conditions, motivation and leadership. Another important influence factor is the *exhaustion* after handling large quantities or high volumes without interruptions. Permanent picking in a commissioning system with dynamic article provision without interruption by travel times leads to decreasing personal availability. The picker can recover during traveling, in phases of low utilization or during waiting times.

17.11.5 Utilization

The productive time of a picker can only be used as far as the picker is not hindered and blocked by *stochastic waiting times*. Waiting times occur in front of the pick places, within the aisles and at the basis due to blockages by other pickers, or are caused by the lack of information or replenishment.

The occurrence and lengths of stochastic waiting times are principally unpredictable. Stochastic waiting times reduce the productive time and affect the commissioning performance. They are taken into account in the performance formula (17.48) by the maximally achievable utilization:

- The *achievable utilization* is the ratio of the position picking time to the sum of the mean waiting times and the position picking time:

$$\rho_{max} = \tau_{pos}/(\tau_{pos} + \tau_{wait}) \quad [\%]. \tag{17.73}$$

Times for regular activities and normal events are not stochastic waiting times but part of the picking time. Waiting times caused by an *underutilization* of the whole commissioning system do not affect the maximal utilization, but determine the actual utilization.

A station which serves a *stochastic flow* λ [Pos/h] with randomly varying service times is a *queuing system*. With a mean service time τ_S [s/Pos] the limit performance of the station is:

$$\mu_S = 3600/\tau_S \quad [\text{Pos/h}] \tag{17.74}$$

Due to the formulas (13.65) and (13.67) of *Sect. 13.5*, the stochastically varying waiting queue in front of a station with the *actual utilization*

$$\rho_S = \lambda_S/\mu_S \quad [\%] \tag{17.75}$$

and *system variability V* causes the *mean waiting time*

$$\tau_{wait} = ((1 - \rho + V \cdot \rho) \cdot \rho/(1 - \rho)) \cdot \tau_S \quad [\text{s/Pos}]. \tag{17.76}$$

The *system variability* is 1 for maximally fluctuating and 0 for clocked inflow and service rates. It can be assumed to be $V = 0.5$ for unknown stochastic fluctuations.

Each *access place* is a service station where the mean service time is the sum of setup time and grip time:

$$\tau_A = \tau_{sup} + \tau_{grip} \quad [\text{s/Pos}]. \tag{17.77}$$

Inserting this into formula (17.74) gives the limit performance of a single access place. The mean inflow of order positions to one of the N_A access places is $1/N_A$ of the position throughput (17.8). If one picker with his picking device occupies N_P access places during picking, the arrival flow of pickers with an order position in this access area is:

$$\lambda_A = N_P \cdot n_{Pos} \cdot \lambda_{Ord}/N_A \quad [\text{Pos/h}]. \tag{17.78}$$

The insertion of the service performance (17.74) with service time (17.77) and of the picker flow (17.78) into formula (17.76) and (17.75) gives the mean waiting time τ_{Await} [s/Pos] of a picker at the access places.

The *basis station* can also be considered as a waiting station where pickers arrive in stochastically fluctuating time intervals and have to wait if other pickers arrived before them. The mean service time per order at the basis is

$$\tau_B = n_{Pos} \cdot \tau_{bas} \quad [s/Ord]. \tag{17.79}$$

The arrival flow of pickers at one of N_{BS} parallel basis stations is

$$\lambda_B = \lambda_{Ord}/N_{BS} \quad [Ord/h]. \tag{17.80}$$

Inserting (17.79) and (17.80) into (17.74), (17.75) and (17.76) gives the mean basis waiting time per order $t_{B\,wait}$, and the division by n_{Pos} gives the mean basis waiting time per position τ_{Bwait} [s/Pos].

In a similar way, the mean waiting times for stochastically fluctuating supplies or randomly provided information can be calculated. The total stochastic waiting time $\tau_{wait} = \tau_{Bwait} + \tau_{Bwait} + ..$ is the sum of the partial waiting times, which gives inserted into (17.73) the *achievable utilization* of the pickers.

For the commissioning system of *Fig. 17.19* with one or two basis stations and the requirements of *Table 17.1*, the calculated dependency of the achievable utilization on the picking demand is shown in *Fig. 17.32*. From this example and the above formula follow the *utilization rules*:

- At low picking demand and system performance, a utilization of 100% is achievable. With higher demand, the utilization decreases first slowly and for further increasing demand it decreases faster and faster.
- The achievable utilization becomes 0% when the demand approaches the limit performance of the *bottlenecks* of the system.
- At low demand, the performance of the pickers does not depend on the throughput whereas at higher demand the performance is continuously reduced by increasing waiting times.
- The necessary number of pickers and picking devices increases for lower picking demand proportional and for higher demand over proportional with the demand.
- As long as the articles are equally distributed in the aisles, the waiting times in front of the basis stations are more critical than the waiting times in front the of access places.
- With fast mover concentration at the entrances of the aisles, the waiting times at the fast mover location can become critical and reduce the effective performance of the pickers.
- The additional waiting times can over-compensate the path-time savings by fast mover concentration.
- The more access places are occupied by the picker, the lower becomes the achievable utilization.

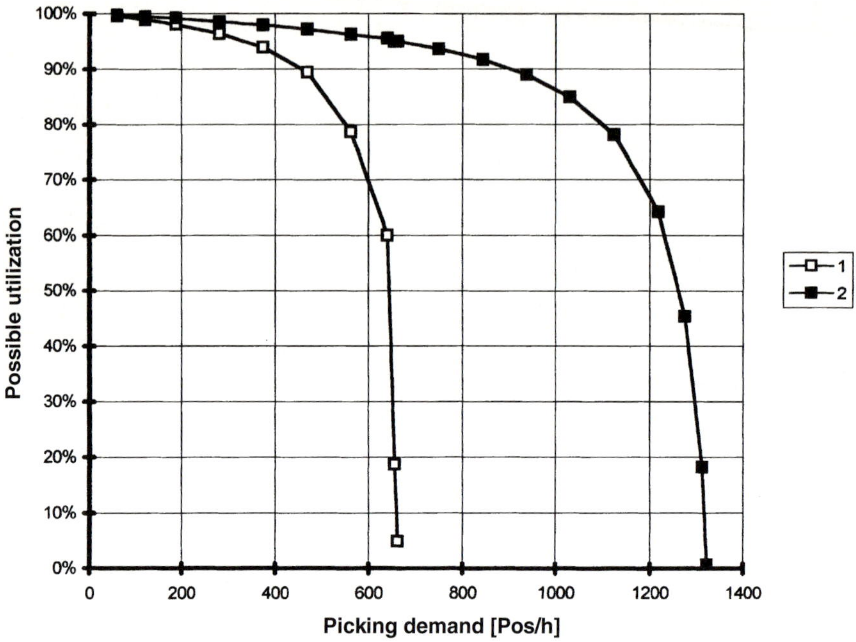

Fig. 17.32 Dependency of the achievable utilization on the picking demand
Parameter: basis stations $N_{BS} = 1$ or 2 other see *Fig. 17.27*

In the example of *Fig. 17.32*, the bottlenecks are the basis stations. With one single basis station in front of 12 aisles, the effective limit performance of the system is 660 Pos/h and achieved with 18 pickers. With two parallel basis stations, the limit performance is 1,320 Pos/h and achieved with 32 pickers.

Consequences of the utilization rules are the *commissioning design rules*:

- High performance commissioning systems should have more than one parallel basis stations in order to avoid long basis waiting times.
- Fast moving A-articles should be equally distributed in the picking aisles and not concentrated in a *fast-mover zone*.

Due to the sensitive dependency of the picking performance on demand, it is essential to calculate the achievable utilization with the given formulas or to determine it by simulation. If in peak times the utilization falls below 80%, the operating time must be extended in order to reduce the hourly demand. Otherwise, the performance of the pickers decreases sharply and the picking costs increase enormously.

17.12 Order Consolidation and Order-Line Reduction

The commissioning orders can be either single *external orders* [EOrd] or internal *series orders* [SOrd] made up by consolidation of a series of s external orders. By consolidation of s external orders with on average n_E external order positions, an *external order flow* λ_{EOrd} [EOrd/h] is converted into the *internal order flow*

$$\lambda_{SOrd} = \lambda_{EOrd}/s \quad [SOrd/h]. \tag{17.81}$$

The number of positions of the internal orders depends on the *execution strategy*. In one-stage serial commissioning, the single external order positions within the series order remain unconsolidated. They are picked separately and deposited into different collection units for the external orders. That means:

- For *one-stage order picking of series orders,* the mean position quantity of the series order m_S [AU/SPos] is equal to the mean position quantity of the external orders m_E [AU/EPos], i.e. $m_S = m_E$, whereas the mean number of *series order positions* is

$$n_S = s \cdot n_E \quad [SPos/SOrd]. \tag{17.82}$$

For two-stage order picking, all positions of the external orders for the same article are consolidated into one *series-order position.* This reduces the positions of a consolidated series order compared with the positions of the unconsolidated series order (17.82). The mean position quantity of the consolidated series order is larger than the position quantity of the unconsolidated series order. This gives the *advantage of two-stage order picking*:

► The execution of consolidated series orders reduces the number of order positions and the *provision demand* for dynamic article provision.

If the external orders are consolidated randomly into series orders as they arrive, the order line reduction depends on the probability that the positions of the external orders request the same article. This probability and the resulting order-line reduction can be assessed by analyzing a large number of external orders. The results of such an analysis differ from project to project.

Since the influence factors and dependencies remain unknown, it is preferable to determine the reduction of order lines analytically with the help of *probability theory*. The result is the *order-line reduction theorem* (Gudehus 1978):

► If an assortment includes a total of $N = N_A+N_B+N_C$ articles with N_A A-articles, N_B B-articles and N_C C-articles, and if the single orders have on average $n = n_A+n_B+n_C$ positions requiring n_A A-articles, n_B B-articles and n_C C-articles, the *mean number of order lines* of the consolidated series orders containing s single orders is:

$$\begin{aligned} n_S = {} & N_A \cdot (1-(1-n_A/N_A)^s) + N_B \cdot (1-(1-n_B/N_B)^s) \\ & + N_C \cdot (1-(1 - n_C/N_C)^s). \end{aligned} \tag{17.83}$$

With the mean position number (17.83) of the consolidated series orders results the *order-line reduction factor*:

$$r_S = n_S/(s \cdot n_E) \tag{17.84}$$

If for example 1,500 single orders with on average 12 positions of an assortment of 30,000 articles with the ABC-distribution of Table 17.1 are picked in 2 consolidated series orders with s = 750 external orders, due to formula (17.83), the series orders have on average 5,179 order lines. The number of external order lines of an unconsolidated series order is 750·12 = 9,000 due to formula (17.82). For this example, the consolidation reduces the number of order lines by the factor $r_s = 5{,}179/9{,}000 = 0.64$. The order line reduction is 36% and the required provision performance is reduced by one third.

The dependency of the order-line reduction factor on the series length s is shown in *Fig. 17.33* for an assortment of 30,000 articles with two different ABC-distributions. From this example and the above formula follow the *order consolidation rules*:

- The order-line reduction by consolidation of external orders to series orders increases with the series length and with the inequality of the assortment.
- The order-line reduction increases with the number of positions of the external orders and decreases with increasing number of articles.
- The series length s is limited by the maximal tolerable order lead time of the external orders.
- If the external orders have a mean position quantity m_E [AU/EPos] and the order line reduction factor is r_S, the mean position quantity of the consolidated series orders is

$$m_S = m_E/r_S \qquad [\text{AU/SPos}]. \tag{17.85}$$

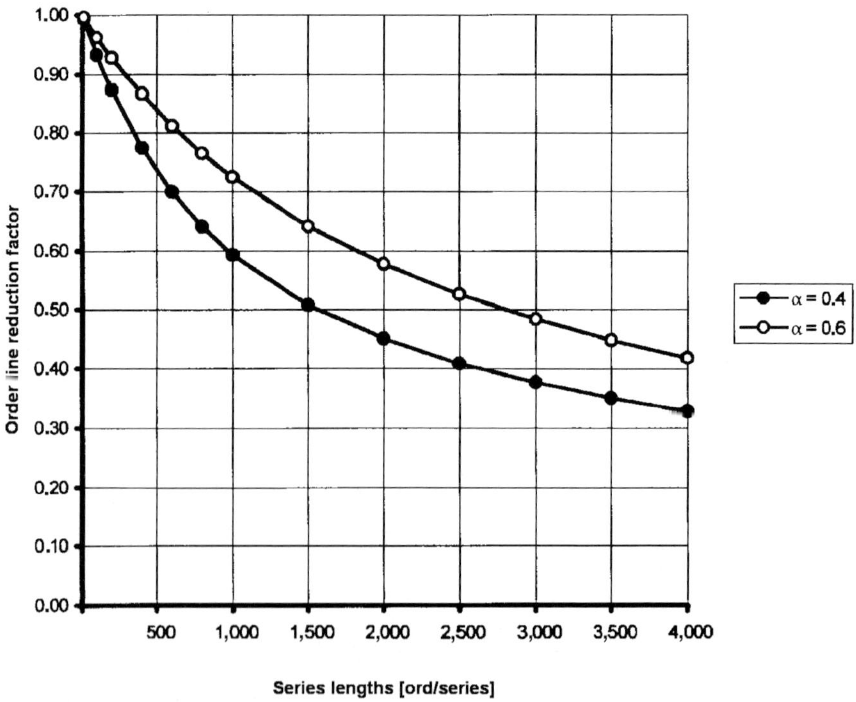

Fig. 17.33 Dependency of the order-line reduction and the series length

Assortment	30,000 articles
Order structure	12 Positions/order
Parameter	Lorenz-asymmetry of the position distribution $\alpha = 0.4$ and 0.6

Under favorable circumstances, the consolidation effect can be improved by intelligent replenishment strategies. For example, if all outlets of a retailer are supplied on 5 different days of the week with the same segments of the assortment, the number of articles required on the same day in the logistic center is reduced by a factor 1/5 and the probability that the same articles are required by the orders from the outlets increases.

The two formulas (17.83) and (17.85) for the structure of the consolidated series orders are fundamental for performance calculations of two-stage order picking systems. They are of particular importance for dimensioning commissioning systems with *dynamic article provision*.

17.13 Dynamic Design of Commissioning Systems

The tasks of dynamic design of a commissioning system are to determine the necessary number of pickers and technical devices, and to design the inward and outward conveyors. The goal is to fulfill the requirements at lowest costs.

The necessary number of pickers for an actual demand $\lambda_{Pos}(t)$ results from the effective limit performance of the single picker (17.48):

- ▶ If a single picker can achieve a picking performance μ_{Pos} [AU/h], the number of pickers necessary to execute a picking demand $\lambda_{Pos}(t)$ [AU/h] is

$$N_P = \{\lambda_{Pos}(t)/\mu_{Pos}\}. \tag{17.86}$$

The curly brackets in the formula indicate to round up to the next integer as the number of pickers is whole-numbered. The active number of pickers changes with the current demand.

For investment planning, the number of picking devices, picking robots and replenishment devices must be calculated for the demand of the expected peak hour. Otherwise, the pickers cannot achieve their limit performance. Additional devices should be available as reserve for emergency cases due to the *experience rule*:

- ▶ The number of picking devices should be about 10%, but at least 1 device higher than the number of pickers in the peak hour of the annual peak day.

Highly mechanized and automated logistic centers require *multi-shift operation*. If the demand fluctuates heavily, even conventional systems have to operate at peak times on flexible *multi-shifts*. Less time-critical orders are postponed to times with lower utilization.

The number of full-time workers needed for commissioning during a year depends on the *work scheduling*. For a highly flexible work model with part-time work, overtime and overtime pools, the necessary number of pickers N_P can be calculated with formula (17.86) for the mean picking demand of the year. With an effective annual *working time* T_{WT} [h/year] and annual *operating time* T_{OT} [h/year], the necessary *number of full time workers* is:

$$N_{FT} = T_{OT} \cdot N_P/T_{WT} \tag{17.87}$$

If the work time is less flexible, the number of full time workers is higher and depends on the peak demand.

The workers, the number of devices and the conveyors for the replenishment and provision of access units (17.10) can be dimensioned by the formulas and algorithms for storage systems of *Chapter 16*.

17.13.1 Dimensioning of Conventional Commissioning Systems

Conventional commissioning systems with static article provision and central deposition can maximally consolidate as many external orders to a series order as collection units for the external orders can be carried on the picking device. This leads to the *dimensioning rule*:

- If an external order requires n_{CU} collection units and the picking device has a capacity for C_{PD} collection units, the maximal series length is

$$s_{max} = C_{PD}/n_{CU} \tag{17.88}$$

The mean number of collection units per order n_{CU} [CU/EOrd] is given by formula (17.50). If for example, the picker can carry 12 cartons for 6 external orders in a honeycomb rack, the proportionate path times per position are reduced by a factor 6 compared to the path times for single order picking.

If the content of the external order fills more collection units than the loading capacity of the picking device, the order is split into two or more *partial orders*.

17.13.2 Dimensioning for Local Commissioning Systems

In order to achieve steady utilization of the picking zones, commissioning systems with local deposition, as shown e.g. in *Figs. 17.3* and *17.15*, generally execute series orders. Based on this consideration the dynamic design of local commissioning systems is an iterative planning process with the steps:

1. For the *initial solution,* the statically dimensioned aisle modules with the access places are separated into as many picking zones as nessessary for the estimated number of pickers in the peak hour.
2. The picking zones are designed as *functional modules*, which are optimized in terms of ergonomics, conveyor technique and IT.
3. For *fixed batch execution*, the operating time is split into fixed time intervals of lengths limited by the buffer capacity of the downstream conveyor system. In *mail order business* these time intervals are called *rhythms.*
4. For *dynamic batch execution*, the series lengths s, i.e. the number of simultaneously processed external orders is set equal to the number of exits of the downstream conveyor system.
5. The batch or series orders are split in so many *partial orders* as picking zones have been assumed for the initial solution. For this picking demand, the necessary number of pickers per zone is calculated by formula (17.86).
6. If the calculated number of pickers is smaller or larger than the assumed number of zones of the initial solution, the number of zones is stepwise reduced respectively increased until the number of pickers matches the number of zones.

If the series order contains only a few external orders, the partial orders are small and vary stochastically in position number and quantity. This increases the local system variability and reduces the *utilization* of the pickers if the local buffer of orders and collection units is insufficient (see *Fig. 17.3*). Therefore, local commissioning is only efficient if long series orders are possible.

17.13.3 Dimensioning for Dynamic Article Provision

The dimensioning steps for *one-stage picking* of external orders in a commissioning system with dynamic article provision, as shown e.g. in *Figs. 17.4*, *17.25* and *17.34*, are:

1. The picking stations are designed as *functional modules* for $n_{Pick} = 1, 2$ or 3 pickers and optimized in terms of ergonomics, conveyor technique and IT.
2. By the number of collection places of a single picking station the *partial series length* s_{PS} is fixed, i.e. the number of orders, which can be processed simultaneously in one station.
3. With the dimensions of the designed station the limit performance per picker is calculated for the external order structure n_E and m_{Pos} with the help of formula (17.48).

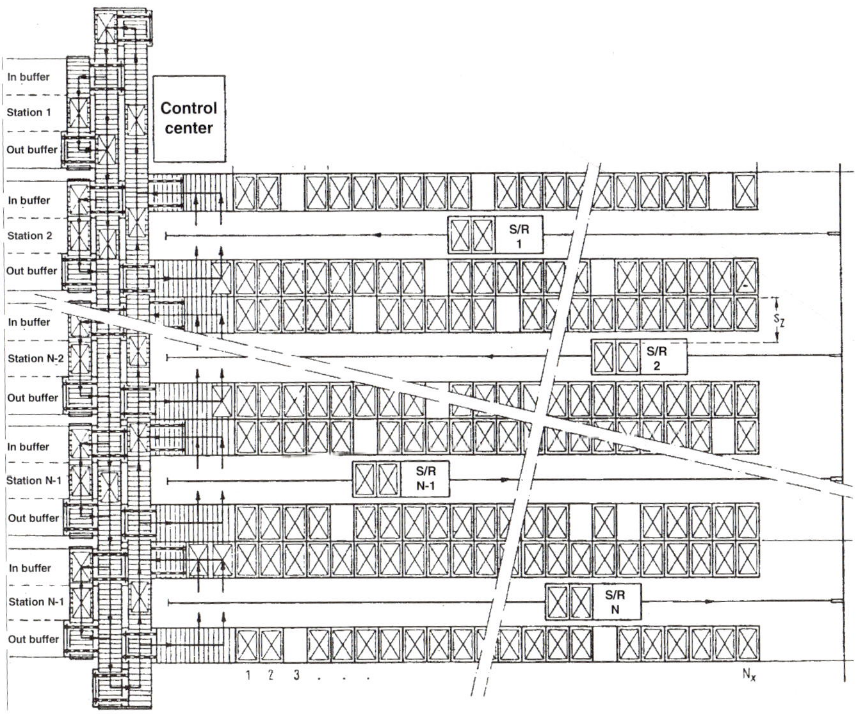

Fig. 17.34 Layout of a mini-load system (MLS) with dynamic article provision for commissioning of small parts

4. For the picking demand in the peak hour, the maximally required number of pickers N_P is calculated by formula (17.86). With n_{Pick} pickers per station the necessary number of picking stations is

$$N_{PS} = \{N_P/n_{Pick}\} \quad (17.90)$$

5. For the *total series length*

$$s = N_{PS} \cdot s_{PS}$$

results with formula (17.83) the mean position number n_S of the series order and with this the *provision demand* for an external order flow λ_{EOrd}:

$$\lambda_P = n_S \cdot \lambda_{EOrd}/s = r_S \cdot n_E \cdot \lambda_{EOrd} \quad [PU/h] \quad (17.91)$$

6. The *storage- and provision-system* for the required provision performance (17.91) is dimensioned by the methods of *Chap. 16.*
7. Finally, the connecting transport system between the provision store and the picking stations is designed and dimensioned for the required throughput. This includes the inward- and outward-conveyors and the buffer places on the feeding and retrieval lines of the stations.

To fulfill, e.g. the external orders of the non-food example of *Table 17.1* in a one shift operation of 8 h, $N_{PS} = 6$ picking stations are necessary each operated by 1 picker and equipped with pallet places for the simultaneous processing of $s_O =$ 4 external orders. The 250 external orders per day with 15 positions are picked in 10.4 series each consisting of $s = 24$ external orders. The order line reduction factor (17.84) is $r_s = 0.79$ and $0.79 \cdot 15 \cdot 250 = 2{,}963$ provisions are required in 8 h, i.e. a provision performance of 370 pal/h.

In case of *two-stage order picking* with dynamic article provision, all access units for one article are conveyed to the same picking station. There the picker grips the required quantity for all external orders of the series order and deposits it on the conveyor or into a collection unit. The dimensioning for the two-stage order picking is similar to the dimensioning of local commissioning systems. Again the series length s is a free parameter and the required provision performance is calculated with the formulas (17.91) and (17.83) from the external order flow and the mean number of order lines.

The success of planning commissioning systems with dynamic article provision critically depends on the correct dimensioning and harmonization of the subsystems. Since many influence factors and design parameters must be considered and quite difficult dimensioning formulas have to be applied, it is advisable to dimension these systems assisted by *planning software*.

17.14 Commissioning Costs

The *specific commissioning costs* or *order picking costs* correspond to the unit costs of production systems. The periodical operating costs K_{com} [€/PE] of a commissioning system depend on the *performance units* or *cost drivers* [CD] *of commissioning*:

$$\begin{aligned}&\text{order [Ord]}\\&\text{position [Pos]}\\&\text{article unit [AU]}\\&\text{provision unit [PU]}\\&\text{dispatch unit [DU].}\end{aligned} \tag{17.92}$$

The corresponding performance rates, demands or throughputs are:

$$\begin{aligned}&\text{order throughput } \lambda_{\text{Ord}} \text{ [Ord/PE]}\\&\text{position throughput } \lambda_{\text{Pos}} \text{ [Pos/PE]}\\&\text{article unit flow } \lambda_{\text{AU}} \text{ [AU/PE]}\\&\text{provision unit throughput } \lambda_{\text{PU}} \text{ [PU/PE]}\\&\text{dispatch unit throughput } \lambda_{\text{DU}} \text{ [DU/PU]}\end{aligned} \tag{17.93}$$

Four of the five flows (17.93) can be calculated from any one of the others by the formulas (17.8), (17.9), (17.10) and (17.11). Therefore, it is sufficient to measure the picking performance by only one of the cost driver (17.92) and to relate the commissioning costs K_{com} [€/PE] to the corresponding performance flow (17.93). Then the specific commissioning costs are simply

$$k_{\text{com}} = \text{K}/\lambda_{\text{com}} \quad [€/\text{CD}]. \tag{17.94}$$

However, the undifferentiated cost rate (17.94) depends of the order structure. To take into account the influence of the order structure and to achieve *use-related cost rates*, it is necessary to split the operating costs as explained in *Sect. 6.7* into the fixed costs K_{fix} and a sum of partial costs, which vary proportionally to the five costs drivers (17.92):

$$\text{k}_{\text{AUeff}} = \text{K}_{\text{fix}} + k_{\text{Ord}} \cdot \lambda_{\text{Ord}} + k_{\text{Pos}} \cdot \lambda_{\text{Pos}} + k_{\text{AU}} \cdot \lambda_{\text{AU}} + k_{\text{PU}} \cdot \lambda_{\text{PU}} + k_{\text{DU}} \cdot \lambda_{\text{DU}}. \tag{17.95}$$

The different *partial cost rates*

$$\begin{aligned}&\text{partial order costs } k_{\text{Ord}} \text{ [€/Ord]}\\&\text{partial position costs } k_{\text{Pos}} \text{ [€/Pos]}\\&\text{partial article unit costs } k_{\text{AU}} \text{ [€/AU]}\\&\text{partial provision unit costs } k_{\text{PU}} \text{ [€/PU]}\\&\text{partial dispatch unit costs } k_{\text{DU}} \text{ [€/DU]}\end{aligned} \tag{17.96}$$

are independent of the order structure. They are determined only by the commissioning method, technique, organization and operating strategies.

In order to calculate the full performance costs for the cost drivers (17.92), it is necessary to allocate due to the rules developed in *Sect. 6.7* the fixed costs dependent on their utilization by the throughputs (17.93). However, the calculation and accounting of commissioning costs in this differentiation is quite complex. In addition, due to the difficult delimitation between the partial functions, it is ambiguous. This holds e.g. for two-stage order picking where the costs of the first stage depend on the internal series orders.

A customer is only interested in the results of the commissioning process. He expects the correct and complete execution of his orders in due time. It is up to the service provider how and by what means he fulfills these requirements. The customer needs prices, which depend on performance units and cost drivers he can influence and control. To achieve a practicable *cost-based* and *use-related pricing*, it is sufficient to differentiate the commissioning costs only with respect to the two main cost drivers. These are the *picks* or *article units* and the *order positions*. In some cases, such as picking onto pallets, the *dispatch units*, i.e. the filled pallets, are a more appropriate cost driver than the positions. The reduction of the cost drivers simplifies accounting, and increases the transparency of *pricing of commissioning services* (see Sects. 6.7 and 7.2).

If the *article unit* [AU] is taken as single cost driver and performance unit, the elimination of the other cost drivers from formula (17.95) by the relations (17.8), (17.9), (17.10) and (17.11) leads to the *specific commissioning costs* or *order picking costs*:

$$k_{\text{AUeff}} = \text{K}_{\text{fix}}/\lambda_{\text{AU}} + k_{\text{AU}} + k_{\text{Ord}}/(\text{n}_{\text{Pos}}\text{m}_{\text{Pos}}) + k_{\text{Pos}}/\text{m}_{\text{Pos}} + k_{\text{PU}}/\text{C}_{\text{PU}} + (k_{\text{DU}}/\text{C}_{\text{DU}}) \cdot (1 + (\text{C}_{\text{DU}} - 1)/(2\text{n}_{\text{Pos}}\text{m}_{\text{Pos}})) \quad [/\text{AU}]. \tag{17.97}$$

Formula (17.97) explicitly shows the following dependencies of the order picking costs on the cost drivers and the order structure:

- Picking costs decrease with increasing throughput and demand as the fixed costs are carried by an increasing number of units (see *Fig. 17.35*).

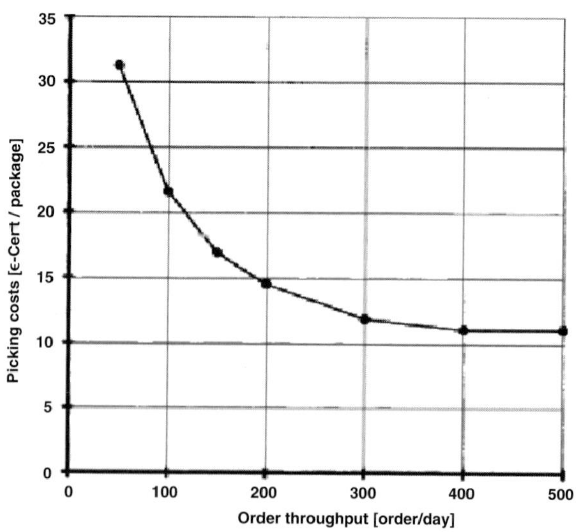

Fig. 17.35 Dependency of the commissioning costs on the order throughput
Parameter: see legend *Fig. 17.28*

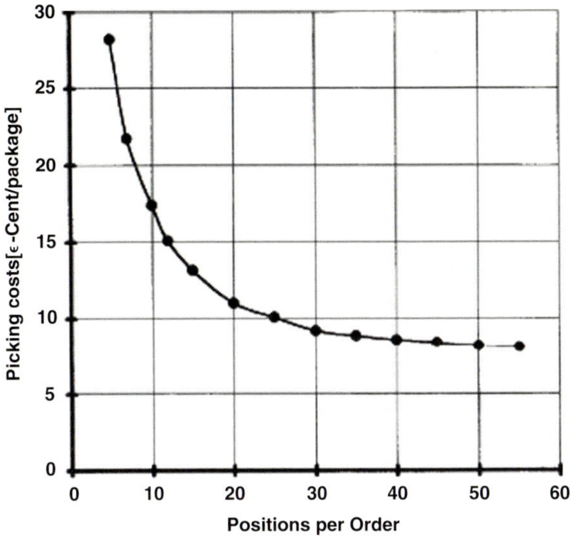

Fig. 17.36 Dependency of the commissioning costs on the positions per order

- ► The picking costs decrease with increasing number of positions per order as the proportionate costs for order processing, generation of collection and dispatch units and - in case of static provision - for the distances between the access places diminishes (see Fig. *17.36*).
- ► The picking costs decrease with increasing quantity per position as the proportionate costs for order preparation, position processing and transport decrease (see Fig. *17.37*).
- ► The picking costs depend on the capacity of the load units used for provision and dispatch (see *Fig. 17.40*).

These general dependencies hold for all kinds of commissioning systems. The economies of scale increase with the share of the fixed costs. This leads to the *application rule*:

- ► Highly mechanized and automatic commissioning systems with fixed costs above 50% are only efficient if continuously used in multi-shift operation.

This holds especially for systems with dynamic order provision, for two-stage order picking with sorters and for fully automatic commissioning systems with robots or mechanical devices.

Formula (17.97) shows that the picking costs decrease with increasing load unit capacity. However, as shown in *Fig. 17.40*, at load unit capacities above a certain *optimal value*, the increasing costs for space, transport and handling compensate against the picking cost reduction.

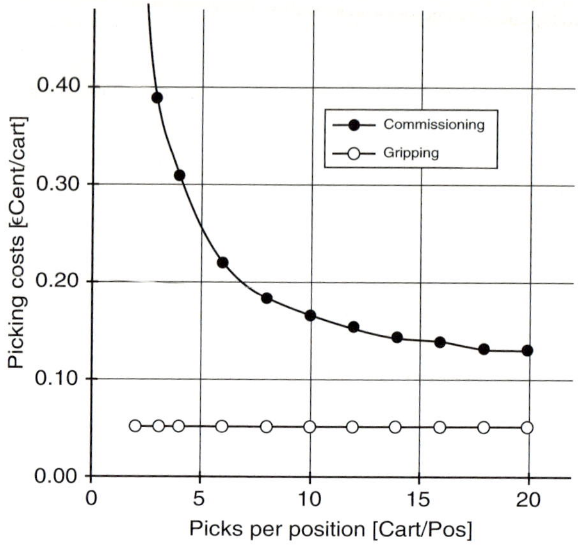

Fig. 17.37 Dependency of the commissioning costs on the position quantity (picks per position)

17.15 Influence Factors on Costs and Performances

Several external and internal factors influence the performance and costs of commissioning systems:

- *External factors* are the order entry flow, order quantities and order structure, the requirements of the customers, the sourcing prices for buildings, equipment and tools, the cost rates for personnel, the interest rates, energy costs etc.
- *Internal factors* are the commissioning system and technique, the design of the function modules and the layout, the dimensioning parameters (17.17) (17.18) and (17.19), the operating strategies, and the work times and operating times.

Planners and operators can change the internal factors to a certain extent but can scarcely influence the external factors. Since they affect the performances and costs in different ways, the internal factors can be used for *system optimization* and for *article and order allocation* most efficiently only with the help of a *commissioning-performance- and costing program*, which operates with the above formulas and algorithms.

An example is the conventional commissioning system shown in *Fig. 17.2* and *17.16*. The requirements are specified in the left column of *Table 17.1*. The key data of the resulting system are given in the legend of *Fig. 17.27*. In this low investment solution, cartons of different sizes are picked manually from pallet to pallet. High picking performances are achieved by optimally positioned access places, empties removal places below the access places, minimal path lengths, optimal aisle layout, provision by dynamic flip-flop and paperless operation of the pickers, who are guided by WMS-instructions on a display.

For the requirements of *Table 17.1* and with German prices and cost rates of 2003 result the project specific data:

- The fixed costs for provision racks, ground space, devices, displays and control are about 40% of the total operating costs of 615,000 € per year.
- The picking costs related to only one of the cost drivers (17.92) are 9.85 €/order, 13 €-cent/carton, 0.66 €-cent/position or 9.85 € per picked pallet.

More important than the project specific data are the *general dependencies*:

- The picking costs for orders with less than 30 positions depend strongly on the number of order positions (see *Fig. 17.36*).
- The picking costs per carton decrease with increasing quantities per position. Even for large position quantities, the costs for gripping are less than 50% of the total picking costs (see *Fig. 17.37*).
- The picking costs for orders with more than 30 positions and over 15 cartons per position are quite independent of the order structure.
- For small pieces up to about 4 *l*, the picking costs are quite independent of the size of the picked pieces, here the cartons, and increase linear with the volume for pieces larger than 6 *l* (see *Fig. 17.38*).
- The picking costs increase linear with the number of articles of the assortment, as the path length gets longer and the required ground area increases (see *Fig. 17.39*).

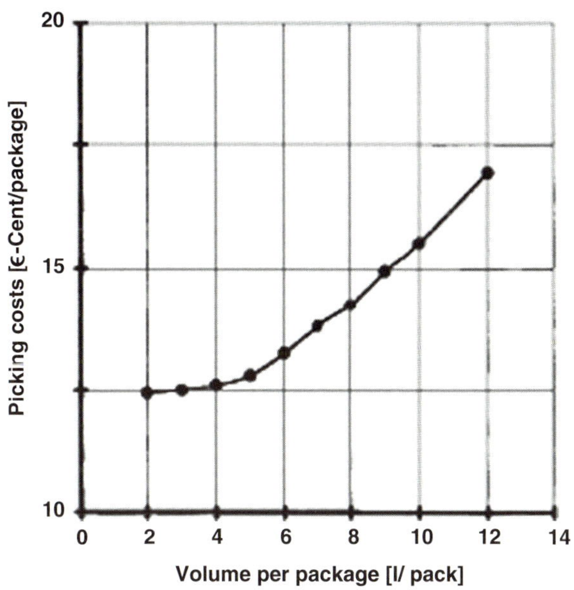

Fig. 17.38 Dependency of picking costs on the size of the picked pieces (cartons)

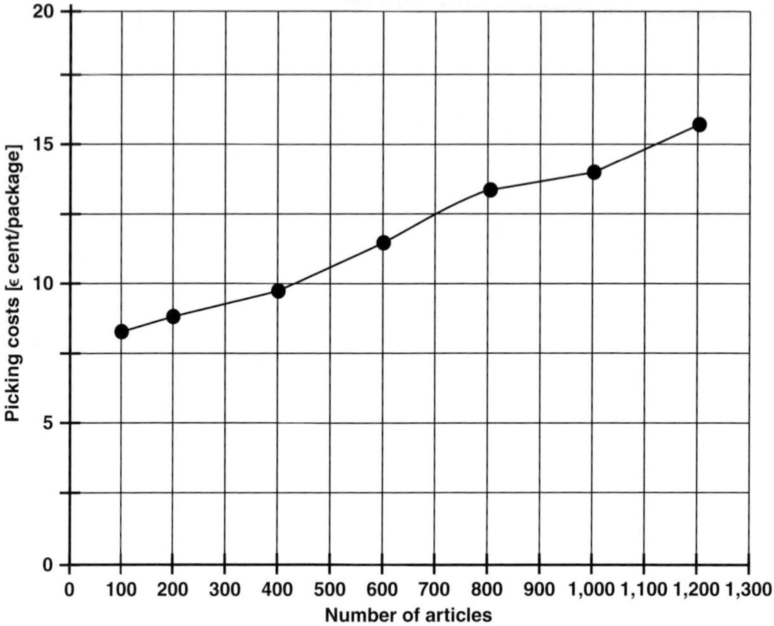

Fig. 17.39 Influence of the number of articles of the assortment on the picking costs

- The influence of the dimensions of the dispatch pallets on the costs illustrated in *Fig. 17.40* shows that the optimal pallet height for this example is between 900 and 1,100 mm.

A general conclusion from these dependencies is:

▸ Commissioning performances and picking costs are only useful *benchmarks* for systems comparisons or for logistic service providers if requirements and frame conditions are the same.

The presented commissioning system for consumer goods has been designed, dimensioned and optimized with the help of a planning program which uses the above formulas and algorithms. The performances and costs achieved during the operation of the realized system agree with the calculated values within $\pm$ 5%. Many other commissioning systems planned with these programs fulfill the expectations and operate successfully in practice.

17.16 Article Allocation and Order Allocation

If different storage and commissioning systems are technically possible or available, the following *allocation questions* have to be answered:

1. Where should which article be provided?
2. Where should which order be picked?

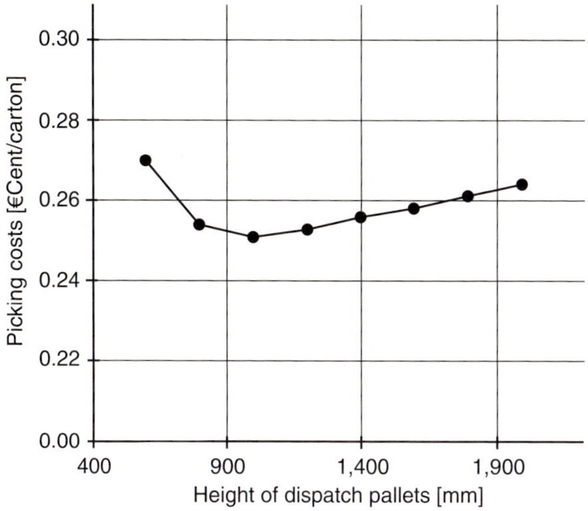

Fig. 17.40 Commissioning costs as function of the height of the dispatch pallets

An example for the first question is the segmentation of an assortment into articles, which can either be provided on pallets or in bins. For the channel-store commissioning system *Figs. 17.19*, *17.20* and *17.21* with the requirements of *Table 17.1*, the dependency of the total operating costs on the share of articles allocated to pallets and to bins is shown in *Fig. 17.41*. The result is that the 35% fastest moving articles are optimally provided on pallets and the other 65% articles on boxes. The area for the optimized solution is 35,000 m^2, the investment is 28 Mio € and the operating costs with 105 fulltime pickers are 6.6 Mio. €/year.

As alternative for the same performance requirements, a commissioning system with dynamic article provision from an automatic high bay store has been planned (see *Fig. 17.25*). For this solution only 60 fulltime pickers are necessary. The ground area is only 8,000 m^2. However, due to the higher investment the alternative solution has only slightly lower operating costs. Despite the advantages of less personnel and smaller area, after longer discussions the channel store-commissioning system was build since this solution offers far higher flexibility. Especially for retailers with fast changing assortment and high seasonal peaks *flexibility* is a key feature.

If a logistic center offers different technically possible *internal logistic chains* for the same articles and several *order chains* for the same orders, *allocation strategies* are necessary to allocate the incoming articles and orders to the cost optimal storage, and commissioning system (see *Fig. 1.12*). The allocation strategies should be *self-regulating* and programmable.

The optimal allocation is the available logistic chain, respectively order chain with lowest process costs. The availability depends on the capacity and limit performances of the storage and commissioning systems within the relevant chains. Therefore, in order to develop optimal allocation strategies, it is necessary to set up an analytical computer model by linking program modules for the performance and

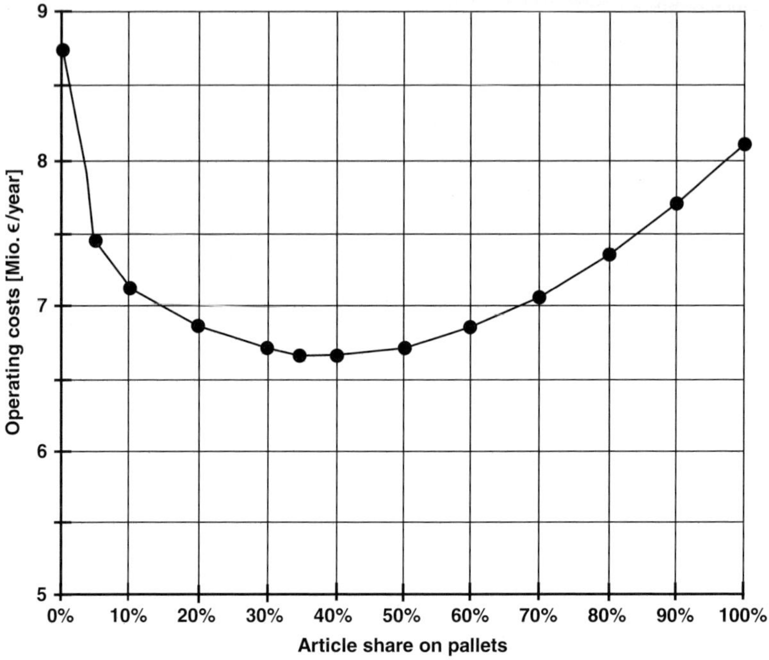

Fig. 17.41 Dependence of the total operating costs for an optimally planned channel-store commissioning system on the share of articles on access pallets (other share in bins)

commissioning system: see *Figs.* 17.19, *17.20* and *17.21*
access units: bins and pallets (see *Fig. 17.20*)

costs calculation of all relevant subsystems. With such a program for each arriving article or order, the costs and the availability for all technically possible logistic and order chains can be calculated and compared.

From *sensitivity calculations* for several logistic centers, result the following general *allocation rules* (see also *Sect. 16.15*):

- The *optimal article allocation* to a technically possible storage and commissioning system is determined by the *size*, the mean *stock* and the *volume flow* of the article.
- The *optimal order allocation* to a technically possible commissioning system depends on *order structure*, *order volume* and the *dispatch units*.

By optimal allocation strategies, the performance of a logistic center with several process chains for the same articles and orders can be improved without additional costs. The operating costs can be reduced significantly by this means. The optimal strategies and their parameters differ from project to project. It is an interesting task for *logistic research* to develop general allocations strategies and decision rules for the different storage and commissioning systems and to derive *rules of thumb* for the strategy parameters. Such allocation strategies and decision rules are of central importance for the *self-regulation* of *dynamic logistic networks*.

Chapter 18
Transport Systems

Transport systems convey goods from the entries and sources to the exits and sinks of a logistic, performance or production network. The *transport good* or *freight* can be *gas, liquids, bulk cargo, single items* or *load units* filled with bulk cargo, general cargo or other goods.

The general task of logistics – provision of the right good in the right time at the right place – encompasses the *general transport task*:

- ► A transport system has to be designed, dimensioned, organized, set up and scheduled to execute transports and freight orders under given spatial, time and technical restrictions at lowest possible costs.

The *design* includes the selection of technically adequate transport means and elements, their connection to a *transport network*, and the arrangement of the paths and tracks within the available space. *Dimensioning* is the determination of transport lengths, capacities and limit performances, and the calculation of the number of transport means. *Organizing* means conception and configuration of the control system. *Scheduling* takes care of the cost optimal execution of current freight orders by a given transport system.

Spatial restrictions are the *locations* $(x_i;y_i)$ of the *stations* S_i, $i = 1, 2, \ldots N_S$, which have to be connected by the transport network, and the available routes, areas and clearance measures. *Time restrictions* are the required pickup times, delivery times and maximal transport times. *Technical restrictions* result from the properties of the freight, the capacities of the transport means and the limit performances of the transport elements (see *Sect. 3.5*).

The main topics of this chapter are *intramodal transport systems*, which operate from the entries to the exits with the same transport technique. After classification of transport systems, the connection between *freight orders* and *transport demand* is explained. Following are a *structure analysis* of transport networks and the development of *network design principles*. After the description of transport control systems, the possible *transport strategies* are developed. Then, the configuration and performance of *conveyor systems* and *vehicle systems* are presented. The next sections develop formulas for transport times and vehicle number and methods to determine *optimal locations* and *transport routes*. After dealing with *transport costs*

T. Gudehus, H. Kotzab, *Comprehensive Logistics*,

DOI 10.1007/978-3-540-68652-1_18,

and *transport pricing*, the chapter closes with a discussion of the relations between *transport* and *traffic*.

Entries and exits of an intramodal transport system are the transitions to transport systems of other technique. The connection of different intramodal transport systems by *transport transitions* and *transfer stations*, to *intermodal transport chains* and *global logistic networks* and the determination of *optimal transport chains* through these networks, are dealt with in *Chap. 21*.

18.1 Classification of Transport Systems

By technical guidelines, such as the German norms DIN 30781 and DIN 25003, transport systems are classified primarily due to technical features, which determine construction and installation. However, for *logistics*, mainly *usability*, *performance*, and *costs* of the transport systems are important.

For the selection, design and dimensioning, transport systems for *continuous transport* and for *discontinuous transport* have to be distinguished. Continuous transport systems are *pipelines* for gases, liquids or mass goods, and *belt conveyors* for coal, ore and other loose goods. Discontinuous transport systems are either conveyor systems or vehicle systems:

- A *conveyor system* moves goods with or without load carrier on a *powered track network* from input stations to output stations.
- A *vehicle system* carries load on powered transport units which move in a *non-powered track network* from loading stations to unloading stations.

Transport units [TU] are the smallest objects, which move separately between the stations of a vehicle system. They are either single vehicles, such as cars, trucks or ships, or *trains* of loaded units and traction units.

Transport quantities are measured in *freight units* or *transport units*. The measure for homogenous freight is a *weight unit*, such as kg or t, or a *volume unit*, such as l or m^3. The measure for discrete freight is a *load unit* [LU], which is defined by its dimensions, volume, weight and *capacity* (see *Sect. 12.3*).

Conveyor systems are generally *open systems* with *input stations* I_i, where *incoming flows* λ_{Ii} of load units enter the system, and output stations O_j, where *outgoing flows* λ_{Oj} leave the system. Intralogistic vehicle systems are generally *closed systems* with a constant number of transport units moving or waiting on the tracks between the stations. Subsystems and extralogistic vehicle systems, such as the *public traffic networks*, are *open systems* with varying numbers of transport units.

For the planning of transport systems, the following general *selection and application rules* hold:

- ▸ *Pipelines* and *belt conveyors* are used for the transport of a continuous flow of gases, liquids and *bulk goods* over short, medium and longer distances between a small number of fixed stations.
- ▸ *Conveyor systems* are best suited for transporting and sorting a permanent flow of *discrete goods* or of standard load units between fixed stations with short distances.

- *Vehicle systems* are optimal for transporting any kind of goods with varying quantity over short, medium and long distances between changing numbers of stations at variable places.

Pipelines supply water, gas, oil and chemicals to and dispose sewage and waste from production plants and households. Belt conveyors are primarily used in mining, raw material industry and power plants. The construction and dimensioning of pipelines for gases and liquids and of conveyors for bulk goods are tasks of *engineering* and will not be discussed in this book.

The most important logistic features of conveyor systems and vehicle systems are compiled in *Table 18.1*. They lead to the *application criteria* of *Table 18.2*. Fully automatic conveyor systems are primarily used for internal transports and sorting. They are generally constructed for one kind of load unit and quite inflexible. Vehicle systems are highly flexible and suitable for internal and external transports.

Table 18.1 Logistic features of conveyor systems and vehicle systems

Characteristics	Conveyor systems	Vehicle systems
Transport units	Unpowered load units pallets, packages, parcels	Powered vehicles, cars, trucks, trains, ships
Transport network	Active conveying elements	Passive traffic lanes and knots
Speed	0.5 to 10 km/h	1 to over 500 km/h
Stations	Fixed installed	Fixed or changeable
Relations	2 to 100	From 10 to more than 1,000
Distances	Quite long	From 10 m to over 1,000 km
Travelling times	Limited	Variable
Functions	Transport Collection and distribution Sorting Buffering	Transport Collection and distribution Pick up Delivery

In addition to the logistic features, *Table 18.1* shows the main differences between conveyor and vehicle systems concerning distances, transport times, limit performances and capacity. Despite these differences, the rules and formulas for the limit performances and queuing effects of *Chap. 13* hold for both classes of

Table 18.2 Application criteria for conveyor systems and vehicle systems

Application criteria	Conveyor systems	Vehicle systems
Freight	Quite uniform	Uniform or different
Transport times	Long	Short or longer
Stations	Fixed	Changing
Relations	Few	Few to many
Distances	Less than 1 km	Up to more than 1,000 km
Troughput	Medium to high Quite continous	Small to high Continuous or varying
Areas of application	Inhouse	Inhouse and Outdoor

transport systems. They are decisive for the design of transport networks, operating strategies, system performance and for the queuing in front of the entrance stations and within the transport network.

Besides the obvious application areas, there are fields where the two classes of transport systems compete. Here, it is necessary to calculate and compare carefully the performances and costs of the suitable systems.

18.2 Transport Requirements

Freight orders [FO] and *transport orders* [TO] initiate the moves of the load units and transport units in a transport system. A freight order contains the information at which *pickup station* S_i and *pickup time* T_{Pi} a *load*, *shipment* or *freight* with *quantity* $M_{FO\,ij}$ [LU/FO] should be collected and to which *delivery station* S_j at what *delivery time* T_{Dj} it has to be delivered. The freight order specifies the freight category:

- *Bulk freight* or *bulk cargo* are unfilled and unpackaged homogenous *loose goods*, such as gases, liquids, granulates and solid material.
- *Piece freight* or *general cargo* are single load pieces and *discrete load units*, such as bottles, boxes, packages, parcels, containers, pallets or ISO-containers with certain dimensions, volume and weight.

Additional requirements hold for goods with specific properties, such as perishables, fire hazardous material, explosives or sensitive, shrinking and valuable goods.

The *scheduler* or *dispatcher* of the sender, or of a logistic service provider, consolidates single upcoming shipments into freight orders and selects the most cost-efficient *freight chain* or *transport chain* due to the rules, which are presented later in *Chap. 21*. Resulting are freight orders with one or several load units from the same or different senders for the same or different customers and destinations. Correspondingly, the freight orders are *single-unit* or *multi-unit orders*, *single* or *mixed shipments* and *single*-or *multi-destination orders*.

From the requested *pickup time* T_{Pi} at station S_i and *delivery time* T_{Dj} at station S_j results the required *transport time* between S_i and S_j

$$T_{Tij} = T_{Pi} - T_{Dj} \quad [\mathrm{PE}]. \tag{18.1}$$

In many cases, in particular for external transports, the pickup and delivery times are not precisely prescribed, but only limited by certain *time-windows*. In other cases, only a *maximal transport time* T_{Tmax} is required.

The *freight order rate* [FO/PE] is the number of freight orders [FO] to be executed per *period* PE [hour or day] on the relation $S_i \rightarrow S_j$. A freight order rate $\lambda_{FO\,ij}$ with mean order quantity $M_{FO\,ij}$ [LU/FO] generates a load flow λ_{ij} [LU/PE]:

- The *load flow* or *freight demand* is the number of load units per period flowing from station S_i to S_j and given by the *freight matrix*

$$\lambda_{ij} = M_{FO\,ij} \cdot \lambda_{FO\,ij} \quad [\mathrm{LU/PE}] \tag{18.2}$$

The elements of the freight matrix (18.2) are the *partial freight flows*. At the *entrance station* S_i the *incoming flow*

$$\lambda_{Ii} = \sum_j \lambda_{ij} \quad [LU/TU] \tag{18.3}$$

enters the transport network to be transported to the stations S_j, $j = 1,2,\ldots N_I$. At the *exit station* S_j the *outgoing flow*

$$\lambda_{Oj} = \sum_i \lambda_{ij} \quad [LU/TU] \tag{18.4}$$

coming from the entrances S_i, $i = 1,2,\ldots N_E$ leaves the transport network. The sum over all partial flows (18.2) gives the *total freight demand*.

For systematically changing demand, the transport systems must to be designed and dimensioned for the required freight matrix during the rush hours of the peak days of the year. For measuring the *peak flows*, generally a period length of an hour [PE = h] is necessary and sufficient. It is long enough to average out *stochastic fluctuations* and short enough to observe with minimal delay the *systematic changes* of the flows (see *Chap. 8* and *Sect. 13.1*).

The freight matrix for which a transport system has to be designed is generally derived by a *demand forecast* based on current freight or traffic flows (see *Chap. 9*). The errors of the measurement and the forecast are in a range of at least $\pm$ 5%. Therefore, calculations of the limit performances and the number of transport units need not to be more accurate than $\pm$ 5%.

A regular freight flow demand from station S_i to station S_j generates a *transport task* $S_i \rightarrow S_j$. If all flows start from one dispatch point DP and run to destinations S_j within a limited area, a *distribution task* DP $\rightarrow S_j$ has to be solved. If all flows come from stations S_i within a certain area and end at one consolidation point CP, a *collection task* $S_i \rightarrow$ CP is given. If an internal load flow collected from different dispatch stations S_i has to be diverted to several destinations S_j, a *sorting task* $S_i \rightarrow S_j$ has to be solved.

18.3 Network Design and System Configuration

Elementary *transport connections* and *transport nodes* span a transport network between the entrances, exits and performance stations of an extended logistic or production system. The *transport system control* triggers, directs and controls the flows of transport units through this network in a way that the freight and transport orders are executed in due time.

If the dispatch and receiving stations are linked by an existing transport net, the transport task is reduced to the execution of the upcoming freight orders within the required transport time at lowest cost. For this purpose, suitable *operating strategies* and a powerful transport control system are required.

If no transport network exists, a new system must designed by selection and connection of appropriate system elements and network modules. General networks consisting of connections and nodes are the topic of *graph theory*. The stations and transport elements of type (n,m) and order n+m of the transport network correspond to the *nodes* of *valence* n+m. The connections between the elements correspond to the edges of a *directed graph*.

Graph theory classifies, analyzes, and quantifies the possible structures and linkages of *networks* (Busacker and Saaty 1968; König 1936, Sachs 1988). The results of graph theory and the limit performance and queuing laws of *Chap. 13* can be used for the systematic design and dimensioning of transport and logistic networks. However, many methods of graph theory and queuing theory are highly mathematical and quite difficult. Therefore, the systematic development of theoretically valid and practically feasible selection and design principles by the investigation of transport network structures are important and widely open tasks for *logistic research* (Domschke 1995, Gudehus 1975/II).

18.3.1 Network Structures

From the simplest possible connections of stations and nodes result the three *elementary network structures* of transport systems:

line structure
ring structure (18.5)
star structure

These elementary structures are shown in *Fig. 18.1*. Dependent on the locations of the stations and the freight demand they can be linked in different ways to more complex two- and three-dimensional networks. Combined *area networks* result from the connection of line, circle and star networks as given in *Fig. 18.2*. Typical examples are *railway networks* and *public traffic networks*.

Spatial networks are area networks in several levels, which are connected by lifts, sloping conveyors or vertical S-conveyors. Such networks serve multi-storey buildings or inner-city urban regions. With transport crossings, connection elements and transfer terminals, the area and spatial nets are linked to inter-modal, local, regional, national and global transport and logistic networks.

Transport systems with line structure consist of a start or entrance station, a chain of connection elements, transport nodes or intermediary stations and a final or exit station. Conveyor systems generally have a line structure or consist of subsystems with line structure. The simplest conveyor systems with line structure are conveyors that supply a chain of workstations or machines.

Also, vehicle systems can have a line structure, e.g. if a number of stations is connected by one transport track. As well part-nets of an extended traffic network have line structure. Examples are *bus-lines* through the public road network and train routes through a railway network.

Figure 18.3 shows different transport *network modules* with line structure built up with the elementary transport elements of *Sect. 13.2*:

- *distribution comb* of connection elements and branching elements
- *consolidation comb* of connection elements and junction elements

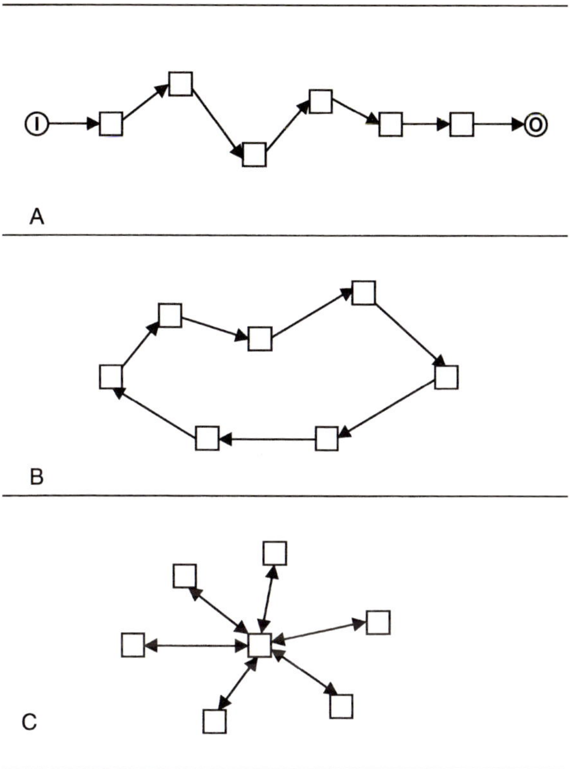

Fig. 18.1 Elementary network structures
A: line structure
B: ring structure
C: star structure
I: input interface
O: output interface
□ stations or transport nodes
→ transport connections

- partly combined *distribution-consolidation comb* of connection, branching and junction elements
- fully combined distribution-consolidation comb of connection, branching and junction elements
- *sorter buffer* with feeder line, branching elements, parallel buffer lines, junction elements and retrieval line

If the exit of a line-net is connected with its entrance, a *ring-net* is generated. Examples for simple ring-systems with *on-line stations* or *off-line stations* as shown in *Fig. 18.4* are rotary chain conveyors, dynamic ring-sorters (see *Fig. 18.14*) and ring-lines of railways or busses.

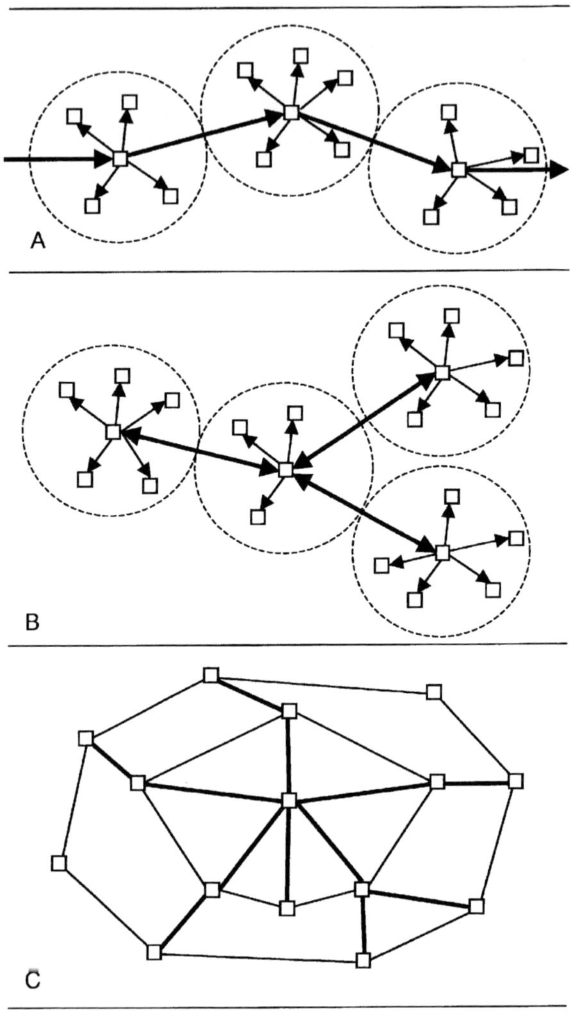

Fig. 18.2 Combined area networks

A: line-star-network
B: star-cluster-network
C: ring-line-network (cobweb)

Figure 18.5 shows how a branching element and a junction element connect a mesh or loop with a transport line. A *mesh* is a sideline running parallel to the main line whereas a *loop* is a circle running opposite to the main line. By meshes and loops, a basic ring net can be extended to a complex *area network* for the connection of many wide spread stations. An example for a meshed and looped ring-network is the track-net of the internal vehicle system of *Fig. 18.21*. Ring-networks are typical

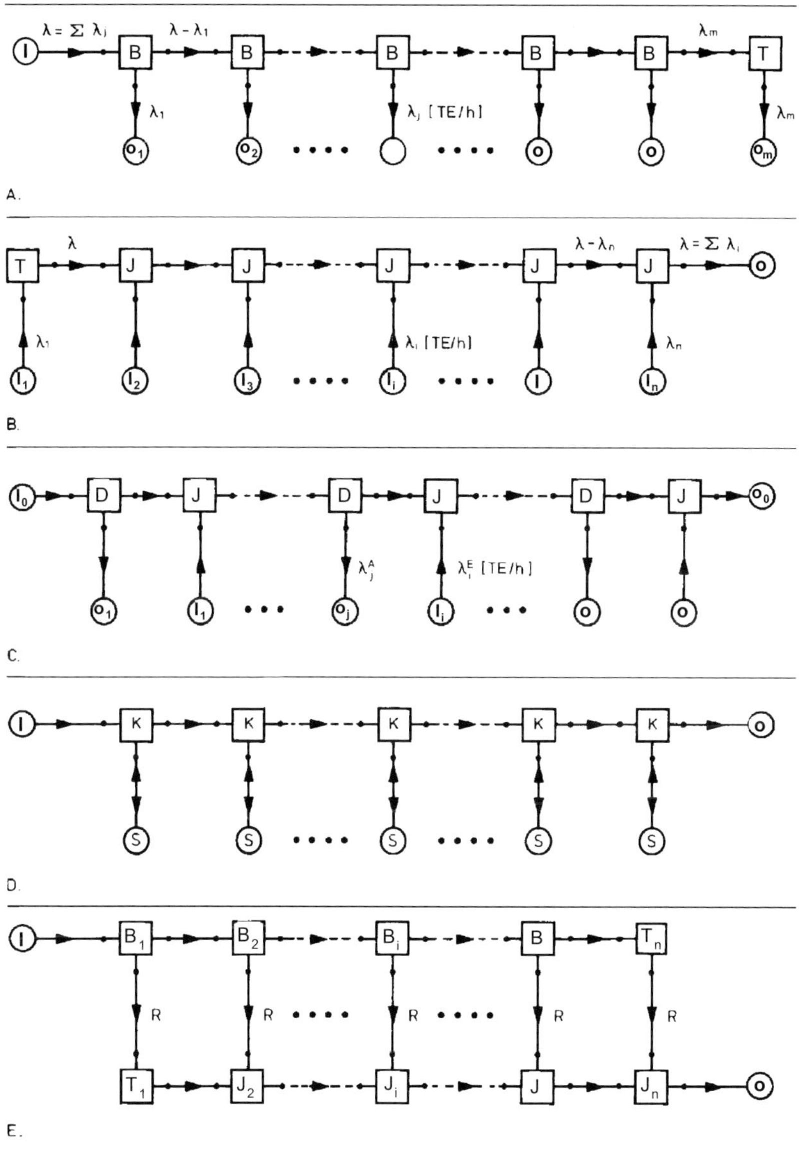

Fig. 18.3 Transport network modules with line-structure

A: single-track distribution comb
B: single-track consolidation comb
C: partly combined distribution/consolidation comb
D: completely combined distribution consolidation comb
E: buffer-sorter with n buffer-lines R transport units

B: branching elements **T**: transfer elements
J: junction elements **K**: connection shuttles

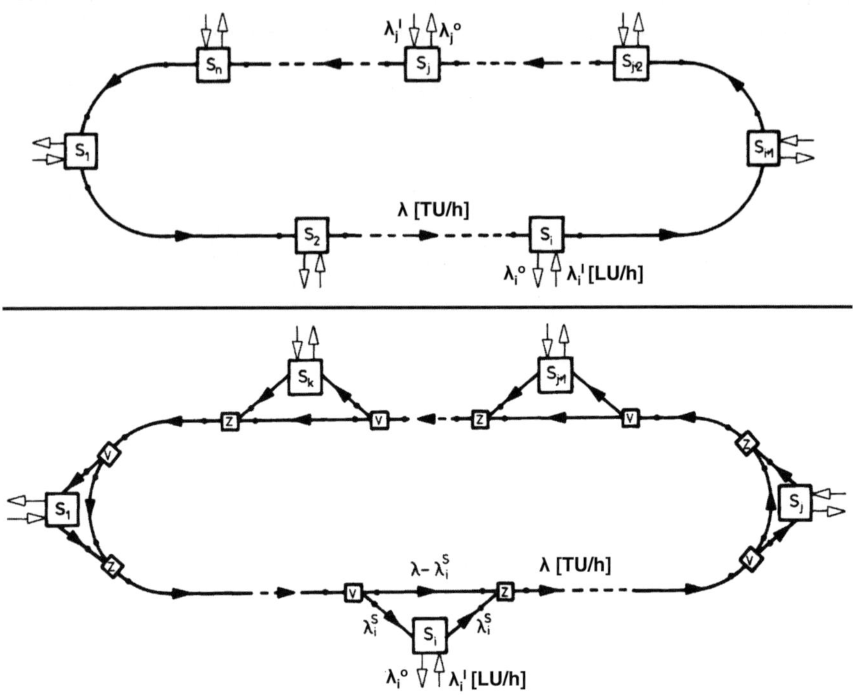

Fig. 18.4 Transport systems with ring-structure

Above: ring-net with online stations
Below: ring-net with offline stations and meshes

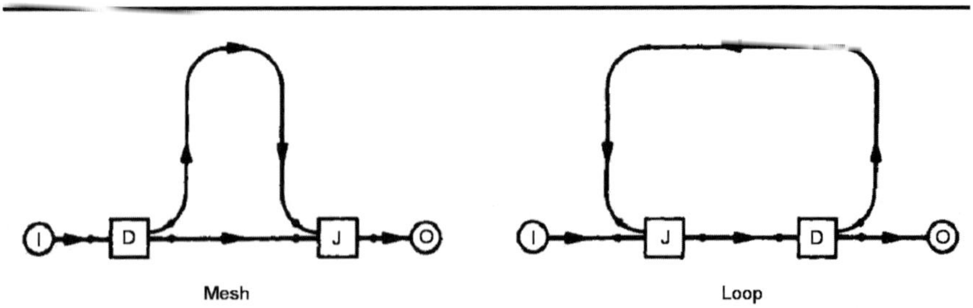

Fig. 18.5 Mesh and loop in a transport net

D: diversion
J: junction

for closed transport systems and for specific transport techniques, such as overhead conveyors, *automatic guided vehicles* (AGV) and railway systems.

18.3.2 Network Length and Distance Matrix

Each transport network consists of a number N_{TE} of transport elements TE_k, k = 1,2, ...N_{TE}, which have the *partial functions* $F_{k\alpha}$, $\alpha = 1,2, \ldots n_k$. Track elements, connection elements, and stations with one entrance and one exit have only one function, i.e. $n_k = 1$. Branching elements with one entrance and two exits and junction elements with two entrances and one exit have two partial functions, i.e. $n_k = 2$. As explained in *Sect. 13.2*, transport nodes of the order o = n+m and stations with n entrances and m exits have $n_k = n \cdot m$ partial functions.

The route lengths for the partial functions $F_{k\alpha}$ through an element TE_k are the *passing lengths* $l_{k\alpha}$. The sum of the passing lengths $l_{k\alpha}$ of all transport elements and stations forming the transport network is the *total network length*:

$$L_{TN} = \sum_k \sum_\alpha l_{k\alpha} \tag{18.6}$$

The total network length and the number and technique of the transport elements determine investment and operating costs of the transport system.

The *shortest path length* l_{ij} between the stations S_i and S_j is the sum of the passing lengths $l_{k\alpha}$ through the shortest chain of transport network:

$$l_{ij} = \sum_k l_{k\alpha} \tag{18.7}$$

The shortest path lengths l_{ij} are the elements of the *distance matrix* between the stations of a transport network.

18.3.3 Selection Rules and Design Principles

The set up of transport systems requires *selection rules* for the transport elements and *design principles* for the network. The design principles for the network depend on the structure of the freight matrix and on the type of the transport system. The selection rules result from the performance demand, the limit performances of the system elements and the buffer capacities of the connection elements.

The maximal throughput of a transport system is determined by the dispatch strategies and the limit performances of the *bottleneck elements* and by the buffer capacity of the connection elements. The bottleneck elements and the buffer capacity can be determined and assessed by a *function and performance analysis* as described in Sect. 13.7. The dimensioning of transport elements with the help of the *limit performance laws* and *queuing laws* has been explained in the *Sects. 13.3, 13.4* and *13.5*.

The throughput demand of conveyor systems, where the transport units equal the load units, is directly given by the freight matrix (18.2). For vehicle systems, where the transport units have a *freight capacity* $C_{TU} > 1$ LU/TU, the *transport matrix* $\Lambda_{ij} = \Lambda_{ij}(C_{TU};\lambda_{ij})$ [TU/PE] must be calculated from the freight matrix λ_{ij} [LU/PE]. The transport matrix for transport capacities $C_{TU} > 1$ LU/TU differs from the freight matrix. It depends on the freight matrix, the freight capacity and the transport strategies and determines the required limit performances of the network elements and the vehicle demand.

Minimal investment and/or lowest operating costs can be achieved by the following basic *design principles for transport networks*:

shortest connection paths
smallest number of nodes
sufficient buffer capacity
few vertical connections
minimal total network length
simplest network structure (18.8)

However, not all of these design principles are compatible. The network with shortest total length and the smallest number of nodes causes the lowest investment and operating costs for the network, but generally does not offer the shortest path lengths between all stations. Therefore, more transport means with higher operating costs are necessary than in a more extended network. The optimal solution is determined by minimizing the total costs, which are the sum of the *network costs* and the *vehicle costs*. Generally, the goal conflicts between the basic design principles (18.8) can only be solved when the specific requirements and restrictions are known.

18.4 Transport Control Systems

The tasks of a transport control system are to initiate and control the moves of the load or transport units from the entrances to the exits and to coordinate and manage the transport flows through the total network. Due to the *organization principles* of *Sect. 2.4*, an efficient and reliable control system of an extended transport system is *hierarchical*.

As shown in Fig. 18.6 , a hierarchical transport control system has several control levels, such as

accompanying controls of the load and transport units
stationary controls at the transport elements
group controls for parts of the transport systems (18.9)
central distance control of the total transport system

In addition, a transfer of the data, information and commands between the single control units and the different control levels is required. This is possible by telephone, *electronic data interchange* (EDI) or by *internet*.

18.4.1 Accompanying Control

The accompanying control of the load units in a *conveyor system* is normally *passive*. It can be an *identification number*, a *barcode label*, a *coding reflector* or a programmable *transponder*, which is read by stationary reading heads or via RFID (Finkenzeller 2002; Scholz-Reiter et al. 2005; Shephard 2004)

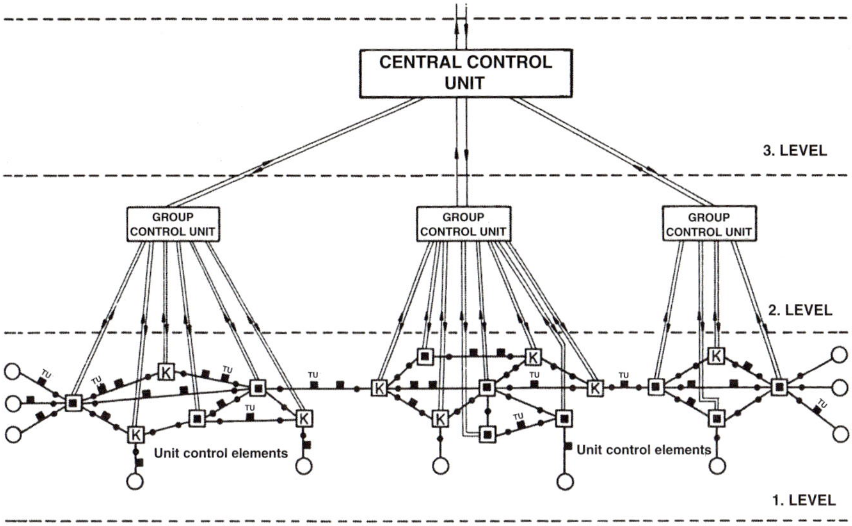

Fig. 18.6 Hierarchical transport control system

K: transport nodes with stationary control

TU: transport units with accompanying control

The accompanying control of a *vehicle system* is generally *active*. Depending on the configuration of the whole transport control system, the accompanying control of the vehicles executes specific sub-functions, such as driving, track and distance control, the transfer through nodes, or tracking the tour through the transport network. For this purpose, each vehicle has sensors, recorders and transfer devices. In addition, it can be equipped with a control system or a *board computer*.

For manually controlled vehicles, a driver performs the steering and control, assisted by displays such as speedometer or mileage counter. Some of the functions can also be performed by a board computer or advised by the commands of a central control system, which are transferred to the driver by displays, head phone or radio. In *automatic guided vehicle systems* (AGV), electronic controls and board computer execute the vehicle and track control completely.

18.4.2 Stationary Control

The stationary control consists of *sensors*, such as contacts, light barriers and reading devices, that register the passing load or transport units, and of *adjusting devices*, that switch between the partial functions of the transport elements.

In extended transport systems, the control of part-nets, sub-systems or longer chains of transport elements is executed by a *group control*. The group control receives input-information and instructions from the central control, a local control unit or from the vehicle control, processes this information, and sends instructions and output-information to the controlled group of local controls, to other group controls and to the central control.

18.4.3 Central Control

Transport systems with an extended network, large number of transport units and many stations require a central control. The central control organizes, manages and coordinates the transport flows depending on the current demand by appropriate *system strategies*. In addition, the central control can perform superior functions such as *tracking and tracing* of freight and transport units, traffic supervision, queuing control or the registration and analysis of failures.

The central transport control receives freight and transport orders either directly from the customers or an operator, or from the ERP system of the company.

18.4.4 Configuration of Transport Control

The allocation of the different tasks to the control levels determines the *configuration* of the control system. The configuration or *system architecture* can be more or less centralized, decentralized or hierarchical:

- In a *decentralized control system,* all functions are executed by the accompanying and the stationary local controls.
- In a *centralized control system,* all control functions are executed by one central control unit. The tasks of the accompanying and local controls are restricted to registration and transfer of information and to the execution of instructions from the central control.
- A *hierarchical control system* has several control levels as shown in *Fig. 18.6*. The different control functions are adequately distributed amongst these levels.

For the design of the transport control system holds the general *decentralization principle* of *Sects. 2.3* and *2.4*:

▸ The control functions should be executed as locally as possible and only as centralized as necessary and efficient.

In addition to this principle, the architecture of the transport control depends on the extension and complexity of the network, on traffic density and transport demand, safety requirements and on the *transport strategies*.

18.4.5 Data Exchange

The configuration of the control system and the distribution of the functions determine frequency and content of the data exchange between the different control units and levels. That necessitates appropriate data exchange systems. Over shorter distances, data are generally transferred via cable, wires or data busses. Extended transport systems require wireless data transmission.

The code and position of the load units in a conveyor system are registered by reading devices at the entrances and exits of the transport elements. They transfer the information to the stationary controls mechanically, optically, inductively or by laser.

Vehicle control systems communicate with stationary controls and central control by long-distance connections, such as radio frequency communication, and by remote data transmission, such as infra-red. The application of the different data exchange techniques depends on the distance between the vehicles, the stationary receivers and the location of the central control.

In external transport systems, cellular radio and satellite communication are used for the information and data exchange between vehicles, central transport control and traffic control systems. The localization of the vehicles is assisted by satellite navigation systems such as the *Global Positioning System* (GPS).

18.5 Transport Strategies

Transport strategies are scheduling and operating strategies for the efficient execution of the current freight and transport orders, for the direction of the transport units on the shortest or fastest routes to their destinations and for the optimal dispatch of the transport units at the stations and transport nodes.

In many cases, good service and high quality at lowest costs can be achieved simpler and quicker by intelligent transport strategies than by high-tech solutions. Transport strategies offer the *optimization possibilities*:

- solution of a transport task with a simpler transport network
- improvement of the performance of an existing transport system
- reduction of the required number of vehicles
- improvement of reliability, availability and traffic safety

Transport strategies can be differentiated in *station strategies*, *routing strategies*, *empty runs strategies* and *traffic strategies*.

18.5.1 Station Strategies

Station strategies regulate the dispatch of the freight orders at the stations and the loading of the transport units with the waiting freight units. Possible *dispatch strategies* for shipments are:

- *Fixed sequence dispatch* (*First-Come-First-Served FCFS*): Incoming freight units are loaded in the sequence of their arrival into the next incoming transport unit, which takes as many load units for the same direction or destination, as its free capacity allows.
- *Optimal sequence dispatch* (*shipment consolidation*): Incoming load units for the same destination or direction are collected for a certain time or until the free loading capacity of a transport unit is reached, loaded irrespective of their arrival sequence and shipped together to the joint destination.

Shipment consolidation improves the utilization of the transport capacities and reduces transport costs. Conditions for consolidation are acceptance of the changed sequences, sorting facilities and sufficient consolidation buffers at the loading station. Possible disadvantages of shipment consolidation are longer waiting times for the single freight orders.

For the filling of the transport units, the following *loading strategies* are applicable:

- *Single-destination loading*: A transport unit is filled only with load units for the same destination.
- *Multi-destination loading*: A transport unit is filled with load units for destinations along the same route.

Conditions for multi-destination loading are either that the load units are filled in the order of their destination or the possibility to restack the load units during the tour or at the destination.

Combined loading and dispatch *strategies* are:

- *Runs without co-loading*: Only completely empty transport units are filled with load units for the same direction.
- *Runs with co-loading*: The free capacity $C_{TUfree} = C_{TU} - M_L$ of partly filled transport units with freight capacity C_{TU} and a load $M_T < C_{TU}$ is filled up.

If runs with co-loading are chosen, a free access to the waiting load units, which are determined for the same destination, is necessary.

Multi-destination loading and runs with co-loading aim at a better filling degree of the transport units. They are only efficient, if the stations offer as many load units for the same destination with appropriate dimensions as fit in the free capacity of the arriving transport unit.

18.5.2 Routing Strategies

Routing strategies regulate the visiting sequence of the destinations of the load and determine the transport route. Possible *routing strategies* are:

- *Minimal route strategies*: Depending on the loads and their destinations the route with minimal length, shortest transport time or lowest transport costs is chosen.
- *Capacity utilization strategies*: Depending on the total freight demand, tours are taken, which lead to maximal utilization of the transport capacity.
- *Freight schedule strategies*: The transports are executed due to a fixed *timetable* that aims at shortest routes and maximal utilization of the capacity for an expected demand of longer time.

For simple transport systems with few connections, the routes with minimal length, shortest travel time or lowest costs can be found quite easily by *full-enumeration* and comparison. The optimal route in more complex transport networks with many connections cannot be found by full-enumeration within limited time, as the number of possible tours increases with the number of connections more than exponentially (see *Sect. 5.2*).

To solve the *optimal routing problem*, *Operations Research* has developed highly sophisticated *heuristics* that generate approximate solutions within acceptable time (Churchman et al. 1961; Laporte 1992; Lawler et al 1985; Müller-Merbach 1970, Domschke 1985; Goczyla and Cielatkowski 1995). Other solutions are *analytical routing strategies*, such as the *stripe strategy* outlined in *Sect. 18.11.2*. They achieve practicable routes close to the optimum and have the advantage that the mean route length can be calculated for a given freight demand.

Capacity utilization and freight schedule strategies aim at minimal operating costs. Their disadvantages are longer waiting and travel times and the necessary buffer capacities in the stations. The required transport times and the available buffer space at the loading stations restrict these routing strategies.

18.5.3 Empty Runs Strategies

Empty runs strategies aim at the optimal use of empty transport units and determine the number of empty or partly filled transport units circulating in a transport network. Possible empty runs strategies are:

- *Single runs*: A transport unit delivers only load units for the same destination and returns empty to the same entrance, as long as there are further load units arriving.
- *Combined runs*: A transport unit delivers load units for the same destination and returns with upcoming *back load* as long as load units arrive at the connected stations.
- *Minimization of empty runs*: After unloading, a transport unit picks up waiting load for any direction or it runs to the closest station where load is waiting for dispatch.
- *Timetable capacity offer*: Independent of the current freight demand, filled, empty or partly filled transport units circulate within the transport network following a fixed timetable for the expected demand.
- *Empty vehicle clearing*: If the freight demand decreases and more transport units are emptied than required, the empty transport units drive to the next free vehicle buffer and wait there until the demand increases.

If the available buffer space at the stations is insufficient, *buffer tracks*, empty rails or *parking lanes* for empty vehicles must be placed within the transport network. They should be located close to the most frequented dispatch stations.

18.5.4 Traffic Strategies

Traffic strategies regulate and control the flows of load and transport units through a transport network. Their objectives are ensuring the required transport times, enabling maximal throughput and preventing collisions. Dependent on the area of influence, traffic strategies can be differentiated into *node strategies*, *partsystem strategies* and *system strategies*. The node and part-system strategies as well as their parameters and effects have already been discussed in *chapter 13*. Advanced system strategies are (Scholz-Reiter et al. 1999, 2005):

- *Adaptive combination strategies*: Dependent on the actual demand, different strategies are combined in order to optimize performance, costs and safety.
- *Self-routing strategies*: The load or transport units are equipped with a programmable transponder and find autonomously the optimal route through the network based on adaptive *self-routing strategies* using locally transferred information about the current traffic situation.

- *Dynamic network strategies*: If stations, connections or nodes are overloaded, underused or defective, *evasion strategies* are applied, such as re-routing, progressive-signal-routes, route-diversion or evacuation.

These system strategies require a central control that observes and controls the whole situation in all parts of the whole transport network. Up to now, the possible system strategies and their parameters, effects and applications have not been fully investigated. This task can only be partly solved by *analytical methods*. Dynamic *OR-methods*, *digital simulation* and other tools are necessary to examine the interactions of single strategies and the effects of system strategies for *complex systems* with fast changing demand (see *Sect. 5.3*).

18.6 Conveyor Systems

The *components of conveyor systems* for continuous transport of discrete goods or load units are:

conveyed load
load carrier
conveyor network (18.10)
material flow control.

The following properties of the goods and load carrier determine their *conveyability* and the conveying technique:

bottom flatness
skid resistance
abrasion resistance (18.11)
stability
stackability

Load carriers, such as containers, trays or pallets, are used for consolidating smaller goods or when the transport good itself is not conveyable. The disadvantages of load units are the costs for the load carriers, the filling and emptying, and the removal and provision of empty load carriers at stations where more respectively less filled load units are arriving than leaving.

For goods with sizes and weights in certain ranges, *standard conveyor techniques* are available, which differ mainly in dimensions and construction:

- *bin conveyor systems* for cartons and small standard containers up to 800 mm and 60 kg
- *pallet conveyor systems* for palletized goods and standard load units up to 1,400 mm and 1,500 kg
- *special conveyor systems* for conveyable goods and load units with larger weight and measures and/or with non-standard shapes

All conveyor systems consist of input and output stations, conveyor lines, connection, junction and branching elements and of transport elements of higher order, such as multi-switches, transfer shuttles and rack feeders. Bin and pallet conveyor systems are generally *modular configurations* of standard conveyor elements with normalized measures, whereas special conveyor systems consist of individually constructed elements, which are designed and manufactured for the single project.

Special conveyor systems are necessary for *baggage transport*, in the production, for car assembling lines and for heavy and bulky goods. Also for *mobile load units*, such as *roller pallets* or *rolling containers*, a special conveyer technique is necessary, such as guiding tracks equipped with pulling or shifting devices (see *Fig. 18.25*). The disadvantages of mobile load units are higher costs, deterioration, and the handling and transport of empties. In most cases, these disadvantages weigh up against the advantages of lighter tracks and a possibly higher speed.

Standard bin and pallet conveyor systems have a *line-net structure* with tracks running in one direction. Circular or endless conveyors have a *ring-net structure* with a rotating chain or endless rope and a central drive. For sorting with high throughput, special sorter techniques are used. Standard conveyors, circular conveyors and sorters can be linked by *transfer stations* to complex multi-task conveyor systems.

18.6.1 Standard Conveyor Systems

Technical realizations of the *continuous connection elements* in standard conveyor systems are:

chutes
belt conveyors
band conveyors
roller conveyors
wheel conveyors (18.12)
chain conveyors
slat conveyors
S-conveyors

The driving force for chutes and downward slope lines is gravity. All other conveyor elements are directly or indirectly powered by electric motors. *Figure 18.7* presents a chain of belt conveyor elements and a chain of roller conveyors whose sections can be switched on and off separately. Both constructions can buffer a queue of load units caused by a *backlog* (see *Sect. 13.5*).

Two technical solutions of *branching elements* in bin conveyor systems are shown in *Figs. 18.8*. *Fig 18.9* presents two different *junction elements*, which join two roller conveyor lines.

An example of a discontinuous connection element between several entrances and exits is the *transfer shuttle* in *Fig. 18.10*. Other elements of container and pallet conveyor systems are shown in *Figs. 13.4, 13.14, 13.16, 16.8* and *17.34*.

By selecting and inserting the appropriate technical elements, the abstract network structures (18.5) with the possible layouts of *Figs. 18.3* and *18.4* are converted into standard system modules. Due to the requirements, the standard modules can

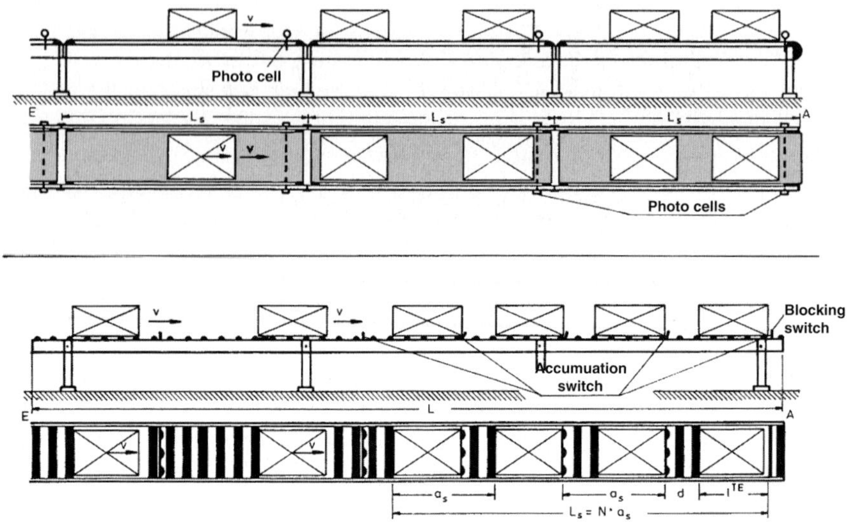

Fig. 18.7 Continuous bin conveyor lines

Above: belt conveyor with separation and buffer elements
Below: roller conveyor with separation and buffer elements

be connected by conveyor elements with the input, output and transition stations in order to set up complete conveyor systems.

For instance, *Fig. 18.11* presents a standard system module designed for the supply and removal of order bins at a *local commissioning station*. In this system module, passing bins can *overtake* stopping bins. Other examples are the in- and outward conveyor systems of an automatic high bay store in *Fig. 16.8* and of a mini load store in *Fig. 17.34*.

Standard conveyor systems for containers and pallets are also applicable for feeding and disposal of workstations in the production, for linking machines, as inward and outward transport systems of automatic stores and in commissioning systems with dynamic article provision or with local picking stations.

18.6.2 Circular Conveyor System

A *circular conveyor* consists of a circulating *pull device*, which can be a chain or an endless rope and is equipped with load carriers. The load carriers are either permanently fixed to the circular chain or rope or can be disconnected for loading, unloading and buffering. If the pull device is moving beneath the floor, the load carriers are roller trolleys linked from above, if it is located overhead, they are hangers or gondolas.

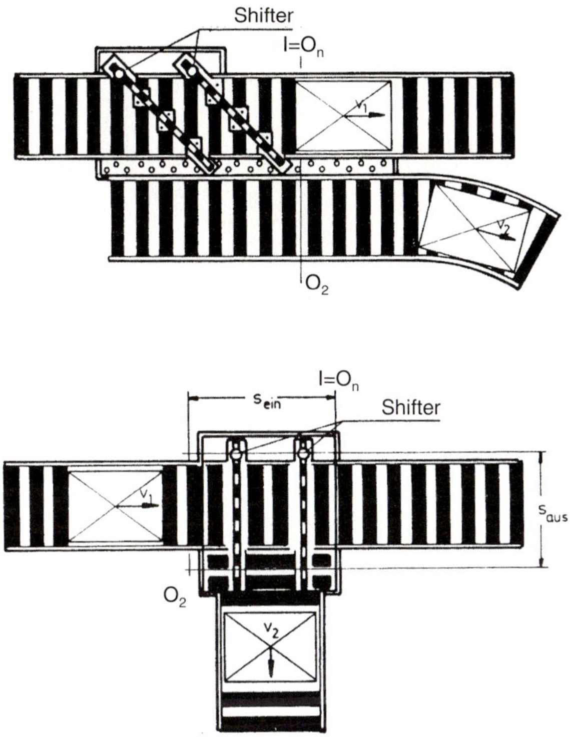

Fig. 18.8 Continuous branching elements in bin conveyor systems
Above: 45°-chain diverter
Below: 90°-chain diverter

Examples for circular conveyor systems are (Arnold 1995; Gudehus 1977):

under-floor chain conveyors
overhead circular conveyors
power & free conveyors
tilt-tray conveyors
cross belt conveyors (18.13)
S-conveyors
ski lifts
cable cars

A ring net structure with online or offline stations as shown in *Fig. 18.4* is, due to the central circular device, typical for endless conveyor systems. *Fig. 18.12* presents an *online loading station* of an overhead chain conveyor with pull chain, slide rail and load hangers. *Figure 18.14* shows a tilt-tray conveyor as central element of a circular sorter system.

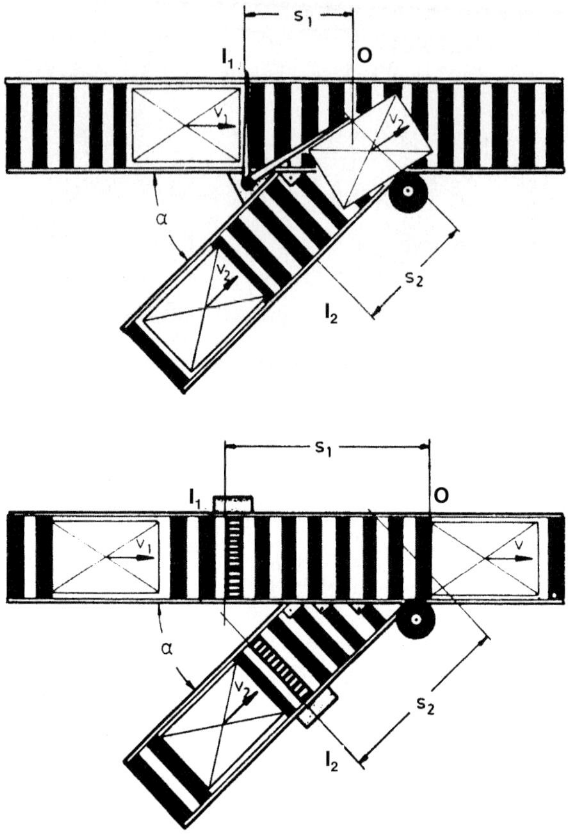

Fig. 18.9 Continuous junction elements in bin conveyor systems

Above: 45°-roller junction with double-track locker
Below: 45°-roller junction with allotment rollers

A circular conveyor is an elementary transport element of type (n,m) with order o = n+m equal to the number n of loading stations plus the number m of unloading stations (see *Fig. 13.9*). The *limit performance* of this steady element is given by relation (13.10). It depends on the speed of the pull device, the minimal distance of the load carriers and on their loading capacity.

The *limit performance* of the central transport element can be fully used only if the limit performances of the loading and unloading stations are sufficient. The stations of a circular conveyor are often transitions to other transport systems. The loading and unloading limit performances of the stations are determined by their construction and location. The stations can be located *on-line* or *off-line*. For off-line-stations, the load carriers are separated from the pull device and loaded or unloaded discontinuously in an idle state. In on-line-stations, as shown in *Fig. 18.12*, the continuous loading and unloading is synchronized with the speed of the circulating pull device.

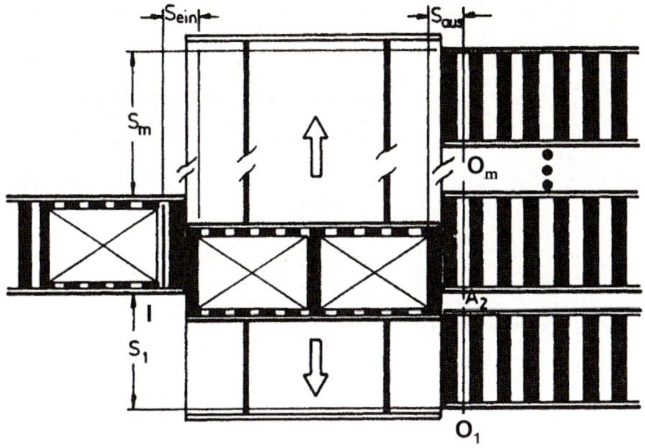

Fig. 18.10 Discontinuous connection element
Roller transfer shuttle

Applications of *overhead circular conveyors* and *power&free conveyors*, which compete with overhead rail systems, are the transport of metal sheets, car bodies or slug material in a series production. They are also used in paint shops, for the dynamic provision of packaging or as consolidation conveyor in two-stage commissioning systems. Floor based chain conveyors compete with internal vehicle systems, especially with *automatic guided vehicles* (AGV). They are particularly suited for *crossdocking* in transit stations and for connecting the receiving docks, storage areas and shipping docks in a logistic center.

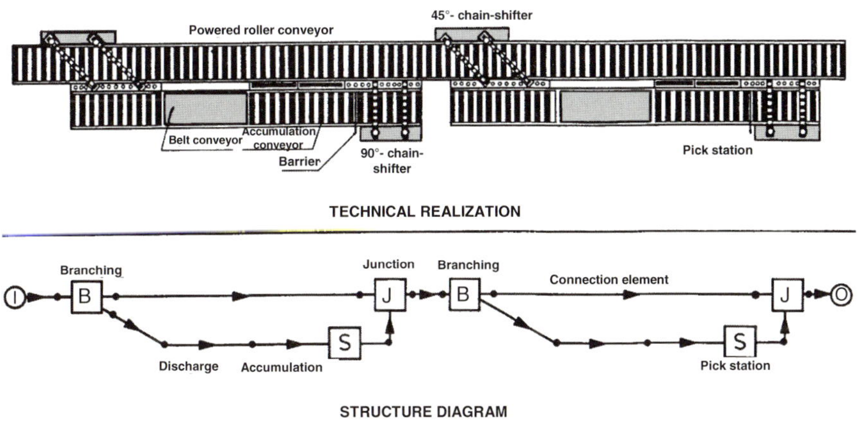

Fig. 18.11 Technical realization and structure diagram of a conveyor module for the order bin provision for local commissioning

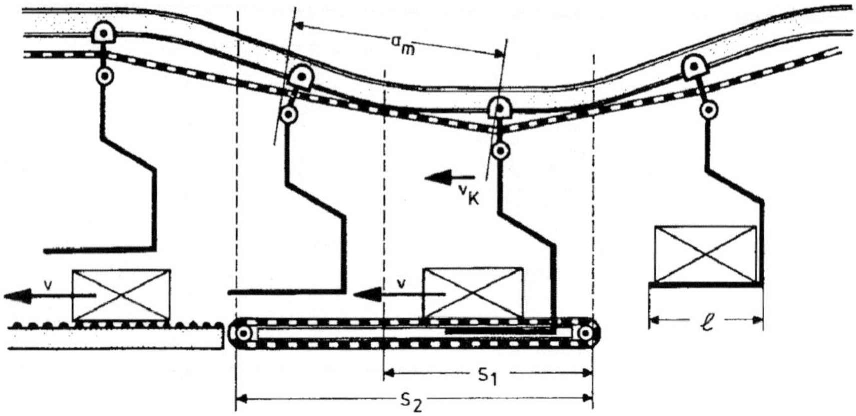

Fig. 18.12 Continuous loading station of an overhead chain conveyor

18.6.3 Sorter Systems

Sorter systems are special conveyors for separation, distribution and sorting due to given *sorting orders*. Depending on the net-structure and the buffer capability, they can be differentiated in *line sorters* and *circular sorters*, with and without intermediate buffer.

Line sorters without intermediate buffers are single-track *distribution-combs* with a line net structure as shown in *Fig.* 18.3A. They consist of a dispatch or incoming station, a chain of line elements and branching elements, and a set of parallel target stations on one or two-sides along the line. The incoming items are identified by the system control at the entrance to the sorter line, and removed by the branching elements at the target stations.

Depending on the required performance and the kind of goods, a line sorter is a roller conveyor line connecting a chain of branching elements, a long belt conveyor with diverters or pushers, or a slat conveyor with movable shoes for *dynamic discharging*. The maximal *sorting performance* of a line sorter without intermediate buffer depends on construction and speed and is restricted by the limit performances of the branching elements. For *sorter goods* [SG] with lengths of up to 600 mm and weights up to 30 kg and sorters consisting of standard conveyor elements, it ranges from 2,000 to 6,000 SG/h, and for special *high performance line sorters* from 8,000 to 13,000 SG/h. Main applications for line sorters without intermediate buffer are hubs and transit stations of *parcel distributors* (see *Fig. 19.2*) and the second stage of commissioning systems in logistic centers of *mail order companies*.

A *sorter buffer* is a *line sorter* with intermediate buffers and a net structure as shown in *Fig.* 18.3E. It consists of a supply line with branching elements, several parallel buffer lines, which end in junction elements, and a retrieval line. Dependent on the sorting orders, the branching elements separate and send the incoming units into the buffer lines, where they are collected. When an order is complete, all its units are conveyed in a single batch through the junction elements via the retrieval line to the destination.

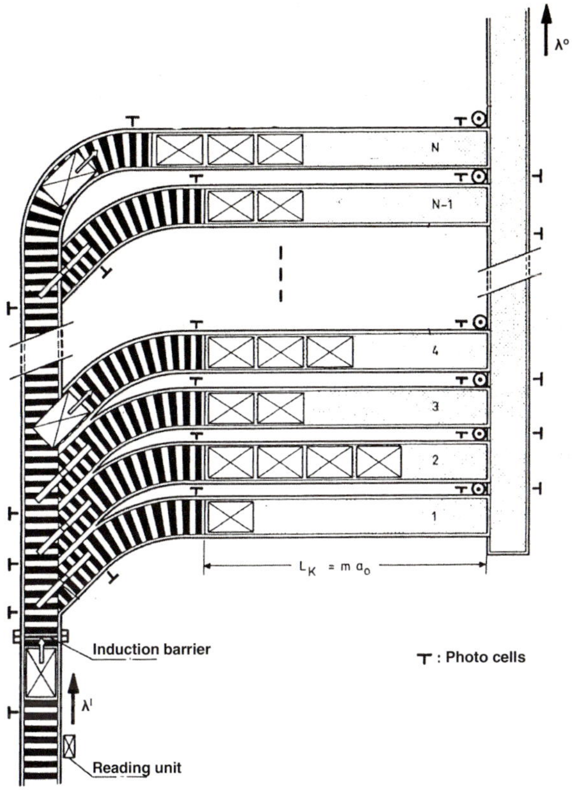

Fig. 18.13 Sorter buffer for bins or cartons

Sorter buffers are generally built up by standard conveyor elements. *Fig. 18.13* presents a sorter buffer that separates mixed cartons of finished goods coming from production in order to palletize cartons with the same article or for the same customer. They are used also in production and logistic sites for order consolidation in the goods dispatch area.

The *limit performance* of sorter buffers depends on the operating strategy, the limit performances of the branching and junction elements, and on the order structure. The queuing capacity of the buffer lines limits the number and content of the sorting orders. Buffer sorters with standard conveyor elements reach performances up to 3,000 SG/h. That is significantly less than the performance of line sorters without intermediate buffer.

Circular sorters without intermediate buffer are special endless conveyors with single accessible load carriers linked in shortest distances to a rotating pull device. They connect several feeding stations with many target stations. The load carriers can be gondolas, tilt-trays as shown in *Fig. 18.14* or small separately powered belt conveyors.

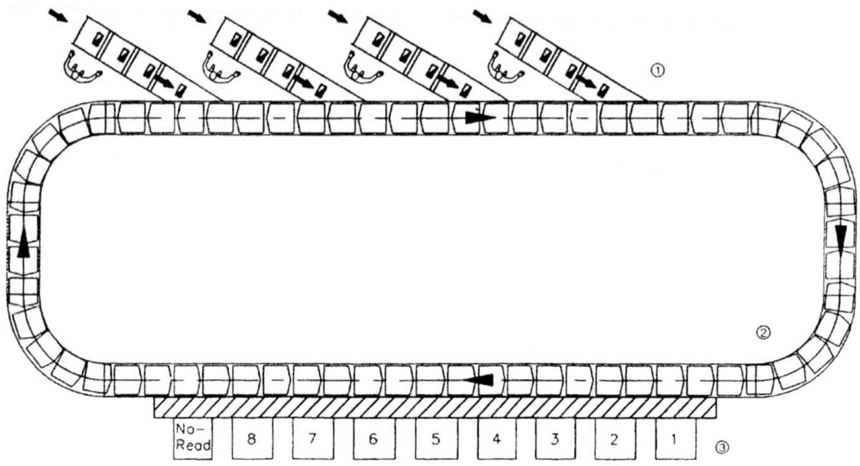

Fig. 18.14 Tilt-tray conveyor

1. In-feed stations
2. circular conveyor with tilt-trays
3. collection chutes

The sorter good is automatically loaded from the feeding lines onto the load carriers and dispatched to the right target station after being identified. Provided loading and dispatch stations operate fast enough, the maximal sorter performance is given by the limit performance of the central rotating conveyor. Nowadays, high performance circular sorters without intermediate buffer reach throughputs up to 15,000 SG/h. They are mainly used as connection between the first and the second stage of *two-stage commissioning systems* especially in logistic centers of *mail order companies*.

Circular sorters with intermediate buffer, also called *dynamic buffer ring*, consist of one or several feeder lines, a long buffer ring of line, curve and branching elements and a number of terminal channels. The incoming items, cartons or parcels rotate in the buffer ring until all units of a sorting order have arrived and discharged to the next available terminal channel.

The performance of a dynamic buffer ring is determined by the limit performance of the bottleneck element. Applications are feeder systems of *mini load stores* and dynamic sorter buffers for parallel workstations.

18.7 Vehicle Systems

The main *components* of a vehicle system are:

transport units
stations
track network (18.14)
transport control
energy supply

The *transport units* are single *vehicles* or *trains* with *tractor* and *trailers*. The *transport control* regulates the moves of loaded and unloaded transport units on the track network. The mobile or stationary *energy supply* stores and supplies the energy for the power engine of the transport units.

18.7.1 Key Features of Transport Units

The following performance and cost features of the *transport units* [TU] determine capacity, vehicle demand, availability and cost efficiency of a vehicle system:

- *Outside dimensions*: outer length l_{TU}, breadth b_{LU} and height h_{TU} [m] of the fully loaded transport unit
- *Loading space dimensions*: inner length l_{TU}, breadth b_{TU} and height h_{TU} [m] of the available loading space
- *Transport capacity*: loading space V_{TU} [m^3/TU], loading weight W_{TU} [kg/TU, t/TU] and load capacity C_{TU} [LU/TU]
- *Driving speed*: *maximal speed* v_{max} and *travel speed* or *effective speed* v_{eff} [m/s; m/min, km/h, miles/h]
- *Acceleration values*: normal acceleration $b^{+}{}_{TU}$ and de-acceleration $b^{-}{}_{TU}$ and emergency break constant $b^{-}{}_{TUem}$ [m/ s^2]
- *Energy supply range*: maximal driveway s_{Emax} [m] with one filled tank or battery
- *Energy consumption*: fuel consumption [l/100 km] or electricity consumption [kW/h]
- *Reliability*: *mean time between failures* (MTBF) or mean drive length between failures [km] and *mean time to restore* (MTTR) (see *Sect. 13.6*)
- *Service life*: maximal operating mileage or hours
- *Investment* for vehicle, trucks and trailers

Capacity, performance and costs are determined by the construction, the power and driving technique, the control and board computer and by other technical features of the transport units. In the *Tables* 18.3 and 18.4 the key features of selected transport units and traffic means are presented.

If, as outlined in *Fig. 18.15*, C_{VH} [LU/VH] is the load capacity of the vehicle, C_{TR} [LU/TR] is the load capacity of the trailers and N_{TR} [TR/TU] the number of trailers per train, the *load capacity* of the whole train or transport unit is

$$C_{TU} = C_{VH} + N_{TR} \cdot C_{TR} \quad [\text{LU/TU}]. \tag{18.15}$$

A conclusion from this relation is:

► The capacity of the vehicles and trailers, and the number of trailers are *optimization parameters* for the planning and *scheduling parameters* for the operation of a vehicle system, which can be used to adjust the capacity of the transport units to the freight demand.

In the simplest case, the number of trailers is $N_{TR} = 0$ and the transport unit is a single loadable vehicle without trailer. Examples are forklift trucks, load trucks, cars, busses, ships and airplanes. The case $C_{VE} = 0$ means that the tractor has no

Table 18.3 Features of selected transport means

Transport Mode		**Net weight**		**Loading Space**				**Load Capacity**	
Transport mean	Transport unit	t	Length m	Breadth m	Height m	Area m^2	Volume m^3	Number	LU
ROAD									
Pick-up van	Loading space	2.6	3.2	2.2	2.5	7	18	5	EUP
Transport truck	Truck body	7.5	7.2	2.4	3.0	18	53	17	EUP
Semi-road trailer	Semi-trailer	27.0	13.6	2.5	3.0	34	102	34	EUP
Swap-body trailer	2 Swapbodies	14.0	7.1	2.5	3.0	18	53	2.17	EUP
RAIL									
Standard wagon	4 or 8 axes	50	14.6	2.6	2.8	38	106	36	EUP
								2	TEU
High-capacity wagon	8 axes	100	26.0	2.6	2.8	68	189	60	EUP
								4	TEU
Part-train	Engine + Wagons	to 1.600	to 500	3.0				12 to 16	Wagons
								60	TEU
Complete train	Engine + Wagons	to 4.000	to 1.000	3.0				17 to 32	Wagons
								120	TEU
WATER									
Europe Riverbarge	Container	3.000	75	10	6.0	750	4.500	to 80	TEU
Feedership small	Container	5.000	≈ 100	≈22				500	TEU
Feedership large	Container	20.000	≈ 150	26 to 28				to 2,000	TEU
Containership large	Container	80.000	≈ 350	28 to 32				>8,000	TEU

EUP: Euro-pallet place TEU: 20" Container

Table 18.4 Performance and cost data of transport means

Transport mode	**Speed**	**Range**	**Milage**	**Investment**	**Fuel**	**Cost rates**		
Transport means	effective	max	total	new	consumption	basic rate	stop rate	milage rate
ROAD	km/h	km/day	M. km	T€/TM	1/100 km	€/trip	€/stop	€/km
Truck small	50	400	0.8	30	14 to 18	11.00	2.50	0.70
Truck large	60	500	1.2	60	20 to 25	21.00	5.10	1.05
Semi-trailer truck	60	800	1.5	130	35 to 40	45.00	17.00	1.20
Swap-body truck	60	800	1.5	140	35 to 40	47.00	22.00	1.20
RAIL				Preparation + building		trip	traction	track
	km/h	km/day	Mio. km	T€/waggon	€/wagon	€/wag-km	€/train-km	€/km
Train+standard waggons	30 to 60	800	3.0	65	35.00	0.16	13.00	5 to 70
Train+high capacity waggons	40 to 80	1,000	3.0	95	45.00	0.22	16.00	5 to 75
WATER	km/h	km/day	Mio. km	Mio. €/ship	1/100 km	€/trip	€/stop	€/km
River ship	15	300	1.0	3 to 5	500			
Feeder ship small	30	650	2.0	10 to 15	2,000	5,000	1,300	20.00
Feeder ship large	35	850	3.0	20 to 25	5,000	21,000	6,600	28.00
Container ship large	45	950	3.0	40 to 50	20,000	32,000	10,000	32.00

Cost rates budget values from model calculations, basis: 2007

Fig. 18.15 Transport train with tug & tows
Vehicle capacity: $C_{VH} = 0$ LU
Trailer capacity: $C_{TR} = 3$ LU
Number of trailers: $N_{TR} = 2$
Transport capacity: $C_{TU} = 0 + 2 \cdot 3 = 6$ LU
Load quantity: $M_{TU} = 5$ LU
Filling degree: 83.33%

load capacity and the capacity of the trailers is $C_{TR} > 0$. Examples are railway trains with a locomotive and wagons, towboats and barges and tug&tows. An example for the general case with $C_{VH} \geq 1$, $N_{TR} \geq 1$ and $C_{TR} > 1$ is a train with a loadable truck and loadable trailers.

18.7.2 Transport Means

For the different transport goods, functions and requirements, a broad variety of transport means is available. They result from the multitude of possible constructions and combinations of the basic *components of a transport mean*:

vehicle
truck
trailers
loading devices (18.16)
track guidance
engine
vehicle control

When no *stationary loading devices* are available at the stations, the transport means must have own *instationary loading devices*. Their technique depends on the load units and on the stationary equipment of the stations. Typical *loading devices* are:

cranes, hooks and chains
spreader and loading gears
pushing, lifting and telescope forks
roller and chain conveyors (18.17)
undercarriages
gripping, pulling and pushing devices

The transport vehicles can be differentiated into *overhead trolleys*, *floor bound vehicles* and *traffic means*.

Overhead trolleys are individually powered movable cranes with hooks, gondolas or other handling devices rolling on or hanging below a *monorail*. Overhead vehicle systems have a permanent power supply and do not occupy floor space.

The permanent power supply ensures an unlimited energy reach length. The floor savings, however, are reduced in many cases by protective grids and guidance tracks. The main application of overhead trolleys – in competition with circular conveyors – is the internal transport of permanent flows of special, heavy or oversized load units between many stations.

Internal transport is also the main application for *floor bound vehicles* such as:

pallet-trucks
forklift trucks
commissioning vehicles
van-carriers (18.18)
load trucks
under-running hauler
tug&tows

An example for a very simple floor bound vehicle is the manual *pallet-truck* shown in *Fig. 18.16*. It is frequently used for the transport over short distances in factories, logistic centers, sales-outlets, especially in the goods entry and the dispatch area for loading, transferring and unloading.

Figure 18.17 presents four different driverless vehicles of an *automatic guided vehicle* system (AGV). They are equipped with loading devices for different load units and applications in internal transport. *Fig. 18.18* shows the transfer of a

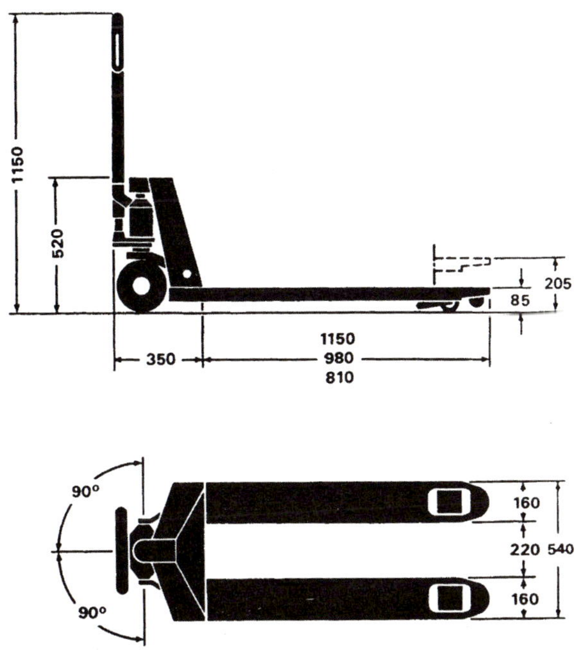

Fig. 18.16 Manual pallet-truck for internal transport

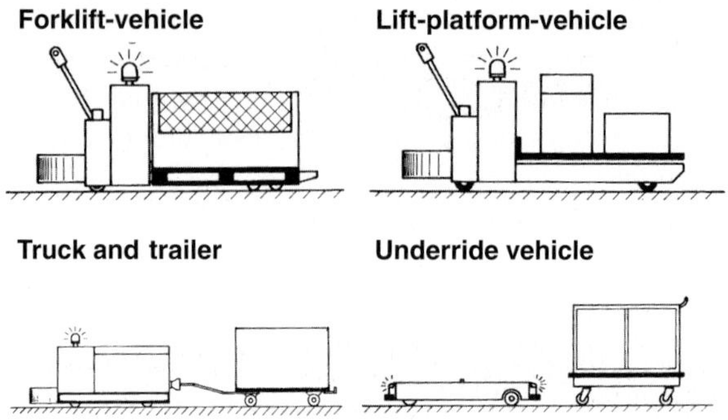

Fig. 18.17 Different types of automatic guided vehicles (AGV)

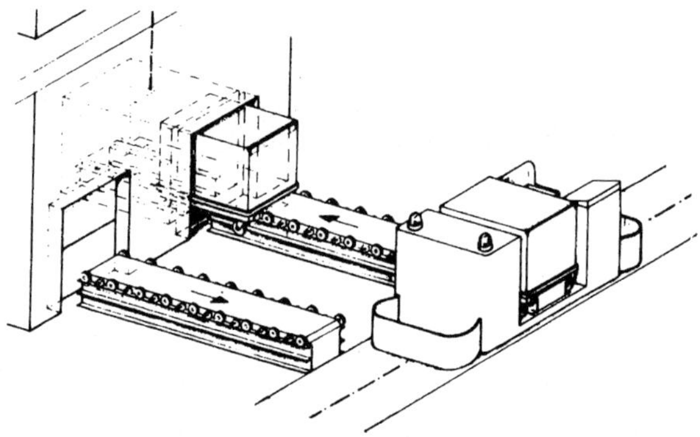

Fig. 18.18 Wheel conveyor transfer of roll-containers between automatic guided vehicle and lift station

roll-container between an AGV equipped with a wheel conveyor and stationary wheel conveyors leading to a lift station.

Automatic guided vehicles compete with manually operated vehicles, overhead trolleys and circular chain conveyors. They are especially efficient for internal transport over longer distances with changing relations and two or three shift operation.

Internal transport is also the main application for *tugs&tows* with driver, as shown in *Fig. 18.15*. They carry pallets and other loads in logistic centers and factories or baggage and freight in railway stations and airports over longer distances. For this purpose, the trucks and trailers are equipped with different loading devices.

Figure 18.19 shows some examples for *traffic means*:

road vehicle
rail-road train
container ship
freight airplane (18.19)

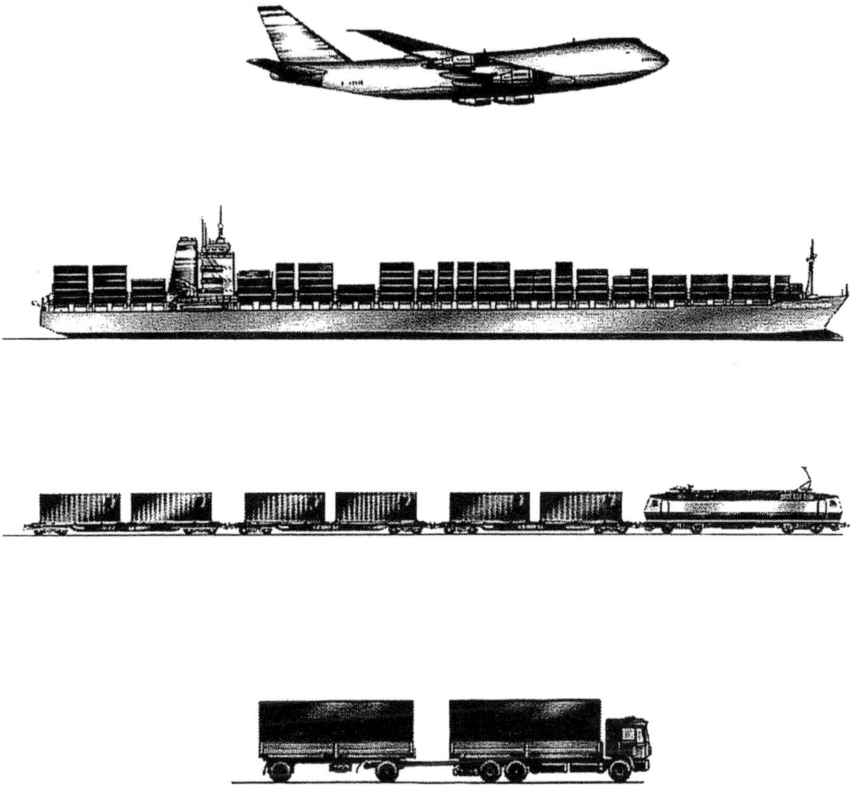

Fig. 18.19 Typical traffic means for road, rail, water and air

Traffic means are mainly used for external transport and operate in the *transport modes* of

road
rail
water (18.20)
air

Figure 18.20 presents an example of an automatic transfer station between an internal conveyor system and an external vehicle system. The vehicle is a semi-trailer truck equipped with chain conveyors for the automatic loading and unloading of pallets. Such a *transfer shuttle* is very efficient for the connection between a production site and a distant store for distances up to 100 km if operated in a multi-shift operation.

18.7.3 Track Guidance and Track Network

The track network, the traffic course and the transport strategies depend on the *track guidance*. The basic possibilities are:

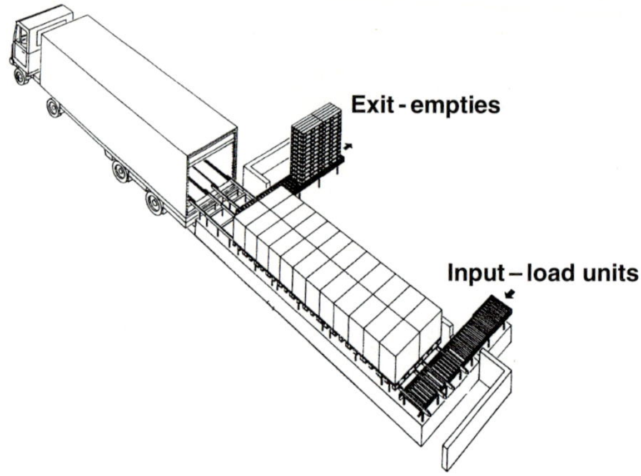

Fig. 18.20 Automatic pallet transfer station with shuttle truck

- *Free track guidance*: The drive way of the transport means can be chosen quite freely within a certain *traffic space*.
- *Fixed track guidance*: The drive way for the transport units is determined by *stationary tracks*.

The traffic space for *free track guidance* is a road, ramp or fairway, a land or water area or the air traffic space. It allows *lane changing* and *overtaking* within the given traffic space. Therefore, vehicles on the same track can drive with different speeds.

With exception of railways, the *public traffic networks*, i.e. the road nets, inland waterway nets, maritime transport nets and air traffic nets are transport networks with free track guidance. They are historically grown, influenced by industrial and population growth, traffic policy, infrastructure and urban planning. The planning and building of new public traffic ways takes many years and does not only depend on the traffic demand, but also on financial feasibility and political circumstances.

The private and commercial user has to take the public traffic network for granted. However, the area-wide traffic networks in industrialized countries offer a large choice of connections and routes. The linked road-, rail-, water- and air-networks around the globe enables the selection and configuration of flexible and temporal logistic networks that use only a small part of the public traffic nets. These are the tasks and options for *network management* and *route planning*.

With *fixed track guidance,* the vehicles or trains can only change tracks at the nodes. Between the nodes, they cannot overtake each other. Hence, the speeds of vehicles using the same track can differ only slightly. The tracks can be mono- or dual-run, guide rails, guiding wires or optical, acoustical or electronic track guidance. Examples for fixed track guidance are railway networks, overhead rails of trolley systems and track networks of automatic guided vehicles.

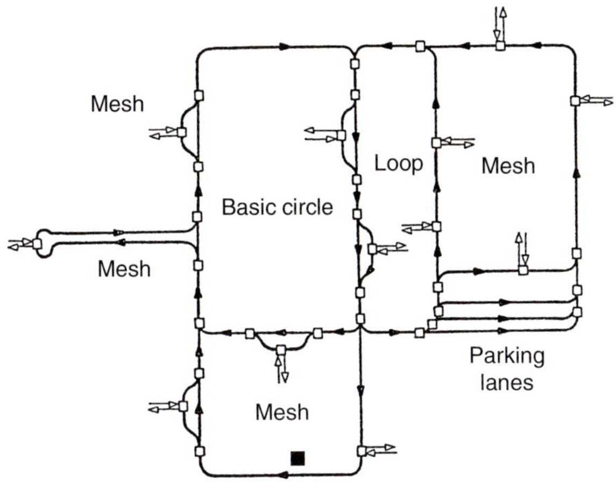

Fig. 18.21 Track Network of a Hospital Service System with AGV

Figure 18.21 shows the underground track network of a *general goods transport* system (GGT) in a *hospital*. On this network run AGV-vehicles as shown in *Fig. 18.18*. They carry roller containers filled with clothes, meals, pharmaceuticals and other material to the lifts to and from the stations on the upper floors of the hospital.

Nowadays, RFID-transponders replace the guiding wires for the AGV. They are installed in a grid of reference points in the ground of the traffic area and detect the positions of the vehicles. This enables *semi-free track guidance*. The central transport computer or the board computer of the vehicle calculates from the actual position and traffic situation, the shortest collision-free path.

The planning of internal vehicle systems, the routing and the design of the networks in plants and sites depend on the transport demand and the project conditions. They are tasks of logistics planning which have to be solved in accordance with factory and operation planning.

18.7.4 Drive Technique and Energy Supply

Nowadays transport units primarily use the following *drive technique*:

electric motors (AC or DC)
petrol, diesel or gas engines (18.21)
jet engines

Former drive techniques, such as wind power or steam engines, have been replaced more or less by diesel engines or electric motors. Other techniques are coming, such as *hybrid engines* for cars and *gyroscope engines* for busses.

Decision criteria for the drive technique of a transport mean are the consumption and costs of the energy resource, the driving power and travel speed, the acceleration and the energy supply. Basically different *energy supplies* of transport means are:

- *stationary energy supply* by a side or overhead conductor or bar or by a tow cable
- *mobile energy supply* from a fuel tank, electric battery or accumulator or by a mechanical energy store, such as a gyroscope or a spring

The advantage of a stationary energy supply is that the driving range of the transport means is unlimited. Their major disadvantages are fixed energy supply lines installed right next to the track. That means:

▸ *Stationary energy supply* enables continuous operation of the transport means but restricts them to fixed tracks.

Stationary energy supply is restricted to track guided vehicle systems. A mobile energy supply is prerequisite for free track guidance. It does not need fixed energy supply lines along the routes and offers more flexibility. However, the mobile energy supply has other *disadvantages*:

- The content of the tank or the energy storage limits the driving range of the transport means.
- Energy storage and energy resource require space and have dead weight, which reduce the load capacity and consume additional energy.
- Fueling or charging stations and energy sources must be installed near to the transport net to reload the transport means. These stations require their own *energy supply logistics*.
- Fueling or charging takes time and reduces the utilization and availability of the transport means.

The consequence is:

▸ *Mobile energy supply* offers higher flexibility of the transport means but limits driving range and availability.

The limited driving range and reduced availability are important *restrictions* for the planning and scheduling of vehicle systems. They also influence the vehicle demand and the routing.

18.8 Transport Matrix and Number of Transport Units

In order to execute a *freight demand* λ_{ij} [LU/h], appropriate *transport flows* λ_{TUij} [TU/h] have to run between the stations S_i and S_j. The *transport matrix* λ_{TUij} determines the *traffic load* of the tracks and nodes and the number of transport units necessary to cope with the freight demand. The transport flow from station S_i to S_j is the sum of a partly or completely *filled transport flow* $\lambda^F{}_{TUij}$ and of an *empty transport flow* $\lambda^E{}_{TUij}$:

$$\lambda_{TUij} = \lambda^F_{TUij} + \lambda^E_{TUij} \quad [TU/h] \tag{18.22}$$

The functional dependency of the filled and the empty transport flows on the freight demand and on the load capacity of the transport units is determined by the *transport strategies*. Explicit formulas for the filled transport flows exist only for basic dispatch strategies. The calculation of the empty flows is only possible for the strategies of single and combined runs. For other empty-run strategies and for system strategies only approximate formulas for the transport matrix and for the number of transport units are known. For complex transport networks and fast changing freight demand, the transport flows and the number of transport units can only be determined by *digital simulation* (see *Sect. 5.4*).

18.8.1 Filled Transport Flows

If each transport unit with load capacity C_{TU} [LU/TU] is fully loaded, it carries C_{TU} load units. This leads to the *1st transport flow law*:

- The *filled transport flow* for serving a freight demand λ_{ij} [LU/h] is *minimal* with 100% used capacity C_{TU} [LU/TU] of the transport units and given by

$$\lambda^F_{TUij} = \lambda_{ij}/C_{TU} \quad [TU/h] \tag{18.23}$$

100% utilization of the load capacity is achievable by maximal *load consolidation, optimal sequence dispatch* and *single-destination runs*. However, maximal load consolidation may cause long waiting times. To keep a *maximal waiting time* T_{Wmax} [h], transports must be executed with a *minimal transport frequency* $\nu_{Tmin} = 1/T_{Wmax}$ [1/h], even if the load for the same destination, which arrives during the maximal waiting time, does not fill up the capacity of a transport unit. A consequence is the *2nd transport flow law*:

- To keep a *maximal waiting time* T_{Wmax} [h] with *optimal sequence dispatch* and *single-destination runs* of transport units with capacity C_{TU} [LU/TU] the *filled transport flows* for serving a freight demand λ_{ij} [LU/h], must be

$$\lambda^F_{TUij} = \text{MAX}\,(\lambda_{ij}/C_{TU}\,;\,1/T_{Wmax}) \quad [TU/h]. \tag{18.24}$$

If the dispatch sequence is fixed by the arrival sequence, a transport unit can be filled for a single run only with load units, which arrive in sequence for the same destination. This number can be calculated for an incoming flow of load units with stochastically mixed destinations with the help of the *sequence probability* (13.46). From this results the *3rd transport flow law*:

- To keep a *maximal waiting time* T_{Wmax} [h] with *fixed sequence dispatch* and *single-destination runs* of transport units with capacity C_{TU} [LU/TU] for serving a freight demand λ_{ij} [LU/h] with stochastically mixed destinations, the *filled transport flows* must be

$$\lambda^F_{TUij} = \text{MAX}\big(((1 - \lambda_{ij}/\lambda_{I\,i})/(1 - (\lambda_{ij}/\lambda_{I\,i})^C)) \cdot \lambda_{ij}\,;\,1/T_{Wmax}\big) \quad [TU/h] \tag{18.25}$$

Here, $\lambda_{I\,i}$ is the total incoming flow (18.3) of station S_i for all destinations S_j. If the load capacity of the transport means is 1, the matrix of the filled transport flows is equal to the freight matrix, i.e. $\lambda^F{}_{TUij} = \lambda_{ij}$ for $C_{TU} = 1$ LU/TU. This holds true for all dispatch strategies.

The filled transport flows for *multi-destination runs* and for *runs with co-loading* are the sums of all required runs, which result from the *route scheduling* as explained in *Sect. 18.11*. Runs with co-loading are possible, if a partially loaded transport unit approaches a station where load units for the same destinations or direction are waiting.

Efficiency conditions for multi-destination and co-loading runs are:

- *Co-loading runs* and optimal sequence dispatch are efficient if the values of the resulting transport matrix are smaller than the values of the transport matrix (18.24) for runs without co-loading.
- *Multi-destination runs* and fixed sequence dispatch are efficient if the values of the resulting transport matrix are smaller than the values of the transport matrix (18.25) for single destination runs.

The *maximal transport flow* of filled transport units is given for optimal sequence dispatch by relation (18.24) and for fixed sequence dispatch by relation (18.25). Hence, if a vehicle system is dimensioned with the respective transport matrices (18.24) and (18.25) for the two dispatch strategies, it will be sufficient for the freight demand even if other transport strategies are applied, such as destination-mixed runs or runs with co-loading.

18.8.2 Transport Matrix for Single and Combined Runs

For single runs from station S_i to station S_j, the returning empty transport flow from S_j to S_i is equal to the filled transport flow from S_i to S_j:

$$\lambda^{E}_{TU\,ij} = \lambda^{F}_{TU\,ij}. \tag{18.26}$$

With this relation, from (18.22) follows the

- *transport matrix for single runs* with empty return

$$\lambda_{TU\,ij} = \lambda^{F}_{TU\,ij} + \lambda^{F}_{TUji}. \tag{18.27}$$

The transport flow for combined runs with *to and fro transports* between a station S_i and a station S_j is the maximum of the forward filled flow and the backward filled flow. This leads to the

- *transport matrix for combined runs* with optimal return

$$\lambda_{TU\,ij} = \mathrm{MAX}\,(\lambda^{F}_{TU\,ij};\quad \lambda^{F}_{TU\,ji}). \tag{18.28}$$

Due to the relations (18.23), (18.24) and (18.25), the filled transport flows in (18.27) and (18.28) and hence the total transport flows depend on the load capacity, the freight demand and the station strategy.

The *utilization of the transport units* is

$$\eta_{TUij} = \lambda_{ij}/(C_{TU} \cdot \lambda_{TU\,ij}). \tag{18.29}$$

It is higher for combined runs than for single runs. With $C_{TU} > 1$, the utilization is better for optimal series dispatch than for fixed series dispatch.

Figure 18.22 presents the dependency of the investment for vehicles of an AGV-system on the load capacity for optimal dispatch and for fixed series dispatch. The dependency results from the opposite effect of the number of vehicles and their utilization on the total investment. This example demonstrates the influence of the dispatch strategy and the existence of an *optimal vehicle capacity*. The optimal capacity depends on the specific project and differs for the two dispatch strategies (Gudehus 1978).

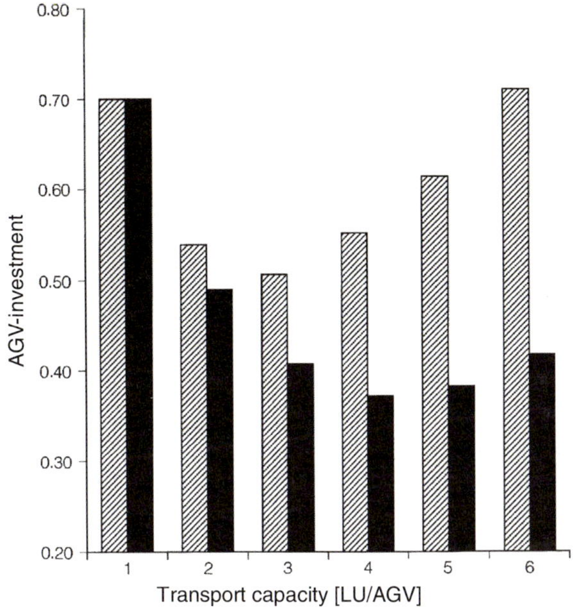

Fig. 18.22 Dependency of total vehicle investment on the load capacity of the AGV-vehicles
Grey bars: fixed sequence dispatch *Black bars*: optimal sequence dispatch

18.8.3 Optimal Flow Allocation for Single and Combined Runs

If the transport matrix (18.27) for single runs, or (18.28) for combined runs is known, the *traffic load* of a transport net results from the *optimal allocation* of the partial transport flows $\lambda_{TU\,ij}$ to the nodes and tracks of the network in a way that they flow on the shortest path from the start S_i to the destination S_j. If NE_k are the *network elements*, i.e. the transport elements and stations, with the partial functions $F_{k\alpha}$ spanning up the network, the optimal flow allocation results in the

- *flow allocation function*

$$\varepsilon_{ij\,k\alpha} = \begin{cases} 1 \text{ if } \lambda_{TU\,ij} \text{ passes element } NE_k \text{ in partial function } F_{k\alpha} \\ 0 \text{ if } \lambda_{TU\,ij} \text{ does not pass element } NE_k \text{ in partial function } F_{k\alpha} \end{cases} \quad (18.30)$$

From the flow allocation function result the

- *partial transport flows* through a network element NE_k using its partial function $F_{k\alpha}$

$$\lambda_{TE\,k\alpha} = \sum_i \sum_j \varepsilon_{ij\,k\alpha} \cdot \lambda_{TU\,ij} \quad (18.31)$$

The sums in (18.31) run over all stations S_i and S_j with shortest connections (18.7) which use the partial function $F_{k\alpha}$ of the network element NE_k.

To allocate the transport flows of a known transport matrix in the described way, one starts with a chart of the network with all stations and connection tracks. The shortest paths between the stations are selected, the transport flows are marked along these paths and all partial flows are added up finally. The result is a quantified map of the transport network with the *traffic load* on all tracks and through all nodes.

18.8.4 Optimal Flow Allocation for Other Strategies

If the complete transport matrix with all loaded and empty flows is unknown, as it is the case for co-loading and multi destinations runs, the traffic load can be determined starting with the filled transport flows (18.23), (18.24) or (18.25) respectively:

1. All filled transport flows between the stations are inserted into the network chart along the shortest paths.
2. For all stations with overflow of empty transport units, the surplus empty flow is directed on shortest paths to the closest stations with demand for empty transport units.

In this way, the empty vehicle demand of all stations S_i with negative *empty vehicles surplus flow*

$$\Delta\lambda^{E}_{TUi} = \sum_{j} (\lambda^{F}_{TUij} - \lambda^{F}_{TUji}) \quad (18.32)$$

is completely supplied from the next stations with positive empties vehicle surplus flow (18.32). The minimal traffic load on the tracks and nodes results, when the strategy of *minimization of empty runs* is consequently applied. For other empty run strategies, the traffic load is higher.

18.8.5 Transport and Forwarding Times

On its way from station S_i to station S_j a transport unit runs successively through N_{ij} network elements NE_k. The total path length l_{ij} is the sum (18.7) of the *passing lengths* $l_{k\alpha}$ through the partial functions $f_{k\alpha}$ of the elements NE_k.

The *passing time* through a steady connection of length l with the effective *travelling speed* v_{eff}, including all accelerations and de-accelerations, is:

$$t_P = l/v_{eff}. \quad (18.33)$$

The *passing time* for an unsteady connection and the *stop time* at a station with *run-in time* t_{in}, *transfer time* $t_{trans}(C)$ for C load units and *run-out time* t_{out} is:

$$t_P = t_{stop} = t_{in} + t_{trans}(C) + t_{out}. \quad (18.34)$$

With the passing times $t_{P\,k\alpha}$ through the network elements NE_k in the partial functions $F_{k\alpha}$, the mean *waiting times* $t_{Wk\alpha}$ in front of the NE_k and the flow allocation function (18.31) follows the

- *transport time* from station S_i to station S_j

$$T_{ij} = \sum_{k}\sum_{\alpha} \varepsilon_{ij\,k\alpha}\cdot(t_{P\,k\alpha} + t_{W\,k\alpha}). \quad (18.35)$$

The first element of the sum is the loading station and the last element is the unloading station. The mean waiting times $t_{W\,k\alpha}$ for given flows and limit performances can be calculated with the queuing formulas (13.67) of *Sect. 13.5*.

If n_{ij} is the *number of stops* on the total path of length l_{ij} and t_{stop} is the *mean stop time* at the intermediate stations, the minimal transport time from S_i to S_j without waiting times is:

$$T_{ij} \approx n_{stop} \cdot t_{stop} + l_{ij} \,/\, v_{eff}. \tag{18.36}$$

The actual transport times exceed the minimal transport times (18.36) by the sum of the waiting times. The elongation of the transport times by the waiting times can be taken into account by a *waiting time factor* $f_{wait}(N_{TE})$ which is 1 for only one transport unit in the network and increases with increasing number of travelling transport units N_{TE} (see e.g. *Fig. 18.23*).

With the partial transport flows $\lambda^F{}_{TU\,ij}$ and the sum of the filled flows $\lambda^F{}_{TU}$ follows from the transport times (18.35) or (18.36) the *mean transport time* of the filled transport units:

$$T^F = \sum_i \sum_j (\lambda^F_{TUij}/\lambda^F_{TU}) \cdot T_{ij}. \tag{18.37}$$

The mean transport time of the loaded transport units determines the *freight forwarding times*, which are critical objectives for many projects. The mean transport time can be minimized by insertion of *bypasses* or *overtaking lanes* or by increasing the travelling speed.

18.8.6 Transport Unit Calculation by Allocation Method

If the transport matrix $\lambda_{TU\,ij}$ and the transport times T_{ij} [h] are known, the required number of transport units can be directly calculated. For a transport flow $\lambda_{TU\,ij}$, the number of transport units travelling from station S_i to station S_j is

$$N_{TUij} = \lambda_{TU\,ij} \cdot T_{ij} \quad [TU]. \tag{18.38}$$

The sum for all transport relations leads to the

► *actual transport unit demand* for a transport matrix $\lambda_{TU\,ij}$ and transport times T_{ij}

$$N_{TU} = \sum_i \sum_j \lambda_{TU\,ij} \cdot T_{ij} = N_{TU\,P} + N_{TU\,W} \quad [TU] \tag{18.39}$$

The transport unit demand is, corresponding to the composition of the transport times (18.35), a sum of $N_{TU\,P}$ progressing units and of $N_{TU\,W}$ of waiting units (see e.g. *Fig. 18.25*).

For transport units with an availability $\eta_{ava} < 1$ which is caused by technical failures, fuelling or battery charging, the

► *effective transport unit demand is*

$$N_{TU\,eff} = N_{TU}/\eta_{ava} \quad [TU]. \tag{18.40}$$

For instance, if due to the charging times, the availability of electric vehicles is only 85%, the number of vehicles increases by 18%. However, if the vehicles are

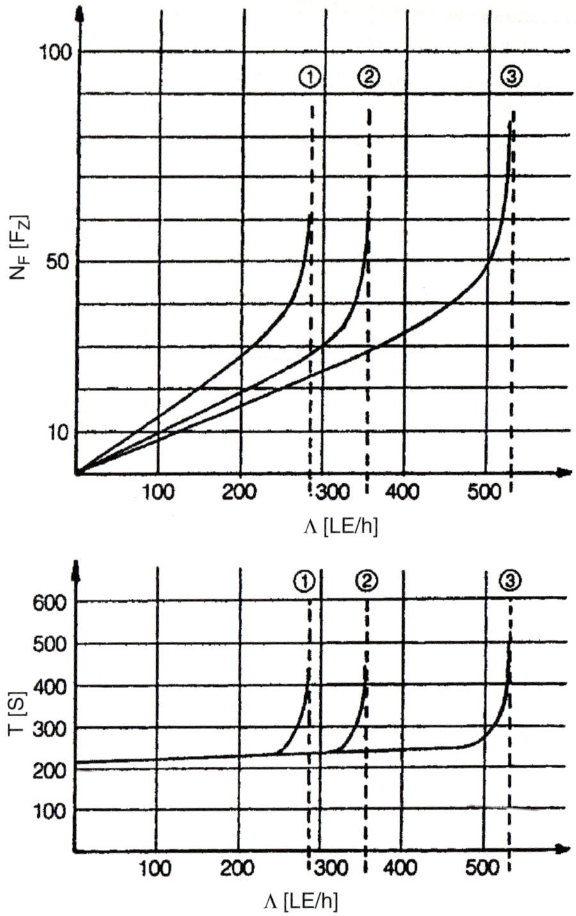

Fig. 18.23 Dependence of the number of vehicles N_V and the mean transport time T on the total freight demand for different empty run strategies

Curve 1: single runs with empty returns
Curve 2: combined runs between stations
Curve 3: optimal tours with minimal empty runs

charged outside of the regular operation times or during periods of low utilization, the availability increases to the technical availability of e.g. 98% and the vehicle demand is lowered by 15%.

18.9 Transport Unit Calculation by Roundtrip Method

The *roundtrip time* T_{RT} [PE/RT] of transport units running from station S_0 back to S_0 along a roundtrip of length l_{RT} with effective travelling speed v_{eff}, n_{stop} intermediate stops and a mean stop time t_{stop} is given by formula (18.36) with $i = j = 0$. In intralogistics, it is called *cycle time* and measured in *seconds* [PE = s] (see *Chap. 7*). In extralogistics, the roundtrip times are generally called *rotation* or *circulation time*. For cars, trucks, trains and airplanes they are normally measured in *hours* [PE = h] and for *sea ships* in *days* [PE = d].

The number of roundtrips, which a vehicle with availability η_{ava} can achieve per period [PE = h, d, week], is the *effective roundtrip limit performance*:

$$\mu_{RTeff} = \eta_{ava} / T_{RT} \quad [RT/PE]. \tag{18.41}$$

It determines the *transport unit demand for roundtrips*:

- The required number of transport units with roundtrip limit performance μ_{RTeff} for a *roundtrip demand* λ_{RT} is

$$N_{TU\,eff} = \{ \lambda_{RT}/\mu_{RTeff} \} \quad [TU]. \tag{18.42}$$

The curly brackets indicating the rounding up to the next integer are only necessary if all transport units must be back in the same start position at the end of the operation period PE. Otherwise they can be omitted.

In intralogistics, the formulas (18.36), (18.41) and (18.42) can be used to calculate the number of forklift trucks, vehicles and storage devices (see *Chap. 16 and 17*). The same formulas are used in extralogistics for the calculation of the number of transport units for systems with *time table operation* (see below).

18.10 Designing and Dimensioning Vehicle Systems

Analytical methods for designing, dimensioning and optimizing transport systems are hardly known. For planning transport systems, often *digital simulation* is recommended. As explained in *Sect. 5.4.3*, digital simulation is useful to examine and improve function and performance of existing systems. However, it is no method to find new solutions. Up to now, not much research has been conducted into the effects of *network parameters*, *vehicle parameters* and *transport strategies* on performance, vehicle demand and operating costs of transport systems.

The *network parameters* for design, dimensioning and optimization of vehicle systems are:

main lines and cross lines
primary and secondary rings
meshes and loops
short cuts
parallel routes
overtaking tracks
counter and bypass rings
branch and bypass lines
positioning of stations
driving direction

(18.43)

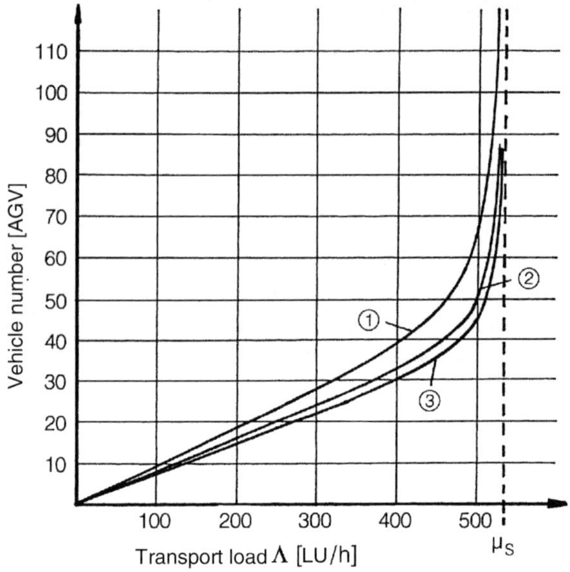

Fig. 18.24 Dependence of the number of vehicles N_V on the total freight demand for different transport networks

Curve 1: initial network without short cuts and with online stations; $L_{TN} = 640$ m
Curve 2: improved network with short cuts and online stations, $L_{TN} = 820$ m
Curve 3: further improved network with short cuts and offline stations, $L_{TN} = 800$ m

The main *vehicle parameters* are

driving speed
acceleration
de-acceleration (18.44)
load capacity
load transfer time

The vehicle parameters are partly interdependent: The speed of vehicles with low capacity is normally higher than for vehicles with high capacity. The load transfer time increases with the load capacity if the transfer technique does not allow simultaneous loading and unloading of several load units.

Two basic analytical methods for designing, dimensioning and optimizing vehicle systems are known: the *ring network method* for ring networks as shown see *Figs. 18.1B, 18.4* and *18.21* and the *line network method* for line networks as shown in *Figs. 18.1A, 18.2A, 18.3B* and *18.3*. These basic methods can be combined for ring-line networks as shown in *Fig. 18.2C*.

For a given freight demand, the stations are optimally located at the sources and sinks with high freight demand or close to the *transport gravity centers* of areas with several sources and sinks with low freight demand (see *Sect. 18.10* and *Chap. 19*).

18.10.1 Ring Network Method

The *ring network method* is applicable for the design of internal vehicle networks and for the route design for public transport services, such as bus ring routes. Its iterative *design steps* are:

1. Stations with permanent freight relation are connected by a minimal number of shortest transport rings considering the spatial restrictions.
2. The driving direction in the transport rings is determined so that the circulating vehicles can execute all freight orders on shortest routes.
3. For the resulting *initial solution*, the transport matrix is calculated using the formulas (18.22) to (18.28) for all adequate transport strategies from the freight matrix with an assumed initial capacity of the transport units.
4. In order to calculate the minimal traffic load on the tracks and nodes, the transport flows are optimally allocated to the tracks and nodes of the transport network.
5. The limit performances and the buffer capacities of tracks and transport nodes are assessed for the calculated traffic load by a *function- and performance-analysis* as explained in *Sect. 13.7*.
6. Bottlenecks and unacceptable queuing effects are eliminated by increasing the limit performance, by deviations or bypasses, or by altered dispatch times (see *Chap. 13*).
7. The transport times are assessed with the relations (18.35) or (18.36), taking into account traffic flow and waiting times.
8. If for a relation $S_i \rightarrow S_j$ the transport time is longer than tolerable, short cuts or bypasses are inserted in the network.
9. The actual and the effective vehicle demand are calculated with formula (18.39) and (18.40) respectively.
10. The vehicle demand is minimized by variation of the network parameters (18.43), e.g. by addition or reduction of shortcuts or overtaking lanes, by changing the driving direction or by shifting of stations.
11. The vehicle demand and the operating costs are further optimized by iterative variation of the vehicle parameters (18.44), for instance by changing the capacity, the travel speed and the load transfer time.

The *Figs. 17.21* and *18.21* show the results of the application of the ring network method for two different AGV-transport networks.

18.10.2 Line Network Method

The *line network method* is applicable for external transport networks, especially for the dimensioning of railway lines and line traffic networks. The design steps of this method are:

1. Pairs of far distant stations with intensive freight relation are taken as preliminary end stations and connected by *main lines* with pairs of tracks for opposite traffic.

2. As far as possible without too much elongation, the main lines are laid along all stations which are located between the pairs of the preliminary end stations.
3. The main lines are eventually extended beyond the preliminary end stations to outer stations with lower freight demand, which now become the final end stations.
4. Further stations which are still unlinked are connected one after the other to the main lines by a minimal number of *feeder lines* or *branch lines*. The feeder lines end at a main line, the branch lines cross the main lines. Both are connected with the main lines by transport nodes or transfer stations.
5. The feeder and branch lines are arranged and extended until all remaining stations are connected and all required freight relations are realized.
6. The traffic flow through the tracks and nodes is derived by optimal allocation of the filled transport flows (18.24) which have been calculated from the freight matrix with a preliminary transport capacity C_{TU}.
7. The minimal necessary transport frequency on a main, feeder or branch line is equal to the frequency of the maximal traffic flow on this line. By using the minimal frequencies, the required number of transport units or trains on the main lines and feeder lines are calculated with the formulas (18.41) and (18.42) from the respective roundtrip times.

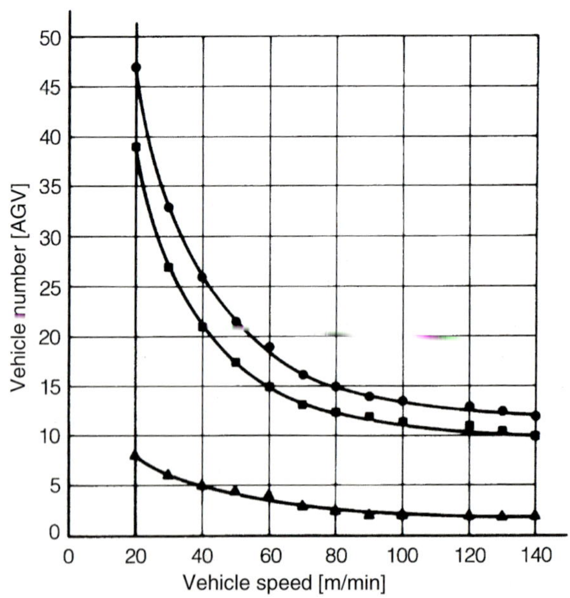

Fig. 18.25 Dependency of the vehicle demand of the overhead trolley system and the vehicle speed

Circles: total number of vehicles
Square: proceeding vehicles
Triangles: waiting vehicles

8. The travel times for the line sections and the transport times between the stations are calculated with the above formulas.
9. If the line system should be operated on a fixed timetable, the departure times from the end stations of the main line are adjusted in accordance with the daily freight demand in order to minimize the waiting and forwarding times.
10. The departure times from the end stations of feeder and branch lines are synchronized with the arrival times of the main line transports at the transport nodes and transfer stations in order to minimize the transfer times.
11. The transport unit demand is minimized by successive variation of the network parameters, for instance by adding direct connections, by reversion of the driving direction on a line or by shifting the stations.
12. By iterative variation of the vehicle parameters, the vehicle demand and the operating costs are further reduced, e.g. by different capacity, travel speed or load transfer times.

The line network method can be improved by additional steps. It has been applied, e.g. to design a network of European piggy-back transports of road semi-trailers, and to plan a distribution network for newly built cars using existing roads, railways and shipping lines.

The above steps of the line and ring network methods describe only roughly the proceedings and the possibilities of network layout. It is a challenging task for *logistic research* to improve these methods and to develop further layout and allocation rules.

18.10.3 Software Tools for Dimensioning Vehicle Systems

The above formulas and algorithms can be used for programming software tools for designing and dimensioning vehicle systems. The inputs of such *vehicle system dimensioning tools* are the freight matrix and the given locations of the stations. The limit performances and the current cost data are provided or calculated for the available transport units and network elements by standard program modules. The program modules can be connected according to the considered network.

The program modules for the network elements include the above formulas for the partial limit performances, throughput times, queue lengths and waiting times for the different dispatch strategies. The user starts with the program module for a central track or node to set up the network program and adds the corresponding program modules for the directly connected tracks and elements and so on. After completing the network program, the partial traffic flows using the partial functions of the network element are inserted.

The finished network program calculates for different input values the limit performances, the throughput times, the mean waiting and transport times, the total number of moving and waiting vehicles, the investment and the operating costs. The consequences of a change of the network or a change of design parameters on the limit performance of the system, the vehicle demand, the costs and on other target values is immediately visible and can be used for the next optimization step.

A vehicle system dimensioning tool of this kind has been used for example to optimize an overhead electric trolley system with 35 stations and an initial network length of 526 m. This system was designed to connect the goods entry with an automatic high bay store and with the commissioning system for palletized good in the European logistic center of a leading *mail order company*. *Figs. 18.23, 18.24* and *18.25* present some of the calculated dependencies of vehicle demand and mean transport time on the freight demand, transport strategies, network structure and travel speed.

For the optimal transport strategy with minimal empty runs, the program calculated for the initial network a demand of 28 trolleys with a speed of 42 m/min. This number was validated by *digital simulation*, which required 29 trolleys. The number of trolleys could be reduced by systematic network improvements from initially 29 to 22. The network length was shortened from 526 to 417 m. Another vehicle system that has been designed and optimized by a similar tool is the AGV-system shown in *Fig. 17.21* for the internal transports in a logistic center of a retail company.

18.11 Optimal Logistic Location

The optimal location of a new logistic station or factory site within a given service area depends on many influence factors: functions of the station, extension of the area, distribution of the sources and sinks, freight flows from the sources and to the sinks, external transport networks, environmental requirements and the general goals of the company (Domschke and Drexel 1990).

The dominating decision factor for the location of a storage center, transshipment point, logistic center and another logistic station are the costs. That means:

- The *optimal logistic location* for serving a given area is the place with the minimal sum of internal operating costs and external transport costs.

An extended area can be served by one or by several logistic stations. In the following, the *one-station problem* for a given area is solved. The *many-stations problem* and the *optimal area division* will be discussed in *Chap. 21*.

The internal operating costs of a logistic station or a factory do not depend as much on the geographic location of the site as the transport costs. As shown in *Fig. 18.27* , the transport costs vary with the location far more in the outer part than in the center of the service area. Due to these dependencies, the optimal logistic location for a given service area with known freight flows can be found quite fast by the following *location optimization procedure*:

1. With the inward and outward freight flows and the locations of the sources and sinks the coordinates of the *transport gravity center* are calculated as shown later.
2. For the resulting location, the external *transport costs* are determined either by model calculations with own transport means or by requests from freight forwarders.
3. The *transport-independent location factors* are evaluated for the location of the transport gravity center and the *internal operating costs* are calculated with the local real estate prices, personal costs and other local cost factors.

4. The *optimal logistic location* is determined by systematic variation of the location in the neighborhood of the transport gravity center and repetition of steps 2 and 3 until the sum of operating and transport costs is minimal and all other criteria are fulfilled.

The advantage of this quite simple procedure, compared with more sophisticated OR methods, is that the consequences for the different influence factors are transparent in each step (Domschke and Drexel 1990). Since many influence factors, e.g. the future freight demand, are not exactly known, more sophisticated solution methods for the optimal location do not necessarily achieve better solutions.

18.11.1 Transport-independent Location Factors

Transport-independent location factors are all requirements, costs, prices and parameters except the external transport costs, which affect the location decision. Such *general location factors* are:

- *land and buildings*: availability and prices of land, buildings, halls and surrounding traffic areas; building and site development potentials; construction regulations.
- *traffic connections*: access to main traffic roads; rail connection; proximity to harbor or airport; driving restrictions, e.g. ban on night driving of trucks.
- *staff and personnel*: local potential to engage adequate work force; salaries, tariffs and wages; vacation regulations; sickness statistics.
- *frame conditions*: public subsidies; taxes and fees; tax allowances and privileges; working time restrictions; approval times.

Missing the minimal requirements of one of these location factors is a *KO-criterium*, also for the transport optimal location. In most cases, additional project specific factors must be considered, e.g. the availability of local freight forwarders or of a qualified logistic service provider for the erection and operation of a logistic station.

18.11.2 Determination of Transport Optimal Location

The costs for the in- and outflow transports are determined by the lengths of the driveways between the sources and the sinks, the freight flows, the capacities of the transport units, the required transport times and by the transport strategies.

The *transport strategy* depends on the size of the freight orders in relation to the load capacity. *Full truck load* (FTL) transports run directly from a single source to the logistic center and from there to a single sink with the option of combined *to and fro runs*. *Less than truck load* (LTL) transports are *collection tours*, *dispatch tours* or *combined* collection and dispatch *tours* as shown in *Fig. 19.5*.

The precise determination of the transport optimal location would require planning, calculating and comparing the optimal pickup and delivery routes between sources, logistic station and sinks. However, since the future freight demand is not exactly known, a precise *route planning*, in order to determine the transport optimal logistic location, is neither adequate nor necessary.

For this purpose, it is sufficient to minimize the weighted mean transport distance for direct runs between the sources and sinks S_i with coordinates (x_i, y_i) and freight demand λ_i [LU/PE] and the logistic station S with the coordinates $(x_s;y_s)$. If $d((x_i;y_i),(x_s;y_s))$ is the driving distance between a sink or source and the logistic station the *mean transport distance* is given by:

$$d_m(x_s;y_s) = \sum_i \lambda_i \cdot d((x_i;y_i),(x_s;y_s)) / \sum_j \lambda_j \tag{18.45}$$

The mean single transport distance in a dense traffic network can be calculated by the approximate formula:

$$d((x_i;y_i),(x_s;y_s)) = f_{dev} \cdot \sqrt{(x_i - x_s)^2 + (y_i - y_s)^2}. \tag{18.46}$$

The *deviation factor* f_{dev} takes into account the mean deviation of the real driveway length from the Euclidian distance. Within a rectangular traffic network, in the best case with linear connection the deviation factor is 1, and in the worst case with rectangular way around corners it is 2. The mean value of these extremes is $f_{dev} = (1.0 + 1.41)/2 = 1.21$. This value complies quite well with the mean deviation factor of the German road network $f_{dev} = 1.23$. The deviation of the single drive way lengths calculated with relation (18.46) using a deviation factor $f_{dev} = 1.21$ from the real length is on average less than 9%.

After inserting (18.46) into (18.45) and equaling the partial differentiation of relation (18.45) with respect to the coordinates x_s and y_s to zero result as solutions of these equations the coordinates of the *transport optimal location*:

$$\begin{aligned} x_s &= \sum_i \lambda_i \cdot x_i / \sqrt{(x_i - x_s)^2 + (y_i - y_s)^2} \Big/ \sum_j \lambda_j / \sqrt{(x_j - x_s)^2 + (y_j - y_s)^2} \\ y_s &= \sum_i \lambda_i \cdot y_i / \sqrt{(x_i - x_s)^2 + (y_i - y_s)^2} \Big/ \sum_j \lambda_j / \sqrt{(x_j - x_s)^2 + (y_j - y_s)^2} \end{aligned} \tag{18.47}$$

Since the denominator in these equations depends on the sought coordinates $(x_s;y_s)$, the optimal coordinates can only be determined by an iterative calculation. The initial solution of the iteration are the coordinates of the *transport gravity center:*

$$\begin{aligned} x_s &= \sum_i \lambda_i \cdot x_i / \sum_j \lambda_j \\ y_s &= \sum_i \lambda_i \cdot y_i / \sum_j \lambda_j \end{aligned} \tag{18.48}$$

The so-called *Miehle-method*, i.e., inserting the initial solution (18.48) into (18.47), gives the second approximation of the transport optimal coordinates and so on (Domschke and Drexel 1990). Model calculations for areas with more than 10 sources and sinks show that the exact solutions deviate only marginally from the transport gravity center. This leads to *optimal location rule*:

- If the freight demand is not exactly known the transport gravity center can be taken for the optimal transport location of a single logistic station.

From relation (18.48) with a freight demand proportional to the population, results that the *transport gravity center of Germany* is located within the city square *Bad Hersfeld-Eisenach-Fulda-Meiningen*. The mean transport distance to the transport optimal location of Germany is 280 km.

Figure 18.27 shows that the mean transport distance (18.45) is only slightly changing for locations in the neighborhood of the optimal location. The dependency on the location increases for deviations of the logistic station from the optimum above 50 km. The small dependency of the mean transport distance on the exact location of the logistic station in the neighborhood of the optimum justifies the described approximation method.

18.11.3 Area-Dependency of Mean Transport Distance

For the calculation of the transport and freight cost rates for a given service area and for the optimization of logistic structures, the explicit dependency of the mean transport distance on the area A must be known. For a circular area with radius R and area $A_s = \pi \cdot R^2$, the mean transport distance of equally spread sources and sinks from the center is given by:

$$d_m(A_s) = f_{dev} \cdot 2/3 \cdot \sqrt{A_s/\pi} = 0.46 \cdot \sqrt{A_s} \qquad (18.49)$$

This is a good *approximate formula* for the mean transport distance from the transport gravity center into a service areas A_s with different shapes as shown e.g. in *Fig. 18.26*, provided they are compact and the distribution of sinks and sources is sufficiently dense and equal. For example, for Germany with an area 358,000 km^2

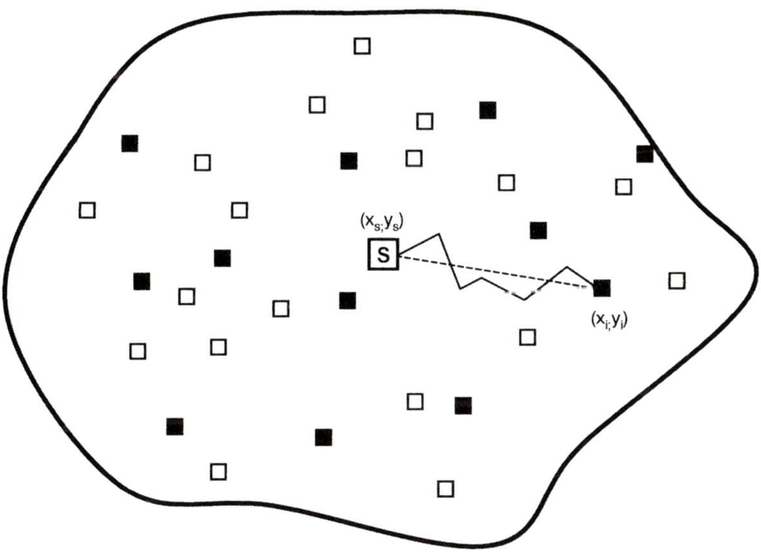

Fig. 18.26 Service area around a logistic station S

■ sources, pickup places □ sinks, delivery places

---- linear distance — driveway

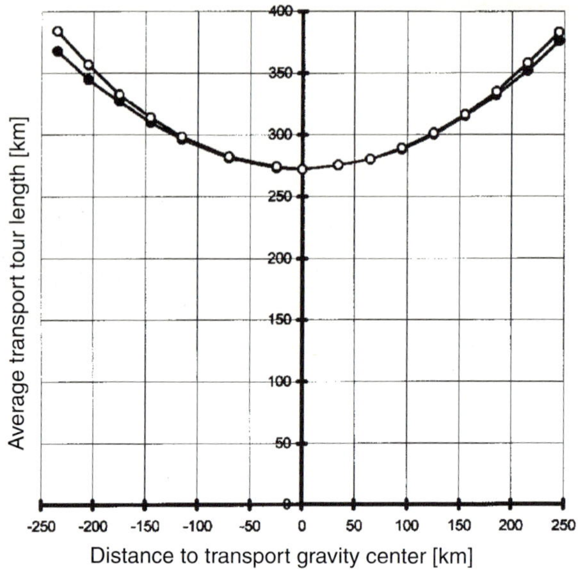

Fig. 18.27 Dependency of the mean transport distance on the deviation from the optimal location

Dots: east-west distance to the optimal location
Circles: north-south distance to the optimal location
Service area: Germany with 358,000 km^2
Freight demand: proportional to population

and a deviation factor $f_{\text{dev}} = 1.23$, results with formula (18.49) a mean transport distance of 277 km to the optimal logistic location, which differs only slightly from the accurately calculated mean distance of 280 km.

18.12 Tour Scheduling

In order to fulfill the freight demand of sources and sinks within a service area as shown e.g. in *Fig. 18.26* from a logistic station S at lowest costs, pickup and delivery tours with minimal number of vehicles and shortest operation time have to be scheduled. In addition, *tour scheduling* or *route scheduling* has to take into account the following *transport restrictions*:

- *Transport capacity*: The load capacity of the available vehicles limits the freight quantities along the runs.
- *Freight goods*: The properties of the freight units restrict the loading possibilities or require a specific loading sequence.
- *Driving times*: The driving time per route should not exceed the allowed driving time for the drivers.
- *Pickup and delivery times*: Prescribed start, pickup, delivery and arrival times must be kept within agreed time windows.

- *Travel speed*: The effective travel speed should take into account speed limits and waiting times caused by high traffic.

In most cases, it is impossible to solve the *tour scheduling problem* with stringent restrictions for many sources and sinks exactly within limited time. However, *Operations Research* has developed different *heuristics* to find approximate solutions for this problem, which are used in many *tour scheduling programs* (Domschke 1985; Gilbert and Miller 1974; Goszyla and Cietlatkowski 1995; Laporte 1992; Lawler et al. 1985). These programs generate within relatively short time quite good results for problems with not too stringent time restrictions. Other routing programs combine OR-search algorithms with *analytical construction methods* for the initial solution.

For routing problems with many transport restrictions, however, most of the standard tour scheduling programs deliver poor results compared with tours scheduled by an experienced dispatcher. Some *dispatchers* refuse to accept the results of a tour scheduling program since they do not understand the algorithms and the mathematical optimization. Others are disappointed that a tour scheduling program does not solve *all* their real-life problems. Generally holds:

- Even the best tour scheduling program needs supervision, assessment and, in exceptional situations, corrections by an experienced dispatcher.

A tour scheduling program can calculate the required number of vehicles and the mean driveway for a given freight demand and service area. However, for the analytical planning and optimization of logistic structures and processes and for the calculation of use-related transport and freight cost rates, the dependency of the *mean vehicle numbers* and the *mean driveways* on the freight demand and the size of the area must be known explicitly.

The mean values and their explicit dependencies can be calculated for solutions generated by *analytical methods* such as the *rotary beam method* for the routing combined with the *segment strategy* for transport paths. These analytical methods result from the basic strategies of logistics, i.e. *clustering*, *sequencing* and *securing* (see *Sect. 5.2)*, and will be described in the following subsections. They give insights to the problems and limitations of route scheduling and path optimization. Further results of the analytical methods are *benchmarks* for the assessment of tour scheduling software.

18.12.1 Analytical Route Scheduling

Full-load orders of sources and sinks can be separately scheduled and executed in direct single or combined runs. Without full load orders, route scheduling is reduced to the generation of optimal *pickup and delivery tours* between sources and sinks with *part-load freight*. This task can be solved after finding a suitable *initial solution*, which can be improved by heuristic methods, in order to fulfill the restrictions.

Initial solutions can be constructed by *cluster strategies*: A number of neighboring sources and sinks with a freight demand which sums up to the transport capacity of one vehicle are clustered and served separately by optimal transport tours. However, there exist many possibilities to select and separate the sources and sinks into

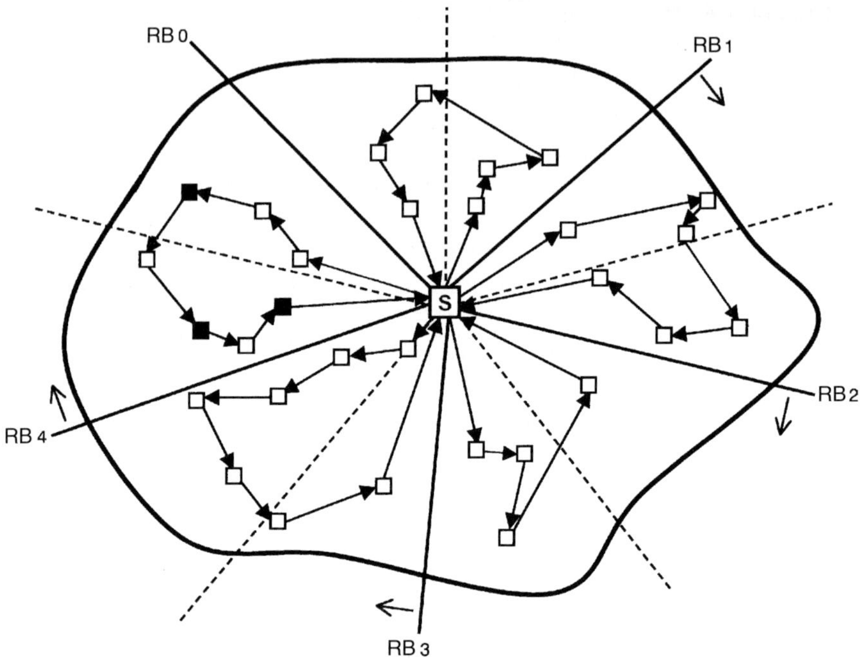

Fig. 18.28 Route scheduling by the rotary beam method and travelling path selection by the segment strategy

□ delivery locations ■ pickup locations
- - - - segment separation
—— RBx rotary beams
→ travelling routes (round trips)

clusters. One quite efficient clustering procedure is the *sweep algorithm* (Gilbert and Miller 1974), another is the *rotary beam method* shown in *Fig. 18.28* (Gudehus 1999/2007).

The steps of the rotary beam method are:

1. Without cutting sinks or sources, a basic beam B0 is drawn from the logistic station through the logistic area.
2. Starting at the position of the basic beam, a first beam B1 is turned clockwise around the logistic station until the *delivery demand* of all *sinks* in the segment between beam B0 and beam B1 for the scheduling period, e.g. for 6 or 12 h, fills between 80% and 90% of the capacity of a first vehicle.
3. Starting from the end position of the first beam, a second beam B2 is turned until the capacity of a second vehicle is filled between 80% and 90%.
4. The last step is repeated with the third, the second and up to the Nth beam until the freight demand of the remaining sinks is smaller than the sum of the unused

capacities of the preceding vehicles. By stepwise adjustment of the N beams, the left over demand is equally distributed to the N vehicles.

5. For the single vehicles, the optimal travelling path to the sinks in the different segments is determined and the roundtrip time is calculated.
6. If the resulting roundtrip time of a vehicle in one segment is longer than permitted, some outer sinks can be attached to neighbored segments where a vehicle is not fully utilized.
7. After the optimal delivery tours have been scheduled, all *sources*, which are located on the way, are visited, if their *pickup demand* fits into the emptied transport capacity.
8. As far as possible, the freight of the remaining sources is allocated to emptied full-load transports on their way back to the logistic station.
9. For the still remaining un-served sources separate roundtrips are scheduled following the steps 1 to 4.

The resulting roundtrips of the rotary beam method or another cluster strategy can be successively improved, e.g. by exchanging predecessor and successor on a route or by transferring stops to neighboring trips, until no further reduction of the vehicles and the travel times is achieved.

18.12.2 Travelling Path Optimization

Travelling path optimization assumes that the capacity of the vehicle is sufficient to serve all destinations in one tour. This condition can be fulfilled by the preceeding route scheduling. To keep the required pickup and delivery time windows and further restrictions is more difficult. With the exception of the full enumeration method, this task is generally postponed until the optimal travelling path without restrictions has been found.

Depending on the objectives, the optimal travelling path is the route with the shortest *length*, the shortest driving *time* or the lowest travelling *costs*. The task to find the *optimal travelling path* for a roundtrip starting and ending at point S_o and visiting n destinations S_i is the classical *travelling salesman problem* of *Operation Research* (Churchman 1961; Domschke 1985, 1995; Laporte 1992; Lawler et al. 19985; Müller-Merbach 1970). For a small number of destinations the optimal travelling path can be found by *full enumeration*:

- For each of the n! roundtrips $S_o \rightarrow S_i \rightarrow S_j \ldots \rightarrow S_r \rightarrow S_o$, which result from permutation of the n destinations S_i, the compliance with the restrictions is examined. An unacceptable path is eliminated. For an acceptable path the travelling length, the driving time or the driving costs are calculated and compared with the best values of the previous paths. By this way, the optimal travelling path is finally left.

As shown in *Fig. 5.3*, the number of travelling paths increases faster than exponential with the number n of destinations. Therefore, by full enumeration the travelling salesman problem cannot be solved within limited time even by powerful computers. Heuristic OR methods aim for approximate solutions for large n which can be achieved within short processing time and deviate only slightly from the optimal solution. Many OR methods, e.g. the *saving method* or the *circle method*, start with an *opening strategy* to construct an *initial solution* which complies with the restrictions and is iteratively improved by different methods (Gilbert and Miller 1974; Vahrenkamp et al. 2000). An example for an opening strategy is the *strategy of the next neighbor*:

- Starting from point S_0, always the station with the shortest distance, shortest driving time or lowest driving costs from the last station is visited next.

In many cases the strategy of the next neighbor generates tours ending somewhere in the service area with long ways back to the starting point. This is avoided by the *segment strategy*[1]:

- By an even number of separation beams the *tour area* is divided into two, four or more segments. Starting at point S_0 the destinations in the first segment are successively visited in the outgoing direction, the destinations of the second segment are visited in the backward direction and so forth until the tour comes back to the start point S_0.

Model calculations as shown in *Figs. 18.30* and *18.31* lead to the *segment scheduling rule*:

▸ Two segments are optimal to serve less than 30 destinations, four segments for more than 30 destinations.

Figure 18.28 shows the roundtrips resulting with a 2-segment strategy applied to the segments of the rotary beam method.

18.12.3 Fulfillment of Time Restrictions

Preserving the scheduled roundtrips as far as possible, many time restrictions, such as required pickup or delivery windows, can be fulfilled by the following *adaptation measures*:

- changing the driving direction of a roundtrip
- adjustment of the start-time of a roundtrip
- advancing the visit of time-critical destinations
- postponing the visit of non-time-critical destinations
- visiting of time-critical destinations of the return trip on the outward trip
- division of the tour in two consecutive roundtrips with shorter driving time
- express tours with smaller vehicles to the most urgent destinations

In ascending order, these measures cause additional efforts and higher costs. Generally holds the *time restriction rule*:

---

[1]The segment strategy for the driveway optimization is an application of the *Miebach-stripe-strategy* described in *subsection 17.6.3* [Miebach 1971].

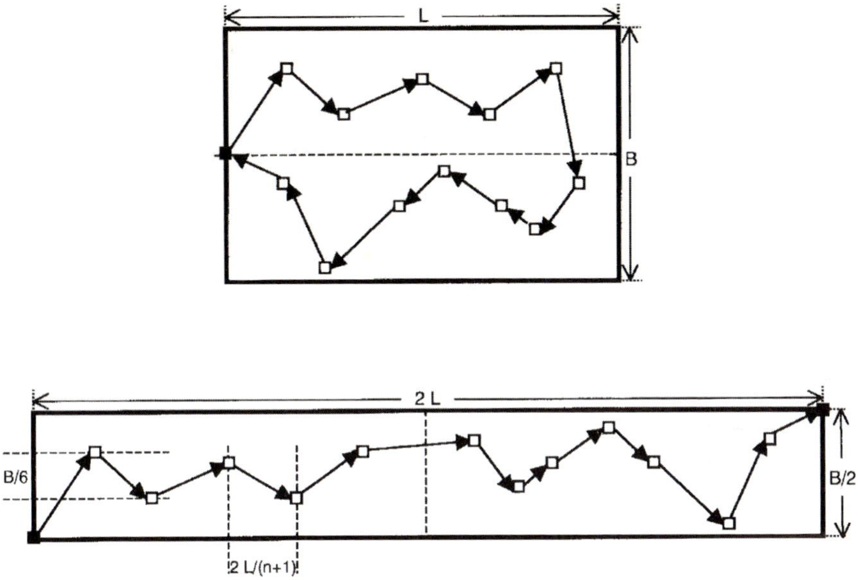

Fig. 18.29 Calculation of the mean route length for the 2-segment strategy for a rectangular tour area
Above: original area *Below*: equivalent area

- The effort to keep time restrictions increases with the number of destinations with narrow time windows for pickup or delivery.

Even the best transport optimization software cannot cope with all possible time requirements and restrictions. For instance, three remote destinations which must be visited *just-in-time* exactly at the same date can only be served in separate runs by three different cars.

18.12.4 Mean Route Length

The mean length of roundtrips executed due to the N-segment strategy in a rectangular area $A = L \cdot B$ of length L and breadth B as shown in the upper *Fig. 18.29* with equally distributed destinations can be calculated explicitly. When the N rectangular stripes of length L and breadth B are unfolded as shown in the lower *Fig. 18.29*, the tour becomes a longitudinal trip in one long stripe of length $N \cdot L$ and breadth B/N. With deviation factor f_{dev} and equally distributed destinations, the mean length of this trip is

$$l_{RT}(n) = f_{dev} \cdot (n+1) \cdot \sqrt{(N \cdot L/(n+1))^2 + (B/3\,N)^2} \tag{18.50}$$

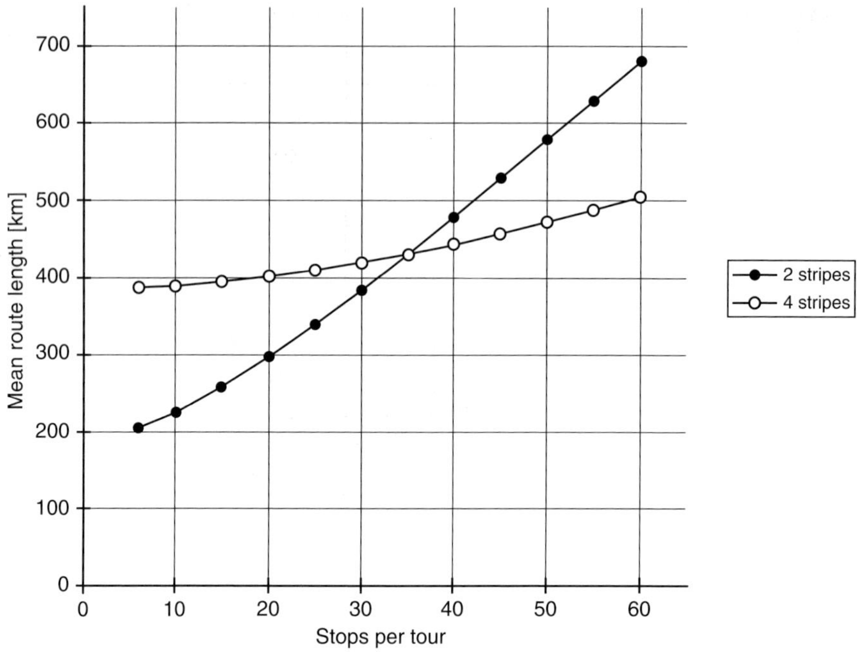

Fig. 18.30 Dependency of the mean route length on the number of stops per roundtrip
Parameter: number of stripes of the travelling strategy
Rectangular tour area: A = 4,000 km^2
Shape factor: f_{shape} = L/B = 1.5
Deviation factor: f_{dev} = 1.24

Figure 18.30 presents the dependency of the mean route length (18.50) on the number n of stops per roundtrip for N = 2 and for N = 4 stripes. This demonstrates the *1st roundtrip rule*:

- For small numbers n of stops per roundtrip, the mean route length increases less than proportional with n; for higher numbers it increases more and more proportional to n.

With the *area* A = L·B and the *shape factor* f_{shape} = L/B, relation (18.50) can be converted into the *approximate route length formula*:

$$l_{RT}(n; A) \approx f_{dev} \cdot \sqrt{A} \cdot \sqrt{f_{shape} \cdot N^2 + (n+1)^2/(9 \cdot f_{shape} \cdot N^2)}. \quad (18.51)$$

Figure 18.31 shows the area dependency of the mean route length (18.51) for three different shape factors. This leads to the *2nd roundtrip rule*:

- The mean route length increases proportional to the square-root of the service area and depends only slightly on the shape of the area.

Due to the small influence of the shape, the route length formula (18.51) holds approximately also for roundtrips in compact service areas of different shapes and

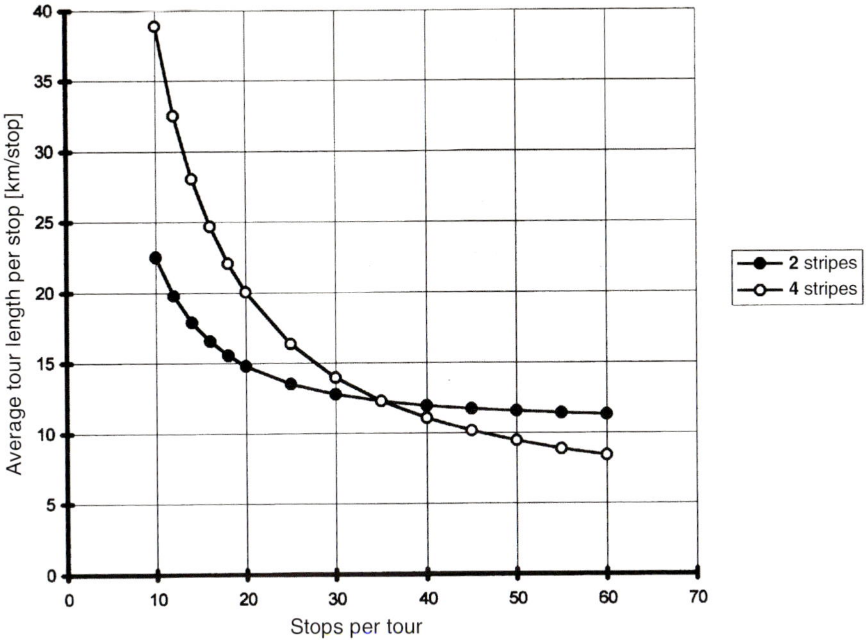

Fig. 18.31 Dependency of the mean route length on size and shape of the service area
Shape factor: $f_{shape} = L/B = 1/\,2/\,4$
Stops per route: $n = 20$
other parameters see *Fig. 18.30*

for the segments of the N-segment strategy. The shape factor for non-rectangular areas is the relation between the maximal length and the mean breadth of the relevant service area.

The division of (18.50) through n gives the *mean travelling length per stop*:

$$s_{stop}(n) = l_{RT}/n = (f_{dev}/n) \cdot \sqrt{(N \cdot L)^2 + ((n+1)(B/3N)^2}. \qquad (18.52)$$

Figure 18.32 shows the dependency (18.52) of the mean travelling length per stop on the number of stops.

Formula (18.52) leads to the *3rd roundtrip rule*:

- The mean travelling length per stop initially decreases rapidly with the number of stops and approaches an asymptotic value for higher numbers.

Figure 18.30 and *18.31* demonstrate the above rule, that with $N = 4$ stripes or segments, the mean travelling lengths per route and per stop are improved compared with $N = 2$ stripes or segments if the number of stops exceeds 30. By further analytical considerations and test simulations result the *master formulas for tour lengths* (Gudehus 2006; somewhat different Daganzo 1999):

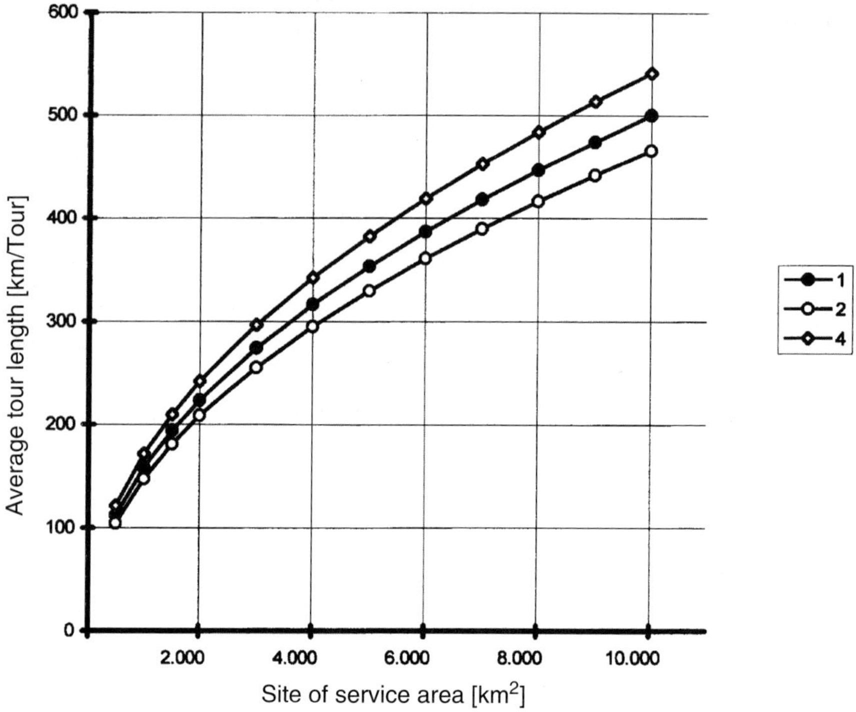

Fig. 18.32 Dependency of the mean travelling length per stop on the number of stops per roundtrip

Parameters: see Fig. 18.30

- The *mean length of optimal n-stop tours* starting and ending in a *distance* d_S from a logistic station and serving an area A with *deviation factor* f_{dev} is

$$l_{RT}(n) \approx 2 \cdot d_s + f_{dev} \cdot (n - 1) \cdot \sqrt{A/n}. \tag{18.53}$$

- The *mean travelling distance between successive stops* of optimal n-stop round tours in a service area A with *deviation factor* f_{dev} is

$$s_{stop} \approx f_{dev} \cdot \sqrt{A/n}. \tag{18.54}$$

These approximate calculation formulas can be used to optimize *freight networks*. They can also be applied to calculate *use-related freight rates*. The explicit calculability of the mean tour is a major advantage of the analytical approach compared to heuristic OR-methods, where many simulations are necessary to determine the mean route length.

18.12.5 Vehicle Demand

The vehicle demand equals the number of simultaneous roundtrips, which are necessary to execute all freight orders keeping time restrictions and other conditions.

If the tours are scheduled due to the rotary beam method, the maximal vehicle demand for a service area is equal to the number of segments. If the driving times of two or more routes given by the relations (18.36) and (18.56) are much shorter than the permitted driving time, one vehicle can execute several roundtrips successively. This reduces the required number of vehicles.

The rotary beam method in combination with the segment strategy can not only be used for tour scheduling, but also for the medium-term transport planning, for the development of fixed route plans and for the approximate calculation of the vehicle demand. For these applications, vehicle demand and tour length are calculated for the mean daily freight demand and a mean distribution of sinks and sources. Qualified route scheduling software for the daily business should be able to generate vehicle numbers and driving times below the benchmark values of the rotary beam method.

18.12.6 Dynamic Tour Scheduling

In order to achieve efficient roundtrips with many stops of highly utilized vehicles of large capacity, batches of freight orders must be collected for a *scheduling period* of sufficient length. *Batch processing* of the freight orders is in effect *periodical scheduling* as described in *Sect. 10.6*. Disadvantages of periodical scheduling are *order waiting times* and *freight forwarding times* which increase with the length of the scheduling period. The goal conflict between effective tours and short forwarding times can be solved only partially by restriction of batch processing to the majority of *standard orders* with longer forwarding times and uncritical time restrictions.

A minority of express orders and special orders with time restrictions, which should not exceed 10%, is scheduled *event-dynamically* and executed with first priority. Whenever an urgent pickup order arrives, the dispatcher or the program checks whether it can be executed by one of the vehicles on a regular tour. Urgent delivery orders are attached to the first vehicle serving the respective area. If by this means the required times cannot be kept, the orders are executed on *extra tours* by express vehicles with smaller capacity. It is an interesting task for *logistic research* to develop further *self-regulating strategies* for *dynamic transport scheduling* (Böhnlein 2008).

18.13 Transport Costs and Pricing

The *operating costs* K_{TS} [€/PE] of a transport system are the sum of the *network operating costs* K_{TN} for the transport network and the *transport unit costs* K_{TU} for the vehicles executing the freight demand:

$$K_{TS} = K_{TN} + K_{TU} \tag{18.55}$$

The main *cost drivers* of a transport system are the transport units, the transport lengths and the number of stops. As explained in *Chap. 6* the transport prices can be calculated from the transport costs rates, i.e. from the partial operating costs divided by the relevant cost driver.

The transport costs and prices together with the capacity and filling degree of the transport units determine the *freight costs*. Their calculation and influence factors will be explained in the *Sects. 21.13, 21.14* and *21.15*.

18.13.1 Network Costs and Network Pricing

The network operating costs consist of:

- depreciations and interests for the network investment
- energy costs for the network operation
- personnel costs for the network operating staff
- network maintenance and repair costs
- costs for traffic control and network safety
- management costs for set up and administration of the network

The network investment depends on the *total network length* $L_{TN} = \Sigma l_{k\alpha}$ and on the network elements TE_k with the partial passing lengths $l_{k\alpha}$ and functions $F_{k\alpha}$. The installed partial limit performances $\mu_{k\alpha}$ of the network elements must be sufficient for the maximal traffic load $\lambda_{k\alpha\,max}$ during the peak operating times.

The division of the *network operating costs* $K_{TN} = K_{TN}(\lambda_{k\alpha};l_{k\alpha})$ [€/PE] by the total traffic load $\Sigma\lambda_{k\alpha}\cdot l_{k\alpha}$ [TU-km/PE] gives the undifferentiated *network cost rate*:

$$k_{TN} = K_{TN} / \sum \lambda_{k\alpha} \cdot l_{k\alpha} \quad [e/\text{TU-km}] \tag{18.56}$$

Except for the costs of wearing, maintenance and repair, the network costs are independent of the traffic load and mainly *fix-costs*. Therefore, the owner or operator of a transport network bears a high *utilization risk*.

If the users are not the owners of the network, such as the participants of a public transport network, they must be charged for the operating costs by *network fees*. The network fees are calculated from the network costs for a planned utilization of the network including an allowance for underutilization. Possibilities to charge the user of a transport network are:

- *direct time-dependent network pricing* per vehicle by a tax, fee or ticket for a certain period, independent of the utilization
- *direct utilization-related network pricing* by tolls, road charges or utilization fees, which depend on driving length and vehicle size and are cashed before or after utilization
- *indirect network pricing* by fees or taxes charged via fuel or energy price

In practice, the different cost pricing possibilities are often combined. Energy consumption pricing is in accordance with the principle of *cost-based and use-related pricing* of *Sect. 7.1*, as fuel or energy consumption are approximately proportional to driving frequency, load capacity and drive length, which are the main cost drivers for wear and tear, repair and maintenance of the network. Further advantages of indirect energy pricing are the low administrative costs since no local toll stations are necessary.

Direct pricing allows the imposition of higher network fees during periods of high utilization and of lower fees during lower utilization in order to force the traffic from rush hours to weak hours. However, this involves unwanted social consequences, for example also workers who cannot avoid driving during rush hours are charged more. Railway network fees, which depend on the network utilization, would be lower for frequently used main lines and higher for less frequented secondary lines. The price difference would shift traffic from the secondary lines to the main lines and further reduce the already low utilization of the secondary lines. A further increase of the network fees for a secondary line due to the reduced utilization starts a vicious circle, which ends with closing down the secondary line. These examples demonstrate some of the still unsolved *problems of network pricing*.

18.13.2 Transport Unit Costs

With the exception of the costs for bins, pallets or containers, the total operating costs of *conveyor systems* are equal to the network operating costs, whereas for *vehicle systems* the transport unit costs contribute considerably to the total transport costs. For all transport systems the *costs for the load carriers* must be taken into account separately as their allocation depends on the ownership. They are only part of the transport costs, if the operator of the transport units is also owner of the load carriers.

The transport unit costs consist of:

- depreciation and interests for transport units
- energy and resource costs for vehicle operation
- personnel costs for the vehicle crew
- maintenance and repair costs for the vehicles
- control costs for vehicle scheduling, control and checkup
- fleet management costs for planning, procuring and administration

Number, type and capacity of the transport units determine investment and interest, and the costs for control and fleet management. The personnel costs depend on the deployment and the staffing of the vehicles. The depreciation for wear and tear, the fuel and energy costs and the maintenance and repair costs vary proportionally with the driving performance.

18.13.3 Transport Cost Rates and Pricing

In practice, many different calculation methods of transport cost rates are in use. Correspondingly, the pricing, which has developed historically in many branches, differs for the transport modes (18.20) and traffic means (18.19). As explained in *Chaps. 6* and *7*, *use-related cost rates* and *cost-based prices* are *self-regulating* and fair for customers and for the operator of a transport system. To calculate them, the total transport operating costs (18.55) are divided up into a sum of partial costs depending on the different cost drivers. The main cost drivers of transport tours are the *basic activities* for empty approach, loading and unloading, the intermediate *stops* and the total *driveway*. The corresponding *transport cost rates* are:

$$\begin{array}{lll} \textit{basic costs} & k_B & [\text{€/TU - run}] \\ \textit{stop costs} & k_{ST} & [\text{€/TU - stop}] \\ \textit{driveway costs} & k_{DW} & [\text{€/TU - km}] \end{array} \tag{18.57}$$

With these cost rates, the *transport costs for a trip* of *driveway length* l_{DW} [km] with n_{ST} stops are:

$$K_{TR}(l_{DW}, n_{ST}) = k_B + n_{ST} \cdot k_{ST} + l_{DW} \cdot k_{DW} \; [\text{€}/TU - trip]. \tag{18.58}$$

With relation (7.1) the corresponding *transport performance prices* can be calculated from the cost rates (18.57) taking into account the required contributions for sales, administration and profit and an allowance for the risks of underutilization. Since a high share of the transport unit costs is variable, the operator of a transport fleet bears a smaller *utilization risk* than the network operator.

For calculating the effects of different influence factors on the transport costs, a *transport cost calculation tool* can be programmed using the above formulas for tour lengths and vehicle demand. With such a program the dependency of the transport costs and cost rates shown in *Figs. 18.33* to *18.37* have been calculated for the *semi-road-trailer* of *Figs. 12.5* and *18.20*.

Figure 18.33 shows that the driveway cost rate k_{DW} [€/TU-km] decreases with increasing travel speed. This demonstrates the significant influence of *speed limits* and *queues* on the transport costs.

From *Fig. 18.34* can be seen how much the *fuel-price* affects the driveway costs. A duplication of the diesel price from 0.75 to 1.50 €/l causes an increase of the driveway costs by 16% from 1.25 to 1.45 €/km.

Figure 18.35 shows the linear dependency of the relation price of a shuttle transport on the distance, calculated with formula (18.58). The relation price depends quite sensitively on the share of *empty runs* of the shuttle trips.

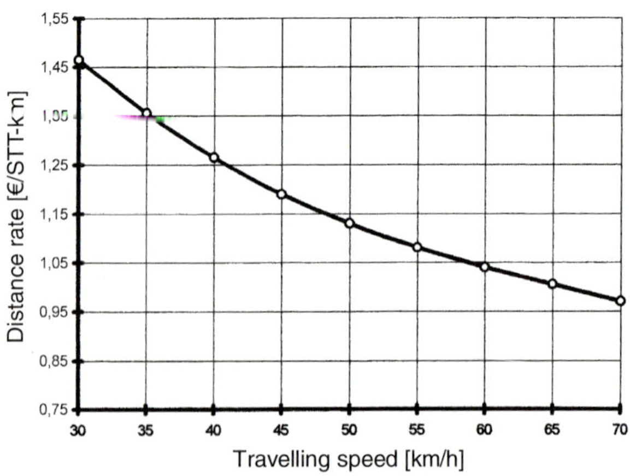

Fig. 18.33 Dependency of the driveway cost rate on the travel speed for a semi-road-trailer

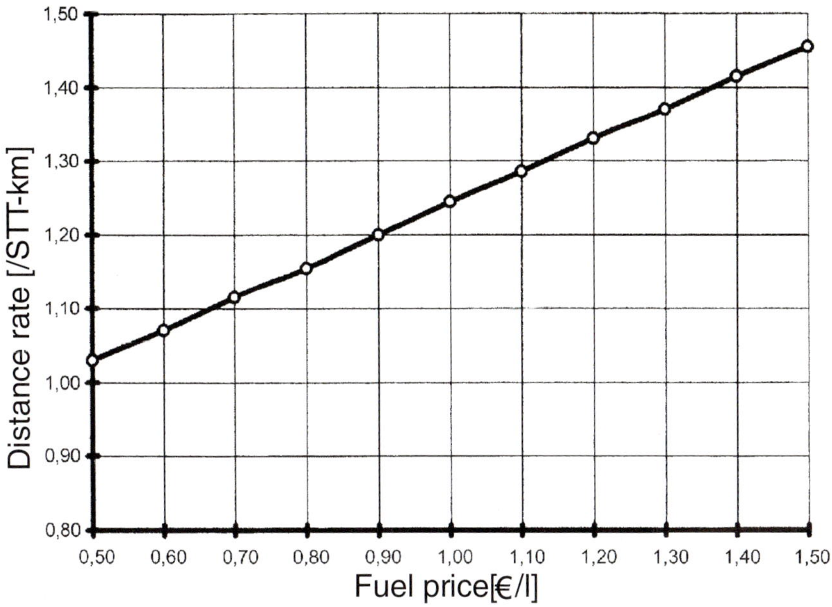

Fig. 18.34 Effect of the fuel price on the driveway cost rate of a semi-road-trailer

18.13.4 Freight Parity

The utilization of a transport system and the costs depend on the parity of the freight flows between the stations. The *freight parity* of the relation $S_i \rightarrow S_j$ is $\lambda_{ji}/\lambda_{ij}$, if the freight flow λ_{ij} from S_i to S_j exceeds the return flow λ_{ji} from S_j to S_j, i.e. for $\lambda_{ij} > \lambda_{ji}$, and $\lambda_{ij}/\lambda_{ji}$ for $\lambda_{ji} > \lambda_{ij}$. Hence, the general definition is:

- The *relation parity* of the freight flows between two stations S_i and S_j is

$$\eta_{\text{pair ij}} = \text{MIN}(\lambda_{ji}/\lambda_{ij};\ \lambda_{ji}/\lambda_{ji}) \quad [\%\ ] \tag{18.59}$$

The parity is 0%, if the freight flows only in one direction, and 100%, if the flows in both directions are equal. The parity is 50% when the return flow is half of the outgoing flow, i.e. for $\lambda_{ji} = \lambda_{ij}/2$. A relation parity $\eta_{\text{pair ij}}$ for shuttle trips leads to the *share of empty runs*:

$$\eta_{\text{ER ij}} = (1 - \eta_{\text{pair ij}})/2 \quad [\%]. \tag{18.60}$$

For instance, with parity 50% the share of empty runs is 25%.

For transport systems with more than two stations and total freight flow

$$\lambda = \sum \sum \lambda_{ij} \quad [\text{LU/PE}] \tag{18.61}$$

the weighted sum of all relation parities is the *system parity*:

$$\eta_{\text{TSpair}} = \sum \sum (\lambda_{ij}/\lambda) \cdot \text{MIN}(\lambda_{ji}/\lambda_{ij};\ \lambda_{ij}/\ \lambda_{ji}) \quad [\%\ ]. \tag{18.62}$$

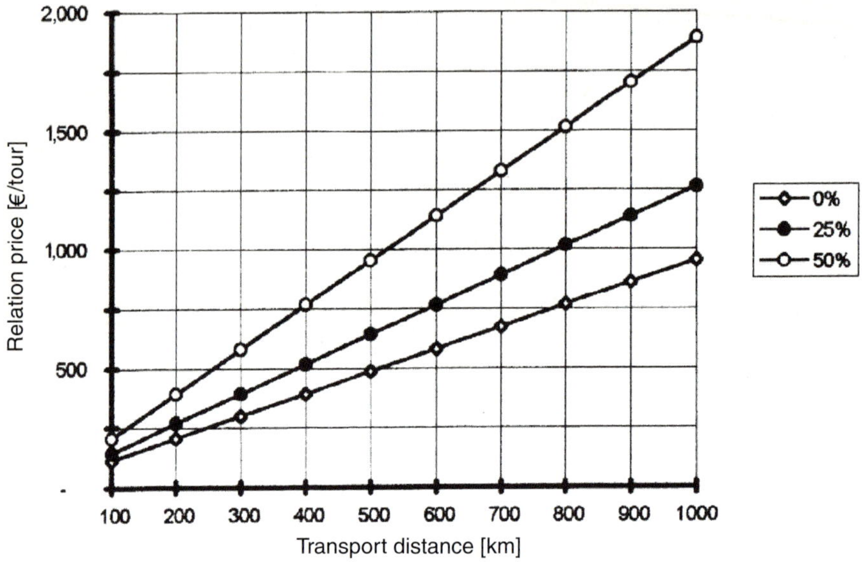

Fig. 18.35 Distance-dependency of relation prices for shuttle transports by a semi-road-trailer

Parameter: share of empty runs of the shuttle trips

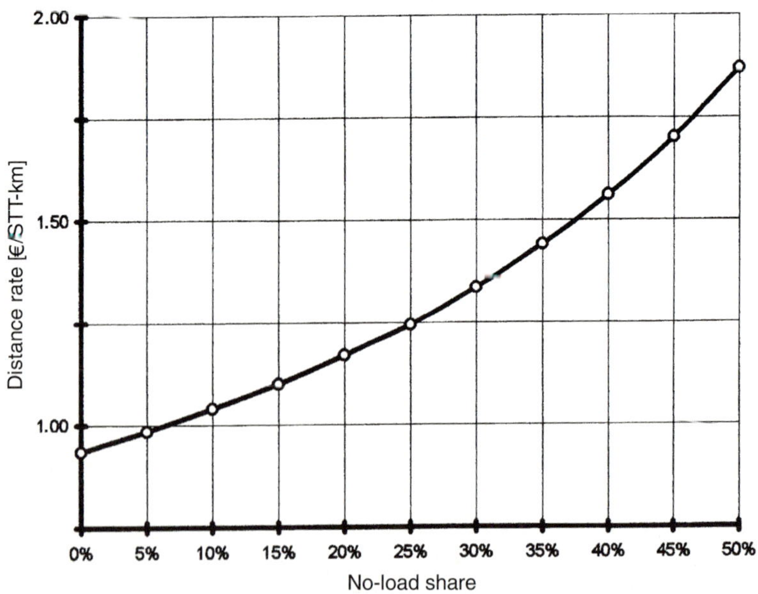

Fig. 18.36 Dependency of the transport cost rate on the share of empty runs

When the sums are restricted to stations in two separate areas, formula (18.62) gives the *area parity* of the freight flows between these areas.

The freight parity is important for the *network design* and the operating strategies of a transport system. If the parity is high and if the freight flow between two stations is sufficient to fill shuttle vehicles, a direct transport connection with outward and return runs is most effective. For low parity and small freight flows between the stations, a network with *ring structure* and roundtrips or a network with *star structure* and broken transport are better solutions (see *Figs. 18.1, 18.21, 19.15* and *19.16*).

The driveway cost rate for the fully used outward runs increases when the return runs are not fully used. For instance, the effective driveway cost rate for 50% empty runs is twice as high as for 0% empty runs. This dependency shows the huge saving potentials by improving the parity. The transport costs between areas with a balanced outward and return freight flows are far lower than between areas with unbalanced freight flows. However, due to the *logic of the market*, the transport prices for the direction with small freight demand are considerably lower than for the direction with high demand (Gudehus 2007).

18.13.5 Distance Cost Rates

Corresponding to relation (18.56) distance cost rates $k_{TR} = K_{TR}(l_{DW})/l_{DW}$ [€/TU-km] are the total transport costs $K_{TR}(l_{DW})$ related only to the transport distance l_{DW}. All-inclusive *distance cost rates* are quite common, although they do not take into account the basic costs for empty approach, loading, unloading and further stops.

However, as shown in *Fig. 18.37* the distance cost rate itself depends on the travelling distance. It decreases with increasing distance as the included basic and stop costs become less important. Therefore, distance cost rates are not in accordance with the principle of cost-based and use-related prices (see *Sect. 7.1*).

18.14 Transport and Traffic

Transport and traffic are two different aspects of the same task. Transport is the *micro-aspect*, traffic the *macro-aspect* of the movement of goods and persons. Hence, the tasks and goals of *transport* and *traffic* differ and are partly incompatible (Ihde 1991).

18.14.1 Transport

Transports of goods and persons are executed due to the orders of individuals, companies and other actors of the society. Hence, transport is a special area of *micrologistics*.

Transport management cares for the individual transports between shippers and receivers. It organizes the transport networks of manufacturing and retail companies and of logistic service providers. Objects of *transport technique* are the transport means, the devices for loading and unloading, the transport tracks and the transport process control.

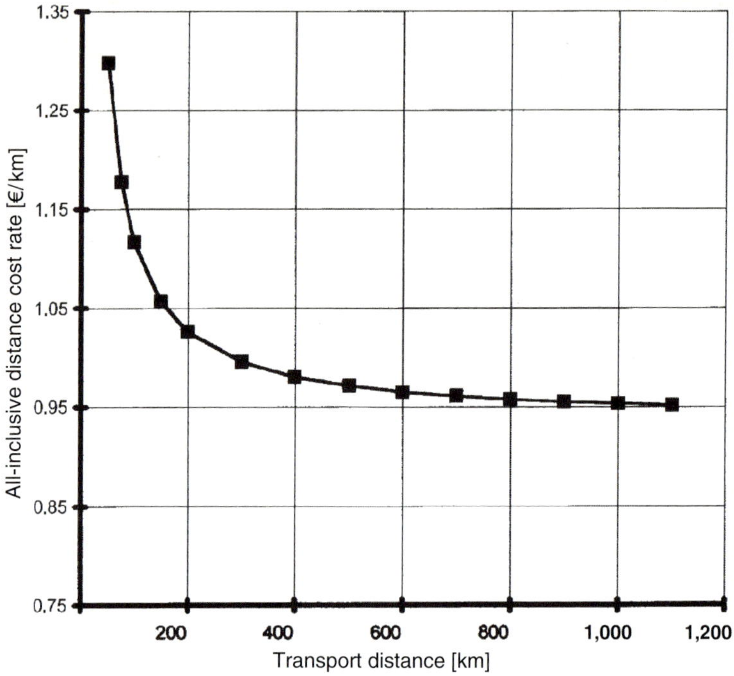

Fig. 18.37 Dependency of the all-inclusive distance cost rate on the transport distance for transports by a semi-road-trailer

Capacity utilization 100%

The main goal of the *transport business* is to fulfill transport and freight orders of people and companies efficiently, safely and reliably. With the aim to reach this objective, *transport economics* investigates the freight processes, the transport chains, and the costs and prices for transport and freight orders.

18.14.2 Traffic

Traffic is a special area of *macrologistics*. *Traffic scientists* investigate the *traffic flows* of goods, persons and transport means between anonymous sources and sinks in a region, of a country or around the globe, which are the sum of all single transport flows between households, companies and other actors of an economy.

Topics of *traffic technique* are development, planning, construction and realization of traffic routes, traffic networks and public transport systems, traffic control, and traffic safety.

Traffic management and *traffic politics* plan and initiate the building of new traffic networks. They care for safe, fast and efficient traffic flows through existing networks. Their goal is to enable and secure the disturbance-free and environmentally safe fulfillment of the transport demand of a region or a country at lowest costs.

Traffic economics deals with the economic aspects of public transport systems and traffic networks. Traffic economists investigate the structure of traffic networks, routes and nodes and the crossings between different *transport modes*. They study costs, prices and pricing-models and the competitors on *transport markets*. Further topics are the structure and directions of traffic flows, the development of traffic within and between regions and countries, the causes of *traffic emergence* and the possibilities of *traffic restriction*.

18.14.3 Goal Conflicts Between Transport and Traffic

Transport and traffic are interdependent:

- Prerequisites for efficient transports between the actors of an economy are safe traffic networks and public transport systems with sufficient capacities, demand-driven traffic control, and cost-based and use-related pricing.
- Prerequisites for investments in traffic networks and public transport systems and for their economic operation are adequate traffic flows respectively sufficient user frequencies.

From the different tasks and the partially deviating interests of the participants result the following *goal conflicts between transport and traffic*:

- *Transport business* cares for the goals of single companies and traffic participants, even if they are inconsistent with the goals of the society.
- *Traffic economy* aims at safe and economic utilization of traffic networks and public transport in the interest of all companies, traffic participants and society, even if some individuals or groups are put at a disadvantage.

These conflicts lead to challenging *tasks for logistic research*:

- ▶ analysis of the organizational, technical and economic options of action to achieve the different goals of transport and traffic
- ▶ development of methods to solve current and future tasks and problems of transport and freight
- ▶ investigation of the transport markets and freight markets, the competitors and the pricing for transport and networks
- ▶ conception of strategies to accomplish or to restrain excessive transport demand and traffic volume
- ▶ proposals for the legislature to regulate goal conflicts between transport and traffic

A fair solution of these tasks requires independency of logistic research from the particular interests of business and daily policy (see *Sect. 7.8 and 23.41*).

Chapter 19
Design of Logistic Halls

In a logistic hall or logistic center, physical goods are handled, stored and transported. Due to the manifold of options, partly incompatible goals and unquantifiable influence factors, the optimal design of logistic halls requires special skills, experience and judgment. It cannot be performed by a computer program alone. However, *digital simulation*, OR-based programs and CAD are useful tools to improve in detail a start solution which has been designed analytically (Armour et Dutta 1963; Arnold 2002; Dangelmaier 1999; Francis et al. 1992; Gudehus 1971; Tompkins et al. 1996).

The most important objectives for designing a logistic hall are *transport optimization* and *area minimization*. They can be achieved by *analytical design* of the hall layout, and by *allocation strategies* for the function modules. These methods are primarily applicable for logistic halls, transshipment stations and logistic centers where transports and movements are the main cost drivers. They are useful also for planning supermarkets, dining halls and office-buildings. The application in factory planning is limited by technical restrictions of the production. The methods are also applicable for designing open-air handling areas and for allocating buildings on a given site.

The following design methods and allocation strategies have been approved in several projects. The derived formulas allow calculating the dependency of the operating costs of a logistic center on throughput and capacity demand. This dependency shows also the limits of *economies of scale in logistics*.

19.1 Requirements and Restrictions

The hall-layout must offer enough *ground area* to allocate the required interior systems and function modules, and should have sufficient *outer length* for the necessary gates. In addition, it has to keep the given *spatial limitations* and *technical restrictions*.

The total ground area of the hall is determined by the sum of the functional areas. For example, the ground area of a *transshipment hall* is determined by the *gate modules* and the buffer places needed for consolidating the incoming goods.

T. Gudehus, H. Kotzab, *Comprehensive Logistics*,
DOI 10.1007/978-3-540-68652-1_19,

The area of a *storage hall* depends on the required storage capacity and the storage technology. A *logistic centre* requires additional areas for commissioning, packaging and other services (see *Sect. 1.6*). The space of a factory hall is determined by the areas necessary for workstations, machines and production systems. In any case, additional space is required for gates and ramps, buffer places and internal transport systems.

The number of gates results from the in- and outflow of goods during peak hours, from the capacity of the external transport units and from the unloading and loading times (see *Sect. 16.3.6*). If designed correctly, during peak times all gates are fully occupied by equally distributed in- and outflows.

External spatial limitations for the hall design are the *maximal length* or *maximal breadth*, the given traffic connections or a *fixed border* determined by the available land and adjacent buildings. These restrictions generally do not exist on a green field. Internal restrictions, which also apply to *green field solutions*, are dimensions of systems and functional zones, the location of their entries and exits and maximal tolerable *escape route lengths*.

Technical restrictions result from the depth, width and *minimal distance* of the *gate modules*. As shown in *Fig. 16.10*, the gate modules are determined by the external and internal transport means, the docking technique, the required buffer places and the size of the load units. Other technical restrictions, such as the *span width* or a standard *upright grid*, result from building technique and *fire sections*.

19.2 Objectives and Design Parameters

The basic objective of hall design is to find a rectangular layout with the required *ground area* A and *gate number* N, which can be built and operated at lowest costs. Main *cost drivers* are the internal transports and the ground area of the hall. Both depend on the layout. Hence, *minimization of transports* and *minimization of ground area* are the most important design objectives.

The transports from and to the gates can be minimized by the hall layout and by the arrangement of the gates. The transports within the building depend on the arrangement of the function modules. The ground space depends on the number and area of the function modules and their allocation.

For a high throughput, the minimization of transports is the most important design objective, as transport costs are significantly higher than area costs. With increasing area for stores, buffer places and other functions, the minimization of the area becomes the second important layout objective. A third objective is in many cases the *expandability* of the whole building or of single function areas. These three design objectives are, however, partly *incompatible*.

The internal transport system is either a *conveyor system* with fixed installed lines or a *vehicle system* with floor bound vehicles, tug and tows or overhead trolleys moving on a track network. The sum of the lengths of all conveyor lines is the total *conveyor network length* (18.6), which determines the investment and the operating costs for a conveyor system. Hence, for a conveyor system, the minimization of transport cost requires a minimization of the total network length. For a vehicle system, the transport costs depend primarily on the number of vehicles, which is determined by the transport effort.

The total *transport effort* is the sum of the products of the *transport lengths* s_{kl} with the *exchange flows* λ_{Ekl} between the function modules FM_k and FM_l of the hall:

$$E_{trans} = \sum_{k,l} \lambda_{Ekl} \cdot s_{kl} = \lambda_T \cdot s_F \quad [TU \cdot m/PE] \tag{19.1}$$

In the sum (19.1) only the relevant exchange flows have to be taken into account, which are for single runs the sum (18.27) and for combined runs the maximum (18.29) of the to-and fro-flows between two stations. The transport effort (19.1) equals to the product of the total sum of the exchange flows

$$\lambda_T = \sum_{k,l} \lambda_{Ekl} \quad [TU/PE] \tag{19.2}$$

with the weighted *mean transport length*

$$s_F = \sum_{k,l} \lambda_{Ekl} \cdot s_{kl}/\lambda_T \quad [m]. \tag{19.3}$$

Hence, by minimization of the mean transport length (19.3) minimal transport costs can be achieved.

The transport lengths from and to the gates are generally far longer than the transport lengths between the function modules, if the latter have been optimally arranged. Therefore, the *mean gate transport length* is the target value to be minimized in order to achieve minimal transport costs.

When the total area requirement A, the number of gates N, the flows λ_i from and to the gate modules GM_i and the function modules FM_k with their exchange flows λ_{Ekl} are given, the *design parameters* and *layout options* for a hall with rectangular ground area are:

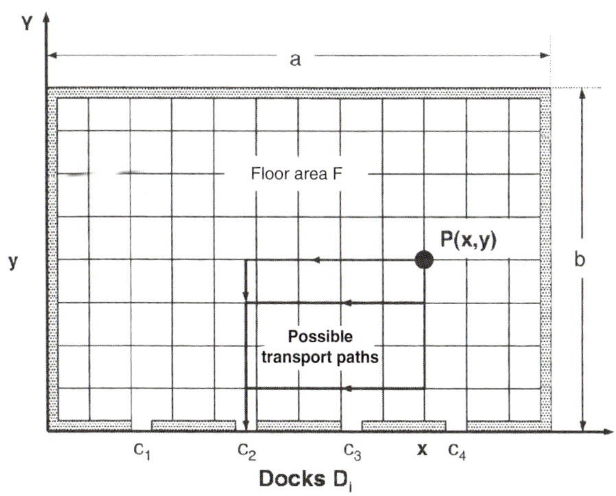

Fig. 19.1 Rectangular hall area with one-sided gate arrangement and transport paths for a rectangular metric

- *gate arrangement* along the sides with the *gate coordinates* $(c_i; 0)$
- *side relation* $f_s = a{:}b$ of the *hall length* a to the *hall breadth* b
- *arrangement* and *coordinates* $(x_k;y_k)$ of the function modules F_k

For a rectangular hall as shown *Fig. 19.1*, with area $A = a \cdot b$ and side relation f_s, *length* and *breadth* are:

$$a = \sqrt{f_s \cdot A} \qquad b = \sqrt{A/f_s} \tag{19.4}$$

For a hall with square ground area, the side relation is $f_s = 1$ and $a = b = A$.

19.3 Mean Transport Lengths

In order to achieve compact areas for the function modules without major *cutting losses*, the conveyor lines and vehicle tracks are normally arranged transversal and parallel to the outside walls of the hall as shown in *Fig. 19.1*. Then the transport length between two points $(x_i;y_i)$ and $(x_j;y_j)$ on the floor is given by the *rectangular metric*:

$$s_{ij} = |x_i - x_j| + |y_i - y_j| \tag{19.5}$$

The shortest way with *Euclidian metric* (18.46) is generally not available for floor transports. For the rectangular metric (19.5), the *mean area transport length* between any two points $(x_i;y_i)$ and $(x_j;y_j)$ within the hall is:

$$s_F = 1/A^2 \int_0^a dx_i \int_0^a dx_j \int_0^a dy_i \int_0^a dy_j \qquad (|x_i - x_j| + |y_i - y_j|) = (a+b)/3, \tag{19.6}$$

if the destinations and sources of the transports from and to the gates are equally distributed over the area. After inserting the relations (19.4) for a and b into (19.6), derivation of the resulting relation $s_F(f_s)$ with respect to $\sqrt{f_s}$, equaling the equation to 0 and solving with respect to $\sqrt{f_s}$ gives the *hall design rule*:

- The *mean area transport length* between any two points within the hall is minimal for a quadratic hall area and given by

$$s_{Fmin} = (2/3) \cdot \sqrt{A}. \tag{19.7}$$

Formula (19.7), the former formulas (18.49) and (18.54) for external transports and the following formulas for internal transports lead to the general *square root rule of transports*:

► The mean transport length is proportional to the square root of the served area.

For halls with optimal layout, the transports within the hall do not contribute as much to the total transport effort as the gate transports. The mean transport length between gate G_i with coordinates $(c_i;0)$ and any point $(x;y)$ within the hall area is (see *Fig. 19.1*):

$$s_{Gi} = (1/A) \int_0^a dx \int_0^a dy \qquad (|x_i - c_i| + |y_i - 0|) = (a+b)/2 + c_i \cdot (c_i - a)/a \tag{19.8}$$

If all gates are equally frequented, the *mean gate transport length* from and to the gates is the mean value of the single mean transport lengths (19.8):

$$s_G = \sum_i ((a + b)/2 + c_i \cdot (c_i - a)/a)/N. \qquad (19.9)$$

The mean gate transport length (19.9) can be minimized by the three *design parameters*: side relation f_s, gate arrangement and gate coordinates c_i.

19.4 Equally Distributed Gates on One Side

When the link to external traffic is only possible from one side, e.g. when three hard borders restrict the accessibility, the gates must be arranged along one side of the hall. If N gates are arranged with equal distance along side a, the gate coordinates are $c_i = i \cdot a/(N+1)$. Inserting this gate coordinates and the relations (19.4) into (19.9) gives after solving the sum:

$$s_G(f_s) = (\sqrt{f_s} \cdot (2N + 1)/(3N + 3) + 1/\sqrt{f_s}) \cdot \sqrt{A/2}. \qquad (19.10)$$

Figure 19.2 shows the dependency of the mean gate transport lengths (19.10) on the side relation f_s for a hall with N = 8 gates. In this case, the mean gate transport length has a minimum for the *optimal side relation* $f_s = 1.5$. It is up to 10% lower than the gate transport lengths for other side relations.

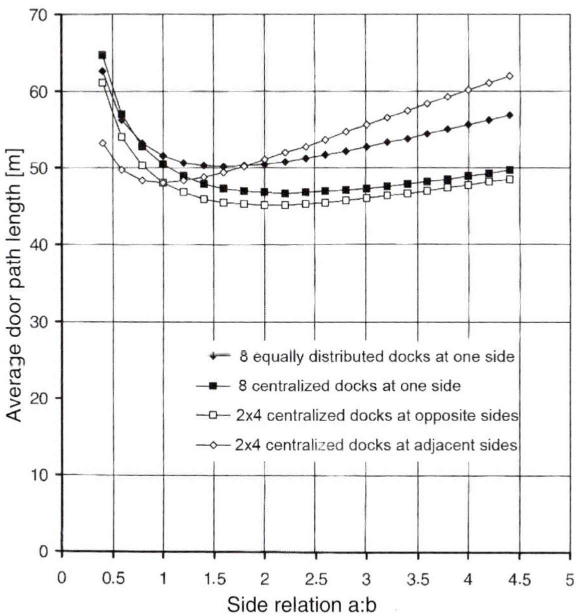

Fig. 19.2 Dependency of the mean gate transport length on the side relation for different gate arrangements

hall area: A = 4,000 m^2 *minimal gate distance*: d = 6 m

By equaling the first derivation of (19.10) with respect to $\sqrt{f}$ to 0 and solving the resulting equation provides the *optimal side relation for equally distanced gates on one hall side* (Gudehus 1971):

$$f_{s\,opt} = (a : b)_{opt} = (3\,N + 3)/(2\,N + 1) \tag{19.11}$$

Inserting this into relation (19.10) and using (19.4) leads to the *minimal mean gate transport length*:

$$s_{Gmin} = \sqrt{(2\,N + 1)/(3\,N + 3) \cdot 2A} \tag{19.12}$$

For the simplest case N = 1 results from (19.11) and (19.12) the *design principle for halls with one gate*:

- ► If only one gate is necessary, it is optimally located in the middle of the longer hall side, which should be twice as long as the other side.

For larger numbers of gates arranged with equal distances along the longer hall side, the optimal side relation is given by relation (19.11) and the mean gate transport length by relation (19.12). With increasing number of gates, the transport optimal side relation approaches the value 3:2 and the mean gate transport length the value 2/3·a.

19.5 Transport Optimal Gates on One Side

Setting the partial derivation of relation (19.10) with respect to the gate coordinates c_i equal to zero, leads to the result that the mean gate transport length is minimal if all gate coordinates are equal $c_i = a/2$, i.e. if all gates would be located in the centre of one side. However, this is not feasible as the minimal gate distance d, i.e. the breadth of the gate module, is finite. Closest to the theoretical optimum is a centered arrangement of the gates with minimal distance d and the gate coordinates:

$$c_i = a/2 + (2 \cdot i - N - 1) \cdot d/2. \tag{19.13}$$

Inserting these coordinates and the relations (19.4) into sum (19.9) leads to the mean gate transport length:

$$s_G(f_s) = \left(\sqrt{f_s} + (2 + (N^2 - 1) \cdot d^2/2F)/\sqrt{f_s}\right) \cdot \sqrt{A/8}. \tag{19.14}$$

The dependency (19.14) of the mean gate transport length on the side relation for a hall with N = 8 gates is shown as well in *Fig.* 19.2. In this case, the optimal side relation with minimal gate transport length is $f_s = 2.2$. The minimal mean gate transport length for centered gates is 8% shorter than for equally distributed gates and about 20% shorter than for other gate arrangements.

Setting the first derivation of (19.14) with respect to $\sqrt{f_s}$ equal to 0, the solution of the resulting equation leads to the *transport optimal side relation for centered gates at one hall side*:

$$f_{s\,opt}(N) = \begin{cases} 2 + (N^2 - 1) \cdot d^2/3A & \text{if } N^2 \le 3A/d^2 - 1/2 \\ N^2 \cdot d^2/A & \text{if } N^2 > 3A/d^2 - 1/2 \end{cases} \tag{19.15}$$

The second line of (19.15) holds for gate numbers and distances, where the narrowest spacing leads to a longer hall length than the first line of (19.15).

Inserting the optimal side relation (19.15) into relation (19.14) gives the *minimal mean gate transport length* for centered gates on one side:

$$s_{G\,min} = \begin{cases} \sqrt{A + (N^2-1) \cdot d^2/6} & \text{if } N^2 \leq 3A/d^2 - 1/2 \\ N \cdot d/4 + A/(4Nd) + (N^2-1) \cdot d/12N & \text{if } N^2 > 3A/d^2 - 1/2 \end{cases} \qquad (19.16)$$

Relations (19.14), (19.15) and (19.16) lead to the *hall design rule*:

- ▸ For gates only at one hall side, *centered arrangement* with minimal distance is transport optimal. The transport optimal side relation for a hall with gates at one side is given by relation (19.15)

Figure 19.3 shows the dependency of the optimal side relation (19.15) on the number of gates. For N = 1, it confirms the above design rule a:b = 2:1.

The optimal side relation changes with increasing number of gates from 2:1 to 3:1, as long as the minimal total gate length N·d required for the N gate modules is shorter than the minimal side length given by the first line of (19.5). Otherwise, the minimal total gate length becomes the optimal side length.

If the minimal total gate length is shorter than the optimal length, further gate modules can be added later on both sides of the existing gates. This *modular*

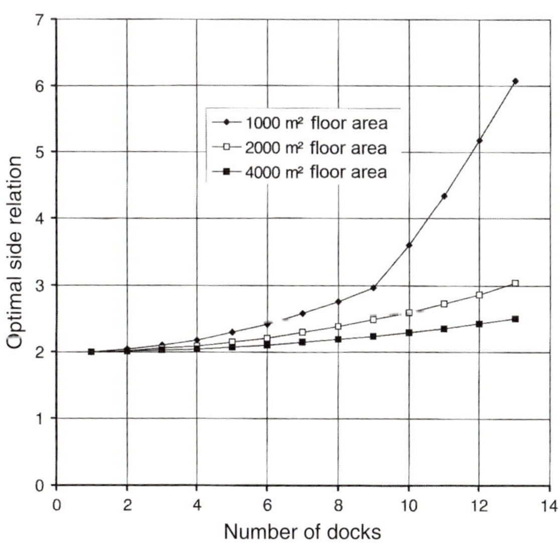

Fig. 19.3 Dependency of the optimal side relation on the number of gates for one-sided centered arrangement

parameter: ground area A = 1,000/2,000/4,000 m^2
minimal gate distance: d = 6 m

expandability is a further advantage of the centered gate arrangement. The dependency of the *mean gate transport time* on the number and location of the gates also leads to the *gate operating strategy*:

- In times of lower throughput, only gates close to the middle of the hall length should be opened, whereas the outer gates can be kept closed.

Experienced managers of logistic halls apply this operating strategy intuitively without knowing the theoretical background.

19.6 Hall Design Principles

If there are no restrictions by traffic connections, building sites or adjoining buildings, the gates of a hall can be arranged on more than one side. For a *gate arrangement on two sides* the number of gates is split up into a sum $N = N_1 + N_2$ of N_1 gates on one side and N_2 gates on a second side. When the gates are arranged on opposite sides, the mean gate transport length is the weighted average:

$$s_G(N_1; N_2; f_s) = (N_1 \cdot s_G(N_1; f_s) + N_2 \cdot s_G(N_2; f_s))/(N_1 + N_2). \quad (19.17)$$

of the mean gate transport lengths $s_G(N_1;f_s)$ and $s_G(N_2;f_s)$, each given by relation (19.14). *Figure 19.2* shows the dependency of the mean gate transport lengths (19.17) on the side relation f_s for a hall with 8 gates, where 4 gates are centered on two opposite sides. In this case, the optimal side relation is slightly larger than 2, and the mean gate transport length is about 3% smaller than for the one-sided centered arrangement, and about 10% smaller than for the equally distributed arrangement on one side.

By setting the partial derivation of (19.17) with respect to the gate numbers N_1 and N_2 equal to zero and solving the resulting equation results that the mean gate transport length is minimal, if the numbers of gates on both sides are the same i.e. for $N_1 = N_2$. This leads to the *design principle for halls with gates on opposite sides*:

▸ If the required number of gates N is even, N/2 gates, if it is odd, N+½ and N-½ gates with minimal distance should be arranged centered on the opposite sides. The optimal side relation of the hall is given by relation (19.15) with N/2 instead of N.

For an arrangement of N_1 and N_2 gates on two neighboring sides, the mean gate transport length is the weighted average of the mean gate transport length (19.14) with side relation $f_s = a/b$ and of the mean gate transport length (19.14) for the inverse side relation $1/f_s = b/a$:

$$s_G(N_1; N_2; f_s) = (N_1 \cdot s_G(N_1; f_s) + N_2 \cdot s_G(N_2; 1/f_s))/(N_1 + N_2). \quad (19.18)$$

The dependency of the mean gate transport length (19.18) on the side relation f_s for a hall with 8 gates arranged on two neighboring sides is shown in *Fig. 19.2*. For this solution, the minimal gate transport length for the centered arrangement at neighboring sides is slightly longer than for the centered arrangement at one side, and significantly longer than for the centered arrangement at opposite sides, each with optimal side relation. The calculation of the optimal side relation leads to the *hall design rule*:

- ► The mean gate transport length for a gate arrangement at neighboring sides is minimal for a quadratic hall with side relation $a : b = 1$.

Analog calculations and considerations can be performed for gate arrangements at three and four sides. It turns out, that by these arrangements no further reductions of the mean gate transport length are achievable.

Theoretically, a *circular ground area* with equally distributed gates around the circle wall is the transport optimal solution. A circle of area A has the minimal mean gate transport length $s_T = 0.63\sqrt{A}$ and wall length $l_{wall} = 2\sqrt{\pi A} = 3.53\sqrt{A}$ compared to the mean gate transport length $s_T = 0.83\sqrt{A}$ and wall length $l_{wall} = 4.0\sqrt{A}$ of a squared hall with the same area and many gates. However, the arrangement of a circular building with other buildings on a site causes space losses. Due this and other disadvantages, it is practically irrelevant. This confirms the *general hall design principle*:

- ► The centered arrangement of half of the required number of gates on two opposite longitudinal sides of a hall with optimal side relation leads to the shortest mean gate transport length.

The optimal side relation is given by relation (19.15) and the minimal gate transport length by relation (19.16), each calculated with N/2 instead of N. But also the opposite gate arrangement restricts the arrangement of the function modules within the hall and the combination with neighboring buildings. The restrictions are even stronger for gates on three or four sides of the hall.

The general hall design principle can be used without any restrictions if a hall, such as the *transshipment station* shown in *Fig. 21.3*, consists only of gate modules and buffer places. The one-sided centered gate arrangement with side relation (19.15) is optimal for small numbers of gates compared with the buffer space requirement. This is normally the case as long as $N < \sqrt{3A}\,/d$. The centered gate arrangement on opposite sides is optimal for a high numbers of gates $N > \sqrt{3A}\,/d$.

For very high gates numbers, i.e. for $N >> \sqrt{3A}\,/d$, also the two-sided gate arrangement leads to extreme long halls with a mean gate transport length which increases linear with the number of gates. This is also the case for non rectangular halls with U-, L-, H- or cross-layout. In these cases, a reduction of the mean transport lengths can be achieved only by two or more separate halls. This indicates the *limits of economy of scale* in logistics.

19.7 Modular Design of Systems and Functional Zones

Before placing logistic systems and functional zones within a hall, it is necessary to design and dimension them separately. After optimal design, they can be arranged due to general *allocation strategies*.

A *logistic system* or *functional zone* is either *indivisible*, such as a mini-load store as shown in *Fig. 17.34* or a production unit, or it is *divisible* into several *function modules* of the same kind and dimensions. For example, the gate area of a logistic center consists of a number of *gate modules*, a store is made up of a number of *aisle modules*, a production zone consists of a number of *production modules*, which can be machines, workstations or workshops, and a packing zone contains several modular *packing places*.

With the number of modules, the *divisibility* and *deformability* of a functional zone and the *changeability* of the material-input- and output-points increase. Divisible, deformable or small systems and functional zones are easy to integrate into a given area, whereas indivisible, un-deformable and large functional zones are most difficult to allocate. In many cases, they determine the total building or allow only a few positions in a given hall. If a large indivisible functional zone is given, optimal allocation reduces to placement of connected smaller function modules in the remaining area.

Examples for *indivisible logistic systems* are flow rack stores, compact stores, automated mini-load stores and high bay stores. Their optimal design and dimensioning has been outlined in *Chap. 16*. The number of aisles results from the throughput demand. The number of shelf racks and the height and length of the aisles are determined by the capacity demand. The location of connected function modules, such as entrances, exits and commissioning stations, is possible in many different ways. Some possibilities are shown in *Figs. 16.1*, *16.8*, *16.9*, *17.34* and *17.42*. Shelf rack stores can be extended in aisle direction if additional capacity is required. Additional aisle modules need to be added if the throughput increases.

The optimal design and dimensioning of *indivisible production units* are tasks of machine construction and *plant engineering*. They must take care of *allocation flexibility*, e.g. by the possibility to bend a production line or to change the input- and output points, and of *extendibility* of the production units.

The design and dimensioning of the *function modules* of a divisible functional zone are tasks of the *logistic planning* in cooperation with plant engineering. It is possible to arrange and combine the modules of a divisible functional zone in different ways. Basic alternatives are *parallel arrangement* and *sequential arrangement*:

- The *parallel arrangement* of function modules for the same purpose ensures good access for material and people.
- The *sequential arrangement* of function modules ensures shortest transport lengths for the passing objects and good accessibility for people and supplies.

Depending on available space and transport links, functional zones or chains of modules are arranged in a *straight line*, single-broken in *L-shape*, twice redirected in *U-shape* or multiple-broken as *meander line*.

Examples for sequential arrangement are *assembly lines* in the production and the picking areas of *local commissioning*. Examples for parallel arrangement are the *aisle modules* of a storage system as shown in *Figs. 16.1* and *16.16* and the *gate modules Fig. 16.10* in a transshipment station or logistic center as shown in *Figs. 19.4* and *21.3*.

When the incoming and outgoing shipments or transport means deviate considerably from each other, different entry and exit gate modules and separate zones for

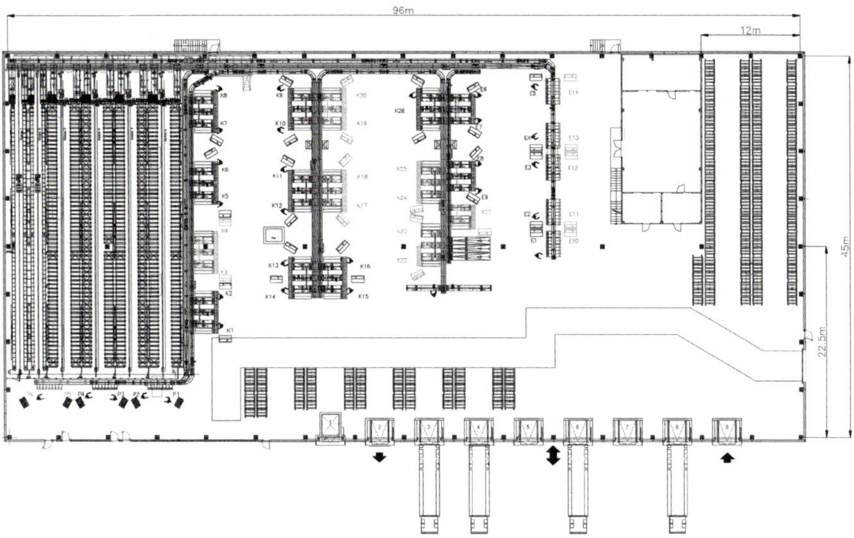

Fig. 19.4 Hall layout of a plant entry and distribution centre (EDC Bosch Rexroth, Lohr)
dispatch gate modules (left of the centre)
entry gate modules (right)
combined goods entry/dispatch gate (centre)
mini load system for bins (close to the left wall)
pallet storage racks (close to the right wall)
small item packaging modules (middle-left)
large item packaging modules (middle)
quality control (back right)
driveways and expansion (free areas)
planning and project management: Reinhard&Ahrens GbR, Berlin

goods entry and goods dispatch are necessary. If the sizes and the transport means are similar, the same gate modules can be used for entry and dispatch. By flexible utilization, *combined gate transports* are possible during times with simultaneous inflows and outflows. Another possibility is the alternative use, only for inflow or only for outflow, during differing peak times.

The N combined gate modules for the joint use by goods entry and dispatch are arranged centered on one side of the hall if the number is small and on two sides if it is large. A goods entry zone with N_E entrance gate modules and a dispatch zone with N_D dispatch gates are optimally arranged side by side on one side, if the total number of gates is small, and on opposite or neighboring sides as shown in *Figs. 17.21* and *21.3*, if the number is large.

The gate transports into and out of the hall or building start and end at the inner side of the gate modules. Therefore, the area for the gate modules A_{gate} must be excluded from the calculation of the optimal side relation, i.e. side relation and internal dimensions should be calculated with the *internal area requirement* $A_{int} = \Sigma A_k - A_{gate}$ for the function modules without the gate areas. The external dimensions and the total area of the hall result after determination of the optimal internal area by adding the gate areas.

19.8 Linking Strategies and Placement Strategies

After the gates are arranged, the inner systems and functional zones must be located. Again, the objectives are minimal transport effort and maximal space utilization. Generally, the solution with minimal ground area differs from the solution with minimal transport effort. Therefore, a compromise between these partly incompatible goals is necessary. The optimal arrangement of the systems and functional zones in a hall depends on:

- exchange flows between the systems and zones
- required ground areas of the systems and zones
- divisibility and deformability of systems and zones
- changeability of number and position of input and output points
- expandability for increasing static or dynamic demand

A minimal transport effort is achieved by the *linking principle*:

▸ The more intensive the relevant exchange flow between two systems or functional zones, the shorter the distance between their entries and exits should be.

However, the linking principle alone does not necessarily lead to the transport-optimal arrangement with lowest costs. This combinatory task can be solved exactly by *full enumeration* of all possible arrangements or with high accuracy by *OR-heuristics* (Arnold 1995, Dangelmaier 1986 and 1999). Approximately minimal transport costs result with the *linking strategy*:

- First, the two systems or functional zones with the most intensive exchange flows are located in adjacent position with entries and exits in shortest distances. Then the system or zone, which is connected with the first two by the next strongest flows, is arranged in adjacent position with shortest length and so forth with the remaining system and zones.

The task to achieve maximal space utilization is equivalent to the task to minimize the *cutting losses* between adjacent function zones. It corresponds to the task of *optimal packaging* with minimal cutting losses as outlined in *Sect. 12.4*. Therefore, similar methods and strategies can be applied to maximize the space utilization of a hall. Function modules with small ground area can be placed in a large area almost without cutting losses. Also modular and deformable functional zones are quite easy to integrate. Large, not deformable systems and functional zones affect the area utilization much more. This leads to the *placement strategy*:

- In order to enable later expansion, the system or functional zone with the largest ground area is located in a corner of the hall with its longitudinal edge parallel to the shorter outside wall, or if impossible, to the longer outside wall. In the remaining area, the next largest system or functional zone is located and so forth until the large systems and functional zones are placed.

If there are no intensive exchange flows between them, the two largest systems or zones can be placed – as shown in *Fig.* 19.4 – in the corners along two opposite outside walls close to the gates to which they have the most intensive transport relations.

The placement strategy can be applied to up to four undividable, not deformable systems or zones with dimensions longer than ½ of the hall side lengths and with ground areas up to ¼ of the hall area. Having arranged the largest systems or zones, the remaining modules are placed transport optimally following the linking strategy. Longer functional chains of small modules fit in after bending the shape or turning the orientation of the chain.

19.9 Efficient Hall Design

The task, to allocate systems and functional zones with given ground area in a rectangular area is comparable with a *puzzle game*, where the final picture is unknown. Similar to the familiar procedure in a puzzle, it is most efficient to start with the corner parts and the wall pieces. For a hall, these are the largest functional zones and the gate modules. After placing them, the modules are sorted by size and transport intensity and, starting with the largest and closest connected units, successively inserted. This consideration leads to the following *procedure of efficient hall design*:

1. Design and dimensioning of systems and functional zones due to the principles of modularity, divisibility and deformability and determination of their base dimensions and connection points
2. Selection of systems with connected function modules, e.g. of a high bay store with in- and out-conveyers, which, due to their technique or dimensions, require a separate hall or a special building.
3. Determination of the external inflows and outflows and of the relevant exchange flows between the functional zones
4. Calculation of the required internal ground area of the hall necessary for all internal functional zones and modules
5. Design and dimensioning of the gate modules, organization of the external inflow and outflow, and calculation of the required number of gates
6. Calculation of the optimal side relation for the internal length and breadth of the hall from the number of gates and from the required internal ground area including an allowance of about 20% for transport systems
7. Drawing the layout of the empty hall with optimal dimensions
8. Optimal allocation of the gate modules on one longitudinal side, for many gates at opposite or neighboring sides
9. Sorting of systems and functional zones by size, divisibility and deformability, and by the intensity of the exchange flows
10. Space optimal allocation of up to four of the largest systems or functional zones on the hall area due to the placement strategy
11. Transport optimal insertion of the deformable, divisible and smaller systems and functional zones due to the linking strategy

12. Insertion of tracks for conveyor and/or vehicle systems between the systems and zones, following the shortest flow connections
13. Determination of the *construction grid* for the hall, which leaves sufficient space for the largest systems and indivisible functional zones. The grid measures should be an integer multiple of the smallest module dimensions

The *construction grid* of a logistic hall or center is determined by the dimensions of the load units, transport means, gate modules and of the storage aisle modules. European *standard grids* for transshipment halls and logistic centers for norm pallets with dimensions 800×1200 and 1000×1200 mm are whole-numbered multiples of 2.5 m, e.g. 12.5, 15.0 or 22.5 m. The *standard height of logistic halls* for pallets ranges from 5 to 15 m depending on the storage type. High bay stores higher than 15 m are separately erected as *silo construction* with wall and roof carrying racks and flanged transversal or parallel to a logistic hall.

The above hall design procedure has been successfully applied in many projects. It leads quite fast to useful results. For example, *Fig. 19.4* presents a realized hall layout, which has been planned by this procedure. Other realized logistic centers planned by this method are shown in *Figs. 17.21* and *20.3*.

A logistic hall can be manually designed and dimensioned with paper, pencil, scissors, ruler and a handheld calculator. For repeated design and for detail planning, computer-based optimization algorithms and CAD-software are helpful. The program calculates transport lengths and transport effort, the occupied area and other target values for each design step.

The result of *layout planning* is the starting point of *detail planning*. As outlined in *Sect. 3.2.2*, detail planning comprises architecture of the building and engineering of construction and technical installations. Further tasks are selecting and dimensioning the transport systems and organization of the operation. By *transport strategies* as described in *Chap. 18*, such as *combined tours* or *fast mover zones*, the transport effort can be reduced further. In order to allow improvements in detail and to take into account technical features or potential obligations, sufficient *allowances* must be left by the layout solution. Therefore, the total ground area should be not too small and the layout must not be very precise. The results achieved by hall design strategies are *benchmarks* for more sophisticated methods, such as OR heuristics and CAD-programs.

19.10 Size Effects of Logistic Centers

The operating costs of a logistic center are caused by *area costs* depending on the storage capacity, *handling and transport costs* determined by the throughput and other costs, which are independent of capacity and throughput.

The required ground area for single place stores increases linear with storeplace demand. The storeplace demand for *transshipment stations* is the product $M_S = T_S \cdot \lambda$ of mean *storing time* T_S [PE] and *throughput* λ [LU/PE]. Handling and internal transports increase proportional with throughput. The mean transport length increases according to the above relations with the square root of the hall area, which is proportional to the throughput λ. That means, the mean transport distance varies proportional to square root $\sqrt{\lambda}$, whereas the transport frequencies vary proportional to throughput λ. Area demand and transport lengths of a commissioning module with static article provision depend also on the number of articles (see *Fig. 17.39* and *Sect. 17.15*).

With the specific cost factors, k_H [€/LU] for handling, k_A [€/m²-PE] for hall area, k_T [€/LU-m] for transports and K_o [€/PE] for all other costs results the general dependency of the operating costs on planned throughput λ and storing time T_S:

$$K_{op}(\lambda, T_{SD}) = K_o + k_H \cdot \lambda + k_A \cdot T_S \cdot \lambda + k_T \cdot c \cdot \lambda \cdot \sqrt{\lambda} \quad [€/PE]. \quad (19.19)$$

The *specific throughput costs* are:

$$K_{op}(\lambda, T_{SD}) = K_{op}/\lambda = K_o/\lambda + k_H + k_A \cdot T_S + k_T \cdot c \cdot \sqrt{\lambda} \quad [€/LU]. \quad (19.20)$$

The dependency of the specific costs (19.20) on the planned throughput can be calculated with a *hall dimensioning program* which applies the above hall design algorithms and calculates the required area and trucks with the help of the formulas of this chapter and of *Sect. 16.6*. Resulting are the dependencies of the throughput costs shown in *Fig. 19.5*.

The diagram illustrates the dependencies of relation (19.20) and confirms the general *cost rules for logistic halls*:

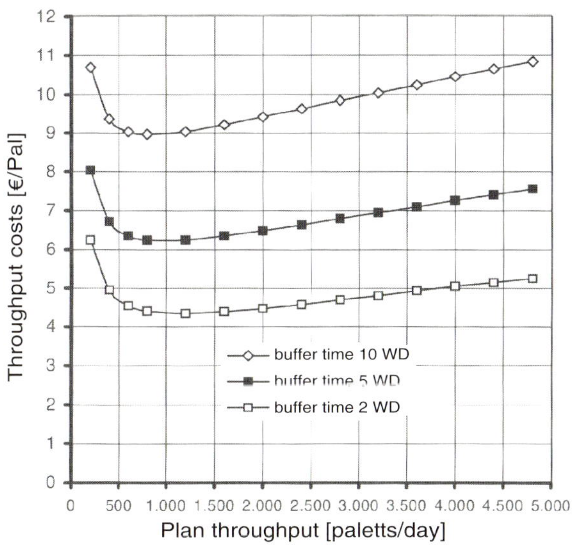

Fig. 19.5 Dependency of the throughput costs of a transshipment hall for standard pallets on the throughput

throughput at 8 h/d operation on 250 days/year
block place store with fork lift service
stacking factor 3 for up to 1000 articles
turnover costs at 100% utilization (cost base 2004)

- The throughput costs of a logistic hall decrease with increasing throughput until they reach a *minimal value* at a critical throughput and increase for higher values due to longer transport paths.
- The *critical throughput* moves to smaller values with increasing buffer time and storing time.

These rules have consequences for business practice: When the throughput exceeds the critical value no further size effects and cost reductions by higher throughput are achievable. The order of magnitude of the critical throughput for *transshipment stations* is 1,000 pallets per day, equivalent to 40 incoming and outgoing semi-road-trailers. When the throughput exceeds this value, it is better to build a second transshipment station and to split the external transport flows.

The calculation of the critical throughput for other logistic stations and storage types is more complex. The design and calculation for multi-functional logistic centers including commissioning require high sophisticated planning tools. It becomes quite complex due to the dependency of the commissioning costs on the number of storekeeping articles and on the size of the handling units (Nowitzky 2003).

The transport costs of a transshipment station with block place storage and truck service make up a share of the operating costs in the range from 25% to 50%, depending on throughput and mean storing time. The area costs contribute 20 to 40% to the operating costs. The share of the transport costs can be reduced by using conveyors or AGV-systems instead of forklift trucks and by automatic storage systems as shown in *Figs. 20.2* and *20.3*. This shifts the critical throughput to larger values, even for high storage capacities.

In the logistic centers of manufacturers and retailers large numbers of *storekeeping articles* are stored which are scheduled on demand. Due to the *square root law of stocks* (11.42) the mean stock of these articles is proportional to the square root of the throughput, i.e. $M_S = A_S \cdot \sqrt{\lambda}$. The storing time reduces inverse proportional to $\sqrt{\lambda}$, i.e. $T_S = c/\sqrt{\lambda}$. Inserting this into relation (19.20) leads to a decrease of the third term with increasing throughput. This causes a shift of the critical throughput for storekeeping logistic centers towards higher values above 3,000 pallets or 100 road semi trailers a day.

The fundamental question of the *limits of economy of scale* in logistics and the development of advanced strategies are interesting tasks for *logistic research* (Nowitzky 2003).

Chapter 20
Production Logistics

Production systems are special performance systems which transform input material into physical goods. They are central parts of the business networks of manufacturing companies. As logistic networks supply the input and distribute the output of production systems, production and logistics are closely interrelated. Production planning without taking into account logistics is as incomplete as logistics without considering production.

In this chapter, the methods of system analysis, the limit performance and queuing laws, and the planning and scheduling strategies, which have been outlined in the first part of this book, are applied to *production logistics*, i.e. the organization, planning and scheduling of production systems (Corsten 2007; Günther and Tempelmeier 2005; Kern et al. 1979; Tempelmeier 1999; Hopp and Spearman 2000; Wagner and Whitin 1958; Ware and Fogarty1990; Vollmann et al. 1997; Zäpfel 1996, 1998).

20.1 Modes and Types of Production

Production systems are multi-stage networks of elementary *production stations* which are either *directly* linked by transport systems or *indirectly* connected via intermediate buffers and storage systems.

A single production station transforms input, such as raw material, parts, preproducts and modules, with the help of people, machines and other means into *products*. The *production units* PU_r and *quantities* m_r [PU_r/PO] of the different *products* P_r, $r = 1,2,\ldots N_P$, and the required completion dates are given by *production orders* [PO]. A *production order rate* λ_{PO} [PO/PE] requiring the mean order quantities m_r of the products P_r causes the *partial production demand* $\lambda_r = m_r \cdot \lambda_{PO}$ [PU_r/PE].

The task of *production planning and scheduling* (PPS) is to enable and to ensure the execution of incoming orders in due time at lowest costs. The execution of this task depends on the production mode and on the type of the production system. Basic *production modes* are:

T. Gudehus, H. Kotzab, *Comprehensive Logistics*,
DOI 10.1007/978-3-540-68652-1_20, © Springer-Verlag Berlin Heidelberg 2009

- *Cyclic production*: In a discontinuous process, discrete product units, which can be single pieces or filled packages, bottles or load units, are generated within certain *cycle times*.
- *Process production*: In a steady process, continuous goods, such as gas, liquids, bulk or long material, are produced during longer time without interruption.

The discrete results of a cyclic production are immediately conveyed to other production stations, directly delivered to customers or kept for short time in a buffer or for longer time in a store. The continuous output of a process production is generally buffered in tanks, silos or stores for mass products, from which succeeding production, consumption, packing or bottling stations are supplied.

The different possibilities of coupling single production stations define the basic *production types* (e.g. Slack et al. 2007):

- In a *workshop production*, the single production stations are separated by order buffers and material buffers. The workstations generate parallel or in series the required goods in isolated part-processes.
- In a *line production*, several production stations which generate products in successive process steps are directly connected in series without buffers. They are supplied from outside with input material via intermediate buffers.
- In a *network production*, a buffer-less main performance chain of stations generates stepwise the final product from input material, modules and parts, which are directly provided by supply chains without intermediate buffers.

Workshop production is more adequate for small orders with changing products, whereas line production becomes opportune with larger orders for the same product. However, by *reduction of setup times*, *standardization* of components and professional *variant management*, a line production can also be made effective for smaller lots of changing products. Hence, the traditional equalization of workshop production with small lots and of line production with large lots does not hold any longer.

Nowadays, many different combinations and hybrid forms of the basic production types can be found in business practice. Some industries, such as the *automotive industry* or *computer assembling*, have aimed for an extended network production with *just-in-time-provision* of parts and modules. However, as explained in *Sect. 8.9*, networks with extreme JIT-production are generally unreliable and not cost-optimal.

20.2 Production Performance

Production planning and scheduling shall ensure that actual and expected customer orders are executed within promised delivery dates at lowest cost. For this purpose, the capability of all stations, which are involved in the order execution process, must be known.

An *elementary production station* PS(n,m), which transforms n *partial input flows* λ_q [MU_q/PE] of *materials* M_q, q = 1,2...n, with material units MU_q into m *partial output flows* λ_r [PU_r/PE] of *products* P_r, r = 1,2...m, with production units PU_r, is an *irreducible performance station* of *type* (n,m) and *order* n+m as defined in *Chap. 13*. Hence, the formula for calculating limit performances, the allocating, operating and scheduling strategies, and the *limit performance laws* and *queuing laws* of *Chap. 13* can be applied as well in production logistics.

In the following, several planning, scheduling and optimization rules for production systems are derived from an analysis of the impact of order scheduling, product changes and process cycle times on costs and utilization. Using these rules, a *standard scheduling procedure* is developed. Combined with other principles and system strategies, the standard procedure can be applied universally in production chains and supply networks of different branches (Gudehus 2005; Günther and Tempelmeier 2005; Tempelmeier 1999).

20.2.1 Production Capability

The input of a production is specified by the bill of material for discrete goods, respectively by a recipe for continuous goods:

- The *bill of material* or *recipe* prescribes the quantities m_{qr} [MU_q/PU_r] of input materials M_q measured by material units MU_q, which are necessary to generate one output unit PU_r of product P_r.

When the bill of material is known, *material requirement planning* (MRP) can calculate the *input flows* λ_q [MU_q/PE] of a production station from the output flows λ_r [PU_s/PE] by the sum:

$$\lambda_q = \sum_r m_{qr} \cdot \lambda_r \quad [MUq/PE] \tag{20.1}$$

The current *output flows* $\lambda_r = \lambda_r(t)$ are the production rates at time t. The future output flows result from the expected customer orders or from the supply orders of succeeding production stations, which are derived from customer orders with corresponding bills of materials.

The production capability, i.e. the maximal periodical output of products P_r, is determined by the limit performances and switch times of the elementary performance stations of the production network:

- The *partial production limit performance* μ_r [PU_r/PE] is the maximal production performance if only product P_r is produced.
- The *product switch time* T_{rs} [PE] is the time needed for changing or switching from product P_r to product P_s.

The product switch times are elements of the *switch time matrix* T_{rs}. They are intermediate *setup times* or *changeover times*, which in some cases can be longer than the pure technical setup times (8.12).

The preparation time until the first start of the station for orders of product P_r is the *initial setup time*. The initial setup time equals the interruption time, that is needed after a maximal output $m_{r\,max}$ in order to clean and maintain the station. It can be taken into account as *eigen-setup time* T_{rr} [PE] and written in the diagonal of the switch time matrix.

The product of the switch time with the cost rate of the station plus the costs for start-up losses, e.g. for waste, gives the *switch costs*, respectively the *setup costs*.

They are of central importance for the optimal scheduling of orders and inventories in supply chains.

Analogous to relation (13.3) for the limit performance of a source, the partial limit performances of a cyclic production are given by:

$$\mu_r = c/\tau_r(c) \quad [PU_r/PE]. \tag{20.2}$$

Herein $\tau_r(c)$ is the *cycle time* for producing or the *service time* for processing a *charge*, *lot* or *batch* of c output units. For batch size $c = 1$, it is a *single-unit production* and for batch sizes $c > 1$, a *batch-wise production*. In a batch-wise production, fixed quantities c of the same product are produced, processed and completed in the same process cycle. The *charge* or *batch size* c is determined by the capacity of the production station, such as a melting furnace, kiln or coating station, and by the capacity of the load carriers.

20.2.2 Limit Performance Law of Production

The total utilization of a production station is the sum of the *productive utilization* ρ_{prod} for generating products and the *unproductive utilization* ρ_{unpr} for setup und switching. The *partial utilizations* of a station with the partial outflows λ_r and limit performances μ_r are $\rho_r = \lambda_r/\mu_r$. A *switching frequency* ν_{rs} with switch times T_{rs} causes a *switching occupation* $\rho_{rs} = \nu_{rs}\cdot T_{rs}$. The sum, including regular interruptions for cleaning and maintenance that happen with cleaning frequency $\nu_{rr} = \lambda_r/m_{r\,max}$ and eigen-setup time T_{rr}, is the *total utilization of a production station*

$$\rho_{tot} = \rho_{prod} + \rho_{unpr} = \sum_r \lambda_r/\mu_r + \sum_{r,s} \nu_{rs} \cdot T_{rs} \quad [\%] \tag{20.3}$$

As total utilization cannot exceed 100%, i.e. due to $\rho_{tot} \leq 100\%$, the relation (20.3) leads to the *limit performance law of irreducible production stations*

$$\rho_{prod} = \sum_r \lambda_r/\mu_r \leq 1 - \rho_{unpr} = 1 - \sum_{r,s} \nu_{rs} \cdot T_{rs}. \tag{20.4}$$

From the limit performance law follow the *availability principles*:

- The maximal *productivity* of a station, i.e. its availability for productive output, decreases with longer switch times and increasing switching frequency.
- If the switch times between products differ, the maximal productivity depends on the *sequence of product changes*.

If the order inflow and/or the cycle times are stochastically fluctuating, an *order waiting queue* builds up in front of the production station. As outlined in *Sect. 13.5*, a *stochastic waiting queue* increases over proportionately with the utilization (20.3) and grows infinitely when the utilization approaches 100%. This causes *waiting times* at the *bottleneck stations* of a production chain or network, and extends the order lead times (see *Fig. 8.1*).

When the total utilization exceeds 100%, a *systematic waiting queue* grows and the order waiting time increases continuously until the total utilization drops again below 100%. During the *overload time*, the respective production station is a *critical bottleneck*.

20.2.3 Order Lead Times

The *minimal lead time* of orders for product P_r with quantity m_r is the sum of switch time and order execution time:

$$T_{LTmin}(m_r) = T_{rs} + m_r/\mu_r \quad [PE]. \tag{20.5}$$

The *switch time* T_{rs} [PE] depends on the last product P_s. For $s = r$ it is the initial setup time. For $s \neq r$ it is the switch time from Product P_s to product P_r. With limit performance μ_r [PU/PE], the *execution time* for a quantity m_r [PU] is m_r/μ_r [PE]. If the production is followed by a ripening process, the lead time is elongated by the *ripening time*. From relation (20.5) follow the *lead time rules*:

- The minimal order lead time increases with growing order quantity and decreases with higher limit performance.
- Long lead times of big orders can be shortened by executing part orders on parallel stations or partly later on the same station.

The minimal lead time (20.5) can be achieved as long as the production station is available and no other orders with higher priority are waiting. If the station is occupied and other orders are waiting, the lead time is elongated by an *order waiting time*, which is the sum of the switch times T_{ss-1} and execution times $T_s = m_s/\mu_s$ of the queuing orders for products P_s with quantity m_s. As ripening does not affect the running production, the order waiting time is not elongated by the ripening times of preceding orders.

The sum of the minimal lead times of the incoming order for product P_r and of all waiting orders with priority is the *actual order lead time* or *delivery time*:

$$T_{LT\,r} = \sum_{s=1}^{r} (T_{s\,s-1} + m_s/\mu_s) \quad [PE] \tag{20.6}$$

The first part of this sum is the *setup time sum*, the second part the *execution time sum*. From relation (20.6) follow the *delivery time principles*:

- The actual delivery time of a production station depends on the number and size of the waiting orders with higher priority.
- The delivery time is elongated by higher switching frequency, i.e. by the number of waiting orders, and by longer switch times.
- With different switch times, the delivery time depends on the *sequence*, in which the waiting orders are executed.

Shorter lead times can be achieved by part orders with smaller production lots. However, part orders increase the switching frequencies and reduce the availability for productive utilization. The opposite influence of smaller lot sizes and higher switching frequencies on lead times and productivity causes the *fundamental goal conflict* between efficient production and short delivery times.

20.3 Production Planning

The goal of *production planning* is the efficient and timely provision of resources and material for *future demand* (see *Chap.* 2). The tasks of *long-term planning* are to set up the production network, to organize and optimize the production processes, to develop *production strategies* and to organize *production scheduling*. The tasks of *medium- and short-term planning* are *advanced resource planning* and *material requirement planning* (Chopra and Meindl 2007; Kern et al. 1979; Stadtler and Kilger 2007; Tempelmeier 1999)

Long-term production planning is performed regularly once a year or as often as required by management. An approved *procedure of long-term planning* is:

1. *Optimization of the existing production network,* respectively *design of a new production network* (see e.g. *Figs. 1.15*, *3.5* and *20.1*), i.e.: Selection of production technique. Design of production stations. Delimitation of parallel and consecutive production areas. Design of production lines and networks, which are directly connected by conveyors and transport means or decoupled by intermediate buffers and storages.
2. *Documentation of network structure*, *specification of limit performances* of available production stations and transport connections, and *determination of capacities* of the necessary buffers and storages.
3. *Segmentation of* the final and intermediate *products* into kanban-storekeeping and order articles and parts (see *Chap. 11*):

   - *Kanban-articles and Kanban-parts* are mass products of low-value with continuous demand. Their consumption must not be registered on a single-item basis. They can flow without IT-assistance due to the self-regulating *pull-principle* from the entrance or production station or from a store directly to the consumption station (see *Sect. 11.11*).
   - *Storekeeping articles and storekeeping parts* are standard products with continuous demand and positive storekeeping profit. They can be produced or procured anonymously and kept on stock. Their production orders result from dynamic inventory scheduling (see *Sect. 11.12*).
   - *Order articles and order parts* are the remaining final and intermediate products that are produced on the basis of customer orders.

4. *Definition of standard production chains and trees*: Selection and design of cost optimal production chains or trees for the different product groups.
5. Specification and documentation of the involved production, performance and storage stations, following the order execution *downstream* from the *order penetration limit* to the exit station (see e.g. *Figs. 3.7* and *8.1*).
6. *Organisation of production scheduling*: Definition of the tasks of production scheduling. Development of production strategies. Synchronization of periods, dates and strategies of production scheduling with periods, dates and strategies of order scheduling and logistic scheduling (see *Sect. 8.2*).

With these steps of long-term planning, the network structure and the process chains of production are established. The results are documented in *structure charts* and in *process plans*. For example, *Fig. 20.1* shows the production structure and *Fig. 20.2* the standard production chains of a metal processing company.

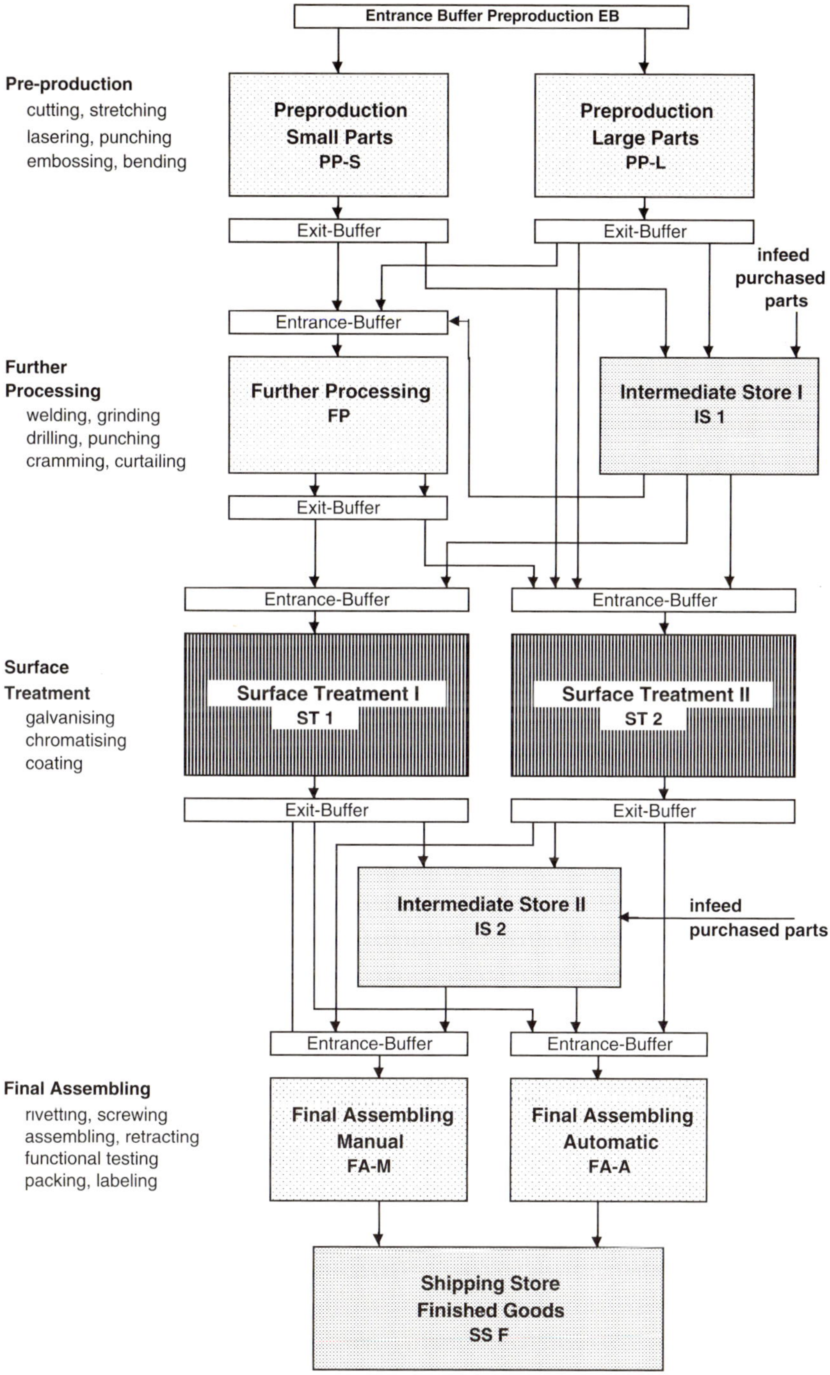

Fig. 20.1 Production network structure of a metal processing company
Bottleneck stations: surface treatment (ST)

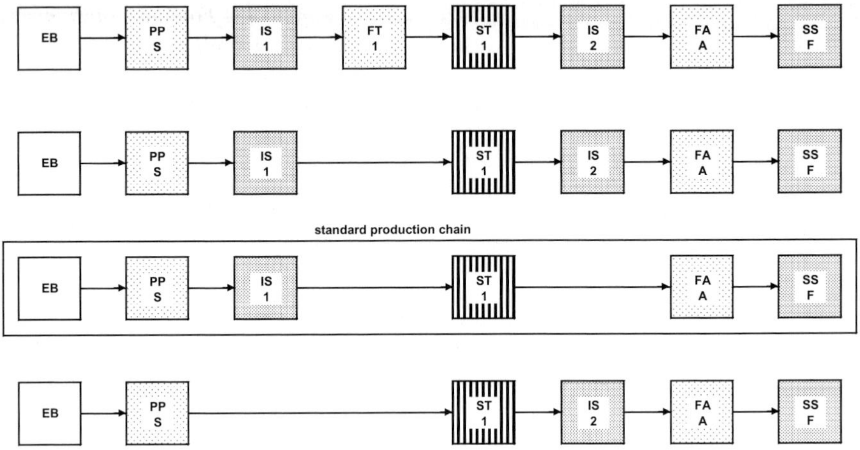

Fig. 20.2 Standard production chains of a metal processing company
Production structure: see *Fig. 20.1*
Bottleneck stations: surface treatment (ST)

A *process plan* determines the sequence and connection of the involved production and performance stations, specifies the technical processes and fixes the *minimal lead times* for representative product groups with standard order quantities. From the partial lead times of the stations, a *standard delivery time* can be calculated for the different process plans taking into account the mean transport and waiting times, which are expected at maximal tolerable utilization. The *punctuality* of production, i.e. the probability to keep standard delivery times, depends on the actual demand and the variability of the order quantities and lead times.

If a certain process step can be executed by several production stations, the corresponding process plan offers different *options*, but leaves open which one should be used. A standard process plan also does not fix the starting date and the finishing date.

Medium- and short-term production planning is performed regularly in fixed *planning cycles*, e.g. weekly or monthly. Approved *steps of short-term planning* are:

1. *Advanced resource planning* (ARP) and *material requirement planning* (MRP): The weekly or monthly demand for final products is forecasted or planned for the next 6 or 12 months. Starting with the outflows of the exit stations and proceeding retrograde, the partial product flows (20.1) of the *upstream* stations are calculated with the bill of material and allocated to the standard process chains or trees.
2. *Announcement and check of resources and material*: The results of ARP and MRP are announced to the production and external suppliers, in order to provide the required resources and capacities.

3. *Identification, adjustment and elimination of bottlenecks*: *Potential bottlenecks* are stations that operate in peak times close to 100% utilization. For *critical bottlenecks*, the expected utilisation exceeds 100% during unacceptable long time. In order to achieve a balanced utilisation, the limit performances of potential bottlenecks are adjusted. Critical bottlenecks are eliminated as far as possible (see *Sect. 13.8*).
4. *Adjustment of operating and scheduling times*: Working hours and operating times of the production stations are adjusted to the expected demand and synchronized.

The selection of the execution stations in case of several options and determining of the execution dates for actual orders are tasks of *production scheduling*. Only for major projects and for pre-production for sales actions and bottleneck phases, production planning fixes starting and finishing dates for the relevant orders.

20.4 Production Scheduling

The main objective of scheduling is to ensure the cost optimal execution of currently *incoming orders* in due time. For this purpose, order scheduling prioritizes incoming *external orders* and disaggregates them due to appropriate *scheduling strategies* into *procurement orders* for suppliers and into internal *production orders* and *delivery orders* (see *Sect. 2.2*). The internal orders are transferred to the production, storage and commissioning systems, where they are executed. Additional tasks are procurement and dispatch scheduling.

The orders can be executed with promised punctuality only, when the production planning has provided the necessary resources and material. Essential for business success is the compatibility of the central order scheduling strategies with the strategies of inventory scheduling, order picking, procurement and production. It is also necessary to adjust the period lengths and to synchronize the starting times of operation and scheduling (see *Sect. 8.3*).

Goals of production scheduling are to ensure the efficient order execution and to keep the agreed lead times. For this purpose, the current orders are allocated – based on adequate *production strategies* – to the available resources of production.

20.4.1 Stock Production or Order Production

Storekeeping articles can be delivered ex stock or produced to order. From the cost opportunity of storekeeping in *Sect. 11.12* result the *delivery rules for storekeeping articles*:

- *Small orders* with quantities less than half of the optimal replenishment quantity and smaller than the current stock should be delivered ex stock
- *Major orders* with quantities larger than half of the optimal replenishment quantity or the current stock should be produced to order if the required delivery time is longer than the lead time of the production
- Major orders with delivery time shorter than the lead time are split into a *make-to-stock-share*, which is half the size of the optimal replenishment quantity and

delivered immediately from stock, and a *make-to-order-share* that is produced to order and delivered later.

In the long run, these delivery rules achieve *self-regulating* a cost-optimal share p of orders for order production and a remaining share 1-p for production on stock (see *Fig. 10.2*).

20.4.2 Dynamic Production Scheduling

Production scheduling can be performed statically in longer fixed *scheduling cycles*, e.g. weekly or monthly, or dynamically depending on the required punctuality or single events, such as arrival of an urgent order or unexpected interruptions. For example, daily scheduling is necessary, if day-accurate delivery is required.

As long as the demand does not exceed the production capacity and the required materials are available in due time, lowest production costs and punctuality are achievable by the following standard *production scheduling procedure*:

1. For all production orders, which have arrived until the scheduling date, the latest *starting date* at the critical *order penetration station* is scheduled backwards from the delivery date with the minimal lead times (20.5) of the involved stations (see *Fig. 8.3* and *Sects. 8.7* and *8.8*).
2. The orders are sorted in *urgency sequence* with ascending starting dates. The orders with *due start date* are scheduled with first priority for the next production period.
3. All due orders concerning the same product P_r are collected in *series-production orders* SO_r with quantities m_r. They generate the *required output flows* $\lambda_r = m_r \cdot \lambda_{SOr}$.
4. The series orders are assigned to the cost optimal standard process chain or process tree in the order of increasing or decreasing order quantities.
5. Starting with the final station and progressing retrograde, *upstream* to the order penetration points, the input flows (20.1) for the single stations of the production chain or tree, which are the output flows of the proceeding stations, are calculated with the help of the bill of material from the respective output flows.
6. For these output flows, the utilizations (20.3) of the production stations are calculated from the limit performances and switch times. In this way, the *potential bottleneck*, i.e. the station with highest utilisation is identified.
7. If the switch times differ for different products, the switch time sum for the potential bottleneck station is minimized by *optimal order sequencing* (see *Sect. 10.4*).
8. If the total utilization of the potential bottleneck station is less than the technical availability, e.g. $\eta_{avail} = 95\%$, steps 3 to 6 are repeated after adding orders with descending urgency until the bottleneck utilization reaches the technical availability or until all orders are scheduled.
9. If the total utilization of the bottleneck station exceeds the availability and its operation time cannot be extended, the station is an *actual bottleneck*, which

prevents the punctual execution of all due orders. In this case, less critical due orders have to be removed one by one, following appropriate priority rules, and the steps 3 to 6 are repeated until the utilization is below availability.

10. With the orders of the bottleneck station, the *upstream stations* are scheduled due to the *pull-principle*, the downstream stations due to the *push-principle*.
11. Finally, the resulting production orders are transferred to the internal production stations and the procurement orders to the suppliers and storage stations.

The steps of the standard production procedure are performed after the last scheduling period has ended and before the next period starts. This can be done by people only in longer scheduling cycles for small numbers of stations, orders and products. For shorter cycles, many orders, longer production chains and extended production networks, suitable *production planning and scheduling software* (PPS) it required.

20.4.3 Additional Rules and Strategies

When implementing dynamic production scheduling, the following rules and strategies must be taken into account:

- The scheduling periods of production should not be longer then the periods of order scheduling. As outlined in *Sect. 8.2*, the period length is determined by the required *punctuality*, e.g. when aiming at day-accurate delivery, daily scheduling is necessary.
- If the limit performances refer to hours and switch times are measured in hours, formula (20.5) gives the lead time in hours.
- For calculating the daily utilization (20.3), the daily production demand and flows have to be divided by the actual number of operating hours per day. This holds for fixed and for varying operating times.
- A *breathing factory* with flexible working hours adapts the operating times dynamically to the current performance demand and can prevent bottlenecks to a certain extend (see *Sect. 10.5*).
- Not yet started and uncompleted orders of the previous day are newly scheduled.

If a production network offers several production stations or chains for the same order with different costs, *minimal costs* are achievable by

- *Cost optimal allocation*: Each order is allocated to the production station or chain, which can execute it at lowest *order costs*.

The order costs are the sum of the quantity-independent *setup cost* and the quantity-dependent *execution costs*. If the order costs for the optional stations are equal or unknown, *optimal utilization of parallel stations* is achievable by

- *Dynamically balanced allocation*: The orders are allocated to the lowest utilized station. When the utilization of all running stations exceeds 90% for a certain time, another station is added, if available. If the overall utilization falls below ((N−1)/N)·80%, the station with lowest utilization is closed.

With dynamic adding and closing of stations, the mean utilization is kept *self-regulating* between 80 and 90% and longer waiting queues and waiting times are avoided.

20.5 Procurement and Dispatch Scheduling

The central tasks of company logistics, *procurement scheduling* and *dispatch scheduling*, are closely connected with production scheduling: Procurement scheduling takes care of the cost optimal and punctual provision of material and parts from external suppliers. It receives orders from production and inventory scheduling. Dispatch scheduling is responsible for the cost optimal and punctual delivery of the finished orders to customers.

The basic strategies of procurement are *procurement on stock* and *procurement on order*, either *advanced* or *just-in-time*. The potentials, risks and costs of these and further procurement strategies have been outlined in the *Chaps. 8* and *11*.

Dispatch scheduling can be performed by a scheduling center, a local dispatch department or a *logistic service provider* who is responsible for storing, commissioning and delivery. In order to avoid self-optimization and to ensure maximal consolidation and cost optimal dispatch, clear dispatch rules should guide the operations of the dispatch department or logistic service provider. Approved *dispatch rules* are:

- Several single customer orders that are due at the same delivery date are consolidated in a single shipment and delivered in the most cost-efficient manner.
- If customers agree upon fixed delivery dates, for example on a delivery on every second day or once a week, all deliveries are collected in an intermediate buffer until the due date is reached and delivered as a whole.
- Further consolidation is achieved by collection of shipments destined for different customers in the same region or country that can be shipped on the same day.

The consolidated shipments are picked up either by a *parcel service provider* or a *freight forwarder* in full or part truck loads and transferred to the next transhipment point. Shipments of sufficient quantity can be transported directly to the transhipment station in the target region. This is especially efficient for shipments to distant countries (see *Chap. 21*).

Order scheduling also decides whether order-produced articles should be shipped directly *ex factory* or via a *dispatch logistic center* (see *Fig. 19.4*). This decision depends on the order content, the logistic costs and on the technical equipment in the production site and the logistic center. If proper equipment is available, standard *rules for ex factory shipment* are:

- Shipment ex factory is necessary for urgent single-item orders or part-order deliveries, if the delivery date can only be kept by direct delivery.
- Direct delivery ex factory is opportune for large, heavy and bulky parts, machines and aggregates, which are packed for shipment at the end of the production.
- Direct delivery ex factory is faster and opportune for large quantities, which can be shipped in full pallets as part or full loads on trucks or trains.

In the last case, continuous production directly into the transport means is optimal. Standard rules for *shipment via logistic center* are:

- Delivery via logistic center is necessary for the complete delivery of *multi-position orders* with storekeeping articles and/or order-articles from several production sites.
- Dispatch via logistic center is cheaper for *small quantity orders* due to more efficient storing, commissioning and packaging.
- A logistic center enables consolidation of shipments to different customers in the same region.

These dispatch rules, as well as the standard production strategies can be implemented by decision-based scheduling software, which proposes the cost optimal execution and delivery mode for each order.

20.6 Bottleneck Strategies

Future peaks in demand and potential bottlenecks can be identified in advance during the medium-term planning for the forecasted demand. The total demand $\lambda_{PS}(t)$ for a production station PS in a future period t is the sum of the demand of all articles or products P_r which will be executed by this station:

$$\lambda_{PS}(t) = \sum_r \lambda_r(t) \quad [PU/PE] \tag{20.6}$$

For the advanced identification of bottlenecks, the *demand structure*, given by the *structural weights* $g_r = \lambda_r/\lambda$, and the mean *switch time loss per production unit* τ_{STm} of the past are assumed to hold also in future. With these values the *effective limit performance* of a production station PS is:

$$\mu_{PS} = \left(\sum_r g_r/\mu_r + \tau_{STm}\right)^{-1} \quad [PE/PU] \tag{20.7}$$

In periods where the total demand does not exceed the effective limit performance, i.e. for $\lambda_{PS}(t) < \mu_{PS}$, all orders can be executed within the standard lead times. If for a *bottleneck phase* of length T_{BN} the demand exceeds the limit performance (20.7), the production becomes a *critical bottleneck*. During the periods t of the bottleneck phase T_{BN} an *order backlog* accumulates which is equal to the *excess demand*:

$$D_{ex} = \sum_{t \epsilon T_{BN}} (\lambda_r(t) - \mu_{PS}) \quad [PU]. \tag{20.8}$$

If the demand in previous periods is less then the limit performance, the order backlog can be reduced or even avoided by *advanced production* of *bottleneck articles*. Suitable scheduling software recognizes and announces bottleneck phases, calculates the accumulating excess demand (20.8) and generates pre-production orders using the excess capacity in periods with low utilization. The disadvantage of *pre-production* for bottleneck phases is a higher stock with all its costs and risks. This leads to the *pre-production recommendation*:

- The scheduler decides on advanced production or procurement in agreement with sales department and management, not the scheduling program.

For different articles that are produced by the same bottleneck station hold the *pre-production selection rules*:

- Articles with the most reliable highest sales frequency should be pre-produced first and kept on stock for the bottleneck phase.
- If two articles have equally high and reliable sales frequency, the one with the lower value should be pre-produced first in order to save inventory interest.

These selection rules prevents the production of non-saleable articles and long storing times in case, that the demand does not reach the expected peak or even falls off. The best selling articles can be produced in cost-efficient lot sizes during periods of low utilization.

If however pre-production is impossible or insufficient and the production demand exceeds the limit performance for more than one period, the respective production station becomes a *critical bottleneck*. During this time, order backlogs and longer delivery times are unavoidable. In such a case, the limited resources should be allocated to the customers by the scheduler, not by the program. In order to avoid conflicts and arbitrary decisions, it is recommendable to agree in advance about appropriate *bottleneck allocation rules*, e.g.:

first come first served
profit margin
absolute profit
urgency (20.9)
customer importance
out-of-stock costs
related to regular demand

The last rule, which distributes the limited production output in relation to demand of the regular customers in normal times, is the fairest allocation.

If production supplies several consumption stations of an extended supply network, *central scheduling* is essential during bottleneck periods in order to keep the allocation rules. Generally holds:

- *Local scheduling* by decentral consumption stations, e.g. by sales outlets or small customers, is satisfactory if the products are distributed via a storekeeping logistic center, as long as the production capacities are sufficient.
- *Central scheduling* is necessary for order-products and if shortages and bottlenecks are expected.

Pre-production is of no help, if the bottleneck phase lasts for a long time. Possibilities in such situation are raising prices, outsourcing of critical production or investment into new production facilities.

20.7 Logistical Optimization of Production

The tasks of *production logistics* are organization, planning and scheduling of production systems, but not to develop new technologies, to improve technical processes or to construct and set up production machines and process plants. These are tasks of *production technology* and *process engineering*.
Production logistics can contribute to optimal production by:

- *Professional long-term production planning* following the above planning procedure
- Computer based *advanced resource planning* and *material requirement planning* with the above strategies
- Mutually agreed and synchronized order scheduling, production scheduling and logistic scheduling based on *optimal scheduling strategies*
- *Dynamic production scheduling* following the above standard scheduling procedure
- *Reduction of setup and switch times*: Shorter setup times and switch times improve productivity, decrease costs and shorten delivery times.
- *Reduction of assortment and variants*: Lowering the number of products and/or variants reduces switch time losses, increases productivity and shortens delivery times.
- *Order consolidation*: Small orders for the same product are consolidated to larger series orders and executed as a whole with low switch time loss.
- *Optimal order execution sequence*: By executing the orders for different products in optimal sequence, the sum of setup times is minimized.

Examples for optimal order execution sequences are the light-dark order sequences, which are customary in bottling operations, printing plants and dyeing mills.

Chapter 21
Optimal Networks and Supply Chains

The tasks of *network management* are the design, organization, set up and operation of logistic networks for the supply of consumers and companies with the required goods in due time at lowest costs. This includes *scheduling* of resources and *replenishment* of stocks in the logistic chains.

A *logistic chain* links a *delivery station* with a *receiving station* by a number of *transport connections* and *intermediate stations*. Depending on purpose and aspect, a logistic chain is called *supply chain, procurement chain, replenishment chain, transport chain, freight chain, carriage chain* or *disposal chain*. In general, different chains are possible for delivering goods from a source to a sink, for the transport of a consignment or for a journey from a starting point to a destination. This leads to the *task of supply chain management*:

- In order to execute *supply orders* under given *restrictions* at *lowest costs*, an efficient logistic network has to be set up, and within this network, the optimal order specific logistic chains have to be selected.

At first sight, this task seems to be quite simple. At closer inspection, it becomes obvious that it covers the whole field of logistics. That means, *logistics* and *supply chain management* deal with the same tasks under different aspects (see *Chaps. 1* and *15*) (Alvarado and Kotzab 2001; Ballou 2004; Bretzke 2008; Bucklin 1966; Christopher 1992; Chopra and Meindl 2007; Cooper and Lambert 1997; Frazelle 2001; Kuhn and Hellingrath 2002; Lambert et al. 1997; Schönsleben 1998; Scott and Westbrook 1991; Simchi-Levi et al. 2008).

A logistic chain is an external or internal chain or a connection of external and internal chains. Accordingly, the tasks of supply chain management arise between companies, suppliers and customers, but also within a production plant or logistic site. *Internal logistic chains* connect the sources and sinks within a plant, site or station. They start at the goods entry or from internal sources and end in internal sinks or at the goods dispatch of the same location. *External logistic chains* connect the goods dispatch of a company, production site, logistic station or supplier with the goods entry of another company, site, station or customer. An *integrated logistic* chain starts at the goods dispatch or from a source and ends after passing several *intermediate stations* at the goods entry of a distant sink.

T. Gudehus, H. Kotzab, *Comprehensive Logistics*,
DOI 10.1007/978-3-540-68652-1_21,

In daily business, supply chain management is confined to the selection of cost-optimal logistic chains for current orders. To remain competitive, the company, which pays the delivery costs, must permanently optimize its supply chains. When demand changes, the supply chains have to be adapted or new logistic structures have to be designed and implemented.

For deliveries *free house* (*franco domicile*), supply chain management is a *delivery task* for the supplier. For deliveries *ex works* (*off factory*), it is a *procurement task* for the recipient. If delivery and procurement are outsourced, supply chain management becomes a *transport, storing* or *logistic task* for the *service provider*.

In this chapter, methods and tools for network design and supply chain management are presented. Although the main focus is on external and integrated logistic chains, these methods are transferable also to internal logistics. Optimal network and supply chains are determined by the *number, locations* and *relations* of customers and suppliers, and by their *performance* and *service demand. Parameters of action* are the number, locations, connections and functions of intermediate stations. The cost optimal utilization of a given logistic network can be achieved by appropriate *supply strategies*.

The most important aim of supply chain management is minimal costs. Hence, the calculation of delivery costs and the development of a general performance cost model are further topics of this chapter. The presented cost-calculation model can be used for optimizing logistic networks, selecting the optimal supply chain, and calculating *freight costs*. Resulting are *design principles* for logistic networks, *selection rules* for supply chains, methods for improving performance and service, and *potentials for cost reductions*.

21.1 Structure Requirements

The *structure requirements* are determined by the *delivery stations* and *receiving stations*. Many structure requirements, for example, the locations of key suppliers and customers, are *fixed points*. Generally the *intermediate stations* and *transport connections* can be changed. Their number, locations and functions are *structure parameters*, which can be used for designing a new supply network and optimizing an existing network.

21.1.1 Recipient Structure

Receiving stations, customers or sinks of the flows of goods and freight are, apart from internal sinks, plants and stores of industry, logistic centers, markets and outlets of retailers and the households of final customers. The *parameters* of the *customer* or *recipient structure* are:

- number N_{RS} of the receiving stations
- locations $(x_j; y_j)$ of the receiving stations RS_j, $j = 1, 2, \ldots\ldots N_{RS}$

Suppliers usually classify their customers based on sales-specific or other non-logistic criteria. However, a sales oriented customer classification should not be binding for logistics. On the contrary:

- Supply chains and distribution networks can only be optimized independent from any sales-oriented and non-logistic customer classifications.

Apart from commercial purchasing activities, within the receiving stations, administrative and operative logistics activities are executed. *Administrative logistic activities* in a receiving station are:

scheduling of stocks and replenishments
release of the required quantities from the suppliers
issue of delivery orders for replenishment ex works
control of delivery dates and quality (21.1)

Operative logistic activities in a receiving station are:

unloading, unpacking, and reception control
break-down of load units in single items
in-storing and keeping of stocks
buffering of goods and consignments
internal provision at the points of consumption and demand
collection and dispatch of empties (21.2)

The *point of consumption*, e.g. of an automotive plant, is the assembly line. The *points of demand* of a retail outlet are the sales counters, and of a self-service shop, the racks and shelves.

The logistic activities (21.1) and (21.2) in the receiving stations generate costs, which depend on the parameters of the supply chains. The most important cost drivers are the *delivery frequencies* and the *delivery mode*. In many cases, the logistic costs of the receiving station make up a considerable part of the total delivery costs and are generally not negligible.

21.1.2 Supplier Structure

The delivery stations or sources of goods for the recipients are production plants or finished goods stores of manufacturers, logistic centers and import stores of wholesalers and retailers, transshipment points of logistic service providers, railway stations, seaports or airports, but also other sites and locations of the own company. The parameters of the *supplier structure* are:

- number N_{DS} of the delivery stations
- locations $(x_i; y_i)$ of the delivery stations DS_i, $i = 1, 2, \ldots\ldots N_{DS}$

Companies cluster their suppliers based on purchasing criteria or other non-logistic aspects into *supplier groups*. As for the customer classification holds for the supplier classification:

- Supply chains and procurement networks can only be optimized independent from purchasing and non-logistic supplier classifications.

Sales activities and commercial order acceptance of a supplier are often located at one point, whereas the goods are shipped from another point or from several delivery stations located elsewhere. For supply chain management, only the location of the delivery stations and their logistic activities are of interest.

The *administrative logistic activities* of a delivery station that are necessary to initiate and to control the shipments are:

acceptance, scheduling and control of orders
scheduling of finished goods stock
placing of production orders
generation of commissioning orders (21.3)
issue of transport and freight orders
control of delivery dates and quality

Standard *operative logistic activities* in a delivery station are:

storing of storekeeping articles
accumulation of customer-specific goods
out-storing, commissioning and provision
filling, bottling and confectioning (21.4)
packing parcels
build-up of load units
provision for dispatch

Depending on the *terms of trade*, additional tasks of a delivery station can be *loading* of consignments into the transport means and *load securing* for shipment and transport. The logistic activities of delivery stations generate costs which are part of the total delivery costs. The main *cost drivers* are the *delivery frequency*, the *delivery mode* and the *transport mode*.

The *internal lead times* of a delivery station contribute considerably to the *delivery time* of a supply chain. The internal lead time for storekeeping articles is the sum of the administrative and operative order execution times within the storage station. For goods and products produced or procured on order, the delivery time increases by the production time respectively the replenishment time.

21.1.3 Intermediate Stations

Between the delivery stations and the receiving stations, the goods and consignments can pass *intermediate stations* IS_k, $k = 1, 2, \ldots N_{IS}$. Their parameters are:

- number N_{IS} of intermediate stations
- locations $(x_k; y_k)$ of intermediate stations
- functions F_k of intermediate stations

These are the most important *design parameters* for logistic networks.

Within an intermediate station, the goods or freight units are unloaded, sorted, stored, buffered, changed and reloaded. Depending on their main function, non-storekeeping *transshipment stations*, *storage stations*, *transformation stations* and multifunctional *logistic center* are differentiated.

The *transfer times* through the intermediate stations of a logistic network increase the total delivery time of a multi-stage delivery chain in comparison to the delivery time for direct delivery. The *minimal transfer time* is the sum of all internal transport and handling times between unloading and reloading. The *effective transfer time* is the sum of the minimal transfer time and the *waiting time* until departure of the next delivery transport or until arrival of the last consignment for the same destination.

21.1.4 Transshipment Stations

In a *transshipment station* for *transit goods* the following handling processes are executed as shown in *Fig. 21.1*:

- *Transfer without load carrier change* or *one-stage crossdocking* (C1): Goods and consignments arriving in destination-mixed load carriers are transferred due to their destination to the dispatch buffer places for the next delivery tour or directly into waiting transport units.
- *Transfer with load carrier change* or *two-stage crossdocking* (C2): Articles and packages arriving in destination-mixed load carriers are *split to zero*, sorted and repacked on destination-pure load carriers, which are transferred to the dispatch buffer for the next delivery tour or directly into a waiting transport unit.

The transfer times of a transshipment station are normally far less than 24 hours.

As an example *Fig. 21.2* illustrates the transshipment station of a *parcel service provider*. Here the arriving parcels, which are normally transported without load carriers, are unloaded onto telescope conveyors, transferred by a *high performance sorter* to the dispatch gates and reloaded directly via telescope conveyors into the waiting transport vehicles.

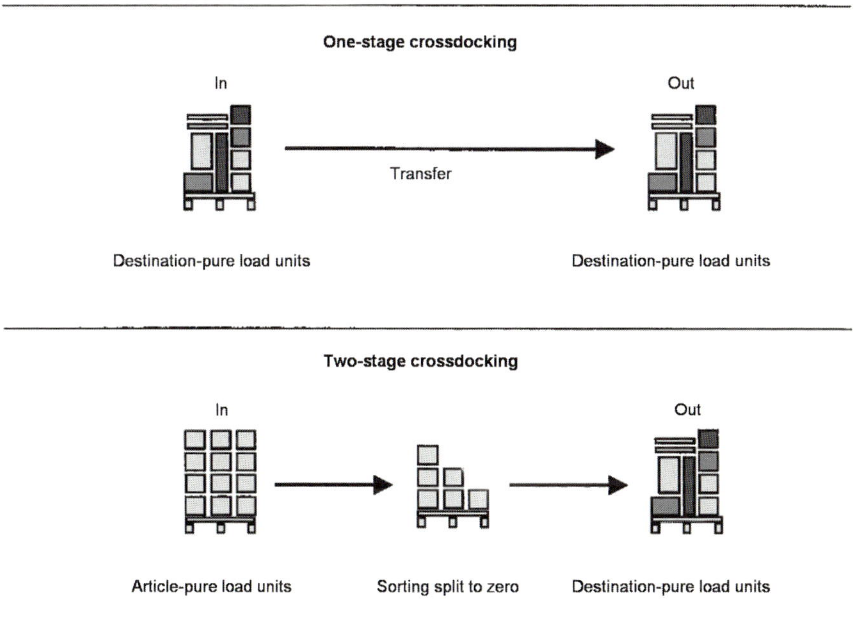

Fig. 21.1 One-stage and two-stage crossdocking of palletized goods

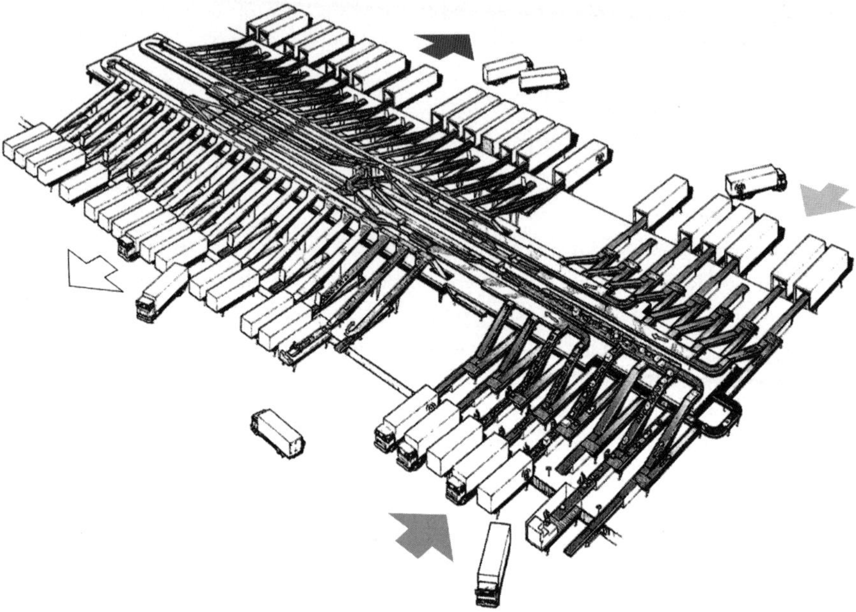

Fig. 21.2 Transshipment station of a parcel service provider
Source: *VanderLande-Industries*

Another example is the transshipment station of a German *DIY-retail chain* shown in *Fig. 21.3*. *Outlet-loaded pallets*, which are destined for the same outlet, are unloaded, checked and transferred to the buffer areas in front of the departure gates, or loaded directly into the waiting transport units. *Article pallets* and consignments with parcels for several outlets are unloaded, split, sorted and commissioned and loaded on *outlet pallets*. This is done in a separate area by *inverse commissioning* (see *Fig. 17.5*) The outlet pallets are also transferred to dispatch buffer places or directly into waiting transport units.

Within smaller *transshipment points* (TSP) of freight forwarders quick-lift and high-speed trucks transfer the pallets from the entrances to the exits. Tug&tow trains, under-floor chain truck conveyors or automatic guided vehicle systems are used in larger logistic centers, for example in *airfreight centers*. The *hubs* of *parcel service providers* operate with fully automated sorters as shown in *Fig. 21.2*. *Container terminals* use cranes, heavy load forklift trucks and Van Carriers. In the most modern terminals, e.g. in the harbors of Rotterdam and Hamburg, 20'- and 40'-containers are carried by *automated guided trucks*.

21.1.5 Storekeeping Intermediate Stations

The dependency on the delivery time of upstream stations, and waiting for supply are reduced or even eliminated, if the intermediate stations keep stocks. That means:

- ▸ *Delivery times* can be reduced by inserting storekeeping stations.

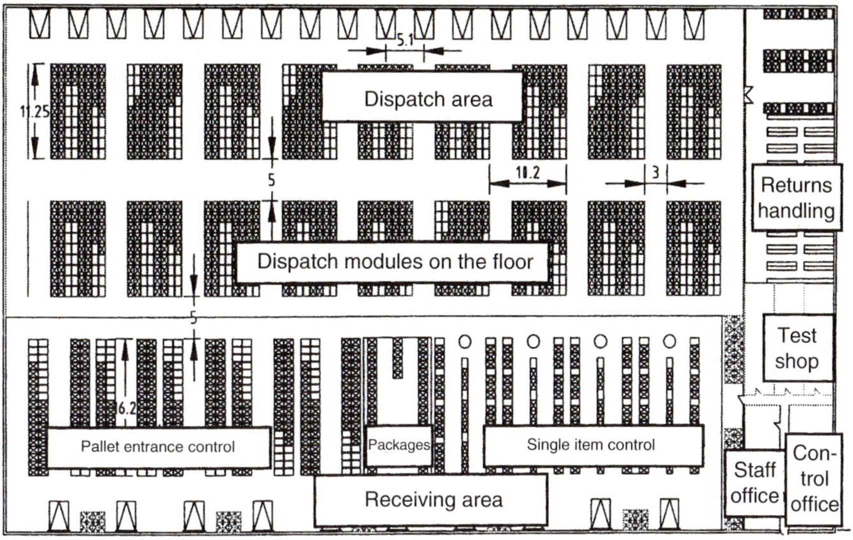

Fig. 21.3 Transshipment station of a DIY retail chain

The delivery time decreases with increasing proximity of stock to the *point of demand*. The disadvantages of shorter delivery are the *storing costs* and the *misallocation risk* of stocks. The closer an anonymous stock, which is not dedicated for a one customer or for one region, the higher the costs and risks of storekeeping (see *Sect. 11.3*).

The internal order lead time for storekeeping articles is the sum of the administrative and the operative order processing time in the storage station. It is independent from the production and supply times, if stock and replenishment are correctly scheduled (see *Chap. 11*).

In storekeeping intermediate stations, the following storing processes are possible:

- *Storing without commissioning*: The arriving article or consignment load units are stored in, kept in storeplaces for a certain time, retrieved due to the incoming orders and dispatched without changing the load carrier.
- *Storing with commissioning*: The arriving article-pure or mixed-consignment load units are stored in, kept in store and disaggregated by commissioning due to the incoming orders. This process changes the load carrier.

Storing with commissioning transforms arriving load units with single articles or mixed consignments into mixed-article or single-consignment dispatch load units respectively. Methods, techniques, dimensioning and examples for storing and commissioning are described in *Chaps. 16* and *17*.

21.1.6 Transformation Stations

Within a transformation station, incoming goods are transformed into logistically different articles are units. According to increasing degree of change, the following *transformation processes* can be differed:

- *Filling, bottling and packing*: Liquids or loose goods are filled into bottles, cans, crates, sacks, bags or other packages. With this process, loose and bulk goods are transformed into packed goods.
- *Cutting and trimming*: Plane material such as plates or panels is cut to required measures; long material such as slug material, cables or strip material is trimmed to a certain length.
- *Co-packing and confectioning*: Different articles or several packages are consolidated, set up and co-packed using packaging material in *displays*, *trays* or customer specific packages.
- *Set up and assembly*: Parts and components are put together and assembled to complete modules and finished products. The involved goods remain physically and chemically unchanged.
- *Production and manufacturing*: A technological process transforms raw material, additives and other supplies into different material or products. The incoming goods are chemically and/or physically modified.

Design and scheduling of assembly, production and manufacturing systems are tasks of *process engineering* and *production logistics* (see *Chap. 20*).

21.1.7 Intermediate Logistic Stations

The basic processes in intermediate logistic stations without assembly, production or manufacturing are:

unloading, reloading, loading
receiving inspection
breaking down and setting up load units
in- and out-storing
buffering and storekeeping (21.5)
bottling, filling and packing
cutting and trimming
confectioning and co-packing
sorting and commissioning

These processes and the design of logistic systems, stations and centers are described and explained in detail in *Sect. 1.6* and in the *Chaps. 16, 17, 18* and *19*. The *transfer costs* for the intermediate stations resulting from handling, storing, commissioning and transport contribute considerably to the total delivery costs.

As shown in *Fig. 1.12,* a multifunctional logistic center offers different internal logistic chains. That means:

- For the same kind of goods or orders, several internal logistic chains are possible which differ in throughput time, performance and operating costs.

Designing, implementing and optimizing internal logistic chains are tasks of *material handling* and *intralogistics*. Design and selection of cost optimal logistic chains through the intermediate stations for the different goods and consignments are important means for the overall optimization of supply chains and networks.

21.1.8 Transport Modes and Transport Means

For the external transport of material, goods and consignments between the stations of a logistic network, the following *transport modes* are available:

road
rail way
inland waterway (21.6)
seaway
airspace

For the individual case, generally only one, two or three of them are relevant. In addition to the transport modes (21.6), liquids and gases, even solids, parcels and letters, can be transported through *pipelines*. Bulk goods and general cargo are also conveyed with cable cars or by belt conveyors. However, pipelines, cable cars and belt conveyors are only efficient for a continuous material flow, which lasts for many years (see *Chap. 18*).

Sections with the same transport mode can be connected by intermediate stations to *intra-modal transport chains*. Different transport modes are combined as shown in *Fig. 21.4* to *inter-modal transport chains* or *freight chains*. This leads to the *options of transport combination*:

- *unbroken transport* with the same transport mean from departure to arrival
- *broken transport* by changing transport means on the same transport mode
- *inter-modal transport* with changing transport modes and transport means

In each transport mode (21.6) different *transport means* are possible (see *Fig. 18.19*). This leads to the *options of transport means*:

- *Road-transport means*: small trucks, delivery vehicles, road semi trailers, trucks with swap body trailers, silo vehicles, road tankers etc.
- *Railway-transport means*: passenger and cargo wagons, platform-, silo- and tank-wagons, wagons groups, linked wagons in part-trains and total trains.
- *Waterway-transport means*: lighters, barges, inland vessels, bulk carrier, container ships, feeder ships and tankers.
- *Air-transport means*: small-, medium- and large-body airplanes, freight and passenger airplanes, balloons and zeppelins.

Each transport mean or *transport unit* [TU] has its specific *transport capacity* or *loading capacity* C_{TU} [m^3, kg, PU or LU per TU], which depends on the dimensions of the loading space and the tolerable net weight. For unpacked and bulk goods, the load capacity is measured in volume units [m^3] or weight units or [t], depending on the specific weight of the load in relation to the specific net weight of the transport unit (see *Sect. 12.5*). For packed goods without load carrier, the transport capacity should be measured in *packaging units* [PU] and for freight on load carrier in *load*

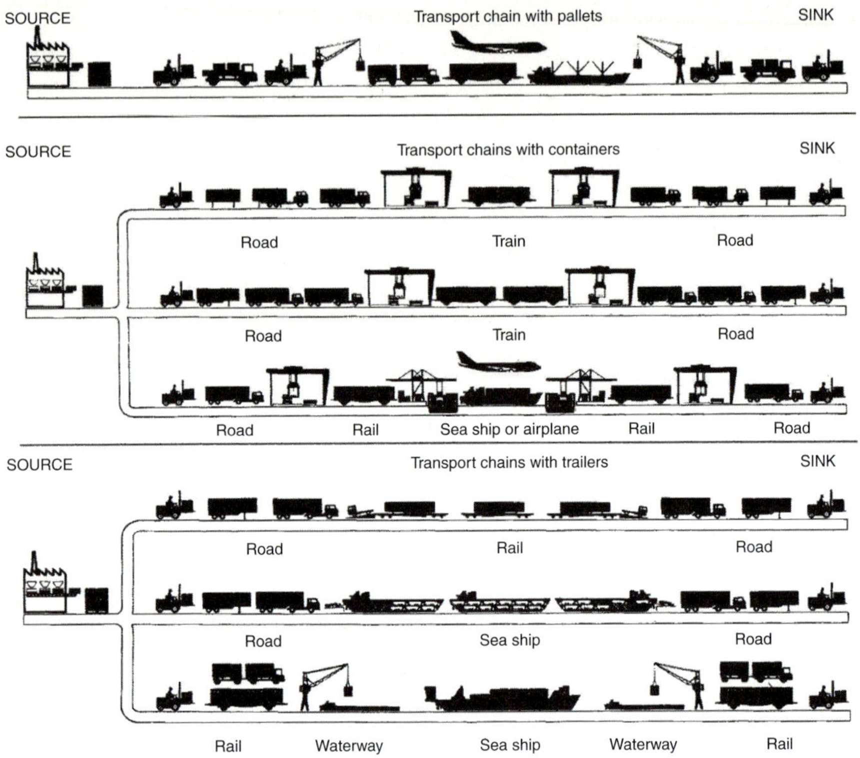

Fig. 21.4 Examples for inter-modal transport and freight chains
Source: SGV *Studiengesellschaft für kombinierten Verkehr* e.V. (German Research Society for Combined Traffic)

units [LU]. Capacities, loading space and net weight of some transport means for road, rail and sea transport are presented in *Table 18.3*. *Table 18.4* contains the specific performances and costs rates of these transport means.

Transport means with great capacity, such as semi-trailers or swap-body trailers on the road, full trains on the rail and big container ships, achieve very low drive costs per load unit, but have high basic costs and stop costs. Big transport means are therefore especially appropriate for the transport of large quantities over longer distances with few stops.

Small transport means, such as *pickup and delivery vans* and *feeder ships*, have far lower basic costs and stop costs but higher drive costs per load unit. They are opportune for the transport of small quantities over shorter distances with many stops.

21.1.9 Transport Tours and Strategies

As explained in the *Sects. 18.5* and *18.11*, it is possible to transport the load from the delivery points to the destination points on different tours following different transport strategies. The basic *transport options* are:

- *Single Runs*: Goods and consignments are transported without intermediate stops in a direct take-over and delivery run from one dispatch station to one recipient station.
- *Mixed Runs*: Goods and consignments are transported in one transport unit with intermediate stops from one or several dispatch stations to one or several recipient stations.

In road traffic, single runs with full trucks are *full load transports*. Mixed runs with part loads for several destinations are *part load transports*.

The task of scheduling single runs is to find the *shortest route* in an existing traffic network. Additional tasks arise for scheduling mixed runs by the *options of route scheduling* as shown in *Fig. 21.5*:

- *Pickup runs* are collection tours for goods from several sources to one collection point or logistic station.
- *Delivery runs* are distribution tours for goods from one distribution point or logistic station to several destinations.
- *Combined runs* are integrated distribution and collection tours, starting and ending at the same transshipment point or logistic station.

The methods and strategies of *transport scheduling* have been described in *Sect. 18.11*.

21.1.10 Transport Organization

The *carriage time* for a shipment is the sum of the *waiting time* until departure, the *loading time*, the *travel time* for the total path to the destination, the *unloading time*, and of the *stop times* at intermediate stations. The travel time depends on the effective speed of the transport mean, the traffic conditions and the path length. The loading times and the stop times are determined by the *drop-quantities* and the loading technique.

The waiting time until departure of the next transport depends on the transport organization. Basic options for the *transport organization* are:

- *Regular transports*: *Planned tours* or *line services* are executed with constant frequency at defined arrival and departure times based on a *time schedule*.
- *Irregular transports*: *Tramp rides* or *demand services* are performed on altering tours, triggered by demand, either if a sufficient load quantity has accumulated or if an urgent consignment must be shipped.

Irregular transports connect stations with *sporadic freight demand* and supplement the regular transports between the scheduled dates for with regular transports.

The travel time of a transport unit for a regular tour is determined by the effective travel speed v_{TU}, the tour length s_{tour}, the number of stops n and the mean stop time t_{stop}. The *mean tour travel time* is:

$$T_{tour} = s_{tour}/v_{TU} + n \cdot t_{stop} \quad [PE/Tour] \qquad (21.7)$$

The stop time is the sum of the waiting time for dispatch and the loading and unloading time for dropping and/or pickup of the load. With a mean tour travel time (21.7),

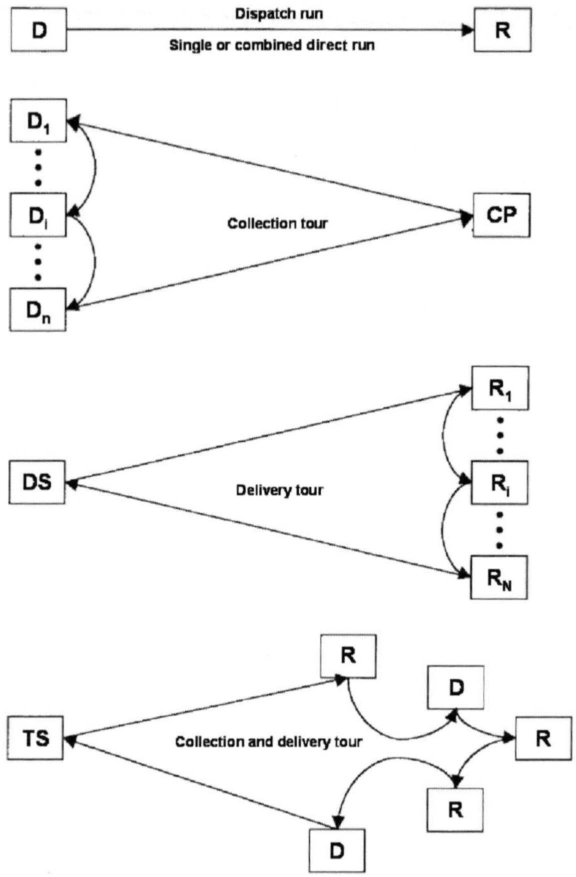

Fig. 21.5 Possible transport tours

D_i = dispatch points
R_i = receiving points
CS = collection station
DS = dispatch station
TS = transshipment station

the *number of transport units* necessary for a *transport demand* λ_{TU} [TU-Tours/PE] is:

$$N_{TU} = \lambda_{TU} \cdot T_{tour} \quad [TU] \tag{21.8}$$

As explained in *Sect. 18.8*, the transport demand $\lambda_{TU} = MAX(\lambda_{LU}/C_{TU}; \nu_S)$. It is determined by the *freight demand* λ_{LU} [LU/PE], the *transport capacity* C_{TU} [LU/TU], and by the required minimal *service frequency* ν_S [1/PE]. For example, for a global container shipping service offering a harbor service frequency of ν_S = 2 ships per week by container ships touring in T_{tour} = 7 weeks around the globe, at least N = 2·7 = 14 ships are necessary.

The number of transport units necessary to cope with a given transport demand determine the *transport costs*, which generally make up the largest share of the total delivery costs or freight costs. The relations (21.7) and (21.8) show that the *transport cost drivers* are freight demand, service frequency, transport distance, travel speed, number of stops and mean stop time (see also *Sect. 18.12*).

21.2 Service and Performance Requirements

In addition to minimal costs, the services offered by the suppliers and the performances required by the customers are the decisive goals of network management. They are determined by the *assortment requirements*, the *service requirements* and the *order requirements*, from which throughput demand and stock levels can be derived (see *Sect. 3.6*).

21.2.1 Assortment Requirements

Key task of *marketing, programme planning* and *inventory policy* is to select the goods and to determine the articles, which are produced on or provided from stock and which are produced or procured on order. The definition of the *storekeeping assortment* in the receiving stations and sinks of the supply network is task of the *assortment planning* of the customers. The storekeeping assortment in the delivery stations of the suppliers results by the *market forces* of bid and demand from the assortment requirements of the customers and the marketing and programme planning of the suppliers (Gudehus 2007).

The resulting *assortment requirements* for a supply network are:

- *number of articles* $N_A = N_{ASK} + N_{ANK}$, of the delivery stations, with N_{ASK} storekeeping articles and N_{ANK} non-storekeeping articles
- *article properties*: loose or packed, shape and bulkiness; durability, danger categories, fire classes; food or non-food; cooled or frozen etc.
- *measuring units* of bulk goods [MU: t, m^3, m^2, m . . .]
- *packing units* of packed articles [PU: barrel, piece, pallet ...] with *dimensions* l_{PU}, b_{PU}, h_{PU} [mm], *volume* v_{PU} [l/PU] and *weight* w_{PU} [kg/PU]

Selection and design of supply chains require a segmentation of the assortment in *logistic article classes* with compatible logistic properties and restrictions concerning handling, storing and transport. *Restrictions* for supply chain optimization are, for instance, prohibitions for an article to be stored or transported together with other articles. Therefore, some goods require separate supply chains, such as *cold chains* for fresh and frozen goods or *safety chains* for valuables or hazardous goods.

As outlined in *Chap. 12,* it is possible to transport the goods and articles within the separate sections of a supply chain in different logistic units. *Table 12.1* summarizes the possible *load carriers* and *logistic units* and the resulting *packaging stages*. The logistic unit in one section of a supply chain can be identical or different from the logistic unit of the next stage. If the logistic units of two subsequent sections are different, it is necessary to build them up, break them down or to repack them in the intermediate station.

In general, the packing units of the sources cannot be changed, whereas the dispatch units, in which the packing units are shipped, and the load units, which contain a consignment of dispatch units for the transport, can be selected and dimensioned within certain restrictions. Also, the load carriers for bulk material such as canisters, tanks, containers or transport silos, and the type and capacity of the transport means, such as silo vehicles or tank wagons, are often free parameters. The selection, dimensions and capacities of the load carriers, load units and transport means are *basic options* of supply chain management (see *Chap. 12*).

21.2.2 Service Requirements

The service requirements concern the *availability to deliver*, the *delivery times* and the *logistic quality* offered by suppliers or requested by customers. For articles produced on order, the availability to deliver is the availability of the necessary production capacity. It is called *availability to promise* or *availability to perform* (ATP) (Bretzke 2008; Straube 2004). For storekeeping articles, the availability is the *delivery capability* of the store. As outlined in *Chaps. 8* and *11*, the required delivery times and the availability to deliver determine the length of the supply chains, the storekeeping articles in the stations and the *safety stocks*.

The *delivery time* T_{Del} of a production station or storekeeping station is the sum of the internal *order execution time* T_{OE} from order entry until the required goods are provided at the ramp, and of the external *shipment time* T_{SH} from the goods dispatch to the entry ramp of the customer:

$$T_{Del} = T_{OE} + T_{SH} \quad [h]. \tag{21.9}$$

To keep a promised *delivery date* t_D by a supply chain, which offers the delivery time T_{Del}, the *latest shipping date* is:

$$t_S = t_D - T_{Del} \quad [d:h]. \tag{21.10}$$

In many cases, delivery times are the most important *selection criterion* for a supply chain. For customers, keeping the agreed delivery dates and arrival times is generally more important than extreme short delivery times. Delivery dates with narrow windows are far stronger restrictions for a supplier than the general promise of short delivery times, like a 24 or 48 h service.

Another important selection criterion for supply chains is the *logistic quality*. As outlined in *Sects. 3.4* and *17.4*, the *logistic quality* of a supply chain is the product of *production availability* or *stock availability*, *punctuality*, and *shipment quality*, which is measures the completeness, intactness and faultlessness of the arriving consignments.

21.2.3 Consignment Requirements

A *consignment* [Con] or *shipment* [Shm] is the quantity of goods, which has to be conveyed or shipped to a specific destination within a required delivery time. A *bulk cargo consignment* consists of a larger quantity of gas, liquid or bulk freight. A *general cargo consignment* contains a number of *freight units* [FU], which can be single packages or load units such as cartons, parcels, bins, boxes, containers and pallets.

A general cargo consignment can contain one complete delivery order, a part of a large *delivery order*, or the quantities of several orders. *Consolidated consignments* consist of several customer consignments accumulated in a collection station for the consolidated transport to a downstream distribution station, where the consolidated consignment is separated again. The single *customer consignments* are destined for receiving stations at the end of supply chains. *Consolidated consignments* and *replenishment shipments* end at an upstream station.

The number and content of the *shipment orders* determine the *consignment requirements*:

- *kind of consignment*: normal, due date or express shipment; hazardous, valuable, safety or cooled consignments
- *shipment times* or *carriage times* [h]
- *departure, pick-up* and *arrival dates* [d:h]
- *consignment content*: bulk good, general cargo, valuables, hazardous goods, cooled goods ...
- *consignment quantity*: total number of freight units m_{Con} [FU/Con], consignment volume V_{Con} [l/Con] and consignment weight G_{Con} [kg/Con]
- *consignment structure*: number of customer orders or order positions per consignment n_{Con} [Pos/Con]; quantity per order m_{Ord} [FU/Ord] or quantity per position m_{Pos} [FU/Pos]
- *consignment flow* λ_{Con} [Con/PE]: Number of consignments per period [PE = hour, day, week, month or year] to be shipped from a dispatch station to a receiving station

The freight units of a general cargo consignment can be classified by weight, dimensions or other criteria. Quite common is the classification due to size and type of the load carrier:

- *small freight* or *mini load* with dimensions up to 600 mm and weights below 30 kg, e.g. packages, parcels, bins and small containers
- *standard freight* or *pallet freight* with length up to 1,400 mm, height up to 2,000 mm and weight up to 1,000 kg, e.g. CCG1, CCG2 and Euro-pallets, industrial pallets, cage pallets, roll containers and other standard containers
- *large freight* or *bulky freight* with dimensions above 1,400 mm, height above 2,000 mm and weights up to 2 t such as large freight cases, long goods cassettes and large load containers
- *heavy freight* with weights above 2 t or *special freight* without load carrier

A *homogenous consignment* consists of freight units of the same type and size, whereas *heterogeneous consignments* or *mixed consignments* contain freight units of different types and sizes. An efficient transport and handling of mixed consignments can be achieved by the *dispatch rules*:

▸ As far as technically possible, a mixed consignment of differing freight units should be transformed into a homogenous consignment using the same load carrier.

▶ If homogenization is not possible, and if the whole consignment cannot be delivered completely by a specialized *combined-freight forwarder*, mixed consignments must be separated into homogenous part-consignments and shipped via different freight chains.

The product of the mean number of consignment positions and the mean position quantity of a consignment is the mean *consignment quantity*:

$$m_{Con} = n_{Con} \cdot m_{FU} \quad [FU/Con] \tag{21.11}$$

The mean freight unit volume v_{FU} [l/FU] gives the mean *consignment volume*:

$$V_{Con} = n_{Cons} \cdot m_{FU} \cdot v_{FU} \quad [l/Con] \tag{21.12}$$

and the mean freight unit weight w_{FU} [kg/FU] the *consignment weight*:

$$W_{Con} = n_{Cons} \cdot m_{FU} \cdot w_{FU} \quad [kg/Con]. \tag{21.13}$$

The efficiency of loading, transport, unloading and internal handling can be improved considerably by the use of *load units* [LU]. The optimal selection, design and dimensioning of the load units and their allocation to the different consignments are important *options for action* to minimize delivery costs (see *Chap. 12*). However, when selecting and dimensioning the load carrier, it is necessary to keep in mind the *disadvantages of load carriers* (see *Sect. 12.1*):

- Rationalization of freight handling by load carriers causes additional efforts for building up and breaking down the load units, a loss of transport capacity by partially filled load units, investments for the load carriers and additional costs for handling the empties.

When filling m_{Con} freight units of a homogenous consignment in a load carrier with capacity C_{LU} [FU/LU] the resulting number of load units is:

$$M_{Con} = \{\, m_{Con}/C_{LU}\} \quad [LU/Con]. \tag{21.14}$$

The curly brackets indicate a rounding up to the next integer as the filling of load carrier generates one partially filled quantity, unless the quantity is a precisely a whole-numbered multiple of the load unit capacity.

The design and optimization of supply chains and freight networks for a continuous freight demand requires the knowledge of the mean consignment structure and its standard deviation, i.e. the mean value and the variance of the number of consignment positions and quantities. When the structure of the single consignments of a freight demand varies strongly, it is necessary to define different *consignment classes* with similar structure in order to handle them separately. Resulting are small quantity and large quantity consignments, single and many unit shipments, or one and many position consignments.

As shown in *Sect. 12.5.3*, averaging of relation (21.14) over many single consignments of the same class leads to the *load unit formula*:

- If the freight units FU of the consignments are filled into load carriers with capacity C_{LU} [FU/LU], the *mean number of load units* per consignment is

$$M_{Con} = \mathrm{MAX}\,(1; m_{Con}/C_{LU} + (C_{LU}-1)/2C_{LU}) \quad [LU/Con]. \tag{21.15}$$

The additional term $(C_{LU}-1)/2C_{LU}$ results from the mean capacity loss per consignment. The capacity loss vanishes for load units which are equal to the freight units, i.e. for $C_{LU} = 1$, and is ½ LU for load units with capacities $C_{LU} >> 1$.

In many cases, it is possible to minimize the capacity loss and the number of load units by the *filling strategies* of *Sect. 12.5*. These are:

- *filling up* or *rounding down* of the consignment quantity to a multiple of the content of completely filled load units
- *compacting* of several partially filled single-consignment load units to mixed-consignment load units

By the rounding strategy, the capacity loss-term in relation (21.15) vanishes. The consolidation of N consignments in mixed load units reduces the loss by a factor 1/N. An example for compacting are *sandwich pallets* which are build up in a *transshipment point* from flat-loaded pallets for delivery to a retail outlet.

21.2.4 Freight Modes

For the different consignment classes, several *freight modes* are possible. Consignments with large and medium quantities can be conveyed in *direct freight mode* as *full-load* or *part-load*:

- *Full-loads* (FL) are consignments with large quantities M_{Con} filling the capacity C_{TU} of a transport unit sufficiently for direct transports. A thumb rule for full loads is

$$f_{FL} \cdot C_{TU} < M_{Con} \leq C_{TU} \quad \text{with } f_{FL} \approx 0.6 \text{ to } 0.9 \tag{21.16}$$

- *Part-loads* (PL) are consignments with medium quantities that fill together with other part loads a transport unit for an efficient direct transport. A thumb rule for part loads is

$$f_{PL} \cdot C_{TU} < M_{Con} \leq f_{PL} \cdot C_{TU} \quad \text{with } f_{PL} \approx 0.1 \text{ to } 0.2 \tag{21.17}$$

By using transport units with smaller capacity, medium size consignments become full-loads. However, this is only opportune to a certain extent, since the transport costs per load unit increase with decreasing transport capacity. On the other hand, with transport units of higher capacity, also larger consignments become part-loads. This shows that the capacities of the transport means which connect the stations of the supply chains are further *design parameters*.

The direct transport of many small consignments as part-loads is opportune only over short distances within a local *pickup and delivery area*. If they fit into the remaining capacity of a load transport, also small consignments can be conveyed directly over long distances most efficiently as co-loads:

- *Co-loads* (CL) are small consignments which are transported directly by using the remaining capacity of a load transport.

Many transport providers and freight forwarders improve their profit by co-loading as it is often charged with the higher rate of an indirect freight mode. An example is the use of the luggage compartment of passenger airplanes for general air cargo. Also *last-minute-passengers* can be seen as a kind of co-load.

If small consignments cannot be conveyed directly as co-loads, an indirect transport via *transshipment points* is more efficient. *Indirect freight modes* are general cargo and parcel cargo:

- *General cargo* (GC) consists of small numbers of load units of medium size. Thumb rule for general cargo is

$$C_{LU} < M_{Con} < f_{PL} \cdot C_{TU} \quad \text{with } f_{PL} \approx 0.1 \text{ to } 0.2. \tag{21.18}$$

- *Parcel cargo* (PC) consists of parcels or packages of smaller size with quantities in the range

$$1 < M_{Con} < f_{PC} \cdot C_{LU} \quad \text{with } f_{PC} \approx 0.1 \text{ to } 0.3. \tag{21.19}$$

The boundaries of the different freight modes (21.16), (21.17), (21.18) and (21.19), and the size of the local pickup and delivery area are additional *design parameters*. The *cost-optimal boundaries* between the different freight modes and the optimal local range depend on the size of the load units, on the transport mode and transport units, and on the freight demand.

For road transport, forwarders generally set the lower limit for full-loads with 10 to 15 t or 20 t 25 pallet-places and for part-loads with 2.5 t or 5 pallet-places per consignment, although the cost-optimal limit between part-loads and general cargo is in many cases significantly below 2.5 t or 5 pallet-places per consignment. Due the sorters, the technical upper limit for parcel cargo is generally 31.5 kg per unit. The economic limit depends on the mean weight and volume of the units and is in a range of 5 to 10 parcels per consignment.

A strategy to reach the *cost-optimal freight mode* is *freight cooperation*:

- If the own freight demand of a company is not sufficient to reach the cost-optimal freight modes, several companies with delivery stations in the same area and freight of similar kind for the same destination areas can reach the cost-optimal boundaries by *freight cooperation*.

The consolidation of the own internal freight demand with the external freight demand of other companies is not only achievable by cooperation but also by a *logistic service provider*. Their core business is the consolidation of the logistic and freight demand of many customers. Also logistic service providers build up *logistic alliances* in order to generate in total a high freight demand, to use the cheapest freight modes and to maximize the utilization of their systems.

21.2.5 Freight Throughput and Demand

A given *consignment flow* λ_{Con} [Con/PE] with the mean consignment volume (21.12) generates the *volume throughput*

$$\lambda_V = V_{Con} \cdot \lambda_{Con} \quad [\text{l/PE}] \tag{21.20}$$

With the mean consignment weight (21.13) the *freight ton throughput* is

$$\lambda_W = W_{Con} \cdot \lambda_{Con}/1000 \quad [\text{t/PE}] \tag{21.21}$$

The mean consignment quantity (21.11) leads to *freight unit throughput*, *freight flow* or *freight demand*

$$\lambda_{FU} = m_{Con} \cdot \lambda_{Con} \quad [FU/PE]. \tag{21.22}$$

Due to relation (21.15), a consignment flow λ_{Con} generates the mean *load unit throughput* or *load unit flow*:

$$\lambda_{LU} = \lambda_{Con} \cdot MAX(1; m_{Con}/C_{LU} + (C_{LU} - 1)/2C_{LU}) \quad [LU/PE]. \tag{21.23}$$

Relation (21.23) and the following relation (21.30) reflect the *filling loss effect*:

- The real flow of load units and transport units is considerably higher than the theoretical flow of completely filled load and transport units due to the filling losses, which increase with the capacities of the load units and transport means and with the transport frequency.

Transport flows measured only in *ton-kilometers* or *ton-miles* neglect the filling loss and underestimate the real transport moves and traffic flows.

Above certain capacities, the cost savings by larger load units and bigger transport means with lower unit costs are overcompensated by the filling losses. Hence, for a low freight flow, smaller load units and transport means are opportune. Theoretically, there is a *cost-optimal load unit* for each consignment class and a *cost-optimal transport unit* for each transport relation.

The design and optimization of supply chains and network structures must also take into account *stochastic fluctuations* and *systematic changes* of the performance requirements and freight demand. Accordingly, transport capacities and limit performances of the intermediate stations need to be flexible.

The regularity and constancy of the freight demand determines also the *transport organization*. *Regular transports* are opportune for steady freight flows. *Irregular transports* with adapted transport means are opportune for sporadic consignments with changing content.

From the aspect of distribution, supply chains and network are designed and optimized for delivering the freight flows λ_j to $N_R > 1$ receiving points R_j out of one source D. From the aspect of procurement, logistic chains and network have to be optimized and designed for the freight flows λ_i from $N_D > 1$ delivery points D_i to one receiving point R. The freight chains for $N_R \cdot N_D$ flows λ_{ij} from N_D delivery points D_i to N_R receiving points R_j must be considered if the return transports and the freight demand of other companies are involved.

In all cases, the required *shipment times* T_{ij} [h] between the delivery and receiving points must be kept. They are determined – as explained above – by the *transport organization* and the *travel distances*. Hence, for network design, it is necessary to analyze the *regional freight structure*, i.e. the locations of the sinks and sources, and their freight demand.

As an example, *Fig. 21.6* shows the regional distribution of more than 2,500 outlets of a German *retail corporation*. Their density is proportional to the regional freight demand. *Figure 21.7* presents the freight distribution of the production quantities from two different plants for *building material* to the two-digit post-code areas of Germany.

Fig. 21.6 Typical location distribution of receipt stations (*outlets*) of a German retailer

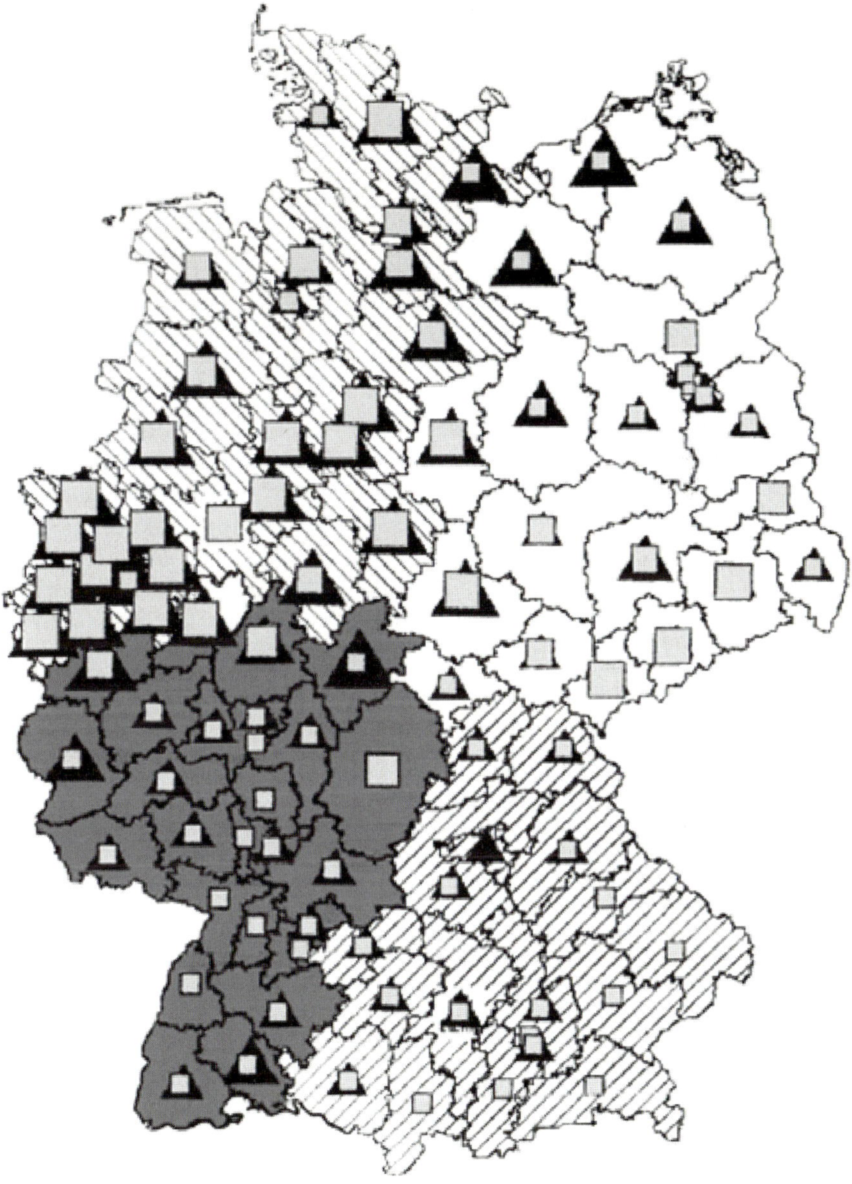

Fig. 21.7 Regional distribution of the dispatch quantities of a German construction material manufacturer

Triangles: quantities from plant 1
Squares: quantities from plant 2

21.2.6 Stock Requirements

Depending on the stocks in the intermediate stations, the supply chains can be differentiated into

- *transport chains* and *freight chains* without storekeeping stations between dispatch and delivery
- *storekeeping supply chains* with article stocks either in the dispatch station and/or in one or several intermediate stations

For storekeeping supply chains, additional optimization parameters are the assortment of storekeeping articles in the different stations and their stock level. As outlined in *Chap. 11*, professional *inventory management* accomplishes not only short delivery times, but also minimal costs.

If the performance costs of the supply chain and the storage costs of the storekeeping station are known, the optimal replenishment quantities for goods with regular demand can be calculated from the scheduled or predicted demand. The *safety stocks*, which are necessary to ensure the required *availability*, are calculated from the stochastic variations of the demand and the supply times of the replenishment. As long as the supply is sufficient to cope with the demand and as long as no *bottlenecks* occur, the independent optimal stock scheduling in the stations of the supply chain, starting from the final receiving station upstream to the first delivery station, results *self-regulating* in optimal *pull-stocks*. Basic *rules for pull-stocks* are:

► A pull-stock in an intermediate station is necessary when the delivery times for direct or transit delivery are longer than acceptable.

► The level of a pull-stock is determined by the required availability, the replenishment time and by the procurement and storage costs

► When scheduled optimally, the mean pull-stock is proportional to the square root of the demand λ_{SU} [SU/PE] of storekeeping units [SU]:

$$M_{BSU} = F_S \cdot \sqrt{\lambda_{SU}} \quad [SU] \tag{21.24}$$

The *inventory structure factor* F_S depends on the scheduling parameters, the required delivery availability, and on the specific cost rates of the storage system. As explained in *Sect. 11.10*, the relation (21.24) is the basis of the *square root law of logistics*, which allows the reduction of logistic costs by *stock centralization*. However, one has to keep in mind:

► The square root law holds only for the pull-stock of articles with regular demand provided that stocks and replenishment are optimally managed.

In addition to pull-stocks also *push-stocks* occur. They result e.g. from a production or procurement in advance, for sales promotions, seasonal peaks or the market introduction of a new product. For both, pull-stocks and push-stock holds the *stock location rule*:

► Stocks of articles destined for customers or consumption stations in different regions should be located close to the production where storage costs and the *misallocation risk* are minimal.

Due to the *pull-principle*, the supply from push-stocks should not be delivered before the articles are ordered.

21.2.7 Extrapolation Factors and Changing Rules

Market developments, technical progress, business policy, change of assortment and altering customer behavior affect the service and performance requirements. Many changes are unpredictable and can only be coped with by *flexibility* and *adaptable systems*.

Some important effects on the logistic network and supply chains can be taken into account by the following extrapolation factors and changing rules, which are derived from probability theory, experience, and the square root laws of logistics (see *Sect. 11.10*):

- A *change of the sales* by the factor f_S leads with probability P to a change of the order or consignment flow by the factor $f_S{}^P$ and with probability $1-P$ to a change of the order or consignment quantities by the factor $f_S{}^{1-P}$. If no change of the scheduling behavior of the customers is expected, the change probability can be assumed with $P = 0.5$.
- *Pull-stocks* with optimal replenishment change by a factor $\sqrt{f_S}$, *push-stocks* change proportional to the *sales change factor* f_S.
- The *mean number of order positions* remains unaffected as long the article assortment does not change, and the customers do not alter the scheduling.
- A *change of the assortment* by factor f_A leads with probability P to a change of the order flow by the factor $f_A{}^P$ and with probability $1-P$ to a change of the number of order positions by the factor $f_A{}^{1-P}$. Again, the change probability can be assume as $P = 0.5$, if no change of scheduling is expected.
- As long as the sales volume remains constant and the scheduling behavior is not altered, an *assortment change factor* f_A causes a change of the quantity per position by the factor $1/f_A$.
- If the customers alter the *replenishment frequency* for the articles by a factor f_R without changing the total demand, the order flow increases with probability P by the factor $f_R{}^P$ and the number of order positions with probability $1-P$ by the factor $f_R{}^{1-P}$.
- A *replenishment change factor* f_R alters the quantity per position by a factor $1/f_R$ when sales and assortment do not change. Without a change of scheduling, $P = 0.5$.

For example, ordering via *internet* has increased the flow of incoming orders for many suppliers in the last years without a relevant change of the total demand. Therefore, the order quantities have decreased whereas the consignment flow increased.

The change of the customer structure and the ordering behavior induced by *e-business* shifted the freight flows from part loads to general cargo, and from general cargo to parcel cargo. Due to this effect, the business of *parcel service providers* has increased.

A reverse consequence is:

- The reduction of the receiving stations of a supply network by a *concentration factor* f_C, e.g. due to a *merger of retailers*, without changing total demand and assortment leads to a reduction of the mean consignment flow by the factor $1/f_C$ and to an increase of the mean consignment quantities by the factor f_C.

For example, a reduction of the receiving stations by a factor of e.g. 2 at constant demand and assortment leads on average to a reduction of the consignment flow by 50% and to a doubling of the delivery quantities.

Another important consequence is the *square root law of stock centralization*:

- A concentration of the *pull-stocks* of N storekeeping stations in a central store without changing sales and assortment causes at the supply stations a reduction of the consignments flow by a factor $1/\sqrt{N}$ and an increase of the mean position quantity by the factor $\sqrt{N}$.

If for example, 9 receiving stations are replenished in future from one central store, the consignment flow decreases by 1/3 while the quantities per position will be tripled. However, the square root rule does not hold for push-stocks.

21.2.8 Handling Units and Load Units

The kind, weight and dimensions of the handling units and load units determine the dimensions and capacities of the logistic stations and transport means and by this the total costs of a logistic network. This leads to the *starting recommendation*:

- At the beginning of a supply chain project, a *synopsis of the logistic units* should be established containing the maximal, minimal and mean dimensions and weights of the smallest handling units in the supply chain, of the available load carriers, and of the resulting load units with the respective capacities.

The next project step is to investigate which load unit parameters are definitely fixed and which can possibly be changed if it is opportune, since they are basic *design parameters*.

21.3 Options for Action and Design Parameters

The supply chains SC_{ij} between delivery stations D_i and receiving stations R_j can be optimized by using the *delivery options, structure options and transport options* and by varying the respective *design parameters*:

- *delivery parameters*: dispatch quantities, delivery frequencies and load units in which the goods are conveyed, buffered and stored
- *structure parameters*: number, locations, functions and stocks of the intermediate stations between sources and sinks
- *transport parameters*: traffic means, capacities and speeds of transport units, transport organization, tours for the transport between the stations.

By selection of the right options and by determining the optimal parameter values, a given freight demand and required services can be theoretically fulfilled at minimal costs keeping all technical and organizational frame conditions. However, due to inaccurateness of the demand, lack of information and changing conditions, it is impossible to solve this task in practice. The most important obstacles are the *goal conflicts* between the different users of the same supply and freight networks. Therefore, supply chain management is based on general *design principles, recommendations* and *thumb rules*, which result from theoretical considerations and have been approved by experience.

21.3.1 Delivery Options

The operative and administrative logistic activities (21.1), (21.2), (21.3) and (21.4) in the receiving and delivery stations determine the dispatch mode, the delivery frequency and the consignment quantities. Depending on the required packaging, two basically different *dispatch modes* are possible:

- *Bulked goods* are shipped loosely in tanks, silos or bulk containers or transported in pipelines
- *Packed goods* are filled in sacks, bags, kegs, bottles, tins or cartons and shipped, stored and transported as *package units* [PU]

For the dispatch via the different freight chains, the package units can be consolidated and transformed into load units by load carriers such as pallets and containers (see *Table 12.2*). This leads to the *load unit option* (Gill and Allerheiligen 1996):

- The selection and dimensioning of load units for bundling and consolidation of goods and consignments are valuable options for the design and optimization of supply chains.

The expensive handling of small package units and the ineffective break down and build up of load units can be avoided by the *load unit selection rules*:

- As far as possible, the same load units should be used in all stations and transport sections of the supply chain.
- Changing of load units should be limited to the build up of larger load units out of smaller load units leaving the content of the smaller units untouched.

For *road transports* over longer distances, packages and parcels are loaded on pallets, which are filled into *trailer trucks, ISO-containers* or *swap bodies*. For transports per *rail* or *sea* over very long distances, the ISO-container and swap bodies are loaded on wagons and ships respectively. Only if freight space is expensive or scarce, like *airfreight space*, it is necessary to achieve high filling degrees by *build up* and *break down* of *destination-mixed load units*.

21.3.2 Structure Options

The structure of a logistic network with given customer and supplier structure is determined by the number N_{IS} and locations $(x_k;y_k)$ of the intermediate logistic stations IS_k, $k = 1,2,\ldots N_{IS}$ and their functions (21.5).

A supplier wants to optimize the delivery to large numbers of receiving stations of his customers from one or a few sources. In this case, the *delivery task* is a *one-to-many* or *few-to-many problem*. It implies the selection of the optimal *delivery chains* within an existing distribution system and the minimization of the operating costs by building a new *supply network*.

The replenishment task for a *manufacturer* or *producing company* with one or a few plants, which are supplied from many delivery stations is a *many-to-one* or *many-to-few problem*. This implies the selection of optimal *procurements chains* and the optimization of a *procurement network*.

Retailing companies with hundreds or even thousands of suppliers and outlets, and *forwarding companies* who deliver daily consignments of many shippers to many customers have to solve a *many-to-many problem*. This task includes the selection of optimal supply chains or freight chains, and the set up and permanent adjustment of a complex *supply network* or *freight network*.

The number of intermediate stations passed by the goods between source and sink determines the *stage degree of the supply chain* (see *Sect. 1.5*):

- An *N-stage supply chain* consists of N transport sections, which are connected by N-1 intermediate stations.

A *single-stage supply chain* is a direct transport connection from delivery station to receiving station without intermediate station. Retailers call this freight mode *direct delivery* or *drop-shipment*. However, they often do not notice that many consignments pass a transshipment point and neglect optimization potentials. The possible delivery and transport options with their parameters lead to a large number of different single-stage supply chains.

In a *two-stage supply chain* as shown in *Fig. 21.8*, the goods are passing one intermediate station. *Mono-functional logistic stations* are:

- *delivery stores* (DS) close to or directly at the supplier
- *consolidation points* (CP) at the transport gravity center of a pickup area
- *transshipment point* (TP) near the transport gravity center of a service area
- *central stores* (CS) near the center of a service area
- *distribution points* (CP) at the transport gravity center of a delivery area
- *regional stores* (RS) near the center of a service area
- *buffer stores* (BS) close to or directly at the recipient

The differentiation between a regional and a central transshipment point as well as between a regional and a central store results only from their distance to the sources and sinks.

With increasing distance from the dispatch stations and decreasing distance to the recipient stations, a consolidation point becomes a transshipment point and a transshipment point becomes a distribution point: CP → TP → DP. Correspondingly, a delivery store becomes a central store, a central store becomes a regional store and a regional store becomes a buffer store: DS → CS → RS → BS.

Mono-functional logistic stations can be combined in many different ways resulting in *multi-functional logistic stations*:

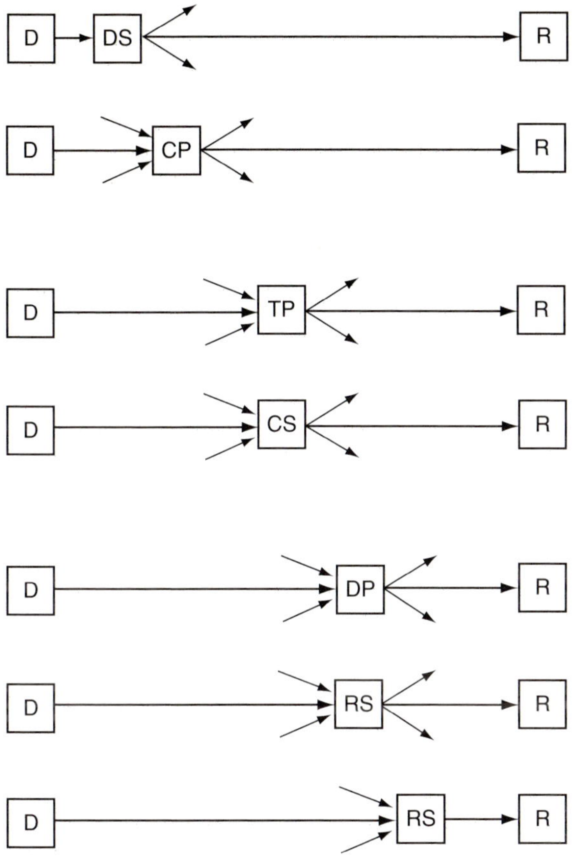

Fig. 21.8 Examples for two-stage supply chains
D: dispatch station
DS: delivery store
CP: consolidation point
TP: transshipment point
CS: central store or logistic center
DP: distribution point
RS: regional store
BS: buffer store
R: receiving station

- Combination of a distribution point and a consolidation point that serve the same area results in a *regional transshipment point* RTP = DP + CP.
- Combination of a distribution point with a regional store gives a *regional logistic center* RLC = DP + RS.
- Combination of a transshipment point with a central store leads to a *logistic center*: LC = TP + CS.

If also bulk goods pass the network, in addition bottling and filling stations are needed. Depending on their location these are:

supplier bottling and filling stations (SBS)
central bottling and filling stations (CBS)
regional bottling and filling stations (RBS)
local bottling and filling stations (LBS)
(21.25)

They can be integrated in intermediate stations or a multi-functional logistic center in many different ways.

A *three-stage supply chain* runs through two intermediate stations, e.g. from a delivery station via a central store and a distribution point to a customer. *Fig. 21.9* shows six common three-stage supply chains. *Four-stage supply chains* have three intermediate stations. Single-, two-, three- and four-stage supply chains of different industrial and retailing companies are shown in *Figs. 21.14, 21.21, 21.23* and *21.25*.

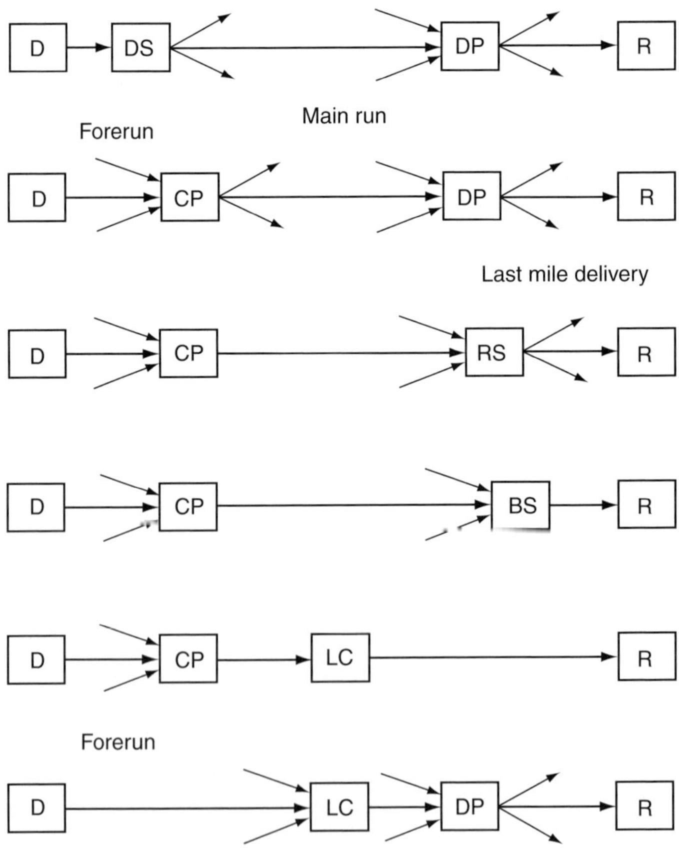

Fig. 21.9 Examples for three-stage supply chains
Notations: see *Figs. 21.7* and *21.8*

By combination with the different delivery and transport options, a huge number of many-stage supply chains can be generated. Most of them are theoretically possible but of no practical use. For the selection holds the *general stage rule*:

- The number of different supply chains increases with the number of stages, whereas their necessity decreases.

Additional options are the possible functions (21.5) of the intermediate stations. Supply chain management must decide in which intermediate stations a stockless transit, with or without a change of the load carrier, should take place and in which stations article stocks are kept and orders are commissioned.

21.3.3 Transport Options

A *transport load* [TL] consists of a number of consignments, which are directly transported either to the same destination or on a transport tour to many receiving points. If for the shipments a *service frequency* ν_S [1/PE] is required, at least ν_S transports must depart from the delivery station. A *load flow* λ_{LU} [LU/PE] resulting by relation (21.23) from the freight demand accumulates until the next regular transport [Trp] departs, on average the *mean transport quantity*:

$$M_{TL} = \lambda_{LU}/\nu_S \quad [\text{LU/Trp}] \tag{21.26}$$

The mean load quantity per transport determines the number and the capacity of the necessary transport units. The exact number of transport units depends on the applied *packing strategies* and *filling strategies*.

As explained in *Chap. 12*, the task of stowing a maximal number of load units with different dimensions and weights in a given transport unit leads to a *three-dimensional cutting problem* (Gilmore and Gomory 1965). Today, powerful *stowing software* is available, which also takes into account restrictions, such as prescribed orientations or stowing sequences of the load units (Isermann 1987) However, these programs do not offer an explicit formula for the mean transport unit demand. For the calculation of *specific freight costs*, it is sufficient to know the mean transport unit demand for a continuous load flow transported with a service frequency ν_S. Due to the explanations of *Sect. 12.5* holds:

- Without filling strategy, the mean number of transport units with *effective capacity* C_{TUeff} [LU/TU] for a *mean transport quantity* M_{TL} [LU/Trp] is

$$M_{TU} = \text{MAX}\,(1; M_{TL}/C_{TUeff} + (C_{TUeff} - 1)/2C_{TUeff}) \quad [\text{TU/Trp}]. \tag{21.27}$$

Relation (21.27) says that at least one transport unit is needed and that several transport units have a *mean filling loss* of $(C_{TUeff} - 1)/2C_{TUeff}$ [TU/Trp].

For equal load units, it is quite easy to calculate the transport capacity. For different load units the transport capacity depends on the size, shape and distribution of the load units and on the packing strategy. The *mean transport capacity* can either be measured in practice or approximately calculated with the mean weight and volume of the load units as explained in *Sect. 12.5*.

The capacity of the transport units is fully usable only, if load units of different consignments can be transported together in one unit. If this is not tolerable, additional filling losses reduce the transport capacity. For *restricted consignment division* holds the rule:

- If consignments with mean quantity M_{Con} [LU/Con] can be loaded in up to N_{CD} different transport units, the capacity C_{TU} [TU/LU] of the transport units reduces by filling losses to the *effective capacity*:

$$C_{TUeff} = \mathrm{MIN}(M_{Con;}C_{TU} - (M_{Con} - 1)/2N_{CD}) \quad \text{for } C_{TU} \geq M_L \geq 1 \quad (21.28)$$

The additional filling loss $(M_{Con} - 1)/2N_{CD}$ vanishes for consignments with quantity $M_{Con} = 1$ LU/Con and is maximal if no *consignment division* is tolerable, i.e. for $N_{CD} = 1$. With unrestricted consignment, i.e. for $N_{CD} \to \infty$, the effective capacity is equal to the full transport capacity. *Figure 21.10* shows the dependency (21.28) of the effective capacity of a transport unit on the mean consignment quantity and on the tolerable consignment division.

By inserting the effective capacity (21.28) into relation (21.27), the mean *filling degree* $\eta_{TU} = M_{TL}/(M_{TU} \cdot C_{TU})$ of the transport units can be calculated. *Figure 21.11* presents the resulting dependency of the filling degree on the load size when loaded

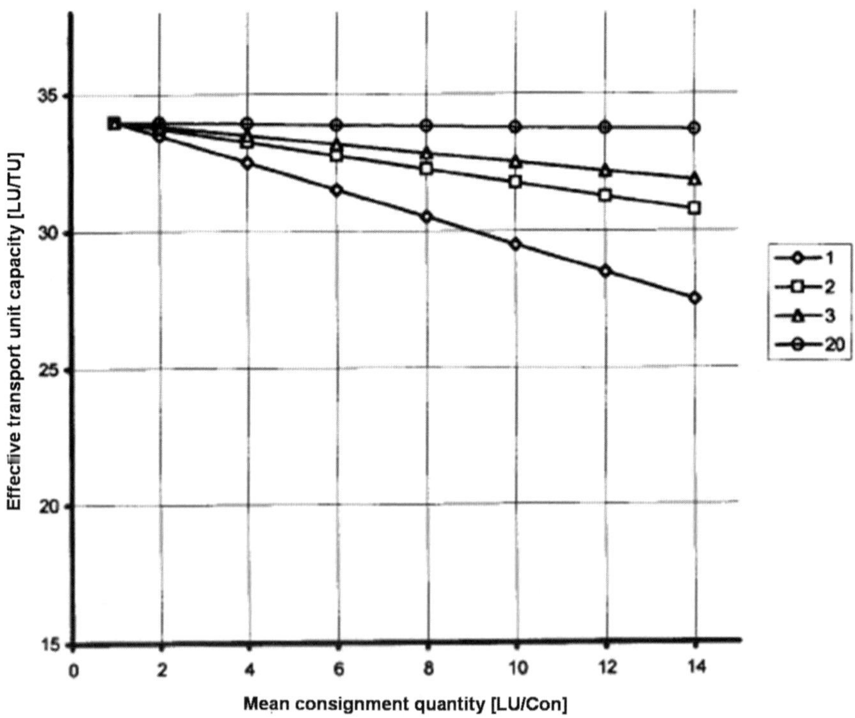

Fig. 21.10 Dependency of the effective transport unit capacity on the mean consignment quantity and on the tolerable separation of the consignments

consignment division: $N_{CD} = 1,2,3,20$ TU/Con
total transport capacity: 34 LU/TU

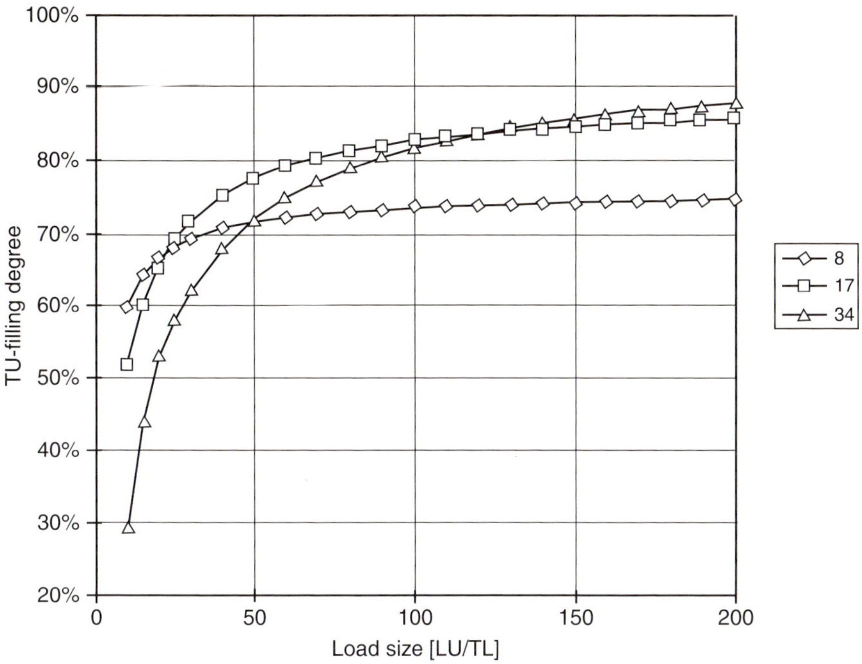

Fig. 21.11 Dependency of the filling degree of transport units on the load size without consignment division

transport capacities: $C_{TU} = 8,17,34$ LU/TU
mean consignment size: $M_{Con} = 5$ LU/Con

without consignment division. *Figure 21.12* shows the dependency for a tolerable consignment division of $N_{CD} = 2$ TU/Con.

The filling losses can be reduced or avoided by the following *delivery strategies*:

- *Postponement of uncritical consignments*: The consignments are loaded according to their urgency. Consignments with later delivery dates are left behind as far as they cause partially filled transport units.
- *Advanced transport of later consignments*: After loading all urgent consignments, the remaining capacity is filled with consignments with later delivery dates.
- *Co-loading of small consignments*: The free capacity of regular load transports is filled with smaller consignments, which otherwise would be conveyed as general cargo.

Some companies in the transport business, e.g. freight and passenger airlines, try to create short-term demand by attractive *last-minute-prices*. However, these *pull-in-step* and *forward buying* strategies bear the risk to create no additional demand but to attract only later regular demand at lower prices.

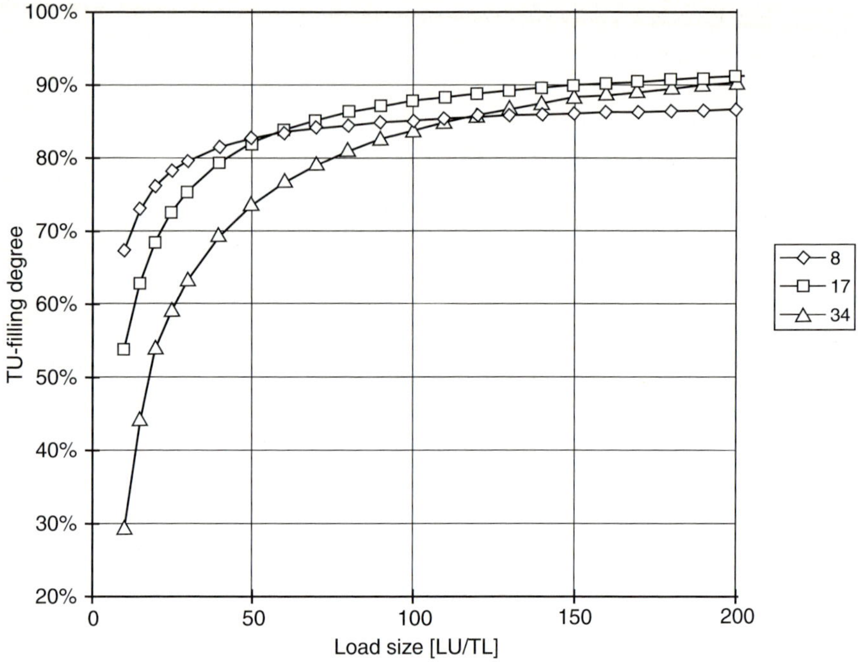

Fig. 21.12 Dependency of the filling degree of transport units on the load size with consignment division

consignment division: $N_{CD} = 2$ TU/Con
other parameters: see *Fig. 21.11*

With a service or delivery frequency ν_S, a transport unit demand M_{TU} per departure results in a transport unit flow:

$$\lambda_{TU} = \nu_S \cdot M_{TU} \quad [TU/PE]. \tag{21.29}$$

By inserting (21.26) in (21.27) and of (21.27) in (21.29) follows:

- A load unit flow λ_{LU} [LU/PE] to be transported with minimal service frequency ν_S [1/PE] generates the *transport unit flow*

$$\lambda_{TU} = \text{MAX}\,(\nu_S; \lambda_{LU}/C_{TUeff} + \nu_S \cdot (C_{TUeff}-1)/C_{TUeff}) \quad [TU/PE]. \tag{21.30}$$

The transport unit flow λ_{TU} on a specific relation is equal to the transport frequency for low load flows and equal to the second term in the brackets of (21.30) for higher load flows. Consequences from relation (21.30) are the *delivery frequency rules*:

- ► Higher load flows offer the opportunity to increase the delivery frequency without lowering the filling degree of the transport units.
- ► For small load flows, the transport unit demand grows with increasing delivery frequency, unless smaller transport units are used.

This leads to the *shipment time dilemma of logistics*:

- Shorter shipment times require higher delivery frequencies and cause smaller loads. Smaller loads lead to lower utilization of the transport units, or to the employment of smaller transport means with higher specific costs. For both solutions, the specific freight costs increase with shorter shipment times.

The dependency of the shipment time and the freight costs on the delivery frequency is shown in *Fig. 21.32* for the distribution of pallet-consignments from a logistic center via regional transshipment points to retail outlets. A consequence of the shipment time dilemma is the *delivery scheduling principle*:

► To operate the supply chains at minimal costs, the delivery frequencies for all transport relations must be made as low as possible without affecting the required shipment times.

Next to the selection of the transport means, the delivery frequencies are the most important transport parameters.

21.4 Delivery Times and Shipment Times

The *delivery time* or *total lead time* for an order consignment, shipped from a delivery station ST_o via the N-1 stages of a supply chain to a receiving station ST_N, is the sum of the *internal transfer times* T_{ITn} of the stations ST_n, the *driving times* T_{DRn} for the transport from one station to the next, the *entry waiting times* T_{EWn} before and in the goods entry of the stations, and of the *dispatch waiting times* T_{DWn} in and behind the stations:

$$T_{TL} = \sum_{n=o} (T_{ITn} + T_{DRn} + T_{EWn} + T_{DWn}) \tag{21.31}$$

The sum (21.31) runs from the first station along the supply chain over N-1 intermediate stations to the receiving station ST_N. The first station ST_o either produces the goods to order or takes them from a finished goods store. Hence, the internal lead time T_{ITo} of the departure station is the sum of the order preparation time and the production time or the commissioning time respectively.

The *forwarding time* or *shipment time* of a consignment is the running time through the transport or freight chain from the ramp of departure to the receiving station. It equals the total lead time (21.3) minus the internal lead time T_{ITo} of the delivery station.

The lead times through the stations are the sum of the process times for the single administrative and operative steps, and the internal waiting times.

The driving times result from travel length and speed of the transport means.

Dispatch scheduling must differ between regular and irregular waiting times:

- *Regular waiting times* result from differences between standard lead times, planned operating times, and departure times for regular transports.
- *Irregular waiting times* are caused by unexpected detentions in the stations and delays on the transport connections.

If internal lead times and operating times are optimally synchronized, the regular waiting time for randomly arriving dispatch orders is for delivery twice a day, i.e. for a delivery frequency $\nu_S = 2$ per 24 h, minimal 0 h, maximal 12 h and on average 6 h.

The reasons for irregular waiting times, which are mostly stochastic, are *bottlenecks* within the stations or overloaded traffic routes. Also, breakdowns of equipment and transport means as well as disturbances and errors of all kinds can cause irregular waiting times. They fluctuate during the day, depend on the season and can only partly be foreseen. In the calculation of standard delivery times, the irregular waiting times are taken into account by *time buffers* resulting from experience. A general *thumb rule for time buffers* is:

- If nothing better is known, the sum of the entry and dispatch waiting times per station can be assumed with 10% of the station lead time, and the transport waiting times with 5% of the regular driving times.

As outlined in *Sect. 8.10*, the aim of the *just-in-time strategy* is to reduce the waiting times by optimal synchronization of internal lead times, operating times and transport frequencies. Ideally, the goods arrive exactly when they are needed. This fails in many cases due to the irregular waiting times. Their effects must be compensated for by sufficient *time buffers* or *material buffers*. Another solution is *postponement* of the last production stage close to point of consumption (Bucklin 1965).

21.5 Delivery Costs

The specific costs $k_{\text{Con ij}}(\text{SC})$ [€/CON] for delivering a consignment from delivery point D_i to receiving point R_j via a *supply chain* SC which passes N_{ij} stations ST_n is the sum of the *proportionate performance costs* k_{STn} within the stations and the *proportionate transport costs* k_{TRn} for the transports between them:

$$k_{\text{Con ij}}(\text{SC}) = \sum_{n=o}^{Nij} (k_{\text{STn}} + k_{\text{TRn}}) \quad [\text{€/Con}]. \tag{21.32}$$

If $\lambda_{\text{Con ij}}(\text{SC})$ [Con/PE] are the consignment flows in the possible supply chains SC between the delivery stations D_i, $i = 1,2,\ldots N_D$, and receiving stations R_j, $j = 1,2,\ldots. N_R$, the total *delivery costs* per period result from summation of (21.32) over all supply chains of the single relations and over all relations:

$$K_D(\lambda) = \sum_{i,j} \sum_{SC} \lambda_{\text{Con ij}}(\text{SC}) \cdot k_{\text{Con ij}}(\text{SC}) \quad [\text{€/PU}]. \tag{21.33}$$

For calculating the total delivery costs per period, all order and consignment flows through the different supply chains must be known. From these, the performance, load and transport flows can be calculated with the help of the above formulas. The delivery costs are the product of the internal flows with the respective performance cost rates in the stations plus the product of the external load and transport flows with the respective freight and transport cost rates.

For system design and optimization, it is sufficient to calculate with target costs rates and performance prices which result from model calculations, estimations or experience. When deciding on the realization of a new logistic concept and when selecting optimal supply chains in daily business, it is necessary to use current costs rates and real prices from biddings, quotations of logistic service providers and internal cost calculations.

Special problems for the supply chain optimization are caused by the dependency of performance prices and costs on the throughput. As explained in *Sects. 16.13, 17.14,* and *19.11* the throughput dependency results from the technical *economies of scale* and the *fixed costs* of the stations. Also, transport prices and freight rates depend – as shown in *Figs. 21.27* and *21.28* – on the freight flows and on the consignment sizes. In addition, the dependency of the prices on the general demand and supply of the *logistic markets*, which are difficult to estimate, needs to be considered.

To solve this dilemma, in the first step of planning and optimization, the dependency of the costs and prices on throughput and utilization is taken into account by calculating the costs for the planned throughput and performance values of a start solution. If the optimization results in throughput and performance values deviating significantly from the initial values, the cost rates and prices are corrected for the next step.

It happens during the planning and optimization process that the throughput of single stations or the freight flow through certain transport connections falls below a *critical value* where the operation of a station or the execution of regular transports is no longer cost efficient. In such a case, the throughput must be shifted to neighboring stations or to other transport connections. Alternatively, neighboring stations with too small throughput or regions with too low freight demand can be combined into a larger station or region respectively. Both actions change the structure of the logistic network.

As long as the own load demand is sufficient, it is generally opportune to use own or rented transport means exclusively for the own consignments. However, when determining the transport cost rates, one should keep in mind the *return load rule* (see *Sect. 18.12*):

- ▸ Own and/or third party *return loads* enable *paired transports* on combined *forth and back trips* and reduce the transport costs considerably.

Return loads are achievable by combining procurement, distribution and reverse transports of only one company, or of several companies with similar freight for neighboring destinations.

If the own load demand is insufficient to fill the transport units at the required delivery frequency, it is unavoidable to ship own consignments together with the consignments of other companies. This is the business of *freight forwarders*. In this case, *cost-based and use-related freight rates* for part loads, general cargo and parcel cargo should be agreed (see *Sect. 21.15*).

For supply network planning and optimization, all performance prices and transport rates must be cost-based and use-related, i.e. they must depend explicitly on the performance and throughput units and on the relevant cost drivers. Logistic service providers and freight forwarders do not always offer transparent cost-based and use-related prices. It is therefore necessary to prescribe the required price structure and price units in the tender document (see *Chap. 22*).

21.6 Order Processes and Information Flows

The provision of the order quantities in the entry station of a supply chain is initiated by a *delivery order* or a *replenishment order*. These orders flow against the consignment flow from the customers to the suppliers. The dispatch of the consignment through a transport or freight chain to a prescribed destination is initiated by a *dispatch order*, *transport order* or *freight order* placed by the supplier, another shipper or the customer. They run ahead, parallel or together with the consignments to the receivers.

The order lead times, the delivery strategies and to some extent also the delivery costs depend on the *administrative order processes* performed in the stations of the supply chain:

order placement
order acceptance
order processing
dispatch scheduling (21.34)
arrival announcement
confirmation of receipt

These processes need time and cause administrative costs. The analysis and design of the administrative order processes are therefore rather important for the optimization of supply chains.

A flow of information, documents and data accompanies the flow of goods through supply chains, in order to control and to track the course of the consignments. Labeling of goods, load units and consignments, generation of accompanying documents, like *shipping documents* or the *bill of loading*, and reading, checking and processing of information in the stations of a supply chain require time and costs, and should not be underestimated. The administrative times influence the delivery times. The administrative costs contribute to the performance prices.

The delivery of a consignment ends when the recipient has acknowledged its reception, completeness, faultlessness and correctness. The notice of receipt is sent as fast as possible to the supplier or shipper in order to close the order and to send the invoice if not paid in advance.

Order processes, information flows and data flows are not an end in itself. They are partly unavoidable due to the requirements of the participants and the *customs*, partly necessary to ensure a high *logistic quality* and required to execute optimal *distribution and supply strategies*. Many supply strategies fail due to insufficient information and communication systems or due to unavailable, faulty and incomplete logistic data. In this field, *transponder* and *RFID* offer new options (Finkenzeller 2002; Shephard 2004).

21.7 Supply Strategies

Most supply strategies are derived from the three basic logistic strategies, *bundling*, *sequencing* and *securing*, or their counter strategies, or are combinations of the basic strategies (see *Sect. 5.2*)

21.7.1 Bundling Strategies

The main *goal* of bundling is to minimize costs without affecting delivery times. Bundling strategies for delivery and their *strategy parameters* are:

- *Batch-Order Processing*: The orders for several customers are bundled to a *series order* and executed together by a producing station or commissioned together in a storekeeping station. Strategy parameter is the *batch length*, also called *lot size* or *series length.*
- *Consignment Bundling*: Several consignments for the same recipient are filled mixed into load units in order to reduce the loading, reloading and unloading effort. Strategy parameter is the *load unit capacity.*
- *Consignment Accumulation*: In a dispatch station or intermediate station, the consignments for the same destination area or tour are accumulated for a certain time in order to achieve larger transport loads. Here the *accumulation time* is the strategy parameter. If λ_S [Con/PE] is the upcoming consignment flow and M_S [LU/Con] the mean consignment size, the mean load quantity accumulating during the T_{Acc} [PU] is

  $$M_{Load} = \lambda_{Con} \cdot M_{Con} \cdot T_{Acc} \quad [LU] \tag{21.35}$$

  Hence, the load quantity grows with increasing accumulation time, whereas the delivery frequency ν_D decreases, as $\nu_D \leq 1/T_{Acc}$, and the delivery time T_D increases, as $T_D \geq \nu_D \geq 1/T_{Acc}$.
- *Source Consolidation*: The consignments from sources in the same collection area destined for the same receiving station are picked up in collection tours. Strategy parameters are the capacity of the transport means, the boundaries of the service area, the service frequency and the lengths of the tours, which are also called *milk runs.*
- *Sink Consolidation*: The consignments from one delivery station destined for sinks in the same distribution area are dispatched together in delivery tours. The strategy parameters are the same as for source consolidation.
- *Transport Consolidation*: In one transport tour, the loads from several sources in a collection area are picked up and carried to a distribution area where they are delivered to several sinks. Strategy parameters are the capacity and speed of the transport means, and the boundaries and distances of the service areas.

Counter strategies to bundling deliveries are:

- *Load Division*: In order to reduce the delivery time, too large or too long accumulated load quantities are divided and sent by smaller transport units.
- *Consignment Separation*: Consignments of different size or with deviating freight units are separated into groups of small, medium, large and special consignments in order to use optimal transport means and freight chains.

Other counterstrategies of bundling are *just-in-time* and *one-piece-flow* with the aim to eliminate waiting times and to achieve minimal delivery times. In the extreme, the accumulation time for both strategies is 0 and the batch size 1. These strategies, however, cause high performance and delivery costs or require a *buffer stock* or spatial *postponement* of the finalization close to the points of consumption.

A bundling strategy which reduces stocks and storing costs in the whole supply network is

- *Stock Centralization*: The stocks of articles destined for many customers or consumption points in different areas are centralized in one store near the production or in a small number of regional stores.

Stock centralization is limited by the required delivery times and the promised availability to deliver. Shorter delivery times and a higher availability can be achieved by the *counter strategy*, i.e. by *stock decentralization* in regional stores or in local buffer stores (see *Sect. 11.10*).

21.7.2 Sequencing Strategies

By sequencing, order lead times and/or costs can be reduced. It is often combined with bundling. Sequencing strategies in the supply chain are:

- *Prioritization*: Rush orders, express consignments and other priority orders are executed immediately, second rank orders and standard consignments are executed as far as the remaining capacities allow.
- *Processing Sequences*: Accumulated orders and consignments are executed in the sequence with the lowest setup costs and switching times.
- *Packing Strategies*: Parcels and freight units are packed into the load units in the sequence and orientation with maximal utilization of the load unit capacity (see *Sect. 12.4*).
- *Stowing Strategies*: Load units and consignments are stowed in the transport units in a sequence and orientation that need minimal loading space and keeps the required unloading sequences (see *Sect. 12.4*).
- *Filling Strategies*: Filling losses in the load units and transport units are minimized by rounding up or down the delivery quantities, by advancing or postponing uncritical consignments, by splitting of consignments or by co-loading (see *Sect. 12.5*).
- *Optimal Transport Sequences*: Transport orders and transport tours are performed in sequences and on paths with lowest costs and shortest times.

If for the same relation, several supply chains or freight chains are available, it is principally possible to calculate for each consignment the most economic chain. If the logistic data and costs rates for the individual calculation are unknown, *allocation strategies* are needed to determine the optimal supply chain for different classes of consignments.

21.7.3 Securing Strategies

Securing strategies aim for keeping promised delivery times, availabilities and logistic quality. Securing strategies in the supply chain are:

- *Safety Stocks*: The storekeeping stations of the supply chain hold safety stocks in order to ensure the availability to deliver in case of fluctuating demand and varying replenishment times (see *Sect. 11.8*)
- *Buffer Places*: They are installed to cope with queues and congestions caused by stochastically fluctuating arrival rates and dispatch times (see *Sect. 13.5*).
- *Time Buffers*: Time buffers of sufficient length are necessary to ensure the uninterrupted operation of the stations, and the keeping of time schedules if irregular waiting times and lead time fluctuations occur (see *Sect. 8.5*).
- *Redundancies*: Breakdowns, disturbances and interruptions of critical stations or linkages in the supply chain are bridged over by *parallel stations* and *bypasses* (see *Sect. 13.6*).

21.8 Specification of Supply Chains

The existing logistic network and supply chains of a company as well as the results of a system optimization or a new network design can be documented in *network structure maps*, *supply chain charts* and *process charts*. As explained in *Sect. 3.8,* rectangular boxes symbolize the operative stations and directed arrows the transport relations. Dotted boxes represent administrative stations. Dotted lines indicate data and information flows. Inscriptions and figures in the boxes and along the lines specify the functions and quantify the flows. Using these symbols *Fig. 21.13* shows

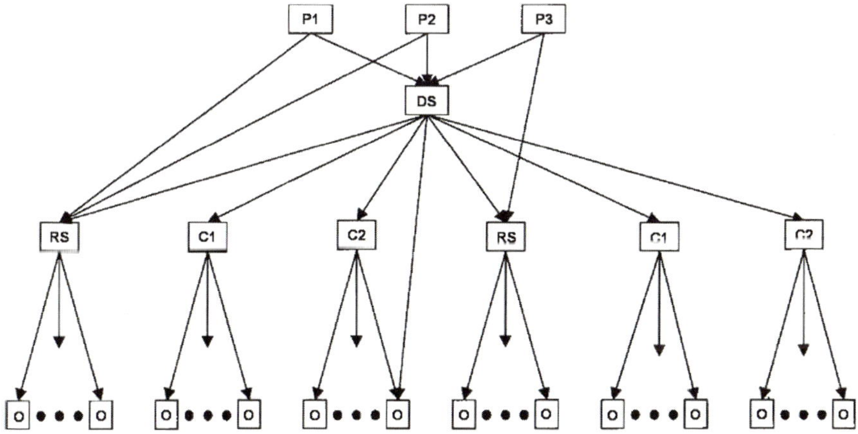

Fig. 21.13 Distribution network structure of a consumer-products manufacturer

Pi: production plants DS: central dispatch store
C1: one-stage crossdocking C2: two-stage crossdocking
RS: regional stores of retailers O: outlets of the retailers

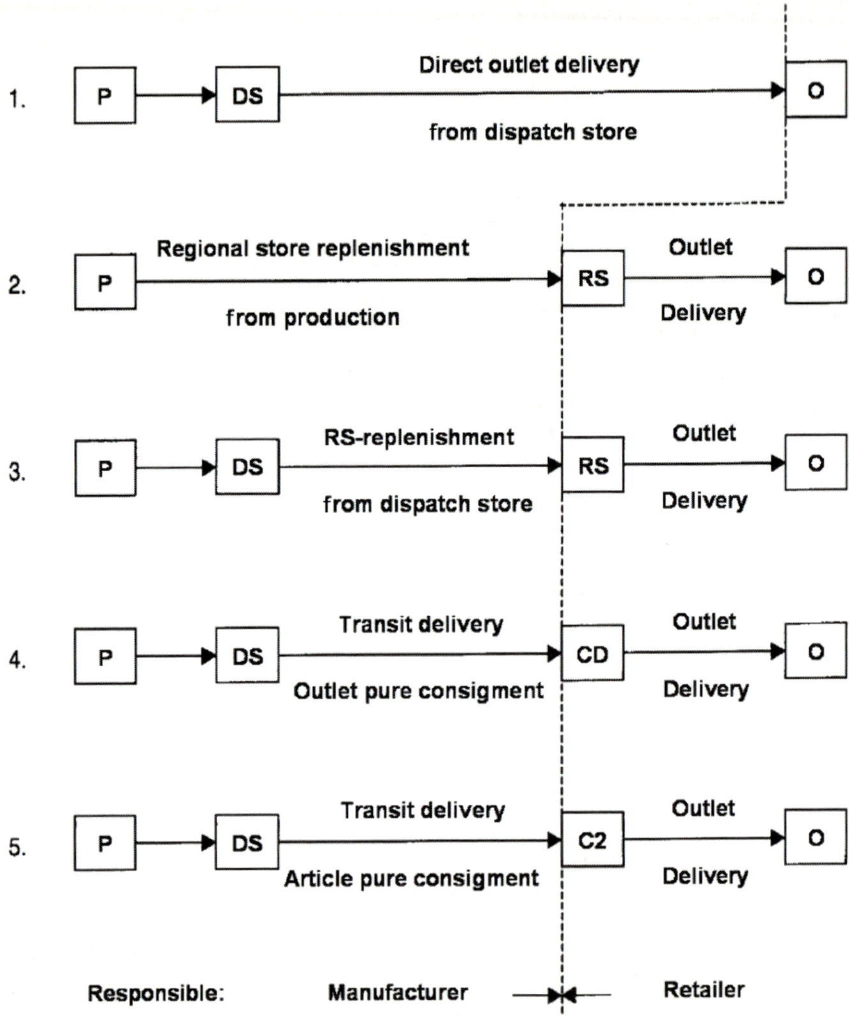

Fig. 21.14 Delivery chains of a consumer-products manufacturer
network structure and symbols: see *Fig. 21.13*
vertical dotted line: border of responsibility between manufacturer and retailer

the distribution network and *Fig. 21.14* the delivery chains of a manufacturer of consumer products, which are send to the outlets of retailers.

Other examples are *Figs. 21.15* and *21.16* presenting two basic freight networks and *Fig. 21.17* showing the most important freight chains through these networks. *Figs. 21.20* and *21.21* show the European distribution network and delivery chains of a German car manufacturer. The supply networks and supply chains of two different retail companies are presented in *Figs. 21.2* and *21.23* and in *Figs. 21.24* and *21.25*.

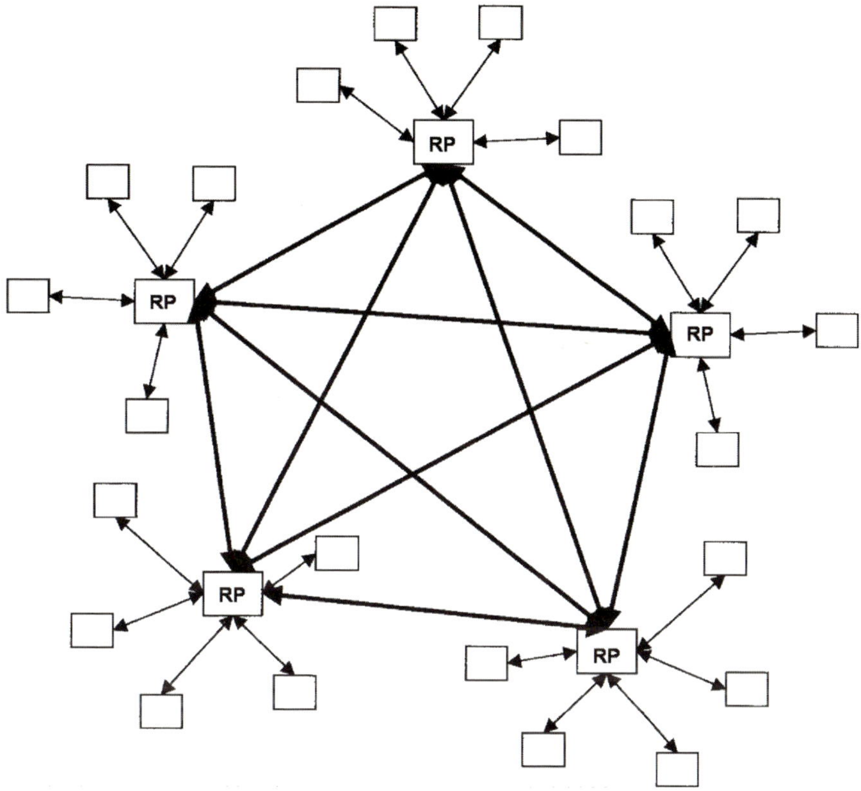

Fig. 21.15 De-central freight network
RP: regional transshipment points (N = 5)

For more detailed information and surveys, it is useful to specify the supply and delivery chains in a *delivery table*.

The *transfer of risks* for the goods between the supply chain actors is marked in the process chain charts and in the delivery table by dotted lines. The *responsibility border* for a delivery *ex works* is the dispatch ramp of the production or store of the manufacturer. For deliveries *franco domicile*, the responsibility border is the goods receiving ramp of a regional store, a crossdocking station or an outlet of the customer.

For the calculation of delivery times and costs, for system optimization, and for dimensioning a new logistic network, a more detailed specification of the delivery and supply chains is necessary. For these purposes, also the orders, the performance flows, the throughput values, and the stocks in the single stages and stations must be specified and quantified. Also the type of delivery, the relevant structure and transport parameters must be quantified. This determines the share of the consignment flow through the different supply chains.

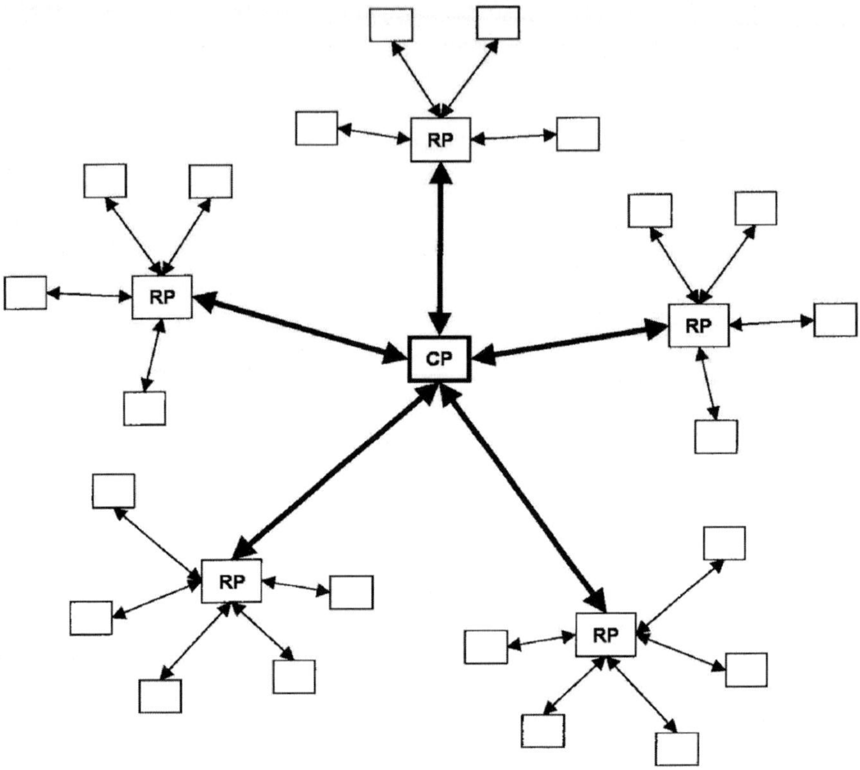

Fig. 21.16 Central freight network

RP: regional transshipment points (N = 5)
CP: central transshipment point (hub)

By incorporating all the necessary data, process times and cost rates and by programming the above formulas, the delivery table of a given supply network can be extended to a *spread sheet program* for the *determination of optimal supply chains* (DOS). Such a DOS-program calculates, for each single stage and station of a supply chain, the partial performance costs generated by the flows. The sum (21.32) of the partial costs over all stages gives the costs for the different supply chains. The cost sum (21.33) over all supply chains results in the total delivery, supply or freight costs per period. From the calculated process, transfer, driving and waiting times, the DOS-program sums the delivery times (21.31) for the different supply chains and the shipment times for the transport and freight chains.

A modularly designed DOS-program consists of several levels of connected *spread sheets*. The single sheets contain the data and formulas for the different stations, connections and subsystems and calculate the partial process times and costs rates. Special DOS-programs are DOF-programs for *determination of optimal freight chains* of transport and freight networks without storekeeping stations.

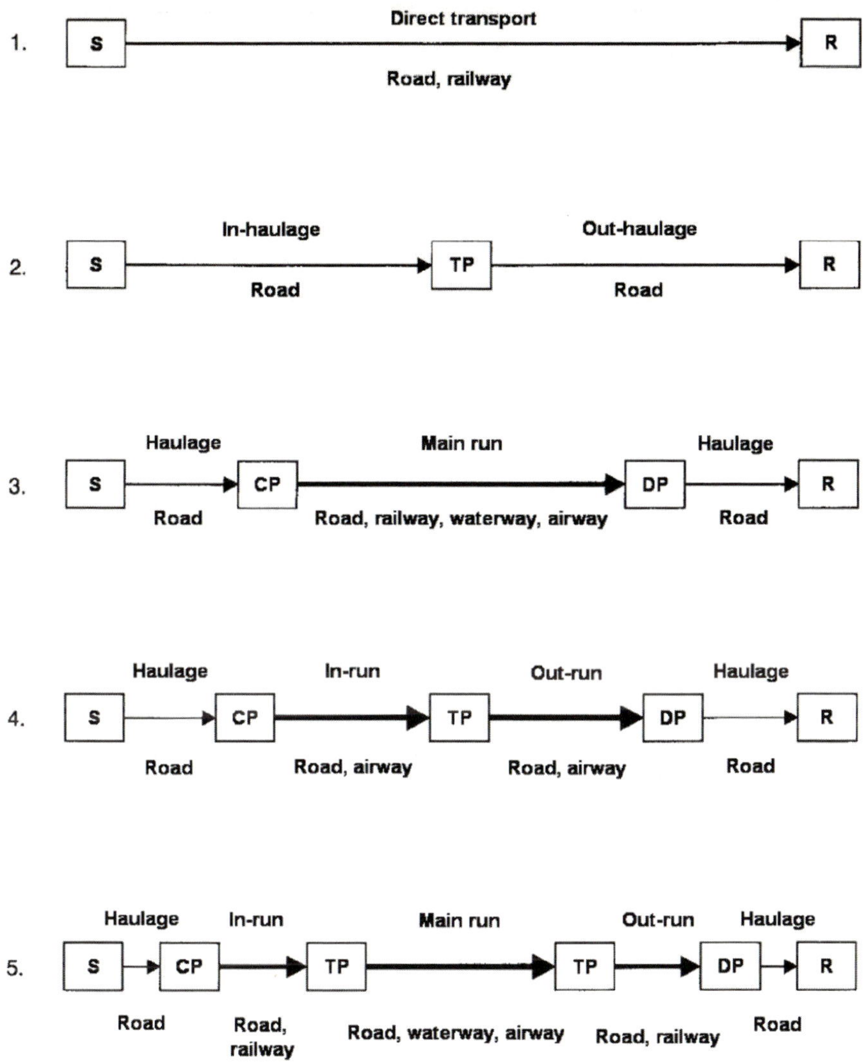

Fig. 21.17 Standard fright chains from supplier (S) to recipient (R)
TP: transfer point CP: collection point DP: distribution point

21.9 Optimization of Logistic Networks

The optimal logistic network of a company should enable the required performances and services at lowest costs. Theoretically, it seems possible to minimize the total costs (21.33) of the logistic network by systematic variation of the design and strategy parameters using the methods of *Operations Research*. However, the number of free parameters and their possible combinations increases more than exponentially with the product $N_D \cdot N_R$ of the number of delivery and receiving stations. Therefore, even powerful computers are not capable to determine by *full enumeration* within limited time the set of parameters for the theoretically optimal logistic network.

The general task of determining the optimal logistic network includes as sub-problems the *OR-standard problems*: *one-* and *multi-stage warehouse location*; *traveling salesman problem*; *optimal tour planning*; *filling, packaging and cutting problems*; one and many stage *lot-size* and *safety-stock problems*; *optimal network design* and others (Armour and Buffa 1963; Churchman et al. 1961; Domschke 1985 and 1990; Goczyla and Cielatkowski 1995; Kirsch et al. 1973; Laporte 1992; Lawler et al. 1985; Prager 1957). In order to make a given problem mathematically solvable, the requirements must often be simplified or restrictions have to be loosened. Some design parameters of logistic networks are neglected by OR-standard methods, e.g. the possibility of several supply chains of different stage degree for the same relation. OR-*heuristics* for finding approximately optimal solutions require suitable *initial solutions* (see *Sect. 5.4*).

The attempt to solve the complex task of determining the optimal logistic network only by mathematical methods is comparable to the attempt to design and construct bridges and buildings only by the methods of *finite elements* and *digital simulation*. However, just as buildings, logistic networks and systems can only be designed, constructed and dimensioned by experienced experts.

In practice, the task of optimizing a logistic network means, to find within limited time and effort a *feasible solution* with total costs significantly lower than the present costs. Far more important than finding the theoretical optimum is to understand which actions enable considerable cost-savings compared to the present situation. An optimized and feasible solution can be designed and dimensioned by qualified logisticians in *iterative steps*, supported by DOS-programs and applying the design principles, construction rules and approximation methods, which are described in the following.

An analytically constructed solution can be used as *initial solution* for further optimization by OR-methods and improved in detail by *digital simulation*. In this way, analytical construction, heuristic OR-methods, and digital simulation complement each another. This holds not only in business practice and consulting but also for the development and test of new design principles, construction rules and methods (see *Sect. 5.4*).

21.9.1 Iterative Network Design and Optimization

Steps of the iterative design and optimization of a logistic network are:

1. Survey, documentation and analysis of the existing *network structure*, *order processes* and *supply chains* including the connected information and data flows
2. Specification of the *structure requirements*, i.e. of the supplier and recipient structure, the intermediate stations and of the available transport relations
3. Determination of the *performance and supply requirements*, i.e. of the assortment, the orders, the service, consignment and throughput demand, the restrictions and the extrapolation factors for the planning horizon

4. *Segmentation of assortment*, *orders* and *consignments* in classes with logistically similar features
5. *Preliminary determination of the load units* and selection of the appropriate load carriers
6. *Determination of the necessary delivery and supply chains* with dispatch modes, structure parameters and transport parameters: First, the single-stage chains, then the two-stage, three-stage and higher-stage chains are defined; chains of higher stages are considered if the requirements cannot be fulfilled by chains with fewer stages
7. *Analysis of the regional distribution* of the delivery and receiving stations, of the consignment demand and of the load flows
8. *Preliminary division of the total service area* into *collection areas* and/or *distribution areas*
9. *Design of an initial network solution*, either starting from the existing network by systematically dropping and adding intermediate stations or constructing a new network with a minimal number of logistic stations in the transport gravity centers of the service areas (see *Sects. 18.10* and *21.10*)
10. *Allocation of the consignments* to the cost-optimal supply chains of the initial solution: Large and urgent consignments are assigned to supply chains with few stages, smaller consignments to supply chains with more stages
11. Preliminary *conception of supply strategies*
12. *Calculation of throughput* and *article stocks* within the stations and of the *load flows* between the stations with the help of the above formulas from the order and consignment demand
13. Determination of the necessary *limit performances*, *buffer places* and *store capacities* within the stations and of the required *transport means*
14. Set up of a project specific *DOS-program* consisting of program modules for the delivery stations, intermediate stations and receiving stations and for the transport relations between the stations
15. *Calculation of the delivery or shipment times* and check of the adherence of the supply requirements
16. *Calculation of the delivery costs* based on preliminary cost rates for the expected throughput and performance demand
17. *Iterative optimization* of the network and supply chains by *adjustment of the supply strategies* and *variation of the free parameters*, such as the structure parameters, the load units, the freight modes, the delivery and replenishment frequencies, and the transport parameters
18. *Design of the order processes* which initiate the supply processes and of the *information and data flows* for control of the processes

If certain requirements or restrictions cannot be fulfilled in one of these design steps, the design process must be repeated at the last or a previous step. By this procedure, the design process becomes iterative.

For a rough outline of the general network structure and for estimation of dimensions and capacities, it is sufficient in step 10 to assign representative mean consignments for each consignment class. For more precise cost calculations, it is necessary to assign the single consignments of a given annual demand one after another to the supply chain with the lowest total delivery costs (21.32) which keeps the required delivery time and all other restrictions.

21.9.2 Consideration of Throughput-Dependent Costs

The summation of the resulting flows for all supply chains gives the total performance demand in the stations, and the total transport and freight demand between the stations. If the resulting throughput deviates significantly from the initially assumed values, the cost rates must be checked and possibly revised.

The throughput for single stations or transport relations may even become lower than the *critical level* where the operation becomes inefficient due to the *fixed-costs dilemma* (see *Sect. 6.8*). If it is not possible to use the respective station or transport relation also for external demand and other freight flows, the optimization process must go back to the step 7 and reduce the number of service areas.

21.9.3 Process Optimization versus Structure Optimization

The logistic network and the supply chains can be optimized by alternating from the *structural aspect* to the *process aspect* and vice verse, until a stable and consistent solution evolves which fulfills all requirements (see *Sect. 1.3*). An optimization of the structure, considering only one supply chain for each relation, is as inadequate as the optimization only of the processes without considering the resulting total utilization of stations and transport relations.

Due to the high number of design parameters, combinations and possibilities, the solution space for the network optimization is huge, although the number of acceptable solutions is reduced by project-specific restrictions. It is quite possible that the iterative optimization process leads to a suboptimal solution and misses the absolute optimum of minimal costs. However, in business practice, the present costs, not the theoretical minimal costs are benchmark for the optimization. For a company, it is important to reduce costs by intelligible means in acceptable time and to adapt performances to the expected demand in feasible steps.

21.9.4 Saving Potentials and Sensitivity Analysis

The present supply chain costs minus the total costs for the new solution are the *saving potential*. The savings must be related to the required *investments* for set up, changes or extension of the network. The *return on investment* (ROI) has to be compared with the ROI-benchmarks of the company (see *Sect. 5.1*).

With the help of a DOS-program it is possible, to perform *sensitivity analyses* for expected changes of the supply requirements or of new constraints. This allows e.g. to quantify the effects of additional or lost orders on the delivery costs.

After a DOS-program has been set up for the logistic network of the company, it is possible to calculate the cost effects caused by a change of supplier to another supply chain or by a change of the *terms of delivery*, e.g. from *free to door* to *ex works*.

It can also be used to calculate order- and article-specific logistic costs. Another application is the calculation of *logistic discounts* for order quantities filling complete load units or transport units.

The DOS-program can be used also in daily business to allocate the upcoming consignments to the optimal supply chains by computer. If no DOS-program is available, the consignments must be assigned to the supply chains on the bases of *allocation strategies* developed and tested by model calculations. Examples for general allocation strategies are the boundaries (21.16), (21.17), (21.18) and (21.19) between full-loads, part-loads, general cargo and parcel cargo.

The described design and optimization procedure for logistic networks and supply chains has been applied successfully in consulting practice to many business cases of different branches. Some cases with their specific aspects and particularities will be presented in the next sections.

Besides adequate methods and computer tools, both, the design of logistic structures and the optimization of supply chains require creativity, experience and knowledge on the options and their effects. Experience and model calculations show that the influence of the various options and parameters on costs differs considerably. Only a few of them enable really interesting cost savings and performance improvements. These need to be identified.

21.10 Transport and Freight Networks

Transport and freight service providers – carriers, forwarders, parcel services, airlines, shipping companies, postal services and railways – regularly execute transport and freight orders for many different customers and shippers. In addition, they rent and charter loading space and transport means.

For their business, the actors use own and hired transport means on a transport or freight network which covers a defined *service area*. They offer the different transport and freight chains shown in *Figs. 21.4* and *21.17*.

21.10.1 Basic Network Structures

A transport or freight network consists of a number of *regional transshipment points* (RP), each located in a *service area* of certain extension. The regional transshipment points are connected by *main transports* which are executed with certain frequency. Depending on the transport connections, two basic *network structures* can be differed:

- A *de-central network,* as shown in *Fig. 21.15*, connects N regional transshipment points directly by N(N-1)/2 transport relations with regular outward and backward transport flows.
- A *central network* or *hub-and-spoke-system*, as shown in *Fig. 21.16*, connects N regional transshipment points indirectly by N transport relations or *spokes* via a *central transshipment point* (CP) or *hub*, where incoming loads from the regions are sorted and consolidated to outgoing loads.

The advantages of de-central networks are shorter mean distances and transport times between the single stations. The main disadvantage is the lower utilization of efficient transport means, if the load flows are low and/or if high service frequencies are required.

In a central network with the same transport frequency, the number of transports as compared to the de-central network is reduced by the factor (N-1)/2. The mean load flows on the transport relations from and to the hub are higher by this factor than the direct load flows. This leads to the *advantages of central networks*:

- Compared to the de-central network, in a central network either larger transport units with high filling degree, or smaller transport units with higher transport frequency can run between the regional transshipment points and the hub.

The positive effect starts with N = 4 and increases linear with the number N of transshipment points. Disadvantages of a central network are the transfer costs, the additional handling in the hub and the longer distances and transport times. This leads to the *freight network design rule*:

► A central network connecting a high number of far distant transshipment points is only opportune, if the direct mutual freight flows are significantly lower than the mean transport capacity for the required delivery frequencies.

Due to differing demand, unequal distribution of freight flows, and deviating service requirements, transport companies and freight forwarders generally operate a *hybrid network*, which is a combination of central and de-central networks. Direct transports are executed between regional transshipment points as long as the freight demand is sufficient to fill the forth and backward running transport means at the required service frequency. Consignments not transported directly are conveyed via the central hub or, if opportune, via another regional transshipment point. The surrounding local area is directly served from the hub.

In regionally structured and densely populated countries as for example Germany the railways, postal services and airlines with area-wide services have more *de-central networks*. In countries with large extension like USA, and in centrally organized countries like France or the UK more *central networks* with one main hub and/or several local hubs are dominant. Worldwide operating *parcel service providers* have *integrated networks* with a few *central hubs*, several *local hubs* and many regional transshipment points.

The integration of the networks of internationally operating freight service providers with the local networks of national transport and logistic service providers leads to *combined networks* with mixed central and de-central structure. In combination with the different traffic-modes result the *inter-modal freight networks* which span the whole globe (see *Fig. 21.4*).

The density and complexity of the global transport networks generate a hardly manageable number of transport and freight chains. They are used simultaneously by industry and trade companies as well as by freight service providers. For both, shippers and service providers, the selection of the optimal transport and freight chains for a given demand is a great challenge. For this purpose, effective *calculation tools* and *allocation strategies* are needed.

21.10.2 Standard Freight Chains

For the development of allocation strategies and the derivation of the optimal strategy parameters like the boundary criteria (21.16), (21.17), (21.18) and (21.19), it is useful to classify the possible freight chains in ascending stage-degree. Resulting are the *standard freight chains* shown in *Fig. 21.17*. Through these standard freight chains flows the greatest share of national and international freight. The standard chains split up in the huge number of existing transport and freight chains, if the possible traffic modes and the means, tours and frequencies for the transports between the stations in these chains are considered.

Standard freight chain 1 of *Fig. 21.17* is the *direct transport* without transit between delivery and receiving station. This transport chain is most efficient and fastest for a sufficient demand of medium-size consignments in a local area, and for larger loads over longer distances. Most direct transports use the road. Regular direct transports on the railway are opportune for large load flows over longer distances, e.g. between factories of the primary industry and the consumer goods industry, in exceptional cases also between the plant of a supplier and the logistic center of a retailer.

Standard freight chain 2 is typical for the distribution of smaller consignments within a regional service area around a transshipment point. The consignments are collected on *pickup tours* or *in-haulages*, consolidated and sorted in the transshipment point and distributed on *delivery tours* or *out-haulages*.

Standard freight chain 3 with *in-haulage* in the service area around a collection point, a *main run* to a distribution point and *out-haulage* in the distribution areas are typical for long distance freight orders of general cargo. The in- and out-haulages are executed by smaller transport means on the road. The main run is performed by semi trailers or swap body trailers on the road, by train or by other traffic modes. Realizations of standard freight chain 3 are three-stage inter-modal road-railway chains, road-airline chains and railway-sea chains within de-central networks.

Standard freight chain 4 is typical for service providers with a central freight network. It consists of an in-haulage to a regional collection point, two main runs to and from a central *transfer point* and an out-haulage from a regional distribution point. Additional transfer points, where other loads and freight flows converge or a change of the transport mean or traffic mode takes place, are useful for inter-modal transports over long distances. For example, the *standard freight chain 5* leads from the sender via a collection point, two central transfer points and a distribution to the recipient.

The freight chains of the service providers and different traffic means compete with each other. The actors must decide, which share of their freight demand should be executed by own transport means and own transfer stations, and for which share it is opportune to rent external transport means or to engage logistics service providers. There is no general answer to these questions. The decision depends on the individual conditions, requirements and business of the companies (see *Chap. 22*).

21.10.3 Zoning Principles for Service Areas

Freight service providers with an own network, and companies with an own procurement and distribution network are faced with the task of dividing the total service area covering all their supply, delivery and receiving points into a number of local *service regions*. This *zoning of the service area* should enable the most cost-effective supply and freight chains, which keep the required delivery times.

For an existing network, the optimal regions can be found by a systematic change of the number and borders of the regions. This is possible with a DOS-program and suitable OR-search algorithms. However, it is quite tedious and implies the danger to miss a structurally different solution with significantly lower costs. In many cases, it is advisable, and for a non-existing network unavoidable, to construct a new network based on general design principles.

Useful *zoning principles* for the division of a total service area are:

- *Principle of the minimal number*: The number of service regions and transshipment points should be as small as possible in order relate low fixed costs to a high throughput, to apply the most efficient transfer technique, and to achieve high load flows for cost-efficient main runs (see *Chaps. 18 and 19*).
- *Principle of the necessary number*: The extension of service regions is limited by the *maximal reach distance* of the pick-up and delivery vehicles, which is determined by the required shipment times, the effective travel speed and the number and duration of the stops. The minimal necessary number of regions results from the complete coverage of the total service area by the *delivery circles* around the locations of the transshipment points.

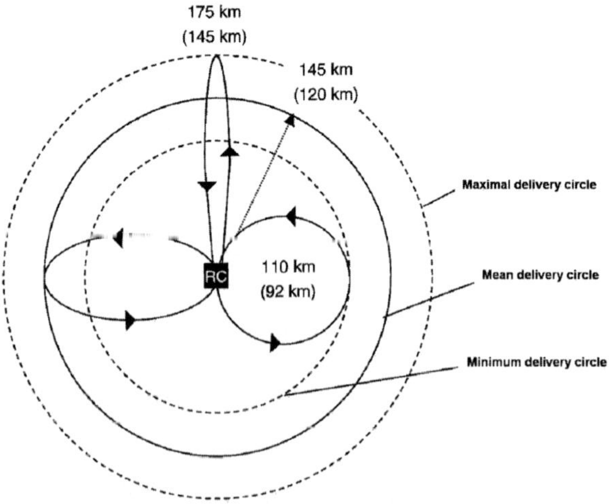

Fig. 21.18 Delivery tours and delivery circle around a regional center

maximal length of delivery tours: 350 km
deviation factor (road to air distance) $f_{dev} = 1.2$
xxx km: road distance (xxx km): air distance

- *Principle of balanced demand*: The single service regions should generate similar load demands in order achieve balanced and paired main runs and equally used transshipment points.
- *Principle of area division*: If the load demand of a regional area exceeds the *critical throughout* of the transshipment station, no further economies of scale can be expected. Then it is advisable to divide the region into two sub-regions with nearly the same load demand and to serve them from two transshipment points (see *Sect. 19.11*).

An application of these zoning principles has been the development of a distribution network for a German wholesaler with the network structure shown in *Fig. 21.24*. In this case, the maximal tour length of a delivery vehicle is 350 km, leading to a *delivery circle* with a mean radius of 120 km and a maximal radius of 145 km as shown in *Fig. 21.18*. *Figure 21.19* shows that 90% of the potential

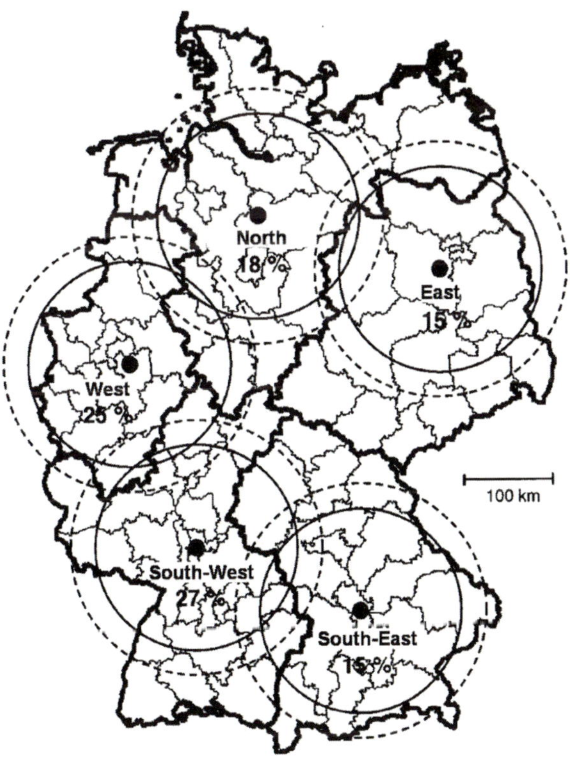

Fig. 21.19 Regional centers and regional service areas with delivery circles of a German wholesaler for installation material

Percentages: shares of total freight flow
Black dots : locations of regional centers
------ : mean delivery circles (120 km)
……… : maximal delivery circles (145 km)

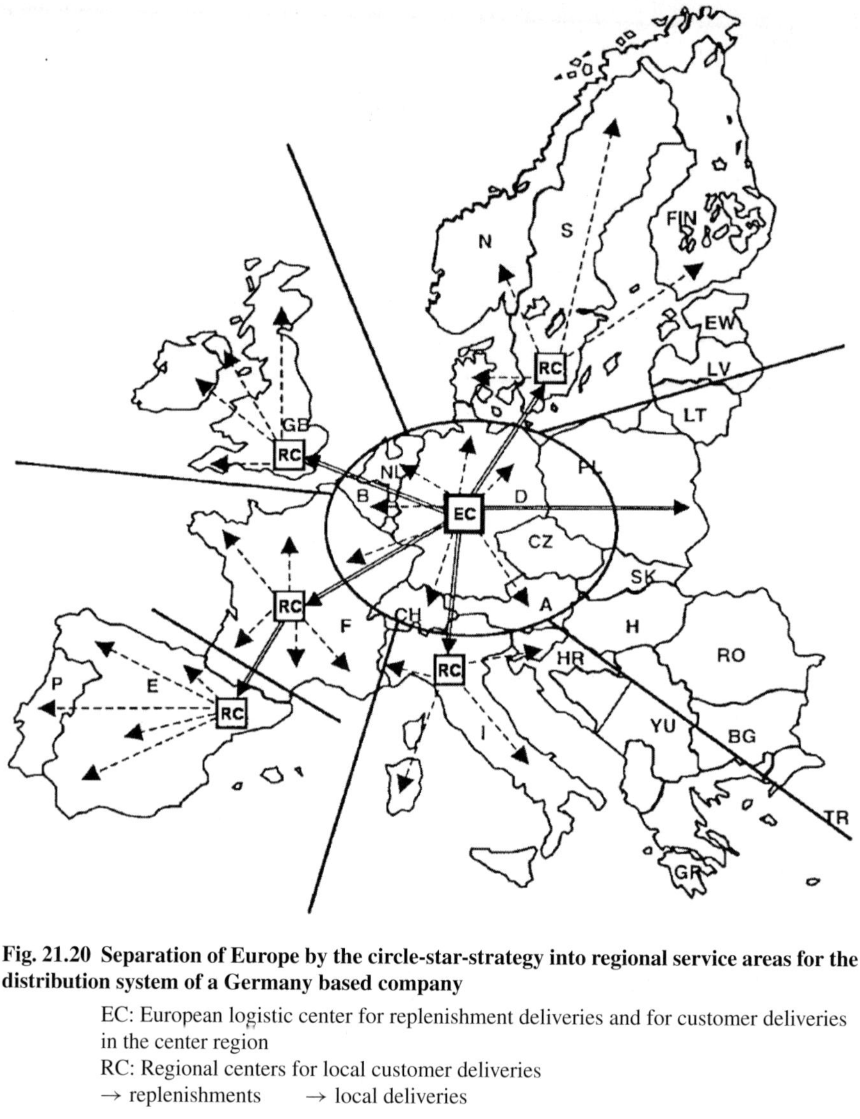

Fig. 21.20 Separation of Europe by the circle-star-strategy into regional service areas for the distribution system of a Germany based company

EC: European logistic center for replenishment deliveries and for customer deliveries in the center region
RC: Regional centers for local customer deliveries
→ replenishments → local deliveries

receiving stations in Germany are covered by 5 delivery circles with a radius of 120 km around 5 regional centers. The left out area in middle-east Germany requires later the addition of a sixth regional center near Erfurt.

The considerations of *Sect. 18.10*, the results presented in *Fig. 18.27* and model calculations show that the allocation of the delivery and receiving stations at the borders to one or another neighboring service region affect the travel distances and delivery costs only marginally. This leads to the

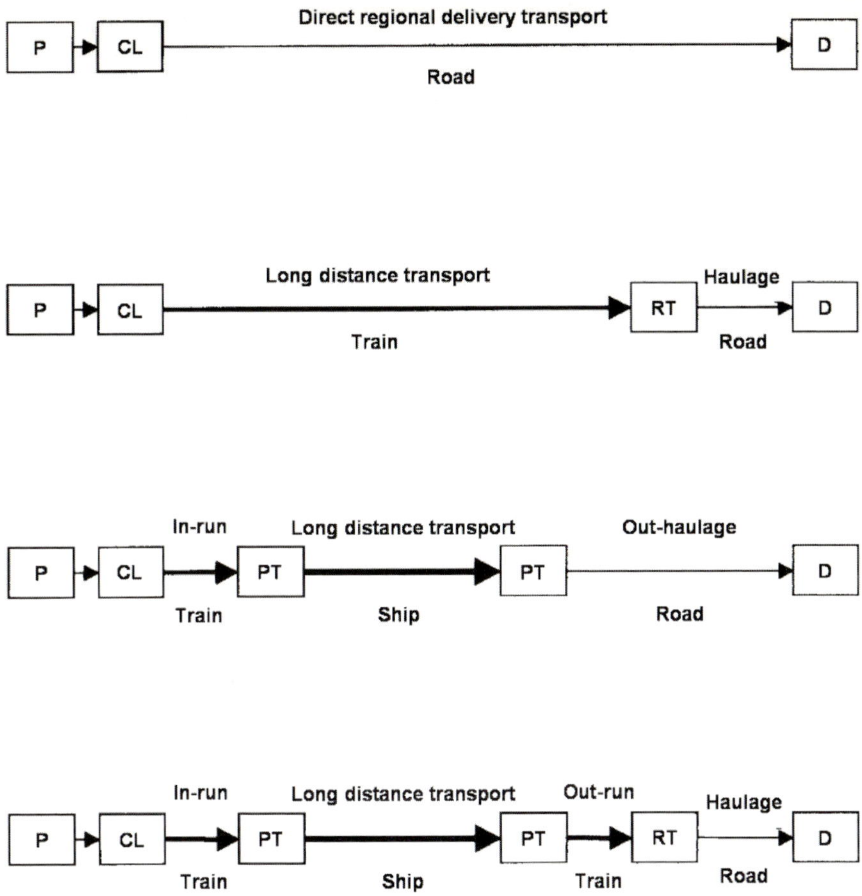

Fig. 21.21 Delivery chains of a German car manufacturer

P: car assembly plant D: car dealers
CL: central loading station with buffer places
PT: port transfer points railway/ship
RT: regional transshipment points railway/road

- *Principle of tolerable simplification*: For business practice, a rough division of the total service area, based e.g. on two-digit post code areas, using the zoning principles and a location of the transshipment points in the vicinity of the transport gravity centers of the regional areas is sufficient.

The exact position of the transshipment points results during project realization from the existing possibilities and restrictions. As explained in *Sect. 18.11*, the precise allocation of delivery and receiving stations at the borders of the regions to the transshipment points is a result of *tour planning* and changes currently.

Another strategy to divide the total service area of a central network for the distribution from a single source, or for the procurement of one sink is the

- *Star-circle strategy*: The local service area around the delivery, receiving or logistic center is determined by the maximal delivery circle around the center. The outside area is separated into a minimal number of service sectors with about the same load demand. If the distances in these sectors are too long for efficient tours, they are further subdivided into smaller local service areas.

The star-circle strategy results from similar considerations as the *rotary-beam method* of tour planning (see *Sect. 18.11*). An application of this strategy is the initial solution shown in *Fig. 21.20* for the *European distribution areas* of a company with a logistic center and plants in Germany. It is quite similar for the European distribution of finished goods, spare parts or cars. *Figure 21.21* shows the European delivery chains of a German car manufacturer. In this case, the final distribution areas have been optimized and adjusted to the available locations of the carriers, the railway and the service providers.

The above zoning principles and design strategies can be applied to find an appropriate initial solution for a logistic network in short time, provided the distribution of the delivery and receiving points and the freight and load flows are roughly known. If a multi-stage logistic system with e.g. two or three regional stores is necessary, the regional areas of the first area division must be subdivided into smaller local service areas using again the design rules.

21.11 Distribution Chains of Consumer Goods

Figure 21.13 shows a typical network structure of the consumer goods industry. The *delivery chains*, which correspond to the different *distribution channels* or *marketing channels*, are presented in *Fig. 21.14*. They replenish the stocks of the retailers and supply the outlets, markets, department stores and smaller shops with the required consumer goods from the factories of the industry (Kotzab and Bjerre 2005).

Large consignments of palletized goods flow through the distribution chains of *Fig. 21.14* from the factory via a finished goods store either directly or via a logistic center or regional store of the retailer to the outlets. Smaller consignments of single pallets, packages or parcels are delivered by *logistic service providers* using the standard freight chains 2 and 3 of *Fig. 21.17*.

Owner and/or operator of the stores, transshipment points and logistic center between production site and *point of sales* can be the supplier, the retailer or a logistic service provider. The ownership of the stations and processes in a supply chain determines – as indicated in *Fig. 21.14* – the *responsibility border* and the *risk transfer* for the goods. If e.g. the manufacturer or on his behalf, a service provider operates the transshipment points, the responsibility border between manufacturer and retailer is shifted to the ramp of the outlets. The supply chains of the manufacturers, service providers and retailers compete with each other. They often differ only in the responsibility for the stations and the transports whereas their structure and processes are more or less the same.

In most cases, the supply chains used by manufacturers and retailers, the number and locations of the factories and the outlets, as well as the number and location of finished goods stores, retail stores and transshipment points, are grown historically. They differ from company to company and are generally not optimal. Therefore, a company must permanently check whether the existing logistic network and supply chains are capable to cope optimally with future market requirements.

This question is crucial for the suppliers of large *retail companies*. Due to the reasons outlined in the next section, more and more retailers have developed own supply strategies, set up own replenishment chains and optimized their logistic networks. This movement has reduced the direct deliveries of small consignments to single outlets and enhanced the delivery of large loads to a small number of stores and transshipment points (Kotzab 1997). The risk transfer has been shifted from the ramp of the outlet to the ramp of a store or transshipment point operated by the retailers themselves or by an assigned logistic service provider. In the extreme, the retailer or his transport provider collects the goods at the ramp of the manufacturer. This enables the retailer to schedule combined pickup and delivery tours, to achieve higher utilization of the transport means and to establish paired transports.

Despite these changes of the networks of big retail companies, the manufacturers must still supply the shops and outlets of the remaining retailers free to the door. The industry delegates this task more and more to specialized logistic service providers (see *Chap. 22*). The logistic changes of the market and the ongoing concentration in retail business, accompanied by a reduction of regional retail stores and a centralization of stocks in a small number of logistic centers, forces the *consumer-goods industry* to adapt and to redesign their distribution networks and chains.

The DOS-program, supply strategies and design rules which are described above have been applied successfully to re-dimension the distribution network and chains of several consumer-goods manufacturers. The possibilities and potentials are exemplified by two *consulting projects*.

The first business case is a German manufacturer of spirits with famous brands. The project was triggered by the merger of three smaller companies. Originally, there have been four plants operating in different locations and eight stores for finished goods. An own fleet and a varying number of forwarders delivered annually 150,000 t of bottled products on 240,000 pallets to about 12,000 customers, who placed about 110,000 delivery orders per year.

The main results of the project have been: closing down two plants; concentration of mass-production in the most modern plant; product development and bottling of special brands in the most traditional plant located in a beautiful old town; closing of all eight stores; delivery of the whole production from only one *logistic center* located close to the main plant; selling of the own fleet and outsourcing of the distribution to three freight service providers. The annual savings only by the redesign of the distribution system were calculated with an DOS-tool to be 3,5 Mio.€, i.e. 18% of the existing costs. Additional savings in the same order of magnitude should come from the consolidation of the production and the centralization of stocks. The calculated cost savings have been achieved within two years after implementation. Furthermore the flexibility has increased, the service was significantly improved, and the negative effects of *self-collection* by the big retailers were reduced.

Similar results have been achieved in the second business case of a leading manufacturer of *home and personal care products*. The company delivered annually

115,000 t finished goods from two German factories via two factory stores to about 2,300 retail outlets and 200 regional stores, partly with crossdocking.

The optimization of the network structure and distribution chains resulted in a delivery from only one optimally located logistic center to be built by a general contractor. The operation was first *outsourced* and later *insourced.* The distribution in Germany was awarded to three forwarders serving different areas. The cost savings in this case were more than 20% of the former distribution costs. The stocks were reduced, the service level was improved and flexibility increased.

21.12 Procurement Chains of Retailers

Several investigations of the business processes in retail outlets and markets have shown that the staff is occupied between 30 and 40%, in some cases even above 50% of the time with logistic activities, such as receipt of goods, inspection, unpacking, labeling, in-storing, rearranging, shelve filling, removal of empties, provision for customer delivery and scheduling of replenishment. Due to the additional occupation by administrative activities, less than 30% of the time is left for selling and customer related activities (Prümper 1998; Kotzab et al. 2007).

A main reason is the high number of uncontrolled incoming consignments of different sizes during the day. This finding and the observation that logistic costs consume up to 50% of the gross margin, has induced many retailers to examine and to redesign their procurement logistics (Bock et al. 1996; Kotzab and Bjerre 2005; Laurent 1996; Prümper 1998, Toporowski 1996).

The *goals of retail logistics* are:

enforcement of the sales activities
reduction of the ramp contacts
relieving the retail outlets from logistic activities
improvement and reduction of scheduling
competitive article availability in the outlets
avoidance of waste and returns
optimal stock levels in the supply chains
minimal total supply costs (21.36)

In addition, retailers must care for an increasing number of customers who expect *home delivery* of goods bought in an outlet, ordered by mail or telephone from a catalogue or purchased via Internet (*E-commerce*).

Also for the optimization and redesign of the *procurement logistic* of retailers the above DOS-program, supply strategies, and design rules have been applied successfully as exemplified by the following two consulting projects.

The first *business case* is a German *Do-It-Yourself-retailer* with an assortment of over 60,000 articles. More than 1,200 suppliers delivered their consignments free to the ramp of 80 DIY-markets. On average, a market received 30–60 deliveries and in peak times, more than 100 deliveries per day. The consignments ranged from one parcel over several pallets to full truck loads. The logistic activities in the markets occupied between 30 and 35% of the time.

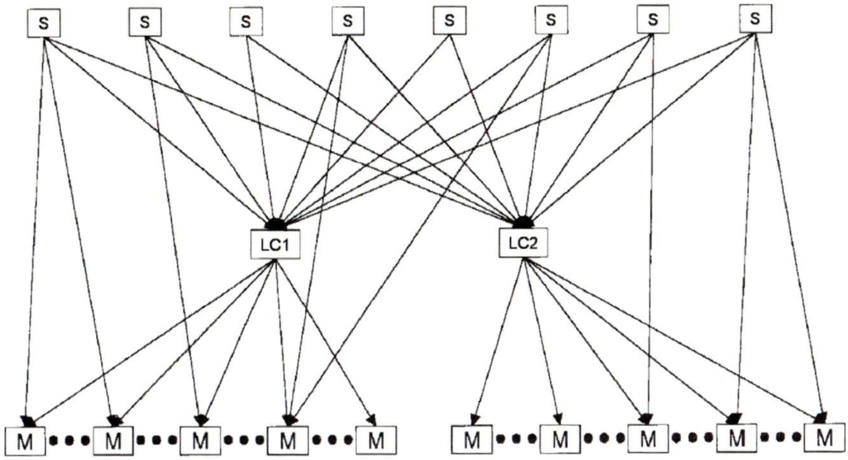

Fig. 21.22 Procurement network of a DIY-retailer

S: suppliers (approx. 1,200) O: Outlets (approx. 80)
M: markets and sales outlets
LC: logistic centers with crossdocking and storing

The redesign of the company logistics resulted in the hybrid network structure shown in *Fig. 21.22* with in the first phase two, and later three logistic centers at optimal locations. The main procurement chains with the borders of responsibility between supplier and retailer are shown in *Fig. 21.23*.

The entry inspection of all goods passing the logistic centers was shifted from the market ramps upstream to the ramp of the logistic center. About 55% of the articles, representing 70% of the volume and less than 50% of the turnover, is shipped as full or part loads via supply chain 1 directly to the outlets. These are mainly consignments with more than five pallets and weight above 1 t, and bulky or large volume items, such as carpets, hazardous goods and incompatible articles such as cement, stones or fertilizers.

About 45% of the articles representing slightly more than 50% of the turnover but only 30% of the volume, are shipped via the own logistic centers. About 50% of the deliveries with more than ½ pallet per market is executed via supply chain 2 with *one-stage crossdocking*. 40% of the deliveries via logistic center with less than ½ pallet per market runs through supply chain 3 with *two-stage crossdocking* (see *Figs. 21.1* and *21.3*). Supply chain 4 with intermediate storing in the logistic center serves about 10% of the total demand and concerns promotions and import articles.

The redesigned procurement logistics of the DIY-retailer has been implemented successfully. The article availability in the markets became better than 98% whereas the stocks in the whole supply chain were reduced significantly. The logistic activities in the markets are less than half as before and the time for sales activities has almost doubled. The procurement costs for the goods passing a logistic center were reduced by more than 15%. At the same time, the preconditions were founded for

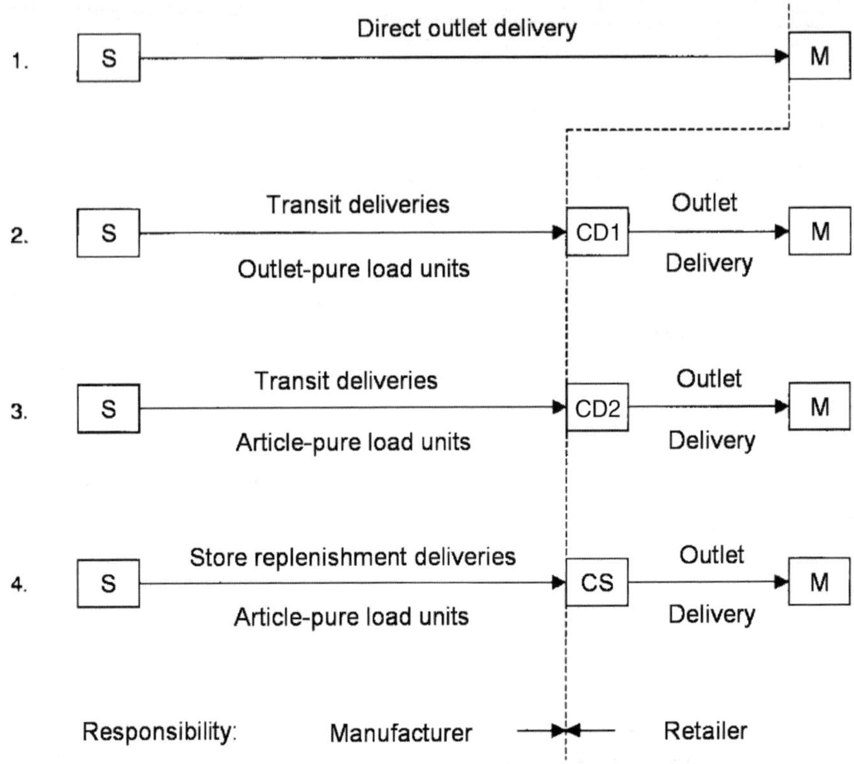

Fig. 21.23 Supply chains of the DIY-Retailer with network *Fig. 21.22*

S: suppliers M: markets with sales stock
C1 and C2: one- and two stage crossdocking
CS: central storing in the logistic center

computer aided scheduling and for further expansion (Makowski 1999). After 10 years of expansion the third logistic center was erected.

The second *business case* refers to a wholesaler for electrical equipment with more than 110 *service branches* located all over Germany. So far, the storekeeping and specifically ordered articles have either been picked up at the branches by the customers themselves or delivered free house. The availability of the storekeeping articles was below 80%, the delivery times for customer-ordered goods too long and unreliable and the procurement costs too high.

The optimization resulted in this case in the distribution structure shown in *Fig. 21.24* with five service areas and regional centers at the locations shown in *Fig. 21.19*. *Figure 21.25* reflects the resulting six standard supply chains and the *responsibility border* between suppliers, retailers and customers.

The regional centers keep stocks of a broad *regional article assortment*, while the sales branches keep stocks of a narrow *local article assortment*. Customer delivery

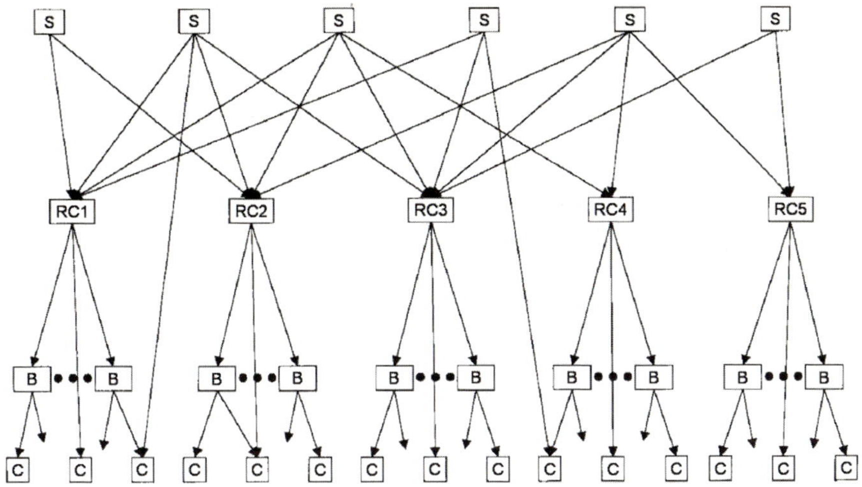

Fig. 21.24 Distribution network structure of a wholesaler for installation material

S: suppliers (approx. 450) B: sales branches (approx. 110)
C: final customers
RC: regional centers with crossdocking and storing

is performed via supply chains 1, 2 and 3. The chains 4, 5 and 6 supply the branches via the regional centers. Up to three different transport and freight service providers execute the full loads, part loads and larger consignments. A parcel service provider delivers small consignments of up to three parcels.

Results of this project were a significant wider assortment of storekeeping articles with availabilities above 95%, reliable short delivery times for customer ordered articles and a reduction of logistic costs by more than 20%. In this case, the logistic project triggered a reorientation of the total business policy with clear distinction between pickup customers and free-house customers and their differing demand.

21.13 Selection of Optimal Transport and Freight Chains

Transport and freight chains are supply chains without storekeeping intermediate stations. *Fig. 21.4* shows common inter-modal transport chains and *Fig. 21.17* several standard freight chains.

The sources, starting points or delivery stations of transport and freight chains are factories or finished goods stores of industry and stores of wholesalers or logistic centers of retailers. The sinks, destinations or receiving stations are plants, companies, supermarkets, outlets, and customers in a service area.

The delivery costs for goods and consignments through transport and freight chains are the *shipment costs* or *freight costs*. For passenger transport, they are the *carriage costs, fare* or *passage price*. The freight costs can be calculated by a *DOF-program* for *determination of optimal freight chains* which is a DOS-program for

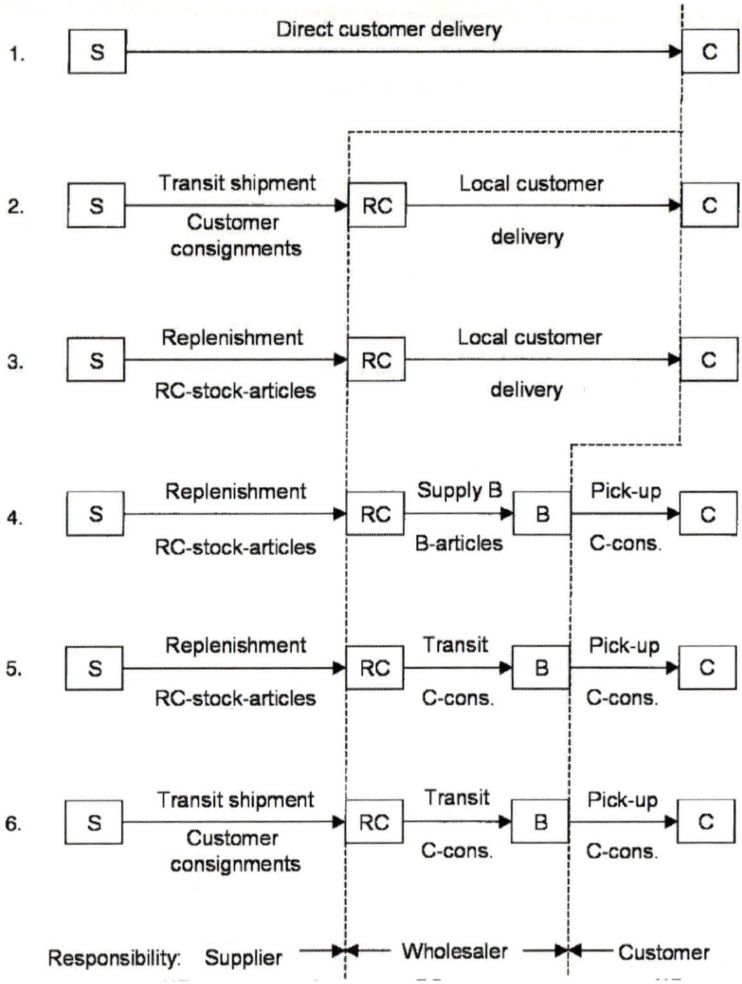

Fig. 21.25 Standard supply chains of the wholesaler with the network *Fig. 21.24*

a network without stocks. The DOF-program can be extended into a *DOS-program* for the *determination of optimal supply chains* by adding further program modules for supply chains with storekeeping stations.

For *return freights*, *empty returns* and *recycling*, either the same freight chains can be used as for the deliveries, or other chains which are mapped by additional program modules. Such *extended DOF-programs* can be used to calculate the optimal share of single and paired transports for the outward and return freight flows and to evaluate further transport cost savings.

The exact transport distances between the stations can be taken from an *electronic map*. From these data the mean distances between and within the service areas are calculated by weighting them with the respective freight flows. Other subprograms calculate the performance costs for internal logistic performances, and the cost rates for the different transport modes and transport means. Instead, also market prices can be used. Further parts of DOF- and DOS-programs are *logistic data tables* with the dimensions and net weights of load carriers, the limit performances and store capacities of the stations and with the travel speeds and capacities of the transport means (see *Sect. 12.6*).

From the dimensions and weights of the basic logistic units which leave the sources, a sub program calculates using the formulas of *Chap. 12*, the effective capacities of the load units and transport units. With these capacities and the relations (21.15) to (21.30) the freight flows, transport flows and performance flows are calculated, and the required numbers of the internal resources and of the transport means are determined.

Input values for the DOF-program are the consignment requirements and the structural parameters. Results are the freight costs per period and the specific freight costs per package unit, per pallet, per load unit and per 100 kg for the different transport and freight chains.

21.14 Influence Factors of Freight Costs

To demonstrate the effects of the various influence factors on the freight costs a series of model calculations have been conducted with a DOF-program. In the *Figs. 21.26* to *21.32* selected results of these calculations are presented.

In this model case, the DOF-program maps six different transport- and freight chains through a distribution network, as shown in *Fig. 21.13*, with one central source, which can be a plant, a finished goods store or a logistic center. The central source supplies the outlets of retailers and delivers final customers in the area of Germany with palletized goods and un-palletized packages. The main runs to the transshipment points and the direct deliveries of full and part loads are carried out by *semi-trailers* and *swap-body trailers*. In the regional areas, the consignments are delivered from the transshipment points to the outlets and other customers by 7.5 t trucks. The legend of *Fig. 21.27* contains the consignment requirements and other parameters for the model calculations. The legends of the other figures indicate the varied parameter. The different dependencies hold *ceteris paribus*.

The dependency of the mean freight costs per pallet on the number of transshipment points is shown in *Fig. 21.26* for *two-stage delivery* of general cargo consignments via one transshipment point with one-stage crossdocking. A similar dependency results for two-stage delivery of parcel cargo. For up to 25 transshipment points, the local haulage costs decrease more than the main run costs. Hence, up to this number the total freight costs go down with additional transshipment points. If the number of transshipment points exceeds 25, the main run costs increase more than the local haulage costs due to the decreasing filling degree of the load transports to the transshipment points. As the transfer costs remain constant, the *optimal number of transshipment points* in this case is 25.

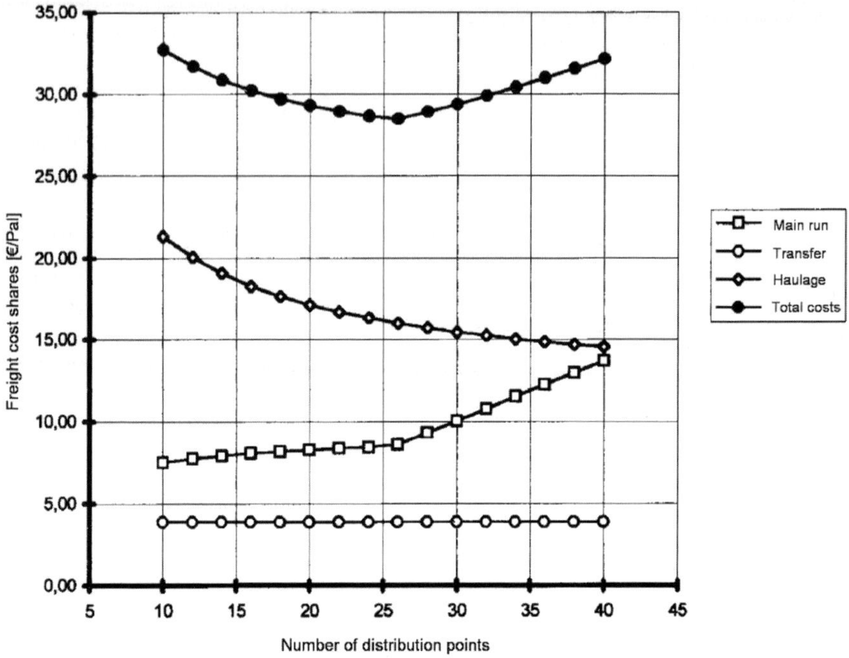

Fig. 21.26 Dependency of the freight costs for general cargo consignments on the number of transshipment points

Freight demand: 30,000 PU/d = 800 Pal/d = 75,000 t/a
Mean package unit [PU]: 12 *l* / 10 kg cartons
Mean consignment [Con]: 50 PU/Con = 2 Pal/Con
Load units [LU]: CCG1-pallets with 49 PU/LU
Mean distance central source [CS] to outlets [O]: 280 km
Delivery chains: two-stages via 1 TP
Transshipment mode: one-step crossdocking

Additional model calculations for general cargo and parcel cargo, and further parameter variations lead to the following *dependencies*.

- In the range of the optimal number, the freight costs change only slightly with the number of transshipment points.
- For a two-stage delivery via transshipment points in the target regions, the optimal number of transshipment points shifts upwards with increasing freight demand and decreasing consignment size and downwards with decreasing freight demand and increasing consignment size.
- If the freight demand becomes lower than a *critical threshold*, a *three-stage delivery* via a consolidation point near the central source and a main run together with third party loads to the regional transshipment point is more cost efficient than the two-stage delivery.

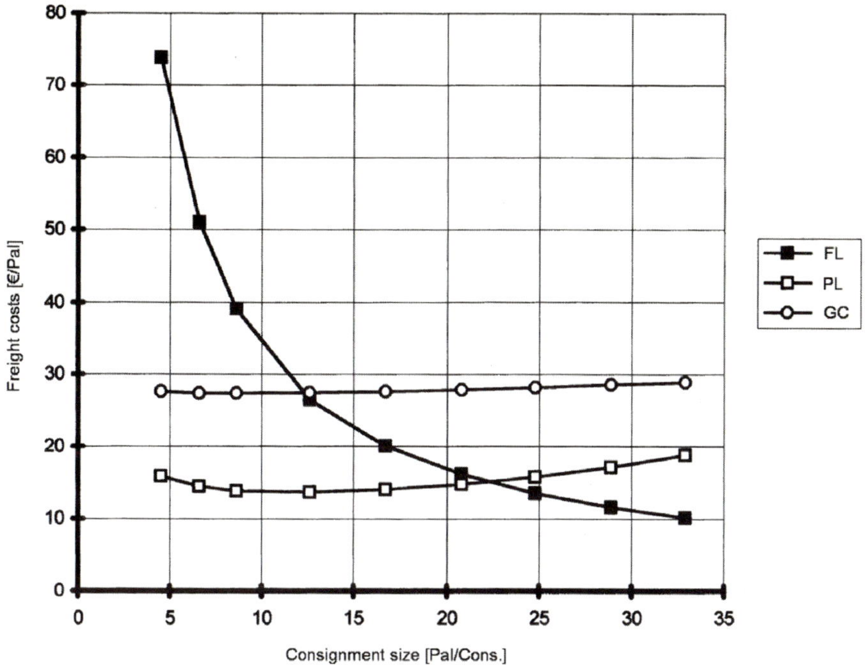

Fig. 21.27 Dependency of the freight costs on the consignment size for full loads, part loads and general cargo

FL: full loads PL: part loads GC: general cargo
Network structure: 20 distribution transshipment points
Other parameters: see *Fig. 21.26*

- The direct main run from the central source to the regional transshipment point is only opportune as long as the loads fill the trucks. Therefore, it is most efficient if the inflow transports to a regional transshipment point can be combined with deliveries of part loads to key account customers.

The calculated dependency of the freight costs on the consignment size is shown for full loads, part loads and general cargo in *Fig. 21.27* and for part loads, general cargo and parcel cargo in *Fig. 21.28*. From these figures and from further model calculations, result the following *principles* and *dependencies*:

- The specific freight costs for all freight modes strongly depend on the consignment size.
- For smaller consignments, the freight costs can decrease by more than 50% if the consignment size is doubled.
- Full-load transport is cheaper than part-load transport for consignments with more than about 22 pallets or 11 t (see *Fig. 21.27*).
- For consignments with more than 3 pallets and a weight above 1,500 kg, part-load transports are cheaper than general cargo, as long as the total consignment demand is sufficient for direct delivery tours (see *Fig. 21.28*).

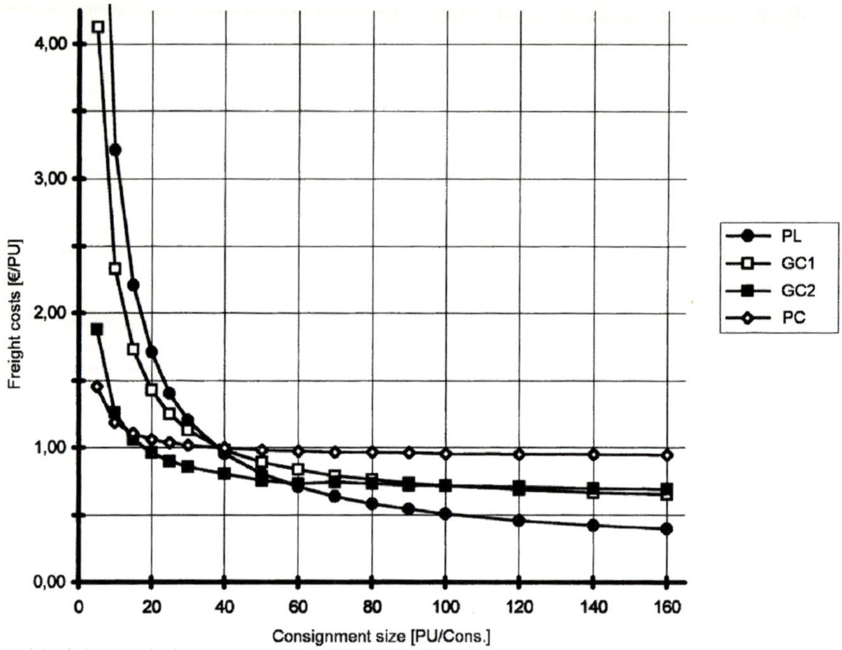

Fig. 21.28 Dependency of the freight costs via 1 TP on the consignment size for part loads, general cargo and parcel cargo

PL: part loads PC: parcel cargo via 1TP
GC1: general cargo with one-stage crossdocking
GC2: general cargo with two-stage crossdocking
Other parameters: see *Fig. 21.26*

- The delivery of general cargo consignments with more than about 1.5 pallets is less expensive with *one-stage crossdocking*; the delivery of smaller consignments is less expensive with *two-stage crossdocking* (see *Fig. 21.28*).
- For consignments with less than 10 packages, parcel freight costs are lower than general cargo freight costs.
- The *opportunity boarders* between the different freight modes depend on the size and on the weight of the packages and pallets, on the total freight demand, and on the mean distance between sources and sinks.

The last five statements approve the *opportunity boarders* (21.16) to (21.19) for the optimal freight chain.

The dependency of the freight costs on the total freight demand and on the stage degree of the freight chain is shown in *Fig. 21.29*. In this example, the mean size of the general cargo consignments with 50 PU/Con differs considerably from the mean size of the parcel consignment with only 3 PU/Con. This dependency, as well as further model calculations, leads to the rules:

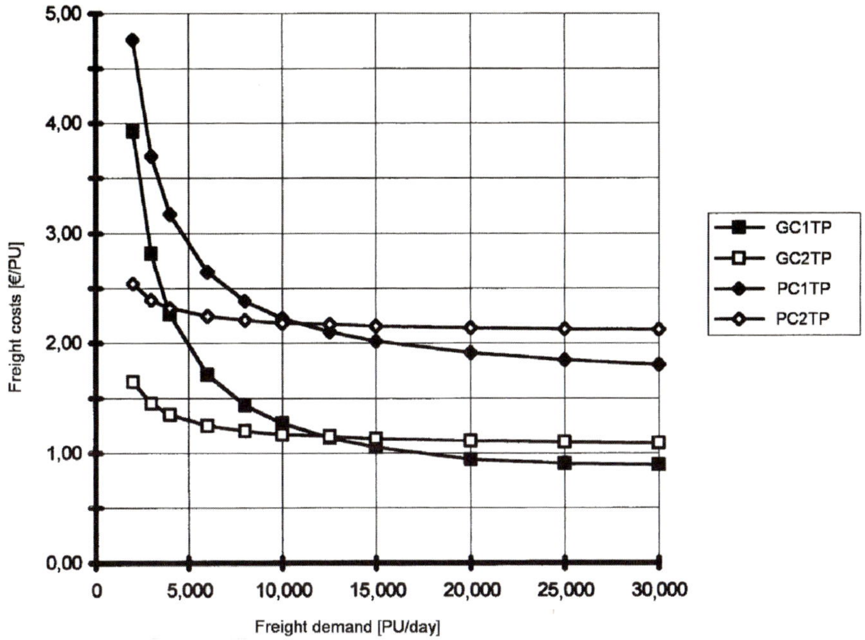

Fig. 21.29 Dependency of the freight costs on the freight demand
GC1TP: general cargo with one-stage crossdocking via 1 TP
GC1TP: general cargo with two-stage crossdocking via 2 TP
General cargo consignment size: 50 PU/Con
PC1TP: parcel cargo via 1 transshipment point
PC2TP: parcel cargo via 2 transshipment points
Parcel consignment size: 3 PU/Con
Other Parameter: see *Fig. 21.26*

- For general cargo as well as for parcel cargo the freight costs decrease with increasing total freight demand due to the better filling degree of the load units and transport units.
- For high freight demand, the freight costs asymptotically reach the lowest values for the maximally used load units and transport units.
- If the freight demand is not sufficient for adequately filled main transports from the source to the transshipment points, delivery via two transshipment points is more efficient than delivery via one transshipment point.
- In this case, the critical value between two-stage and three-stage delivery, which depends on the distances and the transfer costs, is about 12.5 pallets per relation.

The calculated dependency of the freight costs on the transport distance between the central source and the receiving stations is presented in *Fig. 21.30*. This leads to the *dependencies* and *freight rules*:

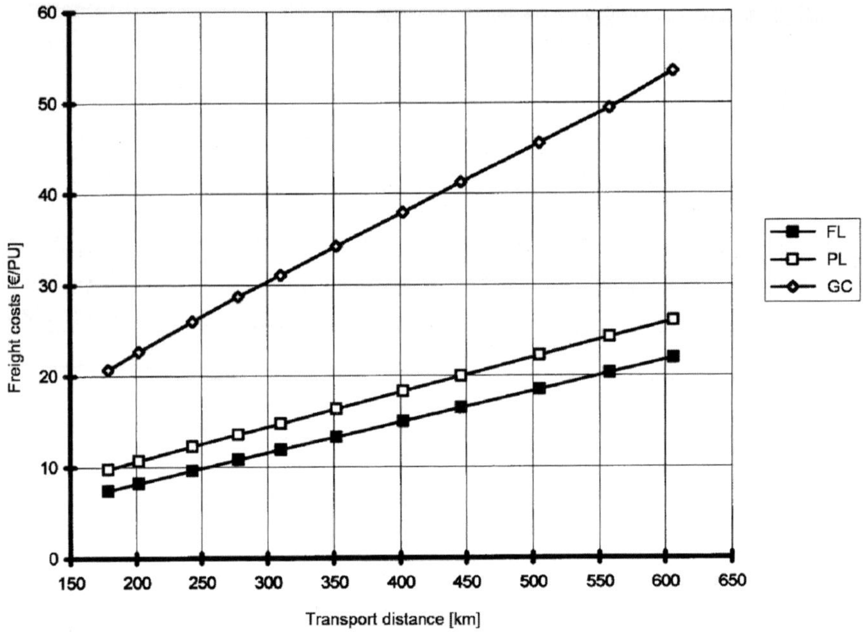

Fig. 21.30 Dependency of the freight costs on the transport distance
FL: full loads with 1,500 PU/Con = 31 Pal/Con
PL: part loads with 600 PU/Con = 12.6 Pal/Con
GC: general cargo with 75 PU/Con = 2 Pal/Con
Other parameters: see *Fig. 21.26*

- The freight costs for all freight modes increase linear with the distance.
- For general cargo and parcel cargo, the distance-related costs increase faster than for full loads and part loads.

A further important influence factor of the freight costs is the size of the freight units. *Figure 21.31* shows the calculated dependency of the freight costs on the package volume for volume determined freight. This diagram and further model calculations confirm the general results of *Sect. 12.5*:

- For *volume determined freight*, the freight costs increase approximately linear with the volume, for *weight determined freight* they increase linear with the weight of the freight units.

For the same freight demand – measured in units per period – big freight units cause a larger volume flow than small units. Hence, above a *critical volume* of the freight units, it is opportune to switch the freight mode and to use general cargo instead of parcel cargo or part loads instead of general cargo.

Finally, *Fig. 21.32* presents the influence of the *delivery frequency* on the freight costs and on the shipment time for general cargo via one transshipment point. In this case, the freight costs increase with higher delivery frequencies due to the

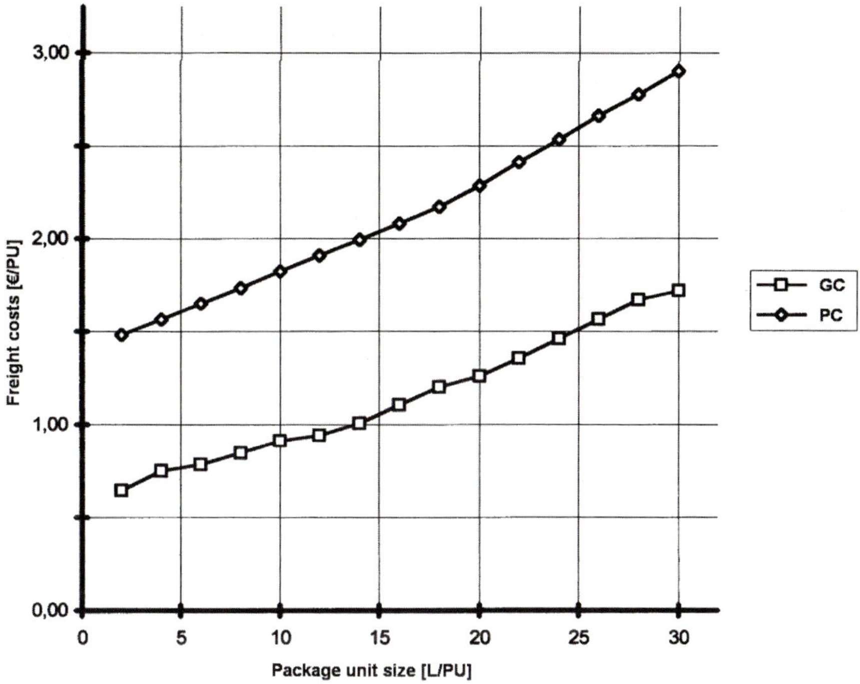

Fig. 21.31 Dependency of the freight costs on the freight size
GC: general cargo with 50 PU/Con
PC: parcel cargo with 3 PU/Con
Freight demand: 20,000 PU/day
Other Parameter: see *Fig. 21.26*

decreasing filling degree of the main transports and to the necessity of using smaller delivery vehicles for the area distribution. This leads to the general *cost-dilemma of short delivery times*:

- A reduction of the delivery times requires higher delivery frequencies and in consequence boosts the freight costs.

The model calculations show that the optimal freight chain with the lowest costs strongly depends on the performance requirements, especially on the freight demand, the freight size and the consignment quantity. Further influence factors are the frame conditions and service parameters, such as the delivery frequencies, the structure parameters, the distances and the service area. Therefore, the comparison of freight costs, delivery times and other *benchmarks* are misleading if the requirements and frame conditions differ or are not completely known. With a DOF-program, *analytical benchmarks* for the freight costs can be calculated. Comparison of the paid freight rates with the analytical benchmarks reveals the saving potentials of optimizing transport and freight chains.

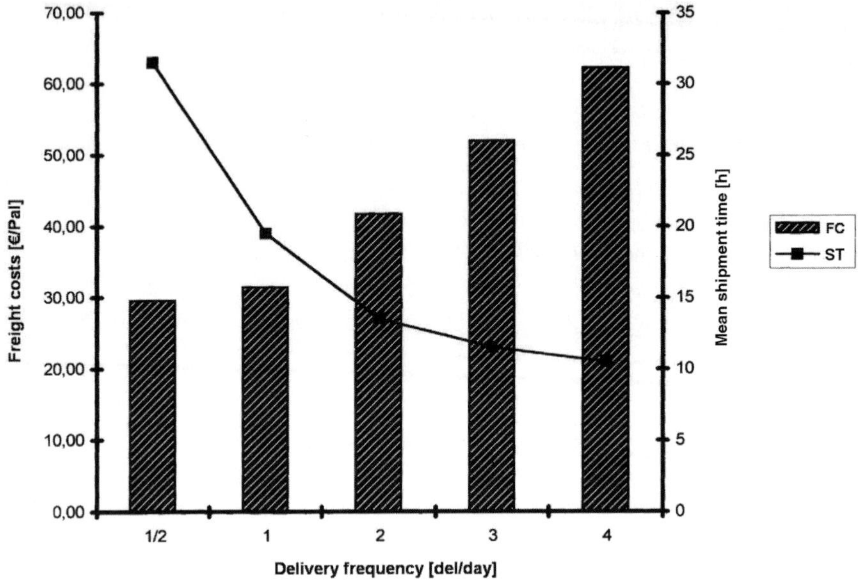

Fig. 21.32 Freight costs (FC) and shipment time (ST) as functions of the delivery frequency
General cargo via one TP with one-step crossdocking
Freight demand: 27 Pal/day and region
Mean consignment: 2 Pal/Cons
Other parameters: see *Fig. 21.26*

21.15 Transport Pricing and Freight Pricing

A company or shipper can operate own, rented or leased transport means or buy transport or freight services on the transport and freight markets. The *transport prices* and *freight tariffs* result by the market mechanism from the *transport* and *freight costs* and the current offer and demand (Gudehus 2007).

21.15.1 Transport Prices

A *transport contractor* or *transport service provider*, e.g. a trucker, taxi driver, barge owner, shipping company or aircraft operator, performs transport tours due the orders and instructions of a customer. The customer can be a private person or a company but also *logistic service provider* (see *Chap. 22*). The contractor provides the transport mean unmanned or manned with a qualified *crew*, including fuel, lubricants, maintenance and repair. The *operating times* are agreed between customer and contractor and the *transport tours* are carried out as required by the customer.

The pricing of transport services differ for the traffic modes and transport means. Price structures and price basis have evolved historically and are not always *cost-based and use-related*. Two basically different transport pricings are possible:

- *Time-charter pricing*: The customer is charged based on *time and effort* with a *charter rate* for the transport mean plus the expenses for fuel, network fees etc. Examples are car rental or ship and aircraft charter.
- *Use-related pricing*: The customer is charged with *use-related transport prices* or *fees* due to utilization of the transport mean. This is typical for taxi rides.

For companies, who want to optimize their supply network and to select the optimal transport and freight chains, use-related pricing is preferable to time charter pricing. From the main costs drivers of transport and the related cost rates (18.57) result the *use-related transport prices*:

$$\begin{aligned} &\textit{basic price} && P_B \quad [€/\text{TU-ride}] \\ &\textit{stop price} && P_{ST} \quad [€/\text{TU-stop}] \\ &\textit{driveway price} && P_{DW} \quad [€/\text{TU-km}]. \end{aligned} \qquad (21.37)$$

A *use-related transport pricing* with these standard prices requires the specification of the type and capacity of the transport mean, and of the loading, unloading and other activities during the stops. With the performance prices (21.37) the price for a transport over a driveway s_{DW} [km] with n_{ST} stops between a start and an endpoint is:

$$P_{TR} = P_B + n_{ST} \cdot P_{ST} + s_{DW} \cdot P_{DW} \quad [€/\text{TU} - \text{ride}] \qquad (21.38)$$

The common pricing of taxi rides is a good example: The basic price – 2004 in a range of 2.00 to 2.50 € per ride – covers waiting, empty rides and provision, and the driveway price – 2004, in a range from 1.30 to 1.50 €/km – charges the carriage of the passengers for each distance-km. In addition, a stop price is charged for an intermediate stop to release a fellow passenger or baggage. Some use-related transport prices for other transport means are listed in *Table 18.4*.

21.15.2 Freight Prices

Freight service providers, like *postal services, general cargo forwarders, parcel service providers*, shipping lines, railway companies, airlines or *public transport services*, pick up, carry and deliver single freight units, loads or passengers and complete consignments. For this purpose, the freight service provider organizes, operates and manages a *freight network* of transfer stations and transshipment points and uses own or rented transport means. Hence, freight service is more than just transport.

For the remuneration of freight services, even more pricing systems are in use than for transport services, e.g. *carriage rates*, *freight tables*, *fares* and different *freight formulas*. They are only partly based on cost drivers, performance and utilization. Many freight rates, as the fares for public transport, are subsidized by the government. For other freight services, the forwarders, e.g. the *shipping conferences* or the *International Air Transport Association* (IATA), agree and control structure and height of the prices.

The above model calculations with the DOF-program have shown that the freight costs for the standard freight chains shown in *Fig. 21.17* depend in a wide range linearly on the volume and weight of the freight units and on the distance between pickup and delivery point (see *Figs. 21.30* and *21.31*). A further cost factor of great importance is the consignment quantity (see *Fig. 21.28*). These dependencies result in the *use-related freight rates*:

$$\begin{array}{lll} \textit{consignment price} & P_S & [€/Con] \\ \textit{quantity price} & P_Q & [€/FU] \\ \textit{distance price} & P_D & [€/FU\text{-}km] \end{array} \qquad (21.39)$$

Quantity price and distance price depend on the *measurement unit* for the consignment quantity and increase with size and/or weight of the freight units.

With the standard freight rates (21.39), the freight price for a consignment with quantity m_S [FU/Cons] carried over a distance d_D [km] is:

$$P_{FR} = P_S + m_S \cdot (P_{QU} + d_D \cdot P_D) \quad [€/Cons] \qquad (21.40)$$

This leads to the freight price per freight unit:

$$P_{QU} = P_{FR}/m_S = P_S/m_S + P_{QU} + d_D \cdot P_D \quad [€/FU] \qquad (21.41)$$

The freight costs per unit decrease, according to *Fig. 21.28*, inverse-proportional with the consignment quantity and increases, according to *Fig. 21.30*, linear with the distance. The freight prices (21.39) are calculated from the corresponding costs rates by adding the costs for sales and general administration and the expected profit rate. The actual prices and profit for the freight service providers depend on *offer and demand* of the transport and freight markets at the time of the inquiry and negotiation. Therefore, the paid freight rates can differ significantly from the calculated costs rates.

21.15.3 Influence of Measurement Units

The freight units and the consignment quantities can be measured in different *quantity units*, like

$$\text{m},\ l,\ \text{m}^3,\ \text{kg},\ \text{t},\ \text{parcel},\ \text{pallet},\ \text{ISO-container or LU}. \qquad (21.42)$$

The freight conditions, the capacity of the transport means and the load handling in the respective freight chain determine the *billing units* for freight rates and for use-related and cost-based prices.

Loading space and *net weight* of the transport means are the most important influence factors of the transport and freight costs. Therefore, the consignment quantities for *volume-determined transports* are measured in volume units, like liter, cubic-meter or load units, whereas for *weight-determined transports* they are measured in weight units, like kilogram or tons (see *Sect. 12.5*). For example, the carriage of *furniture* is charged for load-volume-meters and the carriage of gas for cubic-meters. The transport of bulk cargo, heavy liquids and solids is charged for tons and ton-kilometers.

For general cargo, *freight rates* and *freight tables* based on 100 kg are quite common. The disadvantage of *100-kg-rates* is that they do not directly refer to the size and condition of the freight units, which determine the costs for loading, unloading and handling. If the consignments consist of discrete freight units or load units, such as parcels, containers or pallets with standard dimensions and weights, it is opportune to measure the consignment quantity by the respective *freight units* and to calculate the freight costs and prices with formula (21.40). Parcel service provider invoice typically for standard packages of different size and weight classes. Container shipping lines charge for 20" and 40" ISO-container.

In order to simplify the billing, the basic consignment price is often incorporated in the quantity price. For fixed freight relations and between defined areas also averaged distance prices are used and incorporated in the quantity price to a general tariff per freight unit and distance range. However, this is correct only if the consignment structure, the freight flows and the distances remain the same during the contract period. Averaged prizes are the result of a *mixed price calculation*. As they are not use-related, the prices for small freight units and short distances are generally too high in relation to the costs and subsidize the prices for big units and long distances which are too low in relation to their costs (see *Sect. 7.1*).

Another difficulty is the *rounding problem* which occurs if *freight tables* are used. The price for a consignment with quantity, weight or distance only slightly above a fixed limit of a freight table is far higher than the freight price of a consignment with slightly lower values. Also *smoothing rules* do not solve this problem. It is much simpler and unambiguous to invoice transports and freights based on performance prices by the formula (21.38) and (21.40). Generally holds the *billing rule for transport and freights*:

- The goals of customers and the availability of the data should determine the structure and differentiation of transport prices and freight rates.

More and more companies register the necessary pricing data, e.g. the dimensions and weight of the articles, packages and load units, by computer for inventory scheduling, selection of optimal supply chains and for other purposes. The distances between senders and recipients are available in *electronic maps*. Therefore, the use-related pricing of transport and freight services with the performance prices (21.37) and (21.39) and the formulas (21.38) and (21.40) will become more and more common in future (see also *Sect. 7.6.1*).

21.16 Combined Road-Rail-Cargo Traffic

In the *combined road-rail-cargo traffic* (CRR), trucking units haul *semi-road trailers* and/or *swap-body trailers* (ST) (see *Figs. 12.5* and *12.6*) on the road to a departure rail terminal. There a bridge crane, mobile crane or another transfer device takes the trailers up and loads them on wagons. A train carries the trailers on the rail to a destination terminal, where they are unloaded, picked up by trucking units and hauled to their final destinations.

The CRR-traffic is an example of *inter-modal transport* through a three-stage transport chain as shown *Figs. 21.4* and *21.17*. By CRR-traffic, the long distance and heavy load traffic is shifted partly from the road to the rail and the roads are released. Preconditions for the acceptance of the CRR-traffic by shippers and freight forwarders are *competitive travel times* and *economic attractiveness*.

Today, by professional scheduling, an effective travel speed for the trains of $v_{train} \approx 100$ km/h is achievable whereas trucks on the road reach a mean travel speed of $v_{truck} \approx 60$ km/h provided the traffic is not too heavy. The shorter travel time on the rail is lengthened by the travel time $2 \cdot v_{truck} \cdot d_{stat}$ for the in- and outrun on the road over a mean *station distance* d_{stat} [km] to and from the transit stations, and by the *waiting and loading times* $2 \cdot T_{stat}$ [h] in the stations. Therefore, the travel times of CRR-transports are only attractive, if the total transport distance length l_{Tr} is longer than the *time-critical distance*:

$$l_{Tr\,Crit} = 2 \cdot v_{train} \cdot (d_{stat} + T_{stat} \cdot v_{truck})/(v_{train} - v_{truck}) \quad [km]. \quad (21.43)$$

With the effective speeds $v_{truck} \approx 60$ km/h and $v_{train} \approx 100$ km/h, a mean station distance $d_{stat} = 30$ km and a station time $T_{stat} = 1$ h results a time-critical distance of 450 km. Beyond that distance, CRR-transport offers shorter travel times than direct road transport. For a shorter station distance of $d_{stat} = 10$ km the critical distance becomes 350 km. If the station time can be reduced to $T_{stat} = 0{,}5$ h, the critical distance decreases to 300 km. That means:

▶ The time-critical distance of the CRR-traffic depends strongly on the mean in- and outrun distance and on the mean waiting and loading time.

Not only by shorter station distances and station times but also by a higher reliability and punctuality of the railway the CRR-traffic can be more attractive than direct road transport.

To be economically attractive, the costs for CCR-transports must be lower than for direct road transport. The costs for the road transport are given by relation (18.58) with $n_{ST} = 0$ and a driveway $l_{DW} = l_{Tr}$ for the direct road transport and $l_{DW} = d_{stat}$ for the in- or outrun to the stations. Based on costs and prices of 2000 the basic rate for road trucking of semi-road trailers is $k_{B\,road} = 35.50$ €/ST-ride and the driveway rate is $k_{DW\,road} = 0.85$ €/ST-km. With these costs rates the distance dependency of the direct transport costs for a semi-road trailer is:

$$K_{road}(l_{Tr}) = k_{B\,road} + l_{Tr} \cdot k_{DW\,road} \quad [€/ST] \quad (21.44)$$

The transport costs by train for a driveway $l_{DW} = l_{Tr}$ can also be calculated with relation (18.58) if the respective cost rates for the rail transport are known. Since use-related market prices for rail transports are not available, these cost rates must be determined by model calculations as described in *Chap. 18*.

This was performed in a *consulting project* in 2000, in order to evaluate the CRR-traffic with a new loading technique. For this project, different scenarios, rail connections and time schedules were investigated. The basic cost rate for loading, unloading and waiting turned out to be $k_{B\,rail} = 53.35$ €/ST-ride. The driveway cost rate was assessed with $k_{DW\,rail} = 0.38$ €/ST-km. They result for trains with 39 wagons with an utilization of 75%, when the rail track is charged with a *route price* of 5,00 €/train-km.

With these rates the transport costs for a CRR-transport via two transfer stations are:

$$K_{CRR}(l_{Tr}) = 2 \cdot (k_{B\,road} + d_{stat} \cdot k_{DW\,road}) + (k_{B\,rail} + l_{Tr} \cdot k_{DW\,rail}) \quad [€/ST] \quad (21.45)$$

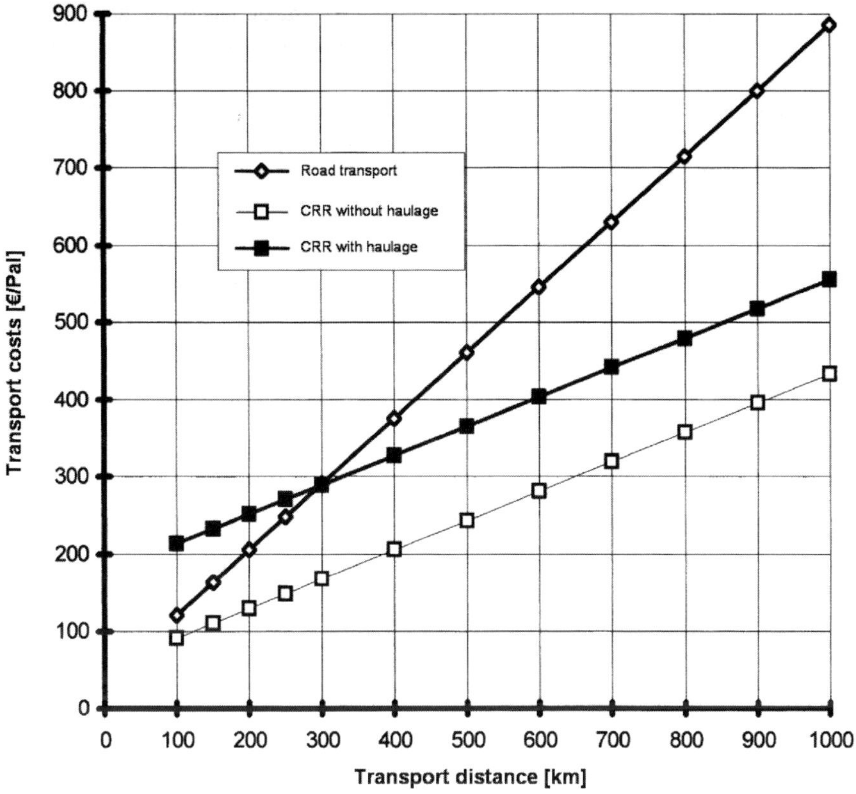

Fig. 21.33 Distance dependency of the transport costs of road transport and CRR-transport of swap-body trailers (ST)

Cost rates	Road	Rail	
Basic costs rates	35.50	53.35	€/ST
Driveway costs	0.85	0.38	€/ST
Mean station distance	30 km		

Figure 21.33 presents the distance dependency of the transport costs for direct transport on the road and for CRR-transport via railway calculated with the formulas (21.44) and (21.45). Without the in- and outrun costs on the road, CRR-traffic would be more economic already for transport distances above 100 km. By the in- and outrun costs of 122 € per semi road trailer the *break-even distance of CRR-traffic* is shifted to values above 300 km.

The model calculations show that the economic break even of CRR-traffic depends critically on several influence factors, especially on the rail-route price, the train capacity and the operating costs for loading, unloading and stations. Further critical success factors for CRR-traffic are the *overhead rates* and the *profit expectations* of the involved companies. As long as railway companies calculate their prices with gross margins far higher than the forwarders on the road, the break-even distance for CRR-traffic shifts to values above 500 km. In this range, however, for most relations in Europe the load demand is insufficient for setting up a cost efficient CRR-traffic.

Analogous to the evaluated project of the road-rail-road transport, the presented methods and tools can be applied for the evaluation of other *inter-modal transport* chains, e.g. of the road-ship-road or the road-air-road transport, and for the calculation of the respective break-even times and distances.

21.17 Consumer Oriented Supply Chain Management

Advocates of *Supply Chain Management* (SCM) and *Efficient Consumer Response* (ECR) propagate to orientate all activities and processes along the supply chains to the requirements and service demand of the consumers (Corsten and Kumar 2005, Kotzab 1999). This begins with the exchange of information and data from the *point of sales* (POS). The sales outlets inform the suppliers immediately on the actual sales data and stock levels by *Internet* or *Electronic Data Interchange* (EDI). Planning and marketing departments of the supplier inform the sales outlets in due time about new products, planned promotions, production changes and the current availability of storekeeping articles.

From the current POS-data, the suppliers are able to forecast the future demand of products with regular demand by the methods outlined in *Chap. 9*. Production orders for the replenishment of stocks and for customer-specific goods are generated automatically by computer from the sales forecast and the current stock levels by the rules, formulas and scheduling strategies outlined in *Chaps. 10* and *11*.

In last consequence, also the development and market introduction of new products and the implementation and operation of *inter-company logistic networks* are tasks of ECR and SCM.

21.17.1 Inter-company Supply Chain Management

At the beginning of the SCM-movement huge potentials and advantages of inter-company supply chain management were announced. In the mean time, many companies found out that it is impossible to reach the expectations, although certain advantages are still possible. The introduction of ECR and SCM requires enormous efforts and changes within the involved companies. The costs for electronic data exchange and the necessary software are generally underestimated. The harmonization of the interfaces and the continuous adaptation of the daily business are more costly than expected.

Within the own network, the customer orientation of logistic chains is principally feasible, although ECR and SCM have been implemented and practiced consequently only by a few companies. Principal difficulties occur if several companies are involved. As long as the single actors – industry, retailer and logistic service provider – try to achieve the advantages only for themselves and shift the efforts and costs to the others, inter-company SCM must fail.

Alternatively, market-dominating companies of the automotive, chemical and consumer goods industry expand the borders of their logistic networks upstream to the suppliers and downstream to the customers in order to become *process owner*, to govern the systems and to implement ECR and SCM. In competition to the attempts of the suppliers, leading retail companies built up and operate their own replenishment network. Logistic service providers expand their networks by horizontal and vertical cooperation.

The ongoing concentration of retailers, and the mergers and alliances of logistic service providers aim for higher transport volumes, consolidation of stocks and the *critical mass* for the cost-efficient operation of the logistic networks. Their main goals are domination of the supply and freight chains and further rationalization.

The potentials of inter-company SCM and ECR can only be achieved if all actors fairly cooperate, no matter who is the process owner or system leader. However, a successful cooperation within the supply chains can only evolve in agreement with the self-interest of the involved actors.

21.17.2 Cooperation, Coordination and Collaboration

Companies participating in the same project or contributing to the production of the same goods and services have always cooperated. Efficient cooperation between the actors is the key for *division of labor* in modern societies. Hence, the voluntary bilateral *cooperation* within the supply networks shown in *Figs. 0.1 and 15.1* is nothing new. The voluntary cooperation is the secret of the superiority of the free market economy. The prices of goods and services result from *offer and demand.* Profits are the result of permanent optimization of products and processes within and between the companies. In their own interest, the actors harmonize interfaces, standardize products, services and processes and set up technical and commercial norms (see *Sect. 3.11*).

Standards and *norms* for products, information and interfaces going beyond more than one stage of the supply networks require inter-company *coordination*. Special institutions like the German DIN, VDI, VDE, CCG (today GS1) and the European FEM (*Federation European Manuentation*) have been established to develop and agree standards and norms. National trade associations and international organizations like WTO and OECD also care for these issues. Neutral norms, standards and behavior rules are the basis for the free trade around the globe.

Some consultants, software companies and other interested groups have suggested a *Supply Chain Collaboration* (SCC) or *Multi Tier Collaboration* (see *Fig. 1.15*). They propose that a *central unit* – generally under control of an OEM or a dominating producer – integrates the supply chains, harmonizes the information flows and schedules the stocks and material flows from the suppliers of the supplier down to the final customers. Claimed advantages are higher *transparency*, minimal stocks, reliable delivery times and lower costs in the total network. However, whether these potentials are really achievable, how they are measured and who takes the final advantage, are open questions.

The price for inter-company supply chain collaboration is the loss of independency, the elimination of competition and in the long run the loss of price information from the free markets (Bretzke 2008; Gudehus 2007). Even for the dominating company, the dependency on other actors reduces their *flexibility*, impedes *innovation* and eliminates the advantages of *outsourcing* which facilitates to break crusted structures.

The failure of totally centralized planning and scheduling has been demonstrated before 1990 by the *planned economy* in socialistic countries. After following for a long time the general trend to *centralism*, many companies in the free-market economies tried to establish *de-central structures*, local decision and fair cooperation with independent partners. However, this was difficult to accomplish as many managers, consultants and theorists still have a preference for centrally controlled structures. They expect new miracles from ERP, APS and SCM and from the propagated *SCM-programs* of dominating software suppliers.

Balancing the obvious disadvantages against the questionable potentials of SCC leads to the *supply chain recommendation*:

- Cooperation and coordination yes, collaboration no!

The successful cooperation of companies with logistic service providers described in the next chapter demonstrates that by qualified organization of business relations and adequate contracts the independency for the involved parties can be kept, and that both sides achieve advantages.

Chapter 22
Logistic Service Providers

A company should regularly check and decide which products should be made and which services should be executed by itself and which should be bought from outside (Coase 1937; Penrose 1959; Williamson 1985). This holds true also for logistic services. The *make-or-buy-decision* for logistic services depends on the goals and core competencies of the company, on the kind and extent of the required services and on the available *logistic service providers* (Barney 1991; Coyle et al. 1996; Peteraf 1993; Van Damme and Amstel 1996).

For a long time, companies were aiming to execute as much as possible *in-house*. Own works fleets accomplished the transports. Own warehouses were planned, built and operated. Internal freight departments organized the shipment of incoming and outgoing goods. Only a few companies – acting against the general trend – had *outsourced* a major part of their transport, freight and warehousing to logistic service providers and focused their resources on *core competencies*. In many cases these companies were more successful.

In the mean time, the trend has reverted. Nowadays, more and more companies assign their logistic demand to service providers. This includes not only freight and external transports, but also warehousing and internal transports up to the provision at the assembly line (Baumgarten 1999). Some companies have *outsourced* the whole *procurement logistics* or *distribution logistics*.

However, outsourcing the company's logistics completely to one single service provider can lead to the loss of own logistic competency and dangerous dependency. In several cases, logistic service providers failed the expectations and disappointed their clients. The *all-inclusive-care-free-package* offered by some *system providers* is not available for free. To plan, organize, set up and manage a customized logistic system is expensive and risky. Therefore, outsourced system services can be more expensive than the sum of the corresponding single services which are executed in-house or purchased seperately.

Outsourcing of logistic services, in particular of sophisticated system services such as the whole procurement or distribution, can only be successful if performed professionally in the *outsourcing steps*:

T. Gudehus, H. Kotzab, *Comprehensive Logistics*,

DOI 10.1007/978-3-540-68652-1_22,

1. conception of *company logistics*
2. delimitation and quantification of *service demand*
3. development and decision of *contracting policy*
4. professional *bidding process*
5. regulation of *control* and *remuneration*

By professional outsourcing to qualified service providers, a reduction of logistic costs up to 20% or more is possible, depending on the present situation. In addition, an improved service quality is achievable, which counts often more than costs savings.

This chapter describes the different kinds of logistic service providers and analyzes their characteristics and application criteria. From this analysis follow the chances and risks and the proceeding for *tendering* and *contracting* of logistic services.

22.1 Conception of Company Logistics

Before the decision about outsourcing, a consistent and future-oriented concept of the whole company logistics should exist. If the competencies within the company are insufficient, the concept can be developed in close cooperation with a logistically experienced *consultant*.

A logistic service provider should not be assigned with this task. It is also risky to develop the logistic concept stepwise during the bidding process. Service providers follow their own interests, which partly deviate from the customer's goals. In conflicts of interests, they tend towards *self-optimization*.

The concept of company logistics implies:

- demarcation and structuring the company's logistic network
- requirements for network management and system control
- benchmarks for transport and freight costs
- benchmarks for costs and prices of internal logistics
- benchmarks for investments and operating costs of logistic centers
- specification and quantification of the required logistic services

An overall logistic concept for the company should be the basis for all investment and outsourcing decisions. It is necessary to regularly revise the concept and to adapt it to changing market conditions.

If the concept requires a new logistic center, a neutral *system planning* is necessary to budget investment and operating costs. The results of qualified planning are needed as *analytical benchmarks* for internal cost rates and external performance prices, and for the evaluation of bids.

It is also advisable to examine whether and to which extent *internal cooperation* with other branches of the own company, or *external cooperation* with other companies, even with competitors, are useful and feasible. By *logistic cooperation*, consolidation effects, a combination of volume-determined or weight-determined freights, increased transport pairing, purchasing advantages and other cost savings can be achieved. For example if several companies located in the same region and serving the same range of customers send out *part-truck loads* separately, jointly they can generate far cheaper *full-truck loads*.

22.2 Service Requirements

The delimitation and specification of the required logistic services, the quantification of the performance demand, an utilization-dependent remuneration concept and a complete specification of the frame conditions are decisive for the outsourcing success. The performance requirements are required for the selection of external service providers as well as for the set up of internal service departments which operate as *profit centers*.

The required services should be separated according to the applied technologies, equipment and resources into transport, handling, warehousing and related *operative services*, in *administrative services* and in *special services*. The connection of several single services results in different *performance chains*. Interlinked performance chains make up integrated *system services*.

The operative services of a performance chain or a system are calculated, measured and priced from the involved single services as explained in *Chaps. 6* and *7*. The related administrative performances can either be calculated and priced individually, or attributed in proportion to the operative services.

22.2.1 Transport Services

Single *operative transport services* are:

internal transports
external part- and full-load transports
delivery drives
distribution and collection tours (22.1)
regular line transports
optional tramp transports
towing and traction

The requirements, the technical solutions and the means of transport have been described in *Chap. 18*. *Transport related administrative services* are:

transport planning and tour scheduling
scheduling of drivers and transport means
tracking and tracing of transport means and load units (22.2)
information about free capacities and state of transport

The *costing and pricing* of operative transport services are content of *Sect. 18.12*. The costs of the related administrative transport services (22.2) are normally included in the performance prices for the operative transport services.

22.2.2 Handling Services

Operative handling services in logistics are:

loading and unloading
transshipment and reloading
build-up and break-down of load units (22.3)
sorting and arranging
packing and unpacking

Related with the handling services are the *administrative services*:

scheduling of packing and filling
order and personnel scheduling (22.4)
scheduling of load carriers and handling devices
set up and management of the handling operations

In general, the operative handling services (22.3) and the related administrative services (22.4) are partial performances of warehousing or of other integrated logistic services. Handling costs are generally included in the freight cost rates as described in *Sects. 21.14* and *21.15*. If the handling services cause a major part of the total costs, it is useful to calculate and to price them separately.

22.2.3 Warehousing Services

In addition to handling services, the *core services of warehousing* are:

in-storing and out-storing
load buffering
storekeeping (22.5)
commissioning
order consolidation

Additional operative services of warehousing are

loading and unloading
quality control
packaging and labeling (22.6)
accumulation of load units
consolidation of shipments

Storekeeping, commissioning and other warehousing services are executed in a logistic hall, a logistic center or other buildings. They require special systems, equipment, buildings and trained personnel.

Related to the warehousing activities are the administrative services:

planning and set up of buildings, systems and equipment
management of warehousing operations
storeplace administration (22.7)
inventory control and scheduling
order processing

The performances and requirements of storage and commissioning systems have been described in *Chaps. 16* and *17*. The calculation of the operating costs and performance prices are presented in *Sects. 16.13, 16.14* and *17.14*.

22.2.4 Special Services

Logistic services are often combined with special non-logistical services, also called *value-added services*. *Operative special services* are e.g.:

filling and bottling
confectioning
setup of displays
cleaning and handling of empties (22.8)
waste removal
repair and maintenance
assembling

They are generally executed at the same location before, after or parallel to the logistic processes. *Administrative special services* are:

billing
encashment
controlling (22.9)
call center services
customs clearance

With an increasing number of value-added services, a logistic system becomes more and more a general *performance system* that competes with the systems of manufacturers and retail companies.

22.2.5 Compounded Logistic Services

By linking single services to a performance chain, *compounded logistic services* are generated, such as:

- *Generation of order-specific shipments*: This is executed successively by in-storing, storekeeping, order processing, out-storing, commissioning, packaging, accumulation and preparation for dispatch.
- *Freight services, forwarding and haulage*: These are operative transport and handling services linked with administrative services.
- *Provision of parts and modules at the point of assembly* (POA): This involves in-storing, storekeeping, out-storing, commissioning, loading, external transport, unloading, buffering, internal transport and handling.
- *Shelf-jobbing of merchandize in sales outlets*: Depending on the range of service, this involves in-storing, storekeeping, out-storing, commissioning, external and internal transports and handling.

Administrative services, necessary for the organization and execution of compounded logistic services, are:

- set up and organization of specific performance chains
- order acceptance and order processing (22.10)
- tracing of shipments and feedback of deliveries

The specifications and constraints of compounded services result from the single services involved. The methods developed in *Chaps. 6* and 7 can be used to calculate performance costs and prices for compounded logistic services. The requirements, costs and prices for delivery and freight services have been described in *Sects. 21.2, 21.14* and *20.15*.

22.2.6 Integrated System Services

Integrated system services are related and connected single services performed by an *integrated logistic system* or *logistic network*, such as:

- *Freight networks* consisting of special transport means, connections and transshipment points.
- *Logistic centers* with special storage, commissioning, sorting and handling systems.
- *Procurement and distribution networks* that are interrelated transport, handling and warehousing chains.

The performance of integrated system services requires logistic centers and logistic networks. The necessary *administrative system services* are:

set up and organization of the freight or logistic network
network management (see *Sect. 1.9)*
set up and organization of logistic stations (22.11)
internal operation management
system management

Specification, costs and prices of integrated performances result from the involved single operative services by the methods of *logistic costing* and *performance remuneration* explained in *Chaps. 6* and *7*. The costs for the administrative system services (22.11) are in a range between 5% and 15% of the operative performance costs. They are either included in the performance prices or invoiced separately as a *basic fee*.

22.3 Logistic Service Providers

Corresponding to the characteristics and features presented in *Table 22.1*, logistic service providers can be differentiated in

single-service providers
compound-service providers (22.12)
system-service providers

Many logistic service providers are specialized on certain *goods* such as

valuables	fresh products	gases	newspapers	
hazardous	chilled goods	liquids	money	(22.13)
furniture	beverages	building material	books	
heavy duty	food	waste	chemicals	

Table 22.1 Characteristics and features of logistic service providers

Characteristics	Single-service provider	Compound-service provider	System-service provider
Competence	**single services**	**compounded services**	**system services**
performances	transport, handling storing, special serv.	linked transport freight services	system operation network services
Resources	**transport means** **logistic operations**	**freight networks** **transfer terminals**	**logistic networks** **logistic centers**
competence	related technical know how, special experience and skills	technology, organization transport scheduling freight forwarding	logistics, IT, SCM system planning project mangm.
Orientation	**service quality**	**service & performance**	**customer service**
	special goods local, regional, national relations	standard freight flexible networks national and global	key accounts branches, funct national, global
Customers	**small, temporary**	**all sizes, varying**	**big, constant**
relations	personal, individual	anonymous order dep.	contractual
binding time	short or varying	up to 1 year	3 up to 10 years
Tendering	**inquiry**	**inquiry/call for bids**	**call for bids**
closing	order placement and confirmation	order and confirmation frame contract	letter of intend service contract

Others concentrate on a specific kind of *load or freight*, such as

general cargo	letters	pallets	animals	(22.14)
bulk cargo	parcels	containers	people	

or on selected *industries and branches*, such as

automotive industry	steel industry	
chemical industry	building industry	(22.15)
beverage industry	raw material industry	
consumer goods industry	retail business	

Passenger service companies specialize on selected people groups, such as vacation travelers, employees, business travelers or patients, and concentrate on local traffic areas or on certain connections. The business activities of logistic service providers are very different. They can be local, regional, national, international and global companies, who act as single, compound or system provider and offer a variety of the services.

Many logistic service providers operate in multiple functions and offer different single, compound or system services. Large logistic groups try to maximize the utilization of their resources, to achieve *economies of scale* and to realize *synergies*. This allows additional cost reductions but implies the danger of scrupulous *self-optimization* at the expense of the customer.

22.3.1 Single-Service Providers

The program of single-service providers is limited to quite simple transport, handling or warehousing services. They often specialize in specific goods, certain types of freight or selected industries and operate in limited areas or on fixed relations. Examples for *logistic single-service providers* are:

- *Transport providers* (*Carriers*), such as haulage contractors, couriers, ambulances, rescue operations, removal companies, inland-waterway and sea-shipping companies, air-transport companies
- *Handling providers*, such as shelf-jobber, cargo-handling operator, stevedore companies, transshipment terminals, railway stations, airport operators
- *Warehousing providers*, such as store, tank and silo operators, staple houses, warehousing companies, parking site and garage operators, archives, depots

Specialized single-service providers often operate on the basis of a long-term agreement for fixed clientele or as *subcontractors* for compound- and system-service providers. Less specialized single-service providers execute orders for a changing clientele on a short-term basis. This holds as well for

- *Special-service providers*, such as filling and bottling operators, co-packers, confectioners, empties-cleaning and maintenance companies, customs-clearance operators and IT-service providers

Transport-service providers execute transport and haulage orders they receive directly from customers or from an internet-based *freight auction* with own, leased or chartered transport means. Also handling-, warehousing- and special-service providers generally operate their own systems and equipment within rented or owned buildings. *Electronic auctions* of freights, transports and other logistic services are confined to *standard-services* and goods packed in normalized packages, load units or containers.

22.3.2 Compound-Service Provider

A compound-service provider links single logistic and other services to a complete service range for anonymous customers. The service range includes single services, related services and integrated system services. Examples of *compound service providers* and *forwarders* are:

letter post
parcel and express services
freight forwarder
container services
railway companies
shipping lines and airlines
operators of logistic centers
empties-handling companies (22.16)

Logistic compound-service providers operate to a certain extent with own systems, logistic centers, transshipment points, transport means, equipment and personnel. Some services are performed by *subcontractors* or in cooperation with *system partners*.

Compound-service providers receive their orders generally on short notice from changing customers. The contract period of agreements with fixed prices is maximally one year.

22.3.3 System-Service Providers

System-service providers or *integrators* develop, set up and operate customer-specific logistic systems. Their integrated services are adapted to the requirements of one or a small number of contract partners. Hence, the distinguishing *features of logistic system-service providers* are:

- *integrated system services* for the special demand of single customers
- *highly reliable*, *superior* and *efficient* execution of customer orders
- completely or partly *dedicated buildings* and *customized systems*
- *overall responsibility* for performance, quality and costs

To some extent, logistic system-providers are comparable to *general contractors* in building industry and plant business. However, the focus of a general contractor is on planning and realization of a system or building, which is handed over ready-to-use to the customer or another operator. A logistic system service provider plans and realizes a system – alone or together with the customer – with the aim to become the system operator.

A system-service provider must be capable of performing a wide range of logistic services on his own. In addition, a system provider must have the *competence* to plan, set up, organize and manage the system due to the demands of the customer. To be competitive, the provider must be significantly better and more cost efficient than a customer, who operates with single-service providers. Otherwise, outsourcing of logistic system services is not attractive. *Key success factors* of qualified system-service providers are:

- competent and trustworthy management
- qualified and skilled staff
- high efficiency by professionalism, experience and specialization
- synergies and saving potentials by joint utilization of resources
- lower staff costs due to low wages and longer working times
- own resources for function-critical services
- capable control, information and communication systems
- balancing potentials and available resources for peak loads
- know how and market power for purchasing single services

A system-service provider operates for single customers based on a *long-term contract* with prices fixed for at least one year. The minimal contract period depends on the investment and the technical life of the customer specific systems, buildings and equipment. The shortest contract times for system services are 3 years, usual contract times are 5 years. In a trusting business relationship, the contracts can last 10 years and longer.

Due to the long-term contracts, the business of logistic system services is called *contract logistics* (Stölzle et al. 2007). System-service providers with own IT systems and logistic resources are called also *third party logistic providers* (3PL) (Halldorsson 2002). System-service providers who have no own logistic resources are named *fourth party logistic providers* (4PL). However, these delusive buzz words offer no real advantage, as long as the customer cannot identify the added value of a 4PL as compared to other service providers. The service of a 4PL is limited to the organization and management of the purchased logistic resources of others. Many 4PL are comparable with the good old *sofa-forwarder* who organizes transports and freight capacities by fax and telephone without own transport means. The temporary fashion for 4PL and other doubtful business models has ceased since the hype in the early 2000's.

22.4 Outsourcing and Contracting Strategies

Not all logistic services should be outsourced. Not every company needs a system-service provider. Without knowing the goals of the company, it is impossible to decide whether and to which extent logistics should be a *core competency*. It is also difficult to state in general whether only parts of company's logistics should be outsourced to single- and compound-service providers or whether it is opportune to transfer the whole logistics to a systems provider.

If logistic services are *key competitive advantages*, it is wrong to outsource them, even if this would reduce costs and avoids own investments. In this case, the company should make logistics its core competency and develop its own competitive and cost-effective logistic system.

Another question arising with the outsourcing of warehousing is the *responsibility for the inventory level*. An independent service provider cannot be alone responsible for scheduling the inventory of customers. He can only execute storage and replenishment orders, which are generated by agreed rules or a program based on algorithms and strategies defined by the client. The goals, strategies and decisions of inventory management are key success factors of the business and should not be delegated to a third party.

Before purchasing logistic services, a *contracting strategy* must be developed in order to achieve the company specific goals. These are:

extension of performance limits
service improvements
cost reductions
concentration on core competencies
release of own resources (22.17)
reduction of staff
use of external competencies
avoiding investments
increased flexibility

A qualified system-service provider simplifies the business of the customer and reduces the complexity of his processes.

22.4.1 Logistic Outsourcing in Industry

The logistic costs of industrial companies are in a range between 5% and 15% of total costs (Baumgarten 1993, 2000). Therefore, most industrial companies do not consider logistics as core competency. Logistics should function but it is not crucial for the business (see *Sect. 21.11*).

Hence, industry, especially the *automotive industry*, has outsourced major parts of their logistics. On the procurement side, system-service providers are responsible for the operation of supply buffer stores and the provision of parts ready for production. Some logistic service-providers even assemble complete modules and become *system suppliers*, who compete with manufacturers that extend their products by logistic services. On the distribution side, system-service providers are responsible for storing finished goods, provision of spare parts and for the delivery to the customers. They have built and operate customized logistic centers and freight networks.

Industrial companies outsource logistic services as far as their core business is not directly affected. On the procurement side, the *outsourcing limit* is given by the necessity of absolute *supply security*, which is required e.g. for expensive production facilities or a continuously operating *process production*. The limit for outsourcing distribution results from the kind of customer contact at the *point of delivery* (POD). It is disadvantageous to outsource deliveries to a third party, if they are inseparably connected with sales, assembly, customer service and advise, which establish a special customer relation or enable to observe the customer and his demand.

Even if outsourcing is highly advantageous, the overall *system management* of logistics should be kept within the company. This requires qualified logistic management and a sufficient number of in-house logistic specialists.

22.4.2 Logistic Outsourcing of Retailers

The logistic costs of retailing companies are in the range from 5% to 25% of the turnover (Baumgarten 1993, 2000; Kotzab and Schnedlitz 1999). They use up between 10% and 50% of the gross margin and are of central importance for the profitability. Next to purchasing, assortment policy, marketing and sales, logistics is a *core competency* for retailers, department stores, purchasing cooperatives and mail-order companies (see *Sect. 21.12*).

In the last few years, many retailers have improved their competitiveness by setting up own logistic networks and loosening the dependency on the distribution of the industry. National and international mail-order and retail groups, such as Sears Roebuck and OTTO, established own logistic companies and parcel services, who also work for external customers. Other retail chains and department stores, e.g. ALDI or the METRO group, have developed their own logistic networks with individually designed supply chains, structures and distribution centers, but use qualified logistic providers for the realization and operation of defined part systems. Some retailer and cooperatives have organized their logistic systems and departments as *profit centers* and operate them like external service providers.

In any case, the management of the logistic network is kept in the company. With an own network, retailer can determine the optimal supply chains themselves and select optimal *delivery conditions*, such as:

- *Free to outlet*: Industry, supplier or their logistic service provider deliver directly to the ramps of the outlets, markets and department stores.
- *Free to logistic station*: Industry or supplier, respectively their carrier or forwarder deliver to logistic centers or transshipment points, which are operated by the retailer or by an assigned logistic service provider.
- *Ex works*: The carrier of the retailer picks up the shipments at the factory ramp or at the store of the supplier.

Retailing companies have achieved significant cost savings by shifting goods entry to an upstream logistic center where *economies of scale* by consolidation, high-tech and automation can be achieved (see *Sect. 1.8*). In addition, the retailer gains back *control of the outlet ramps*. The relief of outlets from *goods entry control* and other logistic activities is not only cost-saving but also *sales supporting*. The personnel can primarily serve customers and not logistics.

However, the positive effects of consolidation cannot be achieved for all goods and shipments. If the quantities shipped directly to neighbored outlets make up already full-truck loads, the costs would increase if they are delivered via logistic center. Only for smaller shipments of *parcels* and *palletized goods* substantial cost reductions and service improvements are possible by *transshipment* and *crossdocking* via own logistic stations. For heavy, bulky, dangerous and special goods and for full-truck loads, the systems of the supplier or of specialized service providers are generally more effective.

22.4.3 Opportunities and Disadvantages

Compound- and system-service providers offer the following *advantages* and *opportunities*:

- higher limit performances and larger store capacities
- improved service quality due to professionalism and specialization
- higher flexibility by extensive resources for several customers
- lower costs due to high efficiency and better utilization
- favorable wages and salary structures due to other tariff agreements

These advantages are of special interest for companies that plan to enter an established market, to develop a new business or to expand an existing business. In cooperation with a logistic system provider, they can avoid investments and surmount the barrier costs of an own logistic system. However, the advantages must be compared with the *disadvantages* and *risks*:

- long-term contractual binding to another company
- critical dependency on the system-service provider
- loss of the own logistic know-how

- potential financial weakness and lack of investment capability
- change of management and lack of skilled staff
- insufficient cost transparency and control
- lack of incentives and decreasing interest
- accumulation of overhead costs and profit surcharges by assignment of sub- and sub-sub-contractors
- no sharing of cost savings by rationalization, increasing throughput, larger quantities and purchasing advantages
- insight of the system-service provider into the business and internal weaknesses of the client
- potential misuse of confidential business data and customer information, especially if the system provider also operates for competitors

By selecting the right partners and a good contract, most of these disadvantages and risks can be limited or avoided. This requires a professional *bidding process* based on complete and qualified *tender documents*.

22.4.4 Contracting Strategies

After the outsourcing decision, it is necessary to fix the number of bidders, to specify the required logistic services and to assign a limited number of providers. By the experience of many tender processes and contract negotiations, the following *contracting strategies* have been approved:

- For transports and shipments to and from the same region, only one carrier or forwarder should be taken.
- In order to participate in the cost savings from *paired transports*, it is opportune to tender truckload transports to and from the same region together and to place the order to one transport service provider.
- Full and less than truckload transports are better tendered separated from parcel, general cargo and special logistic services, as the synergies between these services are minimal, and the bidders are not equally qualified in all areas.
- Also *combined freight*, i.e. shipments containing a mixture of parcels, pallets and other packages for the same destination, is better tendered separately. The *combined-freight providers* with their specialized freight network are more flexible than the highly standardized parcel services and pallet freight forwarders.
- It is opportune to tender a nationwide transport and freight demand to at least five up to maximal ten bidders in order to ensure sufficient competition and to give the single bidder a fair chance.
- If the national transport and freight demand exceeds a critical volume, it is useful to define two or more separate regions and to assign them to different providers in order to have *benchmarks* and the option to change.

- Parcel services, letter mail and combined freight to and from to the same country are better tendered separately and placed each to only one provider for the whole country.
- It is better to tender and negotiate transports and freight separated from warehousing. If these basically different services are given to the same service provider, it is useful to close separate contracts with different *contract periods*, in order to have different *exit options*.
- If warehousing services are needed at separate locations for different regions, they are better tendered independently, as there are normally no synergies. This approach offers benchmarks and separate exit options.
- For warehousing and other internal logistic services which are performed in existing buildings, between 5 and 10 bidders can be asked.
- If new systems and buildings with high investment are necessary, the number bidders should be at least 3, but not more than 5, in order to limit the bidding and planning effort and to motivate qualified bidders.

These contracting strategies are of advantage to the customer, but also fair for qualified service providers. The last point takes into account that the costs for planning and tendering a dedicated project are quite high. The invitation of too many bidders leads to rejections of the most qualified ones. It causes unqualified offers, if bidders recognize too many competitors.

Outsourcing options for warehousing services requiring a major investment for a new building and systems are:

- *Own investment and operation*: The warehouse is realized at own costs and operated by the company itself.
- *Own investment and external operation*: The warehouse is realized by the company itself and operated by a logistic service provider.
- *External investment and operation*: Building and logistic systems are invested, realized and operated by a logistic service provider

Experience has shown that the operating costs without wage differences for a warehouse, which is exclusively erected for one customer and operated by a system service provider, are 10 to 15% higher than the costs for an own optimally designed and organized warehouse. The higher costs are caused by the *overhead and profit surcharge* for management, administration, sales, risks and profit margin. If the warehouse is owned by the service provider, the surcharge is calculated as percentage on full costs including depreciation and interests. For in-house services, it depends only on the operating costs.

The additional costs for the different options between own and external investment and operation should be weighted against the advantages and disadvantages. It is necessary to keep in mind that also own investment and operation bears risks, causes administrative costs and must generate profit on the invested capital. Outsourcing of existing logistic operations causes *transfer costs* for pay-offs of retiring employees and compensations for employees who will be taken over by the service provider and from other legal obligations, as e.g. required by the German BGB §613a.

The contracting strategies should be applied with flexibility, depending on market situation and the behavior of the bidder during the negotiations. E.g., price advantages are achievable by bundling larger contracting volume.

From the general contracting strategies follows, that the assignment of only one *system-service provider* for the total procurement and distribution including warehousing and transports should be the exception and not the rule. This seems to be in contrast to general trend (Baumgarten 2000), but is confirmed by recent tendency to *insource* critical logistic services. More and more *outsourcing-damaged companies*, which are not mentioned in trend reports, have terminated a too tight relation to a single system-service provider and taken back the responsibility for their logistics.

22.5 Tendering and Contracting Logistic Services

As long as the transports, freight or other logistic services do not require customer specific investments by the bidder, a simple *performance request* is sufficient. A complete *letter of request* includes:

- brief description of the service requirements
- tables with the demanded performance quantities
- blank forms for the insertion of standard prices
- purchasing conditions for logistic services

For the purchasing of compounded, system and warehousing services, which require larger investments an *invitation to tender* is necessary.

The effort, time, specialties and risks of tendering logistic services are widely unknown or under estimated. The steps of *logistic tendering and contracting* with required times are shown in *Fig. 22.1*. For volumes up to 5 Mio €/year the tendering does not take longer than 10 to 12 weeks, if it is professionally performed. For larger projects with contract value above 5 Mio €/year it can last from 20 up to 30 weeks.

It is advisable to assign a qualified *tendering team* with specialists from logistics and purchasing and representatives of the operative departments, which are affected by the services to be outsourced. The team selects the bidders, approves the tender documents, assesses the quotations, negotiates with qualified bidders, and proposes the selected bidder and the negotiated contract to management for approval.

Experienced specialists are needed to work out the *tender documents* with specifications, demand figures and price blank forms. Also, explanation of the project to the bidders, answering of questions and analysis of quotations requires qualification and experience. If people with this special know-how and experience are not available within the company, external *consultants* specialized in logistic outsourcing and tendering should be engaged.

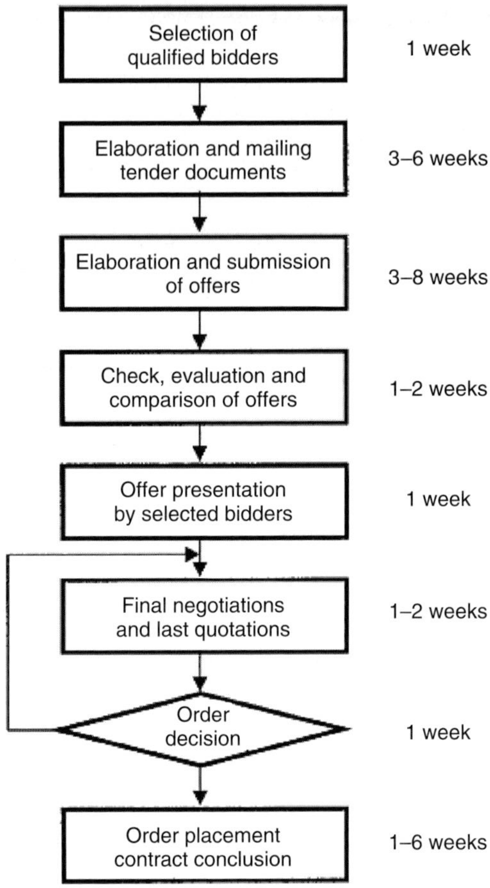

Fig. 22.1 Steps and required times for tendering logistic services
single and compound services below 5 Mio.€/a: 10 to 12 weeks
system services and warehousing above 5 Mio.€/a: 20 to 30 weeks

22.5.1 Tender Documents

Corresponding to the kinds of specification explained in *Sect. 1.1,* i.e. *result specification*, *functional specification* or *technical specification*, different tender documents are possible. For tendering compounded, warehousing and system services, a functional specification is preferable to a technical specification.

The *functional specification* explains basic concept and goals of the project. It specifies and delimitates the required functions and performance demand. For the execution of the performance packages, no technical solution is prescribed. This allows the bidder to develop own ideas for the optimal combination of his resources. It also ensures that the bidder accepts the full *responsibility for functionality*. A complete set of *functional tender documents* includes:

- description of the basic concept and goals of company logistics
- presentation of required services
- explanation of procedure and time table of tendering
- deadline for submission and binding period of the bids
- specifications of service demand and quality requirements
- quantification of performance demand
- delimitation of the service packages to be offered
- required IT and communication systems and interfaces
- general purchasing and awarding conditions
- special contract conditions
- registration, monitoring and remuneration of performance
- intended penalty regulation for quality faults
- blank forms for price quotation
- letter of acceptance to be signed by the contractor

The structure of company logistics, the flow of goods and orders, the procurement, distribution and performance chains, the buffer- and storeplace demand, the IT-frame, the performance demand and the price forms are presented as structure diagrams, process workflows and tables (see *Sect. 3.8*). The bidder has to declare in written form that he has checked the documents, that the requirements are feasible and that he is capable of delivering the required performances under the given circumstances.

In order to prevent misunderstandings and later changes by the bidder, it is useful to specify all cost and performance relevant restrictions and conditions in advance and to require a written confirmation of the completeness, consistency and acceptance of the circumstances and conditions by the bidder. This includes contract duration, cancellation period, compulsory coverage, liability and warranties, regulations for price adjustments, control rights and the legal frame conditions of the contract.

22.5.2 Assessment of Quotations

In order to select the most qualified bidders for the contract negotiations, the submitted quotations have to be assessed and analyzed with respect to *formal, performance* and *commercial criteria*. Already the keeping or missing of the submission deadline gives information about interest and reliability of a bidder. Quotations submitted after *deadline* should be refused if no profound reasons are given. If more than three bidders cannot cope with the deadline, it is probably too short and should be extended.

The *formal assessment* of the offer gives first indications about the qualification. This includes the control for completeness and general appearance. Are the quotation, price forms and approval letter legally signed? Have all price blank forms been completely filled out as required? Are all required specifications and all information included?

Evaluation criteria for quotations are: Clarity and structure; comprehensibility, readability and tidiness; type and frequency of errors; useful diagrams, figures and tables. The quality of the quotation gives a first impression on the service quality to be expected, and about the customer orientation of the bidder.

The *content assessment* includes the examination of the *performance criteria*:

- *System solution*: Is the offered solution sufficient, flexible and reliable to execute the required functions, services and performances?
- *Functional feasibility*: How are the functions and performances realized? What equipment and system technology are proposed?
- *Performance capability*: Are the offered capacities, limit performances and resources sufficient for the demand?
- *Staff qualification*: Experience, reliability, fluctuation and competence of workforce, employees and managers of the bidder
- *References*: Do the named references and experiences fit to the tendered performance range? Are they relevant?

A contractor is unacceptable if not all these criteria are settled. Missing one of the performance criteria is normally a *KO-criteria*. Only if none of the bidders is able to meet a certain requirement, it must be reexamined and eventually revised.

For the assessment of the performance, the methods of *Sect. 3.9* can be applied. From the final *utility value analysis* results a *scoring of the bidders*. *Figure 22.2* shows the result of an assessment of six quotations for the distribution system of a retailer. It includes transports and freight services to outlets and final customers. The best performance quotation according to this synopsis comes from bidder # 4 with score 1.8, followed by the quotation of bidder #6 with score 2.1.

The *commercial assessment* includes a comparison of the following points:

- offered *performance prices* for the different services packages
- annual *operating costs* resulting with these prices for the expected demand
- acceptance or counterproposals for the *terms of payment*
- conditions and penalties of the *quality remuneration* (see *Sect. 7.4*)
- coverage and duration of *warranties* and sum of *liability*

Due to different *market situation*, calculation methods, tactical reasons or cross-subsidies, the offered performance prices for logistic services differ often up to ± 50%, whereas the operating costs do not differ as much. The annual operating costs resulting from the quoted prices for the expected performance demand are the most important commercial selection criteria for a logistic service provider.

In *Fig. 22.3*, the quoted annual costs for freight and transports resulting from the offered prices of the 5 bidders from *Fig. 22.2* are compared with each other and with independently calculated *benchmark* values. In this example, the spread between the quotations with the most favourable and the least favourable prices is 12%. Bidder #4 is most expensive, although – as it is often the case – his performances

Assesment criteria	Weight	Bidder 1	Bidder 2	Bidder 3	Bidder 4	Bidder 5	Bidder 6
Offer quality	10%	3.0	3.5	4.0	2.0	2.5	2.5
References and competence	15%	3.0	1.5	3.0	2.0	3.0	2.5
Coefficient of requirements	30%	3.0	3.5	3.0	1.5	2.5	2.0
IT-systems	20%	2.0	2.0	2.5	2.0	2.0	2.0
Own resources	10%	3.0	2.0	2.0	2.0	4.0	1.5
Personnel qualification	15%	2.0	2.5	2.5	1.5	3.0	2.0
Total score	100%	2.7	2.6	2.8	1.8	2.7	2.1

Fig. 22.2 Performance assessment and utility analysis of quotations for distribution services
Scorings: excellent 1/good 2/satisfactory 3/sufficient 4/insufficient 5

are scored highest. Bidder #6 with the second best performance score offers with 2.9 Mio €/year the lowest operating costs. They are even less than the calculated benchmark costs of 3.15 Mio €/year.

Due to the in-transparency of the *logistic markets*, the offered operating costs for warehousing services differ far more than the operating costs for transports and freights. The differences can go up to ± 25% and more. Therefore, neutral planning and cost budgeting is recommended to *benchmark* the quotations for warehousing, which incorporate investments in new buildings and systems.

22.5.3 Bidder Presentation and Negotiations

At least two and maximally four bidders with the best price-performance relation are invited to present their companies and to discuss the details and terms of their quotations. Together with the invitation and the agenda for the meeting, they should be informed on open questions and relevant price deviations. This saves time and allows the invited bidders to be prepared.

Some logistic service providers, in particular big companies, tend to an exaggerated self-portrayal during the presentation whereas they often miss customer specific issues and questions towards the quotation. No serious customer is distracted by colorful power-point charts and impressed by glossy brochures. Central question for the customer remains: Is the bidder a qualified partner for the tendered logistic services?

After having met the selected bidders and settling all open questions, the assessment of the bidders and their quotations is revised and the actual price offers are

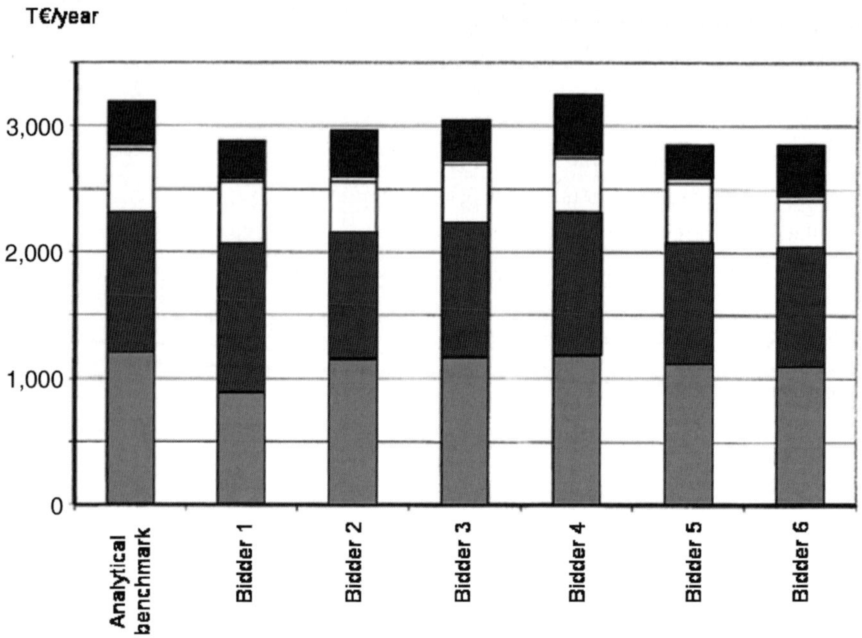

Fig. 22.3 Comparison of transport and freight costs of the quotations assessed in *Fig. 22.2*
Meaning of the bars from bottom up:
Full- and part-truckload transports
Regular distributions tours with 7.5 t trucks
Peak distribution tours with 7.5 t trucks
Transhipment costs
General freight via one transhipment station

compared. This results in a well-founded awarding recommendation to management by the tendering team. The selected bidder is invited to final *contract negotiations*. If several favorable bidders are left, they are asked to revise their prices and to come up with a *final bid*.

The final negotiations and last price biddings require sense of proportion, fairness and experience on both sides. The future business partner must be able to survive in the long run with the contracted prices and the resulting turnover. On the other hand, some service providers try to buy the order with non cost-covering *dumping prices* in the intention to raise prices after the business has started and the client became dependent. It is quite difficult and often very expensive to revise or cancel such a mislead business connection.

22.5.4 Order Placement and Contract Conclusion

Generally, an *order letter* by the client, and a written *order confirmation* by the service provider are sufficient to close the contract. This simple form of contracting is normal for standard transport, freight and logistic services. In most cases, it is also sufficient for warehousing and logistic system services provided the tendering has been performed professionally and well documented.

In order to avoid later disputes, the order letter must confirm the results of the negotiations and the content of the documents in *sequence of importance*:

1. tendering documents and purchasing conditions of the customer
2. final prices and agreements of the assignment negotiations
3. quotation and terms of sales from the contractor

These order statements and their acceptance by order confirmation make the agreed performance prices and conditions as well as the priority of the tender documents legally binding.

If the project requires larger investment and intense pre-work for the contractor, a short *letter of intent* (LOI) of the customer is opportune for both parties. It declares the willingness to conclude the contract after all open questions are settled and all conditions are fulfilled.

If it was not possible to plan buildings, systems and operations during the tendering phase with sufficient accuracy, the order is placed under the reserve of the offered services and costs. For major projects, only a *pre-contract* for *detail planning* should be closed. If the finally planned solution meets the requirements and can be realized within the offered cost frame and time, a *realization contract* or a *project management contract* is concluded:

- *Realization contract*: In case of an external investment, the contract regulates all technical, commercial and legal details for the buildings and systems to be erected and delivered ready to use.
- *Project management contract*: In case of an investment by the customer, the contract regulates the duties and commercial conditions for the project management by the service provider until the buildings and systems are ready to use.

The final step is a *service contract*, respectively a *operating contract* for the operation:

- *Service contract*: In case of a logistic system owned by the service provider, a service contract regulates the performances and duties of customer and contractor as well as frame conditions and remuneration.
- *Operating contract*: If the logistic services are performed within buildings and by systems and equipment owned by the customer, an operating contract regulates the performances and duties of customer and contractor, and the frame conditions and remuneration.

If a service provider takes over the operation of an existing logistic department, the contract needs to regulate the take over of employees according to law, e.g. due to the German BGB §613a.

In order to avoid omissions and changes of the tender conditions and documents, it is advisable that the customer works out the draft agreements and hands them over to the contractor for examination. *Legal departments* and *lawyers* should not be involved before both parties have indisputably agreed upon technical content, services, performances, and commercial conditions.

Trustworthiness, operational readiness and capability of the responsible managers and specialists are decisive criteria for a long-term relation with a system-service provider. Therefore, the bidder must introduce the project manager and people responsible as soon as possible, latest at the presentation of the quotation. During the negotiations and in the planning phase, it becomes obvious, whether the intended cooperation between the people is functioning and the partnership will endure. A final contract for realization and long-term cooperation should only be signed after a trustful relationship between the partners has been established. On the contractor's side, this requires competence, capability and trustworthiness, and on the customer's side, managers, who are open minded, willing to cooperate and prepared to develop common solutions for unexpected problems.

22.6 Performance Control and Remuneration Adjustment

After a *performance and quality remuneration scheme* as described in *Sect. 7.5* has been implemented, the performance control by the customer can be restricted to *check invoices* with respect to content and quantities and to *control service quality* based on error statistics, customer complaints and quality reports. Additional task of the customer's logistic controlling is the observation of *logistic markets* and of the general development of costs and prices.

Controlling the daily operation of the service provider is unnecessary and only causes additional costs. Here holds the *controlling rule*:

- ▸ The logistic service provider, not the customer is responsible for performance and quality and controls his own operations.

A critical test for a longer lasting contractual cooperation is the adjustment of remuneration to changing conditions. Experience has shown that disputes and even a premature cancellation of the contract may happen, if the acceptable reasons and the procedure of *price adjustments* are not regulated in advance.

Two different reasons for adjustments are possible: either changing cost factors or changes of the performance structure. *Cost related price adjustments* are normally negotiated once a year. They are caused by a general increase of wages, material and energy costs and costs of living. Changes of state fees, tolls or taxes lead to exceptional adjustments. The indisputable regulation of cost-related price adjustments requires a contractual documentation of the *calculative weight of the relevant cost factors*, by which the affected prices are adjusted to the different possible cost changes.

Structure-related remuneration adjustments are necessary, if the frame conditions or the performance structure change or if the productivity has been improved. For the indisputable adjustment of remuneration caused by altering structures as well as by changing cost factors, the project specific remuneration scheme described in *Sect. 7.5* is suitable. By this scheme, the consequences of structural and cost-related changes on the performance prices can be calculated comprehensibly for both parties.

Chapter 23
People and Logistics

People are the actors and producers, and at the same time, the customers and beneficiaries of logistics. As producers, they determine the services, performances and costs. As customers, they take the benefits of services and performances, but are also affected by deficiencies and high prices. Most people are not aware of the *goal conflicts* between suppliers and customers and of the influences and consequences of human behavior in logistics. In order to draw the attention to this matter, the last chapter of this book deals with the role of people in logistics, which has been mentioned only briefly in *Sects. 1.10, 2.8, 2.4., 3.1, 3.4* and *13.6.*

In automatic systems, only few persons are employed in the operation. Hence, output and quality depend primarily on the people in the planning and building phase. With increasing number, the people in the operation determine quality and quantity of the output. Correspondingly, at first, the human influences in the *set up phase* and then in the *operating phase* will be investigated. The analysis results in *recommendations*, which help to avoid the negative consequences, and *rules of conduct*, which promote the positive influences of people in logistics. Better understanding and appropriate rules of conduct help to achieve benefits not only for the people but also for the company and the customers (Aggarwal 1985; Ohno 1988; Rinehart and Ragatz 1996; Slack et al. 2004)

Under fair market conditions, the prices for logistic services result due to the *market rules* from *offer and demand* (Gudehus 2007). However, on many *logistic markets*, state intervention, one-sided rules und unfavorable frame conditions foil the fair price formation and cause prices which are not related to costs and utilization (see *Sect. 7.7*). For these *market deficiencies*, managers and politicians, but partly also scientists and teachers of economics can be blamed. They are responsible for the long-term development of the economy and logistics for the benefit of the people, not only of companies.

T. Gudehus, H. Kotzab, *Comprehensive Logistics*,
DOI 10.1007/978-3-540-68652-1_23,

23.1 Human Success Factors

Nature and behavior of the people acting on the strategic, tactical and operative level influence and determine the output, performance and profitability of a company. The dependency of the results of planning, realization and operation on human behavior is obvious. But also the results of management, consultancy, research, education and politics depend on people although this is rarely noticed.

23.1.1 Capability and Willingness to Perform

The *capability* of people is crucial for the *quality* of the work in the set up phase and in the operating phase. The *willingness to perform* is decisive for the output of a plant or a logistic system with many people in the operation.

The human capability depends on the individual *suitability* for the specific task. It results from personal disposition and professional qualification. *Personal disposition* is determined by physical strength and intellectual capability, by the ability to judge and to decide, by communication skills and last but not least, by *character*. For a manager the most important character attributes are self-confidence, understanding and humanity. Damaging attributes are disinterest, conscientiousness and arrogance. *Professional qualifications* are special skills, expertise and know-how. Whereas personal disposition and character can hardly be changed, professional qualifications can be gained by learning and experience, and trained by education and instruction.

The willingness to perform is strongly influenced by *personal state*, such as health and exhaustion, by the *working conditions*, such as heat, chill, atmosphere and monotony, and by *motivation*. Exhaustion and monotony can be reduced by *job rotation* and *job enrichment*. Ergonomic work flows increase performance. Good working conditions improve the personal state and motivation. Distrust and excessive control diminish motivation, whereas trust, appreciation and praise, success and fair remuneration enhance motivation. High motivation is achieved also by involving the employees, i.e. by *kaizen* and *continuous improvement processes* (CIP).

In logistics, the problem of fair, performance dependent remuneration is still not solved. The proven solutions from manufacturing such as *piece wages* cannot be directly transferred to logistics due to the stochastic processes and fast changing requirements.

23.1.2 Self-Interest and Human Weaknesses

Self-interest is the strongest driving force of people: self-interest to satisfy the human needs, such as hunger, thirst, heat, safety and survival, and self-interest to fulfill the human wishes, such as love, appreciation and joie de vivre. Natural and fair self-interest, restrained by law and order, is principally positive. It is the driver of all personal demand and necessary for a functioning economy, which would be useless without natural self-interest. Hence, the economic behavior of people is dominated by the *principle of self-interest*:

- ► The overall goal of people as consumers and as producers is to satisfy their self-interest.

Human self-interest is also the origin of the general *economic principle*:

- ► The goal of economic actions by individuals and companies is to maximize revenues and output at minimal costs and input.

Self-interest and striving for profit become intolerable when pursued ruthlessly to the disadvantage of others. Unfair and ruthless self-interest is dangerous especially in combination with other human weaknesses, such as envy, ignorance, lust for power and self-delusion. Most difficult to defeat is the imaginary or pretended common good, as the opposite of good is not bad but intended good. Recognizable evil can be defeated, whereas against good intentions people are quite helpless.

This holds also in logistics: Companies offer special discounts, rebate systems, mile bonuses and other allowances, pretending they have been developed for the advantage of the customer. In reality, many bonus programs and pricing systems are cheating the customers, especially when they have no choice. Many prices for logistic services are not transparent and comparable. Discriminated customers, such as *seldom travelers*, pay higher prices and subsidize the bonuses for the beneficiaries, such as *frequent travelers*. *Added services of logistics*, e.g. free home delivery, 24-hour-service, extremely short delivery times, just-in-time or tracking and tracing, are questionable offers as long as only some customers really need them, whereas the prices are higher for all customers in order to cover the additional costs.

People know best what is good for themselves and their families. However, they often do not know or ignore what harms them and others. Ruthless self-interest and human weaknesses will always remain, but they can be curbed by law and rules of conduct.

23.2 Recommendations for the Set-up-Phase

The set up phase of a logistic system starts with target planning and ends with the start of productive operation. Results and success of a project depend strongly on the people involved in planning and realization (see *Sects. 3.2* and *3.3*).

Time pressure and stress cause special problems. Experience, common sense and character are needed when working and deciding under pressure. A qualified *project manager* must be able to work efficiently and successfully under time pressure, to execute planning steps as far as possible in parallel, to accelerate the unimportant and to focus on the important without being chaotic, loosing nerves and making mistakes.

23.2.1 Recommendations for Target Planning

In theory, the client or management knows the goals and requirements before starting a project. In reality, the human influence on the success of a project begins already here. Clients, customers or management often have not made up their mind about the goals and priorities. In many cases, they do not know the requirements and performance figures for the planning horizon of the initiated project.

On the other hand, preconceived opinions, general trends and personal interests influence the project and exaggerate expectations. This causes mistakes, defaults and higher costs, which become obvious later and cannot be corrected.

To avoid this, the actors should follow the *recommendations for target planning*:

- No planning without clearly specified goals
- The main goal is to enable the required services and performances at minimal costs with adequate quality
- As far as a project concerns customers, the customers set the goals
- The project goals should be clarified and written down before planning
- All performance, quality and service requirements must be known and quantified
- Before beginning the project, client and project manager must agree on the goals and requirements

As the future can never be predicted precisely, it is impossible to exactly determine the requirements for a planning horizon of 3 to 5 years. This leads to the necessity to plan and decide under *uncertainty*. Due to character or anxiety, some people are unable to decide under uncertainty. They require certainty where there is none, instead of designing flexible and adaptable systems. They do not notice the chances and potentials achievable by *flexible reaction* and *dynamic scheduling* during operation (Gudehus 2002/2007)

23.2.2 Recommendations for System Planning

Success and failure of a project depend on the quality of the system planning. Unsystematic procedure, insufficient knowledge, bias, lack of experience and comprehension, can cause irreparable mistakes. Due to these weaknesses, often the most promising options and chances are ignored.

Recommendations for system planning are:

- Imagination, creativity, expertise and an open mind are decisive for the success of system planning.
- Not benchmarking and imitators, but unprejudiced people find new solutions and innovative breakthroughs.
- Right methods, knowledge and experience are more important than personal interests, tactics, power and hierarchies.
- Knock-out-criteria and constraints must be known completely and considered in advance when selecting solutions and designing the system.

The results of system planning are the basis for detail planning. They should be completely documented and approved by the client. This includes the prospective operating costs, investment budget, a *feasibility analysis* and the *timetable for realization*.

23.2.3 Recommendations for Detail Planning

Unqualified detail planning or insufficient tendering can spoil the best solutions. Important issues are often neglected or insufficiently planned, such as safety and ease of operation, working conditions, or interfaces. Other issues are unnecessarily complicated or delayed by a dominating manager or a powerful department.

Inadequate tender documents without complete specifications of the requirements, appropriate price blank forms and purchasing conditions, jeopardize the success of tendering. Reasons are often time pressure and wrong parsimony, a snappy management or a powerful, but logistically inexperienced purchasing department. Due to the same reasons, often the wrong bidders are preferred and other, more qualified suppliers are ignored.

A help to avoid these situations are the *recommendations for detail planning*:

- The project team for detail planning and tendering should be supervised by an experienced and authorized project manager.
- The project team should be made up by qualified experts of all involved areas and include the responsible manager for the future operation.
- Detail planning and tender documents require sufficient time.
- The results of detail planning and the tender documents have to be checked by the manager responsible for the operation.
- The bidders should be selected based on objective criteria, such as qualification, expertise and references, and not on personal preferences.

23.2.4 Recommendations for Realization

Even after the *point of no return*, when the obstacles of planning are taken, when the management has decided and the orders for realization are placed, people can endanger and affect the project by misbehaving and objections. Quite often opposition and cost-driving requirements come from authorization clerks, shop councils, IT and from people responsible for the operation.

The human influence is especially high just before the operation starts. Human abilities and weaknesses, interests and worries, motivation and enthusiasm decide on the final success and failure. Confusion, lack of communication and misunderstandings between experts of different disciplines, such as architects, engineers, IT- and OR-specialists and economists, can cause failure or delay.

These problems can be avoided or solved by the following *recommendations for realization*:

- Organization of a qualified project team led by an experienced and authorized project manager
- Selection of the right partners for the realization, who are motivated by fair terms, realistic completion dates and sufficient prices to do their best for the project and the client
- Installation of a qualified project management and controlling of mile stones, due dates, performance and costs
- Early selection and involvement of the persons responsible for the operation
- Joint agreement about the terminology of the project in order to avoid misunderstandings between the experts.

- Duly deployment, training and instruction of system operators and workers
- Professional planning, preparation and execution of acceptance tests of part systems and the total system
- Duly organization of preventive maintenance, repair and spare parts
- Commitment of top management in the project, backup for the project manager, acknowledgement of progress and support in difficult situations

If client and management grant sufficient independency and fair working conditions, people will solve most questions and settle disputes on their own. Joy and pride of the common work generate better results than intensive *controlling* and *reporting*.

23.3 Recommendations for the Operating Phase

Output and *quality* of the products and services are key indicators for the capabilities of a plant or a system (Pande et al. 2000). For automatic systems, they depend on the qualification of the operators and the people of quality control, maintenance and repair, who determine *technical availability* and *effective performance*.

The *technical availability* of *logistic systems*, such as automatic high bay stores, sorter systems, mini-load stores and AGV-systems, should exceed 98% as long as the system is properly maintained. Performance losses by non-availability less than 2% are negligible. The technical *failure rate* is below 0.1% and the *technical reliability* above 99.9%.

Output, performance, and quality of *manually operated systems* depend less on technical availability but far more on the *human availability*. Human availability depends on the capability of workers and their willingness to perform, on the competence of the schedulers, and on management. For example, the human availability for manual order picking ranges between 80% and 95%, depending on operating conditions and workload. Also management can affect the human availability significantly. The error rate of human order pickers ranges between 0.5% and 2.0% (see *Sect. 17.4*). These figures demonstrate the general *performance rule*:

► Performance, availability and quality of manually operated systems are significantly lower than for automatic systems.

The quoted figures for the availability and reliability and their differences indicate the improvement potentials of manually operated systems in logistics.

23.3.1 Improvements for the Workers

The availability of an industrial worker is the relation between productive time and total working time, provided sufficient orders are available. The working time is reduced by *personal allowances* for recreation and unproductive activities. Personal allowances increase significantly by exhaustion, distraction, lack of interest and motivation, and by bad working conditions. Means to improve the availability and to increase the effective performance are:

no permanent handling of heavy weights
no excessive work load
right temperature and humidity
good lighting
no distractions
high safety
humane treatment
fair remuneration (23.1)

The *effective human performance* is the product of human availability and personal limit performance, which depends on the cycle times of the work process. The cycle times can be reduced by several measures, such as

ergonomically designed working places
ergonomic process flow
minimal waiting times
instruction and training (23.2)

High performance is only useful if the quality of the resulting products and services is sufficient. Failures and faults can cause more damage than just performance loss, corrections or reconditioning.

A central quality department may register and control the defects. However, the positive effects of *performance and quality controlling* are often over-estimated and the efforts for registration and analysis of faults are not taken into account. The *recommendation* for *quality management* is:

► Avoiding defects is better than measuring and controlling them.

This can be achieved by

continuous quality instruction
quality measurement at the work station
defect reporting by workers
personal marking of results (23.3)

By a *zero-defect-picking-program* it is possible to reduce the error rate of human pickers to less than 0.1%. Another self-regulating solution could be an output related *performance and quality remuneration* of the employees. However, due to the stochastic influences and changing order structure, such a scheme is difficult to establish in logistics.

23.3.2 Influence of Operators and Schedulers

The performance of a logistic system or plant depends not only on the workers but to a great extent on the *operators* and *schedulers*. Their tasks require *attention*, *qualification* and *motivation*. High attention is achievable by optimally designed control devices and displays. Professionalism is ensured by qualified people and improved by instruction and training. The motivation is strongly influenced by the management.

Schedulers decide on the efficient use of the available resources. In many companies, the tasks, qualifications, options and responsibilities of the schedulers are not adequately recognized. Schedulers often operate on experience rules and apply self-developed strategies, which are neither written down nor coordinated.

The results and effects of scheduling are seldom analyzed and assessed. Most schedulers operate behind the scenes. Many are highly motivated and perform better than expected. Others are underperformers. Generally holds:

► Unqualified, overstressed and de-motivated schedulers can cause immense damages.

Therefore, management should recognize and appreciate the work of schedulers. They have to be supported by appropriate software. ERP-software offers many different *algorithms*, *operating* strategies and *free parameters*. However, the software generally does not help in the decision about the application of methods and the adjustment of parameters.

23.3.3 Behavior of Managers

Unqualified management is often the main reason for underperformance and poor quality of a logistic system, plant or company. Ignorance, lack of interest, arrogance and self-overestimation can lead to wrong decisions and misconduct. Subordinated managers and employees sometimes ignore and correct the wrong advices of top management. However, incurable consequences of the misconduct of a manager are frustration and de-motivation of people.

A frequent misconduct of modern managers is their rare presence in the plant and operative departments. Here holds the biblical word: "The masters eye is worth both his hands".

Some managers believe, their tasks can be performed by computers and management systems, delegated to employees or replaced by sophisticated controlling. Anyone, who is unable to make decisions, cannot cope with conflicts, avoids contact with people, or is unable to mediate, should not become manager.

23.4 Outlook

The economy of modern societies is still not based on free markets with fair and effective price formation (Gudehus 2007). This holds true also for *logistic markets*. Interventions of government, missing competition, unfavorable frame conditions, monopolists or obscure pricing foil the fair price formation and lead to misallocation of resources (see *Chap. 7*).

For example, within central Europe, the postal rate for a border crossing letter deviating from standard size can be four times higher than the rate for national delivery even if the distance is shorter. Despite the intention of the *European Union* (EU) to deregulate the service markets, similar distorted price relations can be found for other logistic services.

This leads to the demand on *politicians*:

► In order to improve the efficiency of national and international logistics, improved market rules, laws and frame conditions are needed.

This demand is not a plea for a flood of detailed laws regulating everything. On the contrary, logistics requires general rules and universally applicable laws which avoid special regulations and expensive court decisions. For this purpose, logistic and economic research has to develop practicable solutions, rules and laws.

Historical-descriptive logistics is not sufficient to solve these issues. Also, *knowledge management* does not create solutions. Documentation and administration of existing knowledge neither open new insight, nor do they produce ideas or progress. Nowadays, the creation of new knowledge and the development of innovative solutions are widely neglected.

Analytical logistics as outlined in this book develops theoretically founded and feasible solutions, well-founded proposals for actions and practicable methods. Its results are the basis for the development of general rules, the formulation of effective laws and the improvement of the economic frame conditions. In logistics just as for the whole economy, the overall guide line should be:

► Logistics serves the people, not only companies

No one is free of *self-delusion*. This concerns the own capabilities and is the origin of arrogance and self-overestimation. It also refers to the own knowledge and can lead to the misinterpretation of the risks and consequences of decisions. The greatest danger of self-delusion however are the motives. Pretended common good and altruism quite often conceal pure self-interest.

Self-delusion causes exaggeration and one-sidedness. Often new solutions, strategies or ideas, which after long time of resistance are finally accepted by the majority, are pursued as the one and only possibility. New strategies are often overdrawn, new laws are legislated in an undue haste and regulations are uncritically changed. When negative consequences become obvious, the new strategy is discarded, the new law is changed and the opinion reversed.

An example for such a behavior is the *RFID-hype* in logistics. It started around 2000 with technical enthusiasm and exaggerated expectations. After noticing error rate and costs of the RFID-technique, its limited importance for logistics is generally recognized (Economist 2007). Another example is the endless back and forth between *centralization* and *decentralization* (see *Sect. 2.4*). It happens in logistics, in companies and in the state. Other examples are one-sided therapies and patent solutions, which promise to cure a complex problem.

Helpful against self-delusion, one-sidedness and exaggerations are the following *rules of conduct*:

objective clarification of goals
common prioritization of diverging goals
serious consideration of the relevant aspects
open minds for new ideas, solutions and proposals
systematic development of adequate strategies
critical investigation of compatibility and conflicts
objective analysis of consequences and side effects
pragmatic balancing of advantages and disadvantages

self-critical search for contradictions
respect against well-faunded objections
acceptance and consideration of ambiguities and risks
balancing between freedom and safety
limitation of centralization
fair mediation of diverging interests
balanced combination of old and new

These rules and recommendations are helpful not only in logistics, but also in other areas. As logistics relatively easy to understand, it is a good field to develop, investigate and test strategies, rules and concepts which can be transferred also to other areas of the economy.

The present state of *analytical logistics* has been outlined in this book. The presented strategies, solutions and rules can be improved and extended. Many problems are still unsolved. In some sectors and branches *operative logistics* is years behind the findings and solutions of analytical logistics.

The challenge for the theorists is, to convince managers and practitioners of the value of their ideas and recommendations. The challenge for practitioners and managers is to realize convincing ideas, strategies, recommendations and solutions. This requires acknowledgement of the proposals and results of analytical logistics as well as the will to invest and to take risks.

Bibliography

Given the vast and fast growing number of textbooks, journals, reports, publications and scientific work on logistics and supply chain management, a complete reference of literature to the topics of this book is impossible. In the text all literature is cited from which inspiration, strategies, algorithms, methods, techniques, tools, data, illustrations or examples have been taken or used for this book. In addition, a selection of relevant literature to the specific topics is given, albeit procedures, methods and results may differ from those presented in this book.

Aberle G.: Transportwirtschaft – Einzelwirtschaftliche und gesamtwirtschaftliche Grundlagen, 4th Edition, München, 2003

Ackerman K. et al.: Practical Handbook of Warehousing, Springer, Berlin Heidelberg New York, 1997

Ackerman K., LaLonde B.: Making warehousing more efficient, Harvard Business Review, 58, 2, 94–102, 1980

Ackhoff R.L.: A concept of corporate planning, J. Wighley & Sons, New York, 1970

Aggarwal S.: Materials Management: The Key is People Management, Journal of Operations Management, 6, 1, 43–49, 1985

Alvarado U., Kotzab H.: Supply Chain Management: The Reemergence of Logistics in Marketing. In: Industrial Marketing and Management Special Issue on: Partnering with Resellers in Business Markets: Relationship Marketing and Partnering, 30, 2, 183–198, 2001

Aminoff A., Kettunen O, Pajunen-Muhonen H.: Research on Factors Affecting Warehousing Efficiency, International Journal of Logistics: Research and Applications, 5, 1, 45–57

Andraski J., Novack R.: Marketing logistics value: Managing the 5 P's, Journal of Business Logistics, 17, 1, 23–34, 1996

Apple J.: Materials Handling Systems Design. New York, Ronald Press, 1972

Armour F., Buffa E.: A heuristic algorithm and simulation approach to relative location of facilities, Management Science, 9, 2, 294–309, 1963

Arnold D.: Materialflußlehre, Viehweg, Braunschweig Wiesbaden, 1995, 3rd edition Springer, Berlin, Heidelberg, New York, 2002

Arnold P., Peeters D., Thomas I.: Modelling a rail/road intermodal transportation system, Transportation Research Part E: Logistics and Transportation Review, 40, 3, 255–270, 2004

Arnold D., Rall B.: Analyse des Lkw-Ankunftsverhaltens in Teminals des kombinierten Verkehrs, Internationales Verkehrswesen 6, 1998

Arnold T.: Introduction to Materials Management, 3rd edition, Prentice-Hall, Upper Saddle River, 1998

Arrow K., Karlin S., Scarf H.: Studies in the mathematical theory of inventory and production, Stanford CA, 1965

A.T. Kearney: The efficient unit loads Report, ECR Europe Bluebook Report, ECR Europe, 1997

Axson D., Hermann K.R.: Performance Management and Reporting Best Practice, John Wiley&Sons, 2003

Baker K.R.: Introduction into Sequencing and Scheduling: John Wiley&Sons Inc., 1974

Ballou R.: Business Logistics/Supply Chain Management: Planning, Organizing, and Controlling the Supply Chain, 5.edition, New Jersey, Pearson Higher Education, 2004

Barney J.: Firm Resources and Sustained Competitive Advantage, Journal of Management, 17, 1, 99–120, 1991

Barrett B.: A further digression on the over-automated warehouses: Some evidence. Interfaces, 8, 1, 46–49, 1977

Baumgarten H.: Make-or-Buy entscheidet der Manager: Jahrbuch der Logistik, Verlagsgruppe Handelsblatt, Düsseldorf, 1992

Baumgarten H.: Prozeßanalyse logistischer Abläufe, Fabrik, 43, 14–17, 1993

Baumgarten H.: (Editor), RKW-Handbuch Logistik, Erich Schmidt Verlag, Berlin, 1981–1999

Baumgarten H., Straube F., Wolff S. et al.: Trends und Strategien der Logistik, Analysen-Potentiale-Perspektiven, annual survey in cooperation with BVL (Bundesverband Logistik) Deutscher Verkehrsverlag, since 1993

Beckhard R.: Organization Development: strategies and modules, Reading, Mass. Adison-Wesley series on organization development, 1969

Behrentzen Chr., Reinhardt M.: Kooperation in der Distributionslogistik von Strothmann Spirituosen und Melitta Haushaltswaren, In: Busch A./Dangelmaier, W. (eds). Integriertes Supply Chain Management, Gabler, Wiesbaden, 453–468, 2002

Bellmann R., Glücksberg I., Gross O.: On the Optimal Inventory Equation, Management Science, 2, 1, 83–104, 1955

Bertalanffy L.v.: General system theory: foundations, development, applications, Braziller, New York, 1968

Berry L.L., Yadav M.S., (1997); Oft falsch berechnet und verwirrend – die Preise für Dienstleistungen, HARVARD BUSINESS manager 1/1997, S. 57 ff.

Biggs N.L.: Discrete Mathematics, Oxford Science Publications, revised edition, 1999

Blau P.M., Schoenherr R.A.: The structure of organizations, Basic, New York, 1971

Bode W., Preuß R.W.: Comprehensive Introduction to Intralogistics, A reference book by the STILL Akademie, Wirtschafts-Verlag W.V., Suhl/Thüringen, 2004

Böhnlein D.: Multi Multi-Agent Agent-Based Logistics Management in the Newspaper Industry, Presentation WK Logistik, 12.07.2008

Bock D., Hildebrand H., Krampe K.: Handelslogistik, in Grundlagen der Logistik, hussverlag, München, 2nd ed., p. 233ff, 1996

Bogatschewsky R.: Losgröße, In: Kern W., Schröder, H.-J., Weber, J. (eds). Handwörterbuch der Produktionswirtschaft, 2nd edition, Poeschel, Stuttgart, Sp. 1141–1158, 1995

Bontekoning Y., Macharis J., Trip J.: Is a new applied transportation research field emerging? A review of intermodal rail–truck freight transport literature, Transportation Research Part A: Policy and Practice, 38, 1, 1–34, 2004

Borries v. R., Fürwentsches W.: Kommissioniersysteme im Leistungsvergleich, moderne industrie, München, 1975

Bowersox D., Closs D., Cooper B.: Supply Chain Logistics Management. McGraw Hill, 2nd edition, 2007

Box G., Jenkins M.: Time series Analysis, forecasting and control, Holden Day, San Francisco, 1970

Bozer Y, White J.: Travel time models for automated storage/retrieval systems. IIE Trans., 16, 4, 329–338, 1984

Bozer A., White J.: Design and performance models for end-of-aisle order picking systems, Management Science, 36, 7, 852–866, 1990

Bragg St.M.: Cost Accounting, John Wiley&Sons, 2001

Breiter P.M.: ECR – Efficient Consumer Response, Wer hat was davon?, Distribution, 7/, 12–13, 1996

Bretzke W.-R.: Supply Chain Management, Wege aus einer logistischen Utopie, Logistik Management, 7,2, 21–30, 2005

Bretzke W.-R.: Logistische Netzwerke, Springer, Berlin Heidelberg New York, 2008
Brockmann T.: 21 Warehouse trends for the 21st century, IIE Solutions, July, 36–40, 1999
Brucks W.: Subsidiarität, Definition und Konkretisierung eines gesellschaftlichen Strukturprinzips, Soziologisches Institut der Universität Zürich, Online Publication: http:/deo/t_wbrucks.htm, 1997
Brynzér H., Johansson M., Medbo L.: A Methodology for Evaluation of Order Picking Systems as a Base for System Design and Managerial Decisions, International Journal of Operations & Production Management, 14, 3, 126–139, 1994
Bucklin L.P.: A Theory of Distribution Channel Structure: CA: IBER, Special Publications, 1966
Bucklin L.: Postponement, Speculation and the Structure of Distribution Channels: Journal of Marketing Research, 2, 1, 26–31, 1965
Busacker R.G., Saaty T.L.: Endliche Graphen und Netzwerke, Oldenbourg, München Wien, 1968
Camp R.: Benchmarking: the search for industry best practices that lead to superior performance, Milwaukee: ASQC Quality Press, 1989
Caron F., Marchet G., Perego A.: Layout design in manual picking system: A simulation approach, Integrated Manufacturing Systems, 11, 2, 94–104, 2000
Caron F., Marchet G., Perego A.: Optimal layout in low-level picker-to-part system, International Journal of Production Research 38, 1, 101–117, 2000
Cavonato J.I.: A Total Cost/Value Model for Supply Chain Competitiveness, Journal of Business Logistics, 13, 2, 285–301, 1992
CCG Centrale für Coorganisation GmbH: CCG1 und CCG2, Einheitliche Ladehöhen für EURO-Paletten, 1985
CCG Centrale für Coorganisation GmbH: Der SINFOS-Datenpool, 1993
CCG Centrale für Coorganisation GmbH: EAN 128-Standard, Internationale Codierung zur Übermittlung logistischer Dateninhalte, Coorganisation 1/1994
CCG Centrale für Coorganisation GmbH (CCG): Das EAN-Nummernsystem: Grundlage aller Coorganisation, 1995
CCG today GS1, Standard Knowledge Center: http://www.gs1.org/services/gsmp/kc/, accessed 14-06-2008
Chandler A.: Strategy and structure: chapters in the history of the industrial enterprise, Cambridge, Mass., M.I.T. Press, 1962
Chew E., Tang L.: Travel time analysis for general item location assignment in a rectangular warehouse, European Journal of Operational Research 112, 3, 582–597, 1999
Chisman J.: The clustered traveling salesman problem. Computers & Operations Research, 2, 2, 15–119, 1975
Chopra S., Meindl P.: Supply Chain Management, Strategy, Planning & Operation, 3rd edition, Pearson/Prentice-Hall, Upper Saddle River, NJ, 2007
Coase R.: The Nature of the Firm, Economica, 4, 16, 386–405, 1937
Cooper M., Lambert D., Pagh J.: Supply chain management-more than a new name for logistics, International Journal of Logistics Management, 8, 1, 1–14, 1997
Coyle J., Bardi E., Langley J.: The management of Business Logistics, West Publishing Company, New York, 1996
Christopher M.: Logistics and Supply Chain Management: London, Pitman Publishing, 1992, 3rd ed. 2005, Pearson Education, Edinborough
Christopher M., Peck H.: Marketing Logistics, Butterworth-Heinemann, UK, 2003
Churchman C.W., Ackhoff R.L., Arnoff E.L., Introduction to Operations Research, John Wiley&Sons, New York, 1957, (Deutsch: R. Oldenburg, Wien und München, 1961)
Churchman C.W., Einführung in die Systemanalyse, München, 1970
Collingwood E.F., Lohwater A.J., Bolloba B.: The Theory of Cluster Sets, Cambridge University Press, 2004
Cooper R., Kaplan R.: Profit priorities from activity-based costing, Harvard Business Review, 69, 3, 130–135, 1991
Cooper R., Kaplan R.: Design of Cost management systems, Upper Saddle River, N.J., Prentice Hall, 1999

Cooper M.C., Lambert M.L., Pagh J.D.: Supply Chain Management: More Than a New Name for Logistics, The International Logistics Management, 8, 1, 1997

Cormier G., Gunn E.: A review of warehouse models. European Journal of Operational Research, 58, 1, 3–13, 1992

Corsten H. (2007) Produktionswirtschaft. Einführung in das industrielle Produktionsmanagement, 11th edition, Vahlen, München/Wien

Corsten D., Kumar N: Do Suppliers Benefit from Collaborative Relationships with Large Retailers? An Empirical Investigation of Efficient Consumer Response Adoption, Journal of Marketing, 69, 3, 80–94 (2005)

Coughlan A., Anderson E., Stern L., El-Ansary A.: Marketing Channels, 7th edition, Upper Saddle River N.J., Prentice-Hall, 2006

Coughlan A.: Competition and Cooperation in Marketing Channel Choice: Theory and Application, Marketing Science, 4, 2, 110–139, 1985

Coyle J.J., Bardi E.J., Langley C.J.: The Management of Business Logistics, A Supply Chain Perspective, Thomson Learning, South Western, 2003

Daganzo C.F., Logistic Systems Analysis, Springer, Berlin, Heidelberg, New York, 3rd edition, 1999

Dahlgaard J.J., Kristensen K., Kanij G.P.: Fundamentals of Total Quality Management, Process Analysis and Improvements, Nelson Thomas, 1997

Dangelmaier W.: Algorithmen und Verfahren zur Erstellung innerbetrieblicher Anordnungspläne, IPA Fraunhofer Institut für Produktionstechnik und Automatisierung, Stuttgart, 1986

Dangelmaier W.: Layoutplanung und Standortoptimierung, In: Weber J., Baumgarten H. (eds). Handbuch Logistik, Management von Material- und Warenflußprozessen, Schäfer-Poeschel, Stuttgart, 322–334, 1999

Darkow I.: Logistik-Controlling in der Versorgung, Dissertation TU Berlin, 2001

Darr W.: Integrierte Marketing Logistik, Auftragsabwicklung als Element der marketinglogistischen Strukturplanung, DUV, Wiesbaden, 1992

Dekker R. et al., Reverse Logistics, Quantitative Models for Closed-Loop Supply Chains, Springer, Berlin, Heidelberg, New York, 2004

De Koster R., Le-Due T., Roodbergen K.: Design and control of warehouse order picking: A literature review, European Journal of Operational Research 182, 2, 481–501, 2007

De Koster R., Van der Poort E.: Routing orderpickers in a warehouse: A comparison between optimal and heuristic solutions, IIE Transactions 30, 5, 469–480, 1998

De Wit B., Meyer R.: Strategy: Process, content, context: An international perspective, St. Paul, West Publishing Company, 1994

DIN 66001: Sinnbilder und ihre Anwendung, Standardsymbole zur Darstellung von Programm-, Prozeß- und Funktionsabläufen, Beuth, Berlin-Wien-Zürich, 12/1993

DIN EN ISO 9000ff, Quality Management and Quality Assurance, Beuth, 1994

DIN-, EN- and ISO-Norms for standard load units and load carriers

DIN 55405 Packstücke
DIN 30820 Kleinladungsträger
DIN 15146 Paletten
DIN 15155 Gitterbox
DIN 70013, DIN/EN 284 Wechselbehälter
ISO R 668/830 Überseecontainer

Dittrich J., Mertens P., Hau M.: Dispositionsparameter von SAP R/3-PP, Einstellhinweise, Wirkungen, Nebenwirkungen, Vieweg, Wiesbaden, 2000

Dobler D., Burt D.: Purchasing and supply management, 6th edition, New York, McGraw-Hill, 1996

Domschke W.: Logistik, Rundreisen und Touren, R. Oldenbourg, München Wien, 1985

Domschke W., Drexel A.: Logistik: Standorte, R. Oldenbourg, München Wien, 1990

Domschke W.: Logistik: Transport, R. Oldenbourg, München Wien, 1995

Dorfwirth R.: Wartezeiten und Rückstau von Kraftfahrzeugen an nicht signalgesteuerten Verkehrsknoten. Forschungsarbeiten aus dem Straßenverkehrswesen, Neue Folge, 43, Bad Godesberg, 1961

Drucker P.: The coming of a new organization, Harvard Business Review, 66, 1, 45–53, 1988
Dutta A., Lee H., Seungjin W.: RFID and Operations Management: Technology, Value, and Incentives. Production & Operations Management, 16, 5, 646–655, 2007
Economist: Radio Silence, Wireless technology: It was hailed as breakthrough that would revolutionise logistics, Whatever happened to RFID? 9th June 2007
Ekelund R.: Price discrimination and product differentiation in economic theory: an early analysis, Quarterly Journal of Economics, 84, 2, 268–278, 1970
Elsayed E.: Algorithms for optimal material handling in automatic warehousing, International Journal of Production Research, 19, 5, 525–535, 1981
FEM 9.851: Leistungsnachweis für Regalförderzeuge/VDI 3561: Testspiele zum Leistungsvergleich und Abnahme von Regalförderzeugen, Beuth, Berlin Köln, 1973
FEM 9.831: Berechnungsgrundlagen für die Regalbediengeräte, Toleranzen, Verformungen und Freimaße im Hochregallager, 1995
FEM 9.222: Regeln über die Abnahme und Verfügbarkeit von Regalbediengeräten und anderen Gewerken, Beuth, 1989
Ferschl F.: Zufallsabhängige Wirtschaftsprozesse, Physica Verlag, Wien Würzburg, 1964
Finkenzeller K.: RFID-Handbuch, Hanser, München, 2002
Flores B., Whybark D.: Multiple Criteria ABC Analysis, International Journal of Operations & Production Management, 6, 3, 38–46, 1986
Francis R., McGinnis L., White J.: Facilities Layout and Location: An Analytical Approach, 2nd edition, Prentice-Hall, Englewood Cliffs, NJ, 1992
Frazelle E.H., Supply Chain Strategy, McGraw-Hill, UK, 2001
Frazelle E., Apple J.: Material Handling Technologies, In: Robeson F., Copacino W., Howe E. (eds). The Logistics Handbook, The Free Press/Maxwell Macmillan, New York, 547–603, 1994
Fritsche B.: Advanced Planning and Scheduling (APS), Die Zukunft von PPS und Supply Chain, Logistik Heute, 50–55, 1999
Ford L.R., Fulkerson D.R.: Flows in Networks, Princeton University Press, Princeton, New York, 1962
Forrester J.: Industrial dynamics a major breakthrough for decision makers, Harvard business review, 36, 4, 37–66, 1958
Forrester J.: Industrial Dynamics, MIT Press, John Wighley&Sons, New York, 1961
Förster H.: EDI in Europa, Rollt der EDI-Zug?, Coorganisation 2/95, 14–19, 1995
Foster J., Strasser S.: Carrier/Modal Selection Factors: The Shipper/Carrier Paradox, Journal of the Transportation Research Forum, 31, 1, 206–212
Gardiner I.: War in the Frankincense Mountains, Oman 1973–1975, 2005 (to be published)
Gardner E.: Exponential Smoothing: The State of the Art. Journal of Forecasting, 14, 1, 1–28, 1985
Geiger G., Kanban, Hanser, München 2003
Gertsbaker J., Reliability Theory: With applications to prevent maintenance, Springer, Berlin Heidelberg New York, 2000
Gfrerer H., Zäpfel G.: Hierarchical model for production planning in the case of uncertain demand. European Journal of Operational Research, 86, 1, 142–161, 1995
Gilbert B., Miller L.: A Heuristic Algorithm for Vehicle Dispatching Problems, Operations Research Quarterly, 22, p. 340ff, 1974
Gilmore P., Gomory R.E., Multistage Cutting Stock Problems of Two and More Dimensions, Operations Research, 13, 1, 94–120, 1965
Gill L., Allerheiligen R.: Cooperation in channels of distribution: Physical distribution leads the way, International Journal of Physical Distribution & Logistics Management, 26, 5, 49–63, 1996
Girlich H., Chikan A.: The Origins of Dynamic Inventory Modelling under Uncertainty (The men, their work and connection with the Stanford Studies), International Journal of Production Economics, 71, 1, 351–363, 2001
Gnedenko B.W.: Handbuch der Bedienungstheorie, Akademieverlag, Berlin, 1984

Goetschalckx M., Ratliff H.: Order picking in an aisle, IIE Transactions 20, 1, 53–62, 1988
Goczyla K., Cielatkowski J.: Optimal routing in a transportation network, European Journal of Operational Research, 87, 2, 214–222 (9), 1995
Goldsby T., Martichenko R.: Lean Six Sigma Logistics, Ross Publ., 2005
Goldratt E.M., Cox J.: The Goal, Excellence in Manufacturing, Campus, Frankfurt/New York, 3rd edition, 2002
Grant R.: The Resource-Based Theory of Competitive Advantage: Implications for Strategy Formulation, California Management Review, 33, 3, 114–135, 1991
Graves S., Hausman W., Schwarz L.: Storage retrieval interleaving in automated warehousing systems. IIE Transactions, 23, 9, 935–945, 1977
Grieger M.: Electronic marketplaces: A literature review and a call for supply chain management research. European Journal of Operational Research, 144, 2, 280–294, 2003
Gross D.: Fundamentals of Queuing Theory, John Wiley & Sons, 1998
GS1 Standard Knowledge Center, http://www.gs1.org/services/gsmp/kc/, accessed 14-06-2008
Gudehus T.: Transportwegoptimierung durch optimale Gebäudeauslegung, dhf Nr. 12, 1971
Gudehus T.: Planung von Warenverteil- und Lagersystemen, Betriebsmanagement Service, 1973a
Gudehus T.: Grundlagen der Kommissioniertechnik, Dynamik der Warenverteil- und Lagersysteme, Girardet 1973b
Gudehus T.: Dimensionierung von Durchlauflagern: Industrie-Anzeiger, 1974
Gudehus T.: Engpässe in Fördersystemen, fördern und heben, 25, Nr.3/4, p. 206ff, 1975/I
Gudehus T.: Transporttheorie, Programm einer neuen Forschungsrichtung, Industrie-Anzeiger Nr.64, S. 1379ff, 1975/II
Gudehus T.: Grenzleistungsgesetze für Verteiler- und Zusammenführungselemente, Zeitschrift für Operations Research, Physica Verlag Würzburg, Band 20, 1976 B37-B61, 1975/III
Gudehus T.: Grenzleistungsgesetze für Verteiler- und Sammelelemente, fördern und heben, Krausskopf-Verlag, Mainz, Nr.16, 1975/IV
Gudehus T.: Staueffekte vor Transportknoten, Zeitschrift für Operations Research, Würzburg, B207 – B252, 1976/I
Gudehus T., Grenzleistungen bei absoluter Vorfahrt, Zeitschrift für Operations Research, Physica Verlag Würzburg, Band 20, 1976 B127-B160, 1976/II
Gudehus T.: Zuverlässigkeit und Verfügbarkeit von Transportsystemen, Teil I: Kenngrößen der Systemelemente, f+h 26/10, S.102 6ff, 1976: Teil II: Kenngrößen von Systemen, f+h 26/13, S.1343 ff, 1976: Teil III: Grundformeln für Systeme ohne Redundanz, f+h 29/1, S.23 ff, 1979
Gudehus T.: Kleinbehälter-Bereitstellungs- und Lagersysteme (AKL), Industrie-Anzeiger Nr.72, 1977
Gudehus T.: Die mittlere Anzahl von Sammelaufträgen, ZOR Zeitschrift für Operations Research, Würzburg, Band 22, B71-B78, 1978
Gudehus T.: Strategien in der Logistik, Fördertechnik 9/92, S. 5ff.
Gudehus T.: Logistik, Grundlagen . Strategien. Anwendungen, Springer, Berlin Heidelberg New York, 1999/2000/2003/2005/2007
Gudehus T.: Optimaler Nachschub in Versorgungsnetzen, Logistik Management 3. Jg., Ausgabe 2/3, 2001
Gudehus T.: Dynamische Disposition, Strategien und Algorithmen zur optimalen Auftrags- und Bestandsdisposition, Springer, Berlin Heidelberg New York, 2002/2006
Gudehus T., Kotzab H.: survey of planning reports and consulting projects for companies of industry, trade and logistic service providers in the time 1992 to 2004: 2004
Gudehus T.: Masterformeln der Logistik, Teil II: Qualität, Verfügbarkeit und Tourenplanung, f+h, 9, S. 364–367, 2006
Gudehus T.: Dynamische Märkte, Praxis, Strategien und Nutzen für Wirtschaft und Gesellschaft, Springer, Berlin Heidelberg New York, 2007
Gudehus T., Kunder R.: Kapazität und Füllungsgrad von Stückgutlagern, Industrie-Anzeiger, Nr. 93/74 und Nr. 104, 1972
Günther H.-O., Tempelmeier H.: Produktion und Logistik, 7th Edition, Springer, Berlin Heidelberg New York, 2005

Günthner W.A., Heptner K.: Technische Innovationen für die Logistik, hussverlag, München, 2007
Hadley G., Whitin T.: Analysis of inventory systems, Englewood Cliffs, Prentice-hall, 1963
Halldorsson A.: Third Party Logistics – a Means to configure logistics resources and competencies. CBS Business School, Frederiksberg 2002
Hammer M., Champy J.: Reengineering the corporation – a manifesto for business revolution, London, Nicholas Brealey, 1993
Hammer M., Stanton S: How Process Enterprises, Really Work, Harvard Business Review, 77, 6, 108–117, 1999
Hausman W., Schwarz L., Graves S.: Optimal storage assignments in automated warehousing systems. Management Science, 22, 6, 629–638, 1976
Harders J.: Die Leistungsfähigkeit nicht signalgeregelter Verkehrsknoten. Forschungsbericht, 7, Forschungsgesellschaft e. V. des Bundesverkehrsministe-riums, Bonn, 1968
Hardy G.H., Ramanujan S: Asymptotic formula in combination analysis, Proceedings of the London Mathematical Society, 17, S. 75–115, 1918
Harris F.: How Many Parts to Make at Once, Factory – The Magazin of Management, 10, 2, 135–136, 152, 1913
Harris F.: Operations and Costs (Factory management Series), Chicago, IL, W. Shaw Co, 1915
Hayek F.A.: The Fatal Conceit, The errors of socialism, Routledge, New York, 1988
Hax A., Candea D.: Production and inventory management, Prentice Hall College, 1983
Heskett J.: Cube-per-order index – A key to warehouse stock location, Transportation and Distribution Review 3, 1, 27–31, 1963
Heymann K.: Vernetzte Systeme beherrschen, Netzwerkmanagement – Transport in anderen Dimensionen beherrschen, In: Hossner, R. (ed). Jahrbuch der Logistik, Verlagsgruppe Handelsblatt, Düsseldorf, 35–41, 1999
van Hook R., Harrison A.: Logistic Management and Strategy, 2. ed., Prentice Hall, 2004
Hicks D.T., Activity Based Costing, John Wigley & Sons, 2nd ed. 2002
Holt C. (1957): Forecasting seasonals and trends by exponentially weighted moving averages. ONR Research Memorandum 52 (1957) Available from the Engineering Library, University of Texas at Austin, 2003
Hollingdale S. (ed): Digital Simulation in Operational Research. New York: American Elsevier Publishing Co., 1967
Hopp W., Spearman M.: Factory Physics. New York: McGraw-Hill, 2000
Horváth P.: Controlling, Vahlen, München, 1992
Horváth P.: Brokemper A.: Prozeßkostenrechnung als Logistikkostenrechnung, in Handbuch Logistik, S. 523ff: Schäfer Poeschel, 1999
Hubka V., Theorie der Maschinensysteme, Grundlagen einer wissenschaftlichen Konstruktionslehre, Spinger, Berlin – Heidelberg – New York, 1973
Hwang H., Lee S.: Travel-time models considering the operating characteristics of the storage and retrieval machine. International Journal of Production Research, 28, 10, 1179–1789, 1990
Im J., Schonberger R.: The pull of kanban, Production & Inventory Management Journal, 29, 4, 54–58, 1988
Inderfurth K.: Safety stocks in multistage divergent inventory systems: A survey, International Journal of Production Economics, 35, 1–3, 321–329, 1994
Inderfurth K.: Lagerhaltungsmodelle, In: Weber J., Baumgarten H. (eds). Handbuch Logistik, Management von Material- und Warenflußprozessen, Schäfer-Poeschel, Stuttgart, 349–366, 1999
Ihde B., Lucas G., Merkel H., Neubauer H.: Ersatzteilogistik, hussverlag, München, 1984
Ihde G.B.: Transport, Verkehr, Logistik, Vahlen, München, 1991
Isermann, H.: Ein Planungssystem zur Optimierung der Palettenbeladung mit kongruenten rechteckigen Versandgebinden, 9, 4, 235–249, 1987
Johnson H.T., Kaplan R.S.: Relevance Lost, The Rise and Fall of Management Accounting, Boston 1987
Jomini A.H.: Précis d`art de la guerre (Abriß der Kriegskunst), Berlin, 1881
Kanet J., Ramirez G.: Optimal stock picking decisions in automatic storage and retrieval systems, Omega International Journal of Management Science, 14, 3, 239–244, 1986

Kant I.: Über den Gemeinspruch: "Das mag in der Theorie richtig sein, taugt aber nicht für die Praxis", Königsberg, 1793, Neuausgabe: J. Ebbinghaus V. Klostermann, Frankfurt a.M., 1968
Kaplan R.S., Norton D.P.: The Balanced Score Card, Harvard Business Press, Harvard, 1996
Kern W. et al. (ed): Handwörterbuch der Produktionswirtschaft, Poeschl: Stuttgart, 1979
Kille C., Klaus P.: TOP 100 in European Transport and Logistics Services Market Sizes, Market Segments and Market Leaders in the European Logistics Industry, 2nd edition, Deutscher Verkehrsverlag, Hamburg, 2007
Kirsch W., Bamberger I., Gabele E.: Betriebswirtschaftliche Logistik. Systeme, Entscheidungen, Methoden, Gabler, Wiesbaden 1973
Klaus P., Krieger W. (Editors), Lexikon Logistik Gabler, Wiesbaden, 1998
Klaus P.: Logistik in Nordamerika, In: Klaus P., Krieger W. (eds). Gabler Lexikon Logistik, Gabler, Wiesbaden, 290–296, 1998
Kleiber W., HOAI Honorarordnung für Architekten und Ingenieure, Rehm, München, 1992
Kleinrock L.: Queuing systems 1 and 2, 1975
Kochhar J., Heragu S.: Facility layout design in a changing environment, International Journal of Production Research, 37, 11, 2429–2446
König D.: Theorie der endlichen und unendlichen Graphen, Leipzig, 1936
Kortschak E.B.: Vorsprung durch Logistik. Der Produktions- und Wettbewerbsfaktor Zeit und die Entwicklung der Logistik, Wien, Service Fachverlag, 1992
Kotler P.: Marketing Management, the millennium edition, London, Prentice-Hall International, 2000
Kotzab H.: Neue Konzepte der Distributionslogistik von Handelunternehmen, Deutscher Universitätsverlag, Gabler, Wiesbaden, 1997
Kotzab H.: Improving supply chain performance by efficient consumer response? A critical comparison of existing ECR approaches, Journal of Business & Industrial Marketing, 14, 5/6, 364–374, 1999
Kotzab H.: Der Beitrag integrativer Logistikkonzepte zum Unternehmenserfolg, Kritische Bestandsaufnahme theoretischer und empirischer Befunde, logistik management, 3.Jg., Ausg. 1 S. 17ff, 2001
Kotzab, H.: Conceptual understanding: DSD in the light of supplier-retailer relationships in the CP Industry, In: Otto A., Schoppengerd F.J., Shariatmadari R. (eds). Direct Store Delivery, Springer, Berlin et al., 2008
Kotzab H., Bjerre M.: Retailing in a Supply Chain Management Perspective CBS Press, Copenhagen, 2005
Kotzab H., Reiner G., Teller, C.: Beschreibung, Analyse und Bewertung von Instore-Logistikprozessen, Zeitschrift für Betriebswirtschaft, 77, 11, 1135–1158
Krallmann H., Frank H.F., Gronau F.: Systemanalyse im Unternehmen, Oldenbourg, München-Wien, 2002
Krampe H., Kubat J., Runge W.: Bedienungsmodelle, München Wien, 1973
Kraljic P.: Purchasing must become supply management, Harvard Business Review, 61, 5, 109–117, 1983
Kreyssig E.: Statistische Methoden und ihre Anwendungen: Vandenhoeck & Rupprecht, Göttingen, 1975
Kuhn A., Reinhardt A., Wiehndal H.-P., Handbuch der Simulationsanwendungen in Produktion und Logistik, Vieweg, Braunschweig,Wiesbaden, 1993
Kuhn A., Hellingrath H., Supply Chain Management, Optimierte Zusammenarbeit in der Wertschöpfungskette, Springer, Berlin Heidelberg New York, 2002
Kulick R.: Logistische Aufgaben der Vorbereitung und Abwicklung von Auslandsbaustellen, Bauingenieur, 56, 6, 193, 198, 1981
Kunder R., Gudehus T.: Mittlere Wegzeiten beim eindimensionalen Kommissionieren, Zeitschrift für Operations Research, Bdn. 19: S. B53ff, 1975
Küpper H.-U.: Controlling. Konzeption, Aufgaben und Instrumente, Stuttgart, Schäffer-Poeschel, 1995

Lambert D., Burduroglu R.: Measuring and selling the value of logistics. International Journal of Logistics Research 11, 1, 1–17, 2000
Lambert D., Cooper M., Pagh J.: Supply chain management: Implementation issues and research opportunities, International Journal of Logistics Management, 9, 1, 1–18, 1997
Lambert D., Pohlen T.: Supply Chain Metrics, International Journal of Logistics Management, 12, 1, 1–19, 2001
Lambert D.M., Stockdale J., Ellram L.M., Fundamentals of Logistics, Irwin/McGraw-Hill, Series in Marketing, 1997
Landes D.S.: Revolution in Time, Harvard University Press, 1983
Langford J.W.: Logistic Principles and Applications, McGraw-Hill, London, 1999
Langley C.J., Coyle J.J., Gibson B.J., Novack R.B., Bardi E.J.: Managing Supply Chains, a logistics approach, South-Western Cengage Learning, 2008, 429–434
Lanzendörfer R.: Simulationsmodelle von Transport-, Lager- und Verteilsystemen, Materialflußsysteme II, S. 135ff., Krausskopf, Mainz, 1975
Laporte G.: The traveling salesman problem: An overview of exact and approximate algorithms, European Journal of Operational Research, 59, 2, 231–247, 1992
Laurent M.: Vertikale Kooperation zwischen Industrie und Handel, Neue Typen und Strategien zur Effizienzsteigerung im Absatzkanal: Dt. Fachverlag, Frankfurt a.M., 1996
Lawler E., Lenstra J., Rinnooy Kan G., Shmoys D.: The Traveling Salesman Problem. A Guided Tour of Combinatorial Optimization, Wiley, New York, 1985
Le-Duc T., De Koster R.: Travel distance estimated and storage zone optimization in a 2-block class-based storage strategy warehouse, International Journal of Production Research, 43, 17, 3651–3581
Lee H., Padmanabhan V., Seungjin W.: Information distortion in a supply chain: the bullwhip effect, Management science, 43, 4, 546–558, 1997
Leibfried K.H.J., McNair C.J., Benchmarking, Von der Konkurrenz lernen, die Konkurrenz überholen, Haufe, Freiburg i.Br., 1998
Lenk H., Ropohl G. (Hrsg.), Systemtheorie als Wissenschaftsprogramm, Athenäum, Königstein/Taunus, 1978
Leutzbach W.: Ein Beitrag zur Zeitlückenverteilung gestörter Verkehrsströme, Dissertation, TH Aachen, 1956
Levin R., Kirkpatrick C.: Planning and control with PERT/CPM, New York, McGraw Hill, 1966
Lewandowski R.: Prognose- und Informationssysteme und ihre Anwendungen, de Gruyter, Berlin-New York
Lewis H., Culliton J., Steele J.: The Role of Air Freight in Physical Distribution, Boston, MA, The Alpine Press, 1956
Lock D.: Project Management, 6th edition, Aldershot, Gower, 1996
Lütkepohl H.: Introduction to multiple time series analysis, Berlin Springer, 1990
Maimborg C.: A heuristic model for simultaneous space allocation and block layout planning". International Journal of Production Research, 32, 3, 517–30, 1994
Maister E.W.: Centralization of Inventories and the "Square Root Law", International Journal of Physical Distribution: 6, 3, 124–134, 1976
Makowski E.: Maßstab für mehr Effizienz, Crossdocking bei Hornbach, Logistik Heute, 4 EXTRA, Handelslogistik, 1999
Makridakis S., Wheelwright S.: Forecasting. Methods and Applications, Santa Barbara, CA, Wiley, 1978
Mankiw N.G.: Macroeconomics, Deutsch: Makroökonomik, Schäfer-Poeschel, Stuttgart, 2003
Manrodt K., Durtsche A., Ledyard D. M., Keebler J.: Keeping Score: Measuring the Business Value of Logistics in the Supply Chain, Council of Logistics Management, Oak Brook, Ill., 1999
Malborg C., Krishnakumar B., Simons G.: A mathematical overview of warehousing systems with single/dual order-picking cycles, Applied Mathematical Modelling, 12, 1, 2–8, 1988
Maltz A., Ellram L.: Selling inbound logistics services: Understanding the buyer's perspective, Journal of Business Logistics, 21, 2, 69–88

Mayer R.: Prozeßkostenrechnung und Prozeßkostenmanagement, Vahlen, München, 1991
Maxwell W.: The scheduling of economic lot sizes: Maxwell, W.L. Naval Research Logistics Quarterly: 11 2–3, 89–124, 1965
McGinnis M., John K.: Warehousing, competitive advantage, and competitive strategy, Journal of Business Logistics, 9, 2, 32–54, 1988
Meller R. et al.: Progress in Material Handling Research, The Material Handling Institute, Charlotte/USA, 2004
Mentzer J., DeWitt W., Keebler J., Min S., Nix N., Smith C., Zacharia Z.: Defining supply chain management, Journal of Business Logistics, 22, 1, 1–24
Meyr H.: Supply chain planning in the German automotive industry. OR Spectrum,26, 4, 447–470, 2004
Miebach J.: Die Grundlagen einer systembezogenen Planung von Stückgutlagern, dargestellt am Beispiel eines Kommissionierlagers, Dissertation, TU Berlin, 1971
Mirasol N.: A queuing approach to logistics systems, Operations Research, 12, 5, 707–724, 1964
Monczka R., Trent R., Handfield R.: Purchasing and supply chain management, 2nd edition, Cincinnati, Thomson Learning, 2002
Moon-Kyu L.: A storage assignment policy in a man-on-board automated storage retrieval system, International Journal of Production Research, 30, 10, 2281–2292
Morgenstern O.: Note on the Formulation of the Theory of Logistics, in Naval Research Logistics Quarterly 2, 5, 129–136, 1955
Müller-Merbach H., Optimale Reihenfolgen, Springer-Verlag, Berlin Heidelberg New York, 1970
Müller Steinfahrt U.: Logistik in Deutschland, In: Klaus P., Krieger W. (eds). Gabler Lexikon Logistik, Gabler, Wiesbaden, 283–290, 1998
Murphy P.R., Wood D. F., Contemporary Logistics, Pearson Prentice Hall, 8th ed., 1996/2004
Mulcahy D.: Warehouse distribution and operations handbook, McGraw-Hill Professional, 1993
Naylor J., Naim M., Berry D.: Leagility: Integrating the lean and agile manufacturing paradigms in the total supply chain. International Journal of Production Economics, 62, 1/2, 107–118, 1999
Neumann von J., Morgenstern O.: Theory of Games and Economical Behavior, Princeton, 1994
Nersesian R., Swartz G.: Computer simulation in logistics, Quorum Books, Westport CT, London, 1996
Nicholas J.: Management Business and Engineering Projects: concepts and Implementation, Prentice-Hall, Upper Saddle River NJ, 1990
Nowitzky I.: Economies of Scale in der Distributionslogistik, Dissertation Universität, Institut für Fördertechnik und Logistik, Karlsruhe, 2003
Otto A., Kotzab H.: Does Supply Chain Management really pay? Six perspectives to measure the performance of managing a supply chain, European Journal of Operations Research, 144, 2, 306–320, 2002
Owens G., Mann R.: Materials Handling System Design, In: Robeson F., Copacino W., Howe E. (eds). The Logistics Handbook, The Free Press/Maxwell Macmillan, New York et al., 1994, 519–545
Ohno T.: Toyota Production System: Beyond Large-Scale Production, Productivity Press, Portland Oregon, 1988
Orlicky J.: Material Requirements Planning, New York, McGraw-Hill Book Co, 1975
Pande P., Neuman R., Kavangh R.: The Six Sigma Way, McGraw-Hill, New York, 2000
Park S., Narayan R.: Adaptive scheduling in dynamic flexible manufacturing systems: A dynamic rule selection approach. IEEE Transactions on Robotics & Automation, 13, 4, 486–502, 1997
Penrose E.: The theory of the growth of the firm, Billing & Sons, London, 1959
Perelman M.: Fixed Capital, Railroad Economics and the Critique of the Marketing, Journal of Economic Perspectives, 189–195, 1994
Peteraf M.: The cornerstones of competitive advantage: A resource-based view, Strategic Management Journal, 14, 3, 179–191, 1993
Petersen G., Aase G.: A comparison of picking, storage, and routing policies in manual order picking, International Journal of Production Economics 92, 1, 11–19, 2004

Petersen II G.: An evaluation of order picking routing policies, International Journal of Operations & Production Management, 17, 11, 1098–1111, 1997
Pfohl H.Chr.: Logistiksysteme, betriebswirtschaftliche Grundlagen, 4.Aufl, Springer, Berlin Heidelberg New York, 1990
Pflug G. Chr.: Optimization of Stochastic Models: The Interface between Simulation and Optimization, Springer US/Kluver Academic Publ., 1996
Pinto J., Slevin D.: Critical success factors in successful project implementation" IEEE Transactions, 34, 1, 22–27, 1987
Pohlen T., Lalonde B.: Implementing activity-based costing (ABC) in logistics, Journal of Business Logistics, 15, 2, 1–23, 1994
Poist R.: The Total Cost vs. Total Profit Approach To Logistics Systems Design. Transportation Journal, 14, 1, 13–24, 1975
Popper K.: The Logic of Scientific Discovery, Hutchinson, 5th edition, London, 1968 Deutsch: Logik der Forschung, J.C.B. Mohr, Tübingen, 1971
Porter M.: Competitive Strategy: Techniques for Analyzing Industries and Competitors, New York, Free Press 1980
Pradel U.-H., Süssenguth W. (Editors): Praxishandbuch Logistik, Deutscher Wirtschaftsdienst, 1996–2005
Prager W.: On Warehousing Problems, Operations Research, 5, 4, 504–512, 1957
Probst G.J.B.: Organisation, Strukturen, Lenkungsinstrumente, Entwicklungs-perspektiven, Moderne Industrie (in French Les Edition d'organisation), 1992
Prümper W.: Logistiksysteme im Handel, die Organisation der Warenprozesse in Großbetrieben des Einzelhandels, Thun Verlag, Frankfurt a.M., 1998
Rao A., Rao M.: Solution procedures for sizing of warehouses. European Journal of Operational Research, 108, 1, 16–25, 1998
Reider R.: Benchmarking Strategies, John Wiley & Sons, 1999
Reinhardt M.: Strategien für Planung und Betrieb eines Kommissioniersystems innerhalb eines Warenverteilzentrums, Diplomarbeit, TU Berlin, 1993
Richmond S.: Operations Research for Management Decisions, Ronald Press Company, New York, 1968
Rinehart L., Ragatz G.: Management of facility and human resources in materials and logistics management, Journal of Business Logistics,17, 2, 1–18, 1996
Ritter S.: Warenwirtschaft: ECR und CCG, Dynamik im Handel, 7–97, S. 18ff, 1997
Röder A., Sesterhenn J., Friehmuth U.: Standards für den Logistikvergleich – Benchmarking für Produktionsunternehmen, Logistik Heute, 98, 10, 48–52, 1998
Rosenblatt M., Roll Y.: Warehouse capacity in a stochastic environment. International Journal of Production Research, 26, 12, 1847–185, 1988
Rosenblatt M., Roll Y.: Warehouse design with storage policy considerations. International Journal of Production Research, 22, 5, 809–821, 1984
Ruben R., Jacobs F.: Batch construction heuristics and storage assignment strategies for walk/ride and pick systems, Management Science 45, 4, 575–596, 1999
Saaty T.: Resume of useful formulas in queuing theory, Operations Research, 5, 2, 161–200, 1957
SAP: R/3-System, Software Handbuch, SAP AG, Walldorf, 1994/2000
Sachs H.: Einführung in die Theorie der endlichen Graphen, Hanser, München, 1988
Sargent T., Kay M.: Implementation and utilization of a decentralized storage system. Costing model, International Journal of Operations & Production Management, 15, 9, 210–219, 1995
Sarker B., Babu P.: Travel time models in automated storage/retrieval systems: A critical review, International Journal of Production Economics, 40, 2, 173–184, 1995
Schaßberger R.: Warteschlangen, New York Wien, 1973
Scheer A.-W.: Business Process Engineering (ARIS), Springer-Telos, Berlin Heidelberg, New York, 1984
Scheer A.-W.: Informations- und Kommunikationssysteme in der Logistik, In: Weber J., Baumgarten H. (eds). Handbuch Logistik, Management von Material- und Warenflußprozessen, Schäfer-Poeschel, Stuttgart, 495–508, 1999

Scheutwinkel W.: SCM-Marktübersicht, Anspruch und Wirklichkeit, Logistik Heute, 5, 60–62, 1999

Schmidt H., Unternehmen sparen mit Marktplätzen im Internet viele Milliarden, FAZ , 115, p. 28, 18.5.2000

Schneeweiß Chr.: Modellierung industrieller Lagerhaltungssysteme, Einführung und Fallstudien, Springer, Berlin Heidelberg New York, 1981

Schneider E., Einführung in die Wirtschaftstheorie, J.C. Mohr (Paul Siebeck), Tübingen 1948, 12th edition, 1968

Schniederjans M.J., Olson J.R.: Advanced Topics in Just-In-Time-Management, Greenwood Press, 1999

Scholz-Reiter B., Stahlmann H.-D., Nethe A.: Process Modeling, Springer, Berlin-Heidelberg-New York, 1999

Scholz-Reiter B., Freitag M.: Autonomous processes in assembly systems. In: CIRP Annals, 56, 2, 712–729, 2007

Scholz-Reiter, B., Freitag M., de Beer C., Jagalski T.: Modelling Dynamics of Autonomous Logistic Processes: Discrete-event versus Continuous Approaches, Annals of the CIRP, 55, 1, 413–417, 2005

Schonberger R.: Applications of single-card and dual-card kanban, Interfaces, 13, 4, 56–67, 1984

Schonberger R.: World Class Manufacturing Casebook: Implementing JIT and TQC, New York, 1987

Schönsleben P.: Integrales Logistikmanagement, Planung und Steuerung von umfassenden Geschäftsprozessen, Springer, Berlin Heidelberg New York, 1998

Schulte Chr.: Logistik, 2nd edition, Vahlen, München, 1995

Seifert D.: Collaborative Planning, Forecasting and Replenishment – How to create a Supply Chain Advantage, Bonn, Galileo Press, 2002

Seuring S.: Supply chain controlling: summarizing recent developments in German literature, Supply Chain Management: An International Journal, 11, 1, 10–14, 2006

Shapiro R.: Get leverage from logistics. Harvard Business Review, 62, 3, 119–126, 1984

Shephard S.S.: RFID, McGraw-Hill, 2004

Scott Ch., Westbrook R.: New Strategic Tools for Supply Chain Management, International Journal of Physical Distribution and Logistics Management, 21, 1, 23–33, 1991

Siemens N., Marting C., Greenwood F.: Operations Research, The Free Press, New York, 1973

Silver E., Pyke D., Peterson R.: Inventory management and production planning and scheduling, 3rd edition, Chichester, Wiley, 1988

Simon H.A.: The architecture of complexity, Proceedings of the American Philosophical Society, 106 (1962) 6, 62–76

Simon H., Butscher S., Sebastian K.-H.: Better Pricing Processes for Higher Profits, Business Strategy Review, 14, 2, 63–67, 2003

Simchi-Levi et al.: Designing and managing the supply chain. Concepts, strategies and Case studies, Mc-GrawHill, 2008

Simchi-Levi D., Kaminsky P., Simchi-Levi E.: Designing and managing the supply chain. Concepts, strategies and Case studies, 2nd edition, Mc-GrawHill, Boston et al., 2008

Skjoett-Larsen T.: Third Party logistics – from an interorganizational point of view, International Journal of Physical distribution and Logistics Management, 30, 2, 112–127, 2000

Skjoett-Larsen T., Schary P., Mikkola J., Kotzab H.: Managing the global supply chain, 3rd edition, CBS Press, Copenhagen, 2007

Slack N., Chambers S., Johnston, R.: Operations Management, 5th edition, Harlow et al.: Financial Time Prentice Hall, 2004

Smith A.: The Wealth of Nations, 1.ed. Strahan & Cardell, London, 1776

Southworth F., Peterson B.: Intermodal and international freight network modeling, Transportation Research Part C: Emerging Technologies, 8, 1–6, 147–166, 2000

Speh T., Wagenheim G.: Demand and lead-time uncertainty: the impacts on physical distribution performance and management, Journal of Business Logistics, 1, 1, 95–113, 1978

Stadtler H., Kilger, C.: Supply Chain Management and Advanced Planning: Concepts, Models, Software, and Case Studies, 4th edition, Berlin, Springer, 2007
Stalk G.: Time – The Next Source of Competitive Advantage, Harvard Business Review, 66, 4, 41–51, 1988
Stock J., Lambert D.: Strategic Logistics Management, 4th edition, Boston, Mass., McGraw-Hill/Irwin, 2001
Stölzle W., Weber J., Hofmann E., Wallenburg C.: Handbuch Kontraktlogistik: Management komplexer Logistikdienstleistungen, Wiley-VCH, Weinheim, 2007
Stopford M.: Maritime Economics, London Unwin Hyman, 1988
Straube F.: e-Logistik, Springer, Berlin-Heidelberg-New York, 2004
Straube F., Gudehus T.: Auch die Logistik gehört auf den Prüfstand, Harvard Business Manager, 4, 72–81, 1994
Sugimori Y., Kusonoki K., Cho F.: Toyota production system and kanban system: materialization of just-in-time and respect for human system, International Journal of Production Research, 15, 6, 553–564, 1977
Taniguchi E. (editor): City Logistics, Network Modeling and Intelligent Transport Systems, 2001
Tempelmeier H.: Material-Logistik, Modelle und Algorithmen für die Produktionsplanung und Steuerung und das Supply Chain Management, Springer, Berlin Heidelberg New York, 4th edition, 1999
Tersine R.: Inventory control systems: an overview, Industrial Management, 17, 9, 10–13, 1975
Tompkins J.A., White J.A., Bozer Y.A., Frazelle E.H., Tanchoco J.M.A., Trevino J., Facilities Planning (second ed), Wiley, New York (1996)
Toporowski W.: Logistik im Handel, Optimale Lagerpolitik und Bestellstruktur einer Filialhandelsunternehmeung, Physica-Verlag, Heidelberg, 1996
Trip J., Bontekoning Y.: Integration of small freight flows in the intermodal transport system, Journal of Transport Geography, 10, 3, 221–229, 2002
Van Damme D., Van Amstel M.: Outsourcing Logistics Management Activities, The International Journal of Logistics Management, 7, 2, 85–94, 1996
Van Weele A.: Purchasing and Supply Chain Management, Analysis, Strategy, Planning and Practice, 4th edition, London, Thomson, 2004
VDI 3300, Materialflussuntersuchungen, Beuth, Berlin Köln 1959
VDI- und FEM Richtlinien, Zuverlässigkeit und Verfügbarkeit, Beuth
VDI 3564, Empfehlungen zum Brandschutz in Hochregalanlagen, 3.8. Rettungswege, Beuth, Berlin Köln1999
VDI 4446: Spielzeitberechnungen für Krane, Beuth, Berlin Köln, 1999
VDI 3580: Grundlagen zur Erfassung von Störungen an Hochregalanlagen, Beuth, 1980
VDI 3590: Kommissioniersysteme, Blatt 1 (1976), Blatt 2 (1977), Blatt 3 (1994), Beuth
VDI 3331: Beleglose Kommissioniersystem, Beuth, 1998 (Entwurf)
VDI 3881: Zuverlässigkeit und Verfügbarkeit von Transport- und Lageranlagen, Beuth, 1983
VDI 3619: Sortiersysteme für Stückgut, Beuth, 1983
VDI 3649: Anwendung der Verfügbarkeitsrechung für Förder- und Lagersysteme, Beuth 1992
Vahrenkamp R., Miller L., Eley M.: Logistikmanagement, Oldenbourg, München, 2000
Verpackungsverordnung, Bundesgesetzblatt 1991, Teil 1: Nr. 36, S. 1234ff
Vogt G.: Kommissionier-Handbuch, moderne industrie, München, 1993
Vokurka R., Lummus R.: The Role of Just-In-Time in Supply Chain Management, International Journal of Logistics Management, 11, 1, 89–98, 2000
Vollmann T., Berry L., Whybark C.: Manufacturing planning and control systems, 4th edition, Irwin, Homewood, 1997
Voss C., Robinson S.: Application of Just-in-Time Manufacturing Techniques in the United Kingdom, International Journal of Operations & Production Management, 7, 4, 46–52, 1987
Wagner H., Whitin T.: Dynamic version of the economic lot size model, Management Science, 5, 1, 89–96, 1958

Ware N., Fogarty D.: Master schedule/Master production schedule: the same or different, Production & Inventory Management Journal, 31, 1, 34–38, 1990

Warehouse Logistics: Developments and trends on the WMS market 2007–2008, http://www.logistik-benchmark.org/navigation/default.asp?NAV1=33&NAV2=0&NAV3=0&Sprache_ID=3, 2008

Watson G.: Strategic Benchmarking: How to Rate Your Company's Performance against the World's Best, New York, Wiley, 1993

Weber J. (ed): Praxis des Logistik-Controlling, Schäffer-Poeschel, Stuttgart, 1993

Weber J.: Logistikkostenrechnung, Kosten-, Leistungs- und Erlösinformationen zur erfolgreichen Steuerung der Logistik, Springer, Berlin Heidelberg New York, 2002

Weber M.: The Theory of Social and Economic Organization, Free Press New York, 1947

Wernerfelt B.: A resource-based view of the firm, Strategic Management Journal, 5, 2, 171–180, 1984

Wildemann H.: Das Just-In-Time Konzept. Produktion und Zulieferung auf Abruf, FAZ, Frankfurt am Main, 1987

Winters P.: Forecasting sales by exponential smoothing, Management Science, 6, 3, 324–334, 1960

Williamson O.: The economic institutions of capitalism, The Free Press, New York, 1985

Winston C.: Conceptual Developments in the Economics of Transportation: An Interpretive Survey, Journal of Economic Literature, 23, 1, 57–94, 1985

Whitin T.: The theory of inventory management, Princeton, N.J., Princeton University Press, 1957

Wöhe G.: Allgemeine Betriebswirtschaftslehre, 20th Edition , Vahlen, München 2000

Wöhe G., Döring U: Allgemeine Betriebswirtschaftslehre, 23th Edition, Vahlen, München 2008

Wolf O, Dietze G., Daniluk D.: WMS – heute und morgen, In: Seebauer P. (ed). Software in der Logistik: Markt-Spiegel. Anforderungen, Funktionalitäten und Anbieter in den Bereichen WMS, ERP, TMS und SCM, huss, München, 56–57, 2006

Wolff S.: Zeitoptimierung der logistischer Ketten: Dissertation TU Berlin, huss, München, 1994

Wood D., Johnson J.: Contemporary Transportation, 5th edition, Prentice Hall Upper Saddle River NJ, 1995

Wood D: International logistic channels, International Journal of Physical Distribution and Logistics Management, 20, 9, 3–9, 1990

Woxenius J.: Development of Small-Scale Intermodal Freight Transportation in a Systems Context, Chalmers tekniska högskola, Institutionen för transportteknik, ISSN 0283-3611, no 34, 1998.

Wrennail W., Lee Q.: Material movement and storage, In: Wrennail W., Lee Q. (eds). Handbook of Commercial and Industrial Facilities Management, McGraw-Hill, New York, NY, 1994, 141–99

Zangemeister C.: Nutzwertanalyse in der Systemtechnik, Wittemannsche Buchhandlung, München, 1972

Zäpfel G.: Customer-order-driven production: An economical concept for responding to demand uncertainty? International Journal of Production Economics, 56/57, 3 699–709, 1998

Zäpfel G.: Production planning in the case of uncertain individual demand extension for an MRP II concept. International Journal of Production Economics, 46/47, 3, 153–164, 1996

Zibell R.: Just-In-Time-Philosophie, Grundlagen, Wirtschaftlichkeit, Dissertation, TU Berlin, hussverlag, 1990

Zwehl von W.: Wirtschaftliche Losgrößen, Handbuch der Produktionswirtschaft, Poeschel, 1. Edition, S. 1166ff, 1979

Index

The extensive index makes this book a *dictionary* and *reference book* for modern logistics. Via the index the definition and explanation of the listed terms and keywords can be found in the text: bold numbers indicate pages with definitions and main explanations; the other numbers denote pages with further information and examples.

By unambiguous definition of the *terms of modern logistics* and their consistent use throughout this book misunderstandings should be avoided. Hopefully, this will help also to establish a basis for an international logistic language.